Basic Food Chemistry

Second Edition

BASIC FOOD CHEMISTRY

Second Edition

Frank A. Lee, Ph.D.

Professor of Chemistry, Emeritus
New York State Agricultural
Experiment Station
Cornell University
Geneva, New York

發行人　施　弘　國

發行所　華香園出版社　　電話：761-1001

地　址　台北市松山路287巷11號4樓

郵　撥　111000（三個"1"三個"0"）

華香園出版社帳戶

登記證　局版台業字第1438號

Dedicated to the memory of
Dr. Donald K. Tressler
and Prof. Dr. Reinhold Grau,
both respected friends and colleagues

Contents

Preface xi

1. Photosynthesis **1**

Introduction 1
Role of Chlorophyll 1
Chemistry of Photosynthesis 3
Summary 6
Bibliography 7

2. Water and Solutions **9**

The Structure of Water 9
Solutions 19
Summary 22
Bibliography 23

3. Colloids **25**

What Are Colloids? 25
Behavior of Colloids in an Electric Field 26
Syneresis 27
Imbibition 28
Emulsions 28
Breaking of Emulsions 29
Foams 29
Viscosity 30
Summary 31
Bibliography 32

4. Carbohydrates **33**

Introduction 33
Monosaccharides 34
Structure of Carbohydrates 39
Carbohydrate Reactions 44
Oligosaccharides 53
Polysaccharides 56
Digestion of Carbohydrates 76
Qualitative Tests 76
Quantitative Tests 78

Summary 79
Bibliography 80

5. **Lipids** 87

Simple Lipids 88
Composite Lipids 102
Derived Lipids 107
Oxidation of Lipids 108
Summary 124
Bibliography 126

6. **Proteins** 133

Amino Acids 133
Classification of Proteins 144
Structure of Proteins 146
Properties of Proteins 152
Protein Synthesis 158
Proteins in Foods 165
Summary 172
Bibliography 173

7. **Enzymes** 177

Classification 178
Enzyme Composition 182
Properties of Enzymes 182
Enzyme Reactions 190
Some Enzyme Applications in Food 193
Summary 195
Bibliography 197

8. **The Vitamins** 199

The Water-Soluble Vitamins 199
The Fat-Soluble Vitamins 212
Vitamins and Malnutrition 217
Summary 218
Bibliography 220

9. **Minerals** 225

Occurrence of Minerals 228
Anionic Minerals 231
Minerals in Canned Foods 232
Summary 233
Bibliography 234

10. **Flavor** 237

The Basic Tastes 237
Flavors and Volatiles 241
Flavor Enhancement 253
Flavor Restoration and Deterioration 254

Summary 256
Bibliography 257

11. **Natural Colors** **261**

Chemistry of Natural Coloring Matters 262
Caramels and Melanoidins 272
Color Determination 273
Summary 274
Bibliography 278

12. **Browning Reactions** **283**

Enzymatic Browning 283
Non-enzymatic Browning 288
The Formation of Brown Pigments 297
Inhibition of Browning 300
Summary 301
Bibliography 302

13. **Food Colorings** **307**

Color Safety and Regulations 307
Colors Exempt from Certification 317
Color Analysis and Desired Properties 320
Summary 321
Bibliography 322

14. **Alcoholic Fermentation** **323**

Wine 326
Beer and Brewing 336
Distilled Products 337
Vinegar 337
Summary 338
Bibliography 339

15. **Baked Products** **343**

Flour 343
Leavening 354
Summary 357
Bibliography 358

16. **Milk and Milk Products** **363**

Milk Composition 363
Rancidity and Off-Flavor in Milk 373
Cheese and Cheese Chemistry 375
Fermentation in Milk 384
Milk Products Other than Cheese 385
Determination of Fat in Dairy Products 387
Total Solids Analysis 388
Summary 388
Bibliography 390

17. Coffee **397**

Composition of the Green Coffee Bean 398
Changes During Roasting 401
Roasted Coffee 403
Coffee Products 413
Summary 414
Bibliography 415

18. Tea **419**

Methods of Preparation 419
Chemical Components of Tea 422
Amino Acids 425
Chemistry of Tea Manufacture 428
Tea Aroma 433
Summary 435
Bibliography 435

19. Cocoa and Chocolate **441**

Introduction 441
Manufacture of Cocoa and Chocolate 442
Chemical Composition 444
Changes During Manufacture of Cocoa and Chocolate 447
Volatiles and Chocolate Flavor 456
Chocolate and Cocoa Products 457
Summary 458
Bibliography 459

20. Meat and Meat Products **463**

Muscle Composition 463
Changes in Muscle After Slaughter and During Processing 476
Effect of Ionizing Radiation 486
Summary 492
Bibliography 495

21. Fruits and Vegetables **505**

Ripening and Post Harvest Changes 505
Storage 518
Chemistry Involved in the Texture of Fruits and Vegetables 520
Summary 533
Bibliography 535

Index **546**

Preface 

Food chemistry has grown considerably since its early foundations were laid. This has been brought about not only by research in this field, but also, and more importantly, by advances in the basic sciences involved.

In this second edition, the chapters dealing with fundamentals have been rewritten and strengthened. Three new chapters have been added, Water and Solutions, Colloids, and Minerals. The chapter on Fruits and Vegetables has been expanded to cover texture.

Other chapters discuss flavor and colors, together with one on browning reactions. The last seven chapters give the student a background of the classes of food products and beverages encountered in everyday use. Each chapter includes a summary and a list of references and suggested readings to assist the student in study and to obtain further information.

Basic Food Chemistry is intended for college undergraduates and for use in food laboratories.

The author wishes to express his appreciation to the following people, who reviewed the chapters on their respective specialties: Doctors L.R. Hackler, M. Keeney, B. Love, L.M. Massey, Jr., L.R. Mattick, W.B. Robinson, R.S. Shallenberger, D.F. Splittstoesser, E. Stotz, W.L. Sulzbacher, and J. Van Buren.

In addition, the author wishes to express his appreciation to Dr. H.O. Hultin and Dr. F.W. Knapp for their reviews of the entire original manuscript and for their helpful comments.

The author welcomes notices of errors and omissions as well as suggestions and constructive criticism.

Frank A. Lee

1

Photosynthesis

Introduction
Role of Chlorophyll
Chemistry of Photosynthesis
Summary
Bibliography

INTRODUCTION

The basic source of energy in the world is photosynthesis. This is the result of a series of reactions by which the low energy substances, carbon dioxide and water, are converted into sugars, which are high energy substances. The conversion is accomplished with the addition of a minimal free energy which amounts to +686,000 calories per mole and is obtained from the oxidation of a mole of glucose.

ROLE OF CHLOROPHYLL

The agent for the conversion of carbon dioxide and water into sugars is chlorophyll, the green pigment in plants, and the energy is derived from sunlight. This involves two types of reactions. One takes place in the sunlight, the second takes place in the dark and results in the formation of sugars. Chlorophyll *a* is the agent for solar energy conversion in the primary process, the products of which are enzymatically converted to sugars in the subsequent dark reactions.

Many types of plants have the power to carry on photosynthesis. These include the higher green plants as well as the green, brown, and red algae and many unicellular organisms. Lower forms that carry on photosynthesis include the blue-green algae and purple and green bacteria. Many of these are anaerobes and require H_2S or other compounds of sulfur to complete their activities in this chain of reactions. In such cases the S takes the place of O_2 in the process. These lower forms of life are very important in the photosynthetic process, and it seems likely that a great deal of photosynthesis is carried on by them. It follows that photosynthetic products are produced in great quantities by unicellular forms of life. It is well to note, however, that all forms of life which carry on photosynthesis, except bacteria, make use of water as the hydrogen (electron) donor to reduce carbon dioxide which is the electron acceptor.

Chlorophyll and associated pigments are contained in the plants in bodies known as chloroplasts. Chloroplasts are rather complex structures. They vary in size, being about 1–2 μm in diameter and 4–10 μm long, and are composed of structures known as grana. These latter are composed of lamellae, which, in turn, are made up of sheets of membranes known as quantasomes (Table 1.1).

Chlorophyll is present in the chloroplasts as lipoprotein complexes. However, the physical state of the chlorophyll, which is photochemically active, is not yet understood. This combination is broken by nonpolar solvents, hence the extraction of chlorophyll from plant materials is by means of these solvents. Isolated spinach chloroplasts are often used to study the reactions involved in photosynthesis.

Blue-green algae, and bacteria that are capable of carrying on photosynthesis, do not have chloroplasts, although the blue-green algae have lamellae in the cytoplasm. The photosynthetic bacteria have another arrangement, which is attached to the cell membrane.

Chlorophyll is the important pigment in photosynthesis because of its part in the absorption of light, the energy of which permits the buildup of sugars. The presence of many conjugated double bonds in chlorophyll results in a particular structure resonating at the frequencies of the red and blue range of visible light. Actually, chlorophylls *a* and *b* are involved, usually about 3 parts of *a* to 1 part of *b*, in the higher plants. Chlorophyll *c* is found in brown algae, diatoms, and dinoflagellates. Chlorophyll *d* is found in red algae. Accessory pigments include the yellow carotenoids and the blue or red phycobilins. It is perhaps likely that these accessory pigments have an effect in the part of the solar spectrum that is not in the chlorophyll range. Chlorophyll, depending on its plant source, can utilize light wavelengths from about 400 to almost 900 nm to carry on photosynthesis. For chlorophyll *a*

TABLE 1.1. Approximate Composition of an Average Spinach Quantasome

Component	Molecules per Quantasome	Component	Molecules per Quantasome
Chlorophyll *a*	160	Phospholipids (lecithin, phosphatidyl ethanolamine, phosphatidyl inositol, phosphatidyl glycerol)	116
Chlorophyll *b*	65		
Carotenoids	48		
Quinones			
Plastoquinone A	16	Sulfolipids	48
Plastoquinone B	8	Galactosylglycerides	500
Plastoquinone C	4	Cytochrome b_6	1
α-Tocopherol	10	Cytochrome *f*	1
α-Tocopherylquinone	4	Plastocyanin	5
Vitamin K_2	4	Ferredoxin	5

Source: White *et al.* (1973). Reproduced with permission of the McGraw-Hill Book Company.

from higher plants, the absorption maximum is in the neighborhood of 675 nm while that of chlorophyll *b* is about 650 nm. Accessory pigments such as the carotenoids and other chlorophylls absorb from 400 to about 550 nm.

Evidence has been collected that indicates that two light reactions are involved in the part of the photosynthetic process that evolves oxygen. It has been postulated by Duysens (1964) that photosystem I involves chlorophyll *a* and does not evolve oxygen. However, photosystem I is associated with photosystem II which contains chlorophyll *a* and chlorophyll *b* or another chlorophyll (*c* or *d*) according to the species involved and does evolve oxygen. Other pigments are involved also. There has been much speculation concerning the mechanisms of action of these two systems.

Chlorophylls *a* and *b* have a non-ionic magnesium atom in the structure, which is held by two coordinate and two covalent linkages. When chlorophyll is treated with weak acids, the magnesium is removed from the molecule and pheophytin, the olive green compound, is formed. This is especially important to food chemists because it is the cause of the particular green color found in canned green vegetables that have been processed under pressure.

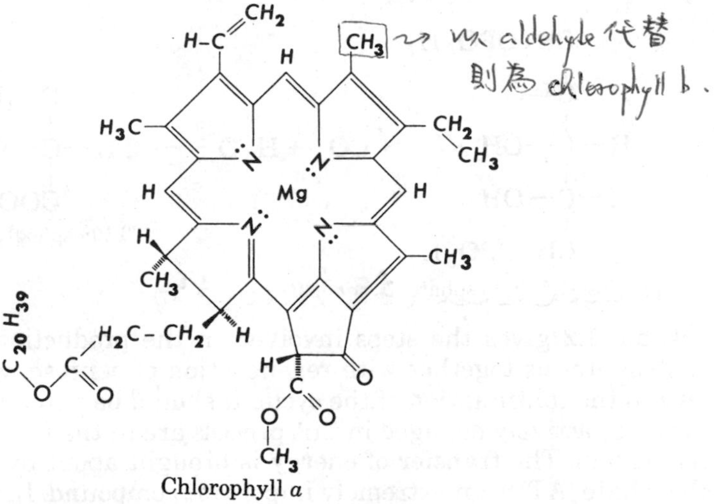

Chlorophyll *a*

The basic structures for chlorophylls *a* and *b* were worked out by H. Fischer (1934, 1937). The structure for chlorophyll *b* is the same as chlorophyll *a* except that a —CHO (formyl) group takes the place of the methyl group at position 3.

CHEMISTRY OF PHOTOSYNTHESIS

For many years it was thought that the path for the conversion of carbon dioxide and water into sugars was through formaldehyde,

CH_2O, the simplest such compound. However, this compound could never be detected. Furthermore, it is toxic to plants if present in more than trace amounts. This theory was abandoned when the work of Calvin showed that a different and complex pathway is the actual course of events. The general equation for this synthesis is

$$6\ CO_2 + 6\ H_2O \longrightarrow C_6H_{12}O_6 + 6\ O_2$$

This may be written in the more general terms as follows:

$$x\ CO_2 + x\ H_2O \longrightarrow (CH_2O)x + O_2$$

These changes were investigated using CO_2 and water with labeled carbon atoms (^{14}C) and suspensions of green algae, followed by two-directional paper chromatography. Finally the spots were demonstrated by placing the chromatograms over photosensitive paper. The first compound to be formed was found to be a 3-carbon compound, 3-phosphoglyceric acid, with the labeled carbon appearing mainly in the carboxyl carbon atom. This compound was formed from a 5-carbon compound, ribulose-1,5-diphosphate rather than a 2-carbon compound. When CO_2 and H_2O react with this compound, two molecules of 3-phosphoglyceric acid result. The enzyme involved in this reaction is diphosphoribulose carboxylase.

$$\begin{array}{c} CH_2OPO_3H_2 \\ | \\ C{=}O \\ | \\ H{-}C{-}OH \\ | \\ H{-}C{-}OH \\ | \\ CH_2OPO_3H_2 \end{array} + CO_2 + H_2O \longrightarrow 2\ \begin{array}{c} CH_2OPO_3H_2 \\ | \\ H{-}C{-}OH \\ | \\ COOH \end{array}$$

Ribulose-1,5-diphosphate

3-Phosphoglyceric acid

Table 1.2 gives the steps involved in the production of sugars by photosynthesis together with regeneration of pentose, which is necessary to the continuation of the cycle. It should be noted that the carbohydrates actively engaged in this process are in the form of phosphoric acid esters. The transfer of energy is brought about by adenosine triphosphate (ATP), an extremely important compound. In reaction 2, the 3-phosphoglycerate from the first reaction is reduced to 3-phosphoglyceraldehyde. The molecules of this compound thus formed are used in 3 and 4 to make fructose-6-phosphate in the amount of five molecules of this compound. As the table shows, one of these molecules is the net gain or final product of the photosynthetic process. The others are changed into ribulose-1,5-diphosphate for use in the next cycle. This set of reactions has been demonstrated by the use of purified enzymes. It is believed that this set of reactions occurs in the chloroplasts. There are, however, reasons to believe the possible existence of other pathways for the production of sugars by photosynthesis.

TABLE 1.2. Hexose Accumulation and Pentose Regeneration in Photosynthesis

Step	Enzyme	Reaction[a]	Carbon Balance
1.	Carboxylation enzyme	6 Ribulose-1,5-diphosphate → 12 3-phosphoglyceric acid	6(5) + 6(1) → 12(3)
2.	Phosphoglyceric acid kinase	12 3-Phosphoglyceric acid + 12 ATP → 12 1,3-diphosphoglyceric acid + 12 ADP	12(3) → 12(3)
	Phosphoglyceraldehyde dehydrogenase	12 1,3-Diphosphoglyceric acid + 12 DPNH + 12 H^+ → 12 3-phosphoglyceraldehyde + 12 DPN^+ + 12 P_i	12(3) → 12(3)
3.	Triose isomeraldelase	5 3-Phosphoglyceraldehyde → 5-dihydroxyacetone phosphate	5(3) → 5(3)
		5 3-Phosphoglyceraldehyde + 5-dihydroxyacetone phosphate → 5-fructose-1,6-diphosphate	5(3) + 5(3) → 5(6)
4.	Phosphatase	5 Fructose-1,6-diphosphate → 5 fructose-6-phosphate + 5 P_i	5(6) → 5(6)
5.	Transketolase	2 Fructose-6-phosphate + 2 3-phosphoglyceraldehyde → 2 xylulose-5-phosphate + 2 erythrose-4-phosphate	2(6) + 2(3) → 2(5) + 2(4)
6.	Transaldolase	2 Fructose-6-phosphate + 2 erythrose-4-phosphate → 2 sedoheptulose-7-phosphate + 2 3-phosphoglyceraldehyde	2(6) + 2(4) → 2(7) + 2(3)
7.	Transketolase	2 Sedoheptulose-7-phosphate + 2 3-phosphoglyceraldehyde → 4 xylulose-5-phosphate	2(7) + 2(3) → 4(5)
8.	Epimerase	6 Xylulose-5-phosphate → 6 ribulose-5-phosphate	6(5) → 6(5)
9.	Phosphoribulokinase	6 Ribulose-5-phosphate + 6 ATP → 6 ribulose-1,5-diphosphate + 6 ADP	6(5) → 6(5)
Net:		6 Ribulose-1,5-diphosphate + 6 CO_2 + 18 ATP + 12 DPNH + 12 H^+ → 6 ribulose-1,5 diphosphate + 1-fructose-6-phosphate + 17 P_i + 18 ADP + 12 DPN^+	6(5) + 6(1) → 6(5) + 1(6)

Source: White *et al.* (1973). Reproduced with permission of the McGraw-Hill Book Company.

[a] Key to abbreviations: ATP = Adenosine triphosphate. ADP = Adenosine diphosphate. P_i = Inorganic orthophosphate. DPNH = Reduced diphosphopyridine nucleotide. DPN = Diphosphopyridine nucleotide.

It has been shown by the use of ^{18}O-labeled water and carbon dioxide that the oxygen formed during the process of photosynthesis comes from the water and not from the carbon dioxide (Kok and Jagendorf 1963). The great force that affects the conversion of light energy is demonstrated in this reaction by the fact that the electrons flow in the direction of the more energy-rich state, which is against the usual flow of electrons. Only a very powerful force could accomplish this.

It was suspected for a long time that two types of reactions were involved in photosynthesis. One of these reactions was thought to require the presence of sunlight to supply the necessary energy, the other could take place in the dark; that is, light was not required. Experimental support for this supposition was first obtained by Hill in 1937. The results of subsequent research have increased the fundamental knowledge of this phase of the photosynthetic process. It was found that a suspension of chloroplasts illuminated in the absence of CO_2 and then placed in the dark with CO_2 added permitted, briefly, the formation of sugars. This indicates that key compounds that can react with CO_2 were formed during the exposure to light.

Further, it was shown that the first step in the photosynthetic process, the one using sunlight as the energy source, reduces $NADP^+$ (nicotinamide adenine dinucleotide phosphate) and phosphorylates ADP, which results in the formation of NADPH and ATP. This first step releases oxygen, but oxygen is not released in the second step. This emphasizes the fact that the oxygen released during photosynthesis comes from water and not from the CO_2. Water, therefore, is the only electron donor necessary. NADPH and ATP are used in the dark reaction to reduce CO_2 to hexoses and other products.

$$2\ H_2O + 2\ NADP^+ \longrightarrow 2\ NADPH + 2\ H^+ + O_2$$

Photosynthetic bacteria use H_2S and other compounds in the first step of photosynthesis, and, therefore, do not release oxygen. However, if H_2S is used by them in this stage of the process, sulfur is released instead of oxygen. In short, the process is fundamentally similar.

Although a great deal has been learned in recent years about the process of photosynthesis and its intricacies, many problems are still unsolved.

SUMMARY

The basic source of energy in the world is the process known as photosynthesis. In this process, carbon dioxide and water are converted into sugars. It is, however, a rather complex process. The agent for the conversion of carbon dioxide and water into sugars is chlorophyll, the green pigment in plants. Chlorophyll is important in photosynthesis because of the part it plays in the absorption of light. The energy thus absorbed permits the buildup of sugars. Calvin (1956; Wilson and

Calvin 1955) showed the complex pathway that is the actual course of events in photosynthesis.

BIBLIOGRAPHY

ARNON, D. I., TSIJIMOTO, H. Y., and McSWAIN, B. D. 1965. Photosynthetic phosphorylation and electron transport. Nature *207,* 1367–1372.

BISHOP, N. I. 1971. Photosynthesis: The electron transport system of green plants. Annu. Rev. Biochem. *40,* 197–226.

CALVIN, M. 1956. The photosynthetic carbon cycle. J. Chem. Soc. 1895–1915.

DUYSENS, L. N. M. 1964. The subcellular localization of the lysosomal enzyme and its biological significance. *In* Progress in Biophysics and Molecular Biology, Vol. 14, J. A. V. Butler and H. E. Huxley (Editors). Pergamon Press, Oxford, England.

FISCHER, H. 1934. Chlorophyll A. J. Chem. Soc. 245–256.

FISCHER, H. 1937. Chlorophyll. Chem. Rev. *20,* 41–68.

GIBBS, M. 1967. Photosynthesis. Annu. Rev. Biochem. *36,* 757–784.

HILL, R. 1965. The biochemists' green mansions: The photosynthetic electron transport chain in plants. *In* Essays in Biochemistry, Vol. 1. P. N. Campbell and G. D. Greville (Editors). Academic Press, New York.

KATZ, J. J. 1972. Chlorophyll function in photosynthesis. *In* Advances in the Chemistry of Plant Pigments, C. O. Chichester (Editor). Academic Press, New York.

KOK, B. 1965. Photosynthesis: the path of energy. *In* Plant Biochemistry, J. Bonner and J. E. Varner (Editors). Academic Press, New York.

KOK, B., and JAGENDORF, A. T. 1963. Photosynthetic Mechanisms in Green Plants. National Academy of Sciences, Washington, DC.

OCHOA, S., and VISHNIAC, W. 1952. Carboxylation reactions and photosynthesis. Science *115,* 297–301.

RUBIN, S., RANDALL, M., KAMEN, M. D., and HYDE, J. L. 1941. Heavy oxygen (^{18}O) as a tracer in the study of photosynthesis. J. Am. Chem. Soc. *63,* 877–879.

SAN PIETRO, A., GREER, F. A., and ARMY, T. J. 1967. Harvesting the Sun. Academic Press, New York.

VERNON, L. P., and AVRON, M. 1965. Photosynthesis. Annu. Rev. Biochem. *34,* 262–296.

VERNON, L. P., and SEELY, G. R. 1966. The Chlorophylls. Academic Press, New York.

WALKER, D. A., and CROFTS, A. R. 1970. Photosynthesis. Annu. Rev. Biochem. *39,* 389–428.

WHITE, A., HANDLER, P., and SMITH, E. C. 1973. Principles of Biochemistry, 5th Edition. McGraw-Hill Book Co., New York.

WILSON, A., and CALVIN, M. 1955. The photosynthetic cycle. CO_2 dependent transients. J. Am. Chem. Soc. *77,* 5948–5957.

2

Water and Solutions

The Structure of Water
Solutions
Summary
Bibliography

Water is an important part of a large number of foods. It is not only an important component of these foods, it is necessary for the digestion of these foods in the body to sustain life. Without water, life as we know it would be impossible. Water is necessary in photosynthesis for the production of carbohydrates by this process. The deterioration of foods, both microbiological and chemical, is affected by the presence of water in the food. Dehydration, which is the elimination of most of the water present, permits food to be stored for relatively long periods of time.

The amount of water found in representative foods is given in Table 2.1.

THE STRUCTURE OF WATER

Water as found under natural conditions is made up mostly of $H_2{}^{16}O$. Very small amounts of other forms of water which contain the ^{17}O and ^{18}O isotopes of oxygen as $H_2{}^{17}O$ and $H_2{}^{18}O$ can be found as well as "heavy water," which contains ^2H (heavy hydrogen, also known as deuterium, D). The formula for the water molecule in general use is H_2O.

The structure of the water molecule accounts for its special behavior. This results from the hydrogen bonding and the tetrahedral arrangement of the electron pairs around the oxygen atom. Its bond angle is 104.5°, and the bond length, the distance between the nuclei of hydrogen and oxygen, is 0.9572 Å.

In the diagram of the water molecule (Fig. 2.1), the dotted lines symbolize the hydrogen-bonding tendency arising from the diffuse cloud of electrons in the direction of the dotted lines. Several molecules held together by the hydrogen bond, as in solid or liquid water, form an assembly of tetrahedral groups; each molecule is bonded to four other molecules, which it closely approaches. An understanding of the princi-

TABLE 2.1. The Water Content of Some Foods, Edible Portion

Food	Percentage Water
Almonds	4.7
Apricots	
raw	85.3
dried	25.0
Asparagus, raw	91.7
Bacon, raw	19.3
Beans	
snap, raw	90.1
common, dried	10.9
Beef, loin, club steak, total edible raw, 36% fat	49.1
Bread, white (1–2% nonfat dry milk)	35.8
Butter and margarine, salted	15.5
Celery, raw	94.1
Cheese, Cheddar, American	37.0
Chicken, light meat	73.7
Cocoa, breakfast, high fat	3.0
Corn, sweet, raw	72.7
Cornmeal, yellow	12.0
Eggs, chicken	
white	87.6
yolk	51.1
Flour, wheat, patent, all purpose	12.0
Grapefruit, Florida	89.1
Jellies	29.0
Lard	0
Lemons, raw, peeled	90.1
Lettuce, raw, Iceberg	95.5
Liver, calf	70.7
Macaroni, dry, uncooked	10.4
Milk	
cow, fresh	
whole	87.4
skim	90.5
canned, evaporated, unsweetened	73.8
Molasses, cane	24.0
Peas	
young, raw	78.0
dry split, raw	9.3
Potatoes, raw	79.8
Rice, white, raw	12.0
Salmon, Chinook	
raw	64.2
canned, solids and liquid	64.4

Source: USDA 1963.

ples of bonding is required to explain the properties of water. These properties are of two types, depending on whether the chemical bon between both the H and O atoms are broken in the action involved whether only the hydrogen bonds are broken, leaving the H_2O molecules intact.

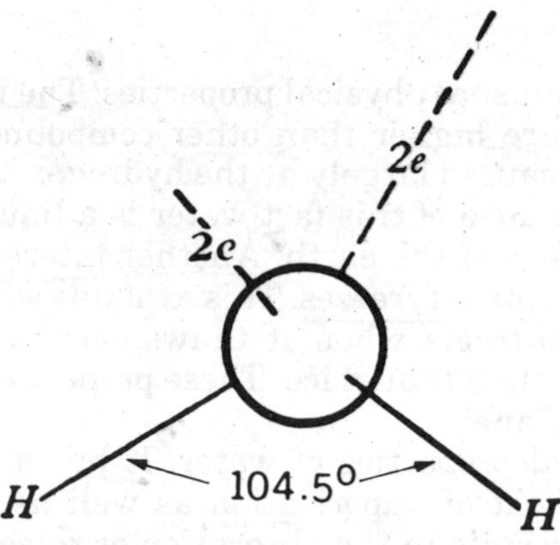

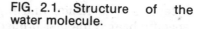

FIG. 2.1. Structure of the water molecule.

Chemical changes like the rusting of iron, the formation of clay in soils, or the splitting of cane sugar in the stomach are of the first type.

Physical changes, such as the melting of ice, evaporation in a boiler or from a lake, or the viscous resistance to flow in a pipe or stream, are of the second type (Hendricks, 1955).

The water molecule is a polar covalently bonded compound which is nonlinear and is made up of two hydrogen atoms which are bonded to oxygen by single covalent bonds. A very important feature of the water molecule is hydrogen bonding. Hydrogen bonding is the result of the dipole moment of the water molecule itself. This leads to a strong interaction between water molecules. The dipole moment is the product of the distance between the two average centers of positive and negative electricity and the electric charge. Hydrogen bonds are not confined to water alone. Compounds other than water molecules are linked with hydrogen bonds. It is likely that a major role is enacted by hydrogen bonding in water–solute interactions in physiological processes.

The two-dimensional associations of water molecules are shown.

```
                                 H
                                 |
H—O····H—O·····H—O
   |            |
   H            H
                ⋮
                O—H
                |
                H
```

Three-dimensional arrangements can also form and coexist.

Physical Properties of Water

Water has a number of rather unusual physical properties. The melting and boiling points of water are higher than other compounds of similar molecular weight. This is caused largely by the hydrogen bonds between the water molecules. Because of this fact water is a liquid in the intermediate temperature areas of the earth. Another interesting property of water is its expansion when it freezes. This explains why ice floats. It also means that ice contracts when it thaws, so that the specific gravity of water is higher than that of ice. These properties are highly important to life on this planet.

These are not the only unusual properties of water. It has a high latent heat of fusion and high heat of vaporization as well as high specific heat. This last property results in the absorption or release of large amounts of heat with a comparatively small change in temperature. Latent heat of fusion and latent heat of vaporization are caused by hydrogen bonding between water molecules.

A change from the solid to the liquid state takes place at the melting point without a rise in temperature. The hydrogen bonds which hold the molecules of water in the crystal lattice are broken by 80 calories per gram, which is necessary to change 1 gram of ice at 0°C to liquid at the same temperature. This is known as the heat of fusion. As the temperature of liquid water rises, this progressive addition of energy causes a decrease in the number of hydrogen bonds, but the water stays liquid up to 100°C. At this stage more heat is necessary to convert it to the gaseous state. This latent heat of vaporization is 540 calories per gram; it is the number of calories needed to convert 1 gram of liquid to vapor at the boiling point.

The increase in density of water as contrasted with that of ice is caused by change in structure to produce a structure of greater density. This condition continues until the temperature of maximum density (4°C) is reached. As the temperature increases above this, the density decreases normally.

Water has a high dielectric constant. In liquid water this is 78.5 at 25°C. This is held to originate from the polarity of the separate molecules and their orientations. The dielectric constant is a measure of the relative effect of the medium on the force with which two oppositely charged plates attract one another.

The dielectric constant (ϵ) is equal to the ratio of the charges of electricity:

$$\epsilon = \frac{C_{\text{liquid}}}{C_{\text{air}}}$$

In this case C_{liquid} is the electrical capacity of the condenser filled with the liquid and C_{air} is that of the condenser filled with air.

Ice

In the crystal structure of ice, each of the oxygen atoms is surrounded in tetrahedral fashion by four additional oxygen atoms. The average distance between oxygen atoms in ice is 2.76 Å. The result of this is an hexagonal crystal lattice structure in ice, as shown in Fig. 2.2.

Ice has a low density because of this open structure. In this structure a specific hydrogen atom is closer to one of the oxygen atoms to which it is connected than to the other. In addition, two hydrogen atoms are bonded to one oxygen atom, and the bonds are strong. The melting of ice results in the breaking of some of the hydrogen bonds. The effect of this is that the water molecules pack together more closely in liquid water, which results in the higher density of the latter. During freezing, water expands almost 9%. However, concentrated sucrose solutions do not show expansion. This is important in the packaging of sweetened fruits for freezing preservation.

Water, as is well known, can exist in three forms, namely solid, liquid, and gas. The basis of freeze drying (lyophilization) depends on the fact that when ice is heated at pressures of less than 4.58 mm Hg, it goes directly into the gaseous form. Under these conditions the material is dried without returning to the liquid state. This technique produces dehydrated products usually of higher quality than those dried by other means. Because of the added cost, however, it is used mainly for the more expensive products.

Supercooling of water is another factor that must be considered. This is the cooling of the water below the temperature at which it normally freezes (crystallizes). The supercooled water immediately crystallizes when a small crystal of ice is introduced. When this happens the temperature rises to 0 °C. The size of the sample influences the temperature at which the water can be held in the supercooled state. Very pure water can be supercooled to −32 °C if the sample size is 1 ml. However, smaller volumes of very pure water can be cooled to somewhat lower

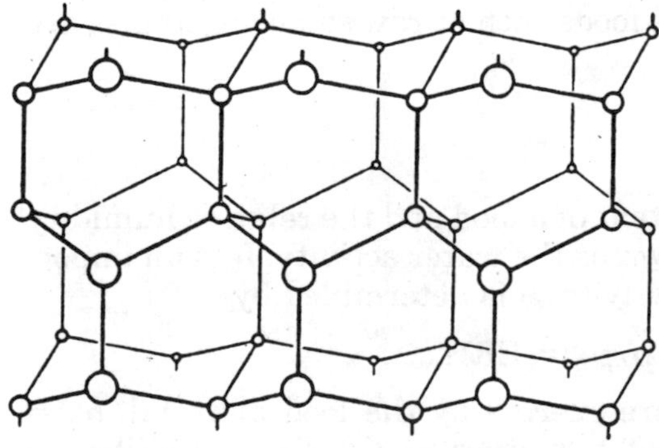

FIG. 2.2. Hexagonal crystal lattice structure in ice.

temperatures, and some experiments have indicated that water can be supercooled down to the range of −40 °C to −50 °C.

It is harder to start crystallization than to keep it going. The rate of the growth of ice crystals decreases with a lowering of the temperature. Rapid freezing results in the formation of small ice crystals while slow freezing produces large ones (Lee *et al.* 1946; Lee and Gortner 1949).

Many solutes have the ability to slow the growth of ice crystals (Lusena and Cook 1954). The reason for this is not clear. The most effective compounds studied that retard ice-crystal growth at 10 °C were alcohols and sugars. It was also found that at low concentrations proteins were equally as effective (Lusena 1955). The reason for this is not clear. Studies have shown (Lusena and Cook 1953) that although membranes may be completely permeable to liquids, they may be completely, partially, or impermeable to growing ice crystals. The increase in the growth of ice crystals in materials of known permeability increased with porosity. In addition, the rate of cooling, the composition of the membrane, and the concentration and properties of the solute in the aqueous phase affect it as well.

Directly related to the ice-crystal size at the end of the freezing period is the number of nuclei. Two types of nucleation are known, homogeneous and heterogeneous. Homogeneous nucleation, according to Matz (1965), is the result of random local fluctuations in the configuration and density of pure water, which leads to the chance spontaneous occurrence of regions having crystal-type orientation and dimensions. Heterogeneous nucleation is the result of the deposition of water molecules in crystalline array on some foreign particle (Matz 1965).

Crystals tend to enlarge in frozen products during storage, especially when storage temperatures fluctuate widely. Slow freezing of food materials such as fruits and vegetables results in the formation of large ice crystals in the tissues, mostly in the extracellular areas. With rapid freezing very small ice crystals are formed that are distributed throughout the tissues.

Many fruits, vegetables and prepared products can be frozen successfully. However, because of deterioration in texture and breakdown of emulsions during thawing some foods such as raw salad vegetables are not satisfactorily frozen.

Water Activity

The ratio of the moisture content of a food and the relative humidity of the air surrounding it is known as the water activity. It is an important characteristic. Water activity (a_w) is determined by

$$a_w = p/p_0 = \text{ERH}$$

where p = water vapor pressure exerted by the food material. p_0 = vapor pressure of pure water at temperature T_0, T_0 = equilibrium temperature of the system, and ERH = equilibrium relative humidity.

When the percentage of moisture is over 50%, the water activity is about 1.0. The water activity is lower than this figure when the moisture content is lower than 50%, and decreases rapidly with the lowering of the moisture content. The data can be prepared in the form of sorption isotherms. For hydroscopic products adsorption isotherms are needed, while in a study of the process of drying, desorption isotherms are necessary.

Figure 2.3 shows the moisture–vapor equilibrium humidity of food substances. A vapor pressure of almost that of pure water is exerted by 80–90% of the total water vapor present in a food. The effect on drying times and the stability of the food is affected by the last 10–20% of the water.

According of Labuza (1968), a sorption isotherm is a "plot of the amount of water absorbed as a function of the relative humidity or activity of the water space surrounding the material."

An adsorption isotherm is compiled from the data collected by placing completely dry samples into atmospheres of increasing relative humidity and noting the gain in weight caused by the water adsorbed.

A desorption isotherm is obtained from data obtained by placing wet samples under these same relative humidities and determining the weight of water lost.

Procedures for carrying out these experiments have been described by Taylor (1961), Karel and Nickerson (1964), and Hofer and Mohler (1962).

According to Labuza (1968), the isotherms in Fig. 2.4 can be shown in regions that depend on the state of water present. Region A corresponds to the adsorption of a monomolecular film of water. The adsorption of additional layers over the monolayer are region B, and the condensation of water in the pores of the material followed by dissolution of the soluble material present is region C. Labuza also noted that

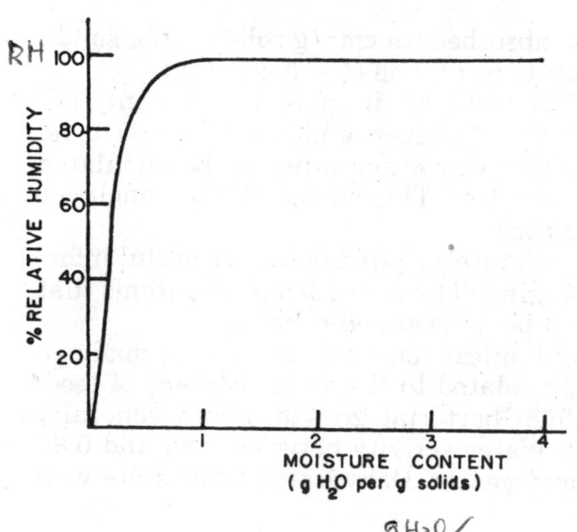

FIG. 2.3. Water activity in foods at different moisture content.
From Labuza *et al.* (1970). Used with permission of the Institute of Food Technologists, copyright owner.

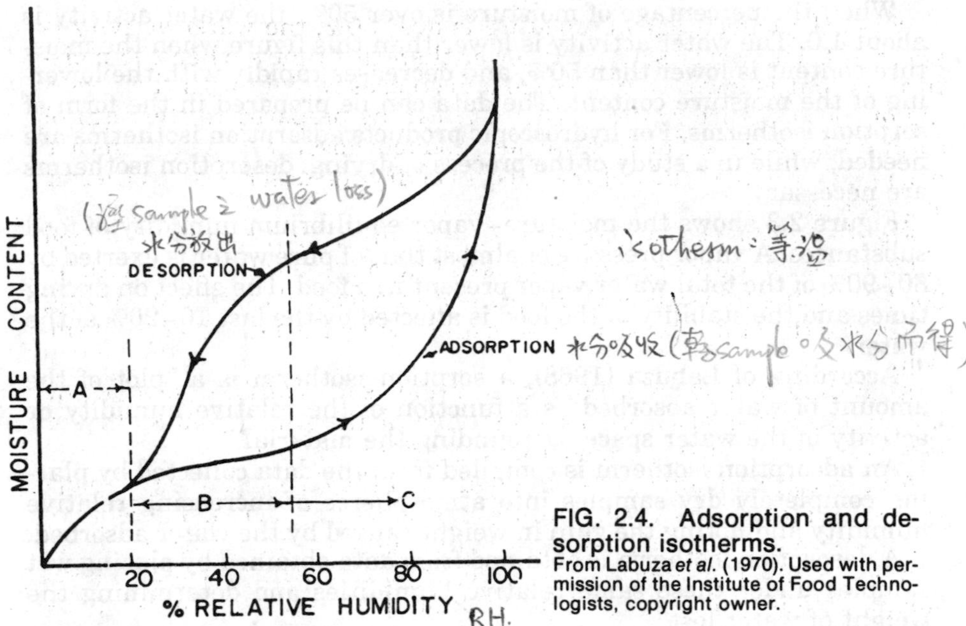

FIG. 2.4. Adsorption and desorption isotherms.
From Labuza *et al.* (1970). Used with permission of the Institute of Food Technologists, copyright owner.

if isotherms are to be used in predicting the drying time of a food substance, it is the desorption branch of the isotherm which is of importance.

The BET isotherm is the one most often used in food work. It was published in 1938 by Brunauer *et al.* The equation for this is

$$\frac{a}{(1 - a)v} = \frac{1}{v_m C} + \left[\frac{a(C-1)}{v_m C}\right]$$

where a = activity, v = volume absorbed (in cm^3/g solid or g/g solid), v_m = monolayer value (same units as v), and C = a constant.

It must be noted that the BET isotherm in general holds only between water activities of 0.1 to 0.5. However, a plot of $a/(1 - a)v$ vs. a gives a straight line. The monolayer coverage value can be calculated from the intercept and slope of this line. This means that a monolayer coverage of water can be calculated.

According to Salwin (1963), moisture-sorption data give useful information for processing and packaging dehydrated foods, and items that are to be packaged together can be determined reliably.

Deterioration as a result of chemical reactions, as well as spoilage from microorganisms, is closely related to the water content of food. According to Loncin *et al.* (1968) bacterial growth occurs generally below a water activity of 0.90. Water activity between 0.85 and 0.80 inhibits the growth of molds and yeasts. Dehydrated foods store well because of the limited amount of water present, although in certain products deterioration does occur slowly as a result of other reasons. As a practical matter dehydrated foods with a water content between 5 and 15% usually have satisfactory storage stability.

Water activity of food products can be lowered by removal of water (dehydration or drying) or by the addition of water-soluble substances such as salt or sugar. Obviously, the agent involved depends on the nature of the original product.

Nonenzymatic browning reactions in foods depend on water activity, and according to Loncin *et al.* (1968) they reach a maximum at *a* values of 0.6–0.7. The browning rate of most foods is slow at low humidities but increases to a maximum in foods in the range of intermediate moisture content. This is illustrated in Fig. 2.5 which shows the browning rate of pea soup mix as influenced by relative humidity (Labuza *et al.* 1970). Figure 2.6 shows the browning in a model system containing sucrose at three relative humidities.

Water activity is of importance in the inactivation of enzymes. When water activity decreases below 0.85, most enzymes are inactivated. These include amylases, peroxidases, and phenoloxidases. According to Loncin *et al.* (1968) lipases may remain active when the water activity values are as low as 0.3 or in some cases 0.1. Acker (1969) gives examples of the effect of water activity on some enzymatic reactions.

Acker (1963) showed that in various foods, some enzymatic reactions take place at very low water activities, but at higher water activities the reaction is accelerated. This is particularly true in the capillary condensation region. Capillary condensation region is one in which water condenses in the porous structure of the food and acts as a solvent for various solutes.

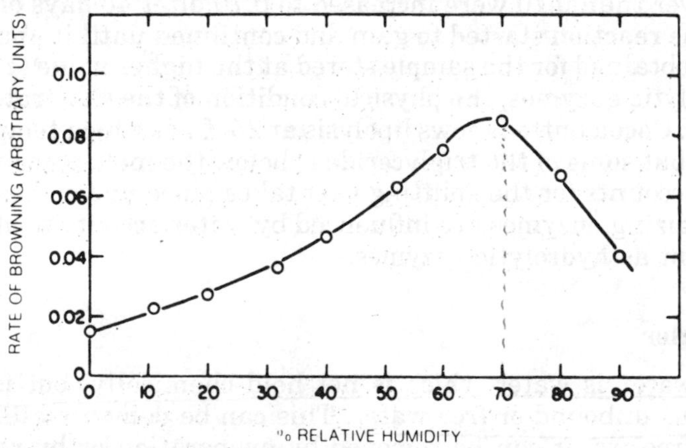

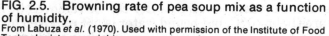

FIG. 2.5. Browning rate of pea soup mix as a function of humidity.
From Labuza *et al.* (1970). Used with permission of the Institute of Food Technologists, copyright owner.

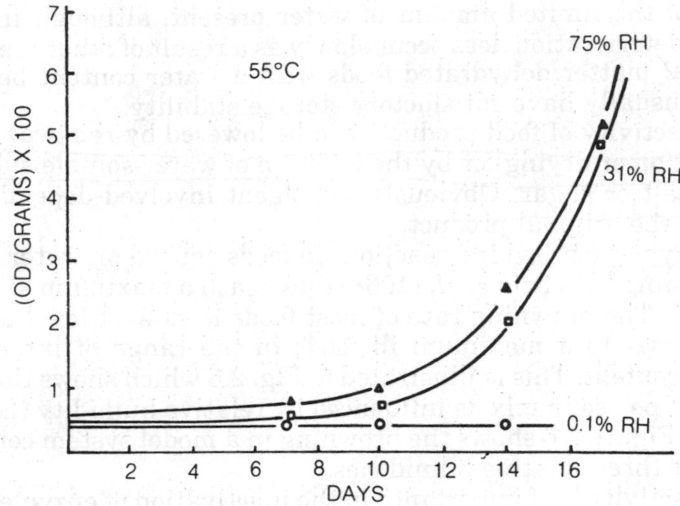

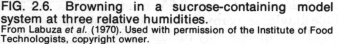

FIG. 2.6. Browning in a sucrose-containing model system at three relative humidities.
From Labuza *et al.* (1970). Used with permission of the Institute of Food Technologists, copyright owner.

The effect of water activity on splitting of lecithin by phospholipases in a mixture of ground barley with 2% lecithin at different water activities at 30°C was studied by Acker and Kaiser (1959). When the water activity was high, the rate of these enzyme activities was enough to effect measurable hydrolysis in hours or days. When water activity was low, hydrolysis did not take place even if the content of water was enough for complete hydrolysis. Hydrolysis at each different level of water activity tends to produce a different final value. When those a values lower than 0.70 were increased to 0.70 after 48 days of storage the enzyme reaction started to gain and continued until it approached the value obtained for the sample stored at the higher value ($a = 0.70$).

For lipolytic enzymes, the physical condition of the substrate is very important. Cocoa butter shows lipolysis at 25°C, at which temperature it is solid, but some of the triglyceride is below the melting point of the fat. This accounts for the splitting that takes place under these conditions. Oxidizing enzymes are influenced by water activity to about the same extent as hydrolytic enzymes.

Bound Water

Bound water is water which is not held chemically, but is not as available as unbound or free water. This can be demonstrated by the fact that free water can be removed by evaporation or by relatively light pressure. Furthermore, free water can be removed by freezing at temperatures below 0°C, but bound water is not removed by this treatment. Bound water is not available as a solvent. Lloyd and Moran

(1934) showed that some water in gelatin can be removed by pressure at 8000 lb, but the remaining (bound) requires 30,000 lb of pressure for removal. Further, the first part can be removed by freezing at −20°C, but not the second part. Gelatin is a protein that can absorb a great amount of water, and when a solution of this protein is allowed to stand it sets to a jelly-like mass. This property is the basis of its importance in foods. Meat proteins are definitely limited as to the amount of water that they can absorb. The formation of gels may be due to the orientation of the molecules so that water is confined in the spaces between the cohering masses of protein.

According to Karmas and DiMarco (1970), Differential Thermal Analysis (DTA) or Differential Enthalpic Analysis (DEA) seem to be uniquely suited for characterization of proteins with respect to water binding or retention. Fourteen amino acids exhibited peaks of bound water beyond the evaporation peak of free water. The acid and basic amino acids indicated no water retention. Exceptional water retention properties, 30–70% of the total water content as measured by peak areas, were exhibited by the nonpolar amino acids: isoleucine, leucine, methionine, and valine. The methyl and methylene groups of these nonpolar structures seem to be responsible for the water retention properties, probably because of hydrate formation. One explanation could be that as the temperature is raised to a critical point, the semicrystalline array around the nonpolar amino acid radicals collapses. Beef muscle tissue and egg albumin showed strong water retention properties.

Higasi (1955) noted that bound water is a quality that is neither well-defined nor exactly measurable. He further noted that the vapor pressure method used in its determination is to be regarded as the most trustworthy. The two methods described are (1) drying over sulfuric acid, and (2) use of a vacuum pump and manometer. The first of these is probably the best, but it takes a long time to run the determination. Because of this, studies on fresh material are practically impossible.

Shanbhag *et al.* (1970) found that bound water at room temperature could be accurately determined by means of nuclear magnetic resonance (NMR). The instrument used is a wideline NMR spectrometer. Hydrogen nuclei from the free water gave a negligible signal at the lowest radio-frequency attenuation, but a strong signal was given by the hydrogen from bound water. This permitted a new definition for bound water and a method for its quantitative determination, both at room temperature. These results are shown in Fig. 2.7.

SOLUTIONS

A solution is a mixture of two or more substances, which is physically and chemically homogeneous. The dissolved substance is in an ionic or

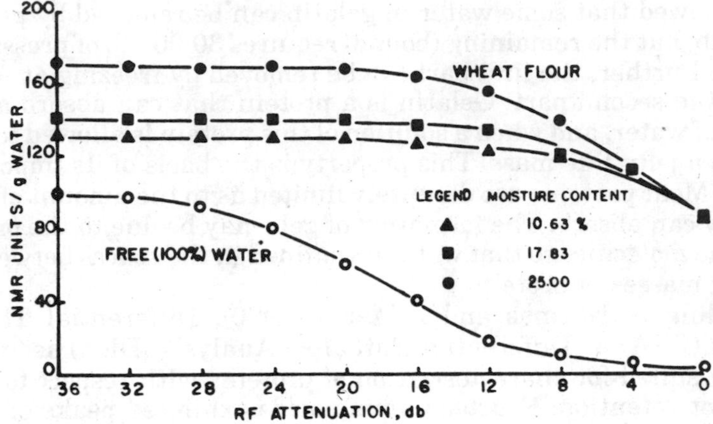

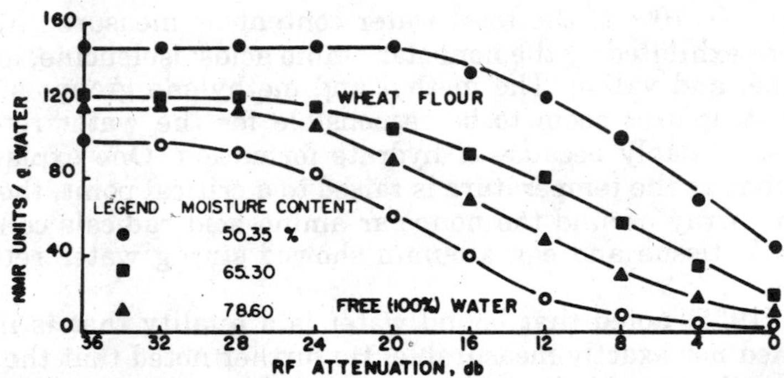

FIG. 2.7. Effect of radiofrequency attenuation on the NMR signal from water in wheat flour containing only bound water *(top)* or both bound and free water *(bottom)*. Curve for distilled water is shown for comparison.
From Shanbhag *et al.* (1970). Used with permission of the Institute of Food Technologists, copyright owner.

molecular state of subdivision. This differentiates the true solution from a colloidal suspension. Colloids will be considered in Chapter 3.

The solution is made up of the solvent and the solute. The food chemist is mainly interested in the following solutions: gas in liquid, liquid in liquid, and solid in liquid.

The tendency of molecules to enter solution governs the solubility of a substance. This tendency is known as solution pressure. When the solution pressure is high, the substance is soluble. Conversely, when the solution pressure is low the solubility of the substance is low. The molecules diffuse in all directions while they go into solution. Some of them go back to the surface of the solute. An increase in the number of molecules in solution causes the number of these molecules returning to the surface of the solute to increase. This tendency is known as diffusion pressure, and it is proportional to the concentration of dissolved molecules. A substance is soluble if the amount of dissolved solute is high relative to the amount of solvent. Most substances, but not all, show an increase in solubility with an increase in temperature. A solution is saturated when a state of equilibrium exists between the rate of return to the solid condition and the rate of solution. It is possible, also, to prepare a supersaturated solution with some substances. These are prepared by saturating at a high temperature and then cooling slowly. However, such solutions are not stable.

The boiling point of a solution of a nonvolatile solute is always higher than that of the pure solvent used. Also, the freezing point of the solution is lower than that of the pure solvent used. Molecular weights of substances can be determined by the lowering of the freezing point and the raising of the boiling point of the solutions as contrasted with the freezing or boiling points of the pure solvent.

It is possible to concentrate a solution by freezing out some of the solvent, which at first separates in the pure condition. However, as the temperature continues to be lowered, some of the solute freezes with the ice, and eventually the entire solution freezes.

Another property of solutions is osmotic pressure. When a solution and a quantity of the solvent are separated by a semipermeable membrane, the solvent passes through the membrane to the solution. The process is known as osmosis. A solution of sucrose in water that contains one gram molecular weight (1 mole) of the solute (sucrose) dissolved in 1000 grams of water would require 22.4 atmospheres of pressure exerted on the solution to affect equilibrium, that is, to stop the passage of water through the membrane to the solution. This is the osmotic pressure of the solution. It can be used to determine molecular weights.

Vapor pressure is another important property of volatile liquids. Vapor pressure increases as a liquid is being heated, and when the pressure is equal to that of the atmosphere above it the liquid boils. Nonvolatile solutes affect a depression of the vapor pressure of a solution. This results in the raising of the boiling point.

Surface Tension

The surface of water has a different behavior from that of the main body of the water, since it behaves as an elastic film which surrounds that body of water. This is known as the resistance of the surface film to rupture, and is called "surface tension." The presence of this surface film explains why a needle can be made to float on the surface of the water if it is carefully placed. However, any disturbance of the surface will cause it to sink.

Water has a high dielectric constant and is therefore a good solvent for salts. This means that the water has a high tendency to oppose the electrostatic forces of attraction between charged particles. This is expressed by

$$F = \frac{e_1 e_2}{Dr^2}$$

where F = the attractive force between two ions of opposite charge, e_1 and e_2 are the charges on the ions, D = dielectric constant, and r = the distance between them.

Basis of Solution

As sugar (a polar covalent compound) goes into solution, hydrogen bonds form between the polar groups of the sugar molecule and the molecules of water with the result that the sugar is held in solution. Sodium chloride acts a little differently. Sodium chloride is an ionic bonded compound. The polar water molecule's electrical attraction to the positive sodium ions is greater than the attraction that holds the ions together. The negative ions of the salt are pulled by the positive ends of the water molecule with the result that the crystal structure is disrupted. This, in turn, destroys the bonding forces between the ions, and the ions go into solution.

SUMMARY

Water is an important part of a large number of foods. It is essential to the very existence of life on this planet. The structure of the water molecule accounts for its special behavior.

A discussion of the physical properties of water is given in which the salient points are presented. This includes the solid state (ice) as well as the liquid.

A considerable discussion is given to water activity. The ratio of the moisture content of a food and the relative humidity of the air surrounding it is known as water activity. It is a very important characteristic. The sorption isotherm is a "plot of the amount of water adsorbed as a function of the relative humidity or activity of the water space surrounding the material." Adsorption and desorption isotherms are considered. The BET isotherm is most often used in food work.

A discussion on bound water includes the methods used for its determination. Bound water is water that is not held chemically, but is not as available as unbound or free water.

A solution is a mixture of two or more substances, which is physically and chemically homogeneous. Solutions and their properties are discussed.

BIBLIOGRAPHY

ACKER, L. 1963. Enzyme activity at low water contents. Rec. Adv. Food Sci. *3,* 239–247.

ACKER, L. 1969. Water activity and enzyme activity. Food Technol. *23,* 1257–1270.

ACKER, L., and KAISER, H. 1959. Influence of humidity on the course of enzymic reactions in low moisture foods II. Z. Lebensm. Unters. Forsch. *110,* 349–356.

BRUNAUER, S., EMMETT, P. H., and TELLER, E. 1938. Adsorption of gases in multimolecular layers. J. Am. Chem. Soc. *60,* 309–319.

CHIRIFE, J., and IGLESIAS, H. A. 1978. Equations for fitting water sorption isotherms of foods. Part I. A review. J. Food Technol. *13,* 159–174.

CHUNG, L., and TOLEDO, R. T. 1976. Predicting the water activity of multicomponent systems from water sorption isotherms in individual components. J. Food Sci. *41,* 922–927.

DUCKWORTH, R.B. 1975. Water Relations of Foods. Academic Press, New York.

EISENBERG, D., and KAUZMAN, W. 1969. The Structure and Properties of Water. Oxford University Press, London and New York.

HÄGERDAL, B., and MARTENS, H. 1976. Influence of water content on the stability of myoglobin to heat treatment. J. Food Sci. *41,* 933–937.

HAMM, R. 1962. The water binding capacity of mammalian muscle. VII. The theory of water binding. Z. Lebensm. Unters. Forsch. *116,* 120–126.

HENDRICKS, S. B. 1955. Necessary, convenient, commonplace. *In* Water, The Yearbook of Agriculture. U.S. Government Printing Office, Washington, DC.

HIGASI, K. 1955. Bound water. Res. Inst. Appl. Elec. Hokkaido Univ. Monograph Ser. *5,* 9–35.

HOFER, A.A., and MOHLER, M. 1962. Establishing sorption isotherms and application in the food industry. Mitt. Geb. Lebensmittelunters. Hyg. *53,* 274–290.

IGLESIAS, H.A., and CHIRIFE, J. 1976. Prediction of the effect of temperature on water sorption isotherms of food material. J. Food Technol. *11,* 109–116.

IGLESIAS, H.A., and CHIRIFE, J. 1976. A model for describing the water sorption behavior of foods. J. Food Sci. *41,* 984–992.

KAREL, M., and NICKERSON, J.T.R. 1964. Effects of relative humidity, air, and vacuum on browning of dehydrated orange juice. Food Technol. *18,* 1214–1218.

KARMAS, E., and DiMARCO, G.R. 1970. Dehydration profiles of amino acids and proteins. J. Food Sci. *35,* 615–617.

KLOTZ, I.M. 1965. Role of water structure in macromolecules. Fed. Proc. Fed. Am. Soc. Exp. Biol. *24,* S24–S33.

LABUZA, T.P. 1968. Sorption phenomena in foods. Food Technol. *22,* 263–272.

LABUZA, T.P. 1977. The properties of water in relationship to water binding in foods: A review. J. Food Process. Preserv. *1,* 167–190.

LABUZA, T.P., TANNENBAUM, S.R., and KAREL, M. 1970. Water content and stability of low-moisture and intermediate-moisture foods. Food Technol. *24,* 543–550.

LABUZA, T.P., ACOTT, K., TATINI, S.R., and LEE, R.Y. 1976. Water activity determination: a collaborative study of different methods. J. Food Sci. *41,* 910–917.

LEE, F.A., and GORTNER, W.A. 1949. Effect of freezing rate on vegetables. Refrig. Eng. *57,* 148–151, 184–187.

LEE, F.A., GORTNER, W.A., and WHITCOMBE, J. 1946. Effect of freezing rate on vegetables. Appearance, palatability, and vitamin content of peas and snap beans. Ind. Eng. Chem. *38,* 341-346.

LLOYD, D.T., and MORAN, T. 1934. Pressure and water relations of proteins. Proc. R. Soc. London Ser. A. *147,* 382–395.

LONCIN, M., BIMBENET, J.J., and LENGES, J. 1968. Influence of the activity of water on the spoilage of foodstuffs. J. Food Technol. *3,* 131–142.

LUSENA, C.V. 1955. Ice propagation in systems of biological interest. III. Effect of solutes on nucleation and growth of ice crystals. Arch. Biochem. Biophys. *57,* 277–284.

LUSENA, C.V., and COOK, W.H. 1953. Ice propagation in systems of biological interest. I. Effect of membranes and solutes in a model cell system. Arch. Biochem. Biophys. *46,* 232–240.

LUSENA, C.V., and COOK, W.H. 1954. Ice propagation in systems of biological interest. II. Effect of solutes at rapid cooling rates. Arch. Biochem. Biophys. *50,* 243–251.

MALONEY, J.F., LABUZA, T.P., WALLACE, D.H., and KAREL, M. 1966. Autoxidation of methyl lineolate in freeze-dried model systems. I. Effect of water on the autocatalyzed oxidation. J. Food Sci. *31,* 878–884.

MARTINEZ, F., and LABUZA, T.P. 1968. Effect of moisture content on the rate of deterioration of freeze-dried salmon. J. Food Sci. *33,* 241–247.

MATZ, S.A. 1965. Water in Foods. AVI Publishing Co., Westport, CT.

MOY, J.H., CHAN, K.C., and DOLLAR, A.M. 1971. Bound water in fruit products by the freezing method. J. Food Sci. *36,* 498–499.

PAULING, L. 1960. The Nature of the Chemical Bond. Cornell University Press, Ithaca, New York.

PIMENTEL, G.C., and McCLELLAN, A.L. 1960. The Hydrogen Bond. W.H. Freeman, San Francisco.

PIMENTEL, G.C., and SPRATLEY, R.D. 1969. Chemical Bonding Clarified through Quantum Mechanics. Holden-Day, San Francisco.

ROCKLAND, L.B. 1969. Water activity and storage stability. Food Technol. *23,* 1241–1251.

ROSS, K.D. 1975. Estimation of water activity on intermediate moisture foods. Food Technol. *29,* no. 3, 26–34.

SALWIN, H. 1963. Moisture levels required for stability in dehydrated foods. Food Technol. *17,* 1114–1123.

SALWIN, H., and SLAWSON, V. 1959. Moisture transfer in combinations of dehydrated foods. Food Technol. *13,* 715–718.

SARAVACOS, G.D. 1967. Effect of the drying method on the water sorption of dehydrated apple and potato. J. Food Sci. *32,* 81–84.

SHALLCROSS, F.V., and CARPENTER, G.B. 1957. X-Ray diffraction study of the cubic phase of ice. J. Chem. Phys. *26,* 782–784.

SHANBHAG, C., STEINBERG, M.P., and NELSON, E.I. 1970. Bound water defined and determined at constant temperature by wideline NMR. J. Food Sci. *35,* 612–615.

SLOAN, A.E., and LABUZA, T.P. 1976. Prediction of water activity lowering ability of food humectants at high a_w. J. Food Sci. *41,* 532–535.

SLOAN, A.E., WALETZKO, P.T., and LABUZA, T.P. 1976. Effect of order of mixing on a_w lowering ability of food humectants. J. Food Sci. *41,* 536–540.

SLOAN, A.E., SCHLUETER, D., and LABUZA, T.P. 1977. Effect of sequence and method of addition of humectants and water on a_w lowering ability in an IMF system. J. Food Sci. *42,* 94–96.

TAYLOR, A.A. 1961. Determination of moisture equilibriums in dehydrated foods. Food Technol. *15,* 536–540.

USDA. 1955. Water. Yearbook of Agriculture. U.S. Government Printing Office. Washington, DC.

USDA. 1963. Agriculture Handbook No. 8. U.S. Department of Agriculture, Washington, DC.

VAN DEN BERG, C., KAPER, F.S., WELDRING, J.A.G., and WOLTERS, I. 1975. Water binding by potato starch. J. Food Technol. *10,* 589–602.

VETROV, A.A., KONDRATOV, O.I., and YUHNEVICH, G.V. 1975. The spectrum of the intermolecular vibration of water. Aust. J. Chem. *28,* 2099–2107.

WARD, A.G. 1963. The nature of the forces between water and the macromolecular constituents of food. Rec. Adv. Food Sci. *3,* 207–214.

3

Colloids

What Are Colloids?
Behavior of Colloids in an Electric Field
Syneresis
Imbibition
Emulsions
Foams
Viscosity
Summary
Bibliography

Colloid science first came into being in 1861 when Thomas Graham published the results of his work on the rate of diffusion of substances in solution. He noted that solutions of substances that could be crystallized diffused rapidly through such membranes as parchment, while substances that did not produce crystals did not diffuse through these membranes at all. The crystallizable substances he called "crystalloids," while those noncrystallizable substances were designated "colloids."

WHAT ARE COLLOIDS?

As a result of his work, Graham concluded that colloids represented a kind of matter. We know now, however, that the term colloid designates a dispersed state of matter rather than a kind of matter. We know further that the size of the particles determines whether they are in a true solution or a colloidal solution. True solutions contain particles less than 1 nm in size. Colloidal solutions contain particles of 1–100 nm (10–1000 Å) in size. The particles of coarser dispersions are larger in size than these particles. The particles in a true or a colloidal solution are too small to be seen with the best microscope.

Two types of colloidal substances are known. One is lyophilic. This describes substances that easily form colloidal suspensions and have affinity for the dispersing medium. If the medium is water, the term hydrophilic is used. Examples of this type are gelatin and glues. The swelling of gelatin in water shows the great affinity between these molecules.

A lyophobic colloid has little affinity for the dispersing medium. If the medium is water the term used is hydrophobic. An example of such a colloid is colloidal gold.

Food chemists are interested mainly in three types of dispersions: solids in liquids; liquids in liquids, called emulsions; and gases in liquids, known as foams.

The two usual phases of the colloidal system are the dispersion medium and the dispersed phase. The particles of the dispersed phase are known also as micelles, or the discontinuous phase, whereas the dispersal medium is known as the continuous phase.

The depression of the freezing point and the increase of the boiling point, as well as osmotic pressure of crystalline substances in solution, were discussed in Chapter 2. These methods do not lend themselves satisfactorily for use on colloidal solutions, although molecular weights of certain proteins, albumin and hemoglobin, have been determined by means of osmotic pressure, but under very carefully controlled conditions. Other methods have been found to be more adapted to these determinations.

Behavior of Colloids in an Electrical Field

The charge on protein particles or molecules in colloidal systems results from the direct ionization of the protein.

Since proteins are amphoteric, they do not possess a net charge at their isoelectric (neutral) point, and, therefore, do not migrate at this point. On the alkaline side of this neutral point they are negative and migrate to the anode. On the acid side they have a positive charge and migrate toward the cathode. This situation can be demonstrated by electrophoresis.

Carbohydrate suspensions are usually neutral or slightly negative. They can, however, capture an ion by adsoprtion from an electrolyte in the liquid medium. Many different substances can acquire a surface charge in this way.

A double layer theory of the charge relations at the surface of colloidal particles was proposed by Helmholtz–Gouy. This theory considers the charged surface of the colloid to be surrounded by a layer of oppositely charged ions. This outer layer is further divided into an immobile layer at the surface of the colloidal particle. Over this is a mobile layer. The outer and inner layers can be oppositely charged. Some properties of colloids are dependent on the electrokinetic potential between the immobile and mobile layers. When this so-called zeta (ζ) potential is lowered to a definite value by the addition of electrolytes, it causes the double layers of the colloid particles to contract, often allowing the particles to aggregate and precipitate.

The precipitation of a hydrophilic colloid can be brought about by the use of electrolytes in high concentration. This is known as "salting-out." In low concentrations the solubility of many proteins can be increased by a process known as "salting-in."

The nature of the anion as well as the cation is important in the salting-out process. Magnesium sulfate ($MgSO_4$) or ammonium sulfate $(NH_4)_2SO_4$ in sufficient concentration are used to salt-out proteins. Because of its solubility $(NH_4)_2SO_4$ is very effective for this purpose. In low concentrations, $(NH_4)_2SO_4$ and $MgCl_2$ are of use in the salting-in process.

Hydrophilic colloids, which are more stable than hydrophobic colloids, can be used to stabilize the latter. A small amount of the hydrophilic colloid is all that is necessary to act as a protective colloid. Gums such as acacia and proteins such as egg albumin and gelatin are used for this purpose.

Gels are semisolid systems that possess high viscosities. They exhibit a behavior similar to that of solids. Gels tend to maintain their form under the stress of their own weight when removed from a mold. Also, they show strain under mechanical stress. Gels will return to the original shape after mild distortion. They show two distinct differences from solids. First, diffusion of substances soluble in the pure solvent takes place at the same rate in the gel as in the pure solvent. Second, chemical reactions can take place in gels at about the same rate as in the solvent. Gels can be disrupted by mechanical action. However, some of them will set again when the agitation stops. Such gels are called "thixotropic." Thixotropy, then, is an isothermal sol–gel interchange that is the result of agitation. In many food products, thixotropy is important. When measurements are made, care must be taken to be certain that results are not affected by thixotropic changes.

Heat-reversible gels are those that return to the solstate when heat is applied. Gelatin solution is such a gel. Sols that form gels which cannot be reversed are known also. Egg white is an example. Coagulation takes place when it is heated.

The bulk of the water in a gel is mechanically immobilized. Since it acts like free water, it can be removed from the gel by means of vacuum or dry air evaporation. As a result of the removal of the water, the gel shrinks considerably, and becomes hard and dry. In this final state it is known as a xerogel. The food gels, gelatin and pectin, produce xerogels that are almost fully reversible in the usual dispersing medium.

Gelatin desserts, pectin in fruit jellies, and modified starches in such products as pie fillings and thickeners in such products as gravies are well-known food uses. Because pectin requires sugar to produce gel, pectin will not work in artificially sweetened jellies, but such substances as agar-agar and certain gums can be used.

SYNERESIS

When a liquid, which is a very dilute solution, exudes, or is released from the surface of a gel, this is known as syneresis. This results from the aqueous phase component of the gel and is independent of the vapor pressure on the system. When syneresis takes place, the volume of the gel shrinks. When a gel is at its isoelectric point, syneresis is maximum, that is, it is influenced by pH. Syneresis is also affected by the nature of the dispersed phase.

IMBIBITION

Imbibition is the absorption of water by a number of substances the result of which is an increase in volume. Some of the factors that influence the process of imbibition are pH and temperature. Slight acidity implements the activity of proteins in this process. When exposed to temperatures over 30°C gelatin and agar gels will swell until they become liquid.

EMULSIONS

An emulsion is a colloidal system made up of two mutually insoluble liquids, one of which is dispersed as droplets in the other. Since such systems tend to be unstable, an emulsifying agent is necessary. This agent holds the droplets apart so that they do not coalesce into a continuous liquid. In short, it insures the permanence of the emulsion.

It is possible to have both oil-in-water and water-in-oil emulsions. Since the adsorption of the emulsifying agent must take place at an interface between the two phases, the resulting surface orientation controls the type of emulsion obtained.

While soap makes an excellent emulsifying agent, it is not, for obvious reasons, used in food products. In foods, the emulsifying agents most likely to be used are such substances as gums and proteins. These emulsifiers are mixtures of several compounds rather than individual substances. The oils in foods, of course, are mixtures, and ordinarily the aqueous phase contains substances such as sugars and/or salts, which may have an effect on the emulsion. Water-soluble substances such as gums, dextrins, proteins, and lecithin favor oil–water emulsions, while oil-soluble substances such as cholesterol favor water–oil emulsions.

The type of emulsion obtained is affected by the ratio of the two phases. When water and oil are shaken together both water–oil and oil–water emulsions result. If the amount of water is much larger than the oil, the oil–water becomes the more stable of the two. Also important is the condition of shaking. Continuous shaking is less effective than intermittent shaking. The period of agitation is necessary to give the emulsifying agent a chance to adsorb on the newly formed surface. Temperature is important also; since higher temperature decreases surface tension and viscosity, emulsification takes place more easily.

Food emulsions are generally complex. However, the theory of emulsions can be illustrated by simply using a soap as the emulsifying agent. Palmitic acid occurs in many fats; the corresponding soap is sodium palmitate ($C_{15}H_{31}COONa$). This soap contains the hydrocarbon radical $C_{15}H_{31}$, which is oil-soluble, and COONa, which is water-soluble. The hydrocarbon radical dissolves in the oil, and the COONa radical goes into the water, ionizing to form $-COO^-$ and Na^+ ions. The result is a negative surface charge in balance with the positive charge of the sodium ions in the water surrounding it. Since the droplets of oil are negatively charged, they repel one another, preventing the droplets from coalescing. This, in turn, stabilizes the emulsion.

BREAKING OF EMULSIONS

It is necessary to know the conditions under which emulsions remain stable, in order to determine how to eliminate a troublesome emulsion.

Methods used to break an emulsion are either physical or chemical. Breaking down an undesirable emulsion is often difficult.

Of the chemical methods, altering the pH and the alteration or destruction of the emulsifier are important. Also, using an oil–water emulsifying agent in a water–oil emulsion will destabilize the emulsion.

Some physical methods include agitation such as churning or stirring, filtration, centrifugation, freezing, heating (to lower the viscosity), addition of a liquid in which both phases are soluble, and addition of two different solids each preferentially wetted by one phase.

FOAMS

Foams are colloidal dispersions of gases in very viscous liquids. There is an interface between the liquid and the gas, similar to that encountered between the liquids in an emulsion. The surface tension at the interface is lowered with an increase in the concentration of dissolved material. If the viscosity at the interface is sufficiently increased by the increase in dissolving material, the foam will be stabilized. Under these conditions, a foam stabilizer is not needed. Colloidal substances, concentrated at the interface, give stable foams, as do proteins.

Whipped cream is a long-lasting foam. Marshmallows and meringues are foams that are stable because of the protein present—gelatin in the case of marshmallows and egg albumin in the meringues. The meringue is further improved by heat.

Foam on beer is stable because of a gum and dextrins extracted from the hops. These substances concentrate at the surface film and yield foam on the release of CO_2.

Pure liquids do not form foams. Foams may be broken by the addition of some of the aliphatic alcohols (C_6–C_9), ether, and toluene.

VISCOSITY

Viscosity, the resistance to flow because of internal friction, is an important property of hydrophilic sols.

Specific viscosity rather than absolute viscosity is what is usually determined. The flow through a capillary tube is measured in terms of time. However, thick emulsions cannot be determined with this equipment, and other methods have been devised. Some of these measure the time it takes for a rotating inner cylinder, or bob, or a rotating forked paddle in the emulsion to make 100 revolutions. This is compared against viscosities of a substance selected as a standard.

Water and other liquids in which the rate of shear and the shear stress are directly proportional are known as Newtonian liquids and a straight-line relationship exists between rate of shear and shear stress. While this is true of dilute suspensions and emulsions, it is not true of many food products and these are considered non-Newtonian. Such products produce "apparent viscosities."

If the flow of a given volume of sample through a tube is used to determine the viscosity of a liquid, the Ostwald viscometer can be used. From the data obtained, one makes use of the Poiseuille equation for the calculation of the results.

$$\eta = \frac{Pr^4\pi}{8\,vl}\,t$$

where η is the viscosity, v is volume of the liquid flowing through a capillary tube of length l and radius r. P is the pressure, and the whole takes place in time t.

It is possible to determine the viscosity of a liquid relative to water by measuring the time of flow of water through the capillary and the time of flow of the other liquid. The following equation will then give the relative viscosity.

$$\frac{\eta_1}{1} = \frac{d_1 t_1}{d_w t_w}$$

η_1 = relative viscosity of the liquid, d_1 = the density, t_1 = time of flow. The viscosity of the water is 1, d_w = its density, and t_w = its time of flow.

To determine absolute viscosity, the relative viscosity can be multiplied by the absolute viscosity of water at the given temperature. Fluidity is the reciprocal of viscosity.

The determination of viscosities of colloidal dispersions (hydrophilic colloids) shows that such dispersions are much more viscous than true solutions. While a 1% solution of sucrose is only 3–4% higher in viscosity than water, a 1% dispersion of starch has a viscosity 50% higher.

Viscosity is affected by the following factors.

1. Viscosity of colloidal solutions decreases with heat (increase in temperature)
2. The viscosity of a hydrophilic colloid increases with an increase in hydration.
3. Colloids with small particles show higher viscosity than one of the same concentration of larger particles. This is called the degree of dispersion.
4. Small amounts of nonelectrolytes may increase viscosity. Viscosity decreases by small amounts of electrolytes. An increase in viscosity usually results from the solution of a large amount of solids.

SUMMARY

The term "colloid" designates a dispersed state of matter. The size of the particles determines whether they are in a true solution or a colloidal suspension. Those less than 1 nm in size are in true solution.

The charge on protein particles in colloidal systems is the result of the direct ionization of the protein. Proteins are amphoteric.

Carbohydrate suspensions are usually neutral or slightly negative. However, they can capture an ion by adsorption from an electrolyte in the liquid medium.

A double layer theory of the charge relations at the surface of colloidal particles was proposed by Helmholtz-Gouy.

Hydrophilic colloids can be precipitated by the use of electrolytes in high concentration. This is "salting out".

Gels are semisolid systems that possess high viscosities. Diffusion of substances in the pure solvent takes place at the same rate in the gel as in the pure solvent. Also, chemical reactions take place at about the same rate as in the solvent.

Heat reversible gels are discussed.

Water in a gel acts like free water and can be removed by means of vacuum or dry air evaporation. As a result the gel shrinks to form a xerogel which is hard and dry.

Syneresis and imbibition are discussed.

An emulsion is a colloidal system made up of two mutually insoluble liquids, one of which is dispersed as droplets in the other. To avoid instability, an emulsifying agent is necessary.

Information is given on the making of an emulsion and in the breaking of a troublesome emulsion.

Foams are dispersions of gases in very viscous liquids. The similarity of foams to emulsions as well as their stabilization are discussed.

Viscosity, its determination, and factors which affect it are given consideration.

BIBLIOGRAPHY

ADAMSON, A.W. 1976. Physical Chemistry of Surfaces. 3rd Edition. Wiley (Interscience) Publishers, New York.

BECHER, P. 1966. Emulsions: Theory and Practice. ACS Monograph Ser. Reinhold Publishing Corp., New York.

BULL, H.B. 1964. An Introduction to Physical Biochemistry. F. A. Davis Co., Philadelphia, PA.

GRAHAM, H.D. (Editor) 1977. Food Colloids. AVI Publishing Co. Inc. Westport, CT.

JERGENSONS, B. and STRAUMANIS, K.E. 1962. A Short Textbook of Colloid Chemistry, 2nd Rev. Edition. Macmillan Co., New York.

SHELUDKO, A. 1966. Colloid Chemistry. Elsevier Publ. Co., Amsterdam.

SIMHA, R. 1940. The influence of Brownian movement on the viscosity of solutions. J. Phys. Chem. *44,* 25–34.

VAN WAZER, J.R., LYONS, J.W., KIM, K.Y., and COLWELL, R.E. 1963. Viscosity and Flow Measurement. A Laboratory Handbook of Rheology. Wiley (Interscience) Publishers, New York.

4

Carbohydrates

Introduction
Monosaccharides
Structure of Carbohydrates
Carbohydrate Reactions
Oligosaccharides
Polysaccharides
Digestion of Carbohydrates
Qualitative Tests
Quantitative Tests
Summary
Bibliography

INTRODUCTION

Carbohydrates are found throughout the world and, of all the biological substances in the plant kingdom other than water, they are present in the largest quantity. They are present in milk, in blood and tissues of animals and are of structural importance in plants and in the shells of such animals as the crab. They are extremely important as a component of foods, in which they are sources of energy, flavor, and bulk. The sugars and starches are sources of energy: sugars provide sweetness, and celluloses and other large molecules contribute to bulk.

Although the name suggests the hydration of carbon, this is not strictly the case. Carbohydrates are compounds of carbon, hydrogen, and oxygen, and usually, but not always, the hydrogen and oxygen are present in the ratio of 2:1, as in water. It is true, however, that crystalline sugars may possess water of crystallization, which is given up on heating, but the carbohydrate part of the molecule gives up water only under drastic heat. Chemically, carbohydrates are polyhydroxyaldehydes or ketones, or condensation products or derivatives of them. Sugars, the simpler carbohydrates, are soluble in water and the solution is sweet in taste. The complex carbohydrates are relatively insoluble in water and are broken down to sugars during hydrolysis by means of acids or enzymes. Polysaccharides, on the other hand, are colloidal and dispersible under the proper conditions in water. Such suspensions are tasteless. Cellulose is not digestible in the human body, but contributes to bulk.

Carbohydrates are formed in plants as a result of the process of photosynthesis. This takes place by the transfer of the energy of the sun through the medium of the catalyst chlorophyll acting on the raw materials, carbon dioxide and water, to produce sugars by a rather complex process discussed in Chapter 1. Conversion into other carbohydrates also occurs and these include monosaccharides, oligosaccharides which include disaccharides and other sugars up to about 10 simple sugars in combination, and large compounds such as hemicelluloses, cellulose, starches, pectins, gums, and mucilages.

The carbohydrates may be organized as follows.

- I. Monosaccharides, the simple sugars.
  - A. Pentoses (arabinose, ribose, xylose)
  - B. Hexoses
    - a. Aldehexoses (galactose, glucose, mannose)
    - b. Ketohexose (fructose)
- II. Oligosaccharides.
  - A. Disaccharides
    - a. Reducing (lactose, maltose)
    - b. Nonreducing (sucrose)
  - B. Trisaccharides
    - a. Nonreducing (gentianose, raffinose)
  - C. Tetrasaccharides
    - a. Nonreducing (stachyose)
- III. Polysaccharides
  - A. Homopolysaccharides (single or one kind of monosaccharide unit)
    - a. Penotsans (arabans, xylans)
    - b. Hexosans
      1. Glucosans (cellulose, dextrin, glycogen, starch)
      2. Fructosan (inulin)
      3. Mannan
      4. Galactan
  - B. Heteropolysaccharides (two or more monosaccharide units on hydrolysis)
    - a. Gums, mucilages, pectins.
  - C. Nitrogen containing (chitin)

MONOSACCHARIDES

Monosaccharides are the simple sugars and are classified according to the number of carbon atoms in the molecule. While the organic chemists are interested in trioses, tetroses, as well as the others, the food chemist is primarily concerned with hexoses (six carbon atoms) and to a lesser extent with the pentoses (five carbon atoms in the molecule). Early research work on the sugars was difficult because sugars are hard to crystallize since they tend to form syrups.

Pentoses

Pentoses are widely distributed as constituents of complex polysaccharides. These polysaccharides are known as pentosans and they yield pentose sugars on hydrolysis. Free pentoses are found only in very small amounts. Pentoses are not fermentable by yeasts and are not of much use as an energy source in the diet.

L-Arabinose. It is found as a constituent of many gums, hemicelluloses, mucilages, and pectins. It has been detected in a variety of fruits, among which are apples, figs, some grapes, and grapefruit. It can be prepared by hydrolysis of gums such as gum arabic and beet pulp with dilute sulfuric acid.

D-Ribose. Ribose is a very important sugar. It is a part of the molecule of riboflavin, which is vitamin B_2. It is also a constituent of nucleic acids and the nucleotide coenzymes. 2-Deoxyribose, a part of deoxyribonucleic acid (DNA), is a derivative of ribose.

D-Xylose. Also known as wood sugar, this sugar occurs in straw, corn cobs, wood gum, and bran as pentosans. It has been found in some of the stone fruits, including cherries, peaches, and plums, as well as in pears.

Hexoses

Glucose and fructose are found in large quantities in the free state under natural conditions. They are also found in various types of combined substances. Some of these sugars are glucose, fructose, galactose, and mannose.

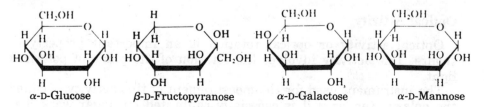

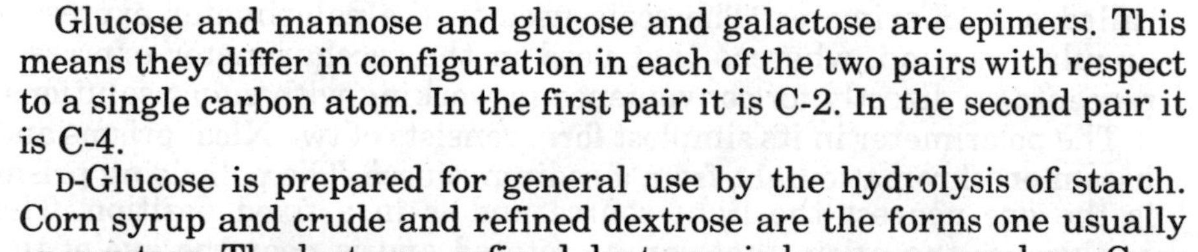

α-D-Glucose β-D-Fructopyranose α-D-Galactose α-D-Mannose

Glucose and mannose and glucose and galactose are epimers. This means they differ in configuration in each of the two pairs with respect to a single carbon atom. In the first pair it is C-2. In the second pair it is C-4.

D-Glucose is prepared for general use by the hydrolysis of starch. Corn syrup and crude and refined dextrose are the forms one usually encounters. The last one, refined dextrose, is known as cerelose. Corn syrup is not, of course, pure glucose, but contains in addition maltose and other oligosaccharides.

Glucose. This sugar occurs as a free sugar in plant leaves, roots, stems, flowers, and ripe fruits. It is present in the sap of the sugar maple from which maple syrup is made. It is found in the blood of most animals. In the combined form it is found in oligosaccharides and polysaccharides such as sucrose, starch, and cellulose.

D-Fructose (Levulose). Gottschalk (1945) found that the β-D-fructopyranose isomer is not fermented by bakers' yeast, while the β-D-fructofuranose is. Fructose in the combined form (i.e., sucrose, inulin, and some phosphate esters) exists as a furanose.

The concentration of fructose in fresh apples and pears was found to be considerably higher than the concentration of glucose. In most other fruits the concentration of glucose is somewhat higher than that of fructose. In a few fruits, namely grapes and strawberries, it is about equal. Fructose is present in honey, molasses, and ripe fruits. It can be prepared by the hydrolysis of inulin, a polysaccharide of fructose that also contains a small amount of glucose.

D-Galactose. Galactose is ordinarily found combined as a constituent of oligosaccharides and polysaccharides. The hydrolysis of lactose, gums, mucilages, raffinose, and stachyose yields galactose. Heating lactose with 2% sulfuric acid yields, under usual conditions, the stable α-D-galactose isomer.

D-Mannose. Mannose is found in the gummy plant exudate known as mannan, a polysaccharide. Mannose is widely distributed in the form of mannosans, a polymeric form. Mannitol is the alcohol derived from mannose. It is found in brown algae, in the bark and leaves of a number of trees, in cauliflower, onion, and pineapple.

Optical Activity

Optical activity or optical rotation is an important property of sugars. This activity involves the rotation of the plane of polarized light.

The instrument used for the measurement of this activity is called the polarimeter, or if it is especially calibrated for sugar work it is called a saccharimeter. The scale used in the polarimeter expresses angular rotation, whereas that used in the saccharimeter expresses percentages directly for convenience in working with sugar solutions.

The polarimeter in its simplest form consists of two Nicol prisms and uses monochromatic light from a sodium source. The polarizing prism is the one nearest the light source and is in a fixed position. The analyzer is the prism that can be rotated and is near the eye of the observer. Essentially all the radiation will pass through if the prisms are arranged so that the optical axes are in the same plane. The radiation will be totally absorbed if the optical axis of the polarizer is at right angles to that of the analyzer. This is known as total extinction. The scale of the instrument indicates the number of degrees the analyzer is rotated. The point where the two Nicol prisms are crossed without a sample in the polarized beam may be used to set the zero point. The plane of polarized light is rotated to the right or to the left when an optically active compound is placed between the prisms. If, in order to obtain total extinction, it is necessary to rotate the analyzer to the left, the compound is levorotatory (−). If it must be rotated to the right the compound is dextrorotatory (+). The angular rotation of the optically active compounds is equal to the angle through which the analyzer must be turned.

The concentration of an optically active compound in solution is directly proportional to the angular rotation of the compound. The angular rotation is proportional also to the rotating power of the substance and to the length of the solution through which the light passes. The specific rotation of a substance $[\alpha]$ is defined as the angular rotation in degrees of a solution which contains 100 g of solute in 100 ml of solution when determined in a true 1 decimeter (dm) long. The result is expressed as

$$[\alpha] = \frac{100 \times A}{l \times c}$$

where A (in degrees) is the rotation (plus or minus), l is the length of the tube in decimeters, and c is the concentration in grams of solute in 100 ml of solution.

Specific rotation is affected by temperature and the wavelength of the light.

Specific rotation is indicated as

$$[\alpha]_{\mathrm{D}}^{20} = \text{rotation value (solvent used)}$$

This is specific rotation at 20°C with D line of sodium.

A dextrorotatory compound is designated by a + or *d*, and a levorotatory one is designated by a − or *l*.

Optical rotation is brought about by carbon atoms which have four different groups attached to them. These are called asymmetric carbon atoms. If two or more groups attached to the carbon atom are the same, the carbon atom is symmetric, and the plane of polarized light will not be rotated. The R_1 CR_2HOH group of the sugars has the asymmetric carbon atom. In the structural formulas these are sometimes designated with an asterisk when asymmetry is being discussed. The open chain formula for glucose has four asymmetric carbon atoms and the ring formula has five.

A mixture of equal parts each of dextro and levo forms of a compound is optically inactive. Each rotates the polarized light the same number of degrees, but in opposite directions. The result is no optical activity. This is known as a racemic mixture. A racemic mixture of this type can be resolved or separated into the two optically active compounds. Another type of compound, the *meso,* cannot be separated because of its structure. The molecule has a plane of symmetry, that is, it can be divided into two mirror image halves. It is, therefore, optically inactive.

The *dextro-* and *levo*-tartaric acids are enantiomorphs, that is, optical isomers, and the molecules have two of the substituents on each asymmetric carbon atom reversed in space and are mirror images. These are able to rotate polarized light in equal but opposite directions. *Dextro-* and *levo*-tartaric acids are enantiomers, *meso*-tartaric acid is diastereoisomeric with these. This discussion is illustrated with the structures of tartaric acid.

```
    COOH                COOH                COOH
    |                   |                   |
H—C—OH             HO—C—H               H—C—OH
    |                   |                   |
H—C—OH              H—C—OH             HO—C—H
    |                   |                   |
    COOH                COOH                COOH
```

meso-Tartaric acid D(−)-Tartaric acid L(+)-Tartaric acid

The small capital letters D and L are enantiomeric designations used in connection with sugars. These do not indicate optical rotation, but instead the position of the hydroxyl attached to the highest numbered asymmetric carbon atom. The compounds designated as D have the hydroxyl group of the asymmetric carbon atom farthest from the aldehyde group projected to the right. Those designated as L have the hydroxyl group projected to the left.

Glyceraldehyde is the simplest aldose, with one asymmetric carbon atom and, therefore, two isomers. It is taken as the reference substance, and the D and L forms are given along with the open chain structures for D-glucose and L-glucose.

```
   CHO                 CHO
   |                   |
  HC*OH           HO—C*H
   |                   |
   CH2OH               CH2OH
```

D(+)-Glyceraldehyde L(−)-Glyceraldehyde

```
   H—C=O           O=C—H
     |               |
   H—C—OH         HO—C—H
     |               |
  HO—C—H           H—C—OH
     |               |
   H—C—OH         HO—C—H
     |               |
   H—C—OH         HO—C—H
     |               |
     CH2OH           CH2OH
```

D-Glucose L-Glucose

According to Van't Hoff (1875) the possible number of isomers is 2^n, where n is the number of asymmetric carbon atoms present in the compound. For glucose, which contains four asymmetric carbon atoms, the number of possible isomers is 16. Of these eight belong to the D-series. All of them have been isolated from nature or synthesized. D-Glucose, D-mannose, and D-galactose have been found in nature. These and the ketohexose, D-fructose, are the only sugars that can be fermented by yeast.

Since the corresponding ketones contain one less asymmetric carbon atom than the aldehydes, fructose contains three asymmetric carbon atoms, and, therefore, has eight isomers. As in the case of the aldoses, half of them belong to the L series and half to the D. D-Fructose is the principal natural ketose.

STRUCTURE OF CARBOHYDRATES

Information on the structure of sugars is available in the book by Shallenberger and Birch (1975) and on carbohydrates in Pigman and Horton (1972). Structure of the molecule is important since it governs the reactions and properties of the compound. Of importance are the number of carbon atoms present, the presence of an aldehyde or a ketone group, and the positions of the hydroxyl groups in the molecule. For the carbohydrates of high molecular weight one must consider the number and kind of units in the chain as well as the position of the linkages connecting the units.

The structures of D-glucose and L-glucose were the first structures drawn for these sugars. This open chain formula does not explain all of the scientifically observed facts. First, it does not explain the failure to obtain a color test with Schiff's reagent. Since this is a test for aldehydes, it indicates an alteration in the aldehyde group. Second, a change takes place in the optical rotation of the freshly prepared solutions of many sugars when they are allowed to stand. This change, known as mutarotation, cannot be explained by the open chain formula. The preparation of the α and β crystalline isomers of glucose produced further evidence.

Mutarotation

Mutarotation, the phenomenon in which a solution undergoes a change in value of the optical rotation when it is allowed to stand after dissolution, has been known for many years. A freshly prepared solution of α-D-glucose has an optical rotation of $[\alpha]_D^{20}$ + 112.2°; a freshly prepared solution of β-D-glucose has an optical rotation of $[\alpha]_D^{20}$ + 18.7°. When either of these solutions is allowed to stand, the reading changes to $[\alpha]_D^{20}$ + 52.7°. This change is the result of the equilibrium between α and β forms of the pyranose ring structure. These are probably in equilibrium with the straight chain form.

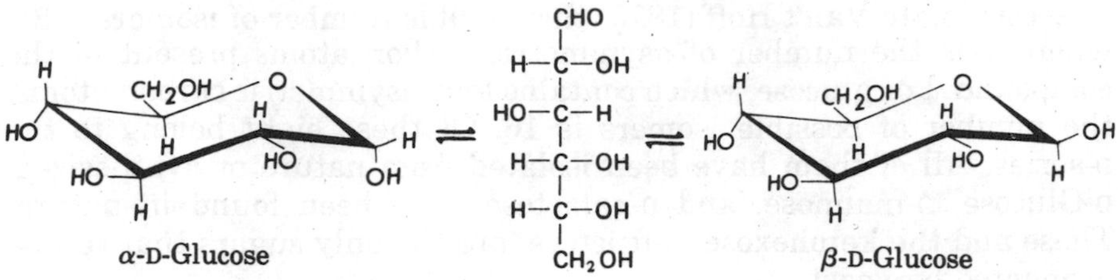

α-D-Glucose ⇌ Open chain form ⇌ β-D-Glucose

CHO
H—C—OH
HO—C—H
H—C—OH
H—C—OH
CH_2OH

Open chain form

When D-glucose mutarotates in water solution a mixture of about 36% α-D-glucopyranose and 64% β-D-glucopyranose, with probably very small amounts of free aldehyde form, is obtained. This is simple mutarotation. Simple mutarotation in sugars, therefore, is usually the interconversion of the α-form of that sugar to its β-form, or the reverse. Complex mutarotation in sugars is usually a transformation of ring forms. Acids and bases catalyze or accelerate the mutarotation of sugars. The enzyme mutarotase acts to catalyze the mutarotation of glucose.

The existence of mutarotation is one of the facts which lead to the Fischer–Tollens formulas. The ring formulas were necessary to explain mutarotation. It should be realized that the aldehyde or ketone groups remain potentially the functional group.

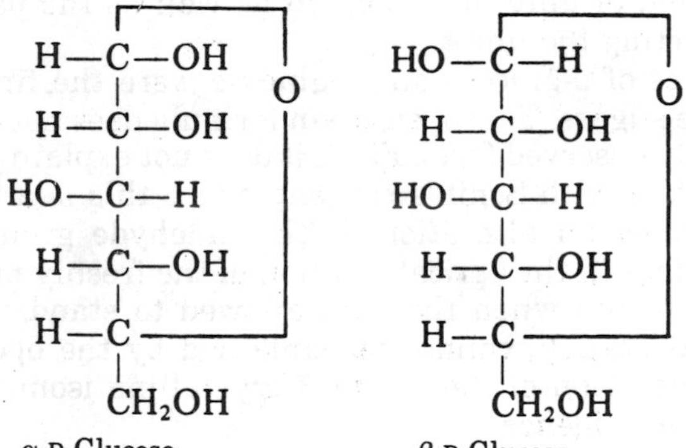

α-D-Glucose	β-D-Glucose
H—C—OH	HO—C—H
H—C—OH	H—C—OH
HO—C—H	HO—C—H
H—C—OH	H—C—OH
H—C (—O— to C-1)	H—C (—O— to C-1)
CH_2OH	CH_2OH

α-D-Glucose
$[\alpha]^{20}_{C} = +112.2$

β-D-Glucose
$[\alpha]^{20}_{D} = +18.7$

From the formulas it is clear that sugars exist in the cyclic form. This appears as the cyclic hemiacetal or hemiketal forms. The six-membered ring is the form in which most of the simple sugars are known.

The stereochemistry at C-1 is the only point of difference in the two cyclic isomers of glucose. The acetal carbon is the former aldehyde carbon. When this takes place the former aldehyde carbon becomes asymmetric. These two cyclic isomers are known as anomers. The acetal carbon is known as the anomeric carbon.

The size of the ring became important with the introduction of this type of formula in sugar chemistry. Carbon-4 or carbon-5 can be involved in the formation of a cyclic hemiacetal. A furanose ring is created with four carbon atoms and one oxygen atom. A pyranose ring is formed from five carbon atoms and one oxygen atom. These names come from furan and pyran, the five- and six-membered cyclic ethers. The pyranose form is ordinarily found in nature and is the more stable of the two. The Fischer–Tollens formula for glucose shows a chain of carbon atoms in which positions 1 and 5 are joined by an oxygen bridge. A long chain of carbon atoms is not feasible for such an oxygen bridge; therefore this molecular configuration cannot be the true one. The C-1 and C-5 atoms must be closer together in an existing configuration. The Haworth formulas correct this problem. In these structures, the plane

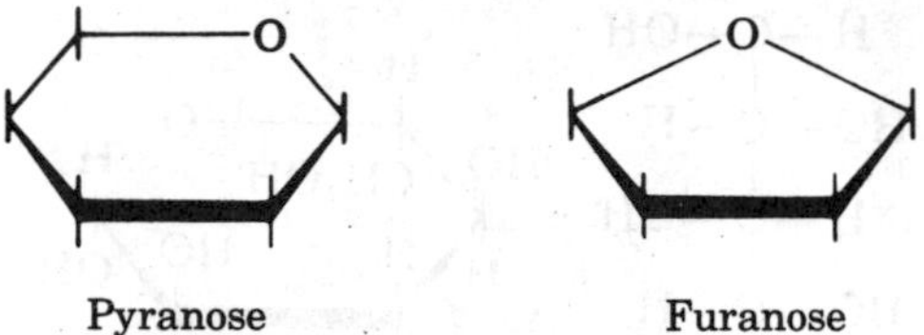

Pyranose Furanose

of the ring is perpendicular to the plane of the page, and the thick lines of the ring are in front of the plane of the page while the thin lines are behind the plane of the page. The difference between the Fischer–Tollens structural formula and the Haworth formula is that any attached group on the right of the carbon chain in the Fischer–Tollens formula is shown below the plane of the ring, while groups to the left are shown above the plane of the ring.

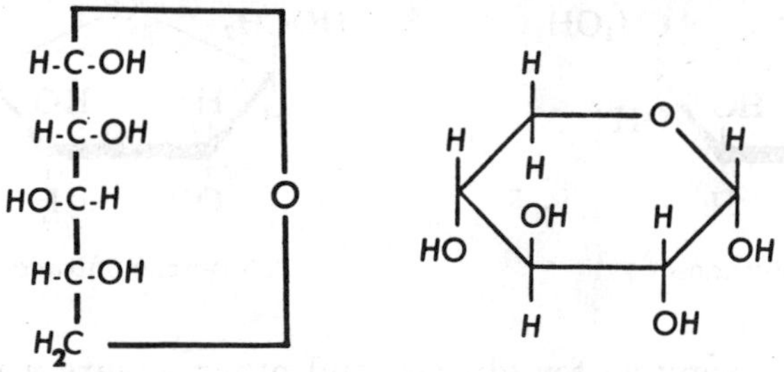

H-C-OH
H-C-OH
HO-C-H
H-C-OH
H_2C
O

α-D-Xylose

α-D-Xylopyranose

It must be determined whether the carbon atoms of a hexose sugar which are not in the ring are above or below the plane of the ring. In α-D-glucose the ring is to the right and the carbon is up. If the ring is to the left, the carbon will be below the plane of the ring.

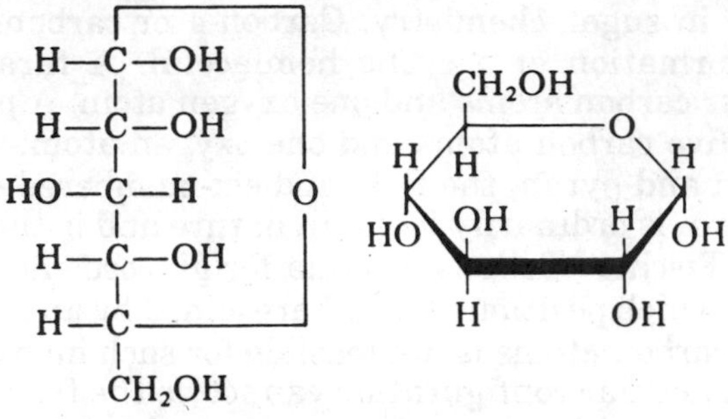

α-D-Glucopyranose

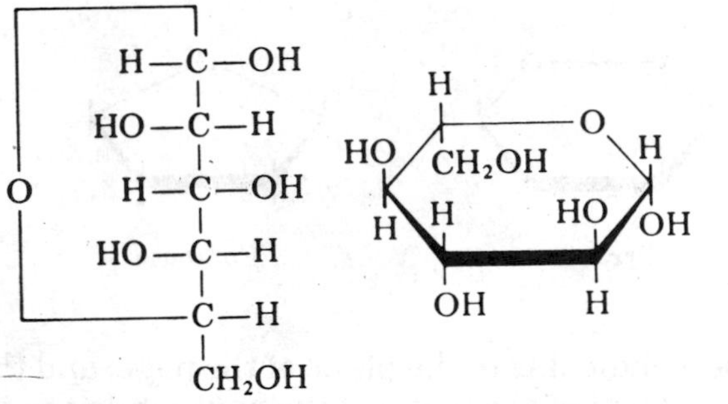

β-L-Glucopyranose

For a ketohexose in the furanose form the position of C-6 is above, but C-1 is above if the formula is in the α form and below in the β form.

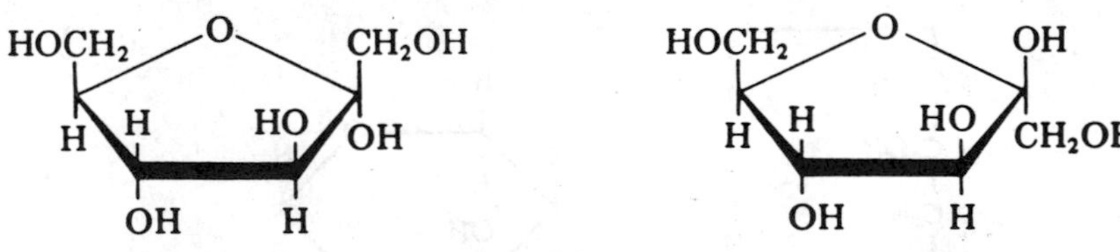

α-D-Fructofuranose

β-D-Fructofuranose

Although the formulas for glucose and other sugars were greatly improved by using the Haworth structure over the Fischer–Tollens structure, further improvement was necessary because the Haworth structure does not give an exact picture of the furanose or pyranose rings. In the Haworth structure all of the atoms are placed on a single plane; this is an oversimplification. The actual molecule cannot exist this way in space because the valence angle of the carbon atom does not permit a stable planar arrangement of the atoms. It was demonstrated that the five-membered furanose ring was nonplanar and strained, but the six-membered pyranose ring was without strain and presented itself in space as either the "chair" or "boat" arrangements. The chair is more stable because it is the structure of lower energy. It is possible to compare the pyranose ring with cyclohexane, which is capable of existing in chair and boat forms.

Chair form Boat form

In work on structural models of cyclohexane it was shown that the arrangement of hydrogen atoms on the ring exists in two ways: those on C–H bonds which are parallel to the plane of the ring (equatorial), and those on C–H bonds perpendicular to this plane (axial). The chair

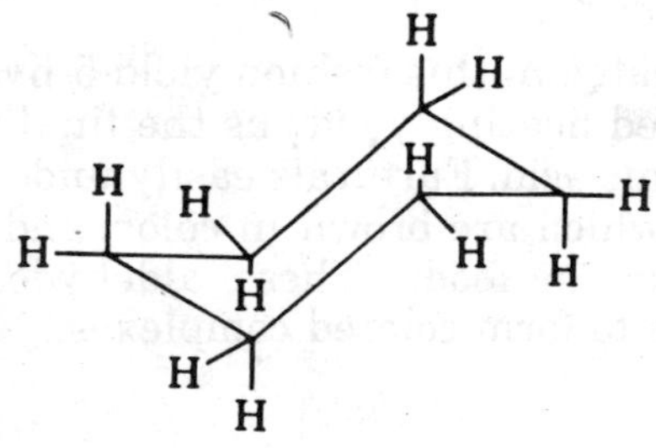

Cyclohexane

form of cyclohexane illustrates this. According to this convention the conformational representation of α-D-glucopyranose is shown as follows.

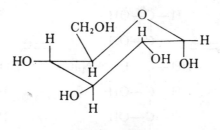

α-D-Glucopyranose

α-D-Glucose and β-D-glucose are stereoisomers and as such are called anomers, because their only difference in configuration involves the steric arrangement of C-1, which is the carbonyl carbon as it appears in the linear formula. This particular carbon atom is known as the anomeric carbon atom.

CARBOHYDRATE REACTIONS

Action of Acids

Dilute mineral acids have little effect on the structure of monosaccharides. However, when a monosaccharide is heated in a strong mineral acid the sugar is dehydrated. Aldopentoses treated in this manner give furfural.

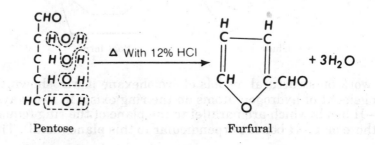

$$\text{Pentose} \xrightarrow{\Delta \text{ With 12\% HCl}} \text{Furfural} + 3H_2O$$

Pentose　　Furfural

Aldohexoses treated in this fashion yield 5-hydroxymethyl furfural. However, continued heating yields as the final product levulinic acid together with formic acid. Furfurals easily undergo further changes to produce humins, which are brown in color, and are involved in some browning reactions in foods. These aldehydes can condense with amines or phenols to form colored complexes.

Action of Alkalies

Reducing sugars undergo tautomerization resulting in an enediol salt 1 formed after standing for several hours in dilute alkaline solution.

$$\begin{array}{c} H{-}C{=}O \\ H{-}C{-}OH \\ HO{-}C{-}H \\ H{-}C{-}OH \\ H{-}C{-}OH \\ CH_2OH \end{array} \rightleftharpoons \begin{array}{c} H{-}C{-}OH \\ \| \\ C{-}OH \\ HO{-}C{-}H \\ H{-}C{-}OH \\ H{-}C{-}OH \\ CH_2OH \end{array} + NaOH \rightleftharpoons \begin{array}{c} H{-}C{-}ONa \\ \| \\ C{-}OH \\ HO{-}C{-}H \\ H{-}C{-}OH \\ H{-}C{-}OH \\ CH_2OH \end{array} + H_2O$$

D-Glucose　　1,2-Enediol　　Enediol salt

Since the formation of the enediol destroys the asymmetry at C-2 and since the last four carbon atoms have an identical configuration, fructose, glucose, and mannose will produce the same enediol salt. Acidification of the enediol salt is necessary to complete the reaction and all three of these sugars will be formed. This rearrangement of related sugars is known as the Lobry de Bruyn-Alberda van Ekenstein transformation. The reaction on glucose gives fructose and mannose.

$$
\begin{array}{ccccc}
\text{H—C=O} & & \text{H—C—OH} & & \text{H—C=O} \\
| & & \| & & | \\
\text{H—C—OH} & & \text{C—OH} & & \text{HO—C—H} \\
| & & | & & | \\
\text{HO—C—H} & \rightleftharpoons & \text{HO—C—H} & \rightleftharpoons & \text{HO—C—H} \\
| & & | & & | \\
\text{H—C—OH} & & \text{H—C—OH} & & \text{H—C—OH} \\
| & & | & & | \\
\text{H—C—OH} & & \text{H—C—OH} & & \text{H—C—OH} \\
| & & | & & | \\
CH_2OH & & CH_2OH & & CH_2OH \\
\text{D-Glucose} & & \text{1,2-Enediol form} & & \text{D-Mannose}
\end{array}
$$

$$
\begin{array}{c}
\downarrow\!\uparrow \\
CH_2OH \\
| \\
\text{C=O} \\
| \\
\text{HO—C—H} \\
| \\
\text{H—C—OH} \\
| \\
\text{H—C—OH} \\
| \\
CH_2OH \\
\text{D-Fructose}
\end{array}
$$

These reactions show the effect that alkali can have on a sugar. Enolization is a general property of aldehydes and ketones. The isomerization reactions, glucose $\rightleftharpoons$ fructose and mannose $\rightleftharpoons$ fructose are enzymatic reactions important to the intermediary metabolism of sugars.

Base cations are important during the action of alkali on sugars. It was found that the initial course of the reaction taking place at 0.5 *N* concentration of alkali was different with monovalent and divalent bases. At temperatures of 35°–37°C both $Ca(OH)_2$ and NaOH easily bring about enolization. However, glucose in $Ca(OH)_2$ after 24 hr yields only mannose, while glucose in NaOH solution yields only fructose. It should be noted that at an alkali concentration of 0.035 *N* no differences in the action of NaOH or $Ca(OH)_2$ on the catalysis of the isomerization of glucose could be detected.

The treatment of reducing sugars with strong alkali produces further isomerization. The result is that enolization along the carbon chain is continued. The substances formed under these conditions could include 1,2-enediol, 2,3-enediol, and 3,4-enediol.

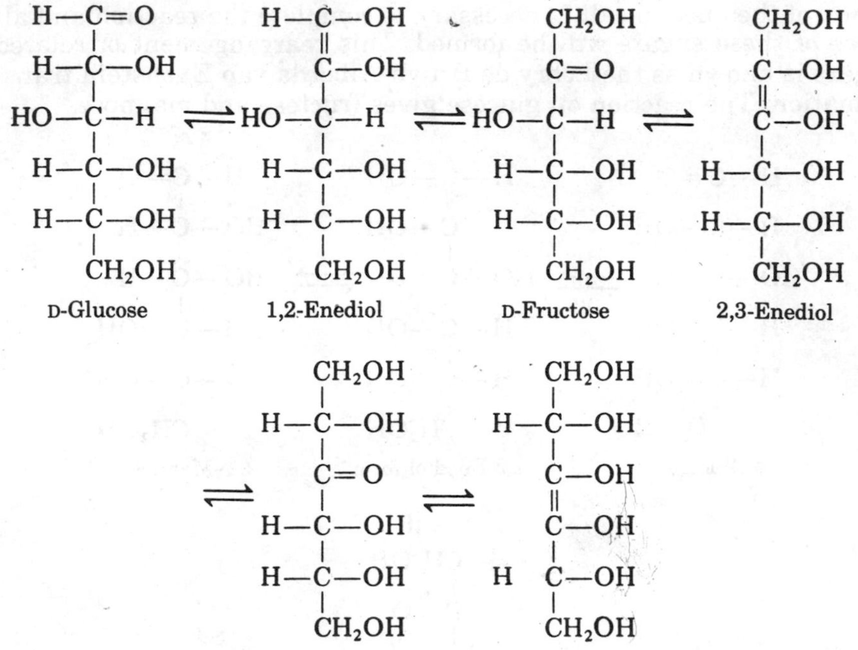

```
     H—C=O             H—C—OH               CH2OH          CH2OH
       |                 ||                   |              |
     H—C—OH              C—OH                 C=O            C—OH
       |                 |                    |              ||
    HO—C—H    ⇌     HO—C—H      ⇌      HO—C—H     ⇌       C—OH
       |                 |                    |              |
     H—C—OH            H—C—OH               H—C—OH         H—C—OH
       |                 |                    |              |
     H—C—OH            H—C—OH               H—C—OH         H—C—OH
       |                 |                    |              |
      CH2OH             CH2OH                CH2OH          CH2OH
   D-Glucose         1,2-Enediol          D-Fructose      2,3-Enediol

                        CH2OH                CH2OH
                          |                    |
                        H—C—OH               H—C—OH
                          |                    |
                          C=O                  C—OH
              ⇌           |          ⇌         ||
                        H—C—OH                 C—OH
                          |                    |
                        H—C—OH               H C—OH
                          |                    |
                         CH2OH                CH2OH
                       3-Ketose            4-Enediol
```

It is quite probable that the major pathway for the formation of brown color in foods is the degradation and dehydration by way of the 1,2-enol forms of aldose o etose amines. This same route by way of the 2,3 and 3,4 forms seems important in the production of flavor.

A mixture of products results when the enediols break at the double bonds. The formation of aldoses by cleavage of the enediol may, in turn, undergo further enolization as well as rearrangement with the formation of a complex mixture of substances. This greatly increases the effective reducing power of a sugar.

One application of the isomerization reactions for use in foods is the creation of fructose. This sugar has higher sweetening power and can be made from dextrose, which is readily available commercially.

In strongly alkaline solutions and in the absence of oxidizing agents, hexoses are changed into carboxylic acids. Starting with glucose, 1,2 enediol is formed. The final result is *meta*-saccharinic acid.

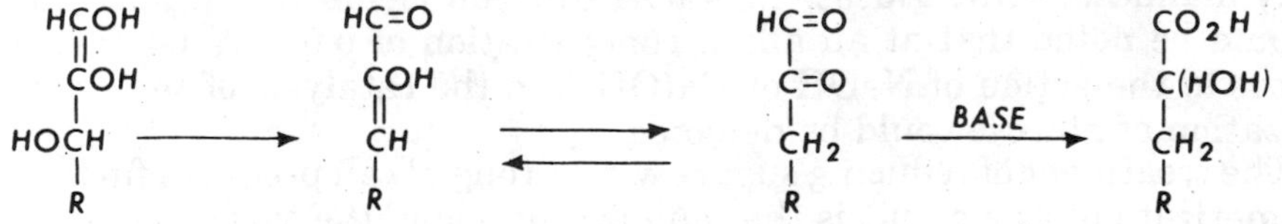

```
  HCOH            HC=O                    HC=O                  CO2H
   ||              |                        |                     |
   COH             COH                      C=O                   C(HOH)
   |               ||         ⇌             |        BASE         |
  HOCH    ——→      CH                       CH2     ——→           CH2
   |               |                        |                     |
   R               R                        R                     R
1,2-Enediol                         3-Deoxy-D-erythro-     meta-Saccharinic
                                       hexosulose                acid
```

The C_6 acids, saccharinic acid, isosaccharinic acid, and metasaccharinic acid, are isomeric, and formation depends on the concentration of alkali. For pentose or higher sugars, the formation of saccharinic acids does not disturb the configuration after C-3. Concentrated alkali favors the formation of the iso- and metasaccharinic acids, while dilute alkali favors the formation of saccharinic acid.

Reducing Action in Alkaline Solution. Reducing sugars are those that have a free aldehyde or ketone group; such sugars are glucose and fructose. These reducing sugars undergo enolization in alkaline solutions. These enediol forms are very reactive and are easily oxidized by oxidizing agents and also by oxygen. Such ions as Ag^+, Cu^{2+}, Hg^{2+}, and $Fe(CN)_6^{3-}$ are easily reduced by reducing sugars, which, in turn, are oxidized to complex mixtures of acids. The most frequently used ion for these determinations is Cu(II) (Cu^{2+}), which is employed both quantitatively and qualitatively. The method was originally worked out by Fehling (1849). The solution used is known as Fehling's solution. The copper ion is used with sodium potassium tartrate and sodium hydroxide. Potassium hydroxide or sodium carbonate may be used also. Some solutions make use of sodium citrate. The tartrate or citrate has the ability to keep the cupric hydroxide from separating out because either one can form complexes with the cupric ions, which are slightly dissociated and soluble.

The sugar fragments which are produced by the alkali and heat are easily oxidized and the Cu(II) is reduced to the Cu(I) (Cu^+) state. This reaction is very complex and proceeds in the following manner.

$$\text{Reducing sugar} + \text{alkali} \longrightarrow \text{enediols and reducing sugar fragments}$$

$$+$$

$$Cu(OH)_2 \longleftarrow \text{copper complex of tartrate}$$

$$\downarrow$$

$$H_2O + Cu_2O \longleftarrow CuOH \xleftarrow{OH^-} Cu^+ + \text{mixture of sugar acids}$$

Reduction is indicated by the appearance of the yellow-orange Cu(I) hydroxide, resulting from the combination of the Cu(I) ions and the hydroxyl ions. This Cu(I) hydroxide loses water from the action of the heat, and the final precipitate is the insoluble Cu_2O.

Oxidation

Monosaccharides can yield a variety of compounds under different oxidizing conditions because of the presence of the aldehyde or ketone group and also because of the hydroxyl groups. The final products of complete oxidation are CO_2 and H_2O.

Oxidation with bromine water or sodium hypoiodite can yield aldonic acids with aldoses but not with ketoses.

Aldonic acids may lose water, the result of which is the formation of lactones. Gluconic acid yields a gamma (γ)- and a delta (δ)-lactone. Of these the γ-lactone is the more stable compound.

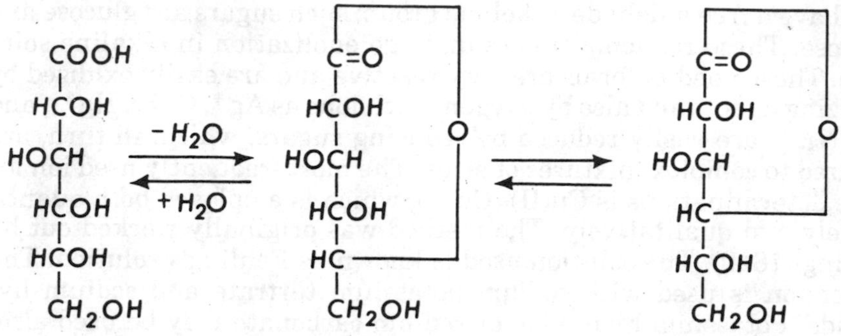

$- H_2O$ / $+ H_2O$

Gluconic acid — Glucono-δ-lactone — Glucono-γ-lactone

Conversion of Ketoses to Aldoses

It has already been noted that a ketose can be reduced to a mixture of two corresponding sugar alcohols. Fructose was shown to yield sorbitol and mannitol. These alcohols are then oxidized to the aldonic acids. The aldonic acids are changed to lactones, which can be cautiously reduced to the corresponding aldoses.

The oxidation of aldoses to aldonic acids is a specific reaction for these sugars and can be used as a method to distinguish them from the ketoses. A simple titration of the excess iodine is all that is necessary to make it quantitative.

When additional calcium is needed in the body, it can be administered orally in the form of the calcium salt of gluconic acid.

Uronic acid is another product of the oxidation of aldoses. In this case the primary hydroxyl group of the sugar molecule is oxidized to a carboxyl group.

Reducing sugars are not used directly for the formation of uronic acids because the carbonyl group is so reactive that it is oxidized first. For this reason natural sources (enzyme systems) are ordinarily used to obtain these compounds.

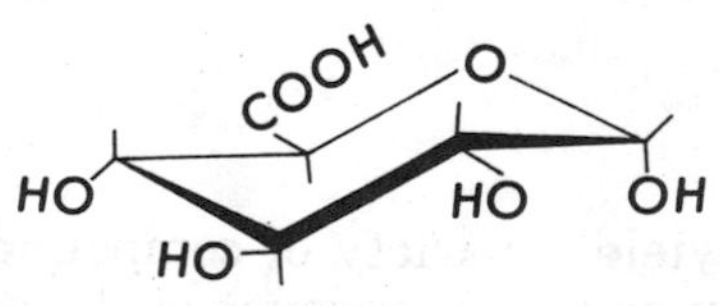

Glucuronic acid

Certain polysaccharides, notably chondroitin, contain glucuronic acid, and various woods have it as monomethyl ethers.

The detoxification of phenol and other such substances in the body by the formation of glucuronides is important.

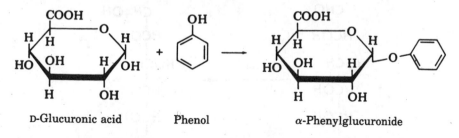

D-Glucuronic acid + Phenol → α-Phenylglucuronide

Strong oxidation by such agents as nitric acid acting on reducing sugars results in the oxidation of the carbonyl group as well as the primary alcohol group. This produces a dicarboxylic acid. Glucaric acid is the result of such oxidation of glucose.

```
  COOH
   |
  HCOH
   |
 HOCH
   |
  HCOH
   |
  HCOH
   |
  COOH
```

Glucaric acid

Pectin, which is an important food polysaccharide, is made up of chains of galacturonic acid residues. Methyl groups esterify part of the C_6 acid groups. This is a constituent also of plant gums and mucilages.

Saccharic acids, also known as aldaric acids, are formed by the oxidation of aldoses with nitric acid. This treatment produces a carboxyl group at each end of the chain. Glucose yields glucaric acid with such treatment.

Reduction

Reduction of free monosaccharides can result from the use of sodium amalgam and water. It can also be done by electrolysis of an acidic solution of the sugar. The result is the corresponding sugar alcohol. Glucose is reduced to D-sorbitol, the corresponding polyol.

The usual practice in naming alcohols is to place the suffix *-itol* at the end of the name of the sugar. Sorbitol does not follow this convention, however; if it did, the name of the compound would be glucitol.

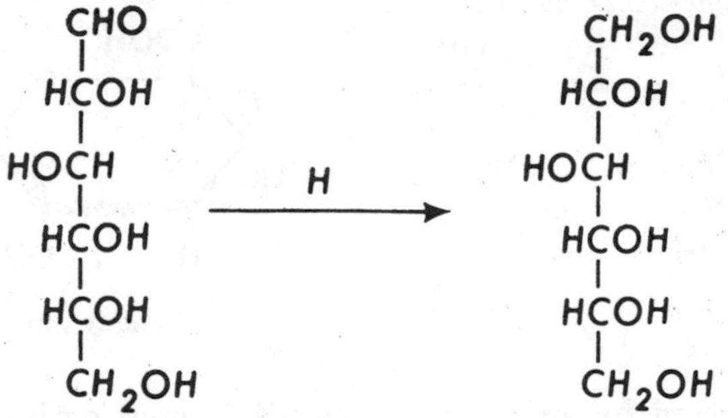

$$\begin{array}{ccc} CHO & & CH_2OH \\ | & & | \\ HCOH & & HCOH \\ | & & | \\ HOCH & \xrightarrow{H} & HOCH \\ | & & | \\ HCOH & & HCOH \\ | & & | \\ HCOH & & HCOH \\ | & & | \\ CH_2OH & & CH_2OH \end{array}$$

Glucose Sorbitol (Glucitol)

During hydrogenation of ketoses a new asymmetric carbon atom forms and two products result. Fructose yields sorbitol and mannitol, which have different configurations at the second carbon atom, hence they are epimers.

The polyhydric alcohols or polyols are crystalline solids, the solutions of which range from strongly to mildly sweet. D-Sorbitol is found in a number of fruits, among which are the drupes (stone fruits) and pomes (apples and pears). Sorbitol as well as mannitol are prepared commercially in large amounts. Sorbitol is used in the manufacture of ascorbic acid and some detergents. It is used also in candy manufacture, and medicinally as a diuretic and cathartic. It is also employed as a sweetener for diabetics.

Condensation Reactions

Aldoses and ketoses react with one molecule of phenylhydrazine to form a phenylhydrazone. The second step is to heat with an excess of reagent to form yellow osazones from the hydrazones. Osazones are bright yellow crystals and can be identified by their decomposition points and microscopic character.

The formation of identical osazones by reactions with glucose, mannose, and fructose show that these three sugars have the same configuration at C-3, C-4, and C-5.

Osazones can be hydrolyzed by dilute acids to osones, which can be reduced to ketones. Thus, a method is provided for the conversion of an aldose to a ketose by way of the phenylhydrazine reaction.

Reactions Involving Hydroxyl Groups

The hydroxyl group in sugars is a very important group, especially if the hemiacetal group is considered.

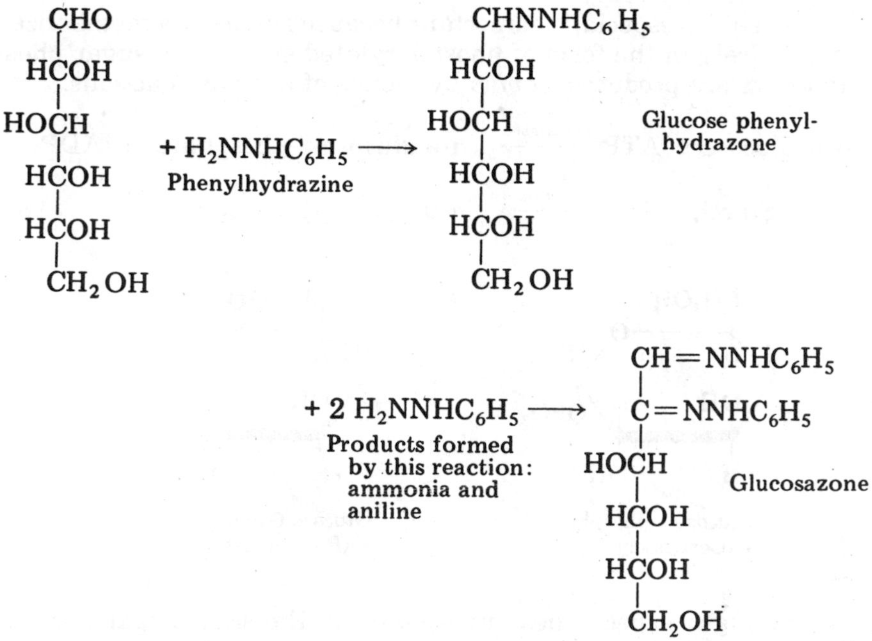

$$\begin{array}{c} CHO \\ | \\ HCOH \\ | \\ HOCH \\ | \\ HCOH \\ | \\ HCOH \\ | \\ CH_2OH \end{array} \quad \begin{array}{c} + H_2NNHC_6H_5 \longrightarrow \\ \text{Phenylhydrazine} \end{array} \quad \begin{array}{c} CHNNHC_6H_5 \\ | \\ HCOH \\ | \\ HOCH \\ | \\ HCOH \\ | \\ HCOH \\ | \\ CH_2OH \end{array} \quad \text{Glucose phenylhydrazone}$$

$$\begin{array}{l} + 2\ H_2NNHC_6H_5 \longrightarrow \\ \text{Products formed} \\ \text{by this reaction:} \\ \text{ammonia and} \\ \text{aniline} \end{array} \quad \begin{array}{c} CH{=}NNHC_6H_5 \\ | \\ C{=}NNHC_6H_5 \\ | \\ HOCH \\ | \\ HCOH \\ | \\ HCOH \\ | \\ CH_2OH \end{array} \quad \text{Glucosazone}$$

Glycosides. A well-known reaction of hydroxyl groups is the formation of glycosides. A glycoside is a sugar derivative in which the hemiacetal (the potential aldehyde or ketone) is replaced with an alkyl or aryl group. This can be done by mixing the sugar with the necessary alcohol and using acid as a catalyst. When glucose reacts with methyl alcohol and HCl under anhydrous conditions, the methyl glycoside or, in this case, methyl glucoside is formed. Because this compound is formed from glucose, it is called a glucoside.

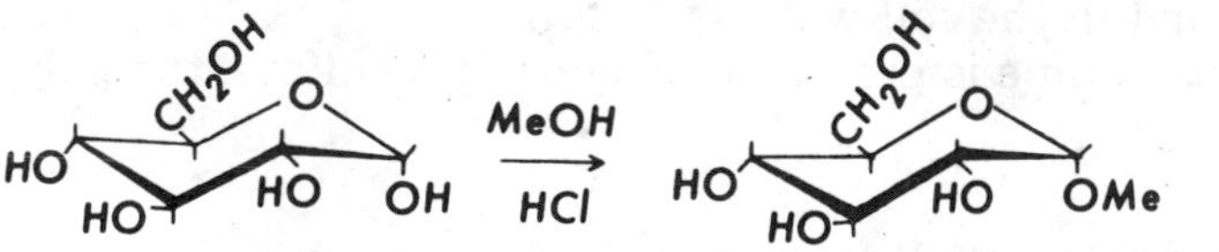

CH$_2$OH, O, HO, HO, HO, OH —MeOH / HCl→ CH$_2$OH, O, HO, HO, HO, OMe

α-D-Glucose   Methyl-α-D-glucose

Sugar acetals have surfactant properties. Some surfactants are solubilizing agents, dispersants, detergents, wetting agents, whipping agents, foaming agents, or defoaming agents. These molecules are useful as emulsifiers, stabilizers, and flavor media. 4,6-Benzylidene methyl-α-D-glucoside is used to develop cherry flavor. It is likely that benzaldehyde is slowly released to bring about this flavor.

Sugar Esters. All sugars are polyhydroxy compounds and, therefore, they can be esterified by acids and anhydrides at the free hydroxyl positions. This is especially important because sugars are metabolized almost entirely in the form of phosphorylated sugar. The sugar phosphate esters are produced *in vivo* by means of enzyme reactions.

$$\alpha\text{-D-Glucose} + \text{ATP} \xrightarrow{\text{hexokinase}} \alpha\text{-D-glucose-6-phosphate} + \text{ADP}$$

$$\text{Starch} + \text{P}_i \xrightarrow{\text{phosphorylase}} \alpha\text{-D-glucose-1-phosphate}$$

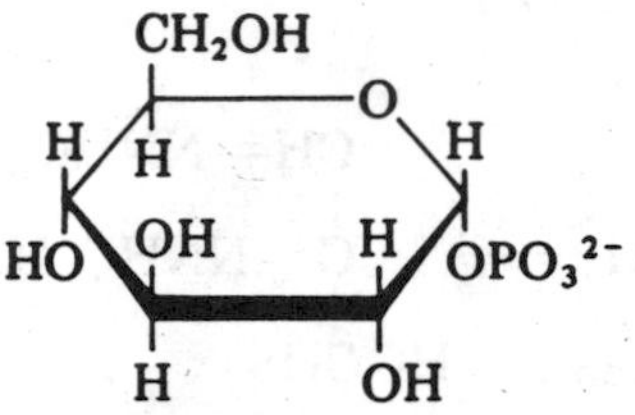

Glucose 1-phosphate
(Cori ester)

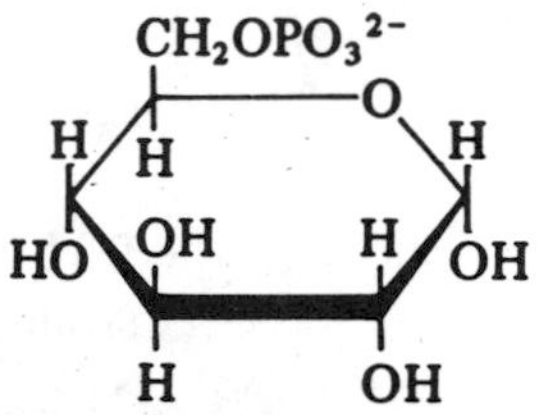

Glucose 6-phosphate
(Robison ester)

As one might expect, these phosphates of the same sugar perform differently in biochemical reactions.

Amino Sugars. The amino sugars are formed by replacement of the hydroxyl group with an amino group at the C-2 atom. Two such sugars often found in organisms are 2-amino-2-deoxy-D-glucosamine and 2-amino-2-deoxy-galactosamine. Their N-acetyl derivatives are found also.

Glucosamine is present in chitin and is formed by hydrolysis of this substance. It is found in mucoproteins and mucopolysaccharides. As chitin it is present in the shells of crustaceans (lobsters, crabs, etc.) and insects and in the cell walls of fungi.

Galactosamine is present in chondroitin sulfate, the polysaccharide of cartilage.

Sialic Acids. Sialic acids are amino sugar derivatives made up of a C_6 amino sugar linked either to lactic or pyruvic acid. Two of these which are frequently found are *N*-acetylneuraminic acid and *N*-acetylmuramic acid.

Deoxy Sugars. These sugars are of particular importance because 2-deoxy-D-ribose is a part of deoxyribonucleic acid (DNA), the sugar ribose being part of ribonucleic acid. DNA is involved in the synthesis of proteins.

Included in the group of deoxy sugars are L-rhamnose, which is 6-deoxy-L-mannose and L-fucose, which is 6-deoxy-L-galactose.

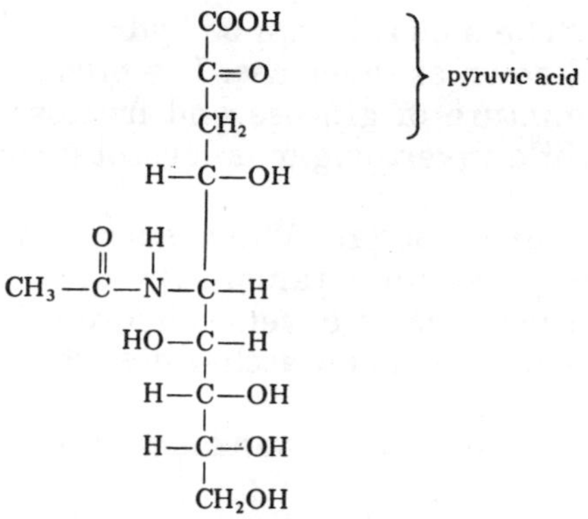

```
              COOH   ⎫
               |     ⎪
              C=O    ⎬ pyruvic acid
               |     ⎪
              CH2    ⎭
               |
          H—C—OH
               |
    O  H       |
    ‖  |       |
CH3—C—N—C—H
               |
         HO—C—H
               |
          H—C—OH
               |
          H—C—OH
               |
             CH2OH
```

N-Acetylneuraminic acid

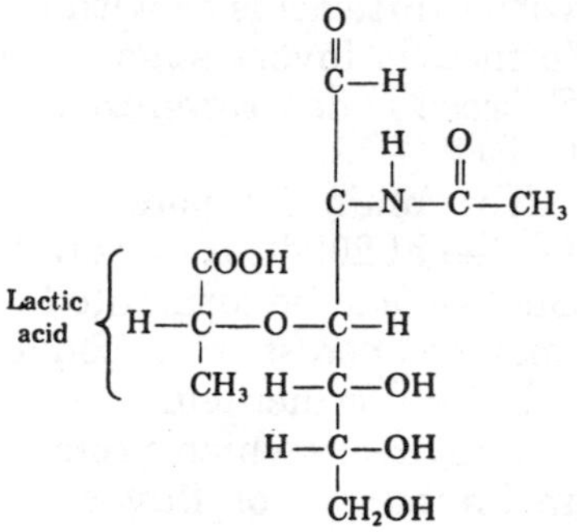

```
                         O
                         ‖
                         C—H
                         |
                         |    H  O
                         |    |  ‖
                         C—N—C—CH3
          ⎧   COOH       |
   Lactic ⎪    |         |
   acid   ⎨ H—C——O——C—H
          ⎪    |         |
          ⎩   CH3  H—C—OH
                         |
                     H—C—OH
                         |
                       CH2OH
```

N-Acetylmuramic acid

OLIGOSACCHARIDES

Oligosaccharides are small polymers formed by the combinations of two to about 10 monosaccharides with the elimination of water. The monosaccharides are linked together by means of a glycosidic bond, which is the result of the condensation of the hydroxyl of the hemiacetal group of C-1 with the hydroxyl of an alcohol.

Disaccharides

A disaccharide is the result of a linkage between two monosaccharides. A disaccharide may be considered a glycoside, the second monosaccharide of which serves as the "aglycone." In this group are sucrose and lactose which occur naturally, and maltose which is formed as a result of the enzymic hydrolysis of starch.

Sucrose. This sugar is available commercially as cane sugar or beet sugar. It is also known as saccharose. The term "sugar" is generally understood to mean sucrose.

Sucrose is a combination of glucose and fructose. It is a nonreducing sugar because in the formation of the glucosidic linkage the anomeric carbonyl groups of both glucose and fructose are used. Because of this

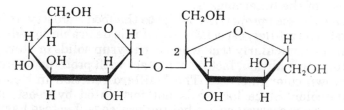

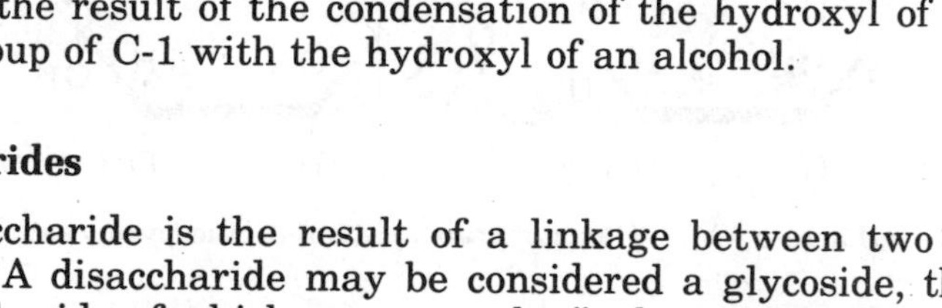

Sucrose (α-D-glucopyranosyl-(1,2)-β-D-fructofuranoside)

linkage, mutarotation is not possible. The molecule can be hydrolyzed with dilute acids or with the enzyme known as invertase. The product formed is invert sugar, which is a mixture of glucose and fructose. Sucrose is dextrorotatory (+66.5°) while invert sugar is levorotatory (−19.8°).

The main component of honey is invert sugar. When sucrose is heated at 200°C it loses water and a brown syrup, caramel, is produced. Sucrose can be fermented by yeast, because yeast develops hydrolytic enzymes necessary to convert disaccharides to monosaccharides which are then fermented.

The theory which explains the sweet taste of a compound is discussed in Chapter 10 on flavor.

Lactose. Lactose is a disaccharide made up of a unit of galactose and a unit of glucose. It is the sugar found in the milk of all animals. Different species have different amounts of lactose in their milk, cow milk contains about 4.5%.

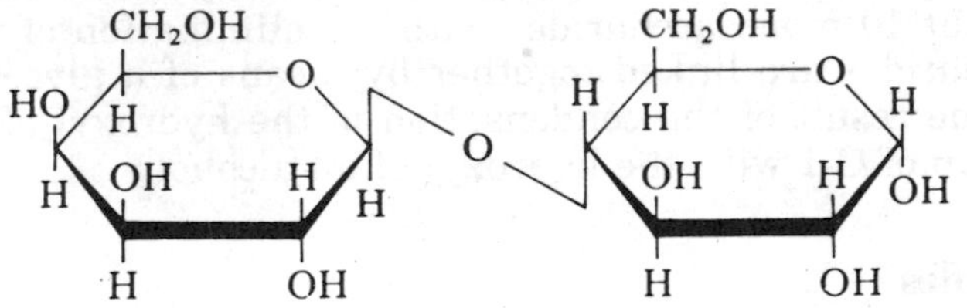

α-Lactose (*O*-β-D-Galactopyranosyl-(1,4)-α-glucopyranose)

Lactose is known and available in two forms, alpha and beta. It is a reducing sugar, and it undergoes mutarotation in solution. Unlike sucrose, lactose is decomposed by alkali. Lactose can be hydrolyzed by the enzyme β-galactosidase, otherwise known as lactase. The usual crystalline form of lactose is α-lactose monohydrate. This hydrate is prepared by crystallizing a supersaturated solution at a temperature below 93.5°C. This hydrate loses water around 130°C and caramelizes between 160° and 180°C. Other crystalline forms of lactose have been prepared, but the presence of water at temperatures below 93.5°C cause these forms to change to the hydrate just described. Lactose is the least sweet of the other sugars.

Lactose possesses properties useful to the food industry. It is used in baked goods as a filler, especially when it is desirable to reduce sweetness. This is particularly true when corn syrup solids or starch are not satisfactory. In addition, lactose can react with proteins to produce the golden brown color in crusts. The Maillard reaction in this case produces the color. Since lactose is not fermented by yeast, its useful properties are preserved as the baking proceeds. Lactose has the property of absorbing flavors, coloring materials, and aromas, hence it is used as a carrying material for flavoring materials and volatile aromas. Lactose used in baked goods functions as a tenderizer. It is also used in infant foods.

Maltose. The name maltose comes from malt liquors, which are prepared from sprouted barley or other cereals. Maltose is formed as a result of the enzymatic hydrolysis of starch. When starch is broken down by amylase approximately 80% maltose is obtained. Maltose is a reducing sugar which shows mutarotation. The enzyme maltase splits maltose into two glucose units. This enzyme is specific for the α-linkage. Maltose, which is prepared by partial hydrolysis of starch with enzymes or acids, is an important component of corn syrups. The commercial starch hydrolysates contain mainly D-glucose, maltose, trisaccharides, and higher saccharides.

These starch hydrolysates enjoy wide application in food products, including soft drinks, confections, infant foods, bread, and coffee substitutes.

Hodge *et al.* (1972) described the technology for making starch hydrolysates that contain 90% or more maltose. The process uses a multiple enzyme process together with β-amylase, that is, isoamylases (amylo-α-1,6-glucosidases), which are able to debranch the amylopectin fraction of the gelatinized starch to linear segments. This has potential as a possible substitute for sucrose as a table sweetener.

Trisaccharides

Raffinose. Raffinose is the trisaccharide (made up of three monoses) of interest to food scientists because it occurs in the free state in the juice of the sugar beet. The three monoses are D-glucose, D-fructose, and D-galactose, and raffinose can be hydrolyzed into these three sugars when strong acids are used. Hydrolysis with weak acids produces the disaccharide melibiose, which comprises galactose and glucose. The enzyme α-glucosidase, also known as maltase, hydrolyses raffinose into galactose and sucrose.

Tetrasaccharides

Stachyose. The tetrasaccharide stachyose occurs in the pea and is made up of α-D-glucose, β-D-fructose, and two α-D-galactose entities.

The α-(1→6) linkage is thought to be a unique feature of the galactosylsucroses. These unusual glycosidic bonds cause special utilization problems when they are ingested in quantity.

According to Olson *et al.* (1975), raffinose and stachyose, both present in dry beans, together with a sugar-free bean residue, seem to be involved in the development of flatulence. This problem is of importance

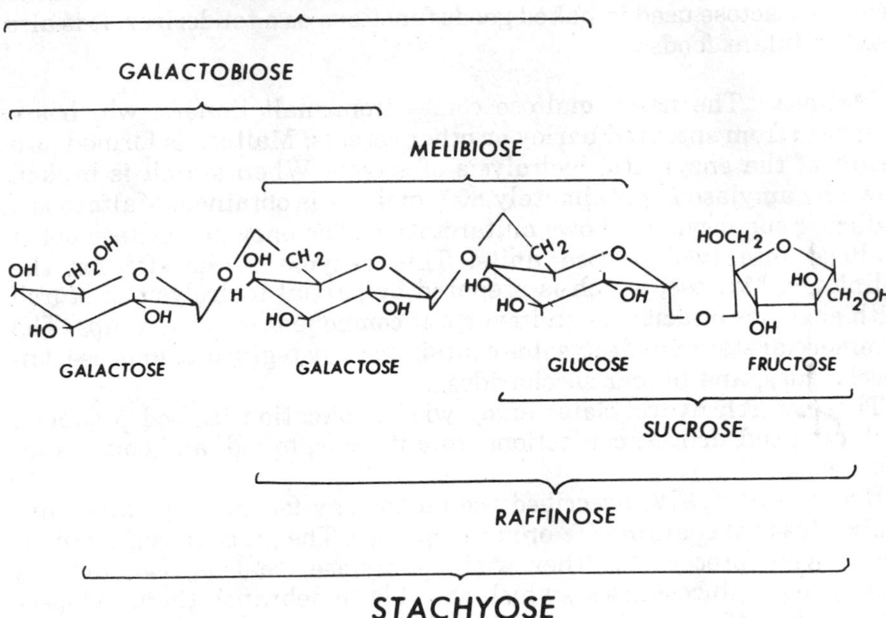

because of the widespread use of beans in the diet. Rackis (1975) reported on this problem in connection with soy products.

POLYSACCHARIDES

Polysaccharides are classed as carbohydrates that contain ten or more carbohydrate units. Most of these substances are large or very large molecules although some have low molecular weights. The larger ones may have thousands of monosaccharide units in the molecule.

While sugars form true solutions, polysaccharides form colloidal solutions, and are, therefore, difficult to purify. The polysaccharides are tasteless and amorphous. The formula $(C_6H_{10}O_5)_x$ is the empirical formula for the hexosans and indicates that the molecular weight is not known.

Polysaccharides are important in structural tissues such as cellulose in plants and chitin in marine life. Muramic acid is found in the cell walls of bacteria. Starch is the storage substance in plants, whereas glycogen is in animals.

Polysaccharides are classified as (a) homopolysaccharides: single monosaccharides linked together—starch, cellulose, and glycogen; (b) heteropolysaccharides; two or more different constituents—hemicelluloses, mucilages, pectins, and resins; and (c) conjugated compounds made up of saccharides, proteins, or lipids.

Starch

Starch is a polysaccharide made up largely of two types of molecules, amylose and amylopectin, usually 1 part of the former to 3 parts of the latter. Starches high in amylose content are known, however. Starch from wrinkled peas, a garden type, contains approximately 70% amylose. About 20% of corn starch is amylose. Wrinkled pea starch does not gelatinize when heated to boiling in water.

The molecules of amylose are relatively small and exist in chains of several hundred glucose units, which are joined by α-D-(1→4) linkage. Amylose is the component of starch that has been found to complex and form a helical structure with iodine, causing the iodine to show a strong absorption of light and a resulting intense blue color; amylopectin does not give a blue color. Amylose can complex with surfactants, fatty acids, and polar agents like butyl and amyl alcohols, and thymol, making possible the separation of amylose from amylopectin, as is described later in this chapter.

Amylopectin, on the other hand, is branched and can have molecular weights ranging from several hundred thousand to millions, which, in turn, amount to thousands of glucose units to the molecule. Several kinds of linkages could exist in the amylopectin molecules, but they are mostly α-D-(1→4) and α-D-(1→6). The α-D-(1→3) linkage is known to be present in amylopectin from waxy maize. The branch point for amylopectin is the α-D-(1→6) linkage. Very small amounts may possibly be in other situations.

The explanation of the α-D-(1→3), (1→4), and (1→6) linkages follows.

(1) D- refers to the group positions which have already been discussed.

(2) The points of union with the numbered carbon atoms in the glucose formula are shown by (1→3), (1→4), and (1→6).

Starch occurs in the form of granules. The immediate precursors of starch are D-glucose-1-phosphate, and uridine diphosphate-D-glucose. These are formed during photosynthesis and will take care of the syn-

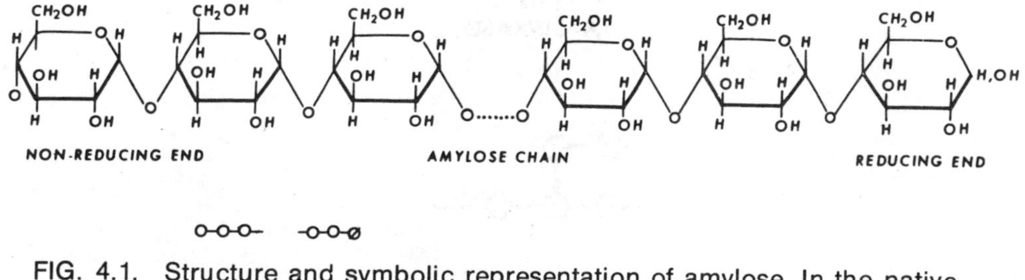

NON-REDUCING END AMYLOSE CHAIN REDUCING END

FIG. 4.1. Structure and symbolic representation of amylose. In the native amylose molecule there may be from a few hundred to 10,000 or more glucose units linked by an α-1,4 glycosidic bond.
From French (1969).

thesis of starch in the leaves. For synthesis elsewhere in the plant, the starch precursors must be converted into sucrose for transport to the storage area, at which point resynthesis of the starch must occur. The synthesis of amylose and amylopectin from sucrose follows pathways of several steps. More steps are required for the production of amylopectin than for that of amylose. Starches are formed and broken down in plant tissues by enzymes. Figure 4.4 shows the pathways for the conversion of sucrose into amylose and amylopectin. These pathways show the current understanding of what occurs, but as research continues they may be subject to change. Sucrose is the starting material for the synthesis of these two components of starch. Sucrose and adenosine diphosphate can undergo conversion to yield adenosine diphosphate D-glucose and D-fructose. This change is brought about by the enzyme sucrose:adenosine diphosphate glucosyltransferase. D-Glucose and D-fructose can also be formed by the hydrolysis of sucrose via a sucrose hydrolase. Following this, the D-glucose and the D-fructose, by the known glycolytic reactions are changed to α-D-glucose-1-phosphate, which through the pyrophosphorylase reaction, reacts with uridine triphosphate to form uridine diphosphate D-glucose. D-Glucosyl units

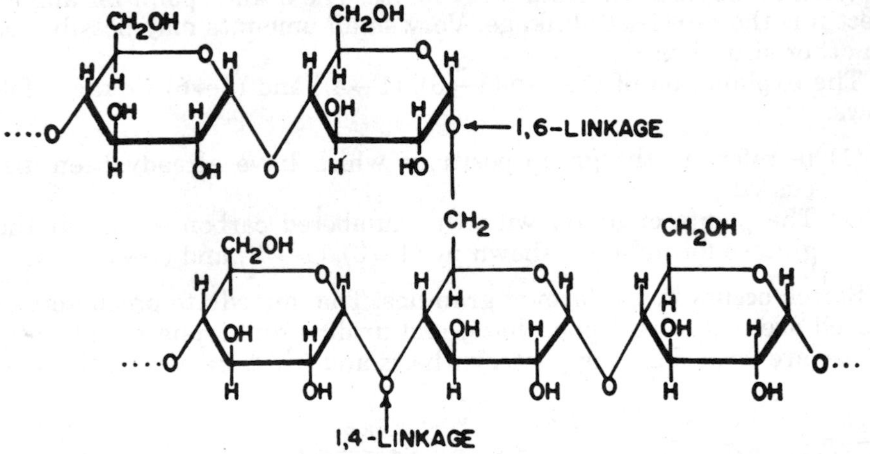

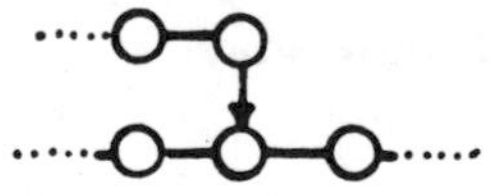

FIG. 4.2. Structure (*top*) and symbolic representation (*bottom*) of an α-1,6 branch point in amylopectin or glycogen. The vertical arrow indicates an α-1,6 bond.

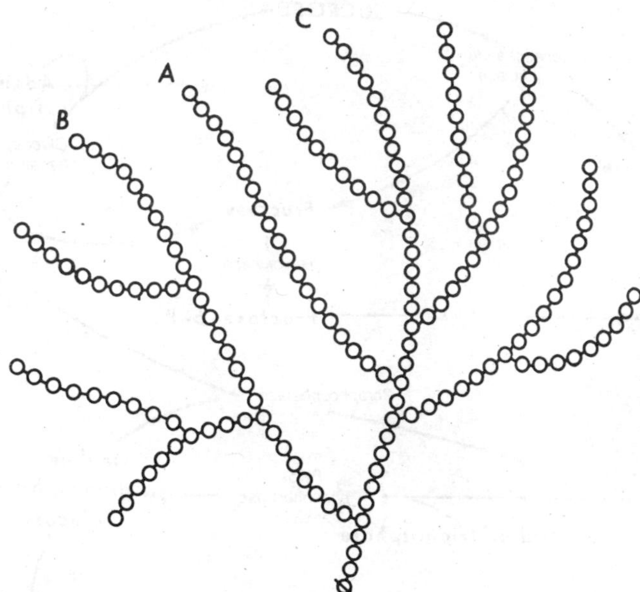

FIG. 4.3. Branching pattern of amylopectin, as proposed by Meyer. The A chains are those which are linked solely at the reducing end by an α-1,6 link to another chain. Each B chain is also linked at the reducing end by an α-1,6 link to another chain, and in addition it is also linked through one or more α-1,6 links to the reducing end or ends of A or B cabins. The C chain carries the reducing group of the molecule. The molecule contains only 11 chains with 170 glucose units. Molecules of native amylopectin range from several hundred to many thousand glucose units in size. One should imagine 5–100; models such as the above joined together to make a single amylopectin molecule.
From French (1969).

required for the synthesis of amylose by the 1,4-α-D-glucan synthetase can be donated by both adenosine diphosphate D-glucose and uridine diphosphate D-glucose. This last action is primarily irreversible since long chains of D-glucose units are formed. These chains are unable to act as substrates for the branching enzyme, probably because of the size of the molecules. The pyrophosphatases in the plant can convert adenosine diphosphate D-glucose and the corresponding uridine compound into α-D-glucose-1-phosphate and the mononucleotides. Phosphorylase synthesizes linear D-glucose polymers from the α-D-glucose-1-phosphate from this last reaction as well as from this compound when it is formed from D-fructose of D-glucose. Because of the degradative effect of phosphorylase, the molecular weight of these linear polymers

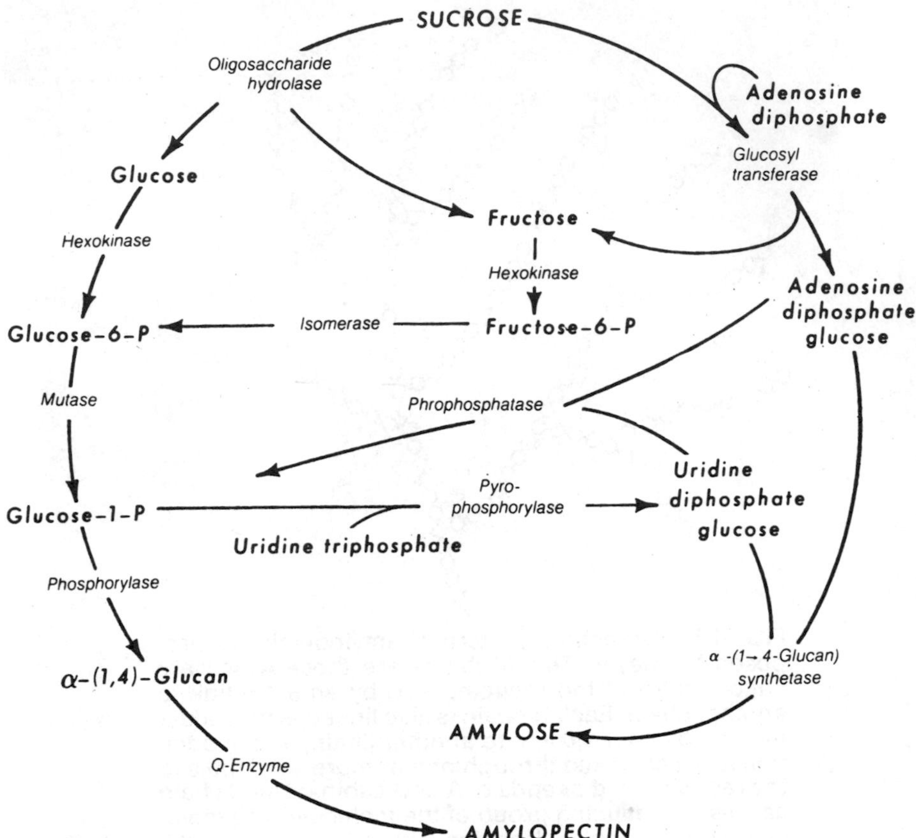

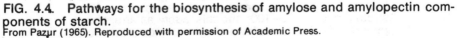

FIG. 4.4. Pathways for the biosynthesis of amylose and amylopectin components of starch.
From Pazur (1965). Reproduced with permission of Academic Press.

is rather low and the Q-enzyme can for this reason convert them to amylopectin. The systematic branching characteristic of amylopectin is brought about by two factors: (1), the phosphorylase reaction is reversible, and (2), more important, the enzyme can redistribute D-glucose units (Pazur 1965).

Starch Hydrolysis. The hydrolysis of starch is accomplished by enzymes which are found in plants, animals, and microorganisms. They are known as alpha-amylases, beta-amylases, glycoamylases, and oligosaccharide hydrolases.

The alpha-amylases are very widely distributed. They bring about a quick breakup of the molecules of starch by splitting the α-D-(1→4) linkages at random, after which the starch is finally but slowly changed to reducing sugars. Maltose is produced together with a little glucose. These enzymes cannot hydrolyze maltose. Amylopectin on hydrolysis gives D-glucose, maltose, and a "limit dextrin," which holds the α-D-(1→6) linkages because these enzymes are unable to hydrolyze them. These polysaccharide fragments are called limit dextrins because the enzyme has reached the limit of its ability to hydrolyze. In addition, it is well to mention that alphaamylases cannot hydrolyze the α-D-(1→3) linkages. The action of plant alpha-amylases on amylopectin produces limit dextrins which are undoubtedly structurally rather similar to those resulting from the action of animal alpha-amylases.

Alpha-amylases of plant origin are important in baking, distilling, and brewing industries because of their ability to form fermentable sugars from starch. They are used in the form of malt enzymes.

Starch can be hydrolyzed by beta-amylase, 1,4-α-D-glucan maltohydrolase, to form maltose and a high-molecular-weight limit dextrin. Beta-amylase liberates maltose of the beta configuration, hence the name given to the enzyme. This enzyme has been prepared in the crystalline form (Balls *et al.* 1946). Beta-amylase attacks the nonreducing ends of the outer chains and in stepwise fashion removes the maltose units. Amylose with an even number of D-glucose units yields only maltose from this activity, but amylose with an odd number of D-glucose units forms not only maltose, but maltotriose, which contains the reducing D-glucose unit present in the original amylose. The maltotriose undergoes slow hydrolysis to D-glucose and maltose if the concentration of the enzyme is high, and the incubation is continued for a period of time. Starting at the nonreducing ends of the outer chains, amylopectin is hydrolyzed in a similar fashion. Because beta-amylase is unable to bypass or hydrolyze an α-D-(1→6) bond, the resulting limit dextrin contains all of the α-D-(1→6) bonds and has a high molecular weight. Therefore considerable amounts of amylopectin remain unhydrolyzed. Beta-amylase is usually associated with alpha-amylase in plant sources. It is found in wheat, barley, sweet potatoes, and in other plant products.

Glucoamylases are found in some species of fungi, and in some yeasts and bacteria. These include species of *Aspergillus* and *Rhizopus* of the fungi, and of the yeast *Saccharomyces diastaticus,* and of the bacterium *Clostridium acetobutylicum*. Experimental data (Pazur and Ando 1959; Pazur and Kleppe 1962) have shown that this enzyme can hydrolyze the three bonds, α-D-(1→3), α-D-(1→4), and α-D-(1→6). Amylopectin, amylose, maltooligosaccharides are hydrolyzed to D-glucose by glucoamylase. It has been noted that glucose of the beta configuration is released by this reaction. This enzyme occurring in animal tissues can convert glycogen by direct action to yield D-glucose.

Starch is degraded not only by enzymes but by the action of acid and heat. Dextrins are partial degradation products of this treatment. Amylodextrin, also known as soluble starch, is one such product; it gives blue color with iodine. If starch is heated dry at 230°–260°C a commercial starch gum results.

Dextrins are soluble in water and can be precipitated from solution by the addition of alcohol. Since they have free carbonyl groups they are able to reduce Fehling's solution. Dextrins are components of corn syrups.

Starch and other similar compounds are hydrolyzed by oligosaccharide hydrolases to oligosaccharides. After this step, other enzymes are necessary to finish the hydrolysis to yield D-glucose. Pazur (1955), using paper chromatography and [^{14}C]D-glucose, determined that oligosaccharide hydrolases are capable of reversing their action. Isomaltose would be formed from glucose. Because of this activity, complete hydrolysis by these enzymes is not possible, and the resulting mixture contains such compounds as isomaltose, isomaltotriose, and other high molecular weight compounds, as well as D-glucose.

Oligosaccharide hydrolases in purified form have been made from molds, yeasts, and animal sources.

Figure 4.5 shows various pathways by which starch may be converted into D-glucose.

A number of routes involving enzymes are possible for the transformation of starch into D-glucose. D-Glucose is the basic compound in the reactions taking place in the chemistry of the cell. It provides the energy and the necessary carbon atoms to form the many complex compounds of the cell. The process of photosynthesis makes this compound available in green plants. Since D-glucose is stored in the plants in the form of starch, D-glucose must be released from this substance to make it available for use.

D-Glucose is released from the nonreducing ends of the starch chains by glucoamylase. In this case both the α-D-(1→4) and the α-D-(1→6) linkages are hydrolyzed, making possible an entire conversion to D-glucose. This is the most direct route.

The phosphorylase pathway is the second route for the conversion of starch to D-glucose. α-D-Glucose-1-phosphate and a limit dextrin, shown as a glucan fragment in Fig. 4.5 are the products of this reaction. R-enzyme hydrolyzes the α-D-(1→6) linkages in this glucan fragment to α-D-(1→4). In the next step, these fragments can be altered to α-D-glucose-1-phosphate and D-glucosyl oligosaccharides by phosphorylase. In the end, phosphatase splits α-D-glucose-1-phosphate to D-glucose and inorganic phosphate, and an oligosaccharide hydrolase hydrolyzes the oligosaccharides to D-glucose.

The beta-amylase pathway is the third of these pathways for the degradation of starch to glucose. The first step is the formation of maltose and a glucan fragment which contains all the α-D-(1→6) linkages and most of the α-D-(1→4) linkages present in the starting material, the starch, by R-enzyme. Beta-amylase continues the hydrolysis of these fragments to maltose. Finally, maltose is hydrolyzed by an oligosaccharide hydrolase to D-glucose.

The alpha-amylase pathway is the fourth main route. D-Glucose, maltose, and low-molecular-weight D-glucosyl oligosaccharides which

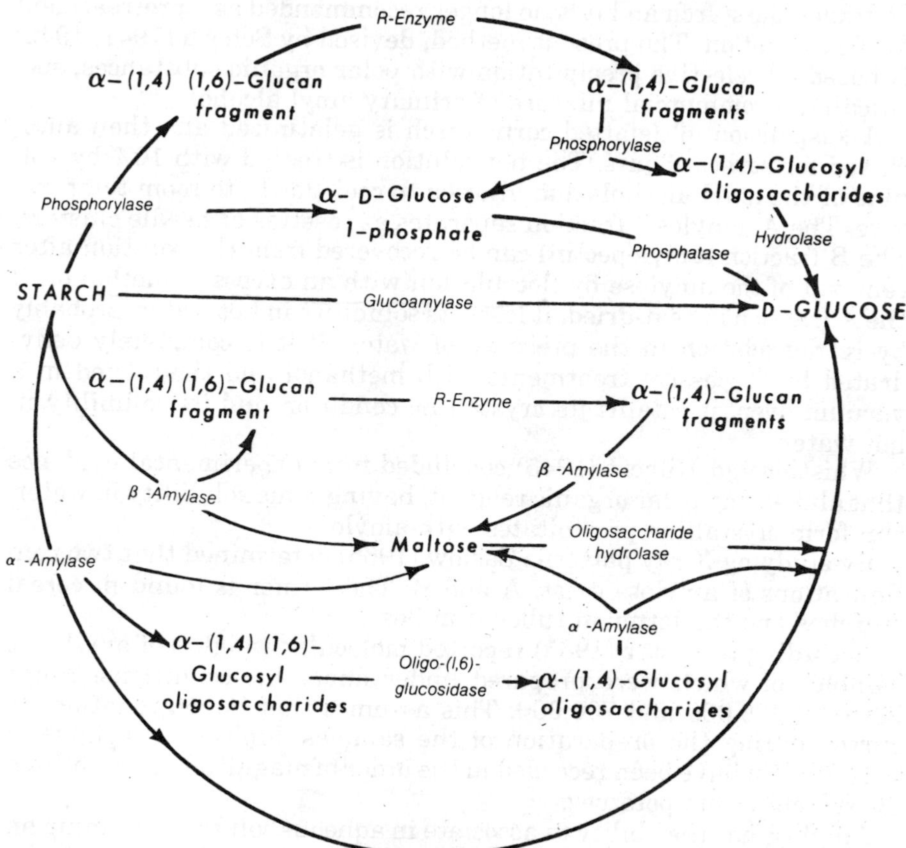

FIG. 4.5. Biochemical pathways for the conversion of starch to D-glucose. From Pazur (1965). Reproduced with permission of Academic Press.

contain α-D-(1→4) and α-D-(1→6) linkages are produced from starch by the action of alpha-amylase. An oligo-(1→6)-glucosidase hydrolyzes the α D-(1→6) linkages in these oligosaccharides, the result of which is the formation of linear glucosyl oligosaccharides. Maltose and D-glucose are formed by the action of alpha-amylase on the linear glucosyl oligosaccharides. Again, an oligosaccharide hydrolase hydrolyzes the maltose to D-glucose, which in this respect is the same as in the beta-amylase pathway (Pazur 1965).

It is important to note that more than one of these pathways may operate at the same time under natural conditions.

Fractionation of Starch. Before starch was found to be heterogeneous, it was thought by some to be a single compound. It was dry-ground in a ball mill, but it has been shown that this mechanical action degrades the starch and it is no longer recommended as a pretreatment for fractionation. The present method, devised by Schoch (1941, 1942) is based on selective precipitation with polar organic substances, specifically, a commercial mixture of primary amyl alcohol.

A suspension of defatted corn starch is gelatinized and then autoclaved for several hours. The hot solution is treated with 10% by volume of the amyl alcohols and allowed to cool slowly to room temperature. The A (amylose) fraction separates as rosettes or needle clusters. The B fraction (amylopectin) can be recovered from the solution after removal of the amylose by flocculation with an excess of methanol. If the A fraction is heat-dried, it loses its solubility in hot water, probably by retrogradation in the presence of water. If it is completely dehydrated by successive treatments with methanol and then dried in a vacuum oven, it retains its crystalline condition and its solubility in hot water.

Whistler and Hilbert (1945) concluded from experimental evidence that almost any polar organic reagent, having some solubility in water, can form crystalline precipitates with amylose.

By studying X-ray patterns Zaslow (1965) determined that two configurations of amylose exist, A and B. The former is found in cereal starches and the latter in tuber starches.

According to Foster (1965) reported molecular weights of amyloses, samples of which were prepared under anaerobic conditions, range between 160,000 and 700,000. This assumes that no degradation occurred during the preparation of the samples. Molecular weights of amylopectin have been recorded in the order of magnitude of 36 million for starches from potatoes.

Amylose has the ability to associate in aqueous solutions, forming an insoluble precipitate. This precipitate results because the linear molecules tend to line up parallel to one another, which causes association through hydrogen bonding, thus decreasing the affinity for water. The aggregate size increases, and a precipitate is formed. The precipitate is called retrograded starch, and the phenomenon is known as retrogradation.

Moisture Absorption by Starch. When held at room temperature, starch equilibrates with the moisture in the atmosphere in which it is held, and reversibly absorbs water. Under normal conditions, this amounts to about 10–17% moisture. The granules possess a limited amount of elasticity which permits this to take place. It has been suggested that water in starch may be held in three ways, namely, water of crystallization, absorbed water, or as interstitial water (Leach 1965).

Gelatinization. Gelatinization is irreversible granule swelling and is brought about in starch suspended in water by heat at a critical temperature, and by certain chemicals at room temperature. Heat gelatinization does not occur all at once at a specific temperature, but over a range of about 10°C. This temperature range varies with starches from different sources. Wrinkled peas yield starch high in amylose and resistant to gelatinization at boiling water temperature because it is made up mainly of highly associated linear molecules. The same is true of high amylose corn starches. Table 4.1 gives gelatinization characteristics of starches. Column 3 gives results on the loss of birefringence, the most accurate method for the measurement of the starting gelatinization temperature of starch. This is done with the Kofler microscope hot-stage (Schoch and Maywald 1958; Watson 1964). Methods measuring increase in optical transmittancy and rise in viscosity lack sensitivity. Granule swelling is the most important event in the gelatinization of starch in an aqueous medium. A known weight of starch is suspended in an excess of distilled water and heated at constant temperature for 30 min with gentle stirring. It is then centrifuged and the supernatant removed so that the swollen sediment can be weighed. The dissolved starch is determined by drying an aliquot of the

TABLE 4.1. Gelatinization Characteristics of Native Starches

Starch Species	Type	Kofler Gel. Temp. Range (°C)	At 95°C		
			Swelling Power	Solubility (%)	Critical Concentration Value
Potato	Tuber	56–66	>1000	82	<0.1
Sago	Pith	—	97	39	1.0
Tapioca	Root	58.5–70	71	48	1.4
Canna	Root	—	72	37	1.4
Arrowroot	Root	—	54	28	1.9
Sweet potato	Root	—	46	18	2.2
Corn	Cereal	62–72	24	25	4.4
Sorghum	Cereal	68.5–75	22	22	4.8
Wheat	Cereal	52–63	21	41	5.0
Rice	Cereal	61–77.5	19	18	5.6
Waxy maize	Cereal	63–72	64	23	1.6
Waxy rice	Cereal	—	56	13	1.8
Waxy sorghum	Cereal	67.5–74	49	19	2.1
Wrinkled pea	Legume	—	6	19	20.0
High-amylose corn	Cereal	—	6	12	20.0
Chick pea (Garbanzo)	Legume	—	13	15	8.3

Source: Leach (1965).

supernatant and weighing the residue to get figures to apply a correction. Calculations from the swelling power data provided the figures for column 6 of Table 4.1, critical concentration values. These figures show the weight in grams of dry starch necessary for use with 100 ml of water to form a paste at 95°C, the swollen granules of which take up almost all the volume—the swollen granules have no free water among them.

The makeup of the micellar network inside the granule controls the swelling activity of starch. This is contingent on the kind and degree of association. The degree of association may be affected by a number of factors, such as molecular weight, distribution of molecular weight, and ratio of amylose to amylopectin. Other factors could be length of outer branches in the amylopectin and degree of branching. The composition, shape, size, and distribution of the micellar areas in the internal lattice could also be affected by these factors. Another vital factor is the presence of naturally occurring impurities of a noncarbohydrate nature.

Weak internal bonding in potato starch is indicated by unusually high swelling. The presence of ionizable esterified phosphate groups assist this swelling by reason of mutual electrical repulsion. Starches with high amounts of amylose show the opposite behavior.

Subjecting an aqueous suspension of starch to heat above the critical temperature, or to appropriate chemicals, weakens the micellar network within the granules by disrupting the hydrogen bonds. Continued disruption of the hydrogen bonds takes place as the temperature is raised further, the swelling of the granules proceeds, and water molecules attach themselves to freed hydroxyl groups. An increase in paste viscosity, paste clarity, and starch solubility result from the swelling of the granules. The patterns of swelling and solubilization are similar for each kind of starch, suggesting a direct relationship between these two functions.

Potato starch starts to gelatinize at about 56°C, whereas corn starch starts at about 62°C. Some chemicals such as sodium sulfate increase the temperature of gelatinization. Sodium nitrate or urea increase swelling or lower the temperature of gelatinization. Because gelatinization is not affected at pH 5–7, in the manufacture of starch products, finishing is done between these two pH values.

Cowie and Greenwood (1957) have shown that no swelling of the granules of potato starch occurred and the birefringent properties were not affected by 0.2 *M* HCl at 45°C. They noted that amylopectin is degraded much more rapidly by HCl than is amylose.

The rate of hydrolysis of wheat starch by acid showed a similar relationship to that of amylose and amylopectin. The rate, however, was in each case 4 to 5 times slower than that for potato starch, indicating a more compact structure for wheat starch than for potato starch (Arbuckle and Greenwood 1958).

Rice starch is said to be more resistant to acid hydrolysis than several other starches.

Swanson and Cori (1948) believe that the α-D-(1→4) glucosidic linkage is less stable to acid hydrolysis than the α-D-(1→6) linkage.

Holló and Szejtli (1961) found that the hydrolysis rates of starch with HCl were markedly greater than when sulfuric acid was used, and that the ratio depends on the acid concentration and the temperature.

Normal sodium hydroxide solution can degrade amylose up to 55–60% under anaerobic conditions. Amylopectin is probably the alkaline-resistant fraction of starch (BeMiller 1965B).

Starch is capable of esterification—either direct esterification with acids or esterification with acid anhydrides and acid chlorides. The latter is the best for most preparations because, in direct esterification, water is a reaction product and some degradation of the starch molecule results from acid-catalyzed hydrolysis of the glucosidic bond. Esterification with formic acid is the main exception to this (Roberts 1965).

Technology of Starch. An understanding of the technical applications of starch is useful because starch is employed extensively in the food industry as a thickener. The ability of starch to thicken solutions and mixtures is a result of heating the starch, which gelatinizes and forms a paste. The interrelationships of other ingredients, such as sugars, fats, acids, and salts, on starches are also of considerable importance in achieving desired results. Starches can be modified to improve their effectiveness as thickeners.

Modified Starches. The Food Chemical Codex of the National Academy of Sciences defines modified food starches as products of the treatment of any of several grain- or root-based native starches (e.g., corn, sorghum, wheat, potato, tapioca, sago) with small amounts of certain chemical agents that modify the physical characteristics of the native starches to produce desirable properties.

Starch is chemically modified by mild degradation reactions or by reactions between the hydroxyl groups of the native starch and the reactant selected. One or more of the following processes are used: mild oxidation (bleaching), moderate oxidation, acid depolymerization, monofunctional esterification, polyfunctional esterification (cross linking), alkaline gelatinization, and certain combinations of these treatments. These methods of preparation can be used as a basis for classifying the starches thus produced.

Modified food starches are usually produced as white or nearly white, tasteless, odorless powders, as intact granules, and if pregelatinized (i.e., subjected to heat treatment in the presence of water), as flakes, amorphous powders, or coarse particles. Modified starches are insoluble in alcohol, in ether, and in chloroform. If not pregelatinized, they are practically insoluble in cold water. Upon heating in water, the granules usually begin to swell at temperatures between 45° and 80°C, depending on the botanical origin and the degree of modification. They gelatinize completely at higher temperatures. Pregelatinized starches hydrate in cold water.

Sugars. Bean and Osman (1959) showed that ten different sugars and syrups slightly increased the high point of hot-paste viscosity and gel strength of 5% corn starch paste. These results were for sugar concentrations up to about 20%. At concentrations of sugar higher than

20%, a decrease in gel strength resulted. Disaccharides at the higher concentrations had greater effect on the inhibition of gelatinization than did equal amounts by weight of monosaccharides. Some confusion is found in the literature concerning the effects of fats. However, Osman and Dix (1960) found that triglycerides lower the temperature at which highest viscosity of a starch paste takes place, while most surfactants raise it.

Acids. Acidity in ordinary food products has little effect on mixtures of starch and water. However, in products with rather low pH values, such as salad dressings, cross-bonded starches are used.

Low concentrations of sodium chloride have been shown to cause a small, lowering effect on the viscosity of potato starch. Low concentrations of calcium chloride showed greater effect in the reduction of the viscosity of potato starch than did sodium chloride, whereas 0.5 to 1 *N* calcium chloride increased the maximum viscosity of corn starch paste slightly.

Acid-modified starch is prepared in an aqueous suspension at subgelatinizing temperature. A mixture of starch and 0.1–0.2 *N* H_2SO_4 between 50° and 55°C is agitated and monitored by in-process controls until the desired change in viscosity is achieved.

Acid-modifed starches have lower hot paste viscosity than regular starches, have a higher ratio of cold- to-hot-paste viscosity, and a higher alkali number. Other differences have also been reported. There are some similarities to the corresponding untreated starches, however, including, among others, approximate solubility in cold water, physical appearance, and birefringence.

Used in the food industry for the manufacture of gum candy, acid-modified starch has advantages over native starch in its ability to yield hot fluid pastes which set to firm gels on standing and are tender and soft. These results are obtained without long cooking and without further use of acid.

Starch Esters. Acetic anhydride or vinyl acetate are used in the manufacture of starch acetates to be used in food products. However, according to the Food Chemical Codex, not more than 2.5% of acetyl groups may be contained in the finished product.

Starch acetates have stability and lend clarity to the finished food product, hence their use. Starches such as waxy corn, which is stable under normal conditions, should be acetylated to prevent "weeping" if they are to be used at low temperatures.

Cross-Linked Starches. These starches can be prepared for food use in several ways, including the treatment of ungelatinized starch with any of the following reagents: acrolein, phosphorus oxychloride, water-soluble metaphosphates, and epichlorohydrin. They can be made also with adipic anhydride not to exceed 0.12% and acetic anhydride in the reaction mixture.

Cross-linking effects the solubility of starch. Intact starch granules can be stabilized by primary bonds. Unmodified waxy corn starch pastes tend to be cohesive and stringy; but, increasing the number of cross-links in the intact starch granules makes these defects tend to disappear. Changing the number of cross-links changes the properties of the products produced, depending on the use to which they will be put—in any case, the thickening effects will be excellent. The products can range from those that are to be subjected to mild cooking at near neutral pH to those subjected to high pressure cooking under conditions of low pH, such as salad dressings and pie fillings.

Starch Phosphates. These comprise another group of starch compounds used to improve products. Starch phosphates are produced by heating a dry mixture of a water-soluble phosphate salt and starch, yielding the products known as starch phosphate monoesters. To produce distarch phosphate, phosphorus oxychloride, not to exceed 0.1%, is used according to the Food Chemical Codex. Sodium trimetaphosphate can be used, but the residual phosphate, calculated as P, must not exceed 0.04%.

Starch phosphates have the property of thickening without gelling and therefore are especially desirable for such products as cream pie fillings, fruit pie fillings, cream soups, baby foods and cream-style corn. Another very desirable property of phosphate starches, especially those with a very low amount of phosphate substitution, is the improvement of freeze–thaw stability of gravies, white sauces, and fillings for frozen pies, since it cuts down the separation of water resulting from freezing and thawing as compared with untreated starches. Albrecht *et al.* (1960B) rated thickening agents in decreasing order of freeze–thaw stability as follows: (1) starch phosphate, (2) ungelatinized regular corn starch (this is limited to products in which solids need not be suspended during freezing and storage and to products which are heated to near boiling point before consumption), (3) pregelatinized, cross-linked waxy maize starch, (4) all-purpose flour, and (5) regular corn starch (which is gelatinized prior to freezing as part of the sauce formulation process).

Cellulose

Cellulose, another type of polysaccharide present in fruits and vegetables and other foods, is of interest to the food chemist because it contributes to bulk in the diet. It is not a nutritious substance because it cannot be utilized by the human body, which lacks the necessary enzymes to digest cellulose. However, when ingested cellulose does contribute to the elimination process because of the bulk. It is insoluble in water.

Cellulose is a polysaccharide made up of units of glucose as building blocks. This cellulose compound is known to be a beta anomer and is beta-linked. Cellulose and its derivatives are split by β-enzymes and not by α-glucosidases. X-Ray evidence also supports the β-structure. Cellobiose, obtained on partial hydrolysis of cellulose, is the simplest of the oligosaccharides and has two glucose units in its structure.

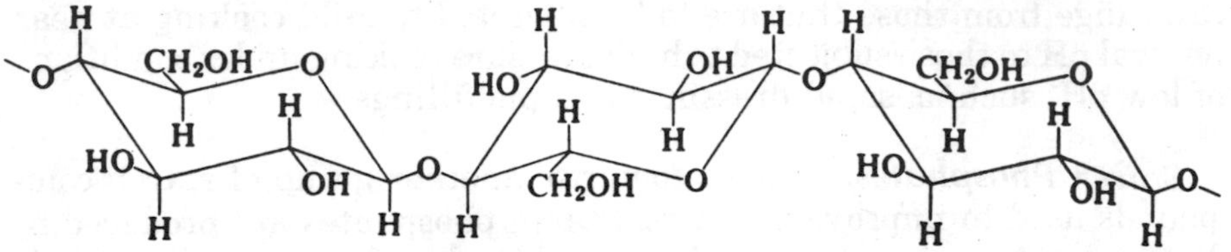

Segment of a cellulose molecule

In the union, C-4 of one glucose is joined to the C-1 of the next one, which gives the 1–4 linkage for the structure of cellulose. Cellulose is a polymer of high molecular weight.

According to Ward (1969) "cellulose is a β-1,4-glucan of considerable size; the macromolecules are held together with hydrogen bonds to form a highly ordered fibrillar structure."

Inulin

Inulin, a homopolysaccharide, is made up of D-fructose units. It is a linear polyfructosan, the fructose units of which are joined by β-(2→1) glucosidic linkages. In addition, research has shown (Palmer 1951) and (Täufel and Steinbach 1959) that small amounts of glucose are connected with the chain. It is readily soluble in hot water, but only sparingly so in cold. It is a white amorphous powder, easily hydrolyzed by acids. Since it is not hydrolyzed by the enzymes of the gastrointestinal tract it has no value as a nutrient.

It is found in the Jerusalem artichoke and in dahlia bulbs. It is present also in garlic and onion, and in dandelion roots.

Glycogen

This is the polysaccharide storage form in animals which corresponds to starch in plants. It is found in the liver and in muscle tissues. Glycogen is branched like amylopectin, but more so, and its molecular weight is higher. Glycogen is made of repeating glucose units joined together by α-(1→4) linkages. The branched points are joined by α-(1→6) linkages. Glycogen has shorter linear chains than amylopectin.

Glycogen can be hydrolyzed to yield glucose, maltose, and dextrins. Glycogen becomes opalescent in solution, which is dextrorotatory. Glycogen gives a violet-brownish red color with iodine.

Glycogen has an important biochemical function. In the liver it is the important reserve for maintaining the level of glucose in the blood. In the muscles it is the source of energy necessary for contraction.

Chitin

Chitin is the polysaccharide that makes up the hard shells of crustaceans. The units of chitin are linked together as they are in cellulose; and the chitin chain is like the cellulose chain except that an $NHCOCH_3$ group at the C-2 atom in every glucose unit takes the place of the hydroxyl group.

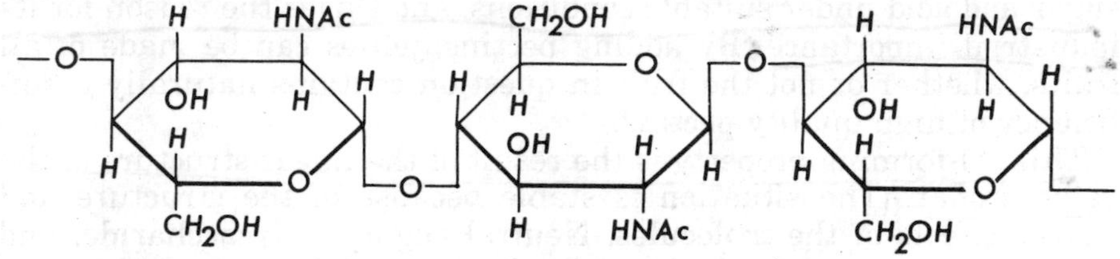

Segment of a chitin molecule

On hydrolysis chitin yields D-glucosamine (2-amino-2-deoxy-D-glucose), an amino sugar.

Pectin

Pectins are polysaccharides that are present in all the higher plants in the intercellular or middle lamella region and in the cell walls.

The building block of the pectic substances is D-galacturonic acid.

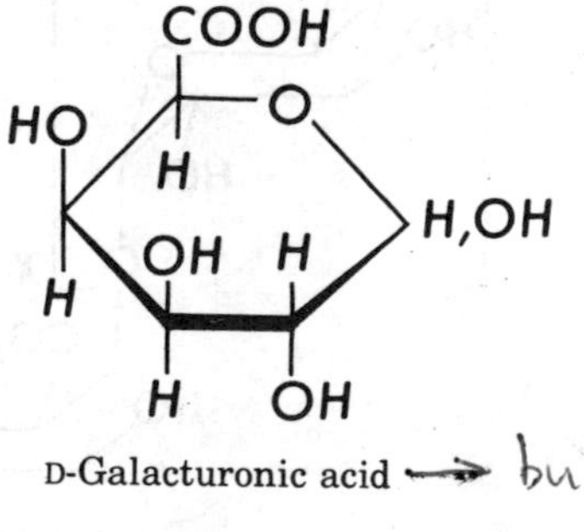

D-Galacturonic acid → building block

Pectins from all sources contain D-galacturonic acid, D-galactose, L-arabinose, and L-rhamnose. These are in the long chains and are known as 1,4-α-D-galacturonan in the partial methyl ester form, a branched L-arabinan, and a 1,4-β-D-galactan. It has been reported that apple pectinic acid contains 87% D-galacturonic acid, 9.3% L-arabinose, 1.4% D-galactose, 1.2% L-rhamnose, 0.9% D-xylose, and traces of fucose, 2-*O*-methylfucose, and 2-*O*-methylxylose (Barrett and Northcote 1965).

Galacturonans (see Fig. 4.6), which on hydrolysis yield only *m*-galacturonic acid, have been obtained. However, separation of the components of pectin is difficult.

Most pectins contain about 9–12% methoxyl ester, but strawberry pectin is much lower, about 0.2%. The highest specific optical rotation seems to be associated with the highest D-galacturonic acid content. Ordinarily the specific optical rotations to the sodium line are +230°–+250°.

Pectin is the substance used in the production of jelly. It reacts with sugar and acid under suitable conditions, and this is the reason for its industrial importance. By adding pectins, jellies can be made of all fruits, whether or not the fruit in question contains naturally a sufficiency of high quality pectin.

This gel-forming property is the result of the linear structure of the galacturonan. The situation is stable because of the structure and charge effects of the molecules. Neutral sugars, polysaccharide, and methylester groups attached to the galacturonan chains keep the poly-

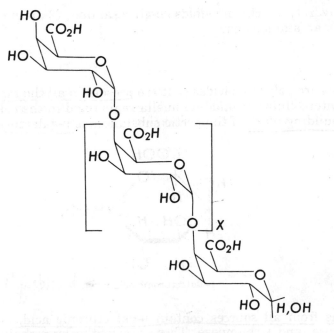

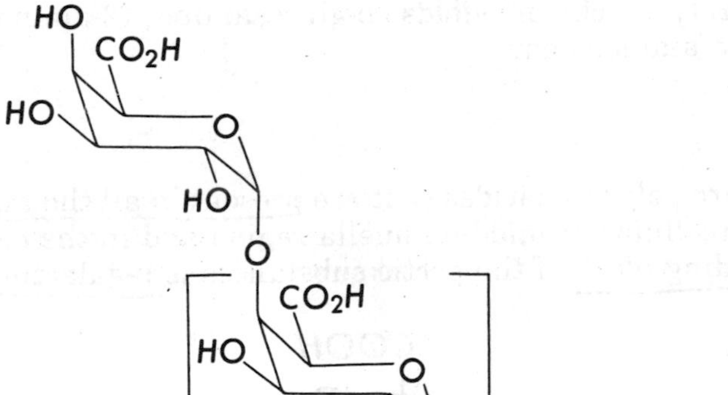

HO
CO_2H
HO
O
HO
O
CO_2H
HO
O
HO
O
X
CO_2H
HO
O
HO
H,OH

FIG. 4.6. Galacturonan from Amabilis fir.
From Whistler (1965).

saccharide chains from an extensive association. This permits the formation of a space that can fill with and hold very large volumes of water and solute molecules of low molecular weight substances. Pectins can be gelled or firmed by means of polyvalent cations such as calcium which bring about intermolecular salt bridging between free carboxyl groups and the polyvalent ions.

Because of the presence of pectin in the intercellular or middle lamella region and to some extent in the cell wall, the texture of fruits and vegetables is materially affected.

In the United States the main sources of pectin are apple pomace and citrus peel (Whistler 1969).

Hemicelluloses

Hemicelluloses are the cell wall polysaccharides other than pectin and cellulose that occur in terrestrial plants. They are alkali soluble, but the solution also extracts some of the pectin materials, and the separation, therefore, is not strictly sharp. Hemicelluloses are classified according to the sugars present in the molecule. While some hemicelluloses are made up of single sugar units, most are made up of two to four different sugar units, and are, therefore, heteroglycans. A mannan is a polymer of mannose units and a xylan is a polymer of xylose units. They are no longer considered intermediates in the biosynthesis of cellulose as they formerly were. They are a separate group of polysaccharides.

A large number of hemicelluloses have been investigated. They are potentially important industrial gums and could find broad application in the food industry if they could be obtained at low cost. A large possible source is corn hulls. Hemicellulose can be extracted with calcium hydroxide, in which the hemicellulose is soluble. This gum is entirely soluble in water and it is molecularly homogeneous, since it is an acidic arabinoglucuronoxylan. Its solutions have many similarities to those of gum arabic (Whistler 1969).

Seaweed Gums

Seaweed gums as a group of hydrocolloids have extensive application in the food industry. Among the most important of these are algin, carrageenan, and agar (Whistler 1969).

Algin. Algin is sodium alginate and is extracted from *Macrocystis pyrifera,* the giant kelp found in the Pacific off the coasts of California and Mexico in shallow water.

The structure of algin is still not fully clear; however, it is a linear polymer of β-D-mannuronic acid units and that L-guluronic acid is present also. Different seaweed sources show variations in the ratio between D-mannuronic acid and L-guluronic acid present in the gum.

Algins behave like pectins in that they are gelled by calcium and other polyvalent cations. It is significant that if these polyvalent cations are present in a solution along with monovalent cations the former will be selectively removed by the algins. Algins gelled by calcium have important food applications. The hydroxylpropyl ester of alginic acid is used in concentrations of 0.1 to 0.5% as an ice cream stabilizer. It should be noted that this particular compound is not precipitated or gelled by polyvalent cations.

Alginic acid precipitates as a gel under acid conditions which repress the ionization of the carboxyl groups.

Carrageenan. Carrageenan is a mixture of two sulfated polysaccharides from the seaweeds known as Irish moss, *Chondrus crispus,* and *Gigartina stellata.* These seaweeds are found along the North Atlantic coasts from Rhode Island to Newfoundland, from Norway to the shores of North Africa, and has been reported near the coast of Chile and also in the South Pacific.

Carrageenans range in sulfate ester content from 20 to 36%, depending on a variety of factors such as source and environmental conditions. Carrageenans were originally held to be two fractions, kappa and lambda, but more recent work has shown that they consist of a series of molecules of different chemical composition and solubility. The two fractions idea came from the fact that κ carrageenans are precipitated from a carrageenan solution with KCl while the λ carrageenans remain in solution. The κ carrageenans and other related polysaccharides, such as furcellaran, are capable of forming insoluble gels in the presence of potassium ions.

Agar. Agar is obtained from different species of *Gelidium* found off the coasts of Japan. Agar is capable of forming gels in concentrations as low as 0.04% and is used as a medium for the growth of bacteria. Agaran (agarose) was isolated from the soluble portion of agar from *Gelidium amansii* and was shown to be an alternating copolymer of 3,6-anhydro-α-L-galactopyranosyl and β-D-galactopyranosyl units joined by 1→3 and 1→4 linkages (Whistler 1969):

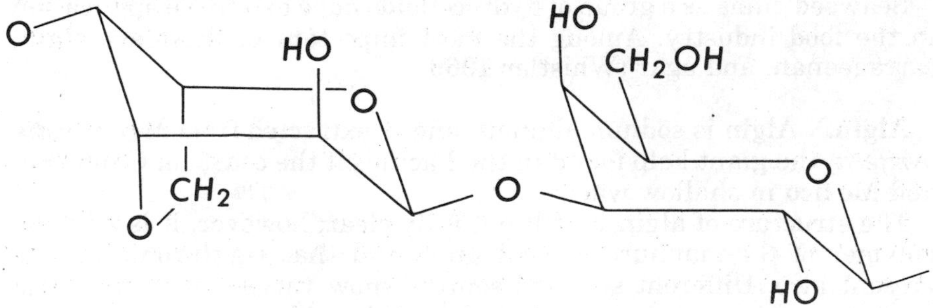

Agaran (agarose) from *Gelidium amansii*

The aqueous gels have a melting temperature that depends on the quality, concentration, and source of the agar used; an aqueous gel of 1.5% strength may melt from 30° to 97°C.

Danish Agar. This comes from a red seaweed, ***Furcellaria fastigiata,*** off the coasts of Denmark and Norway and is known also as furcellaran. At about 40°C the aqueous solutions of this gum gel. The resulting gel is increased in strength by the addition of galactomannans. Furcellaran can be used in the preparation of jellies, jams, and marmalades, as well as in pie fillings, glazes, and icings.

Exudate Gums

Gum Arabic. Gum arabic (acacia) is an exudate from trees of the *Acacia* genus, that grow in tropical and subtropical regions. Gum arabic is one of the oldest known of the commercial gums. Large quantities come from the Sudan, West Africa, Nigeria, Tanzania, Morocco, and India.

Gum arabic is a highly branched and slightly acidic polysaccharide. The main chain of the polysaccharide is composed of D-galactopyranose units connected by β-D-(1→4) and β-D-(1→6) linkages. Side chains made up of D-galactopyranose units are joined usually by β-D-(1→3) linkages. To these side chains L-rhamnopyranose or L-arabinofuranose residues are attached as end units. D-Glucuronic acid units are frequently attached by β-D-(1→6) linkages to D-galactose units and often L-arabinofuranose units are attached to the D-glucuronic acid units by (1→4) bonds. Most of the L-rhamnose is attached to D-glucuronopyranosyl units as 4-D-α-L-rhamnopyranosyl nonreducing terminal units (Aspinall *et al.* 1963; Aspinall and Young 1965). Gum arabic is a complex salt of calcium, magnesium, and potassium with arabic acid.

The molecular weight of gum arabic is in the range of 200,000 to 270,000. It is extremely soluble in water; as a result of its low molecular weight and branch structure it is necessary to use it in higher concentrations than most gums in order to obtain solutions of significant viscosity.

A little over half of the gum arabic imported into the United States is used by the food industry. It is nontoxic, colorless, tasteless, and odorless and does not affect these qualities in other ingredients. Gum arabic influences the viscosity, texture, and body of foods. It improves the quality of these products by preventing crystallization of sugar in confectionery and its thickening power is used in candy glaze and in chewing gums, cough drops, and lozenges. It is used as a stabilizer in such frozen desserts as ice cream, sherbets, and ices.

Tragacanth. Tragacanth is another of the important exudate gums. It is produced in the Near East and is an exudate of a few varieties of small shrub-like plants belonging to the *Astragalus* species. It is one of the oldest of the known gums.

Tragacanth is made up of two fractions: one is soluble in water, the other is only swellable. The structure is not fully known, but is highly branched. Tragacanthic acid has a main chain of 1,4-α-D-galacturonopyranosyl units with side chains attached at C-3. It is used frequently as an additive in salad dressings because of its relative resistance to hydrolysis (Whistler 1969).

DIGESTION OF CARBOHYDRATES

While the saliva can act on starch, it is questionable whether it exerts much influence because of the limited time starchy foods are exposed to it, and also because the acid of the gastric juice has the ability to stop this action. Enzymes concerned with the hydrolysis of starches are not present in the gastric juice.

Starch and glycogen are digested mainly in the small intestine. The major agent of this digestion is pancreatic amylase. This enzyme is most active at pH 7.1, it is stabilized by Ca^{2+}, and the Cl^- ion must be present. The starch granule can be digested by α-amylase whether it is disrupted by previous cooking or whether it is intact. Pancreatic amylase is 1,4-α-D-glucanohydrolase and is capable of endohydrolysis of α-D-(1→4)-glucosidic linkages in polysaccharides which contain α-1,4-linked D-glucose units.

Since the enzymes in the gastrointestinal tract in man are only capable of attacking polysaccharides that have α-1,4 linkages, only such polysaccharides are digestible by man. Under normal conditions carbohydrates in the intestine are converted into monosaccharides, which are absorbed by the intestinal mucosa. This does not seem to be a case of simple diffusion.

QUALITATIVE TESTS

Carbohydrates

A number of color tests and other tests have been developed for the detection of the different types of sugars. Tests are not available, however, for all such compounds.

Anthrone Color Reaction. This is a general test for carbohydrates. It was devised by Dreywood (1946) and depends on the formation of a blue-green or green color when a solution of the carbohydrate is treated with anthrone dissolved in concentrated sulfuric acid. It does not differentiate among the individual compounds.

Molisch Reaction. A test for the presence of carbohydrate, this reaction is given by all members of the carbohydrate group that are able to produce furfural or similar degradation products in trace amounts. The reacting material is naphthol and the color is produced at the junction between a layer of the test solution and concentrated sulfuric acid. The ring produced at the interface is reddish-violet if carbohydrate is present.

Monosaccharides

Fehling's Solution Test. This test is based on the principle of the reduction of the cupric ion to the cuprous. Solutions of cupric sulfate and alkaline sodium potassium tartrate are mixed together and then with the reducing sugar solution and heated. A positive reaction shows the red precipitate of cuprous oxide, or with lesser amounts of reducing sugar, a green or red, or a reddish yellow color. This test works for any reducing sugar, monosaccharide or oligosaccharide. Although the reaction that occurs is rather complex, it can be illustrated as follows (AOAC 1980; Fehling 1849).

$$2\,Cu^{2+} + \begin{matrix} H\diagdown C \diagup\!\!\!\!= O \\ | \\ HC{-}OH \\ | \end{matrix} \longrightarrow \begin{matrix} HO\diagdown C \diagup\!\!\!\!= O \\ | \\ HC{-}OH \end{matrix} + Cu_2O$$

D-Glucose Test. A qualitative test is available for D-glucose which is based on the action of β-D-glucose:O_2 oxidoreductase (1.1.3.4), catalase, and *o*-tolidine. Strips prepared with this mixture are put into the sugar solution, previously neutralized, and in a short time a blue color results. Basically, the glucose oxidase (1.1.3.4) forms D-gluconic acid lactone and liberates H_2O_2, which is acted on by the catalase, liberating oxygen. The *o*-tolidine is acted on giving the blue color.

D-Galactose Test. This test involves a coupled enzyme system. The enzyme galactose oxidase is basic to this system because of its action on D-galactose. Hydrogen peroxide is produced which is broken down to oxygen and water. The leuco compound in the mixture is oxidized to the reddish colored compound which shows the presence of D-galactose in the sample tested.[1]

Seliwanoff Color Test. This test involves heating the sample with hydrochloric acid and resorcinol. A red color results if ketose sugars are present.

---

[1] Galactostat Reagent Set is made by Worthington Biochemical Corporation, Freehold, N.J., and is designed for this test.

Bial Test for Pentoses. The sample is heated for a short time with a solution of orcinol in hydrochloric acid containing ferric ions. If pentoses are present in the sample, a green color results.

Qualitative Paper Chromatography. One method for the qualitative detection and identification of the monosaccharides involves the use of filter paper chromatography (Hodge and Hofreiter 1962; Partridge 1946, 1948). A small drop of a solution of the unknown sugar or mixture is placed a short distance from the end of the filter paper strip. This end is placed into the recommended solvent with the spot above the solvent surface. The solvent then moves by capillary attraction up the paper, and separates the mixture of sugars. After the sheet is dried it is sprayed with a suitable indicator to make the sugar spots visible. Since the sugars move at different speeds, the distance each has moved aids in the identification. Another method involves the use of thin-layer chromatography plates (Hough and Jones 1962).

Oligosaccharides

Sucrose. This substance responds to the Raybin test (Raybin 1933, 1937). This reaction takes place in the cold. Sucrose in solution with sodium hydroxide is shaken with a little diazouracil. When the reagent has dissolved, a blue-green color results. Positive tests are given by certain higher oligosaccharides such as raffinose and stachyose.

Test strips which are commercially available can be used for qualitative work involving maltose, lactose, and sucrose.

As in the case of monosaccharides, paper chromatography can be used for the detection of oligosaccharides. Other types of chromatography can be used also.

QUANTITATIVE TESTS

Quantitatively, sugars are determined by means of optical rotation and by chemical methods involving the reduction of cupric sulfate and weighing the precipitated Cu_2O. From this weight, the equivalent value of glucose, fructose, sucrose, or lactose can be obtained from the Hammond or Munson-Walker tables. Instead of weighing, one or two other methods are employed: titration with sodium thiosulfate and titration with potassium permanganate. In either case standard solutions of these reagents are necessary (AOAC 1980). Other modifications making use of the same basic chemical reaction are used.

Because these methods apply to reducing sugars only, it is necessary to hydrolyze nonreducing sugars like sucrose before making the determination of this substance.

More recent methods involve the use of several forms of chromatography, such as paper chromatography, zone electrophoresis, thin-layer chromatography, column chromatography, and gas chromatography.

The latter has a number of advantages, but requires expensive equipment.

SUMMARY

Carbohydrates are widely spread throughout the world and, of all the biological substances in the plant kingdom other than water, they are present in the largest amount. They are compounds of carbon, hydrogen, and oxygen; usually, but not always, the hydrogen and oxygen are present in the ratio of 2:1, as in water.

Carbohydrates are formed in plants by the process of photosynthesis.

Carbohydrates are classified as follows: monosaccharides; oligosaccharides, which include disaccharides, trisaccharides, tetrasaccharides, and others; and polysaccharides, which include cellulose and starch, among others. Among the monosaccharides, the food chemist is primarily interested in the hexoses and to a lesser extent in the pentoses. Glucose and fructose are found in large amounts in the free as well as in the combined state.

Optical rotation is an important property of sugars. This involves the rotation of the plane of polarized light. Glyceraldehyde is the simplest aldose. It contains one asymmetric carbon atom, and because of this it has two isomers.

The structure of the carbohydrate molecule is important because it governs the reactions and properties of the compound.

Mutarotation is the change which takes place in the optical rotation of a freshly prepared solution of a sugar like glucose when it is allowed to stand. This is the result of the equilibrium between the α and β forms of the pyranose ring structure. The open chain formula alone cannot explain this phenomenon.

Sugars exist in the cyclic form. Considerable discussion is given to the structural formulas of sugars.

A number of interesting reactions result from the treatment of reducing sugars with alkali. In this connection consideration is given to the important Fehling's method for the determination of reducing sugars and the chemical reactions involved.

Because of the importance of the hydroxyl group in sugars reactions involving this group are discussed.

Oligosaccharides are small polymers formed by the combination of two to about ten monosaccharides with the elimination of water. They are linked together with a glycosidic bond. Sucrose is an important disaccharide. It is nonreducing because in the formation of the glucosidic linkage the anomeric carbonyl groups of both glucose and fructose are used.

A number of the oligosaccharides have important industrial uses.

Polysaccharides are large or very large molecules, although some are of low molecular weight. The empirical formula for the hexosans is

$(C_6H_{10}O_5)_x$. These compounds are important in the structural tissues—cellulose in plants and chitin in marine life. Starch is the storage material in plants while glycogen has this function in animals. Starch is made up of amylose, which exists as chains of glucose units while amylopectin has a branched pattern.

The biosynthesis of amylose and amylopectin is described. A rather complete discussion of starch is given together with other polysaccharides.

BIBLIOGRAPHY

ACS. 1954. Natural Plant Hydrocolloids. Adv. Chem. Ser. 11. American Chemical Society, Washington, DC.

ACS. 1963. Rules of carbohydrate nomenclature. J. Org. Chem. *28,* 281–291.

ALBRECHT, J. J., NELSON, A. I., and STEINBERG, M. P. 1960A. Characteristics of corn starch and starch derivatives as affected by freezing, storage, and thawing. I. Simple systems. Food Technol. *14,* 57–63.

ALBRECHT, J. J., NELSON, A. I., and STEINBERG, M. P. 1960B. Characteristics of corn starch and starch derivatives as affected by freezing, storage, and thawing. II. White sauces. Food Technol. *14,* 64–68.

AMARAL, D., BERNSTEIN, L., MORSE, D., and HORECKER, B. L. 1963. Galactose oxidase of *Polyporus circinatus:* a copper enzyme. J. Biol. Chem. *238,* 2281–2284.

AOAC. 1980. Official Methods of Analysis, 13th Ed. Association of Official Analytical Chemists, Washington, DC.

ARAKI, C., and HIRASE, S. 1960. Chemical constitution of agar-agar. XXI. Reinvestigation of methylated agarose of *Gelidium amansii.* Bull. Chem. Soc. Jpn. *33,* 291–295.

ARAKI, C., and HIRASE, S. 1960. Studies on the chemical constitution of agar-agar. XXII. Partial methanolysis of methylated agarose of *Gelidium amansii.* Bull. Chem. Soc. Jpn. *33,* 597–600.

ARBUCKLE, A. W., and GREENWOOD, C. T. 1958. Physicochemical studies on starches. Part XIV. The effect of acid on wheat-starch granules. J. Chem. Soc. 2629–2631.

ASPINALL, G. O. 1969. Gums and mucilages. *In* Advances in Carbohydrate Chemistry and Biochemistry, Vol. 24. M.L. Wolfrom and R.S. Tipson (Editors). Academic Press, New York.

ASPINALL, G. O., and YOUNG, R. 1965. Further oligosaccharides from carboxyl-reduced gum arabic. J. Chem. Soc. 3003–3004.

ASPINALL, G. O., CHARLSON, A. J., HIRST, E. L., and YOUNG, R. 1963. The location of L-rhamnopyranose residues in gum arabic. J. Chem. Soc. 1696–1702.

ATHANASSIADIS, H., and BERGER, G. 1973. γ-Radiation-induced acidity in corn starch. Staerke *25,* 362–367. (French)

AVIGAD, G., AMARAL, D., ASENSIO, C., and HORECKER, B. L. 1962. The D-galactose oxidase of *Polyporus circinatus.* J. Biol. Chem. *237,* 2736–2741.

BALLS, A. K., THOMPSON, R. R., and WALDEN, M. K. 1946. A crystalline protein with β-amylase activity, prepared from sweet potatoes. J. Biol. Chem. *163,* 571–572.

BANKS, W., GREENWOOD, C. T., and MUIR, D. D. 1973. Characterization of starch and its components. 5. Quantitative acid hydrolysis of starch and glycogen. Staerke *25,* 405–408.

BARRETT, A. J., and NORTHCOTE, D. M. 1965. Apple fruit pectic substances. Biochem. J. *94,* 617–627.

BATES, F. J. *et al.* 1942. Polarimetry, saccharimetry, and the sugars. Natl. Bur. Stand. Circ. C440. U.S. Dept. Commerce, Washington, DC.

BEAN, M. L., and OSMAN, E. M. 1959. Behavior of starch during food preparation. II. Effects of different sugars on the viscosity and gel strength of starch pastes. Food Res. *24,* 665–671.

BEMILLER, J. N. 1965A. Acid hydrolysis and other lytic reactions of starch. *In* Starch: Chemistry and Technology, Vol. 1, R. L. Whistler and E. F. Paschall (Editors). Academic Press, New York.

BEMILLER, J. N. 1965B. Alkaline degradation of starch. *In* Starch: Chemistry and Technology, Vol. 1, R. L. Whistler and E. F. Paschall (Editors). Academic Press, New York.

BIRCH, G. G. and PARKER, K. J. (Editors) 1979. Sugar: Science and Technology. Applied Science Publishers Ltd., London.

CALLOWAY, D. H., HICKEY, C. A., and MURPHY, E. L. 1971. Reduction of intestinal gas-forming properties of legumes by traditional and experimental food processing methods. J. Food Sci. *36,* 251–255.

COWIE, J. M. G., and GREENWOOD, C. T. 1957. Physicochemical studies on starches. V. The effect of acid on potato starch granules. J. Chem. Soc. 2658–2665.

DAHLE, L., BRUSCO, V., and HARGUS, G. 1973. Some effects of beta amylolytic degradation of pastes of waxy maize starch. J. Food Sci. *38,* 484–485.

DOESBURG, J. J. 1965. Pectic Substances in Fresh and Preserved Fruits and Vegetables, Communication 25. Institute for Research on Storage and Processing of Horticultural Produce, Wageningen, The Netherlands.

DONER, L. W. 1977. The sugar of honey—A review. J. Sci. Food Agric. *28,* 443–456.

DREYWOOD, R. 1946. Qualitative test for carbohydrate material. Ind. Eng. Chem. Anal. Ed. *18,* 499.

ELBEIN, A. D. 1974. The metabolism of α,α-trehalose. *In* Advances in Carbohydrate Chemistry and Biochemistry, Vol. 30, R. S. Tipson and D. Horton (Editors), Academic Press, New York.

ERLANDER, S. R. 1970. Mechanism for the synthesis of starch and its relation to flagellin and to the newly proposed structural model for DNA. II. Staerke *22,* 393–401.

FEHLING, H. 1849. The quantitative determination of sugar and starch flour by means of copper sulfate. Annu. Chem. Pharm. *72,* 106–113. (German)

FISCHER, E. 1891. On the configuration of grape sugar and its isomers. Ber. Dtsch. Chem. Ges. *24,* 1836–1845. (German)

FOSTER, J. F. 1965. Physical properties of amylose and amylopectin in solution. *In* Starch: Chemistry and Technology, Vol. 1, R. L. Whistler, and E. F. Paschall (Editors). Academic Press, New York.

FRENCH, D. 1966. The contribution of α-amylases to the structural determination of glycogen and starch. Biochem. J. *100,* 2P.

FRENCH, D. 1969. Physical and chemical structure of starch and glycogen. *In* Symposium on Foods: Carbohydrates and Their Roles. H. W. Schultz, R. F. Cain, and R. Wrolstad. AVI Publishing Co., Westport, CT.

FRYDMAN, R. B., and CARDINI, C. E. 1967. Studies on the biosynthesis of starch. II. Some properties of the adenosine diphosphate glucose: starch glucosyltransferase bound to the starch granule. J. Biol. Chem. *242,* 312–317.

GALLANT, D., MERCIER, C., and GUILBOT, A. 1972. Electron microscopy of starch granules modified by bacterial α-amylase. Cereal Chem. *49,* 354–365.

GEDDES, R. 1969. Starch biosynthesis. Q. Rev. (London) *23*(1), 57–72.

GHOSH, H. P., and PREISS, J. 1965. The biosynthesis of starch in spinach chloroplasts. J. Biol. Chem. *240,* PC960–PC962.

GHOSH, H. P., and PREISS, J. 1966. Adenosine diphosphate glucose pyrophosphorylase. A regulatory enzyme in the biosynthesis of starch in spinach leaf chloroplasts. J. Biol. Chem. *241,* 4491–4504.

GIBBS, M., EARL, J. M., and RITCHIE, J. L. 1955. Metabolism of ribose-1-^{14}C by cell-free extracts of yeast. J. Biol. Chem. *217,* 161–168.

GOTLIEB, K. F., and WOLDENDROP, P. 1967. Some properties of cross-linked potato starch. Staerke *19,* 263–271. (German)

GOTTSCHALK, A. 1945. Yeast hexokinase and its substrates D-fructofuranose and D-glucose. Nature *156,* 540–541.

GREENWOOD, C. T. 1956. Aspects of the physical chemistry of starch. *In* Advances in Carbohydrate Chemistry, Vol. 11, M. L. Wolfrom and R. S. Tipson (Editors). Academic Press, New York.

HAWORTH, W. N. 1928. The Constitution of Sugars. Edward Arnold and Co., London.

HAWORTH, W. N., HIRST, E. L., and LEARNER, A. 1927. The structure of normal fructose: crystalline tetramethyl β-methyl-fructoside and crystalline tetramethyl fructose (1:3:4:5). J. Chem. Soc. 1040–1049.

HAWORTH, W. N., HIRST, E. L., and LEARNER, A. 1927. 1:3:4:6-Tetramethyl (γ)-fructose and 2:3:5-trimethyl (γ)-arabinose. Oxidation of *d-* and *l*-trimethyl-γ-arabonolactone. J. Chem. Soc. 2432–2436.

HODGE, J. E., and HOFREITER, B. T. 1962. Determination of reducing sugars and carbohydrates. *In* Methods in Carbohydrate Chemistry, Vol. 1, R.L. Whistler and M.L. Wolfrom (Editors). Academic Press, New York.

HODGE, J. E., RENDLEMAN, J. A., and NELSON, E. C. 1972. Useful properties of maltose. Cereal Sc. Today *17,* 180–188.

HOLLÓ, J., and SZEJTLI, J. 1961. Acid hydrolysis of the glycosidic bond. IV. The influence of the nature and concentration of different acids on the hydrolysis of starch. Staerke *13,* 327–331.

HOUGH, L. and JONES, J. K. N. 1962. Chromotography on paper. *In* Methods in Carbohydrate Chemistry, Vol. 1, R. L. Whistler and M.L. Wolfrom (Editors). Academic Press, New York.

HOUGH, L. and JONES, J. V. S. 1962. Thin-layer chromatography. *In* Methods in Carbohydrate Chemistry, Vol. 1, pp. 21-31, R.L. Whistler and M.L. Wolfrom (Editors). Academic Press, New York.

HUDSON, C. S. 1941. Emil Fischer's discovery of the configuration of glucose. J. Chem. Ed. *18,* 353–357.

HULLINGER, C. H., van PATTEN, E., and FRECK, J. A. 1973. Food applications of high amylose starches. Food Technol. *27,* No. 3, 22–24.

INGLETT, G. E. 1974. Symposium: Sweeteners. AVI Publishing Co., Westport, CT.

ISBEL, H. S. 1944. Interpretation of some reactions in the carbohydrate field in terms of consecutive electron displacement. J. Res. Natl. Bur. Stand. *32,* 45–59.

JOSLYN, M. A. 1970. Methods in Food Analysis, 2nd Edition. Academic Press, New York.

JUNK, W. R., and PANCOAST, H. M. 1973. Handbook of Sugars. AVI Publishing Co., Westport, CT.

KARYAKINA, A. B., LUK'YANOV, A. B., and BEKSLER, B. A. 1971. Effect of the concentration of the starting suspension on the properties of swollen corn starch. Sakh. Prom. *45,* No. 12, 54–57. (Russian)

KEARSLEY, N. W. 1978. The control of hydroscopicity, browning, and fermentation in glucose syrups. J. Food Technol. *13,* 339–348.

KERTESZ, Z. I. 1951. The Pectic Substances. John Wiley and Sons, New York.

LEACH, H. W. 1965. Gelatinization of starch. *In* Starch: Chemistry and Technology, Vol. 1, R.L. Whistler and E.F. Paschall (Editors). Academic Press, New York.

LEE, C. Y., SHALLENBERGER, R. S., and VITTUM, M. T. 1970. Free sugars in fruits and vegetables. New York Food and Life Sci. Bull. No. 1. Dept. of Food Sciences, Cornell University, Geneva, NY.

MacLAURIN, D. J. and GREEN, J. W. 1969. Carbohydrates in alkaline systems. I. Kinetics of the transformation and degradation of D-glucose, D-fructose, and D-mannose in 1 *M* sodium hydroxide. Can. J. Chem. *47,* 3947–3955.

MANNERS, D. J. 1966. The contribution of β-amylase to the structural determination of glycogen and starch. Biochem. J. *100,* 2P.

MARSHALL, J. J. 1974. Application of enzymic methods to the structural analysis of polysaccharides. I. Advances in Carbohydrate Chemistry and Biochemistry, Vol. 30, H. S. Tipson and D. Horton (Editors). Academic Press, New York.

MEYER, K. H., and BERNFELD, P. 1940. Research on Starch. V. Amylopectin. Helv. Chim. Acta *23,* 875–885. (French)

MITCHELL, W. A. 1972. Analyzing the metaphosphate stabilization reaction of starch by acid titration. Food Technol. *26,* No. 3, 34–42, 79.

MONTGOMERY, R. 1969. The chemical analysis of carbohydrates. *In* Symposium on Foods: Carbohydrates and Their Roles. H. W. Schultz, R. F. Cain, and R. R. Wrolstad, AVI Publishing Co., Westport, CT.

MORROW, L., and LORENZ, K. 1974. Some physicochemical properties of starches as affected by changes in atmospheric pressure. J. Food. Sci. *39,* 467–470.

MOUSSERI, J., STEINBERG, M. P., NELSON, A. I., and WEI, L. S. 1974. Bound water capacity of corn starch and its derivatives by NMR. J. Food Sci. *39,* 114–116.

MURATA, T., and AKAZAWA, T. 1969. Enzymic mechanism of starch synthesis in sweet potato roots. II. Enhancement of the starch synthetase activity by malto-oligosaccharides. Arch. Biochem. Biophys. *130,* 604–609.

NATL. ACAD. SCI.–NATL. RES. COUNCIL. 1972. Food Chemicals Codex, 2nd Edition. Committee on Specifications, National Academy of Sciences–Natl. Res. Council, Washington, DC.

NOMURA, T., NAKAYAMA, N., MURATA, T. and AKAZAWA, T. 1967. Biosynthesis of starch in chloroplasts. Plant Physiol. *42,* 327–332.

OKADA, G., and HEHRE, E. J. 1974. New studies on amylosucrase, a bacterial α-D-glucosylase that directly converts sucrose to a glycogen-like α-glucan. J. Biol. Chem. *249,* 126–135.

OLSON, A. C., BECKER, R., MIERS, J. C., GUMBMANN, M. R., and WAGNER, J. R. 1975. Problems in digestibility of dry beans. Nutr. Clin. Nutr. *1* 551–563.

OSMAN, E. M., and DIX, M. R. 1960. Effects of fats and nonionic surface-active agents on starch pastes. Cereal Chem. *37,* 464–475.

PALMER, A. 1951. The specific determination and detection of glucose as a probable constituent radical of certain fructosans by means of notatin. Biochem. J. *48,* 389–394.

PARK, Y. K., and LIMA, D. C. 1973. Continuous conversion of starch to glucose by an amyloglucosidase-resin complex. J. Food Sci. *38,* 358–359.

PARTRIDGE, S. M. 1946. Application of the paper partition chromatography to the qualitative analysis of reducing sugars. Nature *158,* 270–271.

PARTRIDGE, S. M. 1948. Filter paper partition chromatography of sugars. 1. General description and application to the qualitative analysis of sugars in apple juice, egg white, and foetal blood of sheep. Biochem. J. *42,* 238–248.

PAZUR, J. H. 1955. Reversibility of enzymatic transglucosylation reactions. J. Biol. Chem. *216,* 531–538.

PAZUR, J. H. 1965. Enzymes in synthesis and hydrolysis of starch. *In* Starch: Chemistry and Technology, Vol. 1, R. L. Whistler and E. F. Paschall (Editors). Academic Press, New York.

PAZUR, J. H., and ANDO, T. 1959. The action of an amyloglucosidase of *Aspergillus niger* on starch and malto-oligosaccharides. J. Biol. Chem. *234,* 1966–1970.

PAZUR, J. H., and KLEPPE, K. 1962. The hydrolysis of α-D-glucosides by amyloglucosidase from *Aspergillus niger*. J. Biol. Chem. *237,* 1002–1006.

PIGMAN, W., and HORTON, D. (Editors). 1972. The Carbohydrates, 2nd Edition, Vol. 1A. Academic Press, New York.

RACKIS, J. J. 1975. Flatulence problems associated with soy products. Proc. World Soybean Research Conf.

RAYBIN, H. W. 1933. A new color reaction with sucrose. J. Am. Chem. Soc. *55,* 2603–2604.

RAYBIN, H. W. 1937. The direct demonstration of the sucrose linkage in the oligosaccharides. J. Am. Chem. Soc. *59,* 1402–1403.

REYNOLDS, T. M. 1969. Nonenzymic browning. Sugar-amine interactions. *In* Symposium on Foods: Carbohydrates and Their Roles. H.W. Schultz, R.F. Cain, and R. Wrolstad (Editors). AVI Publishing Co. Westport, CT.

REYNOLDS, T. M. 1970. Flavors from nonenzymic browning reactions. Food Technol. Aust. *22,* 610–619.

ROBERTS, H. J. 1965. Nondegradative reactions of starch. *In* Starch: Chemistry and Technology, Vol. 1, R. L. Whistler, and E. F. Paschall (Editors). Academic Press, New York.

SAINI, V. 1968. Ionizing radiation effects on starch as shown by Staudinger Index and differential thermal analysis. J. Food Sci. *33,* 136–138.

SCHOCH, T. J. 1941. Physical aspects of starch behavior. Cereal Chem. *18,* 121–128.

SCHOCH, T. J. 1942. Fractionation of starch by selective precipitation with butanol. J. Am. Chem. Soc. *64,* 2957–2961.

SCHOCH, J. T., and MAYWALD, E. C. 1956. Microscopic examination of modified starches. Anal. Chem. *28,* 382–387.

SHALLENBERGER, R. S., and ACREE, T. E. 1971. Chemical structure of compounds and their sweet and bitter taste. *In* Handbook of Sensory Physiology, Vol. IV, L. M. Beidler (Editor). Springer-Verlag, Berlin and New York.

SHALLENBERGER, R. S., and BIRCH, G. G. 1975. Sugar Chemistry, AVI Publishing Co., Westport, CT.

SIDEBOTHAM, R. L. 1974. Dextrans. *In* Advances in Carbohydrate Chemistry and Biochemistry, Vol. 30, H. S. Tipson, and D. Horton (Editors), Academic Press, New York.

SINGH, D. V., and SHUKLA, R. N. 1973. Phosphorylase in starch metabolism. J. Sci. Ind. Res. *32,* 356–361.

SMITH, F., and MONTGOMERY, R. 1959. The Chemistry of Plant Gums and Mucilages and Some Related Polysaccharides. ACS Monograph Ser. 141. Reinhold Publishing Corp., New York.

SOWDEN, J. C. 1957. The saccharinic acids. *In* Advances in Carbohydrate Chemistry, Vol. 12. M.L. Wolfrom and R.S. Tipson (Editors). Academic Press, New York.

SOWDEN, J. C. and SCHAFFER, R. 1952. The isomerization of D-glucose by alkali in D_2O at 25°C. J. Am. Chem. Soc. *74,* 505–507.

STANEK, J., CERNY, M., KOCOUREK, J., and PACAK, J. 1963. The Monosaccharides. Academic Press, New York.

SWANSON, M. A., and CORI, C. F. 1948. Studies on the structure of polysaccharides. 1. Acid hydrolysis of starch-like polysaccharides. J. Biol. Chem. *172,* 797–804.

TÄUFEL, K., and STEINBACH, K. J. 1959. Glucose as a constituent of inulin. Nahrung *3,* 457–463. (German, English summary)

THOMA, J. A. 1965. The oligo- and megalosaccharides of starch. *In* Starch: Chemistry and Technology, Vol. 1, R. L. Whistler and E. F. Paschall (Editors). Academic Press, New York.

TOLEDO, R., STEINBERG, M. P., and NELSON, A. I. 1968. Quantitative determination of bound water by NMR. J. Food Sci. 33, 315–317.

TOLLENS, B. 1883. On the behavior of dextrose to ammoniacal silver solution. Ber. Dtsch. Chem. Ges. *16,* 921–924.

TSUJISAKA, Y., FUKUMOTO, J., and YAMAMOTO, T. 1958. Specificity of crystalline saccharogenic amylase of moulds. Nature *181,* 770–771.

VAN'T HOFF, J. H. 1875. The formulas and structures in space. Bull. Soc. Chim. Fr. *23,* 295–301.

VOSE, J. R. 1977. Functional characteristics of an intermediate amylose starch from smooth-seeded field peas compared with corn and wheat starches. Cereal Chem. *54,* 1141–1151.

WARD, JR., K. 1969. Cellulose. *In* Symposium on Foods: Carbohydrates and Their Roles. R. W. Schultz (Editor). AVI Publishing Co., Westport, CT.

WATSON, S. A. 1964. Determination of starch gelatinization temperature. *In* Methods in Carbohydrate Chemistry, Vol. 4, R. L. Whistler and M. L. Wolfrom (Editors). Academic Press, New York.

WHISTLER, R. L. 1965. Fractionation of starch. *In* Starch: Chemistry and Technology, Vol. 1, R. L. Whistler and E. F. Paschall (Editors). Academic Press, New York.

WHISTLER, R. L. 1969. Pectins and gums. *In* Symposium on Foods: Carbohydrates and Their Roles. H. W. Schultz, R. F. Cain, and R. Wrolstad (Editors). AVI Publishing Co., Westport, CT.

WHISTLER, R. L., and HILBERT, G. E. 1945. Separation of amylose and amylopectin by certain nitroparaffins. J. Am. Chem. Soc. *67,* 1161–1165.

WHITE, A., HANDLER, P., SMITH, E. L., HILL, R. L., and LEHMAN, I. R. 1978. Principles of Biochemistry. 6th Edition. McGraw-Hill Book Company, New York

WOLFROM, M. L. 1969. Mono- and oligosaccharides. *In* Symposium on Foods: Carbohydrates and Their Roles. H. W. Schultz, R. F. Cain, and R. Wrolstad (Editors). AVI Publishing Company, Westport, CT.

WOLFROM, M. L., and EL KHADEM, H. 1965. Chemical evidence for the structure of starch. *In* Starch: Chemistry and Technology, Vol. 1, R. L. Whistler and E. F. Paschall (Editors). Academic Press, New York.

WURZBURG, O. B., and SZYMANSKI, C. D. 1970. Modified starches for the food industry. J. Agric. Food Chem. *18,* 997–1001.

ZASLOW, B. 1965. Crystalline nature of starch. *In* Starch: Chemistry and Technology, Vol. 1, R. L. Whistler and E. F. Paschall (Editors). Academic Press, New York.

5

Lipids

Simple Lipids
Composite Lipids
Derived Lipids
Oxidation of Lipids
Summary
Bibliography

The term "lipid" is used to describe a number of substances which are soluble in organic solvents (such as diethyl ether, petroleum ether, chloroform, carbon tetrachloride) but usually insoluble in water. A few compounds, such as lecithin, normally classed as lipids are somewhat soluble in water. All of these compounds can be utilized in animal metabolism. Petroleum products cannot be so utilized, and, therefore, are not included here.

Unlike proteins and carbohydrates, which are composed of basic units or "building blocks," amino acids in the former and monosaccharides in the latter, lipids are made up of various chemical substances; also, lipids are heterogeneous in nature and difficult to classify. They are divided into several groups: fatty acids, oils and fats, waxes, phospholipids, sphingolipids, together with such compounds as sterols, hydrocarbons, fat-soluble vitamins, carotenoids, and finally the combined lipids such as lipoproteins and lipopolysaccharides. Some consider fatty acids as one group; neutral fats and phospholipids, both of which contain glycerol, as another group; lipids which do not contain glycerol, such as sphingolipids, aliphatic alcohols, and waxes together with terpenes and steroids, a third group; and combined lipids a final group.

The Bloor classification (Bloor 1926) divides them into (1) simple lipids, consisting of neutral fats and waxes and is subdivided into true waxes, and cholesterol esters and vitamin A esters, and in addition vitamin D esters; (2) The compound lipids comprising phospholipids, which include lecithin, cephalins, and the phosphatidic acids; (3) the sphingolipids cerebrosides, and the sulfolipids; (4) the derived lipids or hydrolytic products, including fatty acids, alcohols of high molecular weight, hydrocarbons, carotenoids, together with vitamin D, E, and K. See Table 5.1 for a more detailed breakdown of the Bloor classification.

Any or all of these substances may be present in foods.

TABLE 5.1. Classification Scheme for the Lipids

1. Simple lipids—compounds containing two kinds of structural moieties
   - Glyceryl esters—these include partial glycerides as well as triglycerides, and are esters of glycerol and fatty acids
   - Cholesteryl esters—esters formed from cholesterol and a fatty acid
   - Waxes—a poorly defined group which consists of the true waxes (esters of long chain alcohols and fatty acids), vitamin A esters, and vitamin D esters
   - Ceramides—amides formed from sphingosine (and its analogs) and a fatty acid linked through the amino group of the base compound. The compounds formed with sphingosine are the most common
2. Composite lipids—compounds with more than two kinds of structural moieties
   - Glyceryl phosphatides—these compounds are classified as derivatives of phosphatidic acid
   - Phosphatidic acid—a diglyceride esterified to phosphoric acid
   - Phosphatidyl choline—more descriptive term for lecithin which consists of phosphatidic acid linked to choline
   - Phosphatidyl ethanolamine—often erroneously called cephalin, a term referring to phospholipids insoluble in alcohol
   - Phosphatidyl serine—also erroneously called cephalin
   - Phosphatidyl inositol—major member of a complex group of inositol-containing phosphatides including members with 2 or more phosphates
   - Diphosphatidyl glycerol—cardiolipin
3. Sphingolipids—best described as derivatives of ceramide, a unit structure common to all. However, as in the case of ceramide, the base can be any analog of sphingosine
   - Sphingomyelin—a phospholipid form best described as a ceramide phosphoryl choline
   - Cerebroside—a ceramide linked to a single sugar at the terminal hydroxyl group of the base and more accurately described as a ceramide monohexoside
   - Ceramide dihexosides—same structure as a cerebroside, but with a disaccharide linked to the base
   - Ceramide polyhexosides—same structure as a cerebroside, but with a trisaccharide or longer oligosaccharide moiety. May contain one or more amino sugars
   - Cerebroside sulfate—a ceramide monohexoside esterified to a sulfate group
   - Gangliosides—a complex group of glycolipids that are structurally similar to ceramide polyhexosides, but also contain 1 to 3 sialic acid residues. Most members contain an amino sugar in addition to the other sugars. However, not all gangliosides contain amino sugars
4. Derived lipids—compounds containing a single structural moiety that occur as such or are released from other lipids by hydrolysis
   - Fatty acids
   - Sterols
   - Fatty alcohols
   - Hydrocarbons—includes squalene and the carotenoids
   - Fat-soluble vitamins, A, D, E, and K

Source: Pomeranz and Meloan (1971).

SIMPLE LIPIDS

Fats and Oils

The largest single group under the lipid classification is made up of fats and oils. These are esters of fatty acids and glycerol. Since glycerol is a trihydroxy alcohol, monoacid, diacid, and triacid esters are known. Most of the naturally occurring fats and oils are composed of triglycerides. Specifically, fats are solid at room temperature while oils are liquid. The term fats can be used as a general term to cover these compounds, both liquid and solid. Triglycerides containing unsaturated fatty acids have lower melting points. The oils are usually higher in unsaturated fatty acids than the fats. Because fats and oils are esters of glycerol and fatty acids, they are composed only of carbon, hydrogen, and oxygen. Most of the saturated fatty acids found in nature are made up of an even number of carbon atoms, from 4 to 24, in straight chains, although small amounts of straight chain acids with an odd number of carbon atoms have been found in fats and oils. In addition, small amounts of branched fatty acids with an even or odd number of carbon atoms have been found in nature. The straight chain unsaturated fatty acids occur mainly as molecules with an even number of carbon atoms. However, those with an odd number of carbon atoms are known. Although fats and oils from plant and animal sources are mainly triglycerides, they are mixtures of similar compounds rather than pure substances. Furthermore, the distribution of the fatty acids in the original molecule can vary considerably. These fats and oils contain small amounts of mono- and diglycerides as well as the triglycerides. Processed fats contain up to 20% mono- and diglycerides.

The newer terminology uses the word triacylglycerols in place of triglycerides. Diacylglycerols is the corresponding term for diglycerides, and monoacylglycerols the term for monoglycerides. While this newer terminology is doubtless more accurate, because of the long use of the older terminology, both will be used in this book.

Fatty Acids

Saturated Fatty Acids. The saturated fatty acids correspond to the general formula $C_nH_{2n+1}COOH$. Unbranched monocarboxylic acids are the usual structure. Some exist which contain cyclic groups and hydroxyl groups. Because of this it is possible to classify according to structure.

The saturated fatty acids C_4 through C_8 are liquid at ordinary room temperature. Those with 10 or more carbon atoms are solid. It can be seen in Table 5.2 that length of chain and unsaturation influence melting point.

Solubility of fatty acids in water decreases with the increase in molecular weight. Butyric acid is miscible with water in all proportions. While caproic, caprylic, and capric acids are only slightly soluble, lauric and those of higher molecular weight are insoluble in water. The solubility is caused by the carboxyl group. The carboxyl group is hydrophilic while the carbon chain of the fatty acid is hydrophobic. Because of this composition, fatty acids have the property of spreading in a thin film, usually of monomolecular thickness, over the water. The carboxyl group is pointed downward, i.e., into the water.

TABLE 5.2. Fatty Acids in Edible Lipid Materials

Acid	Systematic Name[a]	Formula	Number of C Atoms	Melting Point (°C)
Saturated Acids	Saturated Acids			
Butyric	Butanoic	$CH_3(CH_2)_2COOH$	4	−4.3
Caproic	Hexanoic	$CH_2(CH_2)_4COOH$	6	−1.5
Caprylic	Octanoic	$CH_3(CH_2)_6COOH$	8	16.5
Capric	Decanoic	$CH_3(CH_2)_8COOH$	10	31.5
Lauric	Dodecanoic	$CH_3(CH_2)_{10}COOH$	12	44.0
Myristic	Tetradecanoic	$CH_3(CH_2)_{12}COOH$	14	58.0
Palmitic	Hexadecanoic	$CH_3(CH_2)_{14}COOH$	16	63.0
Stearic	Octadecanoic	$CH_3(CH_2)_{16}COOH$	18	71.8
Arachidic	Eicosanoic	$CH_3(CH_2)_{18}COOH$	20	77.0
Unsaturated Acids	Unsaturated Acids			
Palmitoleic	9-Hexadecenoic	$CH_3(CH_2)_5CH=CH(CH_2)_7COOH$ (*cis*)	16	
Oleic	*cis*-9-Octadecenoic	$CH_3(CH_2)_7CH=CH(CH_2)_7COOH$ (*cis*)	18	16.3
Linoleic	9,12-Octadecadienoic	$CH_3(CH_2)_4CH=CHCH_2CH=CH(CH_2)_7COOH$	18	−5.0
Linolenic	9,12,15-Octadecatrienoic	$CH_3CH_2CH=CHCH_2CH=CHCH_2CH=CH(CH_2)_7COOH$	18	−11.3
Arachidonic	5,8,11,15-Eicosatetraenoic	$CH_3(CH_2)_4(CH=CHCH_2)_4(CH_2)_2COOH$	20	−49.5

[a] These acids are straight-chain. The position of the double bond is designated by beginning the numbering of the fatty acid chain with the carbon of the carboxyl group as C-1. The lower number of each pair in the unsaturated linkage is used.

Palmitic acid is the most widely distributed of the saturated fatty acids. However, stearic acid, myristic acid, and lauric acid are often encountered. Stearic acid is found in rather high percentages in animal fats but usually in much smaller amounts in vegetable fats. Butyric acid is found in butter; caproic, caprylic, and capric acids are found in coconut oil. Myristic acid is found in nutmeg butter and arachidic acid is found in peanut oil. All of these are in the combined form as glycerides.

Unsaturated Fatty Acids. Unsaturated fatty acids found in food lipids are generally of the unbranched type, as are saturated fatty acids. A few minor exceptions exist. Unsaturated fatty acids are more reactive than saturated, however. The most common unsaturated acids are palmitoleic, oleic, and linolenic, but others have been found. Palmitoleic acid and oleic acid are widely distributed in animal and plant fats and oils; however, oleic acid is probably the most generally found of all fatty acids. Palmitoleic and oleic acids are monoethanoic acids since each has one double bond. Linoleic acid is widely distributed, and linolenic acid with its three double bonds is found in a number of drying oils. Arachidonic acid occurs in animal fats. Linoleic acid is an essential fatty acid for man and it is the primary dietary essential fatty acid in this case. The 1980 Recommended Daily Allowances of the National Research Council states that there is no evidence of a dietary requirement for linolenic or arachidonic acids in humans (see Holman 1968).

Animal cells cannot synthesize unsaturated fatty acids that have more than one double bond. In the presence of a catalyst such as nickel or platinum, however, unsaturated fatty acids can add hydrogen atoms to form saturated acids. Stearic acid is formed from oleic, linoleic, and linolenic acids. This will be discussed later in this chapter under commercial applications.

Because of the double bonds unsaturated fatty acids can easily be oxidized. When exposed to oxygen, polyunsaturated fatty acids form peroxides, as well as a mixture of volatile aldehydes, ketones, and acids, which is catalyzed by lipoxidase or by the presence of trace metals.

Isomerism. Isomerism is encountered in fatty acids as in other types of organic compounds. There are differences in the way the carbon atoms are linked together, which can cause the properties of the compounds to be significantly different.

In the following discussion of two types of isomerism in the fatty acids are considered: (1) positional isomerism and (2) cis and trans isomerism, also known as geometric isomerism.

Positional isomerism is determined by the position of the double bonds in the carbon skeleton. These usually involve one of two arrangements: (1) the conjugated system in which single bonds alternate with double bonds

$$\text{—CH=CH—CH=CH—}$$

and (2) the nonconjugated arrangement in which two double bonds are separated by one or more methylene groups:

$$—CH{=}CH—CH_2—CH{=}CH—$$

Of these, the conjugated is the more stable form. When a nonconjugated polyunsaturated fatty acid is heated with an alkali it rearranges to yield the conjugated form. This shift also takes place when an unconjugated compound is subjected to high temperatures or is involved in autoxidation.

The position of the double bond can affect the properties of the fatty acid. Both oleic and vaccenic acids have 18 carbon atoms in the chain, but the difference is in the position of the double bond. In the statement of the position of the double bond, one starts with the carbon atom of the carboxyl group at the end of the chain. In oleic acid the double bond is in the 9, 10 position. In vaccenic acid the double bond is in the 11, 12 position. Vaccenic acid is present as a minor component of the fatty acids of butter and tallow.

Under some conditions, the CH_3 carbon, the omega (ω) carbon at the other end of the chain is used. Thus oleic acid is a 9 or an ω-9 acid. This ω terminology is mostly used for biological activity.

Geometric isomerism is caused by the restriction in rotation of two carbon atoms connected by a double bond.

Cis–trans isomerism is possible when two carbon atoms of a compound are attached by a double bond. The carbons of the double bond are not free to rotate on their axis; therefore, the group attached to one of the carbons of the double bond in the cis form can be on the same side of the double bond as the group attached to the other carbon. Trans-isomerism is when the groups are on opposite sides.

The food chemist is interested in the cis-trans isomerism with respect to oleic acid, the cis form of 9-octadecenoic acid. The trans form of 9-octadecenoic acid is known as elaidic acid. This cis to trans isomerization is frequently called elaidinization. In the structural arrangement of oleic acid (cis), the two arms extending from the double bond are folded back together, while in the case of the trans form they are extended, and, therefore, are of maximum length. This relationship is as follows.

$$\begin{array}{rl} CH_3(CH_2)_7 & CH \\ & \| \\ HOOC(CH_2)_7 & CH \end{array}$$

Oleic acid

$$\begin{array}{ll} CH_3(CH_2)_7CH & \\ \quad\quad\quad\quad\| & \\ \quad\quad\quad HC(CH_2)_7COOH & \end{array}$$

Elaidic acid

These formulas show the difference between the cis and trans structure. It is a geometric form of stereoismerism. Trans and cis are from the latin, trans meaning across and cis meaning on this side. While ordinarily the cis acids are found mainly in nature, the body fat of ruminants contains fair amounts of trans acids. It is found also in commercial hydrogenated fats. Several *in vitro* methods can be used to bring about the change from a cis to a trans isomer. Among these the shift of oleic acid to elaidic acid is accomplished with nitrous acid as well as with nitrous oxide. The trans isomer of oleic acid has distinct properties. Oleic acid melts at 13.4°C (a form of oleic acid which melts at 16.3°C is known), while elaidic acid melts at 46.5°C. This indicates that the trans form of a fatty acid has a somewhat higher melting point than the cis form, and this is always the case. Also, when methyl side chains are introduced into a fatty acid, the melting point is always reduced.

The cis form is more soluble, ordinarily, in inert solvents and has a higher energy content, and, consequently, it has a higher heat of combustion than the trans form. It has also a higher ionization constant if the compound is an acid. Usually, the cis form is the less stable of the two.

Four forms (geometric isomers) of linoleic acid are possible. The naturally occurring one is *cis*-9, *cis*-12-octadecadienoic acid. The others have 9-*cis*-12-*trans,* 9-*trans*-12-*trans,* and 9-*trans*-12-*cis* configurations.

It is possible to have eight geometric forms of linolenic acid. The natural form is *cis*-9, *cis*-12, *cis*-15-octadecatrienoic acid.

$$
\begin{array}{l}
CH_3(CH_2)_4CH \\
\qquad\qquad\quad \| \\
\quad HCCH_2CH \\
\quad \| \\
\quad HC(CH_2)_7COOH
\end{array}
$$

Linoleic acid (*cis*-9,*cis*-12-Octadecadienoic acid)

Cyclic Fatty Acids. Malvalic and sterculic acids are cyclopropenoid fatty acids present in cottonseed meal. These must be removed if the meal is to be fed to laying chickens or the eggs will be pink-white.

The crystals of fatty acids are all monoclinic prisms with four molecules to a unit.

Refractive index is easily determined and can be used to determine the purity of fatty acids.

In the study of fatty acids, the absorption spectra have been very important in the determination of the position and number of double bonds in the molecule.

Glycerol. This is a trihydric alcohol which contains primary and secondary alcohol groups and is a constituent of all fats. It gives all the chemical reactions of alcohols. On oxidation it forms such compounds as dihydroxyacetone, glyceric aldehyde, and glyceric acid. The three carbon atoms of the glycerol molecule are labeled alpha, beta, and alpha prime, respectively.

When glycerol is subjected to high heat it loses water and forms acrolein. This compound has a strong irritating odor.

$$\begin{array}{lcl} \alpha\, CH_2OH & & H \\ \quad | & & | \\ \beta\, CHOH & \xrightarrow{\Delta} & C{=}O + 2H_2O \\ \quad | & & | \\ \alpha'\, CH_2OH & & CH \\ & & \| \\ & & CH_2 \\ \text{Glycerol} & & \text{Acrolein} \end{array}$$

Separation and Determination of Fatty Acids. The most effective method for the separation, identification, and determination of fatty acids, both saturated and unsaturated, is by means of gas-liquid chromatography. It has superseded all of the earlier methods. Van Wijngaarden (1967) proposed a modification of the method of Metcalfe and Schmitz (1961) which makes use of BF_3 for the preparation of the fatty acid esters used in this method. Vorbeck *et al.* (1961) prepared the esters by means of diazomethane. Gas-liquid chromatography was used by Lee and Mattick (1961) to show the development of fatty acids in the lipid of unblanched peas held in frozen storage, and in spinach held under similar conditions (Mattick and Lee 1961). Mattick and Rice (1976) used the van Wijngaarden method to determine the fatty acid composition of grape seed oil from native American and hybrid grape varieties.

The earlier method separated saturated and unsaturated fatty acids using the crystallization of the lead salts. The saturated fatty acid salts are less soluble in alcohol (95%) and separate out in crystals first. The lead salts of the unsaturated acids are more soluble.

Stereospecific Analysis. For a complete knowledge of the structure of each fat it is necessary to know the identity of the fatty acids which occupy positions alpha, beta, and alpha prime of the fat molecule. This can be accomplished by the use of the technique known as stereospecific analysis (Brokerhoff 1965). Further work by Brokerhoff and Yurkowski in 1966 showed that in vegetable fats palmitic and stearic acids as well as those of longer chain length are found mainly in the alpha position of the triglyceride molecules. Unsaturated acids are more apt to be found in the beta position.

Di- and Triglycerides. According to Brockerhoff (1965), diglycerides are obtained from triglycerides by means of the lipolitic action of pancreatic lipase. These are then converted to a mixture of D- and L-phosphatidyl phenols. Phospholipase A is then used to hydrolyze L-phosphatide. The result is a lysophosphatide that has a fatty acid in position 1, a free fatty acid from position 2, and D-phosphatidyl phenol, which did not hydrolyze. The identity of the fatty acid in position 3 is determined by calculation.

Triacylglycerols may be made up of three molecules of the same fatty acid or of two or three different fatty acids. These latter are classed as mixed glycerides. A triglyceride made up of three molecules of oleic acid is known as "triolein." A molecule of stearic acid and two of oleic acid is known as dioleostearin or stearodiolein. It is possible for mixed glycerides to exist in isomeric forms. Much has been written about the random distribution of fatty acids in triglyceride molecules, but this idea has not been unaccepted by some (see Kuksis 1972).

The following formulas show possible mixed glycerides.

$$\begin{array}{l} CH_2-O\overset{O}{\overset{\|}{C}}R_1 \\ | \\ CH-O\overset{O}{\overset{\|}{C}}R_2 \\ | \\ CH_2-O\overset{O}{\overset{\|}{C}}R_1 \end{array} \qquad \begin{array}{l} CH_2-O\overset{O}{\overset{\|}{C}}R_1 \\ | \\ CH-O\overset{O}{\overset{\|}{C}}R_1 \\ | \\ CH_2-O\overset{O}{\overset{\|}{C}}R_2 \end{array} \qquad \begin{array}{l} CH_2-O\overset{O}{\overset{\|}{C}}R_2 \\ | \\ CH-O\overset{O}{\overset{\|}{C}}R_1 \\ | \\ CH_2-O\overset{O}{\overset{\|}{C}}R_2 \end{array}$$

Triglycerides can partially hydrolyze to release monoglycerides and diglycerides. This takes place on standing, and the extent of this reaction depends on water and heat, together with an acid or an alkali.

$$\begin{array}{l} CH_2-O-\underset{\underset{O}{\|}}{C}-R_1 \\ | \\ CH-O-\underset{\underset{O}{\|}}{C}-R_2 \\ | \\ CH_2-O-\underset{\underset{O}{\|}}{C}-R_3 \end{array}$$

$\swarrow$ $-R_1-COOH$ $\qquad$ $\searrow$ $-R_2-COOH$

$$\begin{array}{l} CH_2-OH \\ | \\ CH-O-\underset{\underset{O}{\|}}{C}-R_2 \\ | \\ CH_2-O-\underset{\underset{O}{\|}}{C}-R_3 \end{array} \qquad \begin{array}{l} CH_2-O-\underset{\underset{O}{\|}}{C}-R_1 \\ | \\ CH-OH \\ | \\ CH_2-O-\underset{\underset{O}{\|}}{C}-R_3 \end{array}$$

$\downarrow$ $-R_2-COOH$ $\qquad$ $\searrow$ $-R_3-COOH$ $\qquad$ $\downarrow$ $-R_1-COOH$

$$\begin{array}{l} CH_2-OH \\ | \\ CH-OH \\ | \\ CH_2-O-\underset{\underset{O}{\|}}{C}-R_3 \end{array} \qquad \begin{array}{l} CH_2-OH \\ | \\ CH-O-\underset{\underset{O}{\|}}{C}-R_2 \\ | \\ CH_2-OH \end{array} \qquad \begin{array}{l} CH_2-OH \\ | \\ CH-OH \\ | \\ CH_2-O-\underset{\underset{O}{\|}}{C}-R_3 \end{array}$$

When these compounds are hydrolyzed (saponified) with hot alkaline solution, glycerol is released together with fatty acids. The latter, in the form of soaps, are sodium or potassium salts of the fatty acids, depending on the alkali used. This is complete hydrolysis.

During the process of saponification, the following reaction takes place.

$$\begin{array}{l} CH_2OCOR \\ | \\ CHOCOR' \\ | \\ CH_2OCOR'' \end{array} + 3NaOH \rightarrow \underset{\text{Soap}}{RCO_2Na} + \underset{\text{Soap}}{R'CO_2Na} + \underset{\text{Soap}}{R''CO_2Na} + \underset{\text{Glycerol}}{\begin{array}{l} CH_2OH \\ | \\ CHOH \\ | \\ CH_2OH \end{array}}$$

Fats can be hydrolyzed by heating with steam under pressure, releasing fatty acids and glycerol.

The action of animal lipases on triglycerides is shown on p. 97.

The β-monoglyceride remains after this hydrolysis and rapidly isomerizes to form considerable α-monoglyceride. This α-monoglyceride hydrolyzes and the R′ acid is released and glycerol is formed. The acyl group from the β-position is not ordinarily hydrolyzed by most lipases.

Mono- and diglycerides are prepared by the process of interesterification of fats by the use of glycerol and an alkaline catalyst. This will be discussed under commercial applications at the end of this chapter.

Polymorphism

Some triglycerides are known in several different crystal forms. This is called polymorphism. Long chain compounds occur in three different parallel packings of the crystals. Figure 5.1 illustrates this. Table 5.3 gives the melting points of the polymorphic forms of these triglycerides. Triglycerides are quite nonpolar. They behave much like linear hydrocarbons in this respect. In Table 5.3 SSS is the compound of reference. The three forms are α, β', and β. The α form is the least stable, the β form is the most stable, and the more stable can be obtained from the less stable, as follows, $\alpha \rightarrow \beta' \rightarrow \beta$. The fact that the three forms of SSS have three different melting points has been known for over 100 years. Any difference in melting point from one of lower stability to one of higher stability is lower to higher melting point.

Hoerr (1960) described several types of fat crystals as observed magnified and with polarized light as follows. Alpha crystals are fragile transparent platelets which measure about 5 μm. Beta prime crystals are very small delicate needles not much greater than 1 μm long. Beta crystals are coarse and rather large, 25–50 μm long. During long aging these latter can grow to 100 μm in length.

$$
\begin{array}{l}
CH_2-O-\underset{\underset{O}{\|}}{C}-R_1 \\
| \\
CH-O-\underset{\underset{O}{\|}}{C}-R_2 \\
| \\
CH_2-O-\underset{\underset{O}{\|}}{C}-R_3
\end{array}
$$

$-R_1-COOH$ ↙ ↘ $-R_3-COOH$

$$
\begin{array}{l}
CH_2-OH \\
| \\
CH-O-\underset{\underset{O}{\|}}{C}-R_2 \\
| \\
CH_2-O-\underset{\underset{O}{\|}}{C}-R_3
\end{array}
\qquad\qquad
\begin{array}{l}
CH_2-O-\underset{\underset{O}{\|}}{C}-R_1 \\
| \\
CH-O-\underset{\underset{O}{\|}}{C}-R_2 \\
| \\
CH_2-OH
\end{array}
$$

$-R_3-COOH$ ↘ ↙ $-R_1-COOH$

$$
\begin{array}{l}
CH_2-OH \\
| \\
CH-O-\underset{\underset{O}{\|}}{C}-R_2 \\
| \\
CH_2-OH
\end{array}
$$

⇅ isomerization

$$
\begin{array}{l}
CH_2-O-\underset{\underset{O}{\|}}{C}-R_2 \\
| \\
CH-OH \\
| \\
CH_2-OH
\end{array}
$$

↓ $-R_2-COOH$

$$
\begin{array}{l}
CH_2-OH \\
| \\
CH-OH \\
| \\
CH_2-OH
\end{array}
$$

Action of animal lipases on triglycerides.

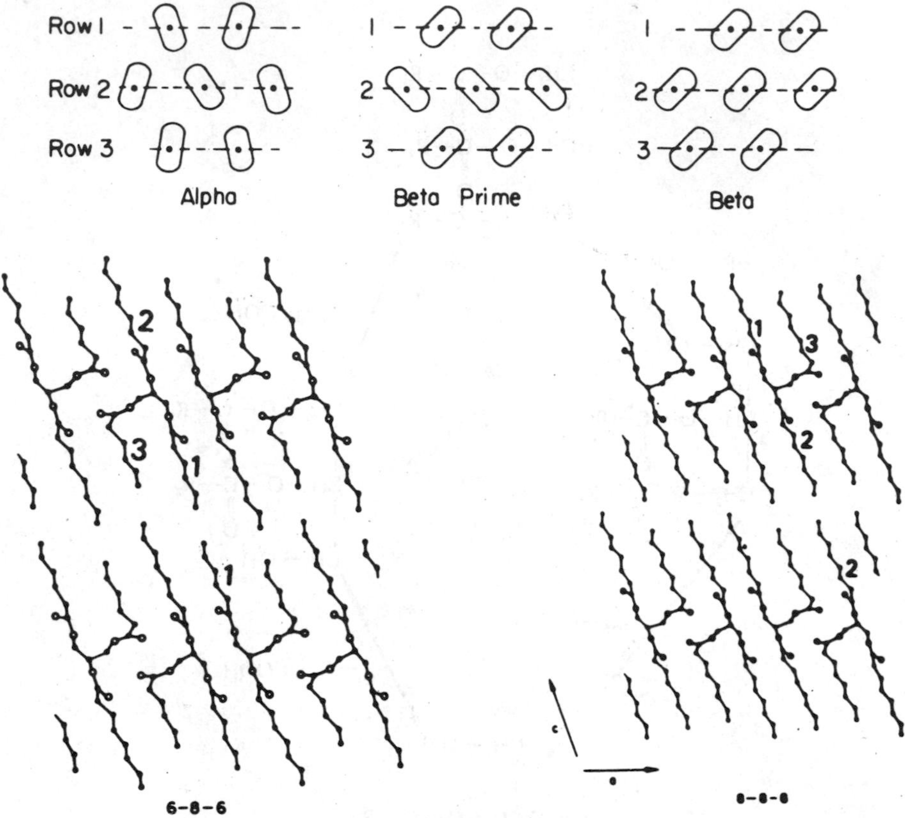

FIG. 5.1. *(top)* Cross-sectional structures of long-chain compounds. *(bottom)* Comparison of postulated 6-8-6 and 8-8-8 triglyceride structures. From Lutton (1972). Used with permission of the American Oil Chemists' Society, copyright owner.

The β' fats are the best for use in shortening since they are able to bring about the inclusion of large amounts of tiny air bubbles. Such fats are best for the making of cakes and icings, producing a tender or short quality.

Grainy or coarse shortening or margarine can be produced from fats which may change into the beta form. This is undesirable.

In the manufacture of shortening, a hardstock, which is basically a combination of fully saturated hydrogenated triglycerides, is used in the mixture to give resistance to heat, that is, warm temperature stability. Soybean hardstock, which contains β-tending SSS, is preferred for a group of solid containing shortenings which can be pumped. The desired component for plastic shortening is β'-tending PSP. This is present in cottonseed hardstock.

Tempering (Controlled crystallization). Tempering is a process used in fat manufacture to put it in the physical state and poly-

TABLE 5.3. Melting Points of SSS, SPS and PSP[a]

Form	SSS	SPS	PSP
α	54.7	51.8	47[b]
β′	64.0	(69.0)[c]	69.0
β	73.3	68.5	(65.5)[c]

Source: Lutton, 1972. Used with permission of American Oil Chemists' Society, copyright owner.
[a] SSS = tristearin; SPS = 2-palmitoyl distearin; PSP = 2-stearoyl dipalmitin.
[b] Softening Point.
[c] Obtained with difficulty.

morphic form in which it is best adapted for the use to which it is to be put. This process consists of holding the product at one or more temperatures for varying periods of time to bring about the formation of different crystal mixes in the fat. In some processes agitation is involved. Different procedures may be necessary for the particular uses of the product.

Bailey (1951) showed that shortening consists of small needlelike crystals enclosing liquid oil.

The most stable crystalline form for hydrogenated vegetable oils is β′, whereas that for interesterified lard in shortenings is β′-2.

Classification of Fats and Oils

Fats and oils can be separated into five groups, based on fatty acid composition and source, that is, plant or animal.

Those of animal origin comprise two groups, those from milkfat and those classed as animal depot fats. Those of plant origin comprise three groups. These include the lauric acid group, which, as the name indicates, are rather high in lauric acid, the oleic–linoleic acid group, and the linolenic acid group.

The milkfat group comes from the milk of the cow and other milk-producing animals. The fat contains a rather large amount of fatty acids up to C_{12} with fair amounts of butyric acid, ranging from 3–15%. A little less than 4% is found in cow's milk. The source of the fat, in general, influences the amount of butyric acid. Palmitic, stearic, oleic, and a large number of other fatty cids also occur.

The animal depot fats are high (up to about 35%) in C_{16} and C_{18} saturated fatty acids and in about 60% oleic and linoleic acids, which are unsaturated. The softness or hardness of these fats is controlled by their composition, which in turn, influences their final use. They are produced from the pig, sheep, and the bovine animals.

The lauric acid group is rather high in lauric acid and is low in unsaturated fatty acids. Fats from this group are obtained from the coconut, the coquilla nut, and seeds of the oil palm.

The oleic–linoleic acid group is made up of fats and oils in which these unsaturated fatty acids comprise the largest amount. They come from the olive and oil palm, and seeds of peanut, sunflower, cotton, corn, sesame, and safflower.

The oils of the linolenic acid group contain fair amounts of linolenic acid, but may have oleic and linoleic acids in their composition. The soybean is the source of the valuable food oil in this group. Linseed oil which is used in paint manufacture contains up to almost 50% linolenic acid, which accounts for its drying qualities. Wheat germ oil and hempseed oil are in this group also.

Another fat which deserves mention but does not fit into these groups is cocoa butter. It is obtained from cocoa beans, is hard at room temperature, and contains palmitic, stearic, and oleic acids in quantity. A small amount of linoleic acid is present also. The amounts of the major fatty acids in cocoa butter are nearly equal—about 26% palmitic, 34% stearic, and 37% oleic.

Preparation of Fats and Oils. Edible oils are now largely commercially extracted by means of solvents. Today, olive oil is made by expression, as it was in ancient times, and the first pressing yields virgin olive oil which brings the highest price and is considered the best and highest in quality. Formerly, many oils were produced by this method. When solvent extraction is used, the solvent is recovered and used again. It is because of this that the cost is reduced and the method is commercially feasible. This process involves countercurrent extraction, that is, the solvent moves in a direction opposite from that of the material being extracted. In this way, the exhausted material meets the fresh solvent so as to extract the most oil or fat possible. Fats can be recovered from animal tissues by rendering. Heat is applied to the material to be extracted. Several types of rendering processes are in use.

Analysis of Fats. ***Saponification Number.*** This is the number of milligrams of KOH necessary to saponify 1 g of fat. When this determination is made in the laboratory a known excess of alkali is used to effect the saponification. The excess of alkali is determined by titration after the reaction has taken place and the value is calculated from the data obtained. In other words, this amount of KOH would neutralize these fatty acids if they were in the free condition. This in turn gives an idea of the average molecular weight of the fatty acids in the fat. The average saponification number of butterfat is 227 while that of mutton tallow has an average value of 194. The molecules of fatty acid in the latter are larger, on the average, and, therefore, it takes less KOH to saponify it than it does for an equal weight of butterfat.

Iodine Number. This is defined as the number of grams of iodine absorbed by 100 g of fat or oil, which amounts to the percentage of iodine absorbed. This determination gives an indication of the degree of unsaturation of the sample. The oleic acid in a sample, for example, would take up to two atoms of iodine to form diiodostearic acid. Ex-

TABLE 5.4. Constants of Edible Fats and Oils[a]

Oil or Fat	Specific Gravity[b]	Refractive Index 40° n/D	Melting Point (°C)	Saponification Number	Iodine Number	Reichert-Meissl Number[c]
Almond oil	0.917[c](15)	1.474[c]	–	192.0[c]	97[c]	0.5
Butter fat	0.911(40/15)	1.4548	32.2	227.0	36.1	29.0
Cocoa butter	0.964(15)	1.4568	34.1	195.0	36.5	0.5
Coconut oil	0.924(15)	1.4493	25.1	257.0	10.4	7.5
Corn oil	0.922(15)	1.4734	–	190.0	122.6	1.9
Cottonseed oil	0.917(25)	1.4735	–	194.3	105.7	0.8
Lard	0.936[c]	1.452[c]	40.5[c]	197.5[c]	57.5[c]	0.4
Lard oil	0.919(15)	1.4615	–	198.5	58.6	0.0
Olive oil	0.918(15)	1.4679	–	192.0	81.1	0.6
Palm oil	0.915(15)	1.4578	35	199.1	54.2	1.4
Peanut oil	0.915(15)	1.4691	–	192.1	93.4	0.5
Safflower oil	0.900(60)	1.462(60°)	–	192.0	145	–
Sesame oil	0.919(25)	1.4646	–	191.5	106.6	1.2
Soybean oil	0.927(15)	1.4729	–	190.6	130	0.7
Sunflower oil	0.923(15)	1.4694[c]	–	188.7	125.5	–
Tallow (beef)	0.948[c]	1.4537[c]	33.5[c]	197.0	49.5	0.4
Tallow (mutton)	0.945(15)	1.4565[c]	32–49[c]	194.0	40	0.3

[a] Most of the data were obtained from Altman and Dittmer (1972), except as noted below.
[b] Specific gravity was calculated at the specified temperature unless otherwise noted in parentheses. Extreme variation may occur, depending on a number of variables such as source, treatment, and age of fat or oil.
[c] Values are typical rather than average, except for almond oil, lard, and some for beef tallow. These latter were obtained from other sources and are averages as were the Reichert-Meissl values.

periments have indicated that the amount of unsaturation is affected by the temperature at the time of biosynthesis. It was shown that linseed oil obtained from seeds produced in a cold climate has a higher iodine value than oil from seeds produced under considerably warmer conditions. It is known that this is a rather general condition, and, therefore, the source of the oil is important. This fact must be borne in mind in evaluating results.

Reichert–Meissl Number. This is an index of the soluble volatile acids, and is determined as the number of milliliters of 0.1 N alkali necessary to neutralize the soluble volatile acids distilled from 5 g of fat or oil. It is most useful in detecting the adulteration of butter, because of all the natural fats butter has the largest amount of steam volatile acids. In making this determination, the procedure must be followed precisely to get satisfactory results. The reason is that complete removal of the volatile acids is not achieved by steam distillation, but comparable values can be had if all the directions, including dimensions of the equipment used, are followed. The acids found in this determination are butyric, caproic, and some of the caprylic acid.

Waxes

True waxes are also classed as simple lipids and are chemically esters of fatty acids and alcohols, both of high molecular weight. The alcohols are long-chained monohydroxy alcohols. In addition, the natural waxes contain secondary alcohols, ketones, hydroxylated and unsaturated fatty acids, and paraffins, all high molecular weight and with rather similar physical properties. Waxes enjoy extensive distribution in both plants and animals. Found in plant cells, and on the surfaces of fruits as well as leaves, they doubtlessly afford protection. The main result of their presence on leaves, usually on the under surfaces, seems to be the conservation of water loss by transpiration. The under surface of the leaves contains the greatest numbers of stomata, the openings through which water is lost.

Beeswax is secreted by honeybees and is used by them to construct combs. It is composed of a variety of substances, the bulk of which are alkyl esters. Of these, myricyl palmitate is present in rather large amounts.

The so-called bloom of fruits such as plums and grapes is made up chemically of waxes. The waxes from the cuticle of the apple and the pear have been investigated. In both cases the main component was found to be *n*-nonacosane, $CH_3(CH_2)_{27}CH_3$, determined by X-ray diffraction, and the melting point (65.1°C). In addition, apple was found to contain the secondary alcohol, 10-nonacosanol, $CH_3(CH_2)_{18}CHOH(CH_2)_8CH_3$, was not found in the wax of the pear; however, some primary alcohols were found (Markley *et al.* 1935).

Waxes from the cherry, grape, cranberry, and citrus fruits were investigated and a variety of compounds were found (Markley *et al.* 1937, 1938; Markley and Sando 1934, 1937).

Waxes generally are solids frequently melting under 100°C, are not so likely to autoxidize, and are harder to saponify than are fats or oils.

COMPOSITE LIPIDS

Phospholipids (Phosphoglycerides, Phosphatides)

While fats and oils, the simple lipids, contain only carbon, hydrogen, and oxygen, the phospholipids contain these elements and nitrogen and phosphorus as well. They are, therefore, compounds of glycerol which are esterified by fatty acids, and by phosphoric acid and basic nitrogen compounds. The nitrogen compounds found are choline, ethanolamine, and L-serine. The structural names of these compounds are phosphatidyl followed by the name of the base. It is well to mention that the sphingomyelins that contain phosphorus are more correctly considered sphingolipids because of the backbone structure connected with the fatty acid. While phospholipids make up about 1–2% of a number of the crude vegetable oils, the amount in animal fats is higher. The phospholipid content of egg yolk amounts to about 20%. During the refining of fats and oils most of the phospholipids are lost, hence only small amounts are found in the finished product. These compounds are involved in some biological functions such as (a) intermediates in the absorption, transport, and metabolism of fatty acids; (b) structural elements in living cells; and (c) involvement in the clotting of blood.

Lecithins. Lecithins are phosphatidyl cholines. They are found in egg yolk, yeast, soybeans, wheat germ, and animal tissues such as liver. The lecithin of egg yolk and liver contains both saturated and unsaturated fatty acids, the saturated is in the alpha position while the unsaturated is in the beta position of the glycerol. It appears unlikely that beta lecithins occur in nature. In the alpha lecithins the choline part of the molecule is attached to the alpha or outside carbon of the glycerol molecule. In the beta lecithins the choline is attached to the beta or middle carbon of the glycerol molecule.

$$\begin{array}{l} H_2C-O-C(=O)-R_1 \\ R_2-C(=O)-O-C-H \\ H_2C-O-P(=O)(O^-)-O-CH_2-CH_2-\overset{+}{N}(CH_3)_3 \end{array}$$

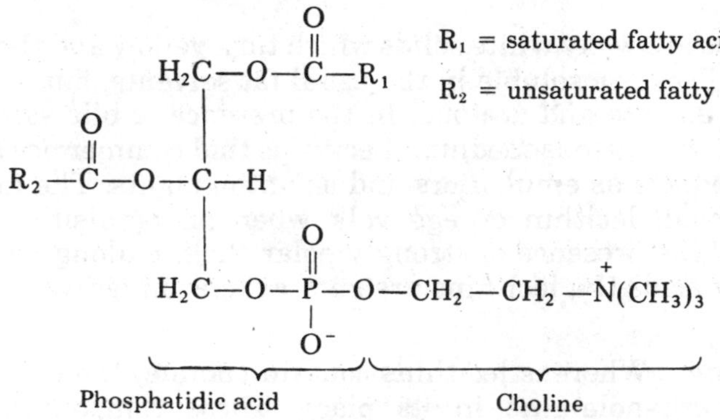

R_1 = saturated fatty acid;
R_2 = unsaturated fatty acid.

Phosphatidic acid | Choline

L-α-Lecithin (Phosphatidyl choline).

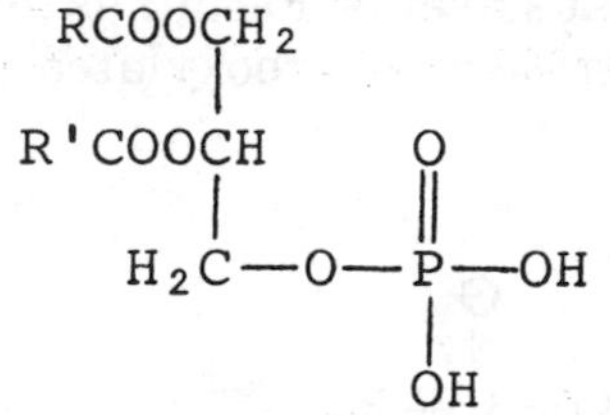

$$\begin{array}{l} RCOOCH_2 \\ R'COOCH \\ H_2C-O-P(=O)(OH)-OH \end{array}$$

L-α-Phosphatidic acid

$$HO-CH_2-CH_2-\overset{+}{N}\equiv(CH_3)_3$$

Choline

Another lecithin, dipalmitoleyl-L-α-glycerylphosphorylcholine is found in brewer's yeast. This is different in that it has two unsaturated fatty acids in its composition.

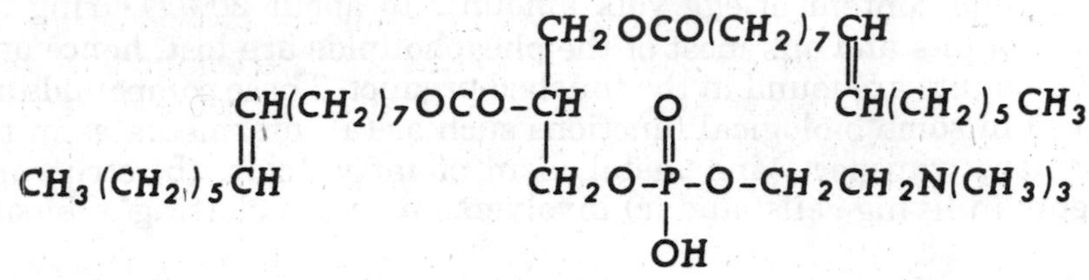

```
                                CH2OCO(CH2)7CH
                                |              ||
          CH(CH2)7OCO-CH    O         CH(CH2)5CH3
          ||                |    ||
CH3(CH2)5CH                CH2O-P-O-CH2CH2N(CH3)3
                                     |
                                     OH
```

Dipalmitoleyl-L-α-glycerylphosphoryl choline

A replaceable hydrogen in the lecithin molecule is in the phosphoric acid part, while the choline acts as a base. However, the basic properties are stronger. Since ethanolamine and L-serine are less basic than choline, the phosphatides (phosphatidyl compounds) containing them are more acid in character than those containing choline.

An absence of choline from the diet results in a fatty infiltration of the liver.

Lecithins are waxy white solids which turn yellow and then brown in the light. They are soluble in the usual fat solvents, but are insoluble in methyl acetate and acetone. In the presence of bile salts they will dissolve in an aqueous medium. Lecithins find commercial application in food products as emulsifiers and as antioxidants. The form used is crude soybean lecithin or egg yolk when an emulsifier is needed. Because of the presence of strongly polar choline along with the nonpolar fatty acids the lecithins are surface tension active.

Cephalins. Whereas lecithins contain choline, the cephalins have serine or ethanolamine in its place. These compounds have the α-L-configuration. They are colorless solids which darken to a reddish brown color when exposed to light and air. They are insoluble in alcohol, but soluble in the usual fat solvents. Formulas for these compounds follow. When phosphatidyl serine is decarboxylated phosphatidyl ethanolamine is the result.

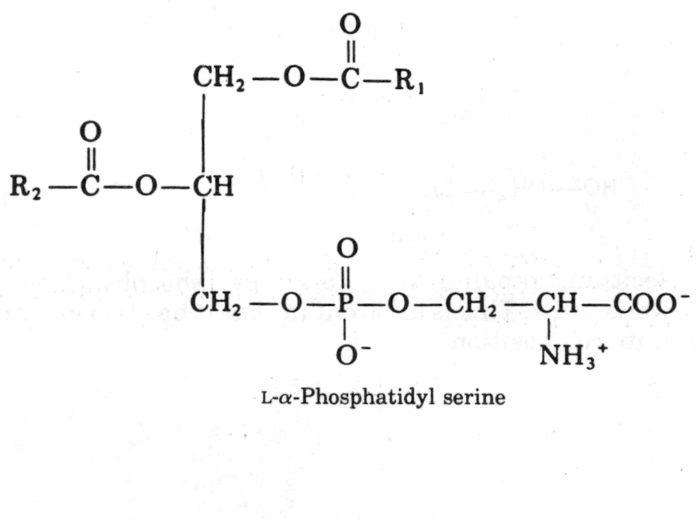

```
                      O
                      ||
               CH2—O—C—R1
                |
       O        |
       ||       |
  R2—C—O—CH
                |
                |       O
                |       ||
               CH2—O—P—O—CH2—CH—COO-
                        |          |
                        O-         NH3+
```

L-α-Phosphatidyl serine

$$
\begin{array}{l}
\qquad\qquad\qquad\qquad\quad\ \ O \\
\qquad\qquad\qquad\qquad\quad\ \ \| \\
\qquad\qquad\quad CH_2-O-C-R_1 \\
\quad\ \ O \qquad\qquad\ | \\
\quad\ \ \| \qquad\qquad\ | \\
R_2-C-O-C-H \\
\qquad\qquad\qquad\ | \qquad\quad O \\
\qquad\qquad\qquad\ | \qquad\quad \| \\
\qquad\qquad\quad CH_2-O-P-O-CH_2-CH_2-NH_3^+ \\
\qquad\qquad\qquad\qquad\qquad\ | \\
\qquad\qquad\qquad\qquad\qquad O^-
\end{array}
$$

L-α-Cephalin (Phosphatidyl ethanolamine). R_1 = saturated acid; R_2 = unsaturated acid.

The cephalins are more acidic than the lecithins. This is because the quaternary ammonium group is more basic than the primary amino group.

Plasmalogens. Plasmalogens are similar to the lecithins and cephalins. The only difference is that an α,β-unsaturated alcohol replaces the fatty acid in the α-position to form an ether-linked compound instead of an ester-linked compound. These compounds are found in the membranes of muscle tissue, heart, and brain.

$$
\begin{array}{l}
\qquad\qquad\quad CH_2-O-CH=CH-R \\
\quad\ \ O \qquad\qquad\ | \\
\quad\ \ \| \qquad\qquad\ | \\
R_2-C-O-C-H \\
\qquad\qquad\qquad\ | \qquad\quad O \\
\qquad\qquad\qquad\ | \qquad\quad \| \qquad\qquad\qquad\qquad\quad + \\
\qquad\qquad\quad CH_2-O-P-O-CH_2-CH_2-N(CH_3)_3 \\
\qquad\qquad\qquad\qquad\qquad\ | \\
\qquad\qquad\qquad\qquad\qquad O^-
\end{array}
$$

A plasmalogen (Phosphatidal choline)

Inositol phosphatides. These compounds contain inositol. This is a cyclic hexahydroxy alcohol which is attached to the phosphate. They can be prepared from soybean oil.

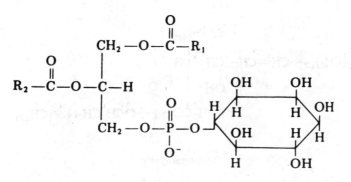

$$
\begin{array}{l}
\qquad\qquad\qquad\qquad\quad\ \ O \\
\qquad\qquad\qquad\qquad\quad\ \ \| \\
\qquad\qquad\quad CH_2-O-C-R_1 \\
\quad\ \ O \qquad\qquad\ | \\
\quad\ \ \| \qquad\qquad\ | \qquad\qquad\qquad\qquad\quad OH \qquad\quad OH \\
R_2-C-O-C-H \qquad\qquad\qquad\quad\ | \qquad\qquad\ \ | \\
\qquad\qquad\qquad\ | \qquad\quad O \qquad\quad H \diagup H \qquad\quad H \diagdown OH \\
\qquad\qquad\qquad\ | \qquad\quad \| \qquad\quad\ \ | \\
\qquad\qquad\quad CH_2-O-P-O-\ \ \diagdown OH \qquad\quad H \diagup H \\
\qquad\qquad\qquad\qquad\qquad\ | \qquad\qquad\qquad\ | \qquad\qquad\ \ | \\
\qquad\qquad\qquad\qquad\qquad O^- \qquad\qquad\quad\ H \qquad\qquad OH
\end{array}
$$

Phosphatidyl inositol

Phosphatidyl glycerols are present in plant and bacterial membranes.

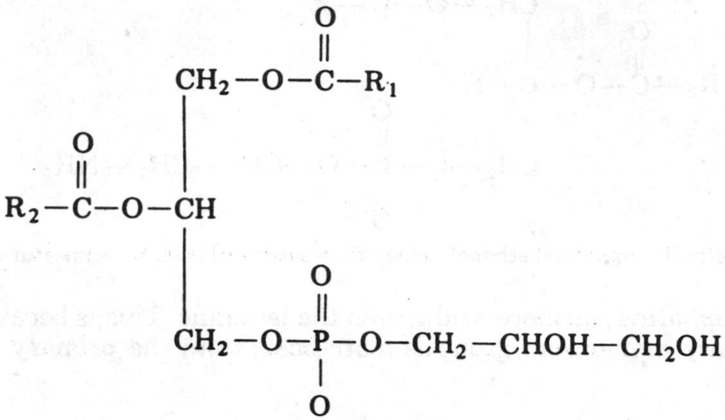

$$
\begin{array}{l}
CH_2-O-\overset{\overset{\displaystyle O}{\|}}{C}-R_1 \\
| \\
R_2-\overset{\overset{\displaystyle O}{\|}}{C}-O-CH \\
| \\
CH_2-O-\underset{\underset{\displaystyle O}{|}}{\overset{\overset{\displaystyle O}{\|}}{P}}-O-CH_2-CHOH-CH_2OH
\end{array}
$$

Phosphatidyl glycerol

Sphingolipids

Sphingolipids are a group of compounds, many of which contain sphingosine, a long chain aliphatic base, which do not contain glycerol. Similar compounds can replace this base in the molecule, hence it is not necessary that sphingosine itself be present.

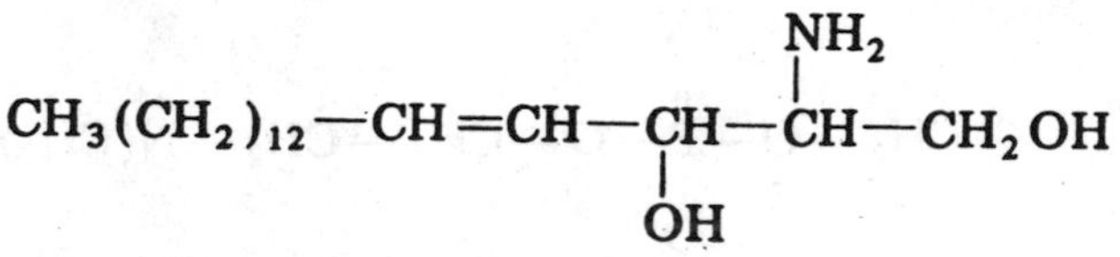

$$
CH_3(CH_2)_{12}-CH{=}CH-\underset{\underset{\displaystyle OH}{|}}{CH}-\overset{\overset{\displaystyle NH_2}{|}}{CH}-CH_2OH
$$

Sphingosine

One group of sphingolipids contain phosphorus and are known as sphingomyelins. They are found mainly in nervous tissue, but also in the blood.

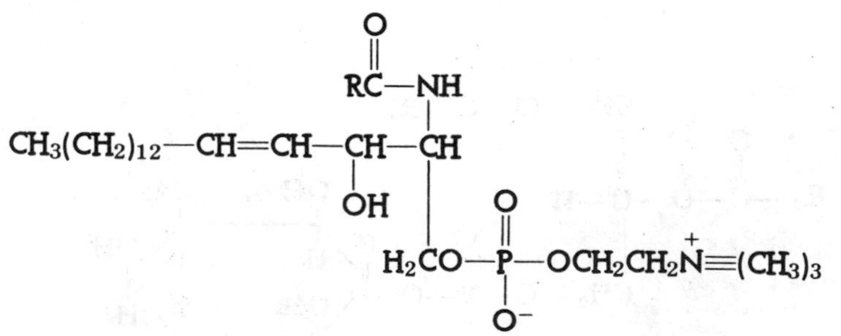

$$
CH_3(CH_2)_{12}-CH{=}CH-\underset{\underset{\displaystyle OH}{|}}{CH}-\overset{\overset{\displaystyle R\overset{\overset{O}{\|}}{C}-NH}{|}}{CH}(-H_2CO-\underset{\underset{\displaystyle O^-}{|}}{\overset{\overset{\displaystyle O}{\|}}{P}}-OCH_2CH_2\overset{+}{N}{\equiv}(CH_3)_3)
$$

Sphingomyelin

DERIVED LIPIDS (UNSAPONIFIABLES)

Terpenes

Of the terpenes of interest is squalene $C_{30}H_{50}$ which is a hydrocarbon and is present in shark liver oil. It is incidentally an intermediate in the biosynthesis of cholesterol. The carotenes, the precursors of the A vitamins, are also in this group.

Steroids

The steroids are another of the groups of compounds that do not have glycerol in their molecules. Those steroids that have eight to ten carbon atoms in the side chain, connected at carbon 17, and with a hydroxyl group at carbon 3 are called sterols. The best known member of this group, having been involved in diet problems, is cholesterol. Cholesterol has eight asymmetric carbon atoms, C-3, -8, -9, -10, -13, -14, and -17.

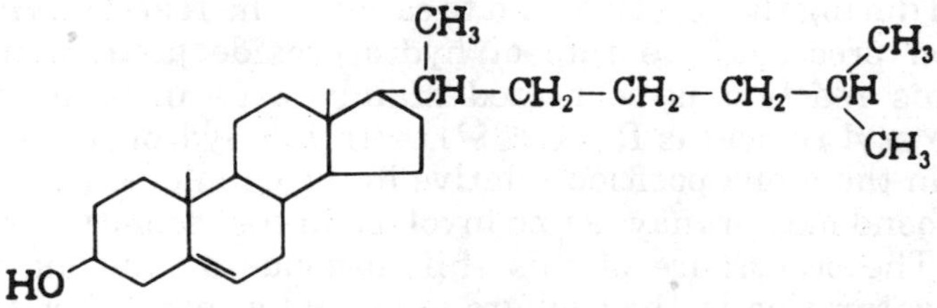

CH_3 — $CH-CH_2-CH_2-CH_2-CH(CH_3)_2$ — HO

Cholesterol

Ergosterol (another sterol which is found in fungi, other plants, and yeast) is well-known also. These sterols have conjugated unsaturation in the B ring and because of this, this ring can be broken by ultraviolet light to form compounds with vitamin D activity. Ergosterol forms vitamin D_2, while 7-dehydrocholesterol yields vitamin D_3.

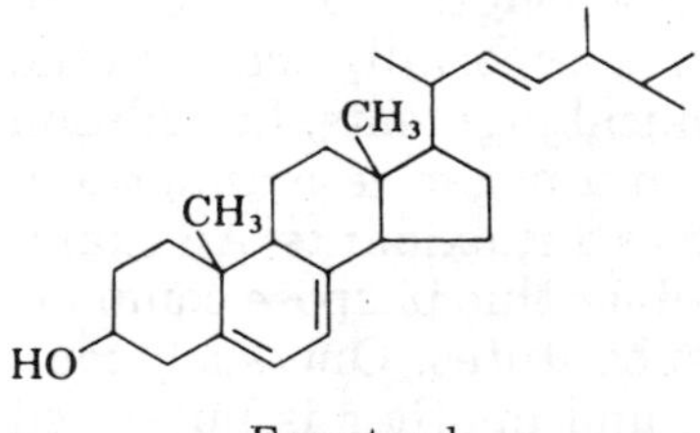

CH_3 CH_3 HO

Ergosterol

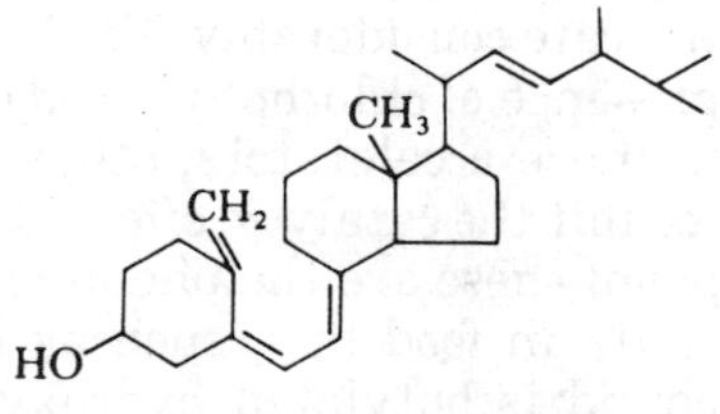

CH_3 CH_2 HO

Vitamin D_2

These compounds will be discussed in the chapter on vitamins.

OXIDATION OF LIPIDS

Lipid oxidation is very important and of much interest to the food chemist because it results in the formation of off-odors and -flavors, reduction or destruction of essential fatty acids, formation of brown pigments, and alteration of pigments and flavors. This is known also as autoxidation. These main reactions are accompanied by secondary reactions which can be of an oxidative or nonoxidative nature. All of these involve reactions with molecular oxygen.

Of these prime reactions the most common type of lipid autoxidation is autocatalytic autoxidation. This occurs in fats and oils and in foods containing these substances, and always occurs in the presence of catalytic substances, examples of which are light, pro-oxidants, and antioxidants. Enzymatic and photochemical peroxidation are the important nonautocatalytic types of autoxidation which are involved in the deterioration of food lipids.

Farmer and Sundralingam (1942) and Farmer *et al.* (1943) showed that hydroperoxides are formed during the usual autoxidation of fats. The rate of an autocatalytic reaction increases with time, because products formed during the reaction tend to catalyze the reaction. Hence, as the reaction proceeds, the rate of hydroperoxide accumulation increases. This reaction is connected mainly with unsaturated acyl groups (the acyl radical is R—C$\leqslant$O), with the hydroperoxide group appearing in the alpha position relative to the double bond. A shift of the double bond may or may not be involved in the formation of hydroperoxides. The occurrence of this shift depends on the quantity of original unsaturation in the acyl group as well as other factors. Also, the rates of autoxidation of the usual fatty acids and derivatives depend particularly on the amount of unsaturation of the fatty acids. Saturated fatty acid derivatives autoxidize very slowly to form hydroperoxides. It is known also that the rate of autoxidation increases exponentially with increasing unsaturation in the case of derivatives of the ordinary fatty acids. The cis isomers of the unsaturated fatty acids are more susceptible to autoxidation than the trans.

The rate of oxidation can be affected by a number of things. Trace metals and biological catalysts including oxidative enzymes, if present, as well as light from short to the longest wavelengths can accelerate the rate considerably. Visible light is, of course, especially active in the presence of chlorophyll and other photochemical pigments. In addition to these accelerators, other substances act in a negative manner and inhibit the catalytic effect. These are known as antioxidants. An example of these are the phenolic substances used for this purpose commercially in food fats, such as lard and lard substitutes. One such compound is butylated hydroxytoluene (BHT), and another is butylated hydroxyanisole (BHA) Stuckey (1962).

It has been found that sometimes a mixture of two antioxidants is more effective than when one is used alone; an example is the use of BHT and BHA.

Lundberg (1962) and Lundberg and Chipault (1947) found that with unsaturated fatty acids, small amounts of conjugated ketones are produced at a speed equal to the production of hydroperoxides.

Figure 5.2 shows the typical curve of hydroperoxide development in fatty materials, an example of uncomplicated autoxidation of a pure fatty material. The slope of the curve shows the rate of hydroperoxide accumulation, which increases as the oxidation progresses.

It is well to note that the hydroperoxides by themselves do not contribute materially to the unwanted odors and flavors of autoxidized food materials. These off-flavors are caused by secondary substances formed during the various reactions and possibly through further oxidation of the peroxides and their degradation products.

The mechanism of this reaction was determined by a study on the autoxidation of ethyl linoleate at various temperatures (Bolland 1946; Bolland and Gee 1946A, B). This work brought to light the following facts: (1) autoxidation is directly proportional to linoleate concentration; (2) above a certain level, oxidation increases in a straight-line relationship with the extent of oxidation. Since at the start practically all the absorbed oxygen appeared as hydroperoxide it was proposed that these hydroperoxides themselves acted as catalysts causing the autocatalytic character of the reaction.

Bolland (1946) found that the rate of oxidation varied with the oxygen pressure. At very low oxygen pressure the rate of oxidation was approximately proportional to the pressure, but at higher pressures the rate was independent of this factor. With ethyl linoleate at 45°C the

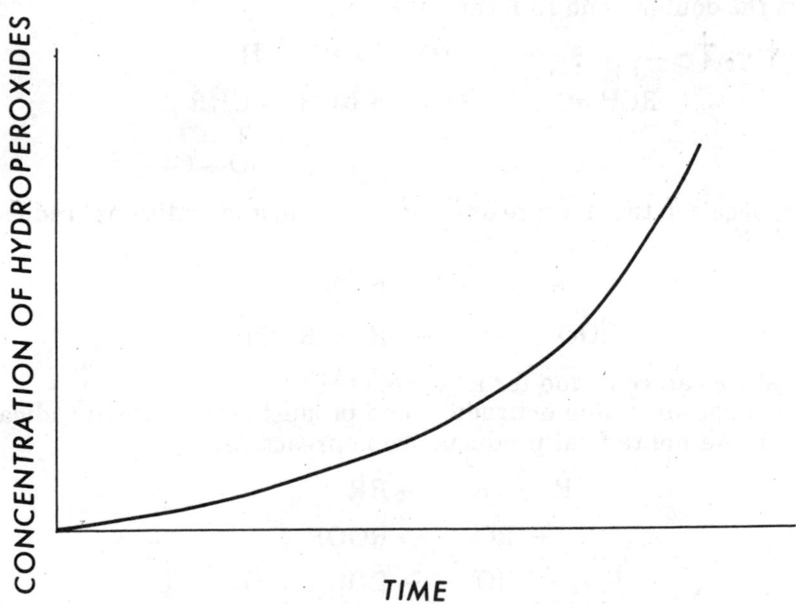

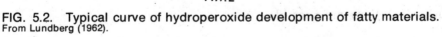

FIG. 5.2. Typical curve of hydroperoxide development of fatty materials. From Lundberg (1962).

rate became independent at a pressure that approximated the partial pressure of oxygen in the atmosphere.

Also, the following observations were considered of importance to the formulation of a mechanism for the autocatalytic autoxidation.

1. In many respects the reaction appeared to be analogous to certain autoxidations in inorganic systems which were known to involve chain reactions.
2. By spectrophotometric methods, it was observed that shifts of double bonds occurred during autoxidation. This, together with the observation that hydroperoxides were formed, suggested that a free radical mechanism might be involved.
3. The autoxidizing linoleate system contained hydroperoxides which decomposed; it was known that the decomposition of other organic peroxides in unsaturated systems initiates polymerization reaction chains. Bolland and Gee (1946A, B) reported that they had investigated the linoleate hydroperoxides and found the decomposition reaction to be bimolecular.

On the basis of all this work, a free radical chain mechanism was proposed which was slightly modified as a result of subsequent work.

The most important cause of the deterioration of lipids or foods which contain lipids is the action of oxygen on unsaturated fatty acids in the lipids.

The oxidation reaction usually occurs in three steps: initiation, propagation, and termination. In the initiation step hydrogen is abstracted from an unsaturated hydrocarbon to form a free radical, and an oxygen adds at the double bond to form a radical.

$$RH \longrightarrow R^{\cdot} + H^{\cdot}$$

$$RCH{=}CHR + O_2 \longrightarrow RCH - \underset{\displaystyle O{-}O^{\cdot}}{\overset{}{\underset{|}{CHR}}}$$

In propagation the chain reaction goes ahead, and additional radicals are formed.

$$R^{\cdot} + O_2 \longrightarrow ROO^{\cdot}$$

$$ROO^{\cdot} + RH \longrightarrow R^{\cdot} + ROOH$$

This stage can continue for a period of time.

In termination, stable nonradical end products result when radicals collide. These nonradical products are nonreactive.

$$R^{\cdot} + R^{\cdot} \longrightarrow RR$$

$$R^{\cdot} + RO_2^{\cdot} \longrightarrow ROOR$$

$$RO_2^{\cdot} + RO_2^{\cdot} \longrightarrow ROOR + O_2$$

Other reactions are possible.

In this reaction mechanism, at low oxygen concentrations, the ratio of $R^{\cdot}$ to $RO_2^{\cdot}$ should be relatively large, and the principal chain-terminating reaction would be the first of the three shown.

If, however, the concentration of oxygen is high, the ratio of $RO_2^{\cdot}$ to $R^{\cdot}$ should also be high, and the third chain-terminating reaction would prevail.

At low total peroxide concentration, the ratio of monomeric to dimeric peroxide is high and the autoxidative chain reaction shown is initiated primarily by a decomposition of monomeric hydroperoxide. At high peroxide concentration, the ratio of dimeric to monomeric peroxide is high, and the chain-initiating reaction involves primarily the decomposition of dimeric hydroperoxide. However, a finite level of monomeric hydroperoxide persists, and, indeed, increases gradually, as the total peroxide concentration increases.

Figure 5.3 shows the initial rates of decomposition of peroxides of methyl linoleate. The samples were autoxidized to various levels at 0°C and then decomposed at 80°C. The initial horizontal portion of the curve is consistent with a unimolecular decomposition of monomeric hydroperoxide, and the rising straight-line portion is consistent with a unimolecular decomposition of dimer hydroperoxide. Thus, the hydroperoxides may be considered to exist in an equilibrium as follows:

$$2\ ROOH \rightleftharpoons (ROOH)_2$$

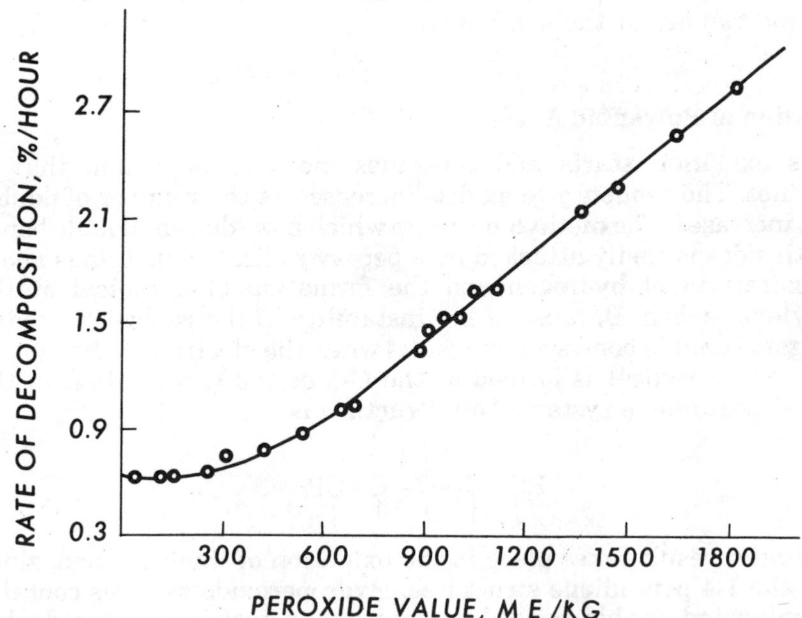

FIG. 5.3. Decomposition of peroxides of methyl linoleate. Autoxidized at 0°C and decomposed at 80°C.
From Lundberg (1962).

Under conditions of normal initial temperature, a rise in temperature affects the rate of autocatalytic autoxidation more than most other chemical reactions. The rising temperature has two effects: it speeds the chain propagation reactions and the decomposition of peroxides, resulting in an increase in the concentration of free radicals available for the start and spread of reaction chains. However, a point is reached where for a given temperature a maximum state of concentration of hydroperoxide is approached. This is the phenomenon of maximum rates (Vorbeck *et al.* 1961). The approach to this condition is speeded by the presence of catalytic factors such as metals and light.

Under usual conditions, however, the only time the phenomenon of maximum rates is likely to be of importance in foods is during deep fat or ordinary frying procedures in metal equipment.

Oxidation of Monoenoic Acids

When oleic acid is involved in such oxidation, four isomeric hydroperoxides can be formed, at four different locations. Abstraction of hydrogen from the carbons α to the double bond occurs during the initiation stage. The result is that two radicals can be produced. Each one of these can assume two forms as a result of resonance. Hydroperoxides are formed at four different positions by the addition of oxygen at each radical site followed by the addition of a hydrogen free radical. Two of these positions are at the carbons of the original double bond and two are at the α carbons.

Oxidation of Polyenoic Acids

This oxidation starts and continues more rapidly than that of monoenes. The tendency to oxidize increases as the number of double bonds increases. The methylene group which has adjacent double bonds on both sides is easily attacked by a peroxy radical. This brings about an abstraction of hydrogen and the formation of a radical at the methylene carbon. Because of the instability of the radical formed, a conjugated double bond system results when the electrons redistribute. In the end a radical is formed at the C-1 or the C-5 position in the original pentadiene system. This structure is

```
          |
RC=C—C—C=CR
 |  |  |  |  |
```

A similar result takes place in the oxidation of linolenic acid, since it has the 1-4 pentadiene structures. Hydroperoxide systems containing conjugated double bonds can form when more than two double bonds exist, as in linolenic acid.

Measurement of Oxidation

The measurement of the oxidized condition tends to be rather uncertain. There are tests that can help in this measurement, including the determination of the peroxide value, particularly at the start of the stage of decomposition of large amounts of hydroperoxides. The use of the thiobarbituric acid test has some value under the right conditions, and finally the determination of the carbonyl compounds as well as the acid determination can help, but interpretation of the data can be difficult.

Secondary Degradation Products

The decomposition of the hydroperoxides that were formed by the action of oxygen on unsaturated fatty acids probably accounts for the degradation products formed. Because of the large number of unsaturated fatty acids found in the fatty materials of foods, and the large number of possible hydroperoxides that can be formed from each, many other compounds can be formed from these hydroperoxides. This is further enlarged by the possible routes of hydroperoxide destruction. Further, oxygen can act on ethylenic bonds to yield other degradation products. In addition to these, alcohols, aldehydes, and other compounds formed during initial degradation are susceptible to further oxidation.

Since these compounds continue to change, analysis by chromatographic procedures show the conditions in the sample at the time of sampling.

Badings (1960) worked on the scheme of Bell *et al.* (1951) and from this some of the basic pathways of hydroperoxide dismutation can be shown.

$$\mathrm{R{-}CH(OOH){-}R \longrightarrow R{-}\underset{\displaystyle \overset{|}{O^{\cdot}}}{CH}{-}R + {}^{\cdot}OH} \tag{1}$$

This shows decomposition to the alkoxy and hydroxy free radicals. The alkoxy radical can react in four ways, as follows.

$$\mathrm{R{-}\underset{\displaystyle \overset{|}{O^{\cdot}}}{CH}{-}R \longrightarrow R^{\cdot} + RCHO} \tag{2a}$$

This illustrates the chain fission which can take place on either side of the radical which results in an aldehyde and a new alkyl-type radical.

$$\mathrm{R{-}\underset{\displaystyle \overset{|}{O^{\cdot}}}{CH}{-}R + R'H \longrightarrow R{-}CHOH{-}R + R'^{\cdot}} \tag{2b}$$

The removal of a hydrogen atom from another molecule can give an alcohol and a new free radical. Both reactions 2a and 2b show the release of free radicals which are able to propagate the autoxidative chain reaction.

$$\underset{\displaystyle \overset{|}{O^{\cdot}}}{R{-}CH{-}R} + R'^{\cdot} \longrightarrow \underset{\displaystyle \overset{\|}{O}}{R{-}C{-}R} + R'H \qquad (2c)$$

$$\underset{\displaystyle \overset{|}{O^{\cdot}}}{R{-}CH{-}R} + R'O^{\cdot} \longrightarrow \underset{\displaystyle \overset{\|}{O}}{R{-}C{-}R} + R'OH \qquad (2d)$$

Reactions 2c and 2d show that interaction between two free radicals ends the chain reaction by the formation of nonradical products: in 2c and alkyl type of free radical is involved, in 2d an alkoxy type is a part of the reaction.

Other secondary reactions are likely to be worthy of consideration in the autoxidation of lipids.

$$ROOH + {-}CH{=}CH{-} \longrightarrow {-}\underset{\diagdown O \diagup}{CH{-}CH}{-} + ROH \qquad (3)$$

$$ROO + {-}CH{=}CH{-} \longrightarrow {-}\underset{\diagdown O \diagup}{CH{-}CH} + RO \qquad (4)$$

These two equations demonstrate the reaction of a hydroperoxide or its radical with a double bond to form an epoxide.

Knight *et al.* (1951) showed that the autoxidation of methyl oleate induced a cis–trans isomerization as one of the changes which took place. Also oxygen added to one of the C of the double bond resulted in the formation of a trans hydroperoxide.

Polymerization is important in the autoxidation of lipids. They may be formed by direct association of alkoxy and alkyl free radicals or as in the following reaction.

$${-}CH{=}CH{-} + HOO{-}R \longrightarrow$$

$${-}CH{-}CH{-}\underset{\displaystyle \overset{|}{H}}{\overset{+}{O}}{-}OR \longrightarrow {-}\underset{\displaystyle \overset{|}{O{-}R}}{CH}{-}CHOH{-} \qquad (5)$$

Carbonyl compounds are of much importance in the autoxidation of fats. Aldehydes result from reaction 2a. A study of such carbonyl compounds was made by Gaddis *et al.* (1961). Since aldehydes are readily susceptible to oxidation to acids, large amounts of them are unlikely to accumulate during the autoxidation of fats. Furthermore, they are susceptible to polymerization and condensation reactions. However, it must be remembered that some aldehydes have very strong flavors and exceedingly small amounts of them can be responsible for serious deterioration of flavor in oil- and fat-containing products (Patton *et al.* 1959).

Bell *et al.* (1951) listed the following primary alkoxy radical reactions.

$$R{-}CH_2O^{\cdot} \longrightarrow CH_2O + R^{\cdot} \quad (6)$$

$$R{-}CH_2O^{\cdot} + R'H \longrightarrow R{-}CH_2OH + R' \quad (7)$$

$$R{-}CH_2O^{\cdot} + R'^{\cdot} \longrightarrow R{-}CHO + R'H \quad (8)$$

$$R{-}CH_2O^{\cdot} + R'^{\cdot} \longrightarrow R{-}CH_2{-}O{-}R' \quad (9)$$

$$R{-}CH_2O^{\cdot} + R'O^{\cdot} \longrightarrow RCHO + R'OH \quad (10)$$

In reaction (6) only formaldehyde is obtained. Reactions (8) and (10) give other volatile aldehydes.

Ellis (1950) reported that the α, β-unsaturated ketone-acids are very reactive with oxygen to form peroxides and that in advance stages of oxidation a decrease in unsaturated ketones is coincidental with a decrease in oxygen absorption. Kummerow (1961) reported that the addition of 12-keto-9-*cis*-octadecenoic acid to corn oil causes a spectacular increase in the peroxide content of the oil.

Light and Radiation

Visible light seems to speed the decomposition of hydroperoxides. This may be the result of absorption of the light by the peroxides or by other compounds that may be present.

As one might expect, the action of ultraviolet light has a greater effect. When the usual polyunsaturated fatty acids are autoxidized, the result is the formation of conjugated unsaturated systems. These in turn absorb the ultraviolet strongly at certain wavelengths. Under these conditions the ultraviolet materially speeds the breakdown of peroxides. It may have an influence also on other autoxidation reactions.

Lundberg (1949) studied the oxidation of methyl linoleate in the presence of chlorophyll and monochromatic light of 600 nm wavelength at 37°C. The results of this work showed that the products of this oxidation were mainly hydroperoxides and that the amount of hydroperoxide produced was directly proportional to the total amount of light absorbed. It was found also that the mechanism is rather different from that involved in the usual autoxidation of linoleate.

Beta, gamma, and other high energy radiations catalyze autoxidation by catalysis of the peroxide decomposition. They effect this also by the production of free radicals from the unoxidized substrate.

Heavy Metals in Lipid Oxidation

Heavy metals, especially those that have two or more valency states with a suitable oxidation–reduction between them (e.g., Co, Cu, Fe, Mn, Ni) speed the oxidative deterioration in food lipids. Actually these metals cut down the time during which no measurable oxidation takes place, and increase the maximum rate of oxidation. This concept is based on the idea that a metal ion initiates a free radical chain by an electron transfer which was introduced by Haber and Willstätter (1931). It has been demonstrated that the basic function of the metal catalyst is to increase the rate of formation of free radicals. In the first instance, the elimination of the catalyst from the substrate during the early stages of the oxygen uptake cuts down the subsequent rate of oxidation (Denisov and Emanuel 1956). In the second instance the addition of a strong free radical inhibitor to an oxidizing system suppresses the chain-propagating steps and further oxidation is completely prevented for lengths of time proportional to the added inhibitor concentration. In metal catalyzed oxidations it has been shown that the rate of chain initiation is considerably greater than the rate in the uncatalyzed oxidations, and at sufficiently low concentrations the rate is proportional to the catalyst concentration.

Traces of heavy metals are usually found in food lipids. Even when purified, such lipid materials hold them in very small amounts. However, prophin complexes of heavy metals are extremely powerful catalysts of autoxidation and they are effective in trace amounts. Incidentally, the pyrrole-containing nucleus, called porphin, is found in many compounds of interest to food chemists, such as chlorophyll, cytochromes, hemoglobin, iron-containing proteins, and myoglobin.

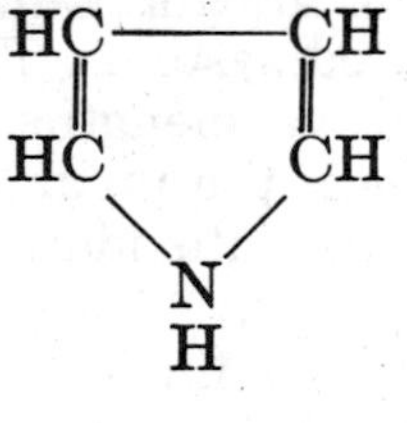

Pyrrole

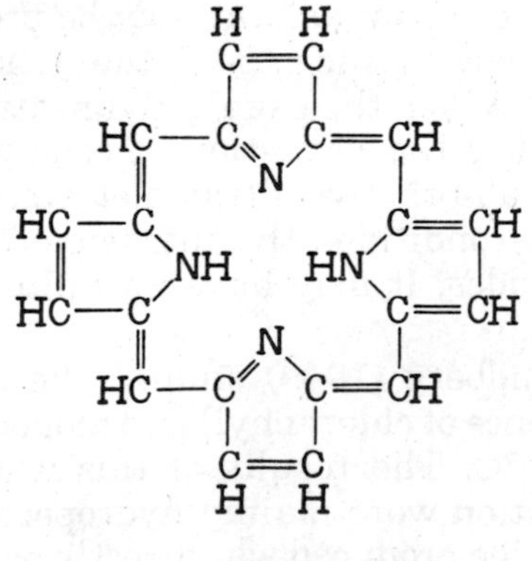

Porphin

The presence of traces of heavy metals in food lipids is without doubt one of the important reasons for their oxidative deterioration.

Hematin Compounds

Of great importance in the peroxidation and, therefore, deterioration in foods containing lipids is the action of hematin compounds and lipoxygenase. Both of these are proteins and their roles of activity are rather well-defined and are more pronounced than other catalysts.

The peroxidations of lipids by hematin compounds is a basic deteriorative reaction. Oxidative rancidity brought about by this type of catalysis limits the storage life of frozen meats, fish, and poultry, and it is also of importance in the deterioration of freeze-dried as well as precooked meats. Experiments resulting in the increase in peroxide value and the determination of the rate of oxygen absorption showed that pork adipose tissue oxidized at a much faster rate than the extracted lard. The study indicated hematin compounds, specifically hemoglobin, myoglobin, and cytochromes, as the important catalysts in *in situ* lipid peroxidation because of their presence in adipose tissue but not in the extracted lard (Tappel *et al.* 1961). Cyanide can be used as a test to inhibit hematin catalysis.

According to Tappel (1962) all the hematin compounds occurring in nature catalyze oxidation of unsaturated lipids and other olefins. Only highly unsaturated lipids have rates of autoxidation similar to hemoglobin-catalyzed oxidation. In all cases, the induction period of hemoglobin-catalyzed reaction is shorter than that for the corresponding autoxidation. These rapid initiation and propagation reactions are characteristic of hematin catalysis. Since the rates of hematin-catalyzed peroxidation of oleic and linoleic are always greater than corresponding autoxidation, the importance of hematin-catalyzed reaction in deteriorative reactions in biological systems containing the usual levels of unsaturation is apparent. Hematin can catalyze linoleate peroxide decomposition when peroxide concentration is as low as 5×10^{-6} *M*. It appears that at low total linoleate concentrations the chain reaction does not occur since oxygen is not absorbed, and higher linoleate concentrations are necessary to be sure of chain propagation.

Hematin-catalyzed lipid peroxidation can have a deleterious effect on other biological compounds near it because of a large steady state production of free radicals made up mainly of lipid peroxy radicals $ROO^\cdot$, lipid oxyradicals $RO^\cdot$, hydroxy radicals OH, and radicals resulting from the splitting of the fatty acid chain. Reacting at random through hydrogen abstraction and various addition reactions, these radicals would damage other lipids, proteins, enzymes, and vitamins. Oxygen-labile compounds such as vitamin A are readily cooxidized. Direct free radical damage to enzymes would be very important. Laboratory studies have been made showing lipid peroxidation damage to proteins. Other studies were carried out on the linolenate peroxidation damage to cytochrome *c*.

Tappel (1962) pointed out that the critical reaction in hematin-catalyzed unsaturated lipid peroxidation is the catalytic decomposition of peroxide into free radicals. Studies of the mechanism of hydroperoxide breakdown showed that cumyl hydroperoxide and linoleate hydroperoxide decompositions were catalyzed by hematin with a hydrogen donor to trap free radicals. The corresponding alcohols that were produced from the hydroperoxides indicated that hematin catalyzes the homolytic cleavage of the O–OH bond of cumyl and linoleate hydroperoxides. This shows that the catalytic homolytic cleavage of the –O–O–H bond of hydroperoxide is the general property of the hematin catalysts. Analytical studies of secondary products from linoleate hydroperoxide decomposition showed that large amounts of oxirane and hydroxyl groups were present. This suggests that a hydroxyoxirane compound forms as a result of the intramolecular attack of the alkoxy radical on the adjacent double bond followed by radical combination between the alkyl radical produced and the hydroxyl radical. This would explain the quantitative loss of conjugated double bonds. Subsequently, the reaction becomes much more complicated.

Antioxidants. Hemoglobin-catalyzed oxidation of unsaturated lipids seems particularly suited for evaluation of antioxidants to be used for the inhibition of oxidative rancidity of fish, poultry, and meats. Lipid peroxides are necessary for the hematin-catalyzed initiation. The formation of these compounds could be retarded by the polyphenolic antioxidants.

Lipoxygenase

Lipoxygenase is an enzyme that catalyzes the direct oxidation of lipids containing *cis-cis*-1,4-pentadiene systems to hydroperoxides. Because of this it has the ability to oxidize linoleic, linolenic, and arachidonic acids, esters, and triglycerides, but not those of oleic acid (Dillard *et al.* 1961). Privett *et al.* (1955) determined that the principal product of the lipoxygenase catalysis is an optically active cis–trans conjugated monomeric hydroperoxide. Polyphenolic antioxidants are important inhibitors of this catalysis. Butylated hydroxyanisole (BHA) and butylated hydroxytoluene (BHT) are typical antioxidants extensively employed in food products. Nordihydroguaiaretic acid was formerly used extensively, but its use is no longer permitted. For a complete list of allowed antioxidants see Food Chemicals Codex, 2nd Edition (Natl. Acad. Sci. 1972).

The baking industry has long used rich sources of lipoxygenase for the bleaching of carotene in dough. Carotene cooxidation can be controlled by the selection of wheat of suitable lipoxygenase and carotene content, by addition of legume flours containing lipoxygenase during dough mixing, and by controlled aeration during the mixing process.

Antioxidants. Antioxidants have already been mentioned in this chapter. These compounds mainly act as free radical acceptors or as hydrogen donors. When they act, they terminate the chain reaction of autoxidation. This is illustrated as follows.

$$ROO^{\cdot} + AH_2 \longrightarrow AH^{\cdot} + ROOH$$

$$AH\cdot + AH^{\cdot} \longrightarrow A + AH_2$$

The synthetic antioxidants are extensively used in food products. These include BHA and BHT which have already been mentioned. BHA is mainly 3-*tert*-butyl-4-hydroxyanisole (3-BHA), but it has varying amounts of 2-*tert*-butyl-4-hydroxyanisole (2-BHA). BHT is butylated hydroxytoluene, 2,6-di-*tert*-butyl-4-methylphenol. In addition to BHA and BHT, propyl gallate (PG), and *tert*-butyl hydroquinone (TBHQ) are used also. The natural tocopherols (α, β, γ, and δ) when present protect the unrefined oils and fats. Since they are heat labile, they are usually destroyed in the refining process of fats and oils. They are also quite easily oxidized. Some compounds, acidic in nature, are known to exert a synergistic effect when used in connection with polyphenolic antioxidants. These include citric, phosphoric, and ascorbic acids.

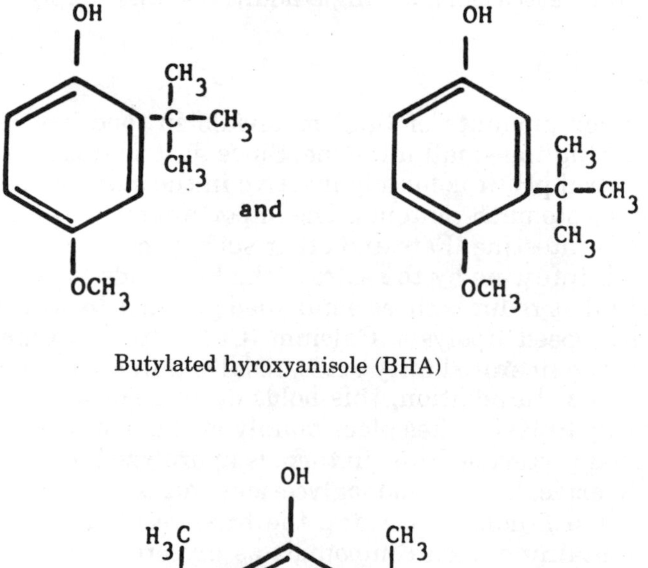

OH CH3 C CH3 CH3 OCH3 and OH CH3 C CH3 CH3 OCH3

Butylated hyroxyanisole (BHA)

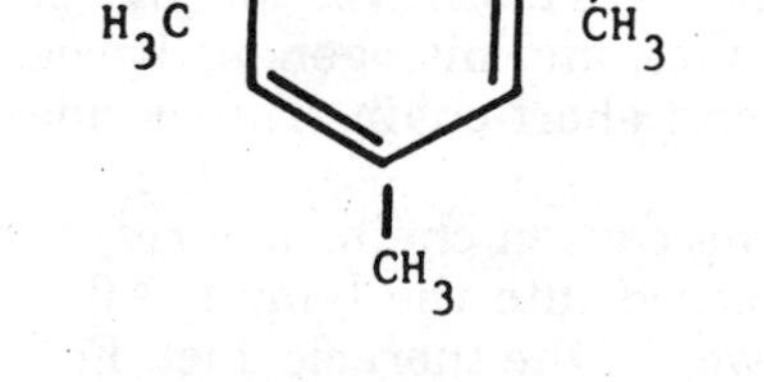

OH H3C H3C C H3C CH3 C CH3 CH3 CH3

Butylated hydroxytoluene (BHT)

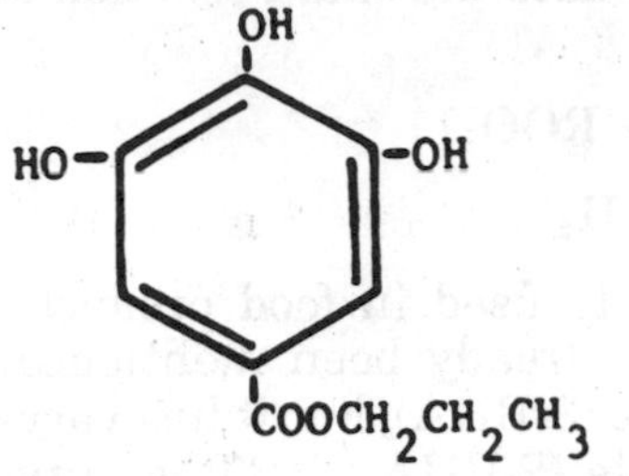

Propyl gallate (PG)

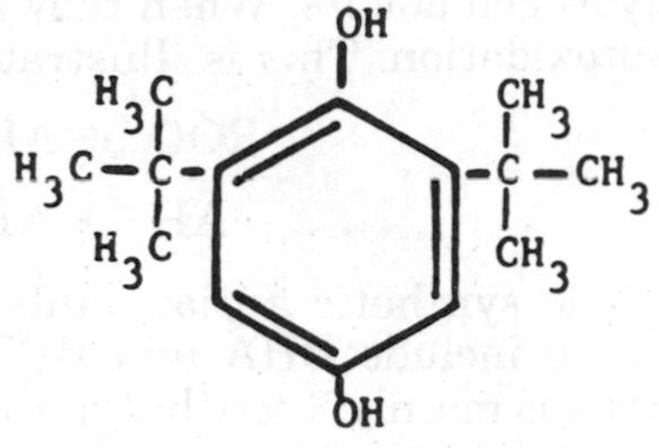

tert-Butyl hydroquinone (TBHQ)

The Use of Antioxidants. A single antioxidant may be used provided that not more than 0.01% is used. This is based on the fat content of the food. If more than one antioxidant is used in the same product the permitted total used must not be greater than 0.02%, and no one antioxidant may exceed 0.01%. If antioxidants and synergists are employed together, the total of all used must not be greater than 0.025%. In this last case, the restrictions of single additives must apply.

Digestion

Only minor amounts of lipid materials in food are broken down before reaching the small intestine. Since gastric lipase is most active around neutral pH, it is largely inactive in the stomach because of the acidity of the stomach contents. The important site of lipid digestion is in the small intestine. Fats and other such compounds are emulsified in the small intestine by the salts of the bile acids to very small globules. Peristalsis reduces these emulsified globules to still smaller globules, which speed lipolysis. Calcium (Ca^{2+}) tends to form insoluble soaps with the liberated fatty acids, resulting in an acceleration of the enzyme action. In addition, this holds down reformation of the glyceride. The hydrolysis takes place mainly at the α or α' positions, forming an α, β-diglyceride. This, in turn, is hydrolyzed to form mainly the β-monoglyceride. Since β-monoglycerides can isomerize to give a mixture of α- and β-monoglycerides, the final result is a heterogeneous mixture containing such compounds as glycerol, fatty acids, di- and triacylglycerols, and other compounds originally present in the fat.

Pancreatic phosphatidases hydrolyze the phospholipids to form fatty acids, glycerol, phosphate, and nitrogenous bases.

Cholesterol esters and short-chain triglycerides are hydrolyzed by pancreatic esterases.

Fatty acids with long carbon chains are resynthesized into triacylglycerols and are released into the lymph. After this they enter the portal circulation by way of the thoracic duct. Fatty acids with a chain length of 10 carbon atoms or less are mainly absorbed in unesterified form and enter the liver by the portal route. The synthesis of triglycerides in the intestinal mucosa occurs in a similar fashion to that which goes on in the liver. The products of these reactions are absorbed through the intestinal wall. It is well to note that complete hydrolysis is not necessary for absorption.

Commercial Applications

Hydrogenation. As already has been pointed out, oils contain more unsaturated fatty acids in their composition than do solid fats. Under natural conditions, some, but not all of the fatty acids present are unsaturated. Oil can be solidifed by a process called hydrogenation. This process adds hydrogen directly to the double bonds of the fatty acids.

$$\underset{\text{Oleic acid}}{CH_3(CH_2)_7CH{=}CH(CH_2)_7COOH} + H_2 \xrightarrow{\text{Ni}} \underset{\text{Stearic acid}}{CH_3(CH_2)_{16}COOH}$$

It is used to modify large amounts of fats and oils. As a result of this treatment a number of changes take place, including alterations in the number as well as the geometry and location of the double bonds. These changes bring about an alteration in the properties of the fat, both physical and chemical. This process is of great importance in producing plastic fats, which have the right consistency for use in margarine and shortening manufacture. At the same time the stability and color of the products are improved.

During this process, deaerated hot oil is whipped with hydrogen gas in a closed chamber in the presence of nickel catalyst, which speeds the addition of the hydrogen to the double bonds. The rate of the reaction is controlled by several factors. Among these are the nature of the fat or oil undergoing treatment, the amount of hydrogen being used, the nature and quantity of the catalyst, as well as the pressure, temperature, and amount of agitation. The agitation results in more effective operation of the catalyst, by keeping it well mixed with the oil under treatment.

Under some conditions very unsaturated fatty acids take up hydrogen more readily than those less saturated acids in a given mixture. This is known as selectivity. Under such conditions, linoleic acid would be hydrogenated to oleic acid, rather than oleic to stearic acid. The ratio of conversion varies according to the conditions of operation.

Isomerization takes place during hydrogenation. The amount of oleic acid that is 9-octadecenoic acid decreases, while the 8, 10, 7, and 11 isomers increase (Allen and Kiess 1955). With the continuation of the hydrogenation, the ratio of 9 to 8 positional isomers nears one. At the same time the ratio of cis to trans isomers comes to an equilibrium value of 1:2. These two types of isomerization proceed simultaneously.

The shift of the double bonds together with the formation of the trans isomers is accounted for by the idea of partial hydrogenation-dehydrogenation. It seems that an equilibrium exists among the several isomers. The explanation of this is the addition of a hydrogen atom to a carbon of the double bond. Following this is the removal of a hydrogen atom to reproduce a double bond. This mechanism is illustrated as follows.

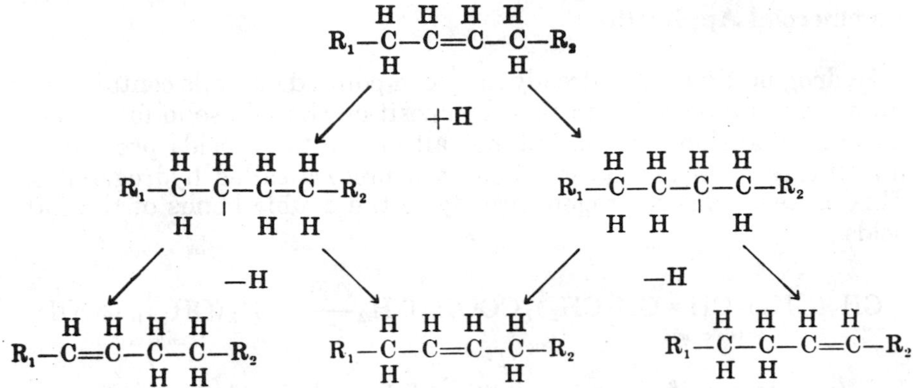

From Allen and Kiess (1955). Used with permission of American Oil Chemist's Society, copyright owner.

Nickel is used extensively as a catalyst for hydrogenation. However, other metals can also be used, including metals of the platinum group, copper, and combinations of copper and chromium. Copper–chromium catalysts have a high selectivity for linolenic acid, and few conjugated dienes result when copper–chromium catalysts are used. Nickel produces more trans isomers than copper–chromium. Of the platinum metal group, palladium is the most effective, and the most likely from the cost point of view to be considered for commercial application. The metals of this group are very sensitive to poisons, such as sulfur-containing compounds, arsenic, and lead. These poisons have the ability to cover or eliminate the active areas of the catalyst (Rylander 1970). This action is not reversible.

Interesterification. This is also known as transesterification or ester interchange. This takes place in triglycerides and is an exchange of acyl groups between or within these triglyceride molecules (Braun 1960). These changes are brought about with the help of catalysts, such as compounds of alkali or alkaline earth metals or such metals as tin, lead, or zinc, or compounds of them. The lower alcoholates are of value at low temperature.

Directed Interesterification. The directed interesterification process has been used for a long time. Hawley and Holman (1956) have applied it as a processing method for lard because untreated lard has a number of disadvantages when compared with hydrogenated vegetable shortenings: among these are grainy, translucent appearance, and rather unattractive texture. Untreated lard is too soft at warm temperatures and too hard at lower temperatures. Furthermore, consistency and flavor vary with such factors as season, area of origin, and rendering methods. Stearic and palmitic acids are the saturated fatty acids found mainly in lard and make up about 37% of the fatty acids.

In directed interesterification, the reaction temperature used is held just below the melting point of the lard. Under these conditions, the trisaturated glycerides, lard substances with the highest melting point, will precipitate, leaving the liquid phase for the interesterification. The reaction is speeded by an active catalyst—sodium–potassium (NaK) alloy, which must be in condition to be rapidly and completely distributed in the fat. The sodium–potassium alloy is pumped into the lard by means of a small continuous mixer and this suspends the NaK throughout the lard in the form of very small particles. The reaction is carried on with gentle agitation. The result is the formation of a larger proportion of the trisaturated glycerides, glycerides with higher melting points, and a smaller amount of the disaturated glycerides which have intermediate melting points. The trisaturated glycerides precipitate out as they are formed. The elimination of these trisaturated glycerides moves the reaction ahead. The reaction is controlled by the temperature used and the length of time of the process. At the end of the reaction, the catalyst is destroyed by pumping in water and CO_2. The water and CO_2 tend to buffer the alkali that has been formed to a lower pH and cuts down the amount of saponification of the lard, as this would reduce the yield of marketable lard if it were allowed to occur.

After the product is washed to remove the soaps, it is dried in a continuous vacuum drier, is then hydrogenated for consistency and stability, and deodorized and plasticized for packaging. Antioxidants can be added to increase the shelf-life.

The improved product brought about by directed interesterification results from the temperature control that permits the precipitation of the trisaturated glycerides as the process progresses. The product which resulted from random interesterification did not solve the plastic range problem because most of the solids in the lard resulting from random interesterification were still in the form of disaturated glycerides.

Margarine. Margarines found on the market in the United States today are made mostly from vegetable oils which have been sufficiently hydrogenated to permit a satisfactory spreading texture. Margarine, like butter, should melt in the mouth when it is consumed. In recent years a great many corn oil margarines have appeared on the market. Sometimes these vegetable oils are blended with small amounts of animal oils, but the total fat present in the finished product must not be less than 80%. The oils used amount to about 100% fat. It is necessary, therefore, to have a water phase in the process of manufacture which contains emulsifiers, salt (unless it is salt-free "sweet" margarine), butter flavor, color, and allowed preservatives, such as sodium benzoate. Actually, the present-day processes for the manufacture of margarine are much the same as those used for the continuous operation for the manufacture of butter. The water phase and the melted oil are pumped into a mixing chamber which is refrigerated and cylindrical in shape. Mixing then distributes the aqueous phase in the form of small droplets throughout the fatty material. Optimum-sized fat crystals, developed by proper temperature control, is important to achieve the proper semiplastic consistency (which means that the oils have crystallized and solidifed with the droplets of water throughout the matrix).

SUMMARY

Lipids are made up of a variety of chemical substances, which are heterogeneous in nature and difficult to classify. They are divided into several groups: fatty acids, oils and fats, waxes, phospholipids, sphingolipids, as well as such compounds as sterols, hydrocarbons, fat-soluble vitamins, carotenoids, and such combined lipids as lipoproteins and lipopolysaccharides. The Bloor classification is one such classification.

Simple lipids are made up of fats, oils, and fatty acids. Monoacid, diacid, and triacid esters are known. Most of the saturated fatty acids found in nature are made up of an even number of carbon atoms, from 4 up to 24 in straight chains. Fats and oils from natural sources are mainly mixtures of triglycerides rather than pure substances. Palmitic acid is the most widely distributed of the saturated fatty acids.

Unsaturated fatty acids found in food lipids are generally of the unbranched type. Oleic acid is probably the most generally found of all fatty acids. Linoleic acid must be included in the diet of man because it is an essential fatty acid, which cannot be produced in the body.

In the presence of nickel or platinum catalysts, unsaturated fatty acids can add hydrogen atoms to form saturated acids. Because of the double bonds, unsaturated fatty acids can be readily oxidized. When exposed to oxygen polyunsaturated fatty acids form peroxides, as well as a mixture of volatile aldehydes, ketones, and acids. This is catalyzed by lipoxidase or by the presence of trace metals.

Position isomerism is determined by the position of the double bonds in the carbon skeleton. There are two possible arrangements: (1) the conjugated system with single bonds alternating with double bonds: $—CH=CH—CH=CH—$, and (2) the nonconjugated arrangement in which two double bonds are separated by one or more methylene groups: $—CH=CH—CH_2—CH=CH—$. The conjugated is the more stable form.

The cause of geometric isomerism or cis and trans isomerism is the restriction in rotation of two carbon atoms which are connected by a double bond.

Stereospecific analysis is a technique used to determine the identity of the fatty acids which occupy the position of α, β, and α' of the fat molecule.

Polymorphism is the existence of several different crystal forms in some triglycerides. These exist in three forms: α, β', and β. The β' are best for use in shortenings because they can include large amounts of tiny air bubbles.

Tempering is controlled crystallization. It is a process which puts the fat in the best condition for the use to which it is to be put.

Oils and fats are classified into five groups, based on fatty acid composition and source. Two of these are from animal fats—milkfat and animal depot fats. Those from plants are the lauric acid group, the oleic–linoleic acid group, and the linolenic acid group.

Waxes are esters of fatty acids and alcohols, both of high molecular weight. The alcohols are long-chain monohydroxy alcohols. The natural waxes contain other compounds also, all of high molecular weight.

Composite lipids contain nitrogen and phosphorus as well as carbon, hydrogen, and oxygen. They are compounds of glycerol esterified with fatty acids, and by phosphoric acid and basic nitrogen compounds. Choline and other such compounds are the nitrogen compounds.

The lecithins are phosphatidyl cholines. The cephalins are phosphatidyl serine or phosphatidyl ethanolamine and are more basic than lecithins.

Sphingolipids do not contain glycerol, but many of them contain sphingosine, a long-chain aliphatic base.

Derived lipids are unsaponifiables and are made up of terpenes and steroids. The best known member of the latter group is cholesterol, since it is involved in diet problems.

Oxidation of lipids is of importance to food chemists because it results in the formation of off-odors and off-flavors, and in reduction or destruction of essential fatty acid, as well as the formation of brown colors.

The most common type of lipid autoxidation is catalytic autoxidation. In the 1940s it was shown that hydroperoxides are formed during the usual autoxidation of fats. The reaction is concerned principally with unsaturated acyl groups, where the hydroperoxide group appears in the alpha position relative to the double bond. The rate of autoxidation increases exponentially with increasing unsaturation in the case of derivatives of ordinary fatty acids.

The off-odors and off-flavors found in autoxidized food materials are not caused by the hydroperoxides, but are secondary substances formed during the reactions.

It has been observed that double bonds shift during autoxidation.

The most important cause of the deterioration of lipids or foods which contain lipids is the action of oxygen on unsaturated fatty acids in the triglycerides.

The oxidation reaction is usually shown as divided into three steps. These include initiation, propagation, and termination. The oxidation of monoenoic and polyenoic acids is discussed, as well as the secondary degradation products. In addition to this, the effect of light and radiation, heavy metals, and hematin compounds are given consideration.

Antioxidants, both natural and synthetic which tend to terminate the chain reaction of autoxidation, are discussed.

The important commercial applications of lipids include hydrogenation, during which process the hydrogen is added to the double bonds, and it can be controlled to yield the hardness necessary for the use to which it is to be put.

Another interesting application is interesterification. This takes place in triglycerides and is an exchange of acyl groups between or within these triglyceride molecules. Catalysts are necessary for these changes to take place. Directed interesterification is used for the improvement of lard.

The manufacture of margarines is discussed.

BIBLIOGRAPHY

ALBRIGHT, L. F. 1963. Mechanism of hydrogenation of triglycerides. J. Am. Oil Chem. Soc. *40,* 16–17, 26, 28–29.

ALBRIGHT, L. F. 1965. Quantitative measure of selectivity of hydrogenation of triglycerides. J. Am. Oil Chem. Soc. *42,* 250–253.

ALBRIGHT, L. F. 1970. Transfer and adsorption steps affecting partial hydrogenation of triglyceride oils. J. Am. Oil Chem. Soc. *47,* 490–493.

ALLEN, R. R. 1956. Isomerization during hydrogenation. II. Methyl *cis*-10, *cis*-12-octadecadienoate. J. Am. Oil Chem. Soc. *33,* 301–304.

ALLEN, R. R., and KIESS, A. A. 1955. Isomerization during hydrogenation. I. Oleic acid. J. Am. Oil Chem. Soc. *32,* 400–405.

ALLEN, R. R., and KIESS, A. A. 1956. Isomerization during hydrogenation. III. Linoleic acid. J. Am. Oil Chem. Soc. *33,* 355–359.

ALTMAN, P. L., and DITTMER, D. S. 1972. Biology Data Book, 2nd Edition. Vol. 1. Federation of American Societies for Experimental Biology, Bethesda, MD.

ANSELL, G. B., and HAWTHORNE, J. N. 1964. Phospholipids: Chemistry, Metabolism, and Function. American Elsevier Publishing Co., New York.

AOAC. 1980. Official Methods of Analysis. 13th Edition. Association of Official Analytical Chemists, Washington, DC.

ARTMAN, N. R. 1969. The chemical and biological properties of heated and oxidized fats. Adv. Lipid Res. *7,* 245–330.

ARTMAN, N. R., and ALEXANDER, J. C. 1968. Characterization of some heated fat components. J. Am. Oil Chem. Soc. *45,* 643–648.

BADINGS, H. T. 1960. Principles of autoxidation processes in lipids with special regard to the development of autoxidation off-flavors. Neth. Milk Dairy J. *14,* 215–242.

BAILEY, A. E. 1951. Industrial Oil and Fat Products. 2nd Edition. Wiley (Interscience) Publishers, New York.

BELL, E. R., RALEY, J. H., RUST, F. F., SEUBOLD, F. H., and VAUGHAN, W. E. 1951. Reactions of free radicals associated with low-temperature oxidation of paraffins. Discuss. Faraday Soc. *10,* 242–249.

BLOOR, W. R. 1926. Biochemistry of the fats. Chem. Rev. *2,* 243–300.

BOLLAND, J. L. 1946. Kinetic studies in the chemistry of rubber and related materials. I. Thermal oxidation of ethyl linoleate. Proc. R. Soc. (London) Ser. A *186,* 218–236.

BOLLAND, J. L., and GEE, G. 1946A. Kinetic studies in the chemistry of rubber and related materials. II. The kinetics of oxidation of unconjugated olefins. Trans. Faraday Soc. *42,* 236–243.

BOLLAND, J. L., and GEE, G. 1946B. Kinetic studies in the chemistry of rubber and related materials. III. Thermochemistry and mechanisms of olefin oxidation. Trans. Faraday Soc. *42,* 244–252.

BRAUN, W. Q. 1960. Interesterification of edible fats. J. Am. Oil Chem. Soc. *37,* 598–601.

BROCKERHOFF, H. 1965. A stereospecific analysis of triglycerides. J. Lipid Res. *6,* 10–15.

BROCKERHOFF, H., and YURKOWSKI, M. 1966. Stereospecific analyses of several vegetable fats. J. Lipid Res. *7,* 62–64.

BROCKERHOFF, H., HOYLE, R. J., and WOLMARK, N. 1966. Positional distribution of fatty acids in triglycerides of animal depot fats. Biochem. Biophys. Acta *116,* 67–72.

CARTER, H. E., JOHNSON, P., and WEBER, E. J. 1965. Glycolipids. Annu. Rev. Biochem. *34,* 109–142.

CHAHINE, M. H., COUSINS, E. R., and FEUGE, R. O. 1958. Positional isomers formed during the hydrogenation of cottonseed oil. J. Am. Oil Chem. Soc. *35,* 396–401.

CHAPMAN, D. 1962. The polymorphism of glycerides. Chem. Rev. *62,* 433–456.

COLEMAN, M. H. 1961. Further studies on the pancreatic hydrolysis of some natural fats. J. Am. Oil Chem. Soc. *38,* 685–688.

CRAIG, B. M. 1957. Dilatometry. Progress in the Chemistry of Fats and other Lipids, Volume 4, pp. 198–226. R.T. Holman, W.O. Lundberg, and T. Malkin (Editors). Pergamon Press, Oxford.

DAHLE, L. K., HILL, E. G., and HOLMAN, R. T. 1962. The thiobarbituric acid reaction and the autoxidations of polyunsaturated fatty acid methyl esters. Arch. Biochem. Biophys. *98,* 253–261.

DENISOV, E. T., and EMANUEL, N. M. 1956. Kinetic characteristics of cyclohexane oxidation in the presence of cobalt stearate. Zh. Fiz. Khim. *30,* 2327–2336. Chem. Abstr. *51,* 9274d.

DEUEL, JR., H. J. 1951, 1955, 1957. The Lipids: Chemistry and Biochemistry, Vols. I, II, and III. Wiley (Interscience) Publishers, New York.

DILLARD, M. G., HENICK, A. S., and KOCH, R. B. 1961. Differences in reactivity of legume lipoxidases. J. Biol. Chem. *236,* 37–40.

DOLEV, A., ROHWEDDER, W. K., and DUTTON, H. J. 1967. Mechanism of lipoxidase reaction. Lipids *2,* 28–32.

DOYNE, T. H. and GORDON, JANICE T. 1968. The crystal structure of a diacid triglyceride. J. Am. Oil Chem. Soc. *45,* 333–334.

ECKEY, E. W. 1956. Esterification and interesterification. J. Am. Oil Chem. Soc. *33,* 575–579.

ELLIS, G. W. 1950. Autoxidation of the fatty acids. 3. The oily products from elaidic and oleic acids. The formation of monoacyl derivatives of dehydroxystearic acid and of α, β-unsaturated keto acids. Biochem. J. *46,* 129–141.

EVANS, C. D., McCONNELL, D. G., LIST, G. R., and SCHOLFIELD, C. R. 1969. Structure of unsaturated vegetable oil glycerides: Direct calculation from fatty acid composition. J. Am. Oil Chem. Soc. *46,* 421–424.

FARMER, E. H., and SUNDRALINGAM, A. 1942. The course of autoxidation reactions in polyisoprenes and allied compounds. I. The structure and reactive tendencies of the peroxides of simple olefins. J. Chem. Soc. 121–139.

FARMER, E. H., KOCH, H. P., and SUTTON, D. A. 1943. The course of autoxidation reactions in polyisoprenes and allied compounds. VII. Rearrangement of double bonds during autoxidation. J. Chem. Soc. 541–547.

FEDELI, E., and JACINI, G. 1971. Lipid composition of vegetable oils. Adv. Lipid Res. *9,* 335–382.

FRANKEL, E. N. 1970. Conversion of polyunsaturates in vegetable oils to *cis*-monounsaturates by homogeneous hydrogenation catalyzed with chromium carbonyls. J. Am. Oil Chem. Soc. *47,* 11–14.

FRANKEL, E. N., THOMAS, F. L., and COWAN, J. C. 1970. Stereo-selective hydrogenation of model compounds and preparation of tailor-made glycerides with chromium tricarbonyl complexes. J. Am. Oil Chem. Soc. *47,* 497–500.

GADDIS, A. M., ELLIS, R., and CURRIE, G. T. 1961. Carbonyls in oxidizing fat. V. The composition of neutral volatile monocarbonyl compounds from autoxidized oleate, linoleate, linolenate esters, and fats. J. Am. Oil Chem. Soc. *38,* 371–375.

GARDNER, H. W., and WEISLEDER, D. 1970. Lipoxygenase free *Zea mays:* 9-D-Hydroperoxy-*trans*-10, *cis*-12-octadecadienoic acid from linoleic acid. Lipids *5,* 678–683.

GROSCH, W., and SCHWARZ, J. M. 1971. Linoleic and linolenic acid precursors of the cucumber flavor. Lipids *6,* 351–352.

GUNSTONE, F. D. 1964. The long spacing of polymorphic crystalline forms of long-chain compounds, with special reference to triglycerides. Chem. Ind. 84–89.

GUNSTONE, F. D. 1967. Introduction to the Chemistry and Biochemistry of Fatty Acids and Their Glycerides. 2nd Edition Chapman and Hall, London.

HABER, F., and WILLSTÄTTER, R. 1931. Unpairedness and radical chains in the reaction mechanism of organic and enzymic processes. Chem. Ber. *64,* 2844–2856.

HANAHAN, D. J., GURD, F. R. N., and ZABIN, I. 1960. Lipid Chemistry. John Wiley & Sons, New York.

HANAHAN, D. J., and JAYKO, M. E. 1952. The isolation of dipalmitoleyl-L-α-glycerylphosphorylcholine from yeast. A new route to (dipalmitoyl)-L-α-lecithin. J. Am. Chem. Soc. *74,* 5070–5073.

HANAHAN, D. J., and THOMPSON, JR., G. A. 1963. Complex lipids. Annu. Rev. Biochem. *32,* 215–240.

HAWLEY, H. K., and HOLMAN, G. W. 1956. Direct interesterification as a new processing tool for lard. J. Am. Oil Chem. Soc. *33,* 29–35.

HAWTHORNE, J. N. 1960. The inositol phospholipids. J. Lipid Res. *1,* 255–280.

HILDITCH, T. P., and WILLIAMS, P. N. 1964. The Chemical Constitution of Natural Fats, 4th Edition. Chapman and Hall, London.

HILL, A. S., and MATTICK, L. R. 1966. The *n*-alkanes of cabbage (var. Copenhagen) and sauerkraut. Phytochemistry *5,* 693–697.

HOERR, C. W. 1960. Morphology of fats, oils, and shortenings. J. Am. Oil Chem. Soc. *37,* 539–546.

HOERR, C. W., and PAULICKA, F. R. 1968. The role of X-ray diffraction in studies of raw crystallography of monoacid saturated triglycerides. J. Am. Oil Chem. Soc. *45,* 793–797.

HOFFMANN, G. 1961. 3-*cis*-hexenal, the green reversion flavor of soybean oil. J. Am. Oil Chem. Soc. *38,* 1–3.

HOFFMANN, G. 1962. Vegetable oils. *In* Lipids and Their Oxidation. H.W. Schultz, E.A. Day, and R.O. Sinnhuber (Editors). AVI Publishing Co., Westport, CT.

HOLMAN, R. T. 1954. *In* Progress in the Chemistry of Fats and Other Lipids, Vol. 2. R.T. Holman, W.O. Lundberg, and T. Malkin (Editors). Pergamon Press, Oxford.

HOLMAN, R. T. 1968. Essential fatty acid deficiency. *In* Progress in the Chemistry of Fats and other Lipids, Vol. 9, Pt. 2, R.T. Holman (Editor). Pergamon Press, Oxford.

JEN, J. J., WILLIAMS, Jr., W. P., ACTON, J. C., and PAYNTER, V. A. 1971. Effect of dietary fats on the fatty acid contents of chicken adipose tissue. J. Food Sci. *36,* 925–929.

JENSEN, R. G. 1973. Composition of bovine milk lipids. J. Am. Oil Chem. Soc. *50,* 186–192.

JENSEN, R. G., SAMPUGNA, J., and PEREIRA, R. L. 1964. Intermolecular specificity of pancreatic lipase and the structural analysis of milk triglycerides. J. Dairy Sci. *47,* 727–732.

KEENEY, M. 1962. Secondary degradation products. *In* Lipids and Their Oxidation. H.W. Schultz, E.A. Day, and R.O. Sinnhuber (Editors). AVI Publishing Co., Westport, CT.

KENDRICK, J., and WATTS, B. 1969. Acceleration and inhibition of lipid oxidation by heme compounds. Lipids *4,* 454–458.

KIMOTO, W. I., and GADDIS, A. M. 1969. Precursors of alk-2,3-dienals in autoxidized lard. J. Am. Oil Chem. Soc. *46,* 403–408.

KNIGHT, H. B., EDDY, C. R., and SWERN, D. 1951. Reactions of fatty materials with oxygen. VIII. *cis-trans* isomerization during autoxidation of methyl oleate. J. Am. Oil Chem. Soc. *28,* 188–192.

KORITALA, S. 1968. Selective hydrogenation of soybean oil III. Copper-exchanged molecular sieves and other supported catalysts. J. Am. Oil Chem. Soc. *45,* 197–200.

KUKSIS, A. 1972. Newer developments in determination of structure of glycerides and phosphoglycerides. *In* Progress in the Chemistry of Fats and Other Lipids, Vol. 12, R.T. Holman (Editor). Pergamon Press, Oxford.

KUKSIS, A., MARAI, L., and MYHER, J. J. 1973. Triglyceride structure of milk fats. J. Am. Oil Chem. Soc. *50,* 193–201.

KUMMEROW, F. A. 1961. Fats and oils. Proc. Flavor Chem. Sym., pp. 109–115. Campbell Soup Co., Camden, NJ.

LABUZA, T. P., TSUYUKI, H., and KAREL, M. 1969. Kinetics of linoleate oxidation in model systems. J. Am. Oil Chem. Soc. *46,* 409–416.

LANDS, W. E. M., PIERINGER, R. A., SLAKEY, P. M., and ZSCHOCKE, A. 1966. A micromethod for the stereospecific determination of triglyceride structure. Lipids *1,* 444–448.

LAW, J. H. 1960. Glycolipids. Annu. Rev. Biochem. *29,* 131–150.

LEE, F. A., and MATTICK, L. R. 1961. Fatty acids of the lipids of vegetables. I. Peas (*Pisum sativum*). J. Food Sci. *26,* 273–275.

LILLARD, D. A., and DAY, E. A. 1964. Degradation of monocarbonyls from autoxidizing lipids. J. Am. Oil Chem. Soc. *41,* 549–552.

LUNDBERG, W. O. 1949. Oxidation of esters of linoleic acid by oxygen. Oléagineux *4,* 86–94.

LUNDBERG, W. O. 1961. Autoxidation and Antioxidants, Vols. I and II. John Wiley & Sons, New York.

LUNDBERG, W. O. 1962. Mechanisms. *In* Symposium on Foods: Lipids and Their Oxidation. H. W. Schultz, E. A. Day, and R. O. Sinnhuber (Editors). AVI Publishing Co., Westport, CT.

LUNDBERG, W. O., and CHIPAULT, J. R. 1947. The oxidation of methyl linoleate at various temperatures. J. Am. Chem. Soc. *69,* 833–836.

LUTTON, E. S. 1971. Postulated scheme for β crystal structures of mixed palmitic-stearic triglycerides. J. Am. Oil Chem. Soc. *48,* 245–247.

LUTTON, E. S. 1972. Lipid structures. J. Am. Oil Chem. Soc. *49,* 1–9.

LUTTON, E. S., and JACKSON, F. L. 1950. The polymorphism of synthetic and natural 2-oleyldipalmitin. J. Am. Chem. Soc. *72,* 3254–3257.

LUTTON, E. S., MALLERY, N. F., and BURGERS, J. 1962. Interesterification of lard. J. Am. Oil Chem. Soc. *39,* 233–235.

MAG, T. K. 1973. Clay-heat refining of edible oils. J. Am. Oil Chem. Soc. *50,* 251–254.

MARION, W. W., MAXON, S. T., and WANGEN, R. M. 1970. Lipid and fatty acid composition of turkey liver, skin, and depot tissue. J. Am. Oil Chem. Soc. *47,* 391–392.

MARKLEY, K. S. 1960–1969. Fatty Acids, 2nd. Edition, Parts 1–5. John Wiley & Sons, New York.

MARKLEY, K. S., and SANDO, C. E. 1934. Petroleum ether- and ether-soluble constituents of cranberry pomace. J. Biol. Chem. *105,* 643–653.

MARKLEY, K. S., and SANDO, C. E. 1937. The wax-like constituents of the cuticle of the cherry, *Prunus avium, L.* J. Biol. Chem. *119,* 641–645.

MARKLEY, K. S., HENDRICKS, S. B., and SANDO, C. E. 1932. Further studies on the wax-like coating of apples. J. Biol. Chem. *98,* 103–107.

MARKLEY, K. S., HENDRICKS, S. B., and SANDO, C. E. 1935. Constituents of the wax-like coating of the pear, *Pyrus communis* L. J. Biol. Chem. *111,* 133–146.

MARKLEY, K. S., NELSON, E. K., and SHERMAN, M. S. 1937. Some wax-like constituents from expressed oil from the peel of Florida grapefruit, *Citrus grandis.* J. Biol. Chem. *118,* 433–441.

MARKLEY, K. S., SANDO, C. E., and HENDRICKS, S. B. 1938. Petroleum ether-soluble and ether-soluble constituents of grape pomace. J. Biol. Chem. *123,* 641–654.

MATTHEWS, R. F., SCANLAN, R. A., and LIBBEY, L. M. 1971. Autoxidation products of 2,4-decadienal. J. Am. Oil Chem. Soc. *48,* 745–747.

MATTICK, L. R., and LEE, F. A. 1961. The fatty acids of vegetables. II. Spinach. J. Food Sci. *26,* 356–358.

MATTICK, L. R., and RICE, A. C. 1976. Fatty acid composition of grape seed oil from native American and hybrid grape varieties. Am. J. Enol. Vitic. *27,* 88–90.

MATTSON, F. H., and VOLPENHEIN, R. A. 1962. Hydrogenation of a fatty acid is not influenced by the position it occupies on a triglyceride molecule. J. Am. Oil Chem. Soc. *39,* 307–308.

METCALFE, L. D., and SCHMITZ, A. A. 1961. The rapid preparation of fatty acid esters for gas chromatographic analysis. Anal. Chem. *33,* 363–364.

METCALFE, L. D., SCHMITZ, A. A., and PELKA, J. R. 1966. Rapid preparation of fatty acid esters from lipids by gas chromatographic analysis. Anal. Chem. *38,* 514–515.

MICHALSKI, S. T., and HAMMOND, E. G. 1972. Use of labeled compounds to study the mechanism of flavor formation in oxidizing fats. J. Am. Oil Chem. Soc. *49,* 563–566.

NATL. ACAD. SCI. 1972. Food Chemicals Codex, 2nd. Edition. National Academy of Sciences, Washington, DC.

NATL. RES. COUNCIL. 1980. Recommended Daily Allowances. National Academy of Sciences, Washington, DC.

NELSON, J. H. 1972. Enzymatically produced flavors for fatty systems. J. Am. Oil Chem. Soc. *49,* 559–562.

OHLOFF, G. 1973. Fats as precursors. *In* Functional Properties of Fats in Foods. J. Solms (Editor). Forster Publishing, Ltd. Zürich, Switzerland (German)

PATTON, S., BARNES, I. J., and EVANS, L. E. 1959. *n*-Deca-2,4-dienal, its origin from linoleate and flavor significance in fats. J. Am. Oil Chem. Soc. *36,* 280–283.

PERKINS, E. G. 1960. Nutritional and chemical changes occurring in heated fats: A review. Food Technol. *14,* 508–514.

POKORNY, J. 1971. Stabilization of fats by phenolic antixoidants. Can. Inst. Food Technol. J. *4,* 68–74.

POMERANZ, Y., and MELOAN, C. E. 1971. Food Analysis: Theory and Practice. AVI Publishing Co., Westport, CT.

PRIVETT, O. S. 1961. Some observations on the course and mechanism of autoxidation and antioxidant action. Proc. Flavor Chem. Symp. Campbell Soup Co., Camden, NJ.

PRIVETT, O. S., and LUNDBERG, W. O. 1954. Paper presented at Fall meeting of American Oil Chemists' Society (Unpublished).

PRIVETT, O. S., and NICKELL, E. C. 1959. Determination of structure and analysis of the hydroperoxide isomers of autoxidized methyl oleate. Fette, Seifen, Anstrichm. *61,* 842–845.

PRIVETT, O. S., and NUTTER, L. J. 1967. Determination of the structure of lecithins via the formation of acetylated 1,2-diglycerides. Lipids *2,* 149–154.

PRIVETT, O. S., NICKELL, C., LUNDBERG, W. O., and BOYER, P. D. 1955. Products of the lipoxidase-catalyzed oxidation of sodium linoleate. J. Am. Oil Chem. Soc. *32,* 505–511.

RAGHUVEER, K. G., and HAMMOND, E. G. 1967. The influence of glyceride structure on the rate of autoxidation. J. Am. Oil Chem. Soc. *44,* 239–243.

RAPPORT, M. M., and NORTON, W. T. 1962. Chemistry of the lipids. Annu. Rev. Biochem. *31,* 103–138.

RAWLS, H. R., and VAN SAMTEN, P. J. 1970. A possible role for singlet oxygen in the initiation of fatty acid autoxidation. J. Am. Oil Chem. Soc. *47,* 121–125.

ROCK, S. P., and ROTH, H. 1966. Properties of frying fat I. The relationship of viscosity to the concentration of non-urea adducting fatty acids. J. Am. Oil Chem. Soc. *43,* 116–118.

RYLANDER, P. N. 1970. Hydrogenation of natural oils with platinum metal group catalysts. J. Am. Oil Chem. Soc. *47,* 482–486.

SCHOLFIELD, C. R., JONES, E. P., NOWAKOWSKA, J., SELKE, E., SREENIVASAN, B., and DUTTON, H. S. 1960. Hydrogenation of linolenate. I. Fractionation and characterization studies. J. Am. Oil Chem. Soc. *37,* 579–582.

SCHOLFIELD, C. R., JONES, E. P., NOWAKOWSKA, J., SELKE, E., and DUTTON, H. S. 1961. Hydrogenation of linolenate. II. Hydrazine reduction. J. Am. Oil Chem. Soc. *38,* 208–211.

SEALS, R. G., and HAMMOND, E. G. 1970. Some carbonyl flavor compounds of peroxidized soybean and linseed oils. J. Am. Oil Chem. Soc. *47,* 278–280.

SHAHANI, K. M. 1966. Milk enzymes: Their role and significance. J. Dairy Sci. *49,* 907–920.

SHERBON, J. W. 1974. Crystallization and fractionation of milk fat. J. Am. Oil Chem. Soc. *51,* 22–25.

SHORLAND, F. B., GERSON, T., and HANSEN, R. P. 1955. Branched-chain fatty acids of butterfat. 7. Investigation of the C_{13} acids. Biochem. J. *61,* 702–704.

SMOUSE, T. H., and CHANG, S. S. 1967. A systematic characterization of the reversion of flavor in soybean oil. J. Am. Oil Chem. Soc. *44,* 509–514.

STUCKEY, B. N. 1962. Antioxidants. *In* Symposium on Foods: Lipids and Their Oxidation. H. W. Schultz, E. A. Day, and R. O. Sinnhuber (Editors). AVI Publishing Co., Westport, CT.

TAPPEL, A. L. 1962. Hematin compounds and lipoxidase as biocatalysts. *In* Symposium on Foods: Lipids and Their Oxidation. H. W. Schultz, E. A. Day, and R. O. Sinnhuber (Editors). AVI Publishing Co., Inc. Westport, CT.

TAPPEL, A. L., BROWN, W. D., ZALKIN, H., and MAIER, V. P. 1961. Unsaturated lipid peroxidation catalyzed by hematin compounds and its inhibition by vitamin E. J. Am. Oil Chem. Soc. *38,* 5–9.

TOBOLSKY, A. V., METZ, D. J., and MESROBIAN, R. B. 1950. Low temperature autoxidation of hydrocarbons: The phenomenon of maximum rates. J. Am. Chem. Soc. *72,* 1942–1952.

VANDER WAL, R. J. 1960. Calculation of the distribution of the saturated and unsaturated acyl groups in fats, from pancreatic lipase hydrolysis data. J. Am. Oil Chem. Soc. *37,* 18–20.

VAN WIJNGAARDEN, D. 1967. Modified rapid preparation of fatty acid esters from lipids for gas chromatographic analysis. Anal. Chem. *39,* 848–849.

VORBECK, M. L., MATTICK, L. R., LEE, F. A., and PEDERSON, C. S. 1961. Preparation of methyl esters of fatty acids for gas-liquid chromatography. Quantitative comparison of methylation techniques. Anal. Chem. *33,* 1512–1514.

WALTKING, A. D., and ZMACHINSKI, H. 1970. Fatty acid methodology for heated oils, J. Am. Oil Chem. Soc. *47,* 530–534.

WATERS, W. A. 1971. The kinetics and mechanisms of metal-catalyzed autoxidation. J. Am. Oil Chem. Soc. *48,* 427–433.

WEISS, T. J., JACOBSON, G. A., and WIEDERMANN, L. H. 1961. Reaction mechanics of sodium methoxide treatment of lard. J. Am. Oil Chem. Soc. *38,* 396–399.

WIEDERMANN, L. H., WEISS, T. J., JACOBSON, G. A., and MATTIL, K. F. 1961. A comparison of sodium methoxide-treated lards. J. Am. Oil Chem. Soc. *38,* 389–395.

6

Proteins

Amino Acids
Classification of Proteins
Structure of Proteins
Properties of Proteins
Protein Synthesis
Proteins in Foods
Summary
Bibliography

Proteins as a group are indispensable components of living matter, particularly as structural components and as the enzymes responsible for the metabolic events of the organism. Some proteins are involved in the contractile process, others in metabolic regulation, while still others serve as antibodies, the defense mechanism against disease.

Plants synthesize proteins from nitrogen, CO_2, and H_2O. Very important to this process are the nitrogen-fixation bacteria found on the roots of certain plants.

Proteins supply amino acids, some of which are necessary to sustain life. Proteins are organic nitrogenous compounds of very high molecular weight and are complex in nature. Not all plant proteins contain all the essential amino acids; therefore a diet based on such foods should be selected carefully to avoid certain deficiency diseases.

Four elements are found in all proteins: nitrogen (16%), carbon (50%), hydrogen (7%), and oxygen (22%). In addition to these, some proteins contain sulfur (3%), and others contain phosphorus. Still others contain such metals as iron, copper, and zinc.

The molecular weights of proteins can range up to many millions. Because of the size, they are colloidal particles, and for this reason are unable to pass through semipermeable membranes. Proteins can act as acids as well as bases, and are called "amphoteric."

AMINO ACIDS

Upon hydrolysis, proteins usually are found to yield 20 amino acids. Two of these are imino acids—proline and hydroxyproline. The primary α-amino acids have the amino and carboxyl groups connected to the same carbon atom.

These α-amino acids are the building blocks of polypeptides and proteins. In proteins the α-amino group of one residue is coupled with the α-carboxyl group of another amino acid residue to form peptide bonds.

$$\begin{array}{c} \quad\ \ \text{H} \\ -\text{C}-\text{N}- \\ \| \quad\quad \\ \text{O} \quad\quad \end{array}$$

Peptide bond

Peptides are the compounds resulting from the formation of peptide bonds. Individual amino acids units of peptides are called residues. Polypeptide chains make up the protein molecule. Each of these chains is composed of about 20 and up to several hundred amino acid residues.

The side chains of the amino acid residues are quite varied, and they contribute to the structural stability of the proteins, as well as to their particular interactions.

By convention, the terminal α-amino group (N-terminus) of the peptide is written to the left while the términal α-carboxyl group (C-terminus) is written to the right.

The reason that protein molecules are easily subjected to modification in solution and are usually labile is that their structure depends on weak secondary (nonconvalent) forces. This modification is brought about by changes in pH, high temperature, organic solvents, and radiation. This situation can cause difficulties during research work on proteins but makes opportunities for creation of special products.

Proteins are hydrolyzed by proteolytic enzymes known as proteases. They are either endopeptidases, which act on the interior peptide bonds of protein substrates (trypsin and pepsin) or exopeptidases, which act on peptide bonds that are adjacent to a carboxyl or amino group. For a more complete discussion see the section "Proteolytic Enzymes" later in this chapter.

Proteins can also be hydrolyzed by acids. However, prolonged heating, together with the use of strong acids (6 *N* HCl at 110°C for 20 or more hours and an anaerobic atmosphere is advisable), is necessary for completion. Some of the serine and threonine residues are destroyed by acid, as is the bulk of the tryptophan. Asparagine and glutamine are changed to the corresponding dicarboxylic acids. Alkaline hydrolysis can be used under certain conditions, but it is usually avoided because of extensive racemization. It causes the destruction of some amino acids.

Amino acids are ordinarily classified according to the number of acid and basic groups present in the molecule. The basic amino acids have an excess of amino groups over the carboxyl groups. The reverse is true of the acid amino acids.

Since α-amino acids with the exception of glycine have one or more asymmetric carbon atoms, they can be found as optically active D- and L- forms and also as the racemic mixtures which show no optical activity. Most of the amino acids found in nature have the same configuration with regard to the asymmetric carbon atom as L-glyceride. It is for this reason that they are classified as L-amino acids.

Essential amino acids must be supplied by the diet because the body cannot synthesize them, at least not in sufficient quantity for use. These are lysine, isoleucine, leucine, methionine, threonine, tryptophan, valine, and phenylalanine. Histidine and arginine are essential in children. The nonessential amino acids are also necessary for nutrition, but the body can synthesize them, and it is not necessary that they be supplied by the diet.

The α-amino acids required for the biosynthesis of proteins are listed in Table 6.1 by structural considerations. pI is the pH at the isoelectric point. The pK value is the pH obtained when half of an equivalent of base is added (Fig. 6.1).

The isoelectric point is the pH value at which the protein or amino acid is electrically neutral. The amino acids behave as dipolar ions and carry both a positive and a negative charge. This is called a zwitterion. The following reactions illustrate this point.

$$\underset{\text{Acid form}}{{}^{+}H_3N\text{—}\underset{H}{\overset{H}{C}}\text{—}COOH} \underset{\bar{O}H}{\overset{H^+}{\rightleftharpoons}} \underset{\text{Zwitterion}}{{}^{+}H_3N\text{—}\underset{H}{\overset{H}{C}}\text{—}COO^-} \underset{\bar{O}H}{\overset{H^+}{\rightleftharpoons}} \underset{\text{Alkaline form}}{H_2N\text{—}\underset{H}{\overset{H}{C}}\text{—}COO^-}$$

Amino acids are amphoteric in character; that is, they have the ability to react with bases and with acids. This is explained by the fact that amino or imino and carboxyl groups occur in the same molecule. According to the Lewis definition, the acid part of the molecule can take up an electron pair to form a covalent bond while the amino group can furnish an electron pair to form a covalent bond. This accounts for their acidity and basicity.

Color Tests for Proteins and Amino Acids

A number of color tests are used to identify proteins.

The Ninhydrin Reaction. This is the most general of these tests. α-Amino acids and protein split products give a blue color when treated with triketohydrindene hydrate, otherwise known as ninhydrin. Free

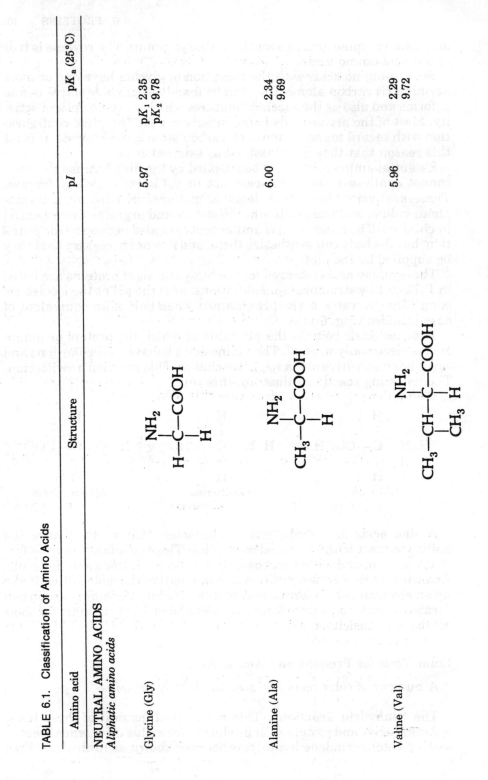

TABLE 6.1. Classification of Amino Acids

Amino acid	Structure	pI	pK_a (25°C)
NEUTRAL AMINO ACIDS			
Aliphatic amino acids			
Glycine (Gly)	$H-C(NH_2)(H)-COOH$	5.97	pK_1 2.35 pK_2 9.78
Alanine (Ala)	$CH_3-C(NH_2)(H)-COOH$	6.00	2.34 9.69
Valine (Val)	$CH_3-CH(CH_3)-C(NH_2)(H)-COOH$	5.96	2.29 9.72

Leucine (Leu)	$CH_3{-}CH(CH_3){-}CH_2{-}C(NH_2)(H){-}COOH$	5.98	2.33 9.74
Isoleucine (Ile)	$CH_3{-}CH_2{-}C(H)(CH_3){-}C(NH_2)(H){-}COOH$	6.02	2.34 9.76
Serine (Ser)	$HO{-}CH_2{-}C(NH_2)(H){-}COOH$	5.68	2.19 9.21
Threonine (Thr)	$CH_3{-}C(H)(OH){-}C(NH_2)(H){-}COOH$	5.60	2.09 9.10

(Continued)

TABLE 6.1. (*Continued*)

Amino acid	Structure	pI	pK_a (25°C)
Sulfur-containing amino acids			
Cysteine (Cys)	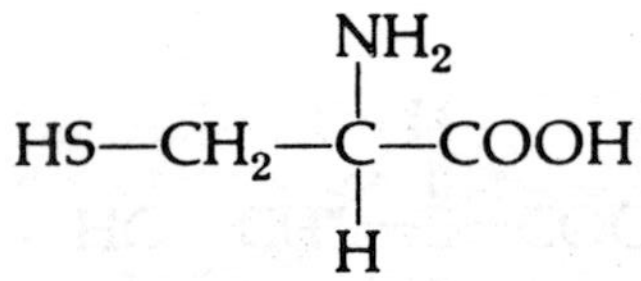 $HS-CH_2-C(NH_2)(H)-COOH$	5.07	1.71 8.39 (sulfhydryl) 10.76 (α-amino)
Cystine (Cys)	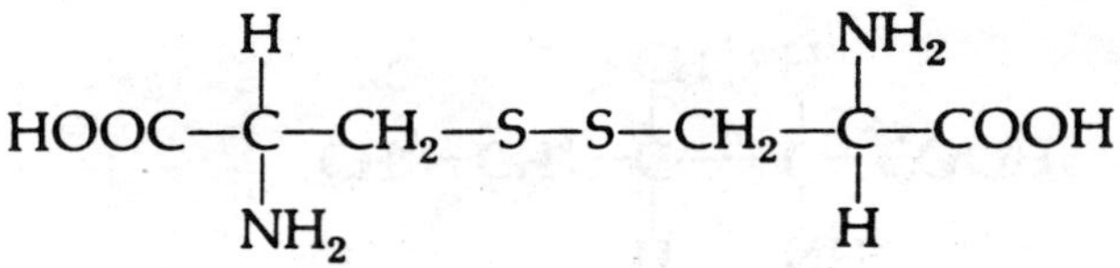 $HOOC-C(H)(NH_2)-CH_2-S-S-CH_2-C(NH_2)(H)-COOH$	4.60	1.65; 2.26 (carboxyl) 7.85; 9.85 (α-amino)
Methionine (Met)	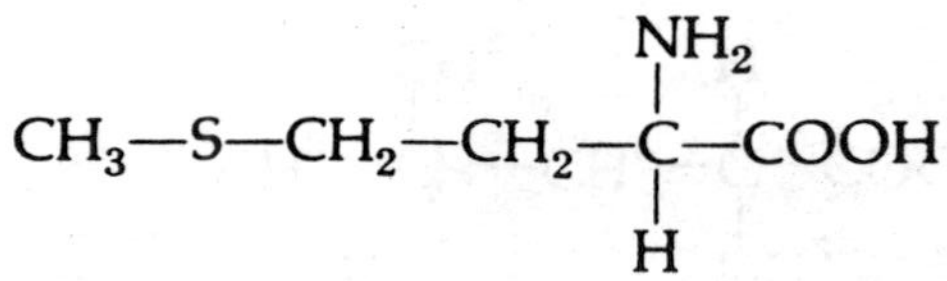 $CH_3-S-CH_2-CH_2-C(NH_2)(H)-COOH$	5.74	2.13 9.28

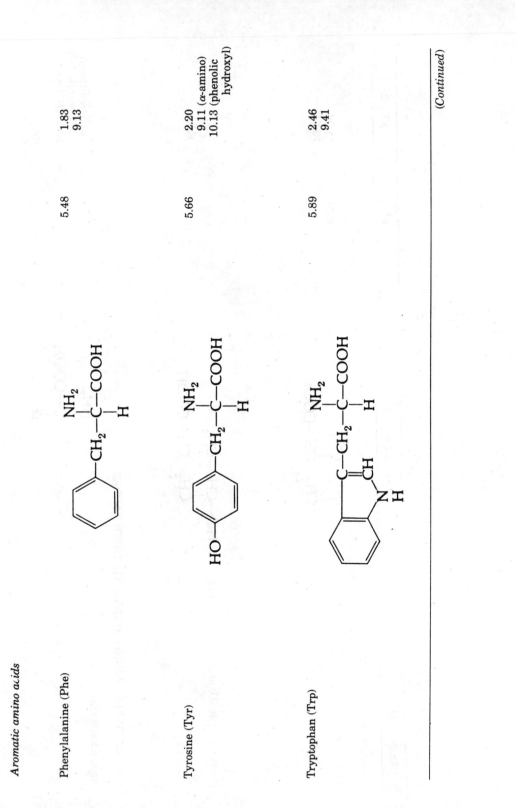

Aromatic amino acids			
Phenylalanine (Phe)	$C_6H_5-CH_2-CH(NH_2)-COOH$	5.48	1.83 9.13
Tyrosine (Tyr)	$HO-C_6H_4-CH_2-CH(NH_2)-COOH$	5.66	2.20 9.11 (α-amino) 10.13 (phenolic hydroxyl)
Tryptophan (Trp)	indol-3-yl $-CH_2-CH(NH_2)-COOH$	5.89	2.46 9.41

(Continued)

TABLE 6.1. *(Continued)*

Amino acid	Structure	p*I*	pK_a (25°C)
Imino Acids			
Proline (Pro)	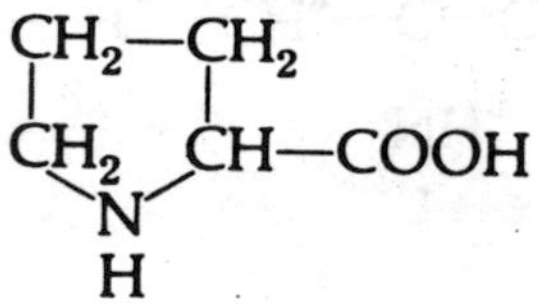 Ring: CH_2—CH_2 / CH_2 / CH—COOH / N—H (pyrrolidine ring)	6.30	1.95 10.64
Hydroxyproline (Hyp)	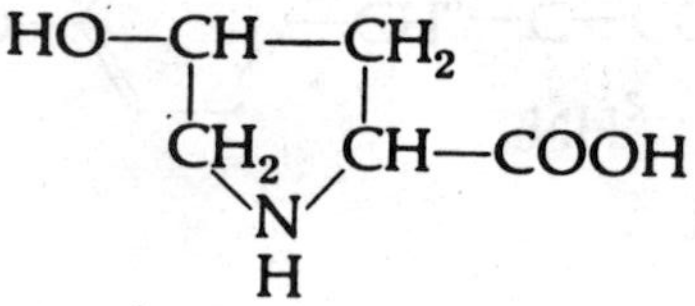 Ring: HO—CH—CH_2 / CH_2 / CH—COOH / N—H	5.83	1.82 9.66
DISCARBOXYLIC AMINO ACIDS AND THEIR AMIDES			
Aspartic acid (Asp)	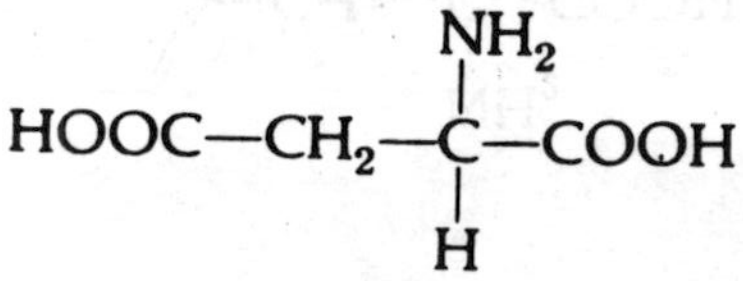 HOOC—CH_2—C(NH_2)(H)—COOH	2.77	2.05 (α-carboxyl) 3.87 (β-carboxyl) 10.00 (NH_3^+)

Asparagine (Asn)	$H_2N-C(=O)-CH_2-C(NH_2)(H)-COOH$	5.41	2.02 8.80
Glutamic acid (Glu)	$HOOC-CH_2-CH_2-C(NH_2)(H)-COOH$	3.22	2.16 (α-carboxyl) 4.27 (γ-carboxyl) 9.36 (NH_3^+)
Glutamine (Gln)	$H_2N-C(=O)-CH_2-CH_2-C(NH_2)(H)-COOH$	5.65	2.17 9.13
BASIC AMINO ACIDS			
Histidine (His)	$HC{=}C(-NH-CH{=}N-)-CH_2-C(NH_2)(H)-COOH$ (imidazole ring)	7.59	1.82 6.04 (imidizole) 9.17 (NH_3^+)

(*Continued*)

TABLE 6.1. (*Continued*)

Amino acid	Structure	pI	pK_a (25°C)
Arginine (Arg)	$H_2N{-}C({=}NH){-}NH{-}CH_2{-}CH_2{-}CH_2{-}CH(NH_2){-}COOH$	10.76	2.17 8.99 (α-amino) 12.48 (guanido)
Lysine (Lys)	$H_2N{-}CH_2{-}CH_2{-}CH_2{-}CH_2{-}CH(NH_2){-}COOH$	9.74	2.18 8.95 (α-amino) 10.53 (ϵ-amino)
Hydroxylysine (Hyl)	$H_2N{-}CH_2{-}CH(OH){-}CH_2{-}CH_2{-}CH(NH_2){-}COOH$		2.13 8.62 (α-amino) 9.67 (ϵ-amino)

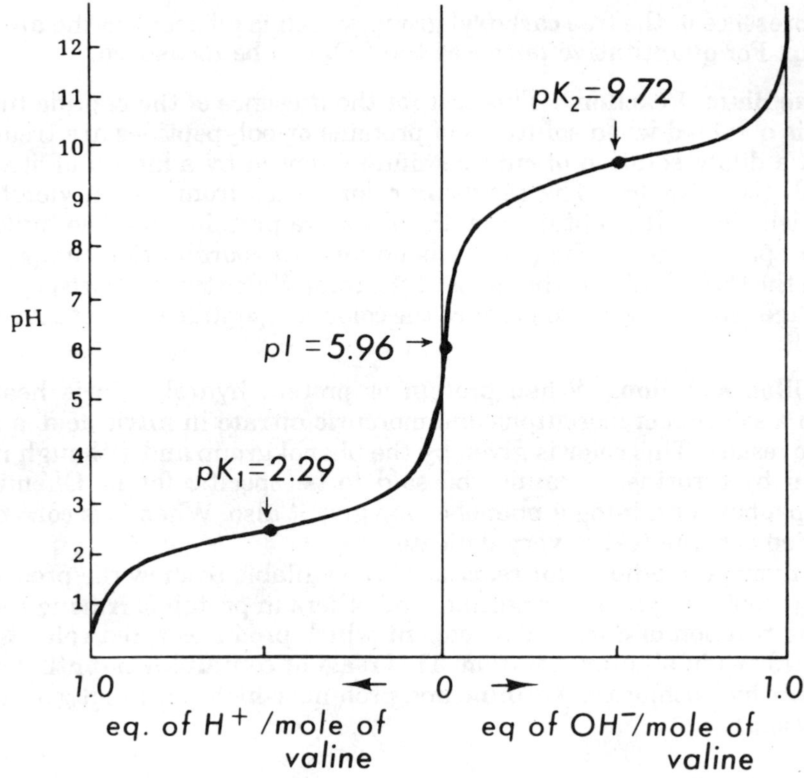

FIG. 6.1. Titration curve of a monoamino-monocarboxylic acid.

amino (—NH_2) and free carboxyl (—COOH) groups must be present to give the test. These conditions exist in the α-amino acids. This color reaction is used with the most modern analytical equipment, which separates the amino acids before the quantities are determined. This is necessary because the depth and shade of the colors are not the same for the different amino acids, and estimation in mixtures would give unsatisfactory results. The CO_2 formed by this reaction is specific for

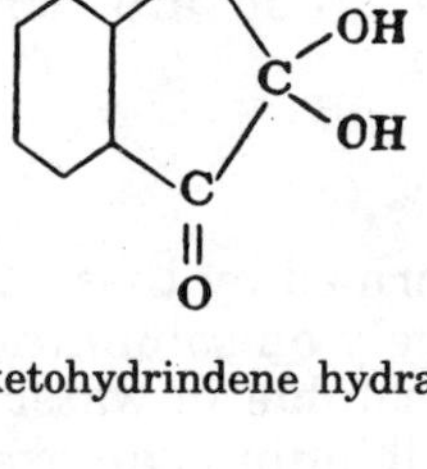

Triketohydrindene hydrate

the presence of the free carboxyl group, which is adjacent to the amino group. For quantitative purposes the CO_2 can be measured.

The Biuret Reaction. This test for the presence of the peptide linkage is obtained when solutions of proteins or polypeptides are treated with a dilute solution of cupric sulfate followed by addition of strong alkali (20% NaOH). The resulting color varies from bluish violet to pinkish violet. It is obtained with all native proteins and the bulk of their split products. The peptide bond forms a coordination compound with the Cu^{2+}, which is the cause of the color. Since two or more peptide linkages are necessary to produce the color, a dipeptide will not give the result.

Millon Reaction. When protein or protein hydrolysate is heated with a solution of mercurous and mercuric nitrate in nitric acid, a red color results. This color is given by the phenol group and although it is given by tyrosine, it cannot be said to be specific for it. Chemical compounds containing a phenol group give it also. When it is correctly carried out, the test is very delicate.

A number of other color reactions are available to show the presence of tryptophan, tyrosine, arginine, and others in proteins. Among these is the reaction of Sakaguchi reagent which produces a red color with guanidines in alkaline solution. This reagent contains α-naphthol and sodium hydrochlorite. Arginine and proteins which contain it give this reaction.

CLASSIFICATION OF PROTEINS

Several general properties are used for the classification of proteins, including solubility, chemical composition, and shape. However, a system now used divides them into three main groups. These include simple proteins, conjugated proteins, and derived proteins. The first two are naturally occurring, whereas the third comprises proteins that have been changed by such things as chemical agents and enzymes.

Proteins can be either fibrous or globular forms. The former are animal proteins and are insoluble. They are the protein of connective tissue, bone, hair, skin, wool, horn, and other such tissues.

Globular proteins are soluble in water, in water solutions containing alkali, acids, or salts, and also in alcohol. Globular proteins in solution frequently occur in the form of ellipsoids or spheroids. Some of the proteins included in this group are hormones, the enzymes, and oxygen-carrying proteins.

Simple Proteins

Only amino acids are formed by these proteins on hydrolysis. This classification is based largely on solubilities.

1. Albumins. They are soluble in water and salt solutions and are coagulated by heat. Egg albumin is an example of this type.

2. Globulins. They are sparingly soluble in water and soluble in salt solutions. Myosin of muscle is an example.

3. Glutelins. These are plant proteins and are soluble in dilute acid or base, but insoluble in neutral solvents. The glutelin of wheat is an example.

4. Prolamines. They are soluble in 50–90% alcohol, insoluble in water and absolute alcohol and neutral solvents. They are found only in plants. Gliadin of wheat and zein of corn are examples.

5. Scleroproteins. They are insoluble in aqueous solvents. These proteins are found only in animals. Examples are keratin and collagen from bones, connective tissue, and hair.

6. Histones. They are basic proteins and are found in animals, are soluble in water and dilute alkalies and acids, but are insoluble in dilute ammonia. They contain much arginine and lysine. Histones are found in the pancreas and thymus of the calf.

7. Protamines. These proteins are of rather low molecular weight and are strongly basic. They are found in large amounts in ripe sperm cells of fish. One of these proteins is salmine which is obtained from the sperm of salmon. Their high nitrogen content is the result of the presence of large amounts of arginine. They are soluble in ammonia and in water and are not coagulated by heat. They form stable salts with strong acids.

Conjugated Proteins

These are somewhat more complex compounds in which the amino acid structure is combined with nonprotein moieties, such as carbohydrates, lipids, and nucleic acids. The important conjugated proteins are as follows.

1. Nucleoproteins. These are a combination of proteins with nucleic acids. Frequently, they are combined with histone and protamine classes of basic proteins. They are found in cell nuclei as chromatin material.

2. Lipoproteins. These are combinations of proteins with lipids, such as lecithin and cholesterol. They are found in milk, egg yolk, and in blood and brain, as well as in other tissues in the body.

3. Mucoproteins. These are proteins combined with carbohydrates with the carbohydrate portion amounting to more than 4%. The mucopolysaccharides are covalently bonded to protein. Heparin and hyaluronic acid are mucopolysaccharides. These are found in connective tissue.

4. Glycoproteins. These are also made up of simple proteins covalently bonded to carbohydrates. These, however, contain smaller amounts of carbohydrate than the mucoproteins. The following make up the main part of the carbohydrate moiety: mannose, *N*-acetylgalactosamine, fucose, glucose, galactose, *N*-acetylneuraminic acid, and *N*-acetylglucosamine. The carbohydrate in gamma-globulin is linked to the protein covalently by the amide bond of asparagine. Other

linkages are known, such as through the hydroxyl group of threonine, hydroxyproline, or serine.

5. Chromoproteins. A colored prosthetic group is the chromophoric group of these compounds. A nonamino acid moiety is called a prosthetic group. Chlorophyll and heme can serve as such groups.

Other types include such combinations as phosphoproteins and flavoproteins.

Phosphoproteins are conjugated proteins in which phosphoric acid is present, combined with hydroxylamino acids. Vitellin of egg yolk and casein of milk belong to this group. Phosphoproteins, acid in character, are present in milk and eggs as the calcium salt.

Derived Proteins

Derived proteins are proteins that have been changed (as a result of hydrolysis) by physical agents such as heat and high hydrogen-ion concentration, as well as by action of enzymes or chemical reagents. They can be divided into two groups which differ by the amount of hydrolysis that has taken place. These are known as primary and secondary derivatives.

Primary Derivatives. These refer to proteins that have been slightly modified by the action of water, enzymes, or dilute acids or alkalies. They are termed proteans and are insoluble in water. The casein of curdled milk and the fibrin of coagulated blood are examples.

Metaproteins are products of further action by acids or alkalies. These are soluble in weak acid or alkaline solutions but not in neutral solvents.

Coagulated proteins are those insoluble proteins which result from the action of heat and such substances as alcohol. The classical example is cooked egg albumin.

Secondary Derivatives. These are made up of proteoses, peptones, and peptides. Proteoses are water soluble but are not coagulated by heat. Saturated solutions of ammonium sulfate will precipitate them. Peptones are simpler products, which are soluble in water, are not coagulated by heat, and are not precipitated by saturated solutions of ammonium sulfate. Peptides are still simpler. They are usually combinations of two or more amino acids.

STRUCTURE OF PROTEINS

Primary Structure

The peptide bond is the basic linkage of protein structure (see Table 6.2). As noted before, this is the bond in which a carboxyl group of one amino acid residue is coupled to the α-amino group of another amino

TABLE 6.2. Structural Forces in Proteins

Type	Mechanism	Energy (kcal/mole)	Distances of Interaction (Å)	Interacting Groups	Example
Covalent bond	Electron sharing	30–100	1–2	C—C, C—N, C=O, C—H	Intra residue bonds
				C—N—C	Peptide
				S—S	Disulfide
Ionic bond	Coulomb attraction between charged groups of opposite sign	10–20	2–3	$—NH_3^+$	α-Amino group; Lysine
				$—C(NH_2)(NH_2^+)$	Arginine
				$>NH^+$	Histidine
				$—COO^-$	Aspartic, glutamic; α-Carboxyl group
Hydrogen bond	Hydrogen shared between two electronegative atoms	2–10	2–3	N—H . . . O=C	Amide-carbonyl group
				OH . . .	Serine, threonine, tyrosine
				NH . . ., NH_2 . . ., NH_3^+ . . ., COO^-	Polar side chains of residues
Van der Waals attractive force	Mutual induction of dipole moments in electrically apolar groups	1–3	3–5	Apolar groups	Apolar side chains
Electrostatic repulsive force	Coulomb repulsion between charged groups of like sign	(q_1q_2/r_2)		Polar groups of like sign	Polar side chains
Van der Waals repulsive forces	Repulsion between apolar groups in close proximity	$(1/r^{12})$		Steric hindrance between side chain groups	All groups

Source: Jones (1964).

acid residue. The order of sequence of the amino acid residues in the peptide chains is an important part of the structure of the protein. This sequence proceeds from the N-terminal position to the C-terminal. Both physical and chemical properties of the protein are governed by the peptide bonds, the side chains, the amino acids, and the sequence of these amino acids in the chains.

After the nature of the amino acids has been determined, the next step is the sequence of the amino acids. It is possible to determine the number of peptide groups in a molecule of protein by the quantitative estimation of the amino or carboxyl end groups. This also gives an indication of the purity of the protein, because a simple integral relationship exists among the number of end groups in a molecule of protein.

The estimation of the N-terminal amino groups was made possible by the introduction by Sanger (1945) of the reagent 1-fluoro-2,4-dinitrobenzene (FDNB). When this reagent reacts with the free amino groups of peptide chains to form dinitrophenyl (DNP) peptides, the α- and ϵ-amino groups give yellow DNP compounds. The polypeptide can be hydrolyzed and the α-*N*-dinitrophenyl derivative is ether-extracted and identified by chromatographic methods.

The Edman degradation (1949) is another method for the determination of the amino end group. This method is extensively employed.

It involves the use of phenylisothiocyanate as the reagent. Two steps are involved in this process. The first step results in the formation of the phenylthiocarbamyl (PTC) peptide, which is accomplished with weak alkali. The PTC peptide is acted on by anhydrous acid. By this treatment the N-terminal residue splits off in the form of a phenylthiocarbamyl amino acid. The remainder of the peptide is unchanged. The phenylthiocarbamyl is then treated with acid which causes it to cyclize to form a phenylthiohydantoin (PTH), which can be identified. Since the remainder of the original peptide is unchanged, it can be further studied by a repetition of this procedure.

In addition to these chemical methods, an enzymatic method, based on leucine aminopeptidase can be used. A terminal free σ-amino acid is necessary for its operation. Since the liberation of the free acids is sequential, the sequence of the amino acids is determined by the rate measurements of their liberation. These results and also those from the Edman method can be of value only when applied to a protein with several identical peptide chains, or to a single peptide chain.

The C-terminal amino acid can be determined by means of hydrolysis with the carboxypeptidases. Carboxypeptidase A does not liberate carboxyl-terminal arginine, lysine, or proline residues, whereas carboxypeptidase B liberates carboxyl-terminal residues of arginine and lysine.

In the determination of the amino acid sequence of a protein made up of a single peptide chain, it is necessary to know the amino acid compo-

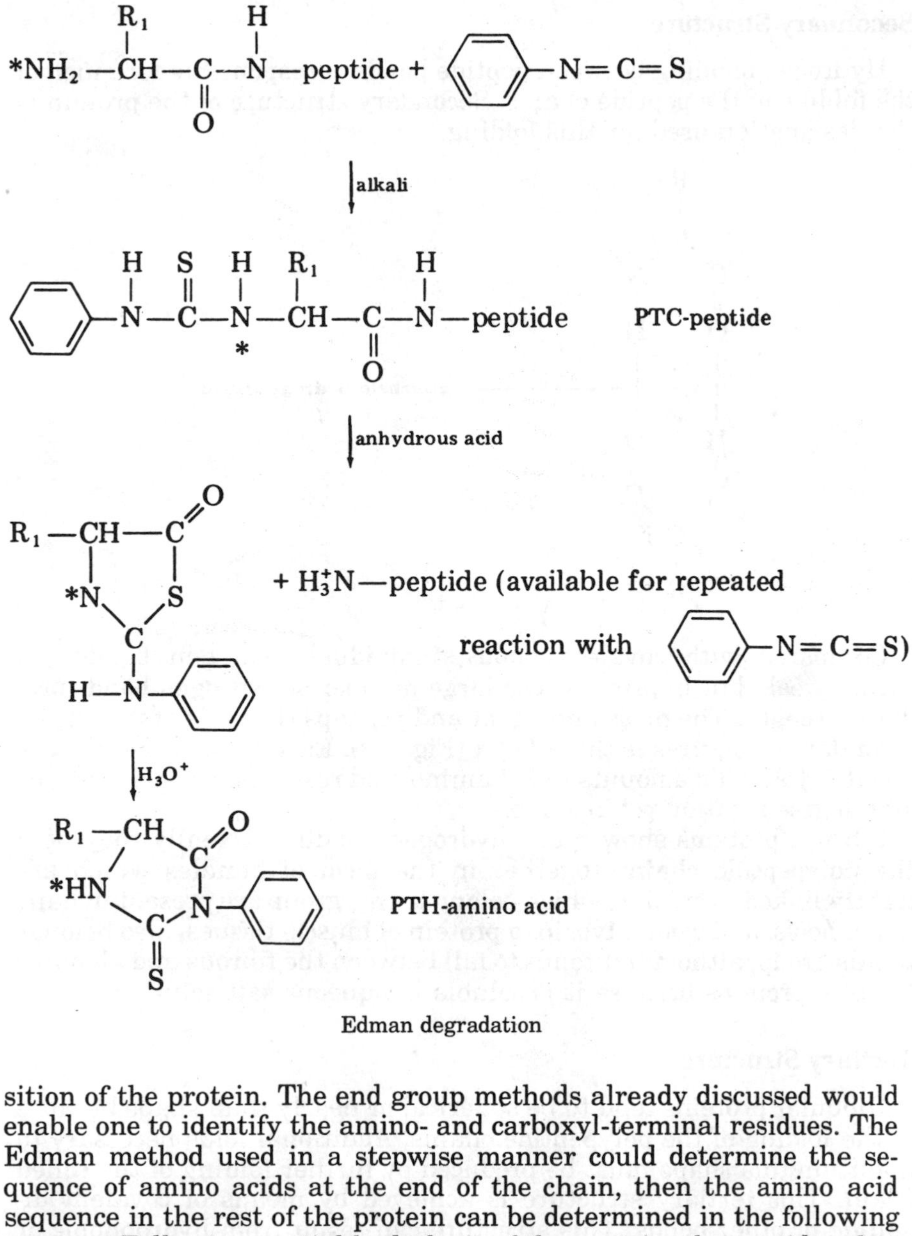

$$*NH_2-CH(R_1)-C(=O)-N(H)-\text{peptide} + C_6H_5-N=C=S$$

$\downarrow$ alkali

$$C_6H_5-N(H)-C(=S)-\overset{*}{N}(H)-CH(R_1)-C(=O)-N(H)-\text{peptide}$$ PTC-peptide

$\downarrow$ anhydrous acid

$$R_1-CH-C(=O)\ (\text{ring: } *N, S, C=N(H)-C_6H_5) + H_3^+N-\text{peptide (available for repeated reaction with } C_6H_5-N=C=S)$$

$\downarrow$ H_3O^+

$$R_1-CH-C(=O)\ (\text{ring: } *HN, C(=S), N-C_6H_5)$$ PTH-amino acid

Edman degradation

sition of the protein. The end group methods already discussed would enable one to identify the amino- and carboxyl-terminal residues. The Edman method used in a stepwise manner could determine the sequence of amino acids at the end of the chain, then the amino acid sequence in the rest of the protein can be determined in the following manner. Following partial hydrolysis of the protein, the peptides that are formed are separated in the pure condition. After this, the sequences of the amino acids in these smaller units are determined, and finally from these data the amino acid sequence of the original protein is deduced.

Secondary Structure

Hydrogen bonding between peptide bonds is responsible for much of the folding of the peptide chains. Secondary structure of the protein is the designation used for this folding.

```
        R
        |
        CH
      /    \
 \  C       N  /
    ||      |
    O       H
    |       |
    |       |  <———————— interchain hydrogen bond
    |       |
    H       O
    |       ||
 /  N       C  \
      \    /
        CH
        |
        R
```

Compared with covalent bonds, individual hydrogen bonds are rather weak, but in proteins the large number of hydrogen bonds produce strength. The most important and perhaps the most stable of the secondary structures is the α-helix (Fig. 6.2). Each turn of the helix is about 5.4 Å. This amounts to 3.7 amino acid residues for each turn, or 100 degree rotation per residue.

Fibrous proteins show much hydrogen bonding. Usually they hold the polypeptide chains together in the form of bundles which are tightly linked. Keratins, which belong to this group are present in hair, nails, hoofs, and wool. Myosin, a protein of muscle tissues, also belongs to this group, although it tends to fall between the fibrous and globular types of proteins because it is soluble in aqueous salt solution.

Tertiary Structure

Globular proteins tend to be spherical or nearly so in shape because of the folding of the polypeptide chains. Additional folds necessary to maintain this shape must be produced by further folding of the coiled chain. This tertiary structure is achieved by means of covalent disulfide or other bonds. This structure can result from hydrophobic or nonpolar bonds, hydrogen bonds, or salt bonds.

Quaternary Structure

Proteins made up of more than a single peptide chain, that is, the association of individual polypeptide chains to produce protein mole-

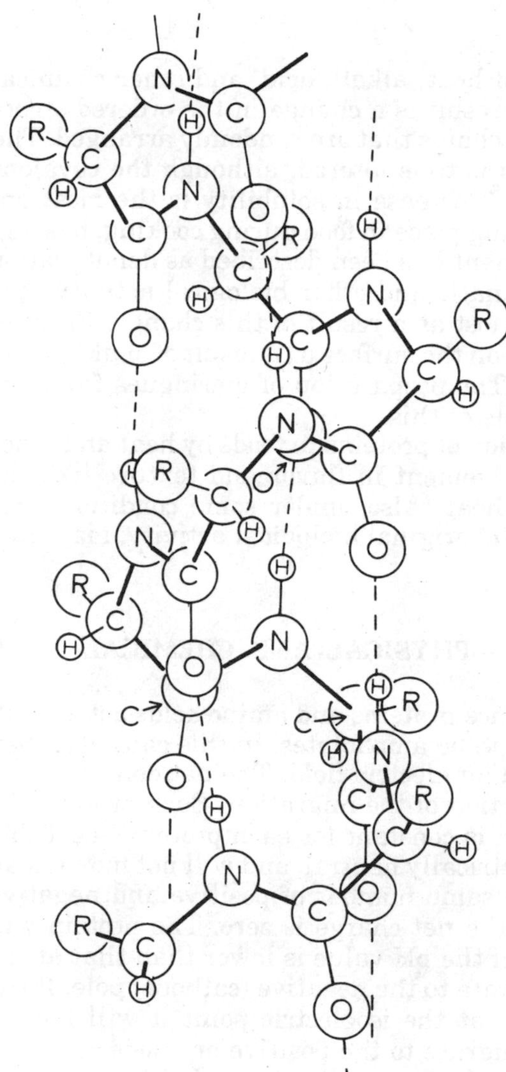

FIG. 6.2. α-Helix form of protein molecule (right-handed). *From Aurand and Woods (1973)*

cules, are said to possess quaternary structure. These are also referred to as oligomeric proteins. Ordinarily, they are not covalently bonded. In most of these proteins, however, the association is extremely close and the complete units act like a single molecule in solution. Hemoglobin is an example of this type of protein.

Secondary, tertiary, and quaternary structures together are known as conformation.

Denaturation

Denaturation is the effect of heat, alkali, acid, and other chemical agents on native protein. The result is a change in the ordered structure of the protein into peptide chains that are randomly arranged. The folded structure of the protein is thus altered, although the covalent peptide bonds are not broken. Decrease in solubility is the most apparent effect. The changes taking place in food during cooking, baking, and other forms of heat treatment has been described as denaturation of the protein involved. Enzymatic and other biological activity that many proteins possess can be lost as a result of this change. Films of denatured proteins are formed on the surface as a result of shaking and foaming of protein solutions. The preparation of meringues from egg white by beating is an example of this.

In some cases the denaturation of proteins in foods by heat and other treatment results in an improvement in flavor and texture like the coagulation of egg white by heat. Also, under some conditions, renaturation, that is, return to its original biological activity, may take place.

PROPERTIES OF PROTEINS—PHYSICAL AND CHEMICAL

As mentioned previously, since proteins and amino acids act as both acids and bases, they are said to be ampholytes. In this capacity, they have the ability to migrate in an electric field. The net charge of the molecule determines the direction of the migration. Each protein has an isoelectric point (p*I*) which is constant for each protein (see Table 6.1). At this pH value it is electrically neutral, and will not move in an electric field. It possesses the same number of positive and negative charges with the result that the net charge is zero. The protein will possess a net positive charge if the pH value is lower than that at the isoelectric point, and will migrate to the negative (cathode) pole. If the pH value is higher than that at the isoelectric point it will have a negative charge and it will migrate to the positive or anode.

Proteins at the isoelectric point are zwitterions. Zwitterions were described earlier in the discussion on amino acids.

Insoluble Protein Salts

A number of ions can act as precipitating agents for proteins because they are able to form insoluble salts with them. These include, among others, the ions of picric, perchloric, phosphotungstic, and trichloracetic acids. When the proteins are on the acid side of their isoelectric points they will form insoluble salts with the anions of any of these acids, removing the proteins from the solution. To determine the amount of protein nitrogen in a solution the Kjeldahl method can be used. This method measures the amount of nitrogen in an aliquot of the solution (AOAC 1980). Another aliquot of the same protein solution is treated with any one of the listed acids to remove the protein. The solution is then assayed for total nitrogen, again by the Kjeldahl method. The difference between the two gives the amount of protein nitrogen in the original solution. The multiplications of this amount by the factor 6.25 gives the amount of protein, which is an approximation because the factor used is not precisely the same for all proteins.

Separation of Proteins

Ammonium Sulfate Fractionation. This is the classical method used to separate proteins in aqueous solution. This salt is of great value because of its high solubility in water and its relatively low temperature coefficient of solubility. Precautions are necessary in this procedure. Specifically, temperature should be kept as low as practical, which in most cases is not far from the freezing point of the solvent used. The pH used should be kept as near the isoelectric point of the protein in question as possible. Control of pH, temperature, and dilution is necessary to limit denaturation. As the concentration of ammonium sulfate increases, different proteins are precipitated and can be collected before the addition of more ammonium sulfate to the solution.

Organic Solvents. Organic solvents can be used for the precipitation of proteins from aqueous solution. Ethyl alcohol is particularly good if proteins are to be separated in large quantity. Close control must be kept on the temperature, pH, dialectic constant, and ionic strength. Variations of these factors have resulted in the separation of many proteins from mixtures.

Chromatography. More recently, chromatography on ion exchange columns has been used. These later methods are largely limited to the more stable compounds of lower molecular weight because of the adverse effect of organic solvents on many proteins. Sometimes, adsorption methods are used to remove unwanted proteins from a mixture, or desired proteins are adsorbed and removed from the mixture with the adsorbent.

Ultracentrifugation and Electrophoresis. The two techniques of electrophoresis and ultracentrifugation have increased our knowledge of proteins a great deal. Formerly, as has already been mentioned, proteins were mainly classified by means of their solubilities. The present techniques have shown that many of these proteins are definitely mixtures rather than pure compounds.

Centrifugal Studies on Proteins. About 1923 Svedberg invented a device known as the ultracentrifuge. This instrument is capable of rotation rates as high as 75,000 rpm, which permit fields up to about

500,000 times gravity. Ultracentrifugation can be employed to determine the molecular weights of proteins. The centrifugation is continued until equilibrium between sedimentation and diffusion is attained. Temperature during the operation must be constant. From the data collected, molecular weights can be calculated. Another method for the determination of molecular weights using this equipment is known as the sedimentation velocity method. This method has been used extensively. It makes use of the rate at which proteins move in centrifugal fields of sufficient intensity, so that the sedimentation is far more rapid than that of free diffusion. Since the protein molecules are more dense than the solvent a rather sharp boundary is formed between the protein solution and the solvent. The refractive index of the liquid shows considerable change at the boundary. A single boundary will be formed if the proteins in a given sample have all the same molecular weight. If several proteins of significantly different molecular weight are present in the mixture, each protein will show a different boundary. This technique has been used to separate proteins (Svedberg 1934, 1937).

Electrophoresis. Another technique for the separation of proteins is the migration of these substances in solution in an electric field. The equipment was invented by A. Tiselius in 1933. Positively charged particles migrate to the cathode and are termed cations, while negatively charged particles migrate to the anode and are called anions. In addition to the isoelectric point mentioned earlier, proteins have an isoionic point. The isoionic point is the pH at which the number of protons dissociated from proton donors is equal to the number of protons combined with the proton acceptors. The isoionic point of a protein is not the point of minimum electric charge. It is, however, the point at which the net charge on the protein is zero. It should be emphasized that the isoelectric point of a protein is not necessarily or usually identical with the isoionic point. When the ionic strength of a protein is changed, the movement of the protein in the electric field will change also. Therefore, since solutions of proteins are buffered for such work it is necessary to consider not only the pH of the solution, but the ionic strength and composition of the buffer mixture as well.

A purified protein which shows only one peak as a result of electrolysis at a given pH should not necessarily be considered a single compound, because at another pH, two or more peaks may show. It is necessary, therefore, before results are considered conclusive, to run this determination at several rather different pH values, at each of which the protein is definitely stable (see Chrambach and Rodbard 1971).

The results of the protein movement are expressed as the distance traveled in unit time in a unit electrical field. This value is constant for the protein under the conditions used (Tiselius and Flodin 1953).

Starch-Gel Electrophoresis. Several other techniques are now extensively employed for the separation and study of proteins. One of these is the starch-gel electrophoresis method. In this technique, strips of starch-gel are connected with a direct electrical current after the sample has been applied to the gel (Smithies 1955). After sufficient time, the current is stopped and the gel is stained to show the location of the bands resulting from this treatment. A number of mixtures, when treated by this method, have shown some interesting new results, and what were once considered to be more or less pure proteins have been shown to be mixtures.

Gel Filtration. This technique, filtration on molecular sieves, makes use of cross-linked dextran gels or gels of polyacrylamide. These gels have pores of fixed small sizes and allow the penetration of the smaller sized molecules but hold out the larger molecules. Gel granules of varying pore sizes are available. Columns made from these gels can be calibrated with known molecular weight substances. When thus calibrated, they can be used to obtain the approximate molecular weight of the unknown sample.

Whitaker (1963) determined the molecular weights of proteins by the use of dextran gels (trade name Sephadex). He found that the linear correlation between the logarithm of the molecular weight of a protein and the ratio of its elution volume V to the void volume V_0 was excellent when a column of Sephadex G-100, a cross-linked dextran, was used. While the ratio of elution volume to void volume (the elution volume of the first peak) was independent of column size, protein concentration, and ion exchange adsorption (at ionic strength of 0.494), it was found to be influenced by temperature for a few of the proteins.

Leach and O'Shea (1965) determined the molecular weights of proteins up to 225,000 by gel filtration on a single column of Sephadex G-200. The temperatures used were 25° and 40°C. Further evidence suggests that proteins of higher molecular weights can be measured.

Blattler and Reithel (1970) determined the molecular weight of proteins using polyacrylamide gels as molecular sieves and for electrophoresis. These authors used catalase, urease, and higher multimers of urease as the protein standards. The gels used ranged in density from 4 to 15%. When urease with a molecular weight of 480,000 was used, it was excluded by gels of 11–12% strength. At a gel strength of 11% the urease entered the gel, but it moved at a very slow rate. This technique is a very important one that is used extensively in protein work at the present time.

Isoelectric Focusing. This is a technique that permits the separation of amino acids and such macromolecular ampholytes as proteins, and is based on differences in isoelectric points. When a protein reaches its isoelectric point it stops moving. Isoelectric focusing is a very valuable research technique because it gives much higher resolution than ordinary electrophoresis. Polyacrylamide gels are used in this test. Isoelectric focusing can also be used to determine isoelectric points. The following articles should be consulted: Awdeh *et al.* (1968), Dale and Latner (1968), Svensson (1961, 1962A, B).

Preservation of Proteins

Proteins are usually preserved by drying. This is done by lyophilization—dehydration in vacuum at very low temperature. The low temperature provided by this method limits deterioration during drying.

Molecular Weight

Proteins are large molecules that tend to be unstable. Therefore methods that are ordinarily used for molecular weight determination such as freezing point and boiling point changes are not applicable to these compounds.

The best method to use is the sedimentation method, which uses the ultracentrifuge, already described.

Several factors determine the rate of sedimentation of a protein from its solution. These include (1) the density, molecular weight, and size and shape of the molecule, (2) viscosity and density of the dispersion medium, and (3) the centrifugal force produced in the ultracentrifuge.

This technique can determine the molecular weight of the protein and its purity. A pure substance will give a sharp boundary, whereas a mixture of compounds varying in particle size will give several boundaries.

The stability of proteins under varying conditions can be determined with the ultracentrifuge. That is, the effect of changes in temperature and pH can be observed by noting the changes in sedimentation constants.

Other methods for use in the determination of molecular weights of proteins include the use of molecular sieves and by calculation from light scattering. The former involves the use of cross-linked dextran gels or polyacrylamide gels. These gels have pores of a size which allow molecules up to a given size to penetrate but the larger ones are not permitted entry. It is possible to separate substances of different molecular weights. These columns are calibrated with substances of known molecular weights.

Calculation by light scattering is based on the opalescence of protein solutions. The intensity of light scattered sidewise by a protein of known weight in solution can permit the calculation of the molecular weight.

The molecular weights of some proteins are given in Table 6.3.

Proteolytic Enzymes

It has already been noted that proteins can be hydrolyzed by acid treatment or by means of enzymes.

These enzymes are known as proteolytic enzymes or proteinases and can be endopeptidases or exopeptidases. Endopeptidases act on the interior and terminal peptide bonds of polypeptides, and such enzymes

TABLE 6.3. Approximate Molecular Weights of Some Proteins

Protein	Molecular Weight
Bovine hemoglobin	66,700
α-Casein (cow's milk)	23,000
β-Casein (cow's milk)	24,000
κ-Casein (cow's milk)	19,000
Catalase (bovine liver)	250,000
Edistin (hemp seed)	310,000
Egg albumin	44,000
β-Lactoglobulin (cow's milk)	35,400
Myosin (muscle)	850,000
Pepsin (pig's stomach)	35,500
Zein (corn)	40,000

as trypsin, chymotrypsin, and pepsin from mammals and ficin and papain from plants are examples. Exopeptidases act only on the terminal peptide bonds with the result that amino acids are sequentially removed. A number of the proteolytic enzymes are quite specific. Tryspin acts specifically to hydrolyze the peptide bonds of the carbonyl group of arginine and lysine residues. Chymotryosin is less specific. It cleaves peptide bonds of carbonyl group of tyrosine, tryotophan, or phenylalanine. The known specificity of some proteolytic enzymes makes them valuable in structural work on proteins.

Denatured proteins are more readily attacked by these enzymes than native proteins.

A number of enzymes in the intestinal tract act on the food proteins in the process of digestion.

Digestion of Protein. Proteins are large molecules that must be broken down into smaller particles, from which the amino acids are released, in order to enter the metabolic pathways.

The digestion of protein starts in the stomach, where the active enzyme is pepsin. Pepsin is secreted in the form of its zymogen pepsinogen, which, in turn, is converted to active pepsin by pepsin at the acid pH of the gastric juice in the stomach. The process is, therefore, autocatalytic. As a result of this activity, 42 amino acid residues are removed as a mixture of peptides from the N-terminal portion of pepsinogen. The hydrolysis of proteins by pepsin starts rapidly. Pepsin attacks preferentially peptide bonds of aromatic amino acids, such as tryptophan, tyrosine, and phenylalanine. It also acts on peptide bonds of leucine and methionine, though less rapidly. Because the process is slow, very few free amino acids are liberated by pepsin; dietary protein is chiefly hydrolyzed to a mixture cf polypeptides.

The digestive process continues in the small intestine. Pancreatic juice, which is secreted into the small intestine, contains a mixture of proteases. These are trypsinogen, chymotrysinogen, two procarboxypeptidases, and proelastase. Enterokinase, an enzyme of the intestine, is able to convert trypsinogen to trypsin. Trypsin in autocatalytic fashion can bring about this change also.

The end products of the action of pepsin in the stomach and pancreatic proteases in the intestine are free amino acids and short peptides, which are absorbed mainly by the small intestine.

Other Enzyme Activity. The activity of enzymes on proteins is also used in a number of other ways. Beef is tenderized by allowing it to stand in cool storage. The cathepsin enzymes present in the muscular tissue are responsible for this action. Preparations of proteolytic enzymes such as bromelain and papain in powder form are used to increase the tenderness of meat.

An important step in the manufacture of cheeses is the formation of *para*-κ-casein as a result of the cleavage of the peptide bond between methionine and phenylalanine in the κ-casein. This releases the insoluble *para*-κ-casein.

PROTEIN SYNTHESIS

Deoxyribonucleic acids (DNA) and ribonucleic acids (RNA) are involved in protein synthesis.

Since the coded information directing the synthesis of protein in higher living organisms comes from the DNA of the nucleus, but the actual synthesis takes place outside the nucleus, i.e., in the cytoplasm, there must be a way for the information to be transferred from the nucleus to the cytoplasm.

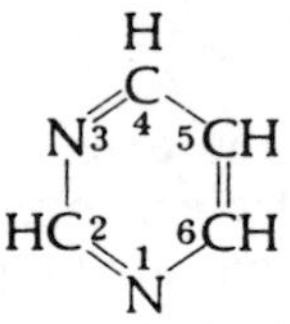

Pyrimidine

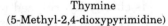

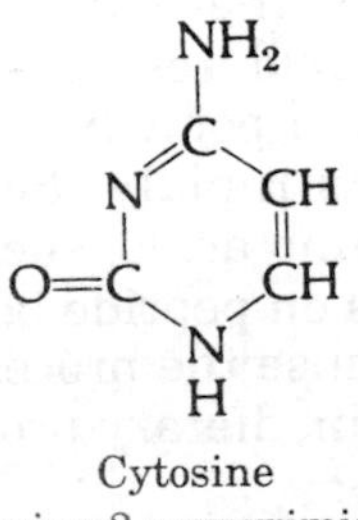

Cytosine (4-Amino-2-oxypyrimidine)

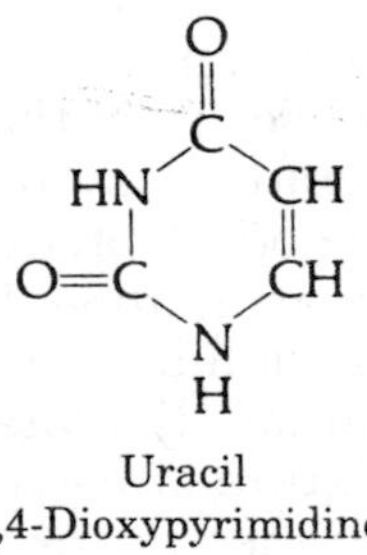

Uracil (2,4-Dioxypyrimidine)

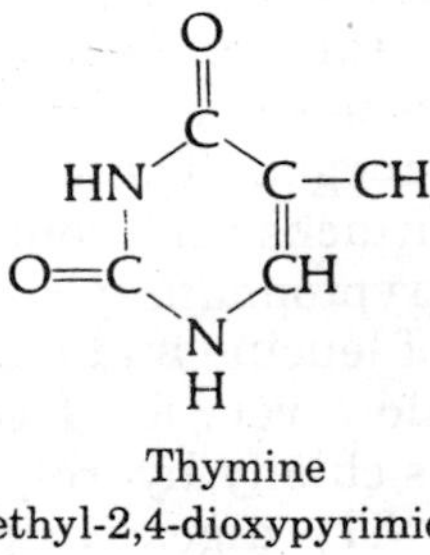

Thymine (5-Methyl-2,4-dioxypyrimidine)

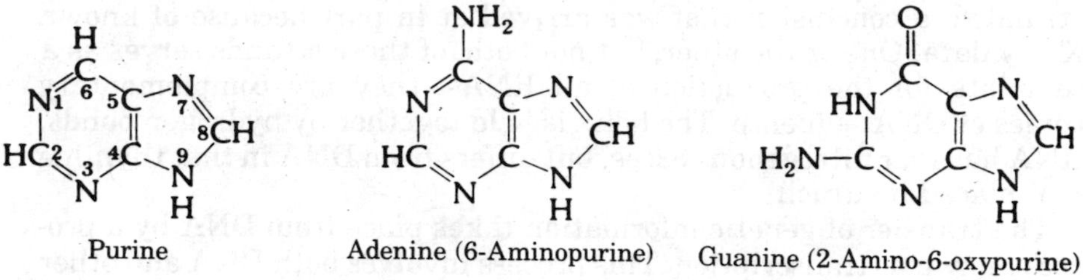

Purine

Adenine (6-Aminopurine)

Guanine (2-Amino-6-oxypurine)

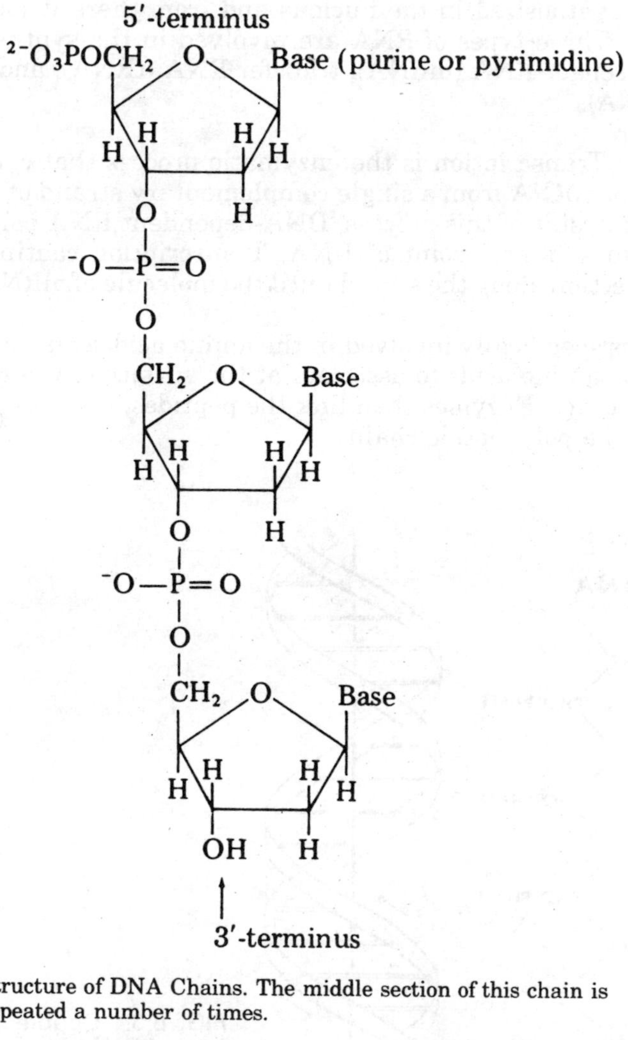

5′-terminus

$^{2-}O_3POCH_2$ Base (purine or pyrimidine)

Base

Base

3′-terminus

Structure of DNA Chains. The middle section of this chain is repeated a number of times.

DNA is the genetic substance. It is made up of the sugar deoxyribose, a pentose; inorganic phosphate, and four nitrogenous bases, adenine

(A), thymine (T), guanine (G), and cytosine (C). DNA is double stranded, a conclusion that was arrived at in part because of known X-ray data. One or the other, but not both, of these strands serves as a template for the production of all RNAs; they are complementary copies of DNA sequence. The helix is held together by hydrogen bonds. RNA has four nitrogenous bases, but differs from DNA in that thymine is replaced by uracil.

The transfer of genetic information takes place from DNA by a process known as transcription. This process involves both DNA and other kinds of nucleic acids, viz., ribonucleic acids.

RNA is synthesized in the nucleus and from there it moves to the cytoplasm. Three types of RNA are involved in the synthesis of proteins: messenger RNA (mRNA), transfer RNA (tRNA), and ribosomal RNA (rRNA).

mRNA. Transcription is the enzymatic process that catalyzes the synthesis of mRNA from a single complementary strand of a length of DNA. As a result of this process, DNA-dependent RNA polymerase is attached to a "start" point of DNA. Transcription continues in the $5' \rightarrow 3'$ direction along the strand until the molecule of mRNA has been created.

mRNA is specifically involved in the amino acid sequence. It causes the various amino acids to associate at the surface of the ribosome in particular order. Enzymes then link the peptides, resulting in the formation of the polypeptide chain.

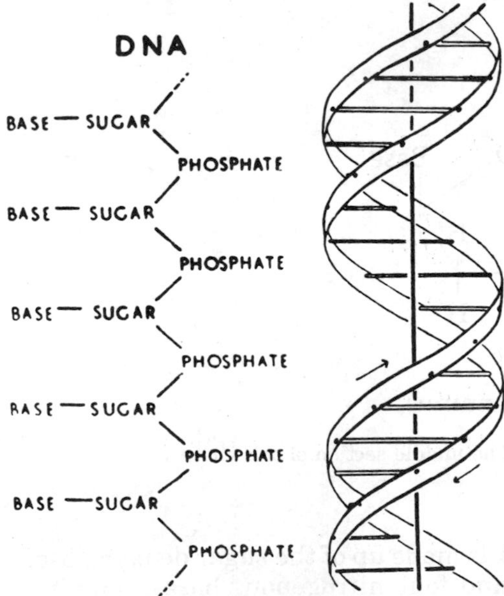

FIG. 6.3. Double helix structure of DNA.
From Watson and Crick (1953). Reproduced with permission of Nature, *copyright owner.*

TABLE 6.4. Genetic Code Codons in mRNA (5′ to 3′)[a,b]

	U		C		A		G	
U	UUU	Phe	UCU	Ser	UAU	Tyr	UGU	Cys
	UUC	Phe	UCC	Ser	UAC	Tyr	UGC	Cys
	UUA	Leu	UCA	Ser	(UAA)		(UGA)	
	UUG	Leu	UCG	Ser	(UAG)		UGG	Trp
C	CUU	Leu	CCU	Pro	CAU	His	CGU	Arg
	CUC	Leu	CCC	Pro	CAC	His	CGC	Arg
	CUA	Leu	CCA	Pro	CAA	Gln	CGA	Arg
	CUG	Leu	CCG	Pro	CAG	Gln	CGG	Arg
A	AUU	Ile	ACU	Thr	AAU	Asn	AGU	Ser
	AUC	Ile	ACC	Thr	AAC	Asn	AGC	Ser
	AUA	Ile	ACA	Thr	AAA	Lys	AGA	Arg
	AUG	Met	ACG	Thr	AAG	Lys	AGG	Arg
G	GUU	Val	GCU	Ala	GAU	Asp	GGU	Gly
	GUC	Val	GCC	Ala	GAC	Asp	GGC	Gly
	GUA	Val	GCA	Ala	GAA	Glu	GGA	Gly
	GUG	Val	GCG	Ala	GAG	Glu	GGG	Gly

[a] Codons in circles are terminals. UUU is the codon for phenylalanine.

[b] U, uracil; C, cytosine; A, adenine; G, guanine.

The unit of the genetic code is a codon. A codon is a triplet, that is, three adjacent nucleotide residues in the mRNA chain. These three residues determine the tRNA species, carrying the activated amino acids that will attach to the mRNA. Three complementary bases in each tRNA species are termed anticodons. When an anticodon associates with a codon, the amino acid is added to the growing polypeptide chain.

tRNA. Transfer RNA has a low molecular weight, that is, not more than 80 ribonucleotides. It is somewhat smaller than mRNA or rRNA. tRNA molecules have a cloverleaf shape (Fig. 6.4). The bottom lobe of the cloverleaf configuration, which contains the anticodon, is probably the cause of the complementary pairing of rRNA with mRNA during protein synthesis.

rRNA. The ribosome is an organelle in the cytoplasm having a diameter of about 20 nm. Ribosomes, which are actively concerned with protein synthesis, occur in groups known as polysomes (polyribosomes). They are held together by messenger RNAs. rRNA is not specific for the sequence of amino acids in proteins.

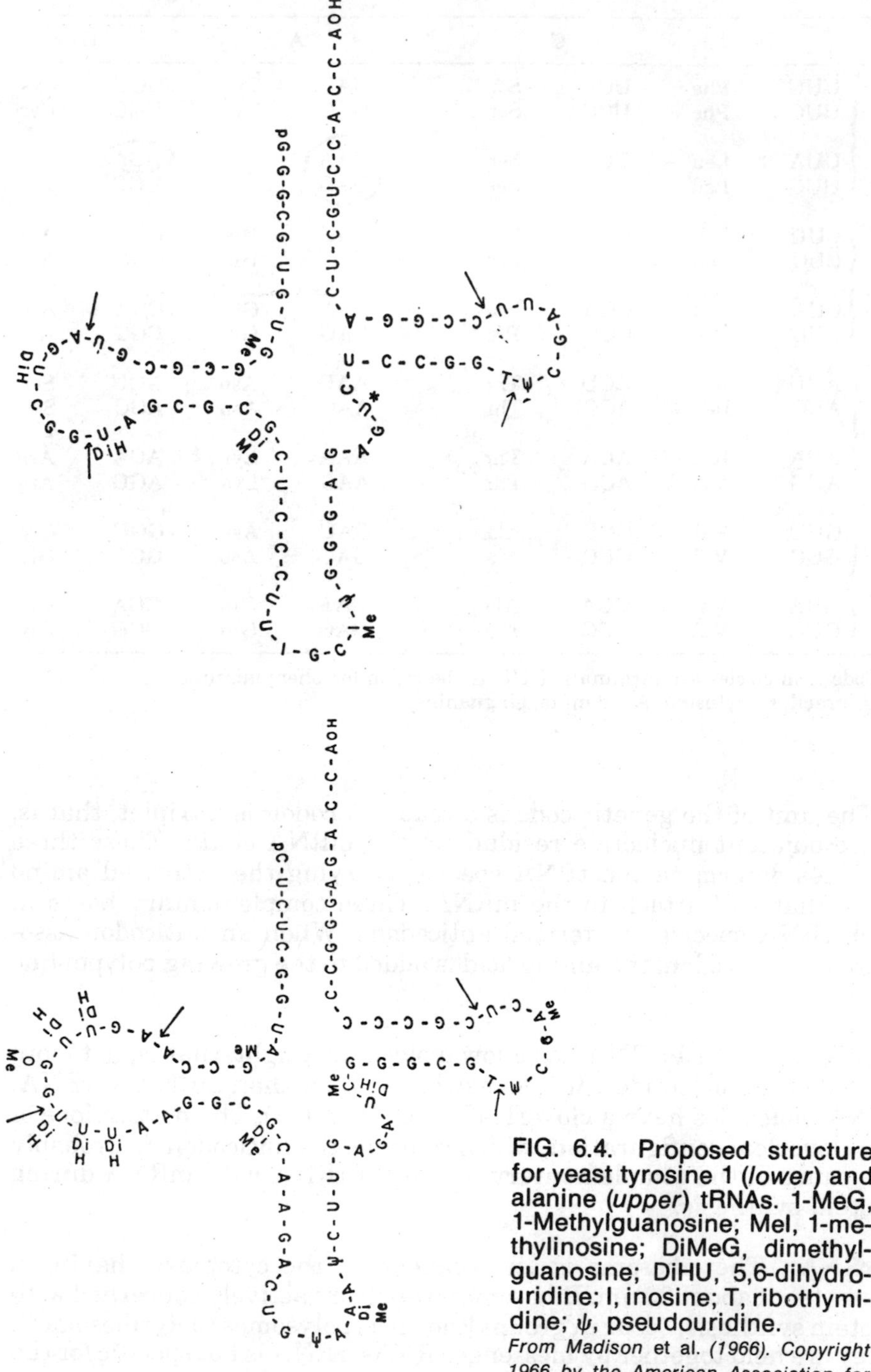

FIG. 6.4. Proposed structure for yeast tyrosine 1 (*lower*) and alanine (*upper*) tRNAs. 1-MeG, 1-Methylguanosine; MeI, 1-methylinosine; DiMeG, dimethylguanosine; DiHU, 5,6-dihydrouridine; I, inosine; T, ribothymidine; ψ, pseudouridine.

From Madison et al. *(1966). Copyright 1966 by the American Association for the Advancement of Science. Used with permission.*

Activation of Amino Acid

Twenty different amino acids require an equal number of amino acyl synthetases for activation. Each enzyme molecule must be able to connect with the corresponding amino acid. In the process of activation, the enzyme, amino acid synthetase, catalyzes the reaction of a given amino acid with adenosine triphosphate (ATP). The amino acyl-adenosine monophosphate (AMP) is the activated amino acid.

$$CH_3{-}\overset{\overset{\displaystyle NH_3^+}{|}}{CH}{-}COO^- + ATP \xrightarrow[\text{Enzyme (Ala specific)}]{Mg^{++}}$$

$$CH_3{-}\overset{\overset{\displaystyle NH_3^+}{|}}{CH}{-}\overset{\overset{\displaystyle O}{\|}}{C}{-}AMP - \text{Enzyme} + PP_i$$

$$\downarrow tRNA_{Ala}$$

$$CH_3{-}\overset{\overset{\displaystyle NH_3^+}{|}}{CH}{-}\overset{\overset{\displaystyle O}{\|}}{C}{-}tRNA_{Ala} + \text{Enzyme} + AMP$$

The formation of amino acyl-tRNA.

After the process of amino acid activation and the binding to tRNA, the complex formed (tRNA–enzyme–amino acid) diffuses to the ribosomes. The tRNA complex arrives at the ribosome and deposits the amino acid into the lengthening chain. Then it returns to the cytoplasm for another amino acid. Three initiation factors are required at this stage for the binding of the amino acyl-tRNA to the ribosome. This stage is known as initiation. The formation into polypeptide chains takes place at the ribosome. Two elongation factors are required, and three release factors. Then the polypeptide is released from the ribosome.

The essential requirements for protein synthesis are amino acids, enzymes, ribosomes, rRNA, mRNA, tRNA, adenosine triphosphate, and additive factors, elongation factors, and release factors.

The synthesis of polypeptides may be summarized as follows. The activated amino acid is coupled to the particular tRNA which then diffuses to the mRNA–ribosome complex. A charged tRNA is temporarily bound to the mRNA site. At this point the tRNA anticodon is complementary to the codon at the mRNA. As a result of the peptide bond formation, the amino acid brought by tRNA is inserted into the lengthening polypeptide chain. At the same time the preceding tRNA is liberated from the surface of the ribosome. Advance relative to the ribosome of mRNA is by one codon. This allows the next arriving amino acid to be added by the same continuing process. The completed polypeptide is released from the surface of the ribosome mRNA complex.

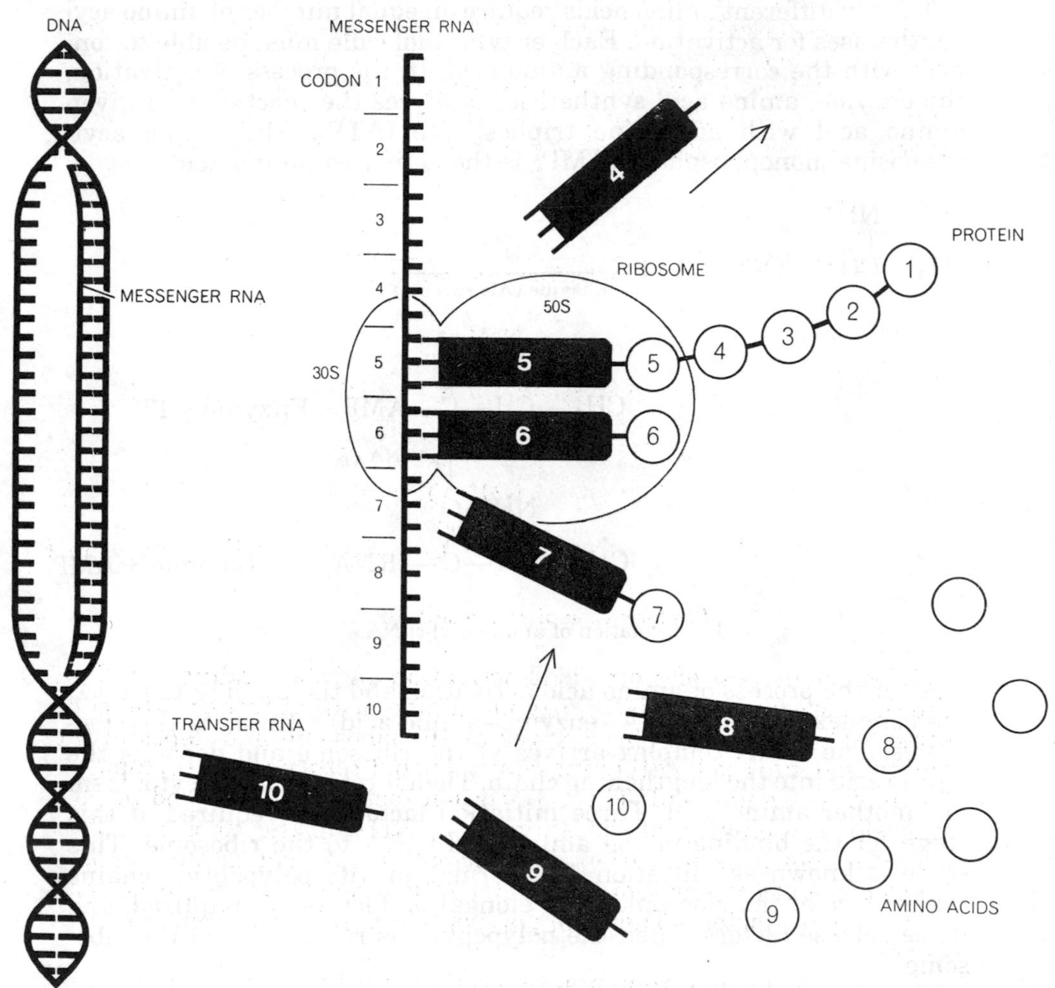

FIG. 6.5. Sequence of protein synthesis. Ribosomes conduct protein synthesis. Genetic information is encoded in the sequence of bases (horizontal elements) in the double helix of DNA (*left*). This information is transcribed into a complementary sequence of RNA bases to form messenger RNA (*shaded dark*). Each group of three bases in the mRNA constitutes a codon, which specifies a particular amino acid and is recognized by a complementary anticodon on a tRNA molecule (*lighter shade*) that has previously been charged with that amino acid. Here amino acid No. 6, specified by the sixth codon, has just been bound to its site on the ribosome by the corresponding tRNA. It will bond to amino acid No. 5, thus extending the growing peptide chain. Then the ribosome will move along the mRNA the length of one codon and so come into position to bind tRNA No. 7 with its amino acid.

From Nomura (1969). Reproduced with permission of Scientific American.

This is a brief resumé of the formation of proteins under natural conditions, which gives an idea of the complexity of the process. It does not pretend to be complete, and the interested student is advised to consult books on genetics for further information.

PROTEINS IN FOODS

Table 6.5 gives the amino acid breakdown of proteins in various foods.

Proteins of Milk

This subject is discussed rather fully in Chapter 13, Milk and Milk Products.

Proteins of Egg

Egg yolk and egg white are considerably different in chemical composition. Over one-half of the solid material in the yolk is lipid, both bound and unbound. The lipoproteins of the yolk are known to be complex mixtures, and are known as lipovitellin and lipovitellenin. A water soluble fraction called livetin is known also. It contains proteins related to blood serum proteins. The important nonlipid phosphoprotein from the egg yolk is known as phosvitin, although its existence as a pure compound has been questioned. It is high in phosphorus, containing 10%. Although not too much work has been done on the proteins of egg yolk, a great deal has been done on the white.

Egg white is composed mainly of a solution of proteins. It also contains a small amount of sugar and salt. Ovalbumin makes up about half of the protein of the total solids of egg white. It was prepared in the crystalline form in 1892. Svedberg (1934) showed that the molecules were almost spherical and had a molecular weight of 40,500 and an isoelectric point of 4.55. Conalbumin has been noted as an antibacterial agent and as an iron-binding protein. It has a molecular weight of about 80,000, an isoelectric point of pH 6.0, and is present to the extent of about 13% of the dry weight. Ovomucoid is high in carbohydrate, containing around 21%. It has rather high heat stability, a molecular weight of about 28,000, and an isoelectric point of about 4.3. About 11% of the dry weight of egg white is made up of this protein. Lysozyme is a protein amounting to about 3.5% of the dry egg white. It has a molecular weight of about 15,000 and an isoelectric point of 10.7. A group of proteins, ovomucin, flavoprotein, avidin, and others are present in rather small amounts.

Proteins of Meat

Meat proteins are discussed in Chapter 20, Meat and Meat Products.

TABLE 6.5. Amino Acid Content of Proteins in Foods

Food		Valine	Leucine	Iso-leucine	Threo-nine	Methio-nine	Lysine	Phenyl-alanine	Trypto-phan
Beef-Veal,	(M)(A)	354	517	337	265	148	523	252	
edible flesh	(CC)(B)	886	852	1435	812	478	1573	778	70
Pork, edible	(M)(A)	324	472	320	307	169	506	261	85
flesh	(B)	616	897	608	583	321	961	496	162
Mutton-Lamb,	(M)(A)	316	481	311	307	153	510	250	79
edible flesh	(B)	790	1203	778	583	383	1275	625	198
Chicken, edible	(M)(A)	318	460	334	248	157	497	250	64
flesh	(B)	1018	1472	1069	794	502	1590	800	205
Crustaceans	(CC)(A)	299	542	291	285	182	493	252	(M) 72
	(B)	765	1388	745	730	466	1262	645	184
Fish, fresh,	(CC)(A)	382	480	299	286	179	569	245	(M) 70
all types	(ep)(B)	1150	1445	900	861	539	1713	737	211
Cow's Milk,	(A)	463	782	399	278	156	450	434	(M) 91
pasteurized	(B)	255	430	219	153	86	248	239	50
Hen's Egg,	(CC)(A)	428	551	393	320	210	436	358	(M) 93
whole	(B)	847	1091	778	634	416	863	709	184
Hen's Egg, yolk	(A)	260	553	321	346	161	480	264	(M) 93
	(B)	998	1370	820	753	364	1202	728	240
Hen's Egg,	(CC)(A)	301	518	321	299	248	415	372	(M) 99
white	(B)	536	922	571	532	441	739	662	176
Wheat, whole grain	(A)	276	417	204	183	94	179	282	(M) 68
	(B)	577	871	426	382	196	374	589	142
Wheat, gluten	(CC)(A)	266	433	258	159	100	89	324	(C) 63
Wheat Flour,	(CC)(A)	258	440	228	168	91	130	304	(M) 67
70–80%	(B)	493	840	435	321	174	248	581	128

Corn, grain	(CC)(A)	303	783	230	225	120	167	305	(C) 44
or whole meal	(B)	461	1190	350	342	182	254	464	67
Oat meal	(CC)(A)	319	454	236	207	105	232	313	79
	(B)	711	1012	526	462	234	517	698	176
Rye, whole	(CC)(A)	279	385	219	209	91	212	276	(M) 46
meal	(B)	561	728	414	395	172	401	522	87
Rice, brown	(CC)(A)	344	514	238	244	145	237	322	(M) 78
	(B)	433	648	300	307	183	299	406	98
Rice, polished	(CC)(A)	361	514	262	207	133	226	303	(M) 84
	(B)	408	581	296	234	150	255	342	95
Coconut, dried	(CC)(A)	339	419	244	212	120	220	283	(M) 68
kernel	(B)	424	524	305	265	150	275	354	85
Peanut	(CC)(A)	261	400	211	163	86	413	363	50
	(B)	1224	1876	990	764	267	1280	1125	155
Soybean, seed	(CC)(A)	300	486	284	241	79	399	309	80
	(B)	1995	3232	1889	1603	525	2653	2055	532
Pea, dry seed	(CC)(A)	294	425	267	254	57	470	287	(M) 56
	(B)	1058	1530	961	914	205	1692	1033	202
Pea, young seed	(CC)(A)	296	435	260	235	58	456	275	(M) 63
	(B)	311	457	273	247	61	479	289	66
Potato, edible	(CC)(A)	292	377	236	235	81	299	251	(M)103
part	(B)	93	121	76	75	26	96	80	33
Sweet potato,	(CC)(A)	283	340	230	236	106	214	241	(M)107
edible part	(B)	59	71	48	50	22	45	51	22
Lettuce	(CC)(A)	338	394	238	256	112	238	319	(M) 50
	(B)	71	83	50	54	24	50	67	10
Apple	(CC)(A)	250	390	220	230	49	370	160	58
	(B)	15	23	13	14	3	22	10	3

Source: Food and Agriculture Organization (1970).

Key: A = mg/100 gm of total nitrogen. B = mg/100 gm of food. Methods of analysis: M = microbiological; CC = column chromatography; C = chemical method. ep = edible part.

Proteins of Fish

A protein similar to myosin is present in fish muscle. Myosin prepared from cod seems to aggregate more rapidly than that prepared from rabbit muscle. Pure fish myosins are very labile and difficult to prepare.

Cod myosin has been isolated by the ultracentrifugation of a solution of the salt-soluble protein in the presence of adenosine triphosphate. However, there have been improvements on this procedure, as follows. Coarsely minced cod muscle is washed to remove sarcoplasmic proteins. It is then extracted with neutral pyrophosphate of ionic strength 0.6. The extract is diluted 10 times with water and the precipitated myosin and actomyosin are rapidly redissolved at an ionic strength of 0.4. The actomyosin is then eliminated by adjusting the ionic strength to 0.23. The molecular weight of the resulting myosin is about 500,000.

Tropomyosins have been extracted from cod and haddock, but are present only in small amounts.

Proteins of Cereals

The classification given for simple proteins at the beginning of this chapter is useful in protein work on cereals. The composition of the proteins varies in the different cereals. Newer techniques have shown that each of these cereal groups contain several or many individual proteins. Starch-gel electrophoresis has been effectively used for this purpose (Smithies 1955). Differences in speed of movement brings about the separation of the several compounds in bands during the electrophoretic process. After the separation the bands are made visible by means of dyeing. Other electrophoretic techniques have been used. See the section "Separation of Proteins" earlier in this chapter for a more comprehensive discussion.

Figure 6.6 shows the distribution of the several classes of proteins in four important cereals. Table 6.6 shows the amino acid contents of proteins of wheat and corn listed by side-chain properties.

The heterogeneity of the albumins and globulins of wheat and of some other grains has been demonstrated by several of the electrophoretic techniques. Analyses of different separated peaks show differences in amino acid composition.

Likewise, prolamines seem to consist of a group of heterogeneous proteins. This conclusion was reached as a result of the use of electrophoretic procedures.

Glutelins are compounds of very high molecular weight which accounts for the fact that they are the least soluble of the cereal proteins. Only after wheat glutelin molecules are fragmented does motion begin in starch-gel electrophoresis.

The cohesive-elastic properties of hydrated wheat gluten proteins are instrumental in the formation of dough which will expand in the pres-

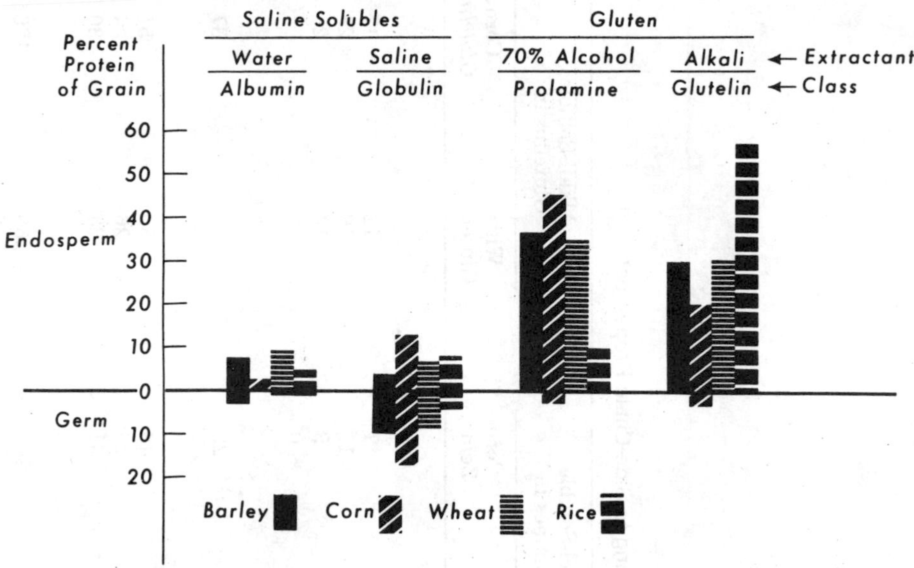

FIG. 6.6. Differences in amounts and distribution of albumins, globulins, prolamines, and glutelins in germ and endosperm of various grains.
From Wall (1964).

ence of leavening agents in the making of such products as bread and cakes.

Changes as a Result of Processing and Cooking. One of the first changes in proteins as a result of heat is the denaturation of the protein. This is more fully discussed in Chapter 20, Meat and Meat Products.

Changes taking place were studied at various temperatures, and it was found that at 65°C most of the globular and myofibrillar muscle proteins are coagulated. Between 70°C and 90°C disulfide bonds are formed by oxidation of the sulfydryl groups of the actomyosin and above 90°C, H_2S splits off from these sulfhydryl groups. According to Hamm (1970), Maillard reactions begin at about 90°C and continue with increasing temperature and time of heating. The Maillard reaction is the reaction between monosaccharides and amino acids which result in the formation of brown colors. This is discussed in Chapter 9, Browning Reactions. Browning of meat is considered to be largely caused by the reaction of carbohydrates in the muscle tissue with the amino acids of proteins. The intensity of browning has been shown to increase with the quantity of reducing sugar in the meat. It is possible that collagen could be converted to gelatin, with a resulting increase in tenderness.

TABLE 6.6 Amino Acid Contents of Proteins of Wheat and Corn Classified According to Side-Chain Properties[a]

Amino Acid	Saline-Soluble Albumins and Globulins in		Alcohol-Soluble Prolamines in		Alkali-Soluble Glutelins in	
	Wheat[b]	Corn[c]	Wheat Gliadin[d]	Corn Zein[c]	Wheat Glutenin[d]	Corn *Glutelin*[c]
Ionizable						
Basic						
Lysine	28	41	4	1	12	17
Arginine	35	42	15	11	19	28
Histidine	23	15	15	9	13	28
Tryptophan	14	—[f]	4	0	8	—[f]
Acidic						
Glutamic acid and aspartic acid[e]	91	124	27	22	35	67
Polar						
Hydroxy						
Serine	48	40	38	48	50	51
Threonine	33	34	18	25	26	34
Tyrosine	19	14	16	31	23	30
Amide						
Glutamine and asparagine[e]	90	64	309	202	266	126

Non-polar						
Glycine	78	81	25	18	78	58
Alanine	51	75	25	120	34	83
Valine	50	33	43	36	41	44
Methionine	13	—[f]	12	3	12	15
Isoleucine	37	20	37	28	28	24
Leucine	58	29	62	166	57	95
Phenylalanine	15	13	38	54	27	30
Sulfhydryl-disulfide						
Half-cystine + cysteine	45	20	12	7	12	16
Secondary						
Proline	70	78	148	101	114	115
Actual analysis in hydrolysate						
Glutamic acid	144	90	317	180	278	143
Aspartic acid	47	98	20	44	23	40
Ammonia	70	64	301	178	240	126

Source: Wall (1964).

[a] In m*M* / 100 gm of protein (16 g N).

[b] Woychik *et al.* (1961).

[c] Woychik and Boundy (1963).

[d] Wu and Dimler (1963).

[e] Based on titration data and ammonia analysis.

[f] Not determined.

While the browning reaction in meat does result in the loss of some amino acids, it seems not to be enough to have a great effect on the nutritional value of the product. However, other monosaccharide–protein systems have shown some reduction in the nutritive value as a result of heating. Furthermore, Hamm (1970) stated that when autoclaved at 112 °C for 24 hr, 45% of the cysteine of pork may be destroyed and other amino acids made unavailable during digestion.

SUMMARY

Proteins are organic compounds of nitrogen of large molecular weight ranging from about 5000 to 1,000,000 and larger. Because of their size, they are colloidal particles.

The α-amino acids are the building blocks of the proteins and these compounds are found to contain usually 20 of these acids. The α-amino group of one residue is coupled with the α-carboxyl group of another amino acid residue to form peptide bonds. Polypeptide chains make up the protein molecule.

Proteins are classified on the basis of solubility, chemical composition, and shape. A system now used divides them into three groups: simple proteins, conjugated proteins, and derived proteins. Two forms of proteins are known, fibrous and globular. Part of the proteins in the latter group are protein hormones, the enzymes, and oxygen carrying proteins.

Primary derivatives of proteins are those which have been subjected to slight modification by the action of water, enzymes, or dilute acids or alkalies. Curdled milk is an example.

Coagulated (denatured) proteins are insoluble proteins, which result from heat treatment or treatment with substances such as alcohol.

The order of sequence of the amino acid residues on the peptide chains is an important part of the structure of the protein. The properties, both physical and chemical, are governed by the peptide bonds and also by the side chains and amino acids, and the sequence of these amino acids on the chains. The various methods used in sequence determination are discussed. The complete amino acid sequence is known as primary structure. Three other structures are discussed.

Proteins act as both acids and bases, and therefore, they can migrate in an electrical field. Proteins, as well as amino acids, have a point at which they are electrically neutral. This is the isoelectric point. At this point they are called zwitterions.

Methods of separating proteins and methods of determining molecular weights are discussed.

Proteolytic enzymes are either endopeptidases, which act on the interior and terminal peptide bonds of polypeptides, or exopeptidases, which act only on the terminal peptide bonds, with the result that amino acids are sequentially removed. Many of these enzymes are quite specific. Denatured proteins are more readily attacked than native proteins. Enzymes in the intestinal tract are importantly involved in the digestion of food.

Subunits called mucleotides make up DNA and RNA. The nucleotides are compounds containing a pyrimidine or a purine base, a sugar, and phosphate. A nucleotide is a phosphate ester of a nucleoside.

In DNA replication, each strand of the double helix has the capacity to act as a template for the replication of the complementary strand. DNA and RNA differ in some of the bases contained, as well as in the sugars. The main classes of RNA are very important in the synthesis of proteins.

Four principal stages are involved in the synthesis of protein. The activation stage, the initiation stage, the elongation stage, and the termination stage.

The proteins in various foods are discussed.

BIBLIOGRAPHY

AHLGREN, E., ERIKSSON, K. E., and VESTERBERG, O. 1967. Characterization of cellulases and related enzymes by isoelectric focusing, gel filtration, and zone electrophoresis. I. Studies on *Aspergillus* enzymes. Acta Chem. Scand. *21,* 937–944.

ANDREWS, P. 1964. Estimation of the molecular weights of proteins by Sephadex gel-filtration. Biochem. J. *91,* 222–233.

AOAC. 1980. Official Methods of Analysis, 13th Edition. Assoc. Official Analytical Chemists, Washington, DC.

AURAND, L. W., and WOODS, A. E. 1973. Food Chemistry. AVI Publishing Co., Westport, CT.

AWDEH, Z. L., WILLIAMSON, A. R., and ASKONAS, B. A. 1968. Isoelectric focusing in polyacrylamide gel and its application to immunoglobins. Nature *219,* 66–67.

BACCANARI, D. P., and CHA, S. 1973. Succinate thiokinase. VI. Multiple interconvertible forms of the enzyme. J. Biol. Chem. *248,* 15–24.

BLATTLER, D. P., and REITHEL, F. J. 1970. Molecular weight determinations and the influence of gel density, protein charges, and protein shape in polyacrylamide gel electrophoresis. J. Chromatogr. *46,* 286–292.

BORGSTROM, G. 1961. Fish as Food, Vol. 1. Academic Press, New York.

BRESSANI, R., and MERTZ, E. T. 1958. Studies on corn proteins. VI. Protein and amino acid content of different corn varieties. Cereal Chem. *35,* 227–235.

CHRAMBACH, A., and RODBARD, D. 1971. Polyacrylamide gel electrophoresis. Science *172,* 440–451.

CRAIG, L. C. 1944. Identification of small amounts of organic compounds by distribution studies. J. Biol. Chem. *155,* 519–534.

CRAIG, L. C. 1950. Partition chromatography and countercurrent distribution. Anal. Chem. *22,* 1346–1352.

CRAIG, L. C. 1963. The isolation of active principles in pure form by partition. Bull. N.Y. Acad. Med. *39,* 686–703.

DALE, G., and LATNER, A. L. 1968. Isoelectric focusing in polyacrylamide gels. Lancet *7547,* 847–884.

EDMAN, P. 1950. Method for determination of the amino acid sequence in peptides. Acta Chem. Scand. *4,* 283–293.

EDMAN, P. 1956. On the mechanism of the phenyl isothiocyanate degradation of peptides. Acta Chem. Scand. *10,* 761–768.

FAO. 1970. Amino-Acid Content of Foods and Biological Data on Protein. Food and Agriculture Organization, Rome, Italy.

FRAENKEL-CONRAT, H., and HARRIS, J. I. 1954. A general micromethod for the stepwise degradation of peptides. J. Am. Chem. Soc. *76,* 6058–6062.

GILBERT, W., and VILLA-KOMAROFF, L. 1980. Useful proteins from recombinant bacteria. Sci. Am. *242,* No. 4, 74–94 (April).

GREWE, E., and LECLERC, J. A. 1943. Commercial wheat germ: Its composition. Cereal Chem. *20,* 423–434.

HAMM, R. 1970. Properties of meat proteins. *In* Proteins as Human Food. R. A. Lawrie (Editor). AVI Publishing Co., Westport, Connecticut.

HARRINGTON, W. F., and VON HIPPLE, P. H. 1961. The structure of collagen and gelatin. *In* Advances in Protein Chemistry, Vol. 16, C. B. Anfinsen, M. L. Anson, K. Bailey, and J. T. Edsall (Editors). Academic Press, New York.

HEIMBURGER, N., and SCHWICK, H. G. 1968. Characterization and isolation of proteins by electrophoretic methods. 4th Int. Symp. Chromatogr. Electrophor., pp. 39–61.

HIGASI, K. 1955. Bound water. Res. Inst. Appl. Elec. Hokkaido Univ. Monograph Ser. *5,* 9–35.

HILL, R. L. 1965. Hydrolysis of proteins. *In* Advances in Protein Chemistry, Vol. 20, C. B. Anfinsen, M. L. Anson, and J. T. Edsall (Editors). Academic Press, New York.

HOLLEY, R. W., APGAR, J., EVERETT, G. A., MADISON, J. T., MARQUISSE, M., MEARILL, S. H., PENSWICK, J. R., and ZAMIR, A. 1965. Structure of a ribonucleic acid. Science *147,* 1462–1465.

HOUSTON, D. F. 1963. Personal communication. U.S. Dept. Agric. Western Regional Res. Lab., Albany, CA.

JONES, D. B. 1931. Factors for converting percentages of nitrogen in foods and feeds into percentages of proteins. U.S. Dept. Agric. Circ. 183.

JONES, R. T. 1964. The structure of proteins. *In* Proteins and Their Reactions. H. W. Schultz and A. F. Anglemier (Editors). AVI Publishing Co., Westport, CT.

KAWATA, H., CHASE, M. W., ELYJIW, R., and MACHEK, E. 1971. A simplified and efficient system for separating proteins by preparative polyacrylamide electrophoresis. With notes on horse heart myoglobin. Anal. Biochem. *39,* 93–112.

KELLN, R. A. and GEAR, J. R. 1980. A diagrammatic illustration of the structure of duplex DNA. BioScience *30,* 110–111.

KOLBACH, P. 1955. The protein modification in barley during malting. Brewers Digest *30,* No. 8, 49–52, 61–62.

KORNBERG, A. 1961. Enzymatic Synthesis of DNA. Ciba Lect. Microb. Biochem. John Wiley & Sons, New York.

LEACH, A. A., and O'SHEA, P. C. 1965. The determination of protein molecular weights of up to 225,000 by gel-filtration on a single column of Sephadex G-200. J. Chromatogr. *17,* 245–251.

LEHNINGER, A. L. 1977. Biochemistry, 2nd Edition, 3rd printing. Worth Publishers, New York.

LLOYD, D. T. and MORAN, T. 1934. Pressure and the water relations of proteins. Proc. R. Soc. London Ser. A *147,* 382–395.

LONGWORTH, L. G. 1942. Recent advances in the study of proteins by electrophoresis. Chem. Rev. *30,* 323–340.

MADISON, J. T., EVERETT, G. A., and KUNG, H. 1966. Nucleotide sequence of a yeast tyrosine transfer. Science *153,* 531–534.

MARSH, R. E., and DONOHUE, J. 1967. Crystal structure studies of amino acids and peptides. *In* Advances in Protein Chemistry, Vol. 22, C. B. Anfinsen, M. L. Anson, and J. T. Edsall (Editors). Academic Press, New York.

MILES, L. E. M., SIMMONS, J. E., and CHRAMBACH, A. 1972. Instability of pH gradients in isoelectric focusing on polyacrylamide gel. Anal. Biochem. *49,* 109–117.

MIRSKY, A. E., and PAULING, L. 1936. On the structure of native, denatured, and coagulated proteins. Proc. Natl. Acad. Sci. U.S.A. *22,* 439–447.

MOUSSERI, J., STEINBERG, M., NELSON, A. I., and WEI, L. S. 1974. Bound water capacity of corn starch and its derivatives by NMR. J. Food Sci. *39,* 114–118.

NEURATH, H., 1963–1966. The Proteins, 2nd Edition, Vols. 1–4. Academic Press, New York.

NEURATH, H., GREENSTEIN, J. P., PUTNAM, F. W., and ERICKSON, J. O. 1944. The chemistry of protein denaturation. Chem. Rev. *34,* 157–265.

NOMURA, M. 1969. Ribosomes. Sci. Am. *221* (4), 28–35.

PAULING, L. 1960. The Nature of the Chemical Bond. Cornell Univ. Press, Ithaca, NY.

PAULING, L., COREY, R. B., and BRANSON, H. R. 1951. The structure of proteins: two hydrogen-bonded helical configurations of the polypeptide chain. Proc. Natl. Acad. Sci. U.S.A. *37,* 205–211.

PENCE, J. W., WEINSTEIN, N. E., and MECHAM, D. K. 1954. The albumin and globulin contents of wheat flour and their relationship to protein quality. Cereal Chem. *31,* 303–311.

PIMENTEL, G. S., and McCLELLAN, A. L. 1960. The Hydrogen Bond. W. H. Freeman and Co., San Francisco, CA.

RADOLA, B. J. 1973. Analytical and preparative isoelectric focusing in gel-stabilized layers. Ann. N.Y. Acad. Sci. *209,* 127–143.

RYLE, A. P., SANGER, F., SMITH, L. F., and KITAI, R. 1955. The disulphide bonds of insulin. Biochem. J. *60,* 541–556.

SANGER, F. 1945. The free amino groups of insulin. Biochem. J. *39,* 506–515.

SANGER, F. 1949. Some chemical investigations on the structure of insulin. Cold Spring Harbor Symp. Quant. Biol. *14,* 153–160.

SANGER, F., and THOMPSON, E. O. P. 1953. The amino acid sequence in the glycyl chain of insulin. 2. The investigation of peptides from enzymic hydrolysates. Biochem. J. *53,* 366–374.

SANGER, F., and TUPPY, H. 1951. The amino-acid sequence in the phenylalanyl chain of insulin. 1. The identification of lower peptides from partial hydrolysates. Biochem. J. *49,* 463–481.

SARGENT, F. 1945. The free amino groups of insulin. Biochem. J. *39,* 507–515.

SCHACHMAN, H. K. 1963. The ultracentrifuge: problems and prospects. Biochemistry *2,* 887–905.

SHANBHAG, S., STEINBERG, M. P., and NELSON, A. I. 1970. Bound water defined and determined at constant temperature by wide-line NMR. J. Food Sci. *35,* 612–615.

SMITHIES, O. 1955. Zone electrophoresis in starch gels: group variations in the serum proteins of normal human adults. Biochem. J. 61, 629–641.

SORM, F., and KEIL, B. 1962. Regularities in the primary structure of proteins. *In* Advances in Protein Chemistry, Vol. 17, C. B. Anfinsen, M. L. Anson, K. Bailey, and J. T. Edsall (Editors). Academic Press, New York.

SPANDE, T. F., WITKOP, B., DEGANI, Y., and PATCHORNIK, A. 1970. Selective cleavage and modification of peptides and proteins. *In* Advances in Protein Chemistry, Vol. 24. C. B. Anfinsen, M. L. Anson, J. T. Edsall, and F. M. Richards (Editors). Academic Press, New York.

STRICKLAND, R. D. 1966. Electrophoresis. Anal. Chem. *38,* 99R–130R.

STRICKLAND, R. D. 1970. Electrophoresis. Anal. Chem. *42,* 32R–57R.

SVEDBERG, T. 1934. Sedimentation of molecules in centrifugal fields. Chem. Rev. *14,* 1–15.

SVEDBERG, T. 1937. Protein molecules. Chem. Rev. *20,* 81–98.

SVENSSON, H. 1961. Isoelectric fractionation, analysis, and characterization of ampholytes in natural pH gradients. I. The differential equation of solute concentrations at a steady state and its solution for simple cases. Acta Chem. Scand. *15,* 325–341.

SVENSSON, H. 1962A. Isoelectric fractionation, analysis, and characterization of ampholytes in natural pH gradients. II. Buffering capacity and conductance of isionic ampholytes. Acta Chem. Scand. *16,* 456–466.

SVENSSON, H. 1962B. Isoelectric fractionation, analysis and characterization of ampholytes in natural pH gradients. III. Description of apparatus for electrolysis in columns stabilized by density gradients and direct determination of isoelectric points. Arch. Biochem. Biophys. Suppl. *1,* 132–138.

TABORSKY, G. 1974. Phosphoproteins. *In* Advance in Protein Chemistry, Vol. 28, C. B. Anfinsen, M. L. Anson, and J. T. Edsall (Editors). Academic Press, New York.

TANFORD, C. 1968. Protein denaturation. A. Characterization of the denatured state; B. The transition from the native to denatured state. *In* Advances in Protein Chemistry, Vol. 23, C. B. Anfinsen, M. L. Anson, J. T. Edsall, and F. M. Richards (Editors). Academic Press, New York.

TANFORD, C. 1970. Protein denaturation. C. Theoretical models for the mechanism of denaturation. *In* Advances n Protein Chemistry, Vol. 24, C. B. Anfinsen, M. L. Anson, J. T. Edsall, and F. M. Richards (Editors). Academic Press, New York.

TISELIUS, A., and FLODIN, P. 1953. Zone electrophoresis. *In* Advances in Protein Chemistry, Vol. 8, M. L. Anson, K. Bailey, and J. T. Edsall (Editors). Academic Press, New York.

TOLEDO, R., STEINBERG, M. P., and NELSON, A. I. 1968. Quantitative determination of bound water by NMR. J. Food Sci. *33,* 315–317.

VALLEE, B. L., and WACKER, W. E. C. 1970. The Proteins, 2nd Edition, Vol. 5. H. Neurath (Editor). Academic Press, New York.

VESTERBERG, O., and SVENSSON, H. 1966. Isoelectric fractionation, analysis, and characterization of ampholytes in natural pH gradients. IV. Further studies on the resolving power in connection with separation of myoglobins. Acta Chem. Scand. *20,* 820–834.

VICKERY, H. B. 1972. The history of the discovery of the amino acids. II. A review of amino acids described since 1931 as components of native proteins. *In* Advances in Protein Chemistry, Vol. 26, C. B. Anfinsen, M. L. Anson, J. T. Edsall, and F. M. Richards (Editors). Academic Press, New York.

WADSTRÖM, T. 1967. Studies of extracellular proteins from *Staphylococcus aureus.* II. Separation of deoxyribonucleases by isoelectric focusing. Purification and properties of the enzymes. Biochem. Biophys. Acta *147,* 441–452.

WALEY, S. G. 1966. Naturally occurring peptides. *In* Advances in Protein Chemistry, Vol. 21, C. B. Anfinsen, M. L. Anson, J. T. Edsall, and F. M. Richards (Editors). Academic Press, New York.

WALL, J. S. 1964. Cereal proteins. *In* Proteins and Their Reactions. H.W. Schultz and A.F. Anglemier (Editors). AVI Publishing Co., Westport, CT.

WATSON, J. D., and CRICK, F. H. C. 1953. Molecular structure of nucleic acids. A structure for deoxyribose nucleic acid. Nature *171,* 737–738.

WATSON, J. D., and CRICK, F. H. C. 1953. Genetical implications of the structure of deoxyribonucleic acid. Nature *171,* 964–967.

WHITAKER, J. R. 1963. Determination of molecular weights of proteins by gel filtration on Sephadex. Anal. Chem. *35,* 1950–1953.

WOYCHIK, J. H., and BOUNDY, J. A. 1963. The effect of reducing agents on corn glutelin. Personal communication. U.S. Dept. Agric. Northern Regional Res. Lab., Peoria, IL.

WOYCHIK, J. H., DIMLER, R. J., and SENTI, F. R. 1960. Chromatographic fractionation of wheat gluten on carboxymethylcellulose columns. Arch. Biochem. Biophys. *91,* 235–239.

WOYCHIK, J. H., BOUNDY, J. A., and DIMLER, R. J. 1961. Amino acid composition of proteins in wheat gluten. J. Agric. Food Chem. *9,* 307–310.

WU, Y. V., and DIMLER, R. J. 1963. Hydrogen ion equilibria of wheat glutenin and gliadin. Arch. Biochem. Biophys. *102,* 230–237.

7

Enzymes

Classification
Enzyme Composition
Properties of Enzymes
Enzyme Reactions
Some Enzyme Applications in Foods
Summary
Bibliography

Enzymes are proteins with the special ability to catalyze specific chemical reactions in living matter. Although they may undergo change during the catalysis, they are unchanged at the end of the reaction. Enzymes are made in living cells, but can act *in vitro*, that is, apart from living material.

The earliest clues and ideas about enzymes were connected with work on digestion and fermentation. In the seventeenth and eighteenth centuries experiments and observations suggested that these two processes were somewhat similar in chemical nature. Early in the 19th century it was shown that starch is converted to glucose in the presence of dilute acid, during which process the acid involved is not altered, and can be recovered unchanged. It was found, also, that something in gastric juice degrades proteins. These and other experiments brought about the idea of catalysis in which the substances acting as agents for the change are in themselves unchanged. It was still thought that life itself in some mysterious way brought about fermentation. Late in the nineteenth century Buchner (1897; Buchner and Rapp 1897), after breaking up yeast cells, showed that a solution made from yeast from which all the cells were removed by special filtration under pressure could cause fermentation of sugar to CO_2 and alcohol. This showed that living cells were not necessary to produce fermentation. The connection between enzymes and fermentation explains why enzymes were formerly called ferments.

Sumner (1926) first isolated an enzyme, urease, in the crystalline form for which he won a Nobel prize. After Sumner's achievement, Northrop prepared other enzymes in crystalline form and extended Sumner's concept that enzymes are proteins. Northrop also was awarded the Nobel prize.

Enzymes are specific in their action; that is, a single enzyme is effective for a particular reaction. Some are highly specific, whereas others may act on a group of related compounds. Furthermore, they are effective at very low concentrations, so low in fact that the usual color reactions for proteins may not be obtained with the dilute solutions still able to show enzyme activity. The material acted on by the enzyme is known as the substrate. After the reaction, the enzyme itself is unchanged, and is therefore a true catalyst. As such, enzymes increase the speed of a chemical reaction.

CLASSIFICATION

The Nomenclature Committee of the International Union of Biochemistry issued Enzyme Nomenclature 1978, Recommendations of the Nomenclature Committee of the International Union of Biochemistry on the Nomenclature and Classification of Enzymes. This edition is a revision of the Recommendations issued in 1972.

The first general rule noted that names of enzymes, particularly those ending in *-ase,* should be used for single enzymes only, single catalytic entities, and should not be used for systems with more than one enzyme. If it is necessary to use these for such a purpose, the name *system* is to be included as part of the name, e.g., *succinate oxidase system*. Furthermore, the suffix *-ase* should not be used in loose and misleading fashion.

The second general rule is that the classification and naming of enzymes is according to the reaction they catalyze, because the chemical reaction that is catalyzed is the particular property that serves to distinguish one enzyme from another.

A condensation of the classification recommended by the Commission including some enzymes of interest to food students is given here and in Table 7.1.

1. Oxidoreductases. To this class belong all enzymes catalyzing oxidoreduction reactions.
2. Transferases. Transferases are enzymes transferring a group from one compound to another.
3. Hydrolases. These enzymes catalyze the hydrolytic cleavage of C–O, C–N, C–O, and some other bonds.
4. Lyases. Lyases are enzymes clearing C–C, C–O, C–N, and other bonds by elimination.
5. Isomerases. These enzymes catalyse geometric or structural changes within one molecule.
6. Ligases (Synthetases). Ligases are enzymes catalyzing the joining together of two molecules coupled with the hydrolysis of a pyrophosphate bond in ATP or a similar triphosphate.

Many older names without the "-ase" terminology have been retained, for example, trypsin, rennin, and ficin.

TABLE 7.1. Condensation of the Classification of Enzymes.

Enzyme Number, Reaction, and Recommended Name	Other Name	Reaction and Comments
1. *Oxidoreductases*		
1.10 Acting on diphenols and related substances as donors		
1.10.3.3 Ascorbate oxidase	Ascorbase	2 L-Ascorbate + O_2 = 2 dehydroascorbate + $2H_2O$
1.11 Acting on hydrogen peroxide as acceptor		
1.11.1.6 Catalase		$H_2O_2 + H_2O_2 = O_2 + 2H_2O$
1.11.1.7 Peroxidase		Donor + H_2O_2 = oxidized donor + $2H_2O$
1.13 Acting on single donors with incorporation of molecular oxygen (oxygenases)		
1.13.11 With incorporation of 2 atoms of oxygen		
1.13.11.12 Lipoxygenase	Lipoxidase	Linoleate + O_2 = 13-hydroperoxy-octadeca-9,11-dienoate
3. *Hydrolases*		
3.1 Acting on ester bonds		
3.1.1 Carboxylic ester hydrolases		
3.1.1.3 Triacylglycerol lipase		Triacylglycerol + H_2O = diacylglycerol + a fatty acid anion
3.1.1.11 Pectinesterase	Pectin methyl esterase	Pectin + n H_2O = n methanol + pectate
3.2 Glycosidases		
3.2.1 Hydrolyzing *O*-glycosyl compounds		
3.2.1.1 α-Amylase	Diastase	Endohydrolysis of 1,4-α-D-glycosidic linkages in polysaccharides containing 3 or more 1,4-α-linked D-glucose units *Comments:* Acts on starch, glycogen, and related polysaccharides and oligosaccharides in a random manner; reducing groups are liberated in the α-configuration

(Continued)

TABLE 7.1. (Continued)

Enzyme Number, Reaction, and Recommended Name	Other Name	Reaction and Comments
3.2.1.2 β-Amylase	Diastase	Hydrolysis of 1,4 α-D-glycosidic linkages in polysaccharides so as to remove successive maltose units from the non-reducing ends of the chains *Comments:* Acts on starch, glycogen, and related polysaccharides and oligosaccharides producing β-maltose by an inversion
3.2.1.3 Exo-1,4-α-D-glucosidase	Glucosamylase Amyloglucosidase	Hydrolysis of terminal 1,4-linked α-D-glucose residues successively from non-reducing ends of the chains with release of β-D-glucose
3.2.1.4 Cellulase		Endohydrolysis of 1,4-β-glucosidic linkages in cellulose, lichenin, and cereal β-glucans
3.2.1.7 Inulinase	Inulase	Endohydrolysis of 2,1 β-fructosidic linkages of inulin
3.2.1.10 Oligo-1,6-glucosidase	Limit dextrinase Isomaltase	Hydrolysis of 1,6-α-D-glucosidic linkages in isomaltose and dextrins produced from starch and glycogen by α-amylase
3.2.1.11 Dextranase		Endohydrolysis of 1,6 α-D-glucosidic linkages in dextran
3.2.1.15 Polygalacturonase	Pectinase Pectin depolymerase	Random hydrolysis of 1,4-α-D-galactosiduronic linkages in pectate and other galacturonans
3.2.1.26 β-Fructofuranosidase	Sucrase Invertase	Hydrolysis of terminal non-reducing β-D-fructofuranoside residues in β-D-fructofuranosides
3.4 Acting on peptide bonds (Peptide hydrolyase)		
3.4.21 Serine proteinases		
3.4.21.4 Trypsin	α- and β-Trypsin	Preferential cleavage: Agr-, Lys- *Note:* β-Trypsin is formed from trypsinogen by cleavage of one peptide bond; single polypeptide chain

Number and name	Other name	Reaction
3.4.21.14 Subtilisin		Hydrolysis of proteins: and peptide amides
3.4.22 SH-Proteinases		
3.4.22.2 Papain	Papainase	Preferential cleavage: Arg-, Lys-, Phe-X-(the peptide bond next-but-one to the carboxyl group of phenylalanine); limited hydrolysis of native immunoglobulins *Comments:* From *Papaya latex.*
3.4.22.3 Ficin		Preferential cleavage: Lys-, Ala-, Tyr-, Gly-, Asn-, Leu-, Val- *Comments: Ficus glabrata* or *Ficus carica latex.* Wide specificity on protein substrates
3.4.22.6 Chymopapain		Specificity close but not identical to that of papain *Comments:* Enzymes with slightly different properties from that of papain have been isolated from papaya latex: Chymopapain A Chymopapain B and payaya 'peptidase'
3.4.23 Carboxyl (acid) proteinases		
3.4.23.1 Pepsin A (Pepsin)		Preferential cleavage: Phe-, Leu- *Comments:* Formed from porcine pepsinogen by cleavage of one peptide bond; single polypeptide chain
3.4.23.2 Pepsin B		More restricted specificity than pepsin A; degradation of gelatin; little activity with hemoglobin as substrate *Comments:* Formed from porcine pepsinogen B
3.4.23.3 Pepsin C		More restricted specificity than pepsin A; high activity towards hemoglobin as substrate

Source: Enzyme Nomenclature. Recommendations (1978) of the Nomenclature Committee of the International Union of Biochemistry. Used with permission of Academic Press, copyright owner.

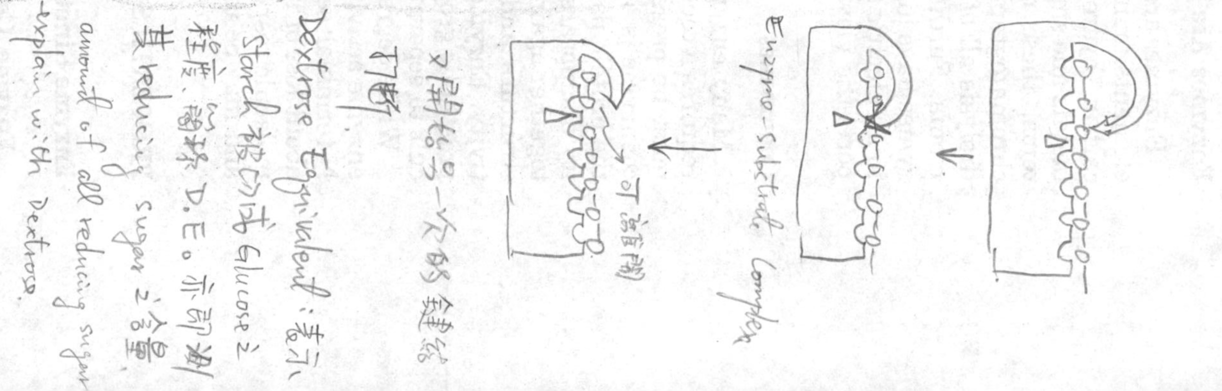

ENZYME COMPOSITION

Enzymes Are Proteins

Enzymes are classed as proteins for several reasons. They contain the elements found in proteins and in the same amounts. These include carbon, hydrogen, nitrogen, oxygen, sulfur, and sometimes certain metals and small amounts of phosphorus. As was discussed under proteins, these compounds yield amino acids on hydrolysis. Indeed, chromatographic studies on pure enzymes have shown that upon hydrolysis all of the nitrogen can be recovered as amino acids and ammonia. Furthermore, studies with the ultracentrifuge show that enzymes have high molecular weights, they act as amphoteric substances in an electric field, and they undergo denaturation—all of which supports the view that enzymes are proteins.

Coenzymes

Many enzymes are simple proteins only. Another type of enzyme requires a coenzyme, that is, a compound, nonprotein in nature, which must be present before enzymatic activity can take place. Since the coenzyme is a much smaller molecule, it can often be separated from the protein part of the enzyme by dialysis. After separation, the activity disappears; but often it can be restored by putting the two parts together again. Other enzymes contain certain metals in their protein structure, and these metals are probably necessary to bring about activity. Enzymes may have covalently bonded organic moieties, called prosthetic groups, which give them the usual activity. They are difficult to separate and if it is done, the activity is lost.

While degradation of the protein molecule usually results in loss of enzyme activity, this does not always happen. The loss of a part of the N-terminal sequence in certain enzymes does not necessarily bring about loss of its enzyme properties. This indicates that the part of the molecule responsible for this activity is located in the area remaining, and may be termed the active center.

PROPERTIES OF ENZYMES

Enzyme Kinetics

Enzyme Concentration. The concentration of an enzyme in an enzyme-catalyzed reaction is directly proportional to the rate of the reaction. Methods used for the determination of enzyme concentration are based on this relationship. A straight line plot results when the amount of enzyme is increased in the presence of an excess of substrate (Fig. 7.1). However, when a large amount of enzyme is added to such a

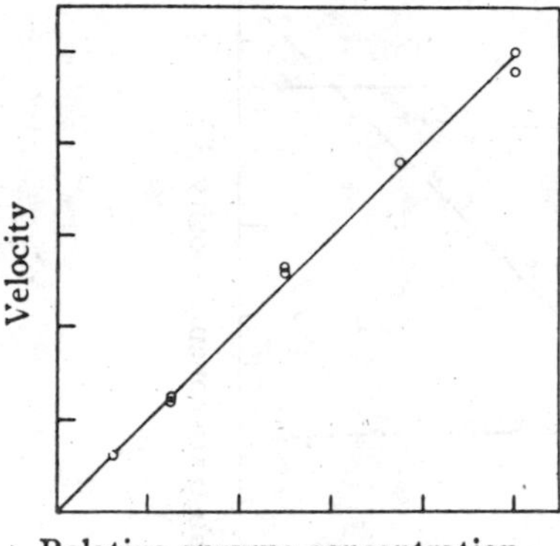

FIG. 7.1. Relative velocity as a function of a carboxypeptidase concentration for the hydrolysis of 0.05 *M* carbobenzoxyglycyl-L-tryptophan. The absence of deviations from a straight-line fit of the data demonstrate a first-order dependence of the substrate inhibition reaction on enzyme concentration. Standard conditions at 25°C.

From Lumpry et al. *1951. Used with permission of the American Chemical Society, copyright owner.*

reaction mixture, the rate of the reaction eventually decreases because the substrate is used up. The linear relationship can be altered by the addition of inhibitors or activators.

If the enzyme concentration is held constant, the resulting initial velocity of the reaction increases as substrate continues to be added, up to a maximum velocity. The resulting curve is not a straight line, but follows a simple hyperbolic function. When the enzyme has completely combined with the substrate, no further increase in the rate of reaction can take place. See Fig. 7.2.

According to the law of mass action, the rate of change in a chemical reaction is proportional to the product of the concentration of the reacting substances. If only one substrate is present in a given reaction, the rate of change is proportional to its concentration. This is a monomolecular or first-order reaction.

This can be expressed mathematically when one uses s as the quantity of substrate at the beginning of the reaction, x as the amount of this substance reacting (or product formed) in time, t. The expression for the remaining substrate is $s-x$. The velocity constant is k. Therefore,

$$dx/dt = k(s-x)$$

$$k = \frac{2.303}{t}\log\frac{s}{s-x}$$

If and when the rate is no longer affected by the substrate concentration, it becomes a zero-order reaction.

The differential equation for such a (zero-order) reaction is

$$dx/dt = k$$

When this is integrated the result is

$$k = x/t$$

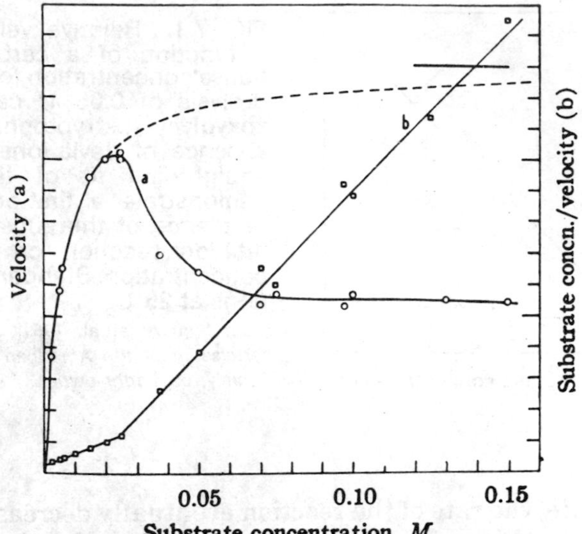

FIG. 7.2. Velocity as a function of initial concentration of carbobenzoxyglycyl-L-tryptophan at 5°C, pH 7.5, and 0.04 *M* phosphate buffer. The dashed line represents the relationship in the absence of inhibition by substrate. The solid segment (*top right*) is the maximum velocity reached at infinite substrate concentration in the absence of inhibitor. (b) The data plotted after the manner of Lineweaver and Burk. *From Lumpry* et al. *(1951). Used with permission of the American Chemical Society, copyright owner.*

Michaelis and Menten (1913) were the first to work out a really useful mathematical analysis of the velocity of enzyme catalyzed reactions as affected by the substrate concentration (Fig. 7.3a). It is assumed that an intermediate enzyme–substrate complex is formed, which is the important feature of this theory. In addition, the assumption is made that the conversion rate of the enzyme–substrate complex to products of the reaction and the enzyme determines the conversion rate of the substrate to the reaction products.

This equation is expressed as follows:

$$v = \frac{V\,[\mathrm{S}]}{K_{\mathrm{m}} + [\mathrm{S}]}$$

where v = initial velocity, V = maximal velocity, [S] = total concentration of substrate, K_{m} = Michaelis–Menten constant.

If one employs the Lineweaver–Burk equation, which is the reciprocal of the Michaelis–Menten equation (Lineweaver and Burk 1934):

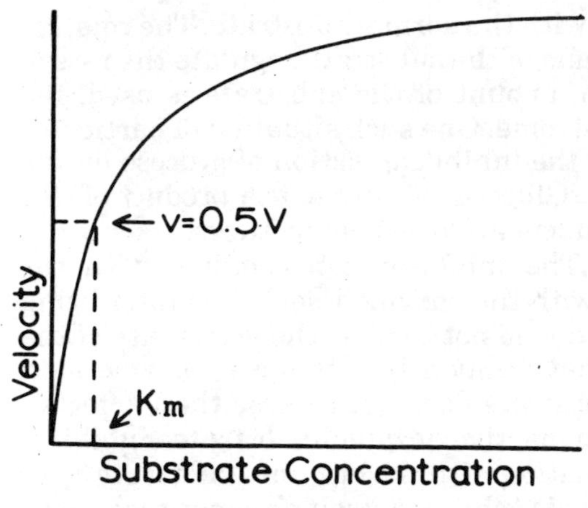

FIG. 7.3a. Michaelis–Menten plot of reaction velocity versus substrate concentration.

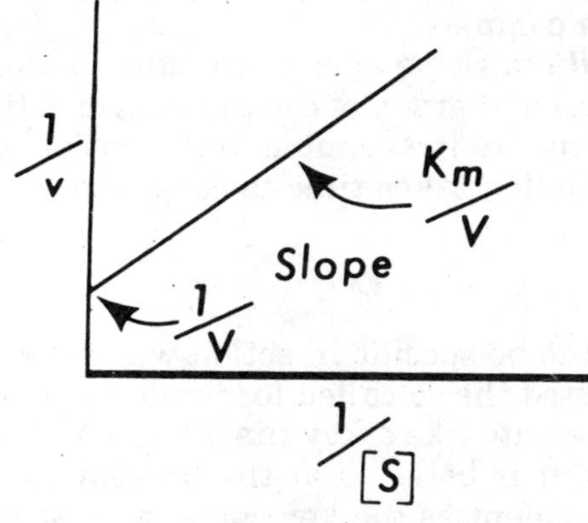

FIG. 7.3b. To calculate K_m plot the reciprocal of velocity versus the reciprocal of substrate concentration.

$$\frac{1}{v} = \frac{K_m}{V}\frac{1}{[S]} + \frac{1}{V}$$

If one plots $1/v$ versus $1/[S]$, K_m/V is the slope, and intercept on $1/v$ axis is $1/V$. One gets a straight line rather than a sigmoid curve as in Michaelis–Menten. It is possible to calculate K_m (Fig. 7.3b).

Enzyme Inhibition. There are three classes of enzyme inhibition. All of these are reversible, and the Michaelis–Menten equation can be used in the analysis of the effects on the reaction kinetics of the enzyme by the inhibitor. The three classes of inhibition are competitive, noncompetitive, and uncompetitive.

Competitive Inhibition. This is the situation in which the enzyme combines with the substrate, but another substance, chemically related to the substrate, is present and competes with the substrate to bind at the active site of the enzyme. This inhibits the activity of the enzyme in the formation of the complex with the normal substrate. The relative concentrations of the inhibitor and of the substrate regulate the rate of the reaction. If a large enough amount of the substrate is used, the effect of the inhibitor can be overcome. One such situation of particular interest in dealing with food is the inhibiting action of glucose on the activity of invertase on sucrose. Glucose, of course, is a product of this enzyme reaction, and such products are often competitors.

Noncompetitive Inhibition. The inhibitor can combine with the enzyme–substrate complex or with the enzyme itself, thus interfering with the activity. These inhibitors do not bind at the active site of the enzyme, but at another site that frequently alters the enzyme as a result. The addition of more substrate does not reverse these effects.

Uncompetitive Inhibition. An inactive enzyme–substrate–inhibitor complex is formed by the combination with the enzyme–substrate complex, which is usually formed. This inhibitor itself does not react with the enzyme to form a complex.

Irreversible Inhibition. Some agents are able to alter permanently and covalently a functional group of certain enzymes that is necessary for catalysis, resulting in irreversible inactivation of the enzyme. Sometimes such a reaction takes time to complete.

Enzyme Specificity

That enzymes tend to be specific in action was noted long ago. Emil Fischer (1894) suggested the so-called lock and key relationship—the enzyme "fits" the substrate like a key fits a lock. Although this is still an acceptable theory, it is believed at the present time that the lock undergoes some adjustment as the key comes to it so that a workable union results, allowing the enzyme to combine with the substrate.

Varying degrees of specificity are known, from highly specific to only moderately so. Urease, highly specific, will act on urea only, and on no other known compound. Some enzymes act on a definite stereoisomeric structure. For example, racemic compounds have been separated by means of enzymes because a specific enzyme was found to act only on the dextro form, leaving the levo form undisturbed. Some enzymes are intermediate in specificity. An example is invertase. It hydrolyzes sucrose and other β-fructosides. The amylases, already discussed, are in this group also, and act on starch. Lipases are enzymes of low specificity, which hydrolyze not only triacylglycerols to fatty acids and glycerol but also simpler esters to alcohols and acids.

Mode of Enzyme Activity

Because of the molecular size of enzymes it seems likely that there is a small area of activity where the action takes place. This is known as the active site. In order for this reaction to proceed asymmetrically, the enzyme and substrate must have a specific spatial relationship.

Three points of interaction between enzyme and substrate are necessary to explain asymmetric reaction on an apparently symmetrical substrate. Experiments by Chance (1951) clearly show the theory of enzyme–substrate combination.

The reaction of enzymes with substrates is shown schematically in Fig. 7.4. The enzyme–cofactor complex that is catalytically active is known as the holoenzyme. The protein remaining after the removal of the cofactor is known as the apoenzyme, which is, by itself, catalytically inactive. However, the holoenzyme has the ability to combine with the substrate, yielding the final products.

Rate of Reaction

The efficiency of an enzyme was formerly expressed as the turnover number. It is now known as molecular or molar activity. This is the

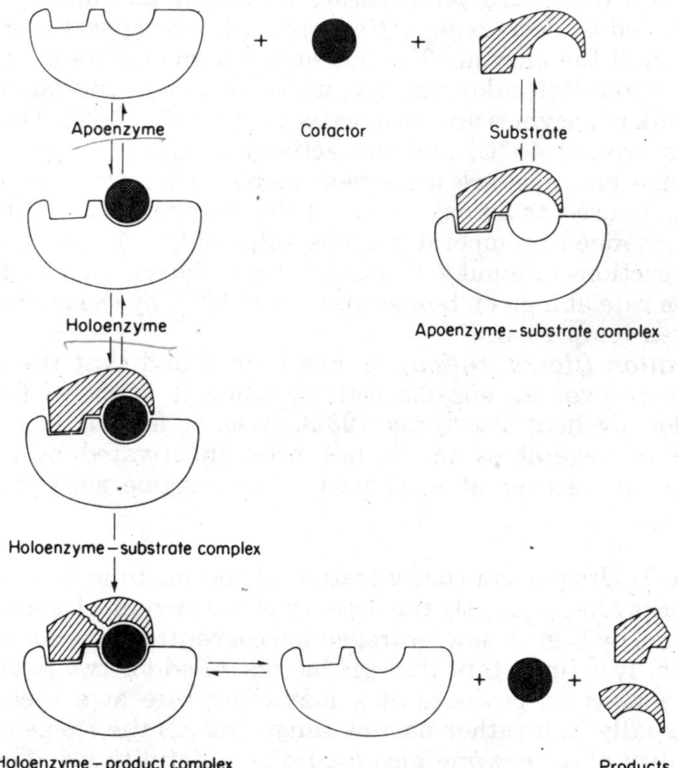

FIG. 7.4. Apoenzyme, holoenzyme, and substrate relationships.

From Whitaker (1972). Used with permission of Marcel Dekker, copyright owner.

number of molecules of a substrate decomposed by a molecule of a given enzyme per minute. Thus, one molecule of catalase decomposes 5,000,000 molecules of H_2O_2 in a minute. During this process, the enzyme combines with the substrate to form the complex compound, and this, in turn, breaks up, yielding the final product and releasing the enzyme for combination with another molecule of substrate to keep the cycle moving. If the molecular weight of the enzyme is not known or if the material employed is not pure, it is internationally accepted that one unit of the enzyme material used will catalyze the reaction 1 μmole of substrate per minute under controlled conditions. Among the factors that influence the rate of an enzyme reaction are the following.

Temperature. Changes in temperature affect the rate of enzyme reactions. It should be remembered also that enzymes, being proteins, are denatured at high temperatures. The usual effect of temperature on reactions involving enzymes is in two stages: (1) the rate of the reaction increases with increasing temperature up to a maximum, (2) this in turn is followed by decreasing activity at higher temperatures, due to denaturation of the enzyme. The increasing temperature at the start produces greater molecular activity, which increases the rate of reaction. The bulk of enzymes are most active around 30°–40°C. Denaturation begins around 45°C, and the activity of the enzyme starts to decline. Some enzymes are more heat stable. The temperature coefficient, Q_{10}, is used as an expression of the change in rate of reaction for a 10°C change in temperature. The value of Q_{10} for many enzyme mediated reactions is about 2. The Q_{10} value is determined by dividing the reaction rate at a given temperature plus 10°C by the reaction rate at that given temperature.

Regeneration (Renaturation). It has been found that the enzyme trypsin can recover its enzyme activity when it is cooled following denaturation by heat (Northrop 1932). Also, it is well known that peroxidase in vegetables which has been inactivated by scalding (blanching) can recover at least part of its enzyme activity during frozen storage.

pH. The hydrogen ion concentration of the medium in which the enzyme works affects greatly the activity of the enzyme. Denaturation of the enzyme by high or low hydrogen ion concentrations can result in inactivation. It is important that pH be controlled for two reasons: (1) the enzyme reaction proceeds at a maximum rate at a specific pH, which is usually in a rather narrow range and (2) the range of maximum stability of an enzyme also occurs at a definite pH. The pH is controlled in an industrial process either to inhibit the enzyme activity or to produce the maximum enzyme activity. The pH can be lowered in fruit products by the addition of such compounds as phosphoric or citric acids.

When the activity of an enzyme is plotted against varying pH values the result is ordinarily a bell-shaped curve. This curve shows the highest activity at the top with rapidly decreasing rates along the sides. This is illustrated in Fig. 7.5.

Activators. These are metal ions which activate or participate in reactions of enzymes. Among these are the following: CA^{2+}, K^+, Fe^{2+}, Cu^{2+}, Zn^{2+}, Co^{2+}, Mg^{2+}, and Mn^{2+}. Pyruvate kinase and enolase are enzymes that require either Mn^{2+} or Mg^{2+}. Either is necessary to form a complex so that binding with a substrate can be accomplished. In this Case Ca^{2+} acts in competition with Mn^{2+} or Mg^{2+} with the result that an inactive complex is formed. In addition, it is necessary that K^+, an alkali metallic ion, be present for pyruvate kinase to act. Two other alkali metallic ions, Rb^+ and Cs^+, are active in this way also.

Papain, bromelin, and ficin, plant proteolytic enzymes, become inactive when exposed to oxygen. These enzymes remain in the active condition if a reductant such as bisulfite ion or cysteine is added. These reducing compounds maintain the sulfhydryl groups in the enzyme protein.

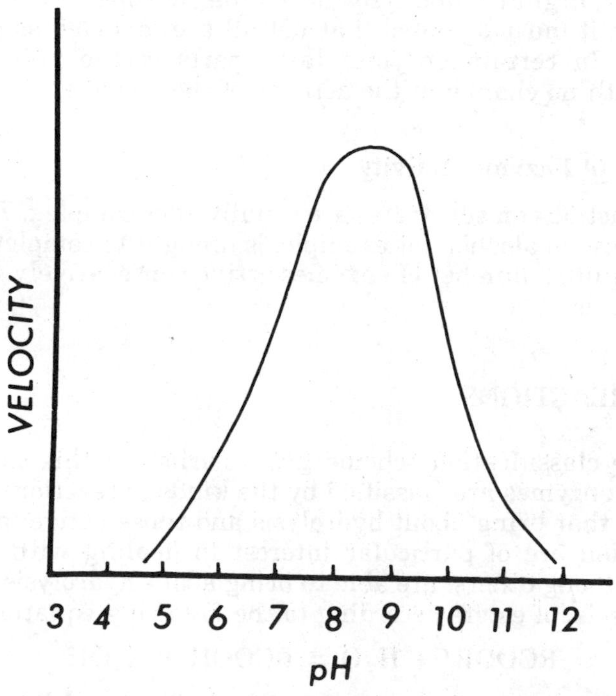

FIG. 7.5. The effect of pH on enzyme reaction velocity.

Active Sites of Enzymes. This is the position on the enzyme at which the catalysis of the substrate with the enzyme takes place. Since many enzymes are solely protein in composition, it follows that an amino acid residue in such enzymes contains the site.

Enzymes have great catalytic efficiency. The large rate accelerations brought about by enzymes seem to be brought about by four factors.

One of these is precisely locating the substrate at the catalytic group of the active site of the enzyme. The second is the formation of a covalent intermediate, which is unstable and changes easily to form the products. Often involved at the active site are residues of serine, cysteine, lysine, or histidine of the enzyme protein. Enzymes that act with the formation of a covalent enzyme–substrate intermediate are classified by the active amino acid residue involved, such as histidine class, etc.

Another of these factors is the susceptible bond in the substrate, which can be more easily broken by the induction of strain. And still another factor involved in catalytic efficiency is that enzymes can be responsible for acid or base catalysis when they can provide groups which can accept or donate protons.

In addition to these, metals and coenzymes are often involved in the catalytic process as part of the enzyme–substrate complex.

Other amino acids found at the active site include glutamic acid, aspartic acid, arginine, and tyrosine. Of the enzymes that contain only amino acids it must be noted that not all are involved as part of the active site. In certain enzymes large parts of the protein can be removed with no change in the activity of the enzyme.

Complexity of Enzyme Activity

Enzyme actions on substrates are usually very complex. The conversion of glucose to alcohol, for example, is brought to completion by the activity of quite a number of enzymes acting consecutively. Figure 7.6 illustrates this.

ENZYME REACTIONS

From the classification scheme given earlier in this chapter it is stated that enzymes are classified by the kinds of reactions they catalyze. Those that bring about hydrolysis and those active in oxidation and reduction are of particular interest in dealing with foods. The esterases, among others, are able to bring about hydrolysis, or reversibly, synthesis, of esters according to the following equation:

$$RCOOR' + H_2O \rightleftharpoons RCOOH + R'OH$$

Similar in reaction to these are the phosphatases and the phosphorylases, which are able to cleave phosphoric acid compounds. Other

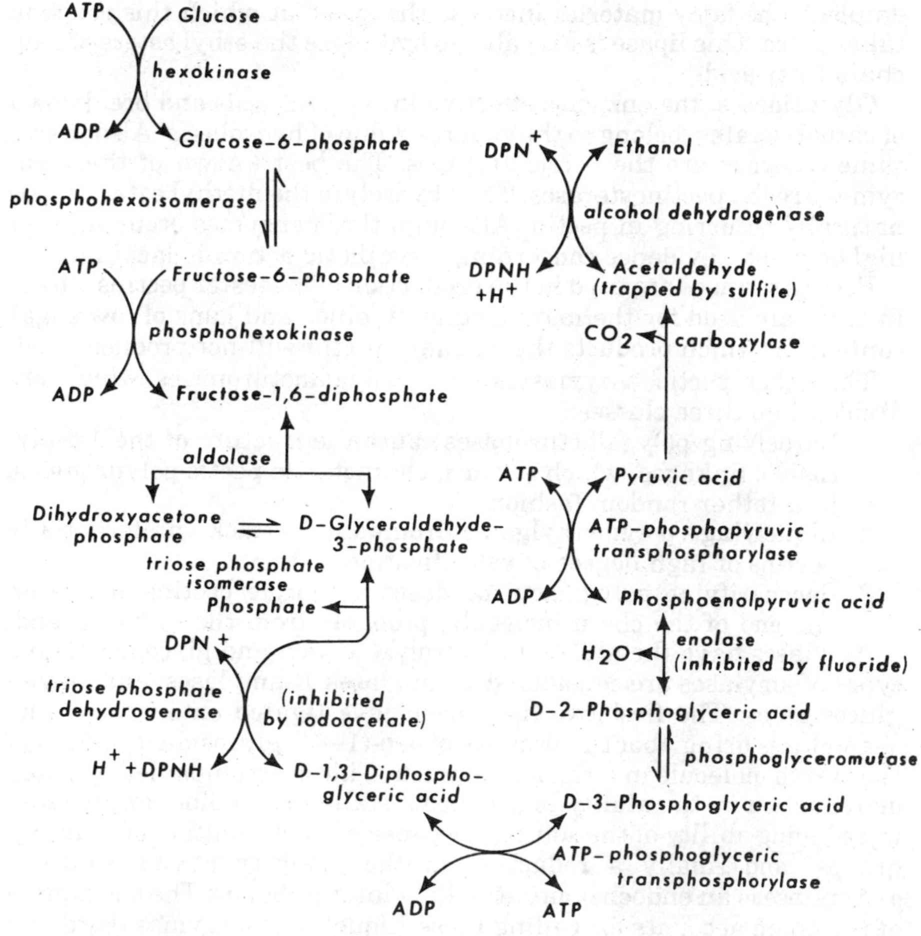

FIG. 7.6 Pathway of anaerobic breakdown of glucose to ethanol and carbon dioxide in yeast.
From Fruton and Simmonds (1958). Reproduced with permission of John Wiley & Sons, copyright owner.

members of this group are the lipases, which have the ability to split fats and nonmineral oils of plant or animal origin, as well as similar compounds prepared in the laboratory. Pancreatic lipase hydrolyzes these fats and oils to glycerol and fatty acids, in a stepwise reaction. The fatty acids are broken off the fat molecule one at a time, to yield first a diglyceride, then a monoglyceride, and finally, free glycerol and third fatty acid. These reactions occur in the gastrointestinal tract during the digestion of foods. Soap and other substances are able to emulsify the fatty material increase the speed at which this reaction takes place. This lipase is also able to hydrolyze the ethyl esters of long chain fatty acids.

Glycosidases, the enzymes effective in the synthesis and breakdown of carbohydrates, belong to the general group of hydrolases. Also in this same category are the pectic enzymes. The best known of these enzymes are the pectinesterases. They hydrolyze the methyl ester groups naturally occurring in pectin. Although these enzymes occur in most higher plants, evidence concerning a synthetic action is lacking.

Pectinesterases are used in the production of low-ester pectins which, in turn, are used for the manufacture of jellies and jams of low sugar content for which products the ordinary pectin will not produce a gel.

The other pectic enzymes are the polygalacturonases which are divided into three classes:

1. Liquefying polygalacturonases cause the fracture of the 1,4-glycosidic linkages, which in turn, characterize pectic polyuronides, in a rather random fashion.
2. Liquefying polymethylgalacturonidases attack preferentially pectins of high degree of esterification.
3. Saccharifying polygalacturonidases hydrolyze pectins only from one end of the chain molecule, probably from the reducing end.

Amylases have the ability to hydrolyze starch and glycogen. Three types of amylases are considered, α-amylases, β-amylases, and amyloglucosidases. The first two have been investigated extensively. The α-amylases bring about hydrolysis of α-D-(1→4)-glucosidic linkages of the starch molecule in a random manner. This is accompanied by a loss or reduction in the ability to produce a stain with iodine, an increase in reducing ability of the solution because of the formation of reducing groups, and finally a reduction in the viscosity of the solution. α-Amylase is an endoenzyme, attacking interior bonds. The shortening of the chain accounts for calling these "liquefying" enzymes (Bird and Hopkins 1954).

β-Amylase acts on the end units of the starch chains; hence, it is an exoenzyme. It removes the maltose units from the nonreducing ends of the chains. It forms a sweet solution, and is known as the sugar-producing enzyme. Limit dextrins are produced by the action of β-amylase on amylopectin because the action of the enzyme stops a couple of glucose units from the branch point.

Phosphatases are able to split phosphoric acid from some types of organic phosphates.

Proteinases, otherwise known as proteolytic enzymes, include papain, ficin, trypsin, pepsin, and others, and have the ability to cleave the CO—NH (peptide) bond.

The peroxidases and catalases inactivated during the blanching of vegetables are important enzymes and contain iron porphyrin. They are oxidoreductases. Peroxidases catalyze reactions in which hydrogen peroxide is an electron acceptor. Thus peroxidase transfers hydrogen as in the following reaction

$$AH_2 + H_2O_2 = A + 2H_2O$$

where A = accceptor and AH_2 may be ascorbic acid. Catalase breaks down hydrogen peroxide as follows.

$$2H_2O_2 = O_2 + 2H_2O$$

Other oxidases such as ascorbic acid oxidase and the polyphenol oxidases contain copper and are found in animal and plant tissues (Nelson and Dawson 1944).

Extracts prepared from plant tissues are known to contain dehydroascorbic reductase, which catalyses the reduction of dehydroascorbic acid to ascorbic acid by glutathione (Crooke and Morgan 1944).

SOME ENZYME APPLICATIONS IN FOODS

The activity of enzymes is of considerable importance in food technology, since many such reactions can have beneficial results and others cause undesirable changes. It is necessary that these facts be taken into consideration in the storage and processing of foods.

Enzymes can be important in the production of flavors in fruits and vegetables. Enzymes are also of importance in the production of tender meat.

Vegetable Processing

Enzymes are present in fresh vegetables. When fresh vegetables are prepared for freezing, it is necessary to inactivate the enzymes by blanching them, thus preventing deterioration during storage. This is accomplished by treatment with steam or boiling water prior to packing and freezing. Lee and Wagenknecht (1951) and Lee and Mattick (1961) showed some of the changes that take place when peas are not blanched and held at −18°C for extended period of storage. The enzymes act on the lipid material resulting in the development of peroxides and the liberation of many fatty acids. In addition, Lee (1958) showed the effect of blanching on the carbonyl content of the crude lipid during the storage of frozen peas. Unblanched peas held in frozen storage yielded fatty materials containing appreciable amounts of unsaturated carbonyl compounds, which were not formed in the corresponding blanched material.

A haylike flavor and aroma develops in dehydrated vegetables unless they are blanched before drying. This is likely the result of enzymatic activity, since blanching would inactivate the enzymes. While low water content protects food from spoilage by microbial action, it does not stop deterioration as a result of enzyme activity. An increase in the moisture content can result in an increase in enzyme activity. Water activity was discussed in Chapter 2. Relative humidity or water activity can affect enzyme activity. If the water activity is low enough, enzyme reactions do not proceed. This conclusion was arrived at as a result of a study of model systems.

Browning as a result of enzyme action is discussed in Chapter 12.

Beer and Wine

These products will be discussed fully in Chapter 14 Alcoholic Fermentation.

Fruit Juices

Pectic enzymes are of considerable value in the production of fruit juices insofar as their use results in considerably better yields of juice. Also, pigments are more completely extracted by the enzyme method.

Meat

Proteolytic enzymes have been known to bring about the tenderizing of meat. Primitive peoples used leaves of the papaya plant to wrap pieces of meat for the purpose of tenderizing it. These leaves contain proteolytic enzymes.

In modern practice this tenderizing is achieved by allowing the carcass to hang at a low temperature (4°C) for several weeks. This improvement in tenderness is the result of enzyme action. Tenderizing can be achieved also by dusting proteolytic enzyme on meat previous to frying or broiling. A third method of tenderizing is achieved by injecting small amounts of proteolytic enzymes into the animal just prior to slaughter. However, it is necessary to hold the meat in cold storage until used, if overtenderization is to be avoided.

Dairy Products

Cheese is another product in which enzymes are of importance in manufacture. Milk is curdled by rennet, which is made from calf stomachs, and, of course, is a crude enzyme. This is discussed fully in Chapter 16.

Isolation of Enzymes

The synthesis of enzymes can take place in different parts of plants. The enzymes may accumulate at the site of formation or elsewhere.

They may also be secreted. Animal organs and muscles are frequent source material for enzymes, as are microorganisms.

To isolate an enzyme, the richest available source should be used in order to reduce the amount of work necessary to prepare the enzyme. The material is comminuted to facilitate the extraction of the enzyme. This step is done at a low temperature to forestall heat denaturation and to held down the action of proteolytic enzymes. In enzyme extraction, it is necessary to observe many precautions, depending on the stability of the enzyme. Extremes of temperature, pH, and ionic strength should be avoided. In some cases a reducing atmosphere is required to prevent oxidation.

The most recent techniques involves the use of isoelectric focusing. Column chromatography making use of such media as Sephadex, cross-linked polyacrylamide, and ion-exchange have been employed.

Enzymes are important in a number of industrial processes. Enzymes can be advantageous because of the following characteristics. (a) They are efficient catalysts. (b) The reaction can be stopped by applying heat sufficient to destroy the enzymes. (c) Temperature, pH, and time can be used to control the reactions. (d) Enzyme activity can be standardized. (e) They are nontoxic and can be left in the product unless it is necessary to destroy them.

SUMMARY

The classification of enzymes is based on the reactions they catalyze.

Among these enzymes are the esterases, which bring about hydrolysis or, reversibly, synthesis of esters. Another member of the group are the lipases, which split fats and nonmineral oils. Pancreatic lipase hydrolyzes the fats and oils to glycerol and fatty acids in a stepwise fashion.

Glycosidases are the enzymes effective in the synthesis and breakdown of carbohydrates and belong to the hydrolases. Also in this group are the pectic enzymes. The pectin esterases hydrolyze the methyl ester groups naturally occurring in pectin. Other pectic enzymes are known.

The three enzymes, α-amylase, β-amylase, and amylglucosidase hydrolyze starch and glycogen.

Proteinases have the ability to split the CO—NH (peptide) bond.

Among the oxidoreductases, catalase and peroxidase are important in vegetable processing. The action of these enzymes on H_2O_2 are shown.

Enzymes are classed as proteins because they contain the elements found in proteins and in the same amounts. Studies on pure enzymes have shown that all the nitrogen can be recovered as amino acids and ammonia.

Although a number of enzymes are simple proteins only, another type of enzyme requires a coenzyme or a metal or both for activity to take place. The coenzyme can often be separated by dialysis, after which the enzyme activity disappears. It can frequently be restored if both parts are reunited.

The concentration of an enzyme in an enzyme-catalyzed reaction is proportional to the rate of the reaction.

The Michaelis and Menten mathematical analysis of the velocity of enzyme-catalyzed reactions as affected by the substrate concentrations is discussed. Curves for the Michaelis and Menten analysis and the Lineweaver–Burk modification are shown. The curve for the latter is a straight line. The assumption is that an intermediate enzyme–substrate complex is formed.

Three classes of enzyme inhibition are known. These are competitive, noncompetitive, and uncompetitive. In competitive inhibition the enzyme combines with the substrate but another substance is present that competes with the substrate to bind with the enzyme at the active site. This can be reversed by the addition of more substrate.

Noncompetitive inhibitors can combine with the enzyme–substrate complex or with the enzyme itself. They do not bind at the active site but at another site.

In uncompetitive inhibition an inactive enzyme–substrate–inhibitor complex is formed by combination with the enzyme–substrate complex. Such competitors do not react with the free enzyme.

Whereas enzymes tend to be specific in action, varying degrees of specificity exist. Some are highly specific, while some others are intermediately so. The lipases have rather low specificity.

Enzymes act by combining with the substrate at a small area of activity which is known as the active site.

The efficiency of an enzyme is known as the molecular activity. This is the number of molecules of a substrate which are decomposed by a molecule of a given enzyme in a minute. Since enzymes are proteins, they are denatured at high temperatures, but at lower temperatures the rate of activity increases with increasing temperature. This in turn decreases as denaturation starts.

It is important to control pH. The enzyme reaction proceeds at a maximum rate at a specific pH, usually rather narrow in range. The range of maximum stability of an enzyme may also be at a definite pH also narrow in range.

Activation is produced by metal ions which activate or participate in the enzyme reactions.

Enzyme action on substrates is usually very complex.

Some enzymes can recover at least a part of their activity following denaturation by heat. One of these is peroxidase which can recover some part of its activity during frozen storage of vegetables.

In the processing of vegetables for freezing it is necessary to blanch them to inactive enzymes if satisfactorily stored products are to result. Likewise dehydrated vegetables which are not blanched prior to drying develop haylike flavor and aroma in storage.

Enzymes are important in a number of industrial processes involving foods.

BIBLIOGRAPHY

BELL, R. M., and KOSHLAND, D. E., Jr. 1971. Covalent enzyme-substrate intermediates. Science *172,* 1253-1256.

BERGMEYER, H. U. 1974. Methods of Enzymatic Analysis, Vols. I, II, III, and IV. Academic Press, New York.

BERGMEYER, H. U., and GAWEHN, K. (Editors). 1978. Principles of Enzymatic Analysis. Verlag Chemie. Weinheim.

BIRD, R., and HOPKINS, R. H. 1954. The action of some α-amylases on amylose. Biochem. J. *56,* 86-99.

BOYER, P. D. 1970-1981. The Enzymes. 3rd. Edition. Vols. 1-14. Academic Press, New York.

BUCHNER, E. 1897. Alcoholic fermentation without yeast cells. A preliminary report. Ber. *30,* 117-124, (German)

BUCHNER, E., and RAPP, R. 1897. Alcoholic fermentation without yeast cells. Ber. *30,* 2668-2678. (German)

CHANCE, B. 1951. Enzyme-substrate compounds. Adv. Enzymol. *12,* 153-190.

COLOWICK, S. P., and KAPLAN, N. O. (Editors). 1955-1982. Methods in Enzymology. Vols. 1-84. Academic Press, New York.

CROOK, E. M., and MORGAN, E. J. 1944. The reduction of dehydroascorbic acid in plant extracts. Biochem. J. *28,* 10-15.

DIXON, M., and WEBB, E. C. 1964. Enzymes. 2nd ed. Longmans, Green, New York.

DOUZON, P. 1977. Enzymology at subzero temperatures. Adv. Enzymol. *45,* 157-272.

ENDO, A. 1964. Pectolytic enzymes of molds X. Purification and properties of endopolygalacturonase III. Agric. Biol. Chem. *28,* 551-558.

ENZYME NOMENCLATURE. 1978. Enzyme Nomenclature Recommendations. International Union of Biochemistry. Academic Press, New York.

ERICKSSON, C. E., and SVENSSON, S. G. 1970. Lipoxygenase from peas, purification and properties of the enzyme. Biochim. Biophys. Acta *198,* 449-459.

EVERSE, J., and KAPLAN, N. O. 1973. Lactate hydrogenases: structure and function. Adv. Enzymol. *37,* 61-133.

FISCHER, E. 1894. Influence of configuration on the action of enzymes. Ber. Deut. Chem. Ges. *27,* 2985-2993.

FRUTON, J. S. 1970. Specificity and mechanism of pepsin action. Adv. Enzymol. *33,* 401-443.

GUTFREUND, H. 1965. An Introduction to the Study of Enzymes. Blackwell, Oxford.

GUTFREUND, H. 1971. Transients and relaxation kinetics of enzyme reactions. Annu. Rev. Biochem. *40,* 315-344.

GUTFREUND, H. 1972. Enzymes: Physical Principles. John Wiley (Interscience), New York.

HALDANE, J. B. S. 1965. The Enzymes. MIT Press. Cambridge, Massachusetts.

HASCHEMEYER, R. H., and de HARVEN, E. 1974. Electron microscopy of enzymes. Annu. Rev. Biochem. *43,* 279-301.

HEWITT, E. J., MACKAY, D. A. M., KONIGSBACHER, K., and HASSELSTROM, T. 1956. The role of enzymes in food flavors. Food Technol. *10,* 487-489.

HULME, A. C. Editor. 1970-1971. The Biochemistry of Fruits and Their Products, Vols. 1 & 2. Academic Press, New York.

JENCKS, W. P. 1963. Mechanism of enzyme action. Annu. Rev. Biochem. *32,* 639-676.

KIRSCH, J. F. 1973. Mechanism of enzyme action. Annu. Rev. Biochem. *42,* 205-234.

KOSHLAND, D. E., Jr. 1973. Protein shape and biological control. Sci. Am. *229* (4). 52-64.

KOSHLAND, D. E., Jr., and NEET, K. E. 1968. The catalytic and regulatory properties of enzymes. Annu. Rev. Biochem. *37,* 359–410.

KUNTZ, M., and NORTHROP, J. H. 1934. Inactivation of crystalline trypsin. J. Gen. Physiol. *17,* 591–615.

LAIDLER, K. J. 1973. The Chemical Kinetics of Enzyme Action. 2nd Ed. Oxford University Press, New York.

LASKOWSKI, M., Jr., and KATO, I. 1980. Protein inhibitors of proteinases. Annu. Rev. Biochem. *49,* 593–626.

LEE, F. A. 1958. The effect of blanching on the carbonyl content of the crude lipid during the storage of frozen peas. Food Res. *23,* 85–86.

LEE, F. A., and MATTICK, L. R. 1961. Fatty acids of the lipids of vegetables. I. Peas (*Pisum sativum*) J. Food Sci. *26,* 273–275.

LEE, F. A., and WAGENKNECHT, A. W. 1951. On the development of off-flavor during the storage of frozen raw peas. Food Res. *16,* 239–244.

LEHNINGER, A. L. 1975. Biochemistry, 2nd Edition Worth Publishers, New York.

LINEWEAVER, H., and BURK, D. 1934. The determination of enzyme dissociation constants. J. Am. Chem. Soc. *56,* 658–666.

LUH, B. S., and DAOUD, H. N. 1971. Effect of break temperature and holding time on pectin and pectic enzymes in tomato pulp. J. Food Sci. *36,* 1039–1043.

LUH, B. S., and PHITHAKPOL, B. 1972. Characteristics of polyphenoloxidase related to browning of cling peaches. J. Food Sci. *37,* 264–268.

LUMPRY, R., SMITH, E. L., and GLANTZ, R. R. 1951. Kinetics of carboxypeptidase action. I. Effect of various extrinsic factors on kinetic parameters. J. Am. Chem. Soc. *73,* 4330–4340.

MASON, H. S. 1955. Comparative biochemistry of the phenolase complex. Adv. Enzymol. *16,* 105–184.

MEISTER, A. Editor. Advances in Enzymology. John Wiley & Sons, New York.

MICHAELIS, L., and MENTEN, M. L. 1913. The kinetics of invertase action. Biochem. Z. *49,* 333–369. (German)

MILDVAN, A. S. 1974. Mechanism of enzyme action. Annu. Rev. Biochem. *43,* 357–399.

NELSON, J. M., and DAWSON, C. R. 1944. Tyrosinase. Adv. Enzymol. *4,* 99–152.

NISHIMURA, J. S. 1972. Mechanism of action and other properties of succinyl coenzyme A synthetase. Adv. Enzymol. *36,* 183–202.

NORTHROP, J. H. 1932. Crystalline trypsin. IV. Reversibility of the inactivation and denaturation of trypsin by heat. J. Gen. Physiol. *16,* 323–337.

RADOLA, B. J. 1973. Analytical and preparative isoelectric focusing in gel-stabilized layers. Ann. N.Y. Acad. Sci. *209,* 127–143.

REED, G. 1966. Enzymes in Food Processing. Academic Press, New York.

SEGAL, H. L. 1973. Enzymatic interconversion of active and inactive forms of enzymes. Science *180,* 25–32.

SUMNER, J. B. 1926. The isolation and crystallization of the enzyme urease. J. Biol. Chem. *69,* 435–441.

VALLEE, B. L., and RIORDAN, J. F. 1969. Chemical approaches to the properties of active sites of enzymes. Annu. Rev. Biochem. *38,* 733–794.

VAN BUREN, J. P., MOYER, J. C., and ROBINSON, W. B. 1962. Pectin methylesterase in snap beans. J. Food Sci. *27,* 291–294.

WHITAKER, J. R. 1972. Principles of Enzymology for the Food Sciences. Marcel Dekker, New York.

WHITAKER, J. R. 1974. Food Related Enzymes. Advances in Chemistry Ser. 136. American Chemical Society, Washington, DC.

8

The Vitamins

The Water-Soluble Vitamins
The Fat-Soluble Vitamins
Vitamins and Malnutrition
Summary
Bibliography

Vitamins are organic compounds of varying composition that are required to maintain health, and in younger animals and adults, growth. Only small amounts of them are necessary for these purposes. In this chapter the chemical composition, properties, occurrence, and use of the several vitamins are considered. The interested student should consult the books and reports given in the Bibliography for nutritional, biochemical, and physiological aspects of these substances. Table 8.1 provides the daily dietary allowances recommended by the Food and Nutrition Board of the National Academy of Sciences—National Research Council (1980).

It was found in the eighteenth century that small quantities of citrus juice would prevent the development of scurvy in seafaring people during long voyages. The vitamin necessary to prevent scurvy is ascorbic acid. Late in the nineteenth century, it was shown that beriberi resulted from the eating of polished rice. The vitamin to prevent beriberi is thiamin, now known to be present in the bran removed from the rice during the milling process.

Early in the twentieth century, it was discovered that for health and normal growth the usual nutrients, proteins, carbohydrates, and fats together with water and some mineral salts will not suffice. Other factors are necessary also. About the same time, these factors were called "vitamines" because of their supposed amino nature. That name has since been shortened to "vitamin." These compounds are of different compositions. Because of these differences in chemical composition, the following classification is used.

THE WATER-SOLUBLE VITAMINS

Vitamins of the B Group

Thiamin (Vitamin B_1). Beriberi, a disease widespread in the Orient, was caused by the extensive use of polished rice. During milling

TABLE 8.1. Recommended Daily Dietary Allowances[a]

	Age (years)	Weight (kg)	Weight (lb)	Height (cm)	Height (in)	Protein (g)	Fat-Soluble Vitamins: Vitamin A (μg RE)[b]	Vitamin D (μg)[c]	Vitamin E (mg α-TE)[d]
Infants	0.0–0.5	6	13	60	24	kg × 2.2	420	10	3
	0.5–1.0	9	20	71	28	kg × 2.0	400	10	4
Children	1–3	13	29	90	35	23	400	10	5
	4–6	20	44	112	44	30	500	10	6
	7–10	28	62	132	52	34	700	10	7
Males	11–14	45	99	157	62	45	1000	10	8
	15–18	66	145	176	69	56	1000	10	10
	19–22	70	154	177	70	56	1000	7.5	10
	23–50	70	154	178	70	56	1000	5	10
	51+	70	154	178	70	56	1000	5	10
Females	11–14	46	101	157	62	46	800	10	8
	15–18	55	120	163	64	46	800	10	8
	19–22	55	120	163	64	44	800	7.5	8
	23–50	55	120	163	64	44	800	5	8
	51+	55	120	163	64	44	800	5	8
Pregnant						+30	+200	+5	+2
Lactating						+20	+400	+5	+3

Source: Natl. Acad. Sci.–Natl. Res. Council 1980.

[a] The allowances are intended to provide for individual variations among most normal persons as they live in the United States under usual environmental stresses. Diets should be based on a variety of common foods in order to provide other nutrients for which human requirements have been less well defined. See text for detailed discussion of allowances and of nutrients not tabulated.

[b] Retinol equivalents. 1 retinol equivalent = 1 μg retinol or 6 μg β-carotene.

[c] As cholecalciferol. 10 μg cholecalciferol = 400 IU of vitamin D.

[d] α-tocopherol equivalents. 1 mg *d*-α-tocopherol = 1 α-TE.

[e] 1 NF (niacin equivalent) is equal to 1 mg of niacin or 60 mg of dietary tryptophan.

[f] The folacin allowances refer to dietary sources as determined by *Lactobacillus casei* assay after

the bran is removed. Eijkman showed in 1897 that fowls fed on polished rice developed beriberi, which, in turn, could be prevented by proper diet. Later Windaus *et al.* (1932) prepared pure vitamin B_1 from yeast and worked out the empirical formula for this compound. In 1936 R. R. Williams and others determined the chemical structure and in the same year synthesized it (Cline *et al.* 1937; Williams and Cline 1936).

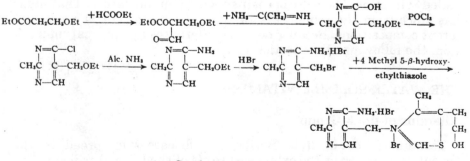

$$EtOCOCH_2CH_2OEt \xrightarrow{+HCOOEt} EtOCOCH(CH{=}O)CH_2OEt \xrightarrow{+NH_2-C(CH_3){=}NH} \text{4-hydroxy-2-methyl-5-(ethoxymethyl)pyrimidine } (N{=}C{-}OH;\ CH_3C;\ C{-}CH_2OEt;\ N{-}CH) \xrightarrow{POCl_3}$$

$$\text{(}N{=}C{-}Cl;\ CH_3C;\ C{-}CH_2OEt;\ N{-}CH\text{)} \xrightarrow{Alc.\ NH_3} \text{(}N{=}C{-}NH_2;\ CH_3C;\ C{-}CH_2OEt;\ N{-}CH\text{)} \xrightarrow{HBr} \text{(}N{=}C{-}NH_2{\cdot}HBr;\ CH_3C;\ C{-}CH_2Br;\ N{-}CH\text{)} \xrightarrow{\text{+4 Methyl 5-}\beta\text{-hydroxy-ethylthiazole}}$$

$$\text{(}N{=}C{-}NH_2{\cdot}HBr;\ CH_3C;\ C{-}CH_2{-}N(Br){-}\text{thiazolium: } C(CH_3){=}C{-}CH_2CH_2OH;\ CH{-}S\text{)}$$

Synthesis of vitamin B_1

Water-Soluble Vitamins							Minerals					
Vitamin C (mg)	Thiamin (mg)	Riboflavin (mg)	Niacin (mg NE)[e]	Vitamin B_6 (mg)	Folacin[f] (μg)	Vitamin B_{12} (μg)	Calcium (mg)	Phosphorus (mg)	Magnesium (mg)	Iron (mg)	Zinc (mg)	Iodine (μg)
35	0.3	0.4	6	0.3	30	0.5[g]	360	240	50	10	3	40
35	0.5	0.6	8	0.6	45	1.5	540	360	70	15	5	50
45	0.7	0.8	9	0.9	100	2.0	800	800	150	15	10	70
45	0.9	1.0	11	1.3	200	2.5	800	800	200	10	10	90
45	1.2	1.4	16	1.6	300	3.0	800	800	250	10	10	120
50	1.4	1.6	18	1.8	400	3.0	1200	1200	350	18	15	150
60	1.4	1.7	18	2.0	400	3.0	1200	1200	400	18	15	150
60	1.5	1.7	19	2.2	400	3.0	800	800	350	10	15	150
60	1.4	1.6	18	2.2	400	3.0	800	800	350	10	15	150
60	1.2	1.4	16	2.2	400	3.0	800	800	350	10	15	150
50	1.1	1.3	15	1.8	400	3.0	1200	1200	300	18	15	150
60	1.1	1.3	14	2.0	400	3.0	1200	1200	300	18	15	150
60	1.1	1.3	14	2.0	400	3.0	800	800	300	18	15	150
60	1.0	1.2	13	2.0	400	3.0	800	800	300	18	15	150
60	1.0	1.2	13	2.0	400	3.0	800	800	300	10	15	150
+20	+0.4	+0.3	+2	+0.6	+400	+1.0	+400	+400	+150	[h]	+5	+25
+40	+0.5	+0.5	+5	+0.5	+100	+1.0	+400	+400	+150	[h]	+10	+50

treatment with enzymes (conjugases) to make polyglutamyl forms of the vitamin available to the test organism.

[g] The recommended dietary allowance for vitamin B_{12} in infants is based on average concentration of the vitamin in human milk. The allowances after weaning are based on energy intake (as recommended by the American Academy of Pediatrics) and consideration of other factors, such as intestinal absorption.

[h] The increased requirement during pregnancy cannot be met by the iron content of habitual American diets nor by the existing iron stores of many women; therefore the use of 30–60 mg of supplemental iron is recommended. Iron needs during lactation are not substantially different from those of nonpregnant women, but continued supplementation of the mother for 2–3 months after parturition is advisable in order to replenish stores depleted by pregnancy.

The thiamin molecule is made up of a thiazole and a pyrimidine derivative. Although thiamin can be synthesized in the laboratory, the knowledge of the exact complete mechanism for the natural synthesis is not entirely known. Thiamin HCl can be prepared in the crystalline form as colorless, monoclinic needles with a melting point of 250° C. It is highly soluble in water, 1 ml of water dissolves 1 g of thiamin. The solution is optically inactive.

Changes in the structure of the thiamin molecule remove the physiological vitamin activity, but several salts of the compound are active. Some forms resulting from the opening of the thiazole ring are also active biologically. Thiamin pyrophosphate (TPP) a coenzyme known also as cocarboxylase and diphosphate ester of thiamin, is the form in which it usually occurs in yeast and in tissues. In this form it is important in the activation and transfer of aldehyde groups and is necessary in the decarboxylation of α-keto acids, such as pyruvic acid.

$$CH_3-\overset{\displaystyle O}{\overset{\|}{C}}-COOH \longrightarrow CH_3CHO + CO_2$$

It is important also in the direct oxidative pathway for the metabolism of glucose. Thiamin is found as free B_1 mostly in plants. Animal muscle contains about half of the thiamin as free thiamin.

Sulfur dioxide and sulfites should not be used in products containing this vitamin because they cause the molecule to break up into its two derivatives, thus destroying its activity.

Of dietary importance is the fact that the enzyme thiaminase can break the thiamin molecule which in turn destroys its vitamin activity. Thiaminase is found in certain raw fish, in crustaceans such as lobsters, and in mollusks such as clams. If such items are consumed as a major part of the diet, thiamin deficiency could result.

Thiamin deficiency results in the loss of appetite and development of nausea and later to a form of neuritis and cardiac difficulties. In addition to these, changes in body tissues, blood, and urine take place.

Thiamin occurs naturally in yeast, in the brans of rice and wheat, in other cereals, and in the seeds of peas and beans, in peanuts and other nuts, in egg yolk and in pork. Fruits and vegetables usually contain small amounts. Thiamin can be determined quantitatively (AOAC 1980) by oxidizing it to thiochrome. Thiochrome gives a blue fluorescence, the intensity of which can be measured. Thiamin can be determined also by the rat growth method.

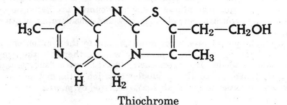

Thiochrome

Riboflavin. This vitamin, formerly known as vitamin B_2, is a growth factor. Karrer and Kuhn established the structural formula for riboflavin, and von Euler *et al.* (1935) worked out its synthesis. This compound is 6,7-dimethyl-9-(1′-D-ribityl)-isoalloxazine. The riboflavin synthesis given in Fig. 8.1 is that of Karrer and Meerwein (1936) combined with the last step from Tishler *et al.* (1947).

Riboflavin, which crystallizes in the form of orange-yellow needles, melts at 280 °C with decomposition. It is only slightly soluble in water, but is very soluble in solutions of alkalies. The alkaline solution (Kuhn and Rudy 1935) is optically active, $[\alpha]^{20}_{D} = -114°$ in 0.1 *N* NaOH. The aqueous solution has a green-yellow color and shows a strong green-yellow fluorescence. The addition of acids or alkalies causes this fluorescence to vanish. In the crystalline form, riboflavin should be protected from light, but it is otherwise stable at room temperature. This vitamin in solution tends to be rather unstable. A bottle of milk exposed to sunlight will lose about 50% of its riboflavin in a half hour.

Riboflavin acts as a coenzyme as well as a vitamin (Warburg and Christian 1932, 1938). In the form of riboflavin-5-phosphate flavin mononucleotide (FMN) it is the prosthetic group of the "yellow enzyme" of yeast. It is known also in the form of a dinucleotide, again as a part of an enzyme. These particular enzymes are known as flavoproteins

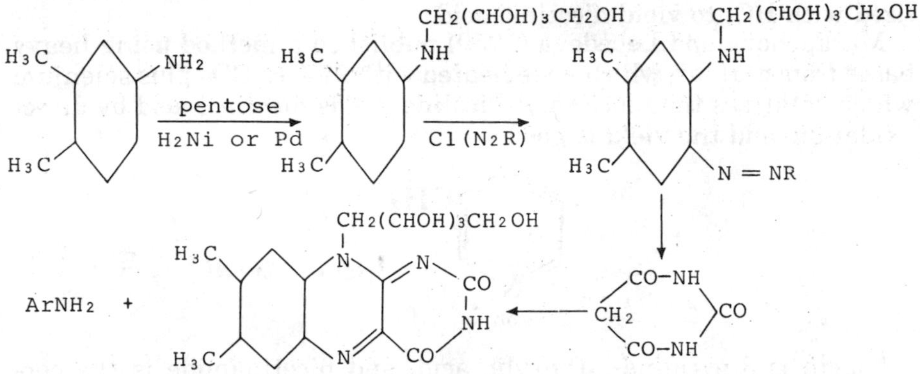

FIG. 8.1. Riboflavin synthesis.

and are involved as catalysts in a number of reactions and act in the transport of electrons. The dinucleotide, flavin adenine dinucleotide (FAD), is a part of the enzymes acyl coenzyme A dehydrogenases, which are involved in the oxidation of fatty acids. In addition to these two compound forms, riboflavin occurs naturally as the free compound.

Riboflavin deficiency in man is responsible for glossitis (inflammation) of the tongue and lips, and also scaliness at the corners of the mouth (cheilosis). Changes in the eye which can involve corneal vascularization are also symptoms of riboflavin deficiency. Riboflavin deficiency in man is associated with deficiency of other B vitamins which would result in a variety of problems.

Liver, kidney, and heart are good sources of riboflavin, as are milk, bran, wheat germ, eggs, cheese, brewers' yeast, and green vegetables. Riboflavin can be determined by measuring the fluorescence of solutions (AOAC 1980).

Niacin, Nicotinic Acid, Nicotinamide. This is otherwise known as the antipellagra factor. Pellagra was known in the eighteenth century but it was not until 1915 that Goldberger proved it was caused by dietary deficiency. Elvehjem and others (1937) found that nicotinic acid acts specifically as a cure for pellagra in the human and black tongue in the dog. Both nicotinic acid (niacin) and nicotinamide have the same vitamin activity.

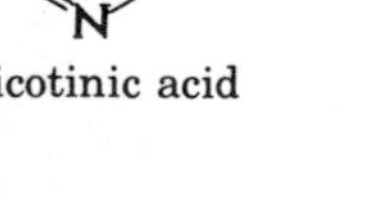

Nicotinic acid

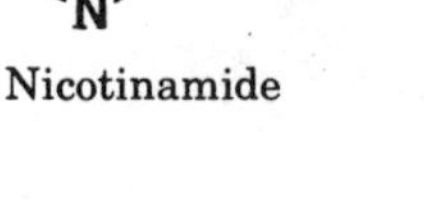

Nicotinamide

A method of synthesis for nicotinic acid, developed by McElvain and Goese (1941), is the result of the direct bromination of pyridine to form 3-bromopyridine. This is heated with cuprous cyanide to form 3-cyanopyridine, which is hydrolyzed with NaOH and finally neutralized with HCl to yield nicotinic acid.

Markacheva and Lebedeva (1950) published a method using heavy bases from coal tar, which were heated with 98% H_2SO_4 plus selenium which catalyzes the reaction. β-Picoline yields nicotinic acid by direct oxidation, and the yield is good.

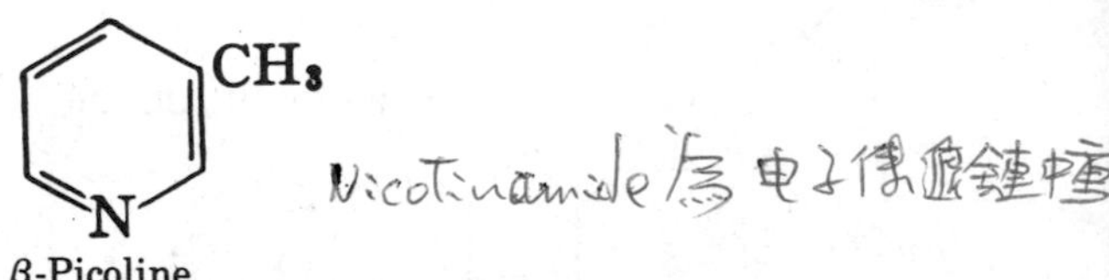

β-Picoline

Niacin is 3-pyridine-carboxylic acid, and nicotinamide is the corresponding acid amide of this compound. Niacin crystallizes in the form of needles from water or alcohol, with a melting point of 236° to 237°C, and sublimes without decomposition. It is slightly soluble in water and alcohol, but soluble in hot solvents. Nicotinamide is very soluble in water and alcohol. It appears as a white powder, but forms needles from benzene, melts at 129°–131°C, and boils under reduced pressure. Niacin is thermostable.

Nicotinamide is a part of two acceptor pyridine protein enzymes which are known as nicotinamide adenine dinucleotide (NAD) and nicotinamide adenine dinucleotide phosphate (NADP). They have also been known as diphosphopyridine nucleotide (DPN) and triphosphopyridine nucleotide (TPN). The first two names and symbols were suggested by the International Commission in 1961 because of their greater accuracy.

These are coenzymes of electron transport. The enzymatic transfer of hydrogen from alcohol to DPN was investigated by Westheimer *et al.* (1951), using alcohol dehydrogenase with deuterium (^2H) as a tracer. Deuterium is heavy hydrogen with an atomic mass of 2.

Nicotinamide adenine dinucleotide, NAD, and its reduced form, NADH, illustrate the part played by nicotinamide in enzyme activity in addition to its function as a vitamin. Other vitamins, of course, participate in enzyme activity.

Tryptophan has been found to be a precursor of niacin in man and many other species. Therefore the daily requirement of niacin is influenced by the dietary protein available.

Pellagra is a deficiency disease which causes skin lesions and an inflamed tongue, as well as diarrhea and dementia. Deficiency in the dog results in black tongue. It has been said that this disease causes the three Ds: dermatitis, diarrhea, and dementia—a fourth may be added: death.

Good sources of niacin are yeast, wheat and rice brans, whole cereals, fish, and milk. Niacin can be determined by chemical and by the microbiological methods. The basis of the chemical method is the reaction of niacin with cyanogen bromide (caution: cyanogen bromide is highly toxic) to produce a pyridinium derivative, which rearranges to form compounds which are able to react with aromatic amines to give colored compounds. If conditions are right, the color density is proportional to the amount of niacin present. Full details of both methods are given by the AOAC (1980).

Vitamin B$_6$. Three compounds have the biological activity of this vitamin—pyridoxine, pyridoxal, and pyridoxamine. Pyridoxal, however, stimulates the growth of some bacteria more effectively. These compounds are able to cure dermatitis in rats. It appears necessary for the maintenance of health in man.

Pyridoxine was isolated from yeast and liver in 1938 and prepared synthetically by workers in the Merck laboratory (Harris and Folkers 1939). Vitamin B$_6$ is found in tissues mainly as pyridoxal and pyridoxamine phosphates. In the liver a specific kinase phosphorylates pyridoxine which in turn is oxidized by a specific flavoprotein to pyridoxal phosphate, a coenzyme form. Pyridoxal phosphate and pyridoxamine phosphate act in many chemical reactions as coenzymes involving amino acids.

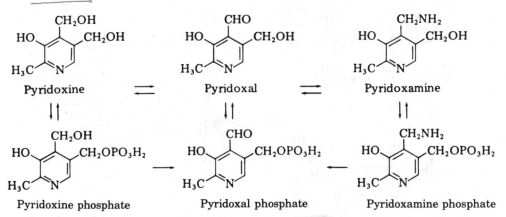

CH_2OH, HO, CH_2OH, H_3C, N — Pyridoxine ⇌ CHO, HO, CH_2OH, H_3C, N — Pyridoxal ⇌ CH_2NH_2, HO, CH_2OH, H_3C, N — Pyridoxamine

⇅ ⇅ ⇅

CH_2OH, HO, $CH_2OPO_3H_2$, H_3C, N — Pyridoxine phosphate → CHO, HO, $CH_2OPO_3H_2$, H_3C, N — Pyridoxal phosphate ← CH_2NH_2, HO, $CH_2OPO_3H_2$, H_3C, N — Pyridoxamine phosphate

Pyridoxine crystallizes in needles from acetone and has a melting point of 160°C. It sublimes and is very soluble in water and soluble in alcohol. Pyridoxine HCl crystallizes in plates from alcohol and acetone. It melts at 206°–208°C, sublimes, and is very soluble in water and slightly soluble in alcohol. These vitamins are sensitive to light.

Deficiency of pyridoxine has caused convulsions in infants, which can be treated by administration of pyridoxine. Deficiency of these vitamin compounds are not known to be connected with any particular disease in adults.

Vitamin B_6 is found in many foodstuffs. Rice bran, yeast, seeds, cereals, egg yolk, as well as meat, liver, and kidney are good sources. Vitamin B_6 is determined quantitatively by the microbiological method (AOAC 1980).

Pantothenic Acid. This compound was found to be a growth factor for yeast by R. J. Williams in 1933. It was later found to be important to animal nutrition.

$$HOCH_2-C(CH_3)_2-CHOH\cdot CO\cdot NHCH_2\cdot CH_2COOH$$

Pantothenic acid [D(+)-*N*-(2,4-Dihyroxy-3,3-dimethlybutyryl)-β-alanine]

This vitamin is a part of the molecule of coenzyme A, and is usually present in animal tissues in this form.

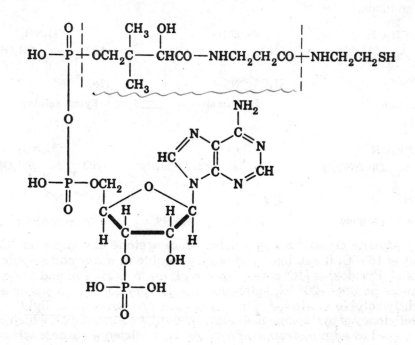

$$HO-P(=O)(-O-)-OCH_2C(CH_3)_2-CH(OH)CO-NHCH_2CH_2CO-NHCH_2CH_2SH$$

Coenzyme A

The part of the coenzyme A formula separated by broken lines is the pantothenic acid part of the molecule. Coenzyme A is vital to the use in the body of fats, carbohydrates, and nitrogen compounds. Pantothenic acid has been prepared synthetically. Several groups of workers were involved in this synthesis (Stiller *et al.* 1940; Williams *et al.* 1950).

Pantothenic acid is a yellow viscous oil, which is dextrorotatory, $[\alpha]_D^{25} = +37.5°$. Calcium pantothenate, the usual commercial form, is a white microcrystalline salt which is also dextrorotatory, $[\alpha]_D^{26} = +28.2°$. The free vitamin is very soluble in water and soluble in ether. The calcium salt is soluble in water and insoluble in alcohol.

The recommended daily allowance is 10 mg a day for adults. A daily intake of 5–10 mg is probably adequate. This vitamin is widely distributed, but the best sources are eggs, liver, and yeast. It is found also in meats and milk. Pantothenic acid is determined by microbiological assay (AOAC 1980).

Biotin. Kögl and Tönnis (1936) isolated the growth factor biotin in crystalline form from died egg yolk. An earlier growth factor had been named coenzyme R. This compound was shown to be identical with biotin.

```
            O
            ‖
          /   \
     H—N       N—H
       |       |
     H—+-------+—H
       |       |
     H—C       C—CH₂—CH₂—CH₂—CH₂—COOH
      H  \   /  H
           S
```

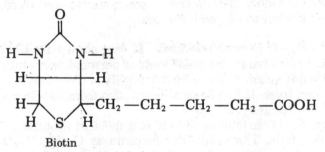

Biotin

Biotin melts and decomposes at 232.3°C, it crystallizes in needles from water and is optically active, $[\alpha]_D^{22} + 92°$. Biotin was synthesized in the Merck Laboratories. Biotin is the prosthetic group in those enzymes having the ability to catalyze the fixation of having CO_2 into organic compounds.

Avidin, a protein in raw egg white, can combine with biotin, thus preventing biotin from being absorbed in the intestine. Cooking the egg white prevents this activity. Raw egg white can bring about a dietary deficiency, but only under experimental conditions.

Man does not usually suffer from biotin deficiency because bacteria in the intestines are able to synthesize this vitamin in the amounts necessary to maintain health. Minimum daily requirement for biotin has not been established, but a daily intake of 150–300 μg is considered adequate. Biotin is widely distributed in nature. Peanuts, egg yolks, beef liver, chocolate, and yeast are good sources of biotin.

Folic Acid. This term is used to cover pteroylglutamic acid and its derivatives which have vitamin activity. It is a nutritional factor. The name, folic acid, comes from the fact that it was prepared from leaves of spinach.

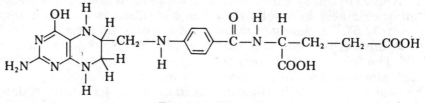

Tetrahydrofolic acid

Folic acid crystallizes as yellow-orange needles from water. It is optically active, $[\alpha]_D^{25} = +23°$, darkens when heated, and decomposes at 250°C. It is slightly soluble in water, but soluble in hot water, and insoluble in ether. It has been separated from liver and from yeast. Deficiency of folic acid results in anemia, leukopenia, and failure of growth. Folic acid is without doubt necessary in human nutrition, the part it plays is not established.

Folic acid is widely distributed in green leaves, and liver, yeast, and kidney are known to be good sources.

Vitamin B_{12}. (Cyanocobalamin) It was shown in 1926 that liver extract was effective in the treatment of pernicious anemia. It was not until 1948 that vitamin B_{12}, which is highly active in this capacity, was isolated from liver. It has been isolated also from *Streptomyces griseus* (Smith 1948, Rickes *et al.* 1948).

Vitamin B_{12} is curious in that it is a natural product that contains the metal cobalt. The empirical formula is $C_{63}H_{90}N_{14}O_{14}PCo$, the probable structural formula is illustrated in Fig. 6.2.

Work on vitamin B_{12} made it obvious that other closely related compounds possessing vitamin B_{12} activity existed. One such compound is cobamide coenzyme. The corrin ring is the central structure and includes divalent cobalt. These cobamide coenzymes contain deoxyadenosine.

The vitamin is deep red in color. The needle-shaped crystals darken to black when heated to 210°–200°C, but do not melt below 300°. The compound is optically active, $[\alpha]_{6563}^{23} = -59° \pm 9°$. It is extremely potent biologically. However, an "intrinsic factor" in gastric juice together with the vitamin B_{12} is necessary to prevent or cure pernicious anemia when orally administered. A patient suffering from pernicious anemia can be successfully treated by intramuscular injection of this vitamin.

Liver and kidney are excellent sources of vitamin B_{12} and it is present also in meat, milk, fish, and poultry. The basic source of this vitamin is soil microorganisms. Vitamin B_{12} is determined by means of the microbiological method. Details for this method are given by the AOAC (1980).

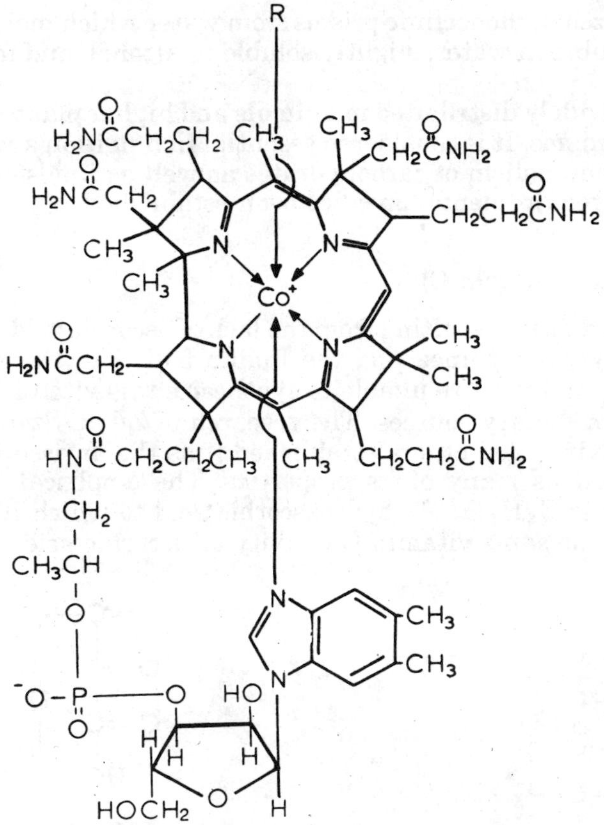

FIG. 8.2. Vitamin B_{12}: R = Cyanide; cyanocobalamin. R = 5′-deoxyadenosyl (via 5′-methylene); coenzyme form of cobalamin.
From Aurand and Woods (1973).

Inositol. This compound occurs in several isomeric forms, but the optically inactive *meso* (or *myo*)-inositol is the only one that is biologically active.

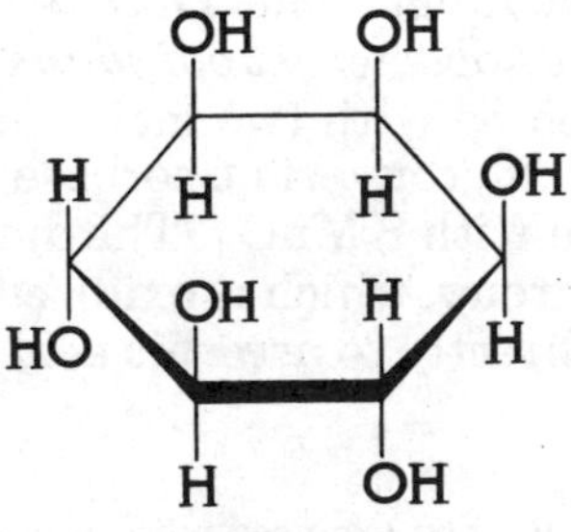

meso-Inositol

It crystallizes in monoclinic prisms from water which melt at 253 °C. It is very soluble in water, slightly soluble in alcohol, and insoluble in ether.

Inositol is widely distributed in animals and higher plants, as well as in microorganisms. It is considered essential in nutrition and has some part in the metabolism of carbohydrates as well as lipids.

Minimum requirements have not been established.

Ascorbic Acid (Vitamin C)

Scurvy is a disease resulting from the lack of ascorbic acid in the diet. Man, monkeys, the guinea pig, the Indian fruit bat, the red vented bulbul (bird), and fish are unable to synthesize this vitamin and must obtain it from dietary sources. The research of Zilva, Szent-Györgyi, Waugh and King, and others established it as the antiscorbutic compound, as well as many of its properties. The empirical formula of ascorbic acid is $C_6H_8O_6$. Dehydroascorbic acid to which it is readily oxidized has the same vitamin C activity as ascorbic acid.

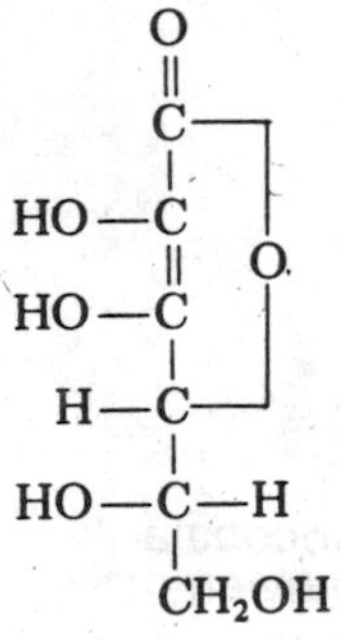

```
        O
        ‖
        C ──┐
        |   |
  HO ── C   |
        ‖   O
  HO ── C   |
        |   |
   H ── C ──┘
        |
  HO ── C ── H
        |
        CH₂OH
```

L-Ascorbic acid

```
  O═C ──┐
    |   |
  O═C   |
    |   O
  O═C   |
    |   |
    HC ─┘
    |
  HOCH
    |
    CH₂OH
```

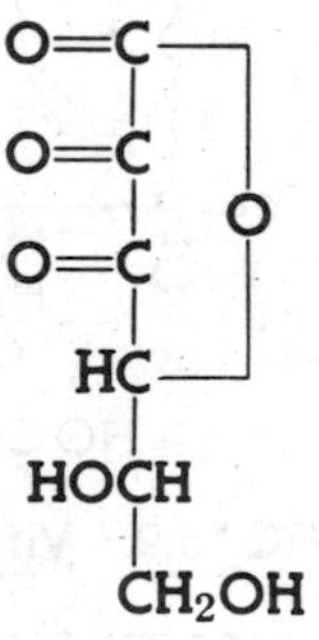

L-Dehydroascorbic acid

In the Reichstein and Grüssner (1934)[1] method for the synthesis of ascorbic acid (Fig. 8.3), glucose is reduced to the corresponding alcohol, D-sorbitol by means of catalytic hydrogenation under pressure. This compound is oxidized by *Acetobacter xylinum* to L-sorbose, which has the necessary structural arrangement. This bacterium is later replaced (Reichstein 1934) by *Acetobacter suboxydans* or other similar organisms. This is then treated with two molecules of acetone which, in the presence of sulfuric acid, converts L-sorbose to diacetone-L-sorbose. The next step is oxidation with $KMnO_4$. The diacetone-L-sorbose molecule has a free hydroxyl group, which is oxidized to the carboxyl group. This latter is converted directly to ascorbic acid, or it is esterified with

[1] Data and equations on this method are used with permission of *Helvetica Chimica Acta*, copyright owner.

methyl alcohol and HCl, which in turn is changed to ascorbic acid by treatment with methyl alcohol and sodium. This method, with some modifications, is used commercially.

Ascorbic acid has strong reducing properties which is used as the basis for its determination. As a matter of fact, Zilva discovered its reducing properties before the identification of this vitamin was completed. It can be reversibly oxidized to dehydroascorbic acid by the removal of two hydrogen atoms. Dehydroascorbic acid can be reduced to ascorbic acid by means of H_2S or by sulfhydryl compounds.

Sometimes, during ascorbic acid oxidation, certain metal catalysts, particularly copper, can cause the release of hydrogen peroxide; copper can cause considerable trouble when ascorbic acid is used as a reducing agent (Maurer and Schiedt 1933; Warren 1943). Coupled reactions are involved. In the beer industry, treatment of beer with ascorbic acid results in considerable increase in stability in most instances.

The acidic property of ascorbic acid is caused by the presence of an enediol group rather than a carboxyl group. It behaves as a monobasic acid and can give salts on reaction with alkalies. Dehydroascorbic acid is chemically a lactone, which is neutral.

Ascorbic acid is very soluble in water, soluble in alcohol, and insoluble in ether. It crystallizes in plates or in monoclinic needles. It melts at 192°C with decomposition. It is dextrorotatory, $[\alpha]_D^{20} = +24°$. The two asymmetric carbon atoms indicate that four stereoisomers are possible, but of these only L-ascorbic acid has important vitamin activity. Isoascorbic acid has minor activity.

Ascorbic acid is widely distributed in the plant kingdom, being especially prevalent in citrus fruits, black currants, strawberries, pineapples, broccoli, and parsley. It is present also in spruce and pine

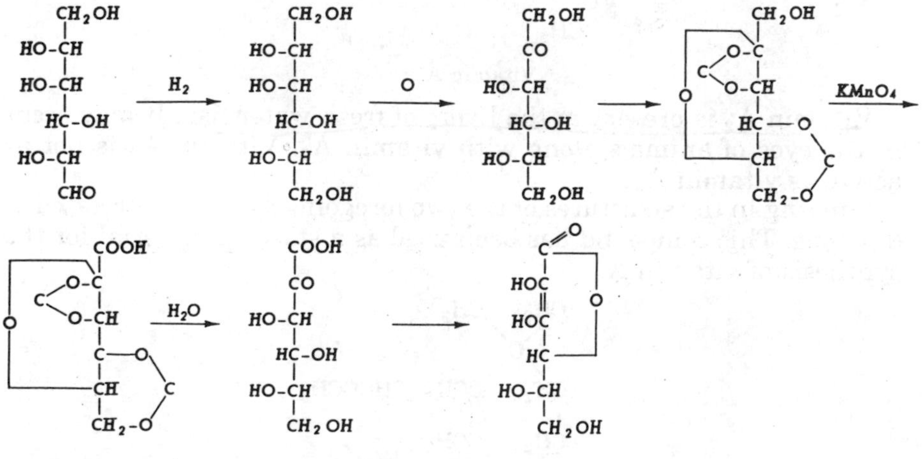

FIG. 8.3. Ascorbic acid synthesis.
From Reichstein and Grussner (1934). Used with permission of Helvetica Chimica Acta, *copyright owner.*

needles, in which case it is the old and darkened needles which contain the larger amount of this vitamin.

Ascorbic acid is determined by the 2,6-dichloroindophenol titration method, in which this dye is reduced by the ascorbic acid, resulting in the disappearance of the color of the dye. It is determined also by the microfluorometric method. In this case, the fluorescence intensity of the compound produced is proportional to the concentration. Both of these methods are given by the AOAC (1980).

THE FAT-SOLUBLE VITAMINS

Vitamin A

This vitamin, also known as retinol, is involved in vision. A deficiency of it is the cause of night blindness, and eventually in the development of the eye disease xerophthalmia. While vitamin A is found in animals, it is derived from the carotenoids of the higher plants and one-celled organisms. The plants, therefore, produce the carotenes, the animals convert them to vitamin A.

The structural formula for vitamin A_1 (Karrer *et al.* 1931, 1933) is the all-*trans* form, and physiologically the most active. Other forms of this vitamin structure exist. The all-*trans* form is the form found in the liver of mammals. The formulas for vitamins A_1 and A_2 differ in only one respect: vitamin A_2 has a second double bond—a conjugated double bond—in the ring.

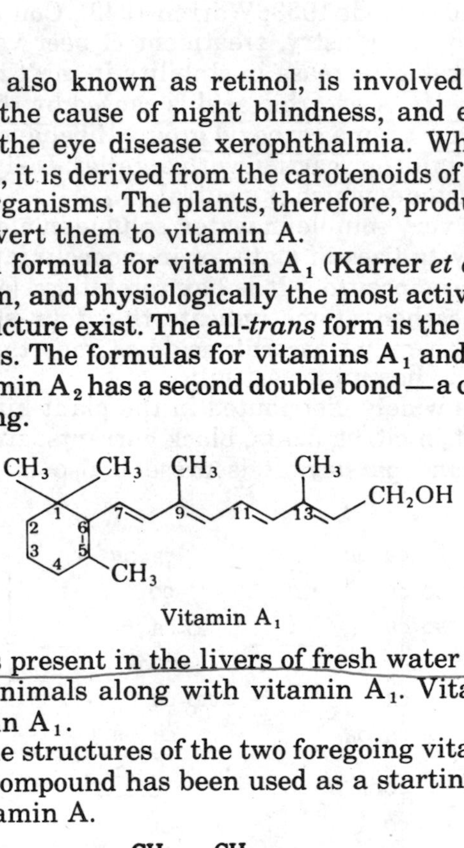

CH_3 CH_3 CH_3 CH_3 CH_2OH 1 2 3 4 5 6 7 9 11 13 CH_3

Vitamin A_1

Vitamin A_2 is present in the livers of fresh water fish. It may occur in the eyes of animals along with vitamin A_1. Vitamin A_2 is not as active as vitamin A_1.

The ring in the structures of the two foregoing vitamins is related to β-ionone. This compound has been used as a starting material for the synthesis of vitamin A.

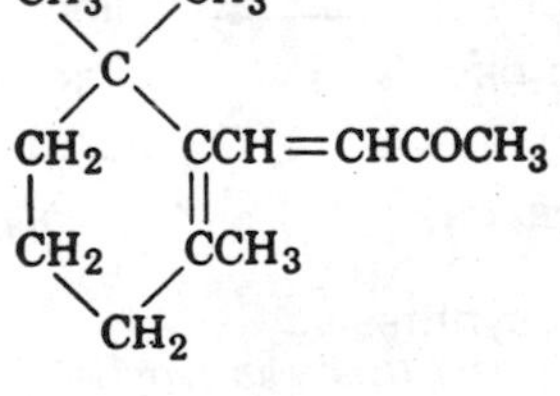

CH_3 CH_3 C CH_2 $CCH{=}CHCOCH_3$ CH_2 CCH_3 CH_2

β-Ionone

The carotenes, as has been already mentioned, are the main source of vitamin A; indeed, they are known as provitamins A. Of these, β-carotene is the most important, but α- and γ-carotenes have part of the activity of β-carotene (Karrer *et al.* 1930; Karrer and Morf 1931).

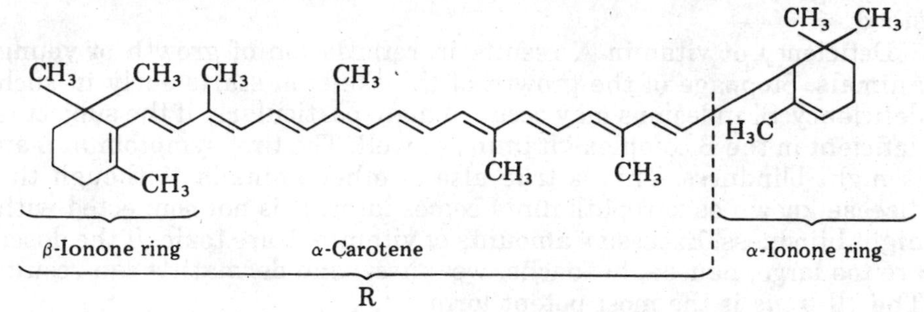

β-Ionone ring α-Carotene α-Ionone ring

R

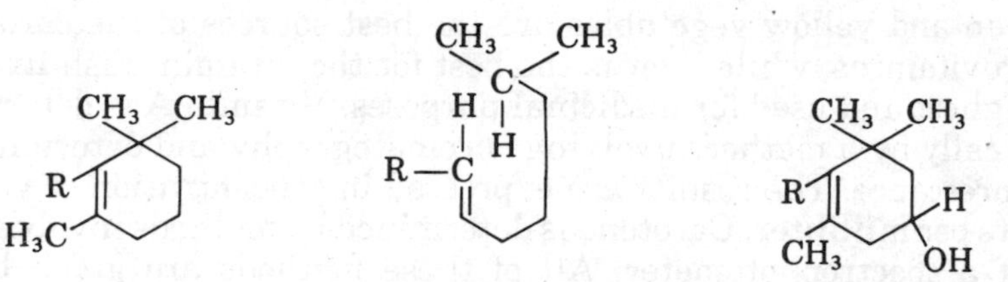

β-Carotene γ-Carotene Cryptoxanthin

β-Carotene has two β-ionone rings. α-Carotene has a β-ionone ring and an α-ionone ring. γ-Carotene has only one (β-ionone) ring with an open structure at the other end. Before the structure of β-carotene was known, workers noted that it was easily oxidized by air with the development of a violet-like aroma. This aroma is accounted for by β-ionone which has this odor.

Vitamin A_1 is insoluble in water but soluble in alcohol and in ether. It crystallizes in yellow plates from petroleum ether, and melts at 63°–64°C.

α-Carotene is insoluble in water, slightly soluble in alcohol, and soluble in ether. It forms red plates or prisms from petroleum ether or benzene-methyl alcohol. It melts at 187.5°C, and is optically active $[\alpha]^{18}_{634.5} = +385°$.

β-Carotene is insoluble in water, slightly soluble in alcohol, and soluble in ether. It forms red-brown hexagonal prisms from benzene-methyl alcohol and red rhombic crystals from petroleum ether. It melts at 184°C.

γ-Carotene is insoluble in water and alcohol and slightly soluble in ether. It forms red prisms from benzene-methyl alcohol and violet prisms from benzene-ether. It melts at 178°C.

The carotenes must be protected from oxidation and light and are sensitive to autoxidation. Vitamin A must be protected from air to avoid oxidation, and it is sensitive to the action of ultraviolet light.

These carotenes are changed to vitamin A_1 mainly in the intestine. If one molecule of β-carotene split it should yield two molecules of vitamin A. However, a vitamin A value of about only one-half of the β-carotene is obtained. People with liver disease have lessened ability to change carotene to vitamin A_1, but the reason for this is not understood.

Deficiency of vitamin A results in retardation of growth of young animals. Stoppage of the growth of the skeleton starts early in such deficiency. Skin lesions may occur in man, particularly if the subject is deficient in the B complex vitamins as well. The first symptom in man is night blindness. This is true also of other animals. Although the disease known as xerophthalmia comes later, it is not connected with night blindness. Excessive amounts of vitamin A are toxic. If the doses are too large, nausea, headache, weakness, and dermatitis can result. The all-*trans* is the most potent form.

Green and yellow vegetables are the best sources of the carotenes, the provitamins, while liver is the best for the vitamin. Fish liver oils are highest and used for medicinal purposes. Vitamin A is determined chemically by a method involving chromatography and determination of fluorescence. The results are expressed in concentration of vitamin A units per milliliter. Carotene is determined by methods involving the use of a spectrophotometer. All of these methods are given by the AOAC (1980).

Vitamin D

The disease of infancy and early childhood known as rickets is a deficiency disease. It is the result of faulty metabolism and faulty deposition of calcium in the bones and teeth.

Steenbock (1924) found that the ingestion of foodstuffs that had been irradiated with ultraviolet light was effective in the cure of rickets. This was traced to the sterol fraction. A substance, listed as vitamin D, was known to be present in fish oils in association with vitamin A. This information explains the fact that people exposed to a great deal of sunlight, and Eskimo children whose diet is rich in fish oils, are free of rickets.

It was found that when ergosterol, a sterol in yeast and in ergot, is irradiated with ultraviolet light, vitamin D_2 results (see Fig. 8.4). This compound is otherwise known as calciferol. The structure of vitamin D_2 was established by the work of Windaus and of Heilbron. A cholesterol derivative, 7-dehydrocholesterol was prepared by Windaus. When this compound was irradiated, vitamin D_3, also known as cholecalciferol, was formed. 7-Dehydrocholesterol has been separated from the outer layers of the skin of the hog. Both ergosterol and 7-dehydrocholesterol are provitamins.

All the provitamins D found in nature are chemically sterols, carbon 3, which has an hydroxyl group attached, also contains the conjugated

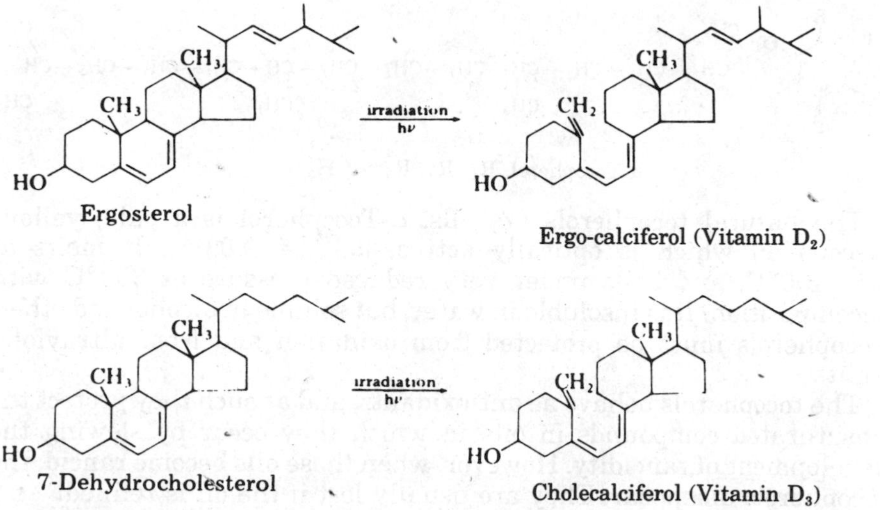

Ergosterol $\xrightarrow[h\nu]{\text{irradiation}}$ Ergo-calciferol (Vitamin D_2)

7-Dehydrocholesterol $\xrightarrow[h\nu]{\text{irradiation}}$ Cholecalciferol (Vitamin D_3)

FIG. 8.4. Vitamin D_2 formation by irradiation of ergosterol.

5,7-diene. Such unsaturation is designated by superscribing the number indicating the lower positioned carbon atom of the two linked together by a double bond, and may be designated also by placing the number of the lower positioned carbon before "ene." During irradiation of the provitamins, ring B splits between carbons 9 and 10.

Calciferol (vitamin D_2) crystallizes as white, colorless prisms from acetone and is opticaly active, $[\alpha]_D^{20} = +102.5°$. It melts at 115°–118°C and sublimes at very low pressures (0.0006 mm). It is insoluble in water, but soluble in alcohol, ether, and acetone. It has one absorption spectrum maximum at 265 nm. Cholecalciferol melts at 82°–83°C.

Since rickets is a disease that affects the growing bone, such conditions as knock knees, bowlegs, and similar problems are noticed in children deficient in this vitamin. Excessive administration of the D vitamins can have a toxic effect. Sufficient vitamin D is obtained for children by exposing the body to sunlight, from irradiated milk, from vitamin D concentrate or cod liver oil. Vitamin D is determined by bioassay (AOAC 1980).

Vitamin E

This factor, now known as vitamin E, was first noticed in the rat, where it affected the sterility of the male and female of this animal.

Actually this factor should be called vitamins E because tocopherol is the basic compound, and seven are known. The alpha form is the most active biologically, and it is the most widely distributed. The biological usefulness of the other tocopherols varies considerably. The *dl*-form of α-tocopherol has been synthesized.

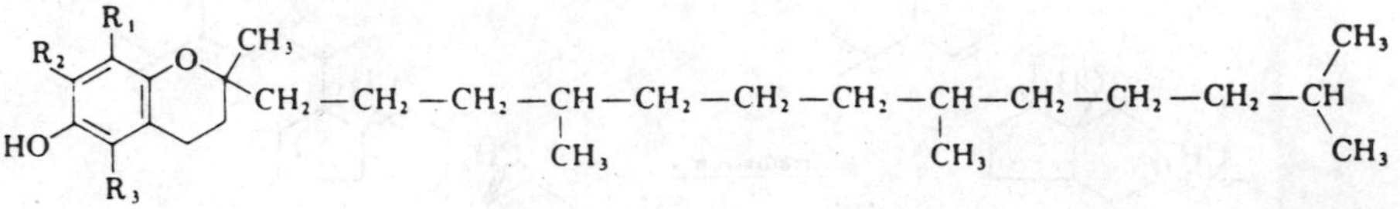

α-Tocopherol. R_1, R_2, R_3 = CH_3

The natural tocopherols are oils. α-Tocopherol is a pale, yellow, viscous oil which is optically active, $[\alpha]_D^{25} = +0.65°$. It melts at 2.5°–3.5°C, and boils under very reduced pressure at 350°C with decomposition. It is insoluble in water, but soluble in alcohol and ether. Tocopherols must be protected from oxidation and from ultraviolet light.

The tocopherols behave as antioxidants, and as such they protect the unsaturated compounds in oils in which they occur by slowing the development of rancidity. However, when these oils become rancid, the tocopherols disappear. They are usually lost if the oil is refined.

The most widely accepted biological role of vitamin E from the physiological point of view is the function of tocopherols as reversible antioxidants. In this capacity they inhibit the oxidation of unsaturated fatty acids in tissues. Numerous studies on vitamin E deficiency in animals have failed to bring forth a unified concept as to its overall function in the living organism.

Sterility occurs in the male rat as a result of tissue damage and failure of movement of spermatazoa, while in the female the lack of this vitamin causes the resorbing of the fetus. Sterility resulting in the male rat is not curable, but in the female the disorder can be cured if the vitamin is included in the food very early in the period of embryonic life.

These vitamins are found in seed germ oils and in some vegetables. Nutritional requirements in the United States are supplied by salad oils, shortenings, and margarine. Fruits and vegetables furnish some, as do grain products.

Vitamin E is determined by methods given by the AOAC (1980).

Vitamin K

Two compounds that have antihemorrhagic properties are known. The first, vitamin K_1 was found in alfalfa by Dam and Karrer in 1939; shortly after, Doisy separated vitamin K_2 from putrefied fish meal. Vitamin K_1 was synthesized by Fieser in 1939.

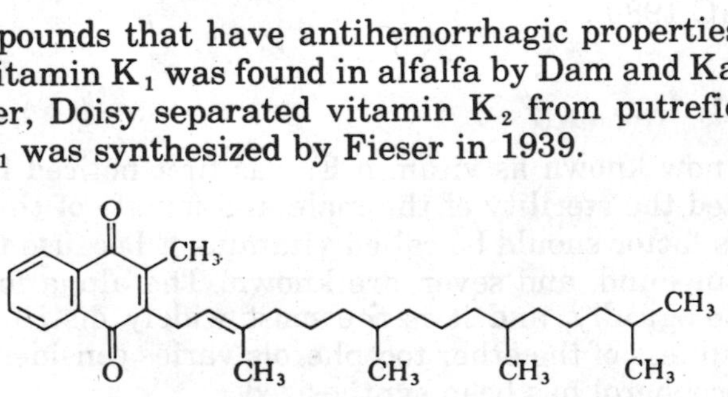

Vitamin K_1 (2-Methyl-3-phytyl-1,4-naphthoquinone)

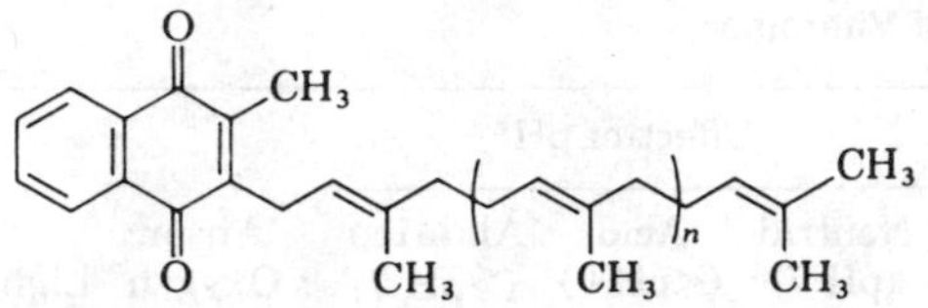

Vitamin K_2 (2-Methyl-3-difarnesyl-1,4-naphthoquinone) $n = 4, 6, 7$ or 8.

Other related compounds are known to have the biological activity of the K vitamins. Among these is menadione (vitamin K_3). Menadione (2-methyl-1,4-naphthoquinone) is a synthetic product and is as active as vitamin K_1.

Vitamin K_1 is a yellow viscous oil which is optically active, $[\alpha]_D^{20} = -0.4°$. It has a melting point of $-20°C$. It is insoluble in water, but soluble in alcohol and ether. Vitamin K_3 crystallizes as yellow needles from alcohol and petroleum ether. It has a melting point of 107°C. It is insoluble in water, slightly soluble in alcohol, and soluble in ether.

If the diet of the chick is deficient in vitamin K, the animal develops a hemorrhagic condition with lengthened clotting time of the blood. Deficiency of vitamin K in the human adult does not occur often because most diets supply sufficient amounts; it is normally produced in the intestine by bacteria as well. A deficiency can occur in newborn infants. This problem results in a hemorrhagic condition which ordinarily disappears when bacteria become established in the intestine. It can be relieved also by administration of vitamin K.

Vitamin K is rather widely distributed, and a sufficiency is obtained from the average diet. Excellent sources are alfalfa, cabbage, kale, spinach, and cauliflower. Cereals and fruits are poor sources.

VITAMINS AND MALNUTRITION

Nutritional science has shown that a balanced diet, even though not entirely adequate in quantity is better than one in which one factor is missing. Should a vitamin be a missing food factor, the deficiency disease with which it is associated will result.

Processing and Holding Losses

Nutritional losses as a result of the processing of foods are caused by one or more of several factors (Table 8.2). Among these are heat, light, the presence of oxygen or oxidizing substances, and pH, as well as enzymes and minute amounts of certain metals. In the case of pH, some nutrients are stable in an acid medium but unstable in an alkaline medium. For others the reverse is true. Thiamin is unstable in neutral or alkaline media, as well as to oxygen and heat. Vitamin C (ascorbic acid) is unstable in neutral and alkaline conditions, as well as to light,

TABLE 8.2. Stability of Vitamins[a]

Vitamins	Effect of pH			Air or Oxygen	Light	Heat	Max Cooking Losses (%)
	Neutral (pH 7)	Acid (<pH 7)	Alkaline (>pH 7)				
Vitamin A	S	U	S	U	U	U	40
Ascorbic acid (C)	U	S	U	U	U	U	100
Biotin	S	S	S	S	S	U	60
Carotene (pro-A)	S	U	S	U	U	U	30
Choline	S	S	S	U	S	S	5
Cyanocobalamin (B_{12})	S	S	S	U	U	S	10
Vitamin D	S		U	U	U	U	40
Folic Acid	U	U	S	U	U	U	100
Inositol	S	S	S	S	S	U	95
Vitamin K	S	U	U	S	U	S	5
Niacin (PP)	S	S	S	S	S	S	75
Pantothenic acid	S	U	U	S	S	U	50
p-Amino benzoic acid	S	S	S	U	S	S	5
Pyridoxine (B_6)	S	S	S	S	U	U	40
Riboflavin (B_2)	S	S	U	S	U	U	75
Thiamin (B_1)	U	S	U	U	S	U	80
Tocopherol (E)	S	S	S	U	U	U	55

Source: Adapted from Harris and Karmas (1975).
[a] S = stable (no important destruction), U = unstable (significant destruction).

heat, and to air (i.e., oxygen). Riboflavin is particularly unstable to light. This instability gets more pronounced as the pH and temperature rise. If milk is held in the sunlight, loss of riboflavin takes place rapidly, and as an effect of this loss, the oxidation of ascorbic acid takes place also. Physicochemical alterations, resulting from evaporation, are more prevalent in milk than nutrient losses in processing.

As far as processing is concerned, washing vegetables prior to blanching does not seem to affect materially such vitamins as ascorbic acid. However, the water-soluble vitamins undergo some losses during the process of blanching before canning or freezing. It is quite obvious that a leafy vegetable such as spinach, which exposes a great deal of surface, will lose more than those that are more compact. Steam blanching, in general, brings about smaller losses of water-soluble vitamins than water blanching because of the much smaller contact with water during the time the vegetable is being blanched. Ascorbic acid is better retained in vegetables if a short, high temperature blanch is employed. A patent was issued recently to Smith and Hanscom (1974) for the dry blanching of vegetables. The process makes use of a recirculating oxygen-free gaseous heating medium operating at >100°C with some steam to supply moisture. A blanching system of this sort should cut down many of the losses of water-soluble vitamins in such vegetables as spinach.

During the preparation of foods on a large scale for immediate consumption, a fair amount of the soluble nutrients may be washed out. These losses can be cut by the use of smaller amounts of cooking water, or by the use of steam, or pressure cooking. The actual amounts of the losses vary, from small amounts to very large losses. Ascorbic acid can vary from losses of about 10% to as much as 80%. Harris and von Loesecke (1960) and Harris and Karmas (1975) should be consulted for further information.

SUMMARY

Vitamins are organic compounds of varying composition which are essential for the maintenance of health. Only small amounts are necessary for this purpose. They exist in two groups, water soluble and fat soluble. Those vitamins that are not synthesized by the organism must be supplied in the diet.

Many of the coenzymes contain water-soluble vitamins. These compounds are necessary in metabolic pathways.

Thiamin (Vitamin B_1) is a part of thiamin pyrophosphate, which is a coenzyme necessary for the decarboxylation of α-keto acids such as pyruvic acid. Thiamin deficiency results in the disease known as beriberi.

Riboflavin (Vitamin B_2) is the prosthetic group of the "yellow enzyme" of yeast. It is a part of the coenzyme flavin mononucleotide (FMN) and a part of flavin adenine dinucleotide (FAD).

Niacin (Nicotinic acid) absent from the diet causes pellagra. Its absence also causes black tongue in the dog. Nicotinamide is a part of two acceptor pyridine protein enzymes known as nicotinamide adenine dinucleotide (NAD) and nicotinamide adenine dinucleotide phosphate (NADP). They are involved in electron transport.

Vitamin B_6 (Pyridoxine) is necessary for the formation of pyridoxal phosphate, which is a coenzyme. It is involved in many chemical reactions connected with amino acids. The vitamin seems to be necessary for the maintenance of health.

Pantothenic acid is important in animal nutrition. It is a part of the molecule of coenzyme A.

Vitamin B_{12} (Cyanocobalamin) is effective in the treatment of pernicious anemia. It contains cobalt and is made up of a corrin ring system. Several related compounds possess vitamin B_{12} activity. The cobamide coenzymes are known to contain deoxyadenosine.

Ascorbic acid (Vitamin C) deficiency leads to a disease called scurvy. Its acidic property is caused by the presence of the enediol group rather than a carboxyl group. Ascorbic acid has strong reducing properties. This was discovered before the compound itself was isolated.

Vitamin A (Retinol) is involved in vision. Although it is found in animals, it is derived from the carotenes, mainly β-carotene, found in plants. Carotenes must be protected from oxidation and light and are sensitive to autoxidation. Carotene is changed to vitamin A mainly in the intestine. Vitamin A is stored in the liver.

Vitamin D (Ergocalciferol). When ergosterol, a sterol in yeast and in ergot, is irradiated with ultraviolet light, vitamin D_2 results. Rickets, a disease which affects growing bone, is the result of deficiency of this vitamin.

Vitamin E. Tocopherol is the basic compound and seven are known. α-Tocopherol is the most active biologically. Vitamin E acts as a reversible antioxidant. A number of symptoms result from deficiency of vitamin E.

Vitamin K is the antihemorrhagic vitamin.

Nutritional losses as a result of processing of foods are caused by one or more of several factors. Light, heat, the presence of oxygen or oxidizing substances, and pH, as well as enzymes and minute amounts of metals. Thiamin is unstable in oxygen and heat and in neutral or alkaline media. Vitamin C is unstable in neutral or alkaline conditions as well as to light, heat, and to air (oxygen). Riboflavin is unstable to light.

There is some loss of the water-soluble vitamins during blanching of foods previous to canning or freezing. In general, steam blanching results in smaller losses because less water comes in contact with the foods during the processing.

BIBLIOGRAPHY

ANON. 1971. Fish require dietary vitamin C. Nutrition Rev. *29,* 207–210.

AOAC. 1980. Official Methods of Analysis, 13th Edition. Assoc. of Official Analytical Chemists, Washington, DC.

ASSOC. OF VITAMIN CHEMISTS. 1966. Methods of Vitamin Assay, 3rd Edition. John Wiley & Sons, New York.

AURAND, L. W., and WOODS, A. E. 1973. Food Chemistry. AVI Publishing Co., Westport, CT.

AXELROD, A. E., and MARTIN, C. J. 1961. Water-soluble vitamins. Part I. Ascorbic acid, biotin, inositol, nicotinic acid, pyridoxin group. Annu. Rev. Biochem. *30,* 383–408.

BARRON, E. S. G., DE MEIO, R. H., and KLEMPERER, F. 1936. Studies on biological oxidations. V. Copper and hemochromogens as catalysts for the oxidation of ascorbic acid. The mechanism of the oxidation. J. Biol. Chem. *36,* 625–640.

BAUERNFEIND, J. C., and PINKERT, D. M. 1970. Food processing with added ascorbic acid. *In* Advances in Food Research, Vol. 18, C. O. Chichester, E. M. Mrak, and G. F. Stewart (Editors). Academic Press, New York.

BAUMANN, C. A. 1953. Fat-soluble vitamins. Annu. Rev. Biochem. *22,* 527–544.

BESSEY, O. A., LOWE, H. J., and SALOMON, L. L. 1953. Water-soluble vitamins. Annu. Rev. Biochem. *22,* 545–628.

BLAXTER, K. L. 1957. The fat-soluble vitamins. Annu. Rev. Biochem. *26,* 275–306.

BORENSTEIN, B., and BUNNELL, R. H. 1966. Carotenoids: properties, occurrence, and utilization in foods. *In* Advances in Food Research, Vol. 15, C. O. Chichester, E. M. Mrak, and G. F. Stewart (Editors). Academic Press, New York.

BOYER, P. D. 1955. Fat-soluble vitamins. Annu. Rev. Biochem. *24,* 465–496.

BRIGGS, G. M., and DAFT, F. S. 1955. Water-soluble vitamins. Part I (vitamin B-12, folic acid, choline, and *p*-aminobenzoic acid). Annu. Rev. Biochem. *24,* 339–392.

BRINK, N. G. WOLF, D. E., KACZKA, E., RICKLES, E. L., KOMINSZY, F. R., WOOD, T. R., and FOLKERS, K. 1949. Vitamin B-12. IV. Further characterization of vitamin B-12. J. Am. Chem. Soc. *71,* 1854–1856.

BRO-RASMUSSEN, F., and HJARDE, W. 1961. Fat-soluble vitamins. Annu. Rev. Biochem. *30,* 447–472.

BROWN, G. M., and REYNOLDS, J. J. 1963. Biogenesis of the water-soluble vitamins. Annu. Rev. Biochem. *32,* 419–462.

BURNS, J. J. 1957. Missing step in man, monkey, and guinea pig required for the biosynthesis of L-ascorbic acid. Nature *180,* 553.

BURNS, J. J., PEYSER, P., and MOLTZ, A. 1956. Missing step in guinea pigs required for the biosynthesis of L-ascorbic acid. Science *124,* 1148–1149.

CHAPON, L., and URION, E. 1960. Ascorbic acid and beer. Wallerstein Lab. Commun. *23,* No. 80, 38–44.

CLIFCORN, L. E. 1948. Factors influencing the vitamin content of canned foods. *In* Advances in Food Research, Vol. 1, E. M. Mrak and G. F. Stewart (Editors). Academic Press, New York.

CLINE, J. K., WILLIAMS, R. R., and FINKELSTEIN, J. 1937. Studies of crystalline vitamin B-1. XVII. Synthesis of vitamin B-1. J. Am. Chem. Soc. *59,* 1052–1054.

DAVIDSON, S., PASSMORE, R., and BROCK, J. F. 1972. Human Nutrition and Dietetics, 5th Edition. Williams & Wilkins, Baltimore.

ELVEHJEM, C. A., MADDEN, R. J., STRONG, F. M., and WOOLLEY, D. W. 1937. Relation of nicotinic acid and nicotinic acid amide to canine black tongue. J. Am. Chem. Soc. *59,* 1767–1768.

FIESER, L. F. 1939. Identity of synthetic 2-methyl-3-phytyl-1,4-naphthoquinone and vitamin K. J. Am. Chem. Assoc. *61,* 2561.

FOLKERS, K., and WOLF, D. E. 1954. Chemistry of vitamin B-12. *In* Vitamins and Hormones, Vol. 12, R. S. Harris and K. V. Thimann (Editors). Academic Press, New York.

FRIED, R., and LARDY, H. 1955. Water-soluble vitamins. Part II. Vitamin B-6, niacin, biotin, ascorbic acid. Annu. Rev. Biochem. *24,* 393–418.

FRIEDEMANN, T. E., and FRAZIER, E. I. 1950. The determination of nicotinic acid. Arch. Biochem. *26,* 361–374.

GLOOR, U., and WISS, O. 1964. Fat-soluble vitamins. Annu. Rev. Biochem. *33,* 313–330.

GOLDBERGER, J. 1916. Pellagra causation and a method of prevention. J. Am. Med. Assoc. *66,* 471–476.

GOLDBERGER, J. 1922. Relation of diet to pellagra. J. Am. Med. Assoc. *78,* 1676–1680.

GOODWIN, T. W. 1963. The Biosynthesis of Vitamins and Related Compounds. Academic Press, New York.

GREENBERG, L. D. 1957. Water-soluble vitamins. Part II. Biotin, pyridoxin group, nicotinamide, ascorbic acid. Annu. Rev. Biochem. *26,* 209–242.

GROLLMAN, A. P., and LEHNINGER, A. L. 1957. Enzymic synthesis of L-ascorbic acid in different animal species. Arch. Biochem. Biophys. *69,* 458–467.

HARRIS, R. S. *et al.* (Editors). 1943–1982. Vitamins and Hormones, Vol. 1–39. Academic Press, New York.

HARRIS, R. S., and KARMAS, E. 1975. Nutrition Evaluation of Food Processing, 2nd Edition. AVI Publishing Co., Westport, CT.

HARRIS, R. S., and VON LOESECKE, H. 1960. Nutritional Evaluation of Food Processing. John Wiley & Sons, New York.

HARRIS, S. A., and FOLKERS, K. 1939. Synthesis of vitamin B-6. J. Am. Chem. Soc. *61,* 1245–1247, 3307–3310.

HENZE, M. 1934. The oxidation of homologous pyridines and quinolines with selenium dioxide. Chem. Ber. *67,* 750–757. (German)

JOHNSON, B. C. 1955. Water-soluble vitamins. Part III. Thiamine, lipoic acid, riboflavin, pantothenic acid, inositol, and miscellaneous factors. Annu. Rev. Biochem. *24,* 419–464.

KARRER, P., and MEERWEIN, H. F. 1936. An improved synthesis of lactoflavin and 6,7-dimethyl-9-(1′-arabityl)-isoalloxazine. Helv. Chim. Acta *19,* 264–269. (German)

KARRER, P., and MORF, R. 1931. Plant pigments. XXXV. The constitution of β-carotene and β-dihydrocarotene. Helv. Chim. Acta *14,* 1033–1036. (German)

KARRER, P., and MORF, R. 1933. Synthesis of perhydrovitamin A. Purification of the vitamin A preparations. Helv. Chim. Acta *16,* 625–641. (German)

KARRER, P., HELFENSTEIN, A., WEHRLI, H., and WETTSTEIN, A. 1930. Plant pigments. XXV. On the constitution of lycopene and carotene. Helv. Chim. Acta *13,* 1084–1099. (German)

KARRER, P., MORF, R., and SCHÖPP, K. 1931. To the knowledge of vitamin A from fish oils. II. Helv. Chim. Acta *14,* 1431–1436. (German)

KARRER, P., MORF, R., and SCHÖPP, K. 1933. Synthesis of perhydrovitamin A. Helv. Chim. Acta *16,* 557–561. (German)

KARRER, P., SCHÖPP, K., and BENZ, F. 1935. Synthesis of flavins. Helv. Chim. Acta *18,* 426–429. (German)

KARRER, P., BECKER, B., BENZ, F., FREI, P., SALOMON, A. A., and SCHÖPP, K. 1935. To the synthesis of lactoflavin. Helv. Chim. Acta *18,* 1435–1448. (German)

KODICEK, E. 1956. Fat-soluble vitamins. Annu. Rev. Biochem. *25,* 497–536.

KÖGL, F., and VAN HASSELT, W. 1936. Plant-growth substances. XXI. Isolation of bios I (meso-inositol) from yeast. Z. Physiol. Chem. *242,* 74–80. (German)

KÖGL, F., and TONNIS, B. 1936. Plant growth substances XX. On the bios problem. Isolation of crystalline biotin from egg yolk. Z. Physiol. Chem. *242,* 43–73. (German)

KUHN, R., and RUDY, H. 1935. On the optical activity of lactoflavin. Chem. Ber. *68,* 169–170. (German)

KUHN, R., and WEYGAND, F. 1935. 6,7-dimethyl-9-1-araboflavin. Chem. Ber. *68,* 166–169. (German)

LAMBOOY, J. P. 1951. Activity of 6,7-diethyl-9-(D-1′-ribityl)-isoalloxazine for *Lactobacillus casei.* J. Biol. Chem. *188,* 459–462.

LEE, F. A., BEATTIE, H. G., ROBINSON, W. B., and PEDERSON, C. S. 1947. The preservation of apple juice by freezing. Fruit Prods. J. *26,* 324–326.

MARKACHEVA, T. M., and LEBEDEVA, G. N. 1950. The production of nicotinic acid from heavy bases of coal tar. J. Appl. Chem. USSR *23,* 313–319. Chem. Abstr. *45,* 1593g, 7571g (1951).

MAURER, K., and SCHIEDT, B. 1933. The preparation from glucose of an acid, $C_6H_8O_6$ which in its reducing power resembles ascorbic acid (preliminary communication). Ber. *66,* 1054–1057.

McELVAIN, S. M., and GOESE, M. A. 1941. The preparation of nicotinic acid from pyridine. J. Am. Chem. Soc. *63,* 2283–2284.

MERCK AND COMPANY. 1954. Vitamin B-6. Merck Service Bulletin, Merck and Co., Rahway, N.J.

MERCK AND COMPANY. 1955. Vitamin B-1. Merck Service Bulletin, Merck and Co., Rahway, N.J.

MERCK AND COMPANY. 1956. Vitamin C. Merck Service Bulletin, Merck and Co., Rahway, N.J.

MERCK AND COMPANY. 1958. Vitamin B-12. Merck Service Bulletin, Merck and Co., Rahway, N.J.

MISTRY, S. P., and DAKSHINAMURTI, K. 1964. Biochemistry of biotin. *In* Vitamins and Hormones, Vol. 22, R. S. Harris and I. G. Wool (Editors). Academic Press, New York.

MOORE, T. 1950. Fat-soluble vitamins. Annu. Rev. Biochem. *19,* 319–338.

NATL. ACAD. SCI.—NATL. RES. COUNCIL. 1980. Recommended Daily Dietary Allowances, 9th Edition. Food and Nutrition Board, National Academy of Sciences—National Research Council, Washington, DC.

NOMENCLATURE POLICY: 1980. Generic Description and Trival Names for Vitamins and Related Compounds. J. Nutr. *110,* 8–15.

NOVELLI, G. D. 1957. Water-soluble vitamins. Part III. Pantothenic acid, thiamine, lipoic acid (thioctic acid), riboflavin, and inositol. Annu. Rev. Biochem. *26,* 243–274.

PFIFFNER, J. J., and BIRD, O. D. 1956. Water-soluble vitamins. Part I. vitamin B-12, folic acid, choline, and *p* amino-benzoic acid. Annu. Rev. Biochem. *25,* 397–434.

PITT, G. A. J., and MORTON, R. A. 1962. Fat-soluble vitamins. Annu. Rev. Biochem. *31,* 491–514.

PLAUT, G. W. E. 1961. Water-soluble vitamins. Part II. Folic acid, riboflavin, thiamine, vitamin B-12. Annu. Rev. Biochem. *30,* 409–446.

PLAUT, G. W. E., and BETHEIL, J. J. 1956. Water-soluble vitamins. Part III. Pantothenic acid, inositol, riboflavin, thiamine, lipoic acid (thioctic acid), unidentified factors. Annu. Rev. Biochem. *25,* 563–596.

REICHSTEIN, T. 1934. *l*-Adonose (*l*-erythro-2-keto-pentose). Helv. Chim. Acta *17,* 996–1002. (German)

REICHSTEIN, T., and GRÜSSNER, A. 1934. A high-yield synthesis of *l*-ascorbic acid (vitamin C). Helv. Chim. Acta, *17,* 311–328. (German)

REICHSTEIN, T., GRÜSSNER, A., and OPPENAUER, R. 1933. Synthesis of *d* and *l*-ascorbic acid (vitamin C). Helv. Chim. Acta *16,* 1019–1033. (German)

RICKES, E. L., BRINK, N. G., KONIUSZY, F. R., WOOD, T. R., and FOLKERS, K. 1948. Crystalline vitamin B_{12}. Science *107,* 396–397.

ROY, R. N., and GUHA, B. C. 1958. Species difference in regard to the biosynthesis of ascorbic acid. Nature *182,* 319–320.

SMITH, E. L. 1948. Purification of anti-pernicious anemia factors from liver. Nature *161,* 638–639.

SMITH, T. J., and HANSCOM, G. I. 1974. Dry blanching process. U.S. Pat. 3,801,715.

SNELL, E. E., and METZLER, D. E. 1956. Water-soluble vitamins. Part II. Vitamin B-6, nicotinic acid, biotin, ascorbic acid. Annu. Rev. Biochem. *25,* 435–462.

SNELL, E. E., and WRIGHT, L. D. 1950. The water-soluble vitamins. Annu. Rev. Biochem. *19,* 277–318.

STEENBOCK, H. 1924. The induction of growth promoting and calcifying properties in a ration by exposure to light. Science *60,* 224–225.

STEENBOCK, H. and BLACK, A. 1924. Fat Soluble Vitamins. XVII. The induction of grwoth promoting and calcifying properties in a ration by exposure to light. J. Biol. Chem. *61,* 405–422.

STILLER, E. T., HARRIS, S. A., FINKELSTEIN, J., KERESZTESY, J. P., and FOLKERS, K. 1940. Pantothenic acid. VIII. The total synthesis of pure pantothenic acid. J. Am. Chem. Soc. *62,* 1785–1790.

STOKSTAD, E. L. R. 1962. The biochemistry of the water-soluble vitamins. Annu. Rev. Biochem. *31,* 451–490.

STROHECKER, R., and HENNING, H. M. 1965. Vitamin Assay—Tested Methods. Verlag Chemie, Weinheim/Bergstr., Germany.

THEORELL, H. 1935. The yellow oxidation enzyme. Biochem. Z. *278,* 263–290. (German)

TISHLER, M., PFISTER, K., III, BABSON, R. O., LADENBURG, K., and FLEMING, A. J. 1947. The reaction between *o*-amino-azo compounds and barbituric acid. A new synthesis of riboflavin. J. Am. Chem. Soc. *69,* 1487–1492.

TOTTER, J. R. 1957. Water-soluble vitamins. Part I. Vitamin B-12, folic acid, choline, and *p*-aminobenzoic acid. Annu. Rev. Biochem. *26,* 181–208.

UL HASSAN, M., and LEHNINGER, A. L. 1956. Enzymatic formation of ascorbic acid in rat liver extracts. J. Biol. Chem. *223,* 123–138.

VON EULER, H. V. *et al.* 1935. Synthesis of lactoflavin (vitamin B-2) and other flavins. Helv. Chim. Acta *18,* 522–535.

WAGNER, F. 1966. Vitamin B-12 and related compounds. Annu. Rev. Biochem. *35,* 405–434.

WARBURG, O., and CHRISTIAN, W. 1932. On the new oxidation enzyme and its absorption spectrum. Biochem. Z. *254,* 438–458. (German)

WARBURG, O., and CHRISTIAN, W. 1938. Isolation of the prosthetic group of the *d*-amino acid oxidase. Biochem. Z. *298,* 150–168. (German)

WARBURG, O., and CHRISTIAN, W. 1938. Observations on the yellow enzyme. Biochem. Z. *298,* 368–377. (German)

WARREN, F. L. 1943. Aerobic oxidation of aromatic hydrocarbons in the presence of ascorbic acid. Biochem. J. *37,* 338–341.

WESTHEIMER, F. H., FISHER, H. F., CONN, E. E., and VENNESLAND, B.

1951. The enzymatic transfer of hydrogen from alcohol to DPN. J. Am. Chem. Soc. *73,* 2403.

WILLIAMS, R. J. 1939. Pantothenic acid—a vitamin. Science *89,* 486.

WILLIAMS, R. J., LYMAN, C. M., GOODYEAR, G. N., TRUESDAIL, N. J., and HOLADAY, D. 1933. Pantothenic acid, a growth determinant of universal biological occurrence. J. Am. Chem. Soc. *55,* 2912–2927.

WILLIAMS, R. J., MITCHELL, H. K., WEINSTOCK, H. H., JR., and SNELL, E. E. 1940. Pantothenic acid VII. Partial and total synthesis studies. J. Am. Chem. Soc. *62,* 1784–1785.

WILLIAMS, R. J., EAKIN, R. E., BEERSTECHER, E., JR., and SHIVE, W. 1950. The Biochemistry of B Vitamins. ACS Monogr. No. 110. Reinhold Publishing Corp., New York.

WILLIAMS, R. R., and CLINE, J. K. 1936. Synthesis of vitamin B_1. J. Am. Chem. Soc. *58,* 1504–1505.

WINDAUS, A., TSCHESCHE, R., RUHKOPF, H., LAGUER, F., and SCHYLTZ, F. 1932. The preparation of crystalline antineuritic vitamin from yeast. Preliminary communication. Z. Physiol. Chem. *204,* 123–128.

WISANSKY, W. A., and ANSBACHER, S. 1941. Preparation of 3,4-dimethyl-aniline. J. Am. Chem. Soc. *63,* 2532.

WOHL, M. G., and GOODHART, R. S. 1968. Modern Nutrition in Health and Disease. Lee and Febiger, Philadelphia.

WOOLLEY, D. W., WASIMAN, H. A., and ELVEHJEM, C. A. 1939. Nature and partial synthesis of the chick antidermatitis factor. J. Am. Chem. Soc. *61,* 977–978.

9

Minerals

Occurrence of Minerals
Anionic Minerals
Minerals in Canned Foods
Summary
Bibliography

Mineral substances play an important part in the nutritional value of foods. They are usually present in small amounts, frequently in traces, but their importance is well recognized. Minerals are present in the form of salts of metals or in combination with organic compounds such as phosphoproteins and enzymes containing metals.

The major elements include potassium, sodium, calcium, magnesium, chlorine, sulfur, and phosphorus. Trace elements include iron, copper, iodine, cobalt, fluorine, and zinc. The role of selenium in the human is not completely established. Others of unknown nutritive value include aluminum, boron, chromium, nickel, and tin. Toxic elements include arsenic, cadmium, mercury, lead, and antimony. Food materials are the important source of the major and minor elements. For recommended daily allowances see Table 8.1.

A number of the elements present in foods in very small amounts have no known nutritional function, but it is possible that some or many of them will be found to be significantly useful. Not long ago several metals, present in very small amounts, were found to be of importance physiologically. These include Co, Cu, Mn, and Zn. Mineral materials contribute to structure and are of value physiologically. Table 9.1 shows some of the minerals found in a variety of foods.

In the past, knowledge concerning such elements present in very small quantities was limited by the fact that satisfactory analytical methods for the detection and determination of these elements were lacking. New methods such as spark-emission spectroscopy, neutron activation, flame photometry, and the very important atomic absorption spectroscopy have done much to assist and make possible the study of these substances.

Metal ions become involved in nutrition as a result of their ability to associate with various molecules. Such molecules are called ligands. Metals vary in thier ability to form complexes, called chelates. A

TABLE 9.1. Minerals in Representative Foods[a]

Food	Ash (g)	Calcium (mg)	Magnesium (mg)	Phosphorus (mg)	Iron (mg)	Sodium (mg)	Potassium (mg)
Almonds, dried	3.0	234	270	504	4.7	4	733
Apricots, raw	0.7	17	12	23	0.5	1	281
Bananas, raw common	0.9	8	33	26	0.7		533
Beans, lima immature seeds boiled and drained	1.0	47	62	121	2.5	1	422
Beef, sirloin broiled	1.1	10	21	191	2.9		
Cabbage, raw	0.7	49	13	29	0.4	20	233
Carrots, raw	0.8	37	23	36	0.7	47	341
Chicken, fryer flesh only, fried	1.2	13		257	1.6	78	381
Lemon Juice, raw	0.3	7	8	10	0.2	1	141
Lettuce, iceberg raw	0.6	20	11	22	0.5	9	175
Macaroni, dry enriched	0.7	27	48	162	2.9	2	197
Ocean perch, Atlantic, fried	1.9	33		226	1.3	153	284
Oranges, peeled raw	0.6	41	11	20	0.4	1	200
Peanuts, roasted with skins	2.7	72	175	407	2.2	5	701
Peas, green raw, immature	0.9	26	35	116	1.9	2	316
Potatoes, raw	0.9	76	34	53	0.6	3	407

Source: Watt and Merrill (1963).
[a] In 100 g, edible portions.

chelate is a particular coordination compound in which a metal or other central atom is connected with two or more other atoms of one or more different ligands with the result that one or more heterocyclic rings are formed with the central atom as part of each ring. The derivation of the word chelate is a Greek word meaning lobster's claw. The reason is that two or more bonds hold the metal to the ligand. While four-membered chelate rings are known, five- and six-membered rings are the most stable. If the ligand is relatively more basic, the chelate is more stable.

Relatively high levels of organic acids are found in fruits. These may combine with metal to form chelates (Fig. 9.1). Vitamin B_{12}, hemoglobin, and myoglobin are metal chelates. Metals play an important role in a number of physiological and biochemical reactions. One such reaction is that in which 2-phosphoryl-D-glycerate undergoes dehydration. An enolase in the presence of either Mg^{2+} or Mn^{2+} forms phosphoenolpyruvate. This reaction is reversible, and it is one of the reactions in alcoholic fermentation.

Another is that of phosphorylenolpyruvate in the presence of phosphoenolpyruvic transphosphorylase and Mg^{2+} and K^{+} to form pyruvate. In this reaction ADP is changed to ATP. The Mg^{2+} is essential for the enzyme to work. It is inhibited by Ca^{2+}.

Still another is the reaction of hexokinase on fructose, also in alcoholic fermentation. In this case, fructose is acted on by hexokinase in the presence of Mg^{2+} and ATP to yield fructose 6-phosphate plus ADP.

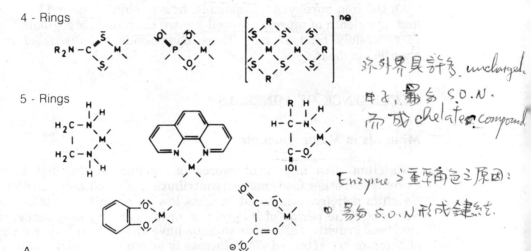

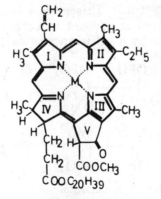

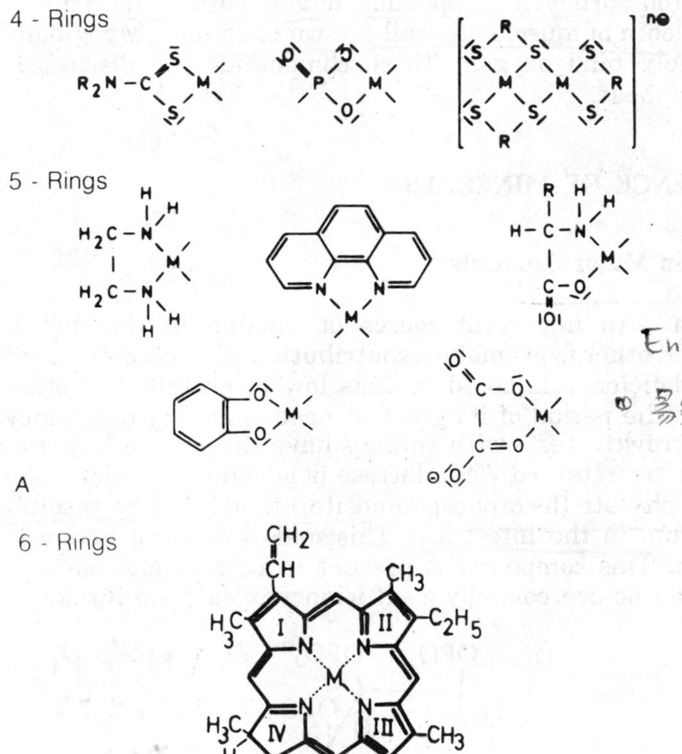

FIG. 9.1. Examples of metal chelates. Only the applicable parts of the molecules are shown. The chelate formers are (A): thiocarbamate, phosphate, thioacid; diamine, *o*-phenantrolin, α-amino acid, *o*-diphenol, oxalic acid; (B) chlorophyll *a*.

From Pfeilsticker (1970). Used with permission of Forster Publishing Co., copyright owner.

In order for the body to use glucose the glucose enters the blood and from there, hexokinase converts it into glucose 6-phosphate. This reaction also takes place in alcoholic fermentation.

Metalloproteins containing the more common as well as the trace metals are widespread. The enzymes ascorbic acid oxidase, δ-aminolevulinic acid dehydrase, tyrosinase, dopamine hydroxylase, and laccase need the presence of Cu^{2+} in order to function.

Carboxypeptidase A is a zinc-containing metalloenzyme that hydrolyzes a peptide bond adjacent to the terminal free carboxyl group.

Chlorophyll is a metalloporphyrin compound which contains the ion Mg^{2+}. It is present in green plants. See Fig. 9.1.

Of the iron porphyrin compounds, hemoglobin of the red blood cells, and myoglobin of muscle are well-known examples. Myoglobin is able to reversibly bind oxygen. These compounds are discussed in the chapter on meat.

OCCURRENCE OF MINERALS

Minerals in Major Amounts

Calcium. An important source of calcium in the diet is milk. Cereals and other foods make a contribution of calcium to the diet also. Calcium deficiency is noted in diets low in protein and other major nutrients. The period of its greatest need is during pregnancy, lactation, and growth. Tests with animals have shown that larger amounts of calcium are retained when lactose is added to the diet.

Sodium phytate (hexaphosphoinositol) tends to form insoluble salts with calcium in the intestines. This makes calcium unavailable for absorption. This compound is present in cereals and soybeans. The problem can be overcome by a sufficiency of calcium intake.

Hexaphosphoinositol

Calcium salt in small amounts is frequently added to tomatoes during the canning process to increase the firmness. The compound formed is calcium pectate (Loconti and Kertesz 1941).

Magnesium. All plants and animals require magnesium. Chlorophyll supplies the largest amount of magnesium in the adult diet.

Infants are given supplementary amounts of magnesium because milk does not have an adequate supply.

Magnesium is important to cardiovascular function, including involvement in blood pressure, myocardial contraction, and myocardial conduction and rhythm. It is important in oxidative phosphorylation, enolase, and glucokinase.

Sodium. Sodium chloride as well as sodium present in foods of animal and vegetable origin is the source of this element in the diet. Water supplies contain considerable amounts of sodium but not as the chloride.

Salt is important in the diet. However, it is thought by some that the average intake of salt in the United States is too high. Sodium is important in the preservation of pH of the body fluid and in body fluid volume.

Potassium. Potassium is present in plant materials and ingestion of these plant materials is a source of potassium in the diet. Potassium is important in the maintenance of the pH of body fluids, as well as in water and electrolyte balance in these fluids. K^+ has a role in enzyme systems.

The infant protein malnutrition condition known as kwashiorkor has been found to affect the K^+ levels. This involves a deficiency of dietary amino acids. The result is disturbances of electrolyte and water balances and a loss of potassium from the cells. Such a loss is made worse by the diarrhea which accompanies it.

Under experimental conditions rats fed diets deficient in potassium showed slow growth with renal hypertrophy and with other symptoms. The final result was death.

Minerals in Minor Amounts

Trace elements are found everywhere and are present in all natural foods. Under ordinary circumstances the quantity of a trace element in a given food is related to the amount of this element found in the environment. It is obvious that these elements arise from the soil in the case of plants and from feed in the case of animals. These elements can also get into foods from the processing equipment. Some animals such as swordfish have the capability to absorb mercury from the sea water. Such absorption has made the flesh of this species undesirable for food when the mercury level is too high.

Iron. Foods which supply large amounts of iron include liver, kidney, and heart, as well as egg yolk, shellfish, cocoa, parsley, and molasses. Muscle meats, green vegetables, nuts, poultry, wholewheat four, and bread supply smaller amounts.

Iron is ingested ordinarily in the form of iron porphyrin or iron–protein complexes. The use of iron salts in the diet has long been practiced for the beneficial effect. The hemoglobin of red cells, cytochromes of respiration, and myoglobin of heart muscle require iron for normal operation. Iron is also involved in the activity of the enzymes catalase and peroxidase. The complexing of iron by the porphyrin nucleus concerns the bulk of these situations. Increased assimilation of iron is helped by diets high in ascorbic acid and low in phosphates.

Copper. It has been only during recent years that copper has been found to be involved in human normal and abnormal physiology. Other metals and anions influence fluid retention and metabolism. In addition, copper is a component of a number of enzymes. A dietary balance of copper and other minerals is necessary.

Good dietary sources of copper include liver, kidney, shellfish, including oysters and crustaceans, nuts, cocoa, peaches, and grapes. Fresh fruits, nonleafy vegetables, and refined cereals contain small amounts of copper.

Cobalt. Vitamin B_{12} contains cobalt. This is discussed in Chapter 9 Vitamins. This vitamin is primarily connected with pernicious anemia. A patient suffering from this disease fails to absorb the vitamin from the intestinal tract. It is not just dietary inadequacy. The cause is the lack of an intrinsic factor in the gastric juice. This factor is a mucoprotein which binds with a molecule of vitamin B_{12}. It can then be used by the body. Vitamin B_{12} is not synthesized by the body, but is of microbial origin. Vitamin B_{12} is found in protein foods of animal origin of which liver and kidney are the best sources.

Manganese. The role of manganese in essential enzyme reactions is well known. Enzymes which require Mg^{2+} can usually act with Mn^{2+}. Pyruvate kinase functions with either Mg^{2+} or Mn^{2+}. It forms a complex with either of these metals before it binds with the substrate. A deficiency of this metal has not been known in humans. Liver, heart, and kidney contain relatively high amounts of manganese. Nuts and cereals are highest. Meranger and Somers (1968) give much information on the manganese values of seafoods. Other metals are included in this report also.

Zinc. A fair amount of zinc is present in the human body. It is involved in the healing of wounds, and there seems to be a connection between the levels of zinc and the production and action of insulin. It seems likely that zinc is essential for a large number of physiological processes.

It is present in some enzymes. Alcohol dehydrogenase of yeast and glutamic acid dehydrogenase of beef liver contain zinc in the protein molecule.

Bran and wheat germ, as well as oysters, are high in zinc. The former contain 40–120 $\mu g/g$, while oysters contain 1000 $\mu g/g$. White bread, meat, fish, leafy vegetables, nuts, eggs, and whole cereals contain moderate amounts of zinc.

Acid foods should not be stored in galvanized containers because it is possible to dissolve enough zinc to cause poisoning.

Molybdenum. Molybdenum seems to have an effect on dental caries. It was found that molybdenum plus fluoride is more effective than fluoride alone.

The metabolic enzyme xanthine oxidase contains molybdenum in the molecule. It catalyzes the oxidation of xanthine to uric acid.

Molybdenum is found in large amounts in liver, kidney, legumes, leafy vegetables, and cereal grains. Cereal grains contain from 0.12 to 1.114 μg of molybdenum while dried legume seeds contain 0.2–4.7 $\mu g/g$. Fruits, meats, root and stem vegetables, and milk are poorest in this metal.

ANIONIC MINERALS

Of this group, chlorides, phosphates, and sulfates are found in the largest quantities. Selenium and iodine are present in minor amounts. They are supplied in the diet in the form of various salts. In addition, phosphates are contributed by the sugar phosphates such as glucose 1-phosphate, fructose 6-phosphate and others. Further, the nucleotides which include ATP and others supply phosphorus.

Sulfur is available from both organic and inorganic sources. Sulfur-containing amino acids and proteins supply much of it. Most of the organic sulfur comes from such compounds as glutathione, cysteine, cystine, methionine, chondroitin sulfate, and the bile salts. Glutathione is a tripeptide. It is γ-glutamyl-cysteinylglycine.

Chloride. Chloride is widely distributed and as a result there is no lack of it in the diet. Sodium chloride is the best source of chloride.

Chloride is important in maintaining the water and electrolyte balance, and maintaining the pH of extracellular fluids and HCl in the gastric juice. Vegetables and fruits contain less Cl^- than do meats and animal products.

Iodine. This element is important because it is part of the thyroxine molecule. This and other derivatives are found in the thyroid gland.

$$HO-C_6H_2I_2-O-C_6H_2I_2-CH_2-CH(NH_3^+)-COO^-$$

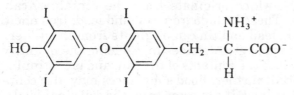

Thyroxine (3,5,3′,5′-tetraiodothyronine)

Insufficient intake of iodine produces the condition known as simple goiter. This illness results in an enlargement of the thyroid gland. It is controlled by the addition of small amounts of iodine in food, usually by iodized salt.

Phosphorus. Phosphorus is contained in both inorganic and organic compounds. The former are phosphoric acid compounds while the latter include the sugar phosphates and nucleotides. Among the sugar phosphates are compounds such as glucose 6-phosphate, fructose 6-phosphate, and glucose 1-phosphate. Among the nucleotides are included ADP and ATP.

Phytates are found in soybeans and about 70–80% of the phosphorus in soybeans is present as phytic acid. This compound is mentioned earlier in this chapter. It occurs in foods which are of plant origin.

Phosphorus is highly important to bone and tooth formation. A low intake of phosphorus interferes with calcification with an effect on both teeth and bones. This situation is known as rickets. Treatment involves the use of more vitamin D in the diet.

Selenium. Animal studies have shown that selenium protects against liver necrosis. However, work on the value of selenium in the human is not complete. Animal tests have shown that selenium and vitamin E have some connection in a number of disorders.

It is quite probable that a simple role for selenium does not exist. Kiermeyer and Wigand (1969) found loss of some selenium in drying milk.

Fluorine. This element is of great importance in reducing dental caries. When used in small amounts it tends to prevent this problem. However, if large amounts are ingested, mottled teeth are the result. It was stated before that this effect is strengthened by the use of molybdenum together with fluorine. Fluorine is now used safely in municipal water systems.

MINERALS IN CANNED FOODS

It is known that foods packed in cans may cause internal corrosion of the cans with the result that metals released will appear in the food. Three metals, which originate from the container, can be found in canned foods. These include iron, tin, and lead. Iron and tin come from the can, while lead and tin can originate from the solder. The canning industry has made great strides in reducing the quantity of lead that finds its way into the contents of the can, and is continuing work in this area to reduce it further. Food substances may also contain some lead before being canned. It can come from the soil since lead is found in the crust of the earth, in crop application of municipal sewage sludge, from industrial sources, and as fallout on crops from automobile exhausts. In addition to lead, municipal sewage can contain cadmium, another undesirable metal. Obviously, metals from these sources will be present in fresh vegetables and fruits.

Some metal contaminants of canned food can come from the processing equipment. The release of such metals as chromium and nickel from stainless steel surfaces of equipment used in the processing of foods depends on a number of factors such as the presence of organic chelating components of the food product, its pH, the area of metal exposed, its temperature, length of time of contact, the amount of agitation, and the particular stainless steel alloy used in the equipment. It has been found that soluble chromium salts orally administered to animals resulted in complete and rapid excretion. Nickel also has been found to be largely excreted after oral administration (Conn *et al.* 1932; Stoewsand *et al.* 1979).

During the manufacture of cans, a layer of tin forms a layer of alloy with the iron, also known as the iron–tin couple. Cans formerly were made by the hot dipped tinplate; today they are coated by the electrolytic process.

When food is preserved in this type of can, almost all of the oxygen is eliminated during the canning process. In a process called detinning the tin coating of the can erodes slowly, with the resulting shelf life of about 2 years, depending on the nature of the food canned. The tin very slowly goes into solution and hydrogen gas is very slowly evolved. This process protects the steel base of the can by electrochemical activity. Normal detinning takes place when the tin is just anodic enough so that any still-exposed steel, the result of discontinuities in the tin coating, is protected.

Internal corrosion can cause several types of can failure. One of these is the rapid production of hydrogen gas, which, in turn, causes some swelling of the can. Another type of failure is the perforation of the end or body of the can. While lacquer provides protection under many conditions, faults in the lacquer such as minor scratches and pores can permit contact of the packed food with the underlying metal.

Staining in canned foods is another problem. It is caused by the formation of iron sulfide (FeS) and tin sulfide (SnS) immediately after processing. The sulfide can come from the sulfur-containing amino acids, cysteine, cystine, and methionine in foods.

Other problems concerning discoloration in canned vegetables may occur. For further information see Piggott and Dollar (1963), Van Buren and Dowining (1969), and Lueck (1970) and other references listed in the Bibliography at the end of this chapter.

SUMMARY

Minerals are usually present in small amounts in foods, but their importance is well recognized. They occur in major and trace amounts.

A number of elements found in foods have no known nutritional value, and some are toxic.

The knowledge concerning the elements present in small amounts, formerly limited by the lack of satisfactory analytical methods, has been greatly increased by modern analytical methods.

Some metals are present in foods in the form of chelates. In such complexes the metal is combined with one or more molecules or atoms which are called ligands. Vitamin B_{12}, hemoglobin, and myoglobin are metal chelates.

A number of enzymes require the presence of such metallic ions as K^+, Mg^{2+}, and Mn^{2+} in order to function. A number of the enzyme reactions in alcoholic fermentation require metal ions of this sort.

Chlorophyll is a metalloporphyrin compound which contains the metal Mg^{2+}. Hemoglobin and myoglobin are iron porphyrins.

The major metals, Ca, Mg, Na, and K are discussed. The trace metals which are of importance in nutrition also are discussed. These include Fe, Cu, Co, Mn, Zn, and Mo.

The anionic elements are discussed. They include Cl, P, I, S, F, and Se.

Minerals in canned foods are given consideration. The sources of these minerals are those present in the raw material before being packed, those which come from the processing equipment, and those which come from the metals of the cans. Iron, tin, and lead come from the cans.

Normal detinning of the can permits an average 2 year shelflife of the can. Internal can corrosion can produce several types of can failure. Among these are swelling and perforation of the can. Another problem is the production of stain. The source of the sulfur is discussed.

BIBLIOGRAPHY

ALBU-YARON, A., and SEMEL, A. 1976. Nitrate-induced corrosion in tinplate as affected by organic acid food components. J. Agric. Food Chem. *24,* 344–348.

BAKAL, A., and MANNHEIM, H. C. 1966. The influence of processing variants of grapefruit juice on the rate of can corrosion and product quality. Isr. J. Technol. *4,* 262–267.

BJERRUM, B., SCHWARZENBACH, G., and SILLEN, L. G. 1957. *In* "Stability Constants Part I. Organic Ligands" Spec. Publ. 6, p. vii. The Chemical Society, London.

BRITTON, S. C., and BRIGHT, K. 1957. An examination of oxide films on tin and tinplate. Metallurgia *56,* 163–168.

CONN, L. W., WEBSTER, H. L., and JOHNSON, A. H. 1932. Chromium Toxicology. Absorption of chromium by the rat when milk containing chromium was fed. Am. J. Hyg. *15,* 760–765.

FISHMAN, M. J., and MIDGETT, M. 1968. Extraction techniques for the determination of cobalt, nickel, and lead in fresh water by atomic absorption. *In* Trace Inorganics in Water. R. A. Baker (Editor) American Chemical Society, Washington, DC.

GOMEZ, M. I., and MARKAKIS, P. 1974. Mercury content of some foods. J. Food Sci. *39,* 673–675.

HARTWELL, R. R. 1951. Certain aspects of internal corrosion in tin plate containers. Adv. Food Res. *3,* 327–383.

HOAR, T. P. 1934. The electrochemical behavior of the tin-iron couple in dilute acid media. Trans. Faraday Soc. *30,* 472–482.

KAMM, G. G., and WILLEY, A. R. 1961. Corrosion resistance of electrolytic tin plate I. Electrochemical studies of tin, iron-tin alloy, and steel in air-free acid media. Corrosion (Houston) *17* (2), 77t–84t.

KAMM, G. G., WILLEY, A. R., BEESE, R. E., and KRICKL, J. L. 1961. Corrosion resistance of electrolytic tin plate Part II. The alloy-tin couple test—A new research tool. Corrosion (Houston) *17* (2), 84t–92t.

KIERMEIER, F., and WIGAND, W. 1969. Selenium content of milk and milk powder. Z. Lebensm. Unters. Forsch. *139,* 205–211. (German)

LAMBETH, V. N., FIELDS, M. L., BROWN, J. R., REGAN, W. S., and BLEVINS, D. G. 1969. Detinning by canned spinach as related to oxalic acid, nitrates, and mineral mineral composition. Food Technol. *23,* 840–842.

LOCONTI, J. D., and KERTESZ, Z. I. 1941. Identification of calcium pectate as the tissue-firming compound formed by treatment of tomatoes with calcium chloride. Food Res. *6,* 499–508.

LUECK, R. H. 1970. Black discoloration in canned asparagus. Interrelations of iron, tin, oxygen, and rutin. Agric. Food Chem. *18,* 607–612.

McKIRAHAN, R. D., CONNELL, J. C., and HOTCHNER, S. J. 1959. Application of differentially coated tin plate for food containers. Food Technol. *13,* 228–232.

MERANGER, J. C., and SOMERS, E. 1968. Determination of the heavy metal content of sea foods by atomic absorption spectrophotometry. Bull. Environ. Contam. Toxicol. *3,* 360–365.

MERKEL, R. A. 1971. Inorganic constituents. *In* The Science of Meat and Meat Products. J. F. Price and B. S. Schweigert (Editors). W. H. Freeman and Co., San Francisco.

MONTY, K. J., and McELROY, W. D. 1959. The trace elements. *In* Food, The Yearbook of Agriculture. A. Stefferud (Editor) U.S. Department of Agriculture, Washington, DC.

MORRIS, V. C., and LEVANDER, O. A. 1970. Selenium content of foods. J. Nutr. *100,* 1383–1388.

MORRIS, V. H., PASCOE, E. D., and ALEXANDER, T. L. 1945. Studies on the composition of the wheat kernel. II. Distribution of certain inorganic elements in center sections. Cereal Chem. *22,* 361–372.

PFEILSTICKER, K. 1970. Food components as metal chelates. Lebensm. Wiss. Technol. *3,* 45–51 (German)

PIGGOTT, G. M., and DOLLAR, A. M. 1963. Iron sulfide blackening in canned protein foods: Oxidation and reduction mechanisms in relation to sulfur and iron. Food Technol. *17,* 481–484.

PRICE, W. J., and ROOS, J. T. H. 1969. Analysis of fruit juice by atomic absorption spectrophotometry. I. The determination of iron and tin in canned juice. J. Sci. Food Agric. *20,* 437–439.

SCHRENK, W. G. 1964. Minerals in wheat grain. Tech. Bull. 136. Kansas State University Agric. Exp. Station, Manhattan, Kansas.

SCHWEIGER, R. G. 1966. Metal chelates of pectates and comparison with alginate. Kolloid. Z. *208,* 28–31.

SEILER, B. S. 1968. The mechanism of sulfide staining in tin foodpacks. Food Technol. *22,* 1425–1429.

SHERLOCK, J. C., and BRITTON, S. C. 1972. Complex formation and corrosion rate for tin in fruit acids. Br. Corros. J. *7,* 180–183.

SHERLOCK, J. C., HANCOX, J. H., and BRITTON, S. C. 1972. Rate of dissolution of tin from tin plate in oxygen-free citrate solutions. Part I. Assessment of polarisation measurements. Br. Corros. J. *7,* 222–226.

SHERLOCK, J. C., HANCOX, J. H., and BRITTON, S. C. 1972. Rate of dissolution of tin from tinplate in oxygen-free citrate solutions. II. Effect of coating porosity. Br. Corros. J. *7,* 227–231.

STEVENSON, C. A., and WILSON, C. H. 1968. Nitrogen closure of canned apple sauce. Food Technol. *22,* 1143–1145.

STOEWSAND, G. S., STAMER, J. R., KOSIKOWSKI, F. V., MORSE, R. A., BACHE, C. A. and LISK, D. J. 1979. Chromium and nickel in acidic foods and by-products contacting stainless steel during processing. Bull. Environ. Contam. Toxicol. *21,* 600–603.

VAN BUREN, J. P., and DOWNING, D. L. 1969. Can characteristics, metal additives, and chelating agents. Effect on the color of canned wax beans. Food Technol. *23,* 800–802.

VANDER MERWE, H. B., and KNOCK, G. G. 1968. In-can shelf-life of tomato paste as affected by tomato variety and maturity. J. Food Technol. *3,* 249–262.

WARREN, H. V. 1972. Geology and medicine. West. Miner Sept. 34–37.

WATT, B. K. and MERRILL, A. L. 1963. Composition of Foods. USDA Handbook No. 8. Revised ed. Department of Agriculture, Washington, DC.

WILLEY, A. R. 1972. Effect of tin ion complexing substances on the relative potentials of tin, steel, and tin-iron alloy in pure acid and food media. Br. Corros. J. *7,* 29–35.

ZOOK, E. G., and LEHMANN, J. 1968. Mineral composition of fruits. J. Am. Diet. Assoc. *52,* 225–231.

10

Flavor

The Basic Tastes
Flavors and Volatiles
Flavor Enhancement
Flavor Restoration and Deterioration
Summary
Bibliography

Flavor is a very important characteristic in food or food products, and as such, is of special interest to students of food chemistry. One of the most important quality standards in a food product should be flavor, of far greater importance than appearance—a cosmetic effect—although food manufacturers are unfortunately more concerned with the latter.

Flavor is often confused with taste. While they are related, it must be realized that flavor is the end result, as taste contributes to flavor. The two main contributors to flavor are taste and odor. Flavor, both quantitatively and qualitatively, is difficult to measure. Certainly, it does not lend itself to determination in the way that one can measure the quantity of acid in a sample by titration. Most of the tests are highly subjective and are not likely to have the precision that direct quantitative chemical analysis affords.

THE BASIC TASTES

Four basic tastes are generally recognized: sweet, sour, salty, and bitter. Two other tastes are known: metallic and alkali. To be tasted, a substance must be soluble in water. The sensation of sweet taste is detected at the tip of the tongue, sour at the edges, bitter at the rear, and salty at the tip and edge. If a physical condition deadens the appreciation of odor, the effect of volatiles on flavor disappears, but not the basic tastes.

As far as chemical structure is concerned, sour and salty tastes were in the past better correlated than were bitter and sweet tastes. Recently, however, a theory has been proposed (Shallenberger and Acree 1967) to correlate sweet taste with chemical structure. It is possible, also, that modified structures could explain bitter taste.

Salty Taste

The only substances that show a salty taste are those compounds that are chemically salts. Sodium chloride is, of course, the prime example of a compound with a "salty" taste. Salts of relatively low molecular weight have a salty, but not precisely the same, taste. As the molecules get heavier the predominately salty taste tends to lessen and bitterness develops. Potassium iodide has a bitter taste, which predominates. Salts having a mainly salty taste include NaCl, KCl, NH_4Cl, and Na_2SO_4. Magnesium sulfate among other salts is basically bitter.

Sour Taste

The hydrogen ion is responsible for the sour taste. The reason acids have a sour taste is the production of hydrogen ions in solution in water.

Pangborn (1963) found no relation among pH, total acidity, and relative sourness.

Sour flavor may be affected by several factors, namely, pH, titratable acidity, effect of buffers, the nature of the acid group, and the presence of compounds such as sugars.

Sweet Taste

Many workers have thought that chemical structures determine sweet and bitter taste. According to Cohn (1914), saporous groups ordinarily occur in pairs and sweet taste is produced by a number of OH groups in the molecule. This would explain the sweetness of sugar molecules. Five years later Oertly and Myers (1919) expressed the opinion that a compound tastes sweet because it contains an auxogluc and a glucophore. A glucophore is a group of atoms or a radical which is considered to be responsible for the sweet taste in compounds containing such groups. It has the power, therefore, to form sweet compounds by its presence in otherwise tasteless molecular groups. An auxogluc when present in such compounds has the ability to increase the sweet taste. An example of a glucophore is $CH_2OHCHOH$—. This appears in $CH_2OHCHOH$ (+H), glycol, which is sweet. Another example is —CO—CHOH—(H). This glucophore yields with 2 H atoms glycolaldehyde, the simplest sugar.

Tsuzuki (1948) noted that closely related compounds with lowest resonance energy tasted bitter, while those with the highest resonance energy were sweet.

Shallenberger and Acree (1967) proposed that in order to possess a sweet taste, a compound must have an electronegative atom A which is usually oxygen or nitrogen. To this atom is necessarily attached a proton by a single covalent bond. AH can represent OH, an imine or an amine group or perhaps given the right conditions, a methine group. A second electronegative atom B must be within a distance of about 3 Å of the AH proton, usually O or N. This accounts for the sweet taste of seemingly unrelated compounds (Fig. 10.1).

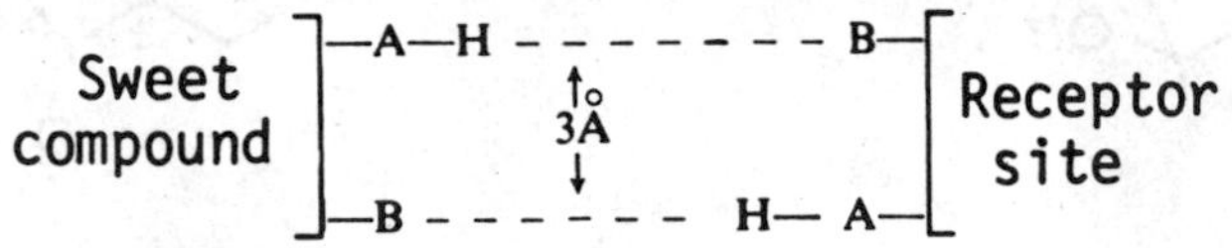

Reproduced with permission of *Nature*

Simple sugar analogues may provide excellent models for sensory evaluation. Most of these compounds are sweet.

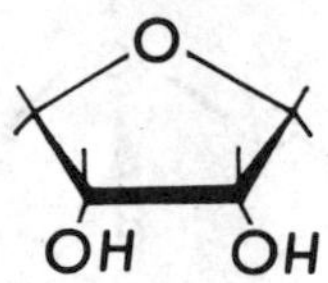

1,4-Anhydroerythritol (not sweet)

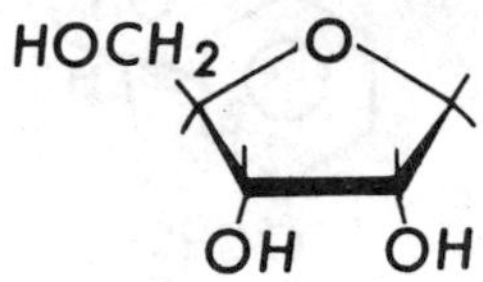

1,4-Anhydroribitol (sweet)

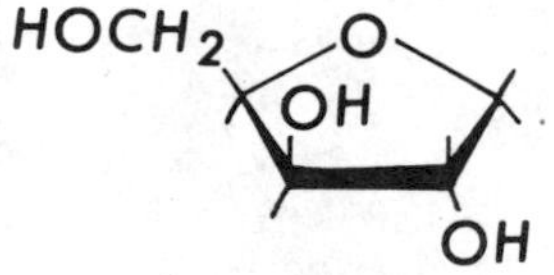

1,4-Anhydroxylitol (sweet)

It has been shown that L-sugars as well as D-sugars are sweet. This is in line with the AH,B theory. Also it has been suggested that the π bonding cloud of the benzene ring could serve as a B moiety, explaining why anti-anisaldehyde oxime is sweet, but syn-anisaldehyde oxime is tasteless. In the tasteless isomer the AH is farther removed from B than 3 Å, and, therefore, no taste results.

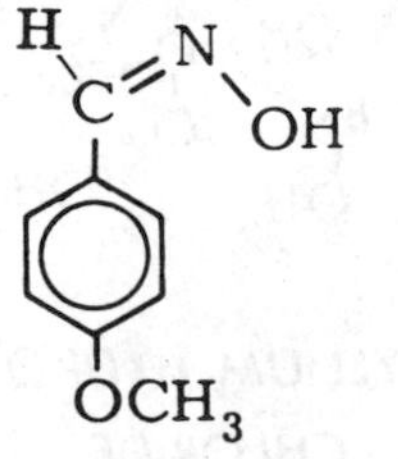

anti-Anisaldehyde oxime

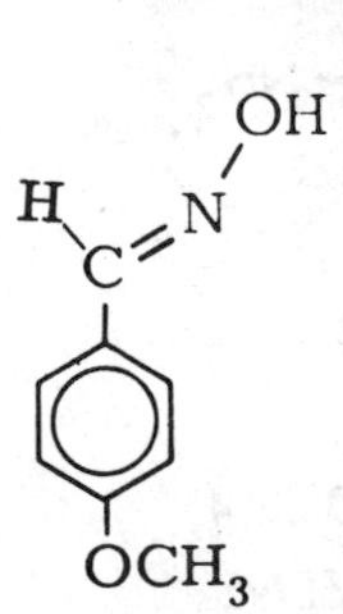

syn-Anisaldehyde oxime

Saccharin is extremely sweet—about 400 times as sweet as sucrose. According to the AH,B theory, the reasons are as follows (Shallenberger 1974).

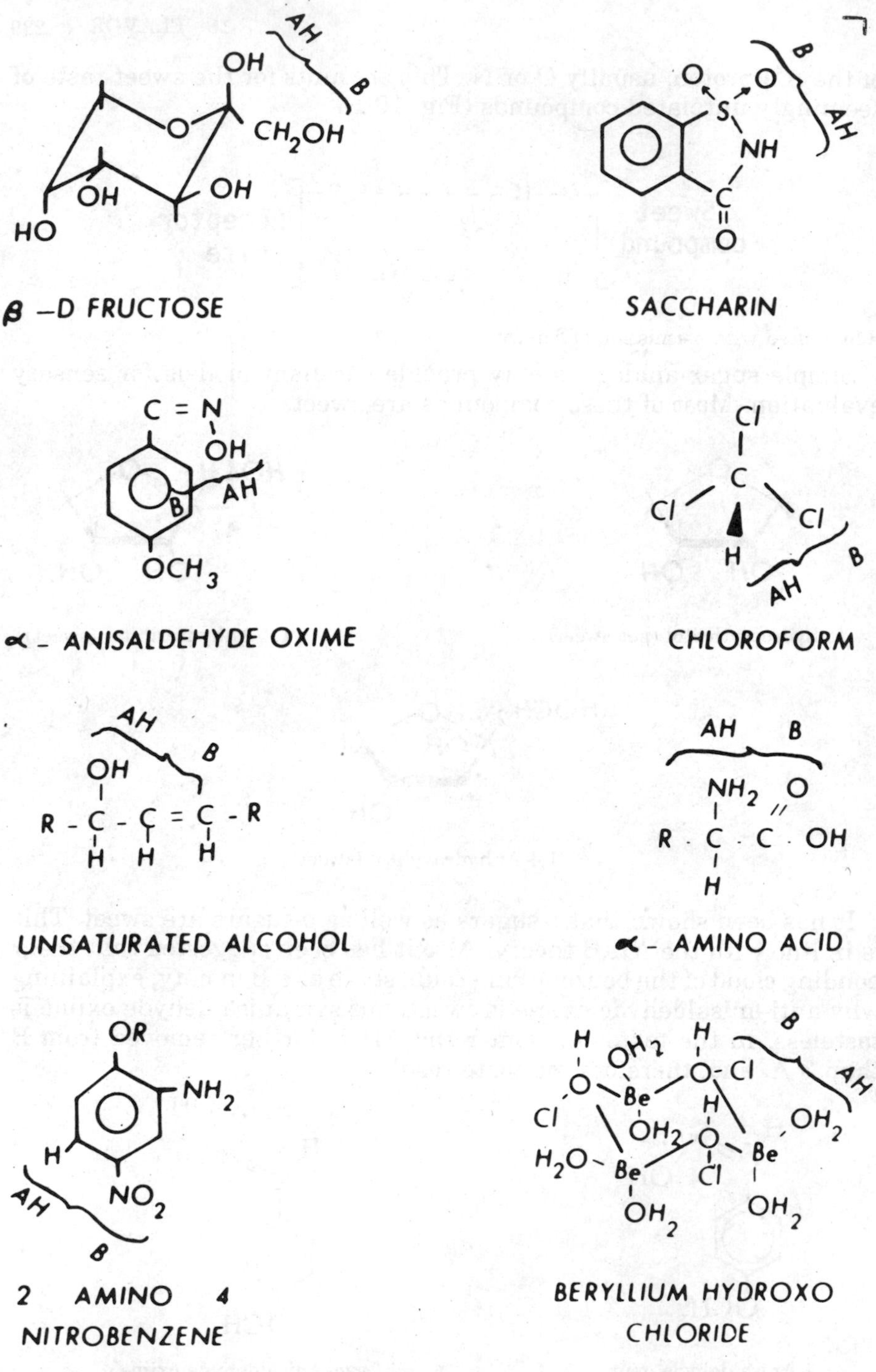

β –D FRUCTOSE

SACCHARIN

α – ANISALDEHYDE OXIME

CHLOROFORM

UNSATURATED ALCOHOL

α – AMINO ACID

2 AMINO 4 NITROBENZENE

BERYLLIUM HYDROXO CHLORIDE

FIG. 10.1. Representative compounds that taste sweet and the AH,B unit common to all of them.

From Shallenberger and Acree (1969). Reproduced with permission of the American Chemical Society, copyright holder.

1. The AH,B of this compound is rigid, exactly 3 Å.
2. This compound has lipoid solubility and, therefore, easy access to the reception site.
3. Electron distribution over the molecule is such that the AH proton has greater acidity than sugar AH proton.

To effect the sweet taste, it is necessary that an analogous unit be present at the taste bud site and between the two an intermolecular hydrogen bond is formed. This situation is supposed by theory to produce the sensation of sweetness. The idea of a loose combination between the sweet compound and the taste bud receptor site was proposed by Beidler (1954). Dastoli and Price (1966) substantiated this idea by preparing proteins from the papillae from the tip of bovine tongues and fractionating the resulting solution with ammonium sulfate. They got a significant response to sweetness with the protein fraction which was soluble in 20% ammonium sulfate, but insoluble in 40%.

Bitter Taste

The fourth of the important tastes includes those substances which produce the sensation of bitterness. Not a great deal of information has been collected as to the exact cause of this taste sensation. Recently, Kubota and Kubo (1969), working with diterpenes obtained from plants, found that the common group needed to produce the bitter taste is the AH,B unit which produces the sweet taste. However, the difference is that the AH proton to the B orbital distance is 1.5 Å. More work must be done on this phase of the subject of taste.

FLAVOR AND VOLATILES

There has been considerable speculation as to the reason the volatile substances responsible for the odor part of flavor possess their particular odor qualities. One of the more valid speculations is the stereochemical theory. Amoore (1967B) stated "Considering this value of P (*probability*) achieved or bettered for all five graphs attempted, the odds are astronomical that a strong correlation does exist between molecular size and shape and odor quality." The basic idea is that the molecules of a number of compounds fit into the receptive sockets or sites at the olfactory nerve endings and produce the odors. This, however, is not entirely accurate, and much work is being done to validate the concept. It is quite likely, however, that this is basically the explanation of odor, and is a similar idea as the one given earlier in this chapter for basic taste sensations.

The qualitative and quantitative determination of the volatile substances active in the flavor components of foods has been greatly improved recently by two very important techniques: gas chromatography and mass spectrometry. At the present time, combinations of these two techniques are increasing the knowledge of these compounds.

The gas chromatography part of this equipment is effective in separating the volatile components. The individual compounds eluted from the gas chromatographic column are detected electronically and are graphically presented as peaks on a strip chart recorder. These compounds are passed directly to the mass spectrometer for analysis to determine the chemical structure of the compounds. In cases in which two components are eluted at approximately the same retention time, information concerning the structures of the two compounds can be obtained by sampling at the beginning or ascending part of the peak. Obviously the reliability of the identification is considerably increased if the sample is known to consist of one component. The reason these two systems are so compatible is that any sample that can be separated by gas chromatography can be expected to be run satisfactorily in a mass spectrometer. The carrier gas used in this equipment is helium.

The mass spectrometer must be operated at very low pressures (10^{-7} torr). The effluent from the gas chromatograph is at atmospheric pressure. In order to operate the gas chromatograph–mass spectrometer combination a method of providing a pressure drop from atmospheric conditions must be employed. This is accomplished by using an easily diffusible carrier gas such as helium. Interfacing in this manner also serves the purpose of concentrating the eluted compound by eliminating large quantities of helium.

During the development of techniques of gas chromatography the problem of peak identification was a major consideration. The important methods were elution time relationships and comparison with known compounds. Also, the retention time of a compound on column substrates of different affinities for the compound aided in identification. However, in many cases, the results were not always reliable. If the first method was used, the possibility that a peak contained more than one compound was a hazard. Because of these difficulties more reliable methods of identification were necessary. The method involving the use of the mass spectrometer provided the answer. Further convenience was the result of incorporating the two techniques into one complete unit. The mass spectrometer because of its sensitivity, speed, and specificity was considered to be the best of the several analytical tools available (infrared, ultraviolet, nuclear magnetic resonance, and mass spectrometry).

The vapor phase chromatography (VPC) detector gives the quantitative data and the mass spectrophotometer the qualitative data as well as quantitative information on the unresolved peaks.

However, speed alone is not the only consideration in peak identification. The continuous monitoring as afforded by the VPC effluent in the combined instrument has certain limitations which must be considered as well as the advantages of peak trapping for subsequent batchwise analysis. Some advantages of batchwise analysis follow.

1. Low voltage ionization and determination of the apparent peak, which requires several scans under different operating conditions, enable the operator by using the batch system to bleed in more sample repeatedly for relatively long periods. This in turn, yields multiple spectra, which can be useful.
2. If compounds producing peaks are of low volatility they may not be pumped out of the mass spectrometer prior to the emergence of succeeding peaks, thus causing confusion.
3. Slower scanning speeds resulting from batch introductions can give higher resolution, and better resolution of smaller peaks.
4. Correlation with spectral data in the literature is improved. Continuous monitoring instruments sometimes produce very sharp VPC peaks which may distort the spectra.
5. If only a few peaks in a chromatographic sample are of interest, the instrument can be used for a greater number of chromatograms in a given time.

Regardless of the advantages of the trapping of peaks, there are some important advantages of direct monitoring, as follows.

1. Samples can be conveniently handled.
2. It is possible to identify the profile of the peak.
3. If the quality of material is limited, the improvement of sample utilization is important.
4. High speed characterization of effluent from chromatograph columns.
5. It can be used in pilot plants to monitor process chromatograph effluents.

Usually it takes 1–30 sec to record a spectrum when a vapor phase column monitoring is used.

It has been found that for practical purposes the combination (VPC and MS) generally gives the best results for the elucidation of the components of mixtures of flavoring compounds present in foods. It is usually more satisfactory for this purpose than the batch method; however, batch samples can be introduced into the combined unit at the direct outlet, if necessary.

In cases where further work is required, infrared spectroscopy can be used for determining the functional groups present. Another acceptable technique is nuclear magnetic resonance.

Theory of Odor

Over the years a number of theories have been proposed to explain the odor part of the flavor sensation. The purpose of each of these theories is to understand and explain how odorivectors and odorivector mixtures are differentiated by the chemoreceptors in the nose.

The objectives of the olfaction theories were further subdivided by Dravnieks (1967). They must outline (1) how odorivector molecules cause electrical changes in the chemoreceptor neutrons; (2) why different odorivectors produce different response patterns from the assembly of chemoreceptors; (3) what properties of odorivectors and their mixtures determine their olfactory thresholds, quality, and intensity; and (4) how odors could be classified and precisely described. He listed also a fifth objective which is indeed very important from the flavor chemist's point of view: to arrive at devices that can measure odors objectively without the use of olfactory panels.

Of the many theories proposed to explain the phenomenon of odor, three seem to have the best possibilities. The first of these is the vibrational theory of Wright (1954, 1957, 1964; also Wright and Michels 1964). In this theory it is postulated that the characteristic odor of a compound has a connection with the particular vibrational frequencies of parts of the molecule or of the entire molecule. These frequencies can be measured by Raman spectroscopy. Evidence is offered for recurring frequency distributions among compounds of the same odor class. However, not enough evidence has been presented to permit its acceptance at the present time.

The second is the adsorption theory of Davies (1962, 1965; also Davies and Taylor 1959). This theory basically explains the intensity of the odor rather than the quality. It fits in easily with current biochemical concepts. According to this theory, the polarized neuron cell wall supplies the energy needed. Odorivectors trigger collapse of the electrical field that exists across the membrane wall to provide the energy needed. However, this theory has not been proved and it is backed only by circumstantial evidence. This theory must be studied in considerable detail, and must include confirmation.

The stereochemical theory of olfaction was proposed by Amoore (1952). It is based on size and shape of the molecule and views the molecules of primary odors fitting into the receptor socket of the nerve ending to bring about the sensation of odor. However, this idea of "lock and key" site fitting assessed only half of the molecule. In order to take all aspects of the molecular surface into consideration, the shadow matching idea was adopted. In this way, very significant correlations were obtained between the odor qualities and the molecular sizes and shapes. This theory sets out to explain odor quality, not intensity. It is likely that many more than the original seven postulated primary odors will be found before work on this theory can approach completion. Figure 10.2 shows the concept of shadow matching.

Several classes of organic compounds are responsible for different types of odors, and many of these compounds are used in synthetic flavors. Esters, which are compounds of acids and alcohols from which water has been eliminated, comprise an important group. While the organic fatty acids are strong, harsh, and sour, particularly in the lower molecular weights, the corresponding esters, especially of acids

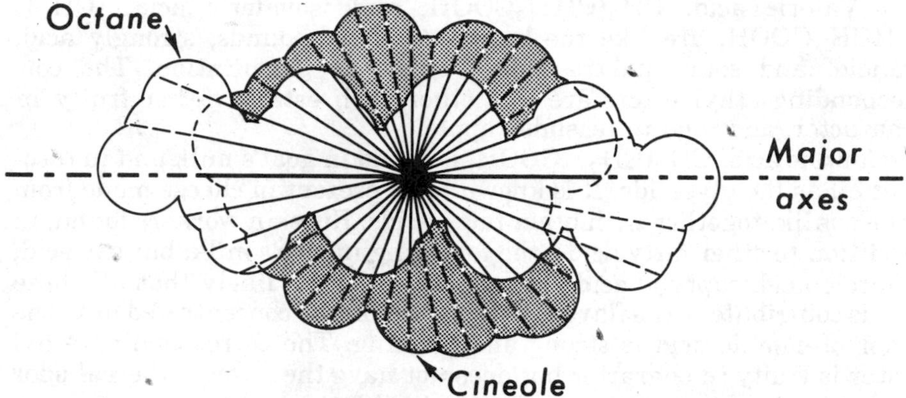

FIG. 10.2. The "shadow matching" method.
From Amoore (1967).

and alcohols of the lower molecular weights, are usually fruity and pleasant.

Fatty Acids

Formic acid, HCOOH, the organic acid of the fatty acid group of lowest molecular weight, is pungent and strongly irritating in aroma. However, the corresponding ethyl ester tends to be somewhat fruit-like in aroma.

Acetic acid, CH_3COOH, the next higher in the group in number of carbon atoms and in molecular weight, has a pentrating and sour odor, and is the acid of vinegar. It is the dominant volatile acid in cheddar cheese (Patton 1963). The corresponding ethyl ester, the compound of acetic acid and ethyl alcohol, is definitely fruity in aroma.

Propionic acid, CH_3CH_2COOH, is sour and rancid in character. This acid is present in aged Swiss (Emmentaler) cheese (Patton 1964) and is produced in the cheese by propionic acid bacteria, *Propionibacterium shermanii.* It has an important role in the flavor of Swiss cheese. This is an example of an important flavor principle: in concentrated form substances can have very undesirable odors and flavors, but in dilute form they can be very acceptable. Ethyl proprionate, the corresponding ethyl ester is a liquid, as are the other esters in the flavor series, and is strongly fruity in character.

n-Butyric acid, $CH_3CH_2CH_2COOH$, and isobutyric acid, $(CH_3)_2CHCOOH$, possess very strongly sour and rancid odors. *n*-Butyric acid is present in rancid butter, having been derived from glyceride, and is responsible for its unpleasant odor. The ethyl ester, ethyl *n*-butyrate has a rather strong fruit-like odor which tends to be suggestive of pineapple.

n-Valeric acid, $CH_3(CH_2)_3COOH$, and isovaleric acid, $(CH_3)_2CHCH_2COOH$, are like the butyric acid compounds, strongly acid, rancid, and sour, and resemble somewhat perspiration. The corresponding ethyl esters are like other such esters, rather fruity in character, and tend to resemble apple.

Caproic acid, $CH_3(CH_2)_4COOH$, is found in goat's milk and in coconut oil as the glyceride. It is known to be present in cheese made from ewe's milk, together with other fatty acids. Russian workers found, in addition to other fatty acids, the following in ewe's milk: butyric acid, caproic acid, caprylic acid, and capric acid. It is likely that all these acids contribute to the flavor of the cheese. In the concentrated form the odor of caproic acid is strong and piercing. The corresponding ethyl ester is fruity in character but does not have the strong ethereal odor found in the esters of the shorter chained fatty acids.

n-Heptylic acid, $CH_3(CH_2)_5COOH$, is not so sour in odor as the fatty acids of lower molecular weight, but it is piercing. It is known also as oenanthic acid, and the commercial form of the ethyl ester has other esters as impurities and has an odor resembling wine. The weakening of intensity of sourness and odor shown by heptylic acid continues in trend as the number of carbon atoms in the molecule increases. Likewise, the higher molecular weight acids and esters show increasing weakness in flavor and pungency.

Fatty acids can be altered in odor as in other properties by the addition to or subtraction from the molecule. The replacement of the methyl group by a carboxyl group in acetic acid forms oxalic acid, HOOC—COOH, which is odorless, toxic, nonvolatile, and very much in contrast in these respects with the two-carbon acid, acetic acid, which contains only one carboxyl group. Equally important changes in properties can be noticed with replacements, additions, or subtractions in other compounds.

Alcohols

The changes in odor and other properties can be obtained in alcohols and, of course, in other classes of compounds. Chemical changes (that is, replacement of groups) can alter the properties of all compounds. The lower molecular weight alcohols are soluble in water and have spirit-like odors. These characteristics tend to disappear as the molecules increase in size, and they become oily liquids. *n*-Decyl alcohol, n-$C_{10}H_{21}OH$ has an odor rather similar to orange flowers.

Esters

The esters show a similar picture. As molecular size of the alcohol group increases, the strength of odor of the corresponding ester decreases. However, specific compounds have specialized odors and flavors. Butyl acetate has a fruity flavor. Isoamyl acetate is well-known for its banana-like aroma; amyl acetate has this character also, as does amyl butyrate. The acetates of the higher alcohols, i.e., octyl, nonyl, and decyl acetates, tend to be citrus-like and somewhat less pungent than those of the lower alcohol esters. In general, the higher the molecular weight of the esterifying alcohol, the less pronounced and piercing the odor becomes.

Other esters of different organic acids have interesting flavor characteristics and are employed extensively in this way. Methyl salicylate (b.p. 223 °C) is synthetic oil of wintergreen and is the chief component of natural oil of wintergreen. The acid of this compound, salicylic acid, is an aromatic acid in chemical structure, and is odorless.

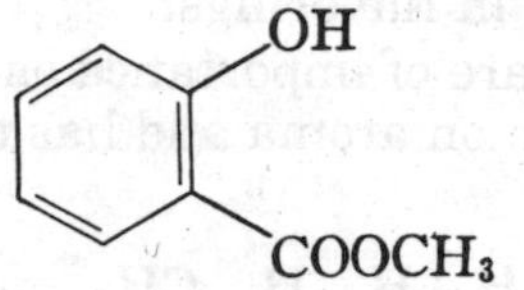

OH

$COOCH_3$

Methyl salicylate

Salicylic acid is the ortho hydroxy compound of benzoic acid. The methyl ester of this compound, as described, has a rather intense odor. The methyl ester of benzoic acid, however, is much smoother and rather minty in aroma. It would seem that this difference is brought about by the hydroxy group in the molecule since this is the only structural difference.

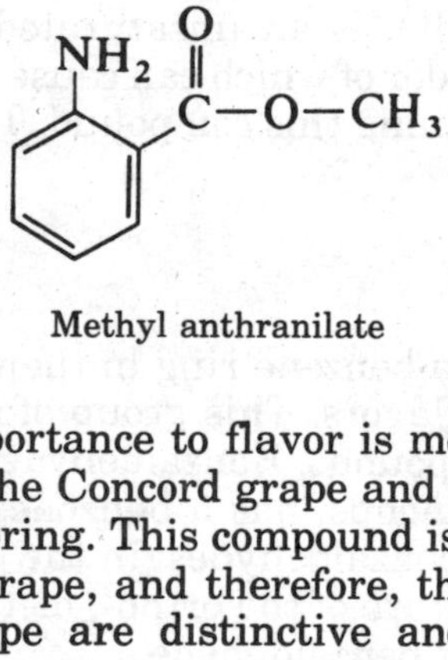

NH_2 O ‖ $C-O-CH_3$

Methyl anthranilate

Another ester of importance to flavor is methyl anthranilate. This compound is found in the Concord grape and for this reason it is used in artificial grape flavoring. This compound is not present in the *Vitis vinifera* or European grape, and therefore, the wines and juice made from the Concord grape are distinctive and not pleasing to most European palates.

Aldehydes

Aldehydes are another group of compounds that are important to flavor. Of the three groups of organic compounds (alcohols, carboxylic acids, and aldehydes) the aldehydes have the lowest boiling points.

They possess the double bond oxygen

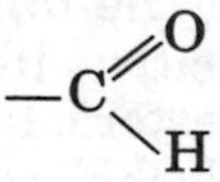

$$-C\overset{\displaystyle O}{\underset{\displaystyle H}{\lessgtr}}$$

which makes them more reactive than the corresponding alcohols. These are two important factors in flavor production. Acids are formed by oxidation of aldehydes, whereas alcohols are formed by reduction of aldehydes. The lower fatty aldehydes are rather pungent in odor, and disagreeable. As the number of carbon atoms in the molecule increases to 10, 12, and 14 in the saturated straight chain aldehydes, odors are formed which can be used in flavorings.

Unsaturated aldehydes are of importance as flavors. Citral, present in lemon oil, has strong lemon aroma and has two double bonds in the molecule.

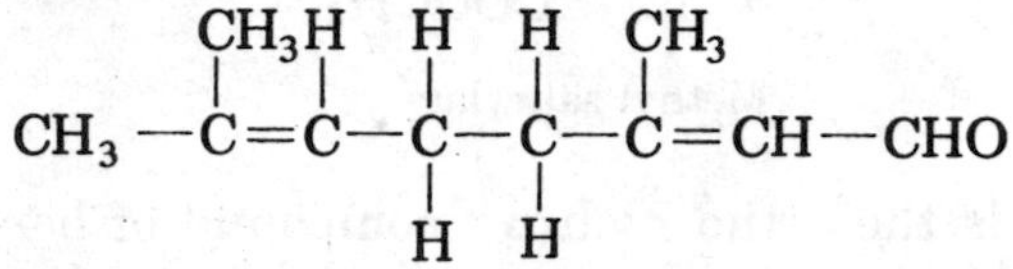

$$CH_3-\overset{CH_3}{\overset{|}{C}}=\overset{H}{\overset{|}{C}}-\underset{H}{\underset{|}{\overset{H}{\overset{|}{C}}}}-\underset{H}{\underset{|}{\overset{H}{\overset{|}{C}}}}-\overset{CH_3}{\overset{|}{C}}=CH-CHO$$

Citral

This compound is harsher than the corresponding compound with one double bond, which in turn, is harsher than the corresponding saturated compound.

Acrolein, $CH_2=CHCHO$, is an unsaturated low-molecular-weight aldehyde, the irritating odor of which can cause adverse flavors in foods prepared with fats containing this compound. This is especially true of fried foods.

Ring Compounds

Aldehydes containing a benzene ring in their structure are particularly important in food flavors. This group of aldehydes contains the following important compounds. Benzaldehyde, the flavor-active compound in oil of bitter almonds, has a benzene ring with an aldehyde group attached. Methyl benzaldehydes (in the plural because as found on the market it is a mixture of the ortho-, meta- and paraforms) have a strong odor resembling benzaldehyde.

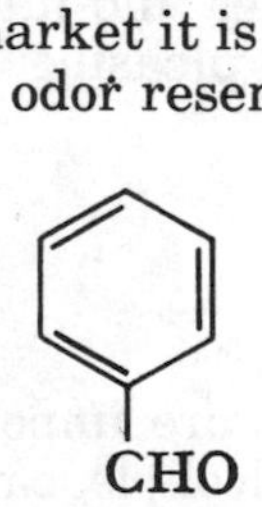

CHO

Benzaldehyde

CHO CH₃

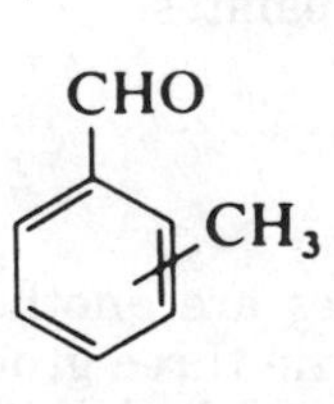

Methyl benzaldehyde (mixed *o*-, *m*-, and *p*-)

Vanillin, the basic flavoring component of the vanilla bean, belongs chemically to the aldehydic derivatives of polyhydric phenols and is chemically 3-methoxy-4-hydroxybenzaldehyde. A basic flavoring compound rather widely distributed in nature, vanillin is used also as an adjunct in other flavors, such as in milk chocolate and other forms of chocolate.

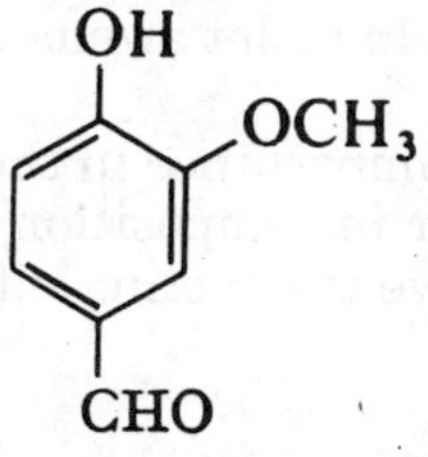

OH
OCH_3
CHO

Vanillin

Anisaldehyde, *p*-methoxybenzaldehyde, is present in trace quantities in vanilla beans and is a sweet aromatic compound. It has a pungent flavor resembling anise.

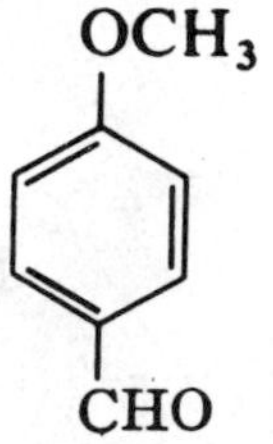

OCH_3
CHO

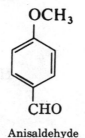

Anisaldehyde

Another aromatic compound is cinnamaldehyde or cinnamic aldehyde. It is present in oil of cinnamon and oil of cassia. Its chemical formula is $C_6H_5CH=CHCHO$, b.p. 127 °C at 15 mm Hg. At atmospheric pressure it boils at 252 °C. Like all the compounds in the series starting with benzaldehyde it is aromatic, pungent, and spicy in character. Many other compounds in this series are known, and are used in flavoring mixtures. Cinnamic acid, $C_6H_5CH=CHCOOH$, and cinnamic alcohol are the source of a number of esters which are of considerable importance in the making of flavoring materials.

Ketones

A number of ketones are of much value as flavoring agents. Ketones are carbonyl compounds with two alkyl groups joined to the carbonyl group. The lower molecular weight ketones are of little significance from the flavor point of view, but those with seven carbon atoms or more are important for this purpose. One of these compounds is methyl amyl ketone, $CH_3CO(CH_2)_4CH_3$, which has a blue cheese aroma. Another group of these compounds are the diketones. Diacetyl,

$CH_3COCOCH_3$, b.p. 87°–88°C, is present in cultured butter, sour cream, buttermilk, and cottage cheese, and is important in the flavor of these dairy products. Diacetyl is formed during the manufacture of such products.

The ionones are another type of ketones. They are used as fruit and berry flavorings. β-Ionone is often used in artificial raspberry flavor. The ionones are related also to violet aroma. The formula for β-ionone is given in Chapter 8.

Ketone compounds are of importance in the mint flavor group. Menthone, b.p. 210°C, is similar in composition to menthol, m.p. 42.5°C, b.p. 216°C, but does not have the cooling effect of menthol.

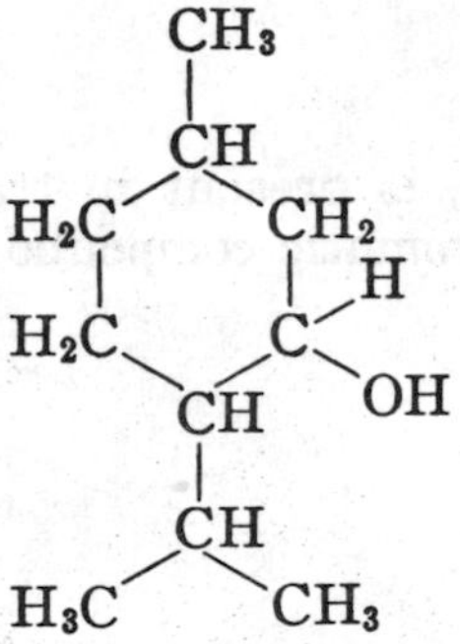

1-Menthol

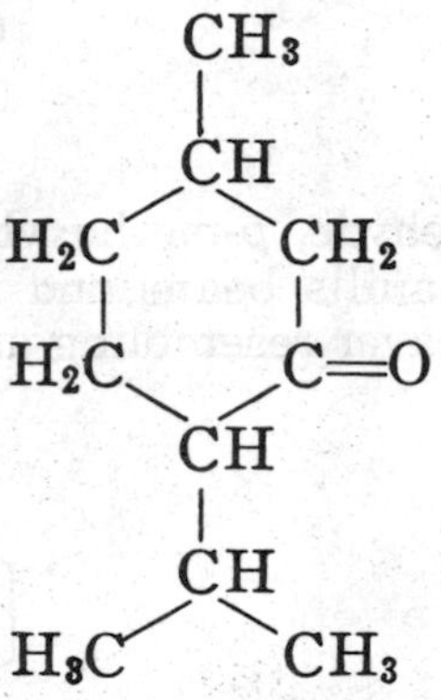

1-Menthone

Terpene Alcohols

Alcohols derived from acyclic terpenes are important as flavoring compounds and are found as such in volatile (essential) oils. Geraniol is a trans isomer. Nerol is the cis isomer of the same compound. These compounds are basically related to isoprene, although the latter compound is not known to occur in nature.

Geraniol and nerol are diunsaturated terpene alcohols, while citronellol is monounsaturated. Geraniol is found in a large number of essential oils including those of lemon and orange. Citronellol occurs in plants of the family Rosaceae and in a large number of volatile oils. Nerol is found in oil of lemon and of sweet orange and in many other essential oils. It has been found in the volatile substances from the currant. Geraniol and nerol are used in flavoring mixtures and tend to impart a floral and fruity character. Linalool is heavier and more spicy in its flavoring capabilities. Citronellol is naturally present in many fruits and spices. Esters of these alcohols are used as aromatic materials in flavor preparations and, as with other esters, the lower molecular weight compounds tend to be more volatile and sharp, the higher molecular weights are less sharp and smoother.

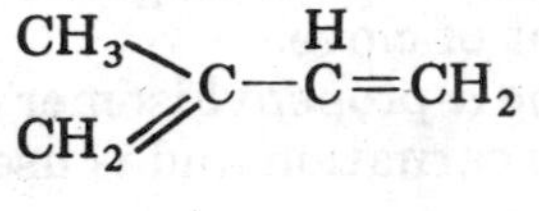

$$CH_2{=}C(CH_3){-}CH{=}CH_2$$

Isoprene

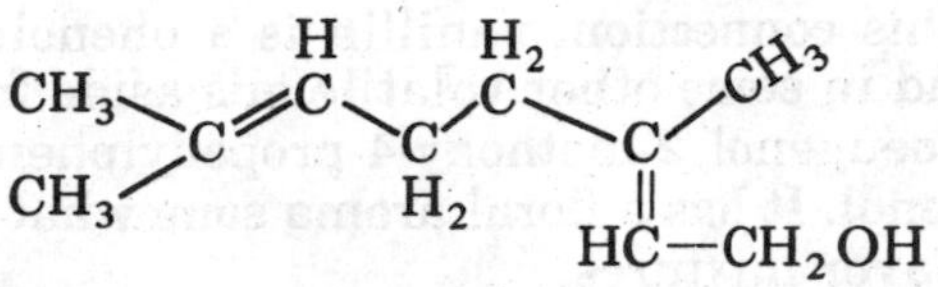

$$(CH_3)_2C{=}CH{-}CH_2{-}CH_2{-}C(CH_3){=}CH{-}CH_2OH$$

Geraniol

$$(CH_3)_2C{=}CH{-}CH_2{-}CH_2{-}CH(CH_3){-}CH_2{=}CH_2OH$$

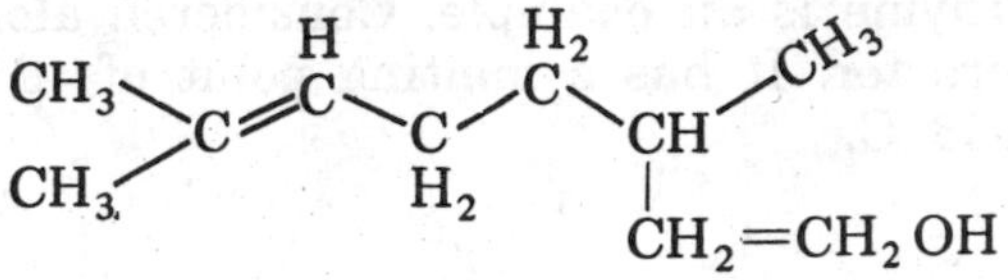

Citronellol

Lactones

Some flavoring compounds belong to the class known as lactones. The organic chemists classify these as alpha, beta, gamma, and delta lactones, depending on the number of members in the ring system. Among compounds of this class in use in flavor work are the γ and δ lactones. The γ-lactones are formed by treating the γ-hydroxy acids with sulfuric acid. Under these conditions a molecule of water is eliminated and a five-member ring is formed. These γ-lactones are stable substances and are neutral. However, the ring can ordinarily be split by means of warm alkali. The misnamed aldehyde C-14 is really γ-undecalactone with 11 carbon atoms in the molecule and has a peach-like character. The misnamed aldehyde C-18 is also a lactone structure, γ-nonalactone with nine carbon atoms, and has on dilution an odor somewhat like coconut. It can be prepared by heating ζ-oxypelargonic acid with 50 vol % of H_2SO_4.

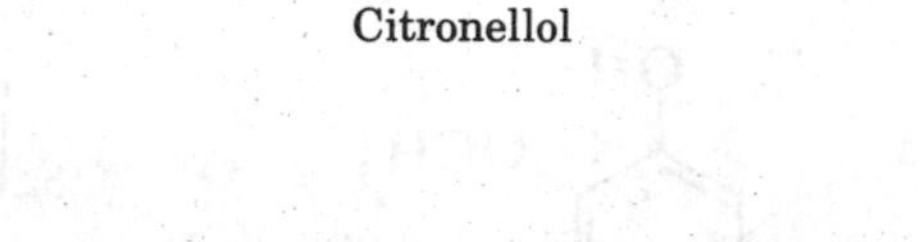

$$CH_3{-}(CH_2)_4{-}CH{-}(CH_2)_2{-}C{=}O \quad (\text{CH and C=O joined through } {-}O{-})$$

γ-Nonalactone

δ-Lactones are important in butter and milk products. They are found in butter and are added to margarine. δ-Decalactone and δ-dodecalactone have been found in heated and dried milk products. Mattick *et al.* (1959) indicated that 5-hydroxy acids are precursors of these compounds which contribute to dry whole milk off-flavor.

Phenols

Some monohydric phenols are present in spices, and are important in flavoring.

Eugenol, 2-methoxy-4-allylphenol, is the important ingredient of oil of cloves and has a pleasant odor of cloves, together with a sharp, burning, clove flavor. It can be used for making vanillin synthetically.

In this connection, vanillin is a phenolic compound also. Eugenol is found in some other volatile oils aside from that of clove.

Isoeugenol, 2-methoxy-4-propenylphenol, is the α-propenyl isomer of eugenol. It has a floral aroma somewhat akin to carnation and is used in flavor mixtures.

Thymol is another member of this group. It is a component of several spices of which thyme is an example. Considered alone, its odor is medicinal in character. It has a melting point of 48°–51°C and a boiling point of 233°C.

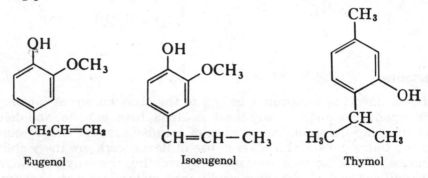

Eugenol: ring with OH, OCH_3, $CH_2CH{=}CH_2$

Isoeugenol: ring with OH, OCH_3, $CH{=}CH{-}CH_3$

Thymol: ring with CH_3, OH, $CH(CH_3)_2$ (H_3C–CH–CH_3)

Eugenol   Isoeugenol   Thymol

Sulfur Compounds

A number of sulfur-containing compounds are of considerable importance as flavoring substances. Compounds of this character are responsible for the odors and flavors of such vegetables as onions and garlic, the condiment horseradish, and the spice mustard. Oil of mustard contains the compound allyl isothiocyanate, $CH_2{=}CHCH_2N{=}C{=}S$. Sinigrin is the compound which occurs in black mustard seed and is hydrolyzed by the enzyme myrosinase as follows.

$$C_6H_{11}O_5S{-}\underset{\text{Sinigrin}}{\overset{\overset{\displaystyle OSO_3K}{|}}{C}{=}NCH_2CH{=}CH_2} + H_2O \longrightarrow$$

$$C_6H_{12}O_6 + KHSO_4 + \underset{\text{Allyl isothiocyanate}}{CH_2{=}CHCH_2N{=}C{=}S}$$

Sinigrin is also present in horseradish.

Several sulfur-containing compounds are present in the volatile substances obtained from coffee; among them are methyl mercaptan, furyl mercaptan, and dimethyl sulfide. An important property of mercaptans is their behavior as reducing agents. Also, they are acidic and are able to form salts with alkalis.

Allyl sulfide (allyl disulfide), $(CH_2{=}CHCH_2)_2S_2$, together with allyl propyl disulfide and other products make up oil of garlic. The corresponding compound formed in onion is $C_3H_5SSC_3H_7$, allyl propyl disulfide. These and other compounds present in onion and garlic have been determined by gas chromatography (Saghir *et al.* 1964). Allyl disulfide is not present in intact garlic, but when the bulb is crushed, enzymatic action produces allicin, $C_3H_5—S—S—C_3H_5$, and this breaks down to form allyl disulfide. Garlic preparations in which the enzyme has been destroyed do not possess the usual odor and flavor.

FLAVOR ENHANCEMENT

There are compounds that have the ability to improve the flavor of foods. The compound known for the longest time is monosodium glutamate, $NaOOCCH_2CH_2CH(NH_2)COOH$. Although it exists in both D- and L- forms, only the L- form, which is the naturally occurring form, has the flavor-enhancing activity. It is a crystalline white powder which is very soluble in water but only slightly soluble in alcohol. This compound can be prepared from wheat gluten, as a by-product in the manufacture of beet sugar, and from casein. The Japanese chemist Ikeda, who first discovered the flavor-enhancing property of monosodium glutamate, prepared it from a seaweed. Its principal use is to improve the flavor of meats, gravies, bouillons, sauces, and foods rich in protein. Under some circumstances it has been used in higher quantities than desirable.

Monosodium glutamate was, for many years, the only compound with flavor-enhancing capability; however, other compounds are now known to have this property. These include disodium 5′-guanylate and disodium 5′-inosinate. Three isomers of inosinic acid exist, but only 5′-inosinic acid is active. On a qualitative basis, the effect of disodium 5′-guanylate and disodium 5′-inosinate are the same. The disodium 5′-guanylate was found, however, to be 3.8 times as active. The 5′ nucleotides are synergistic with monosodium L-glutamate. The general structure of the nucleotides with flavor activity is shown in Fig. 10.3.

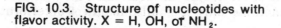

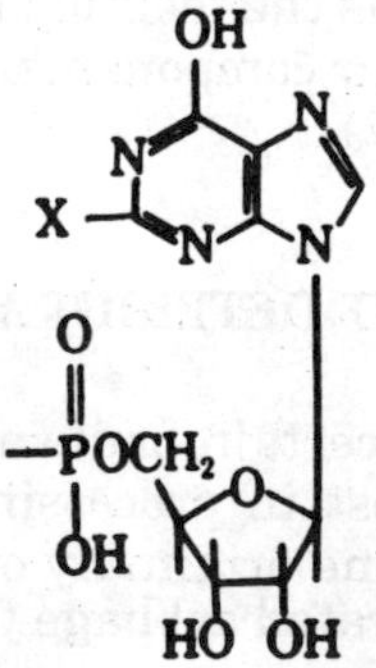

FIG. 10.3. Structure of nucleotides with flavor activity. X = H, OH, or NH_2.

Other flavor enhancers are maltol and isomaltol, which enhance sweetness. Maltol can also cause a sensation that is said to be velvety. Ethyl maltol is active as a sweetness enhancer, also, and is said to be 4 to 6 times as effective as maltol.

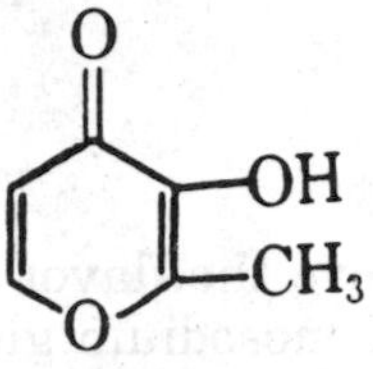

Maltol

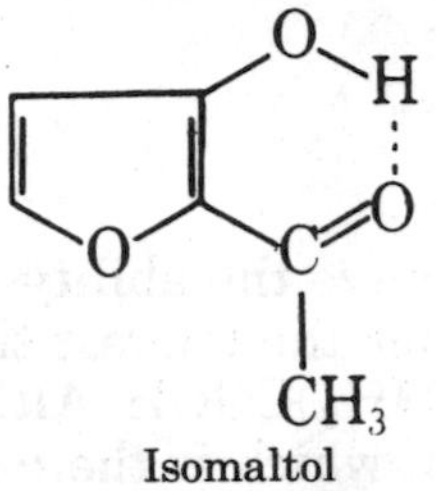

Isomaltol

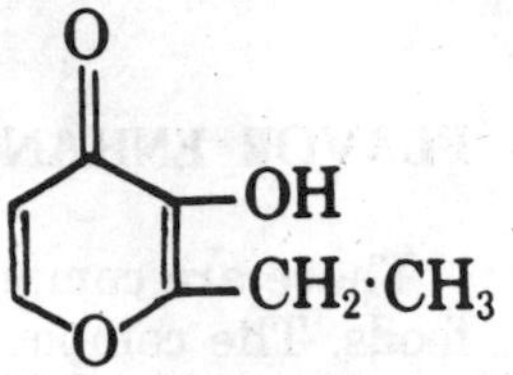

Ethyl maltol

Maltol has been found in roasted coffee beans, soybeans, cacao, cereals, and overheated skimmilk. It is produced during the heating process. It has been found to result from the caramelization of maltose. It is formed in the crust of bread during the baking process. Maltol has a caramel-like odor and dilute solutions of this compound have burnt and fruity characteristics. Maltol enhances the flavor of soft drinks, cakes, ice cream, fruit juices, and other such products.

Certain parts of two tropical plants have the ability to effect changes in taste. One of these is the tropical fruit *Synsepalum dulcificum,* called the miracle berry. It contains a basic glycoprotein with a probable molecular weight of 44,000 that has the ability to modify taste (Kurihara and Beidler 1968, 1969). When it is placed on the tongue the taste of a sour substance becomes sweet. The African natives use it to improve the flavor of some of their foods.

The leaves of another tropical plant *Gymnema sylvestre* contain a compound, gymnemic acid, which is able to bring about the disappearance of sweet taste. In addition, it is able to reduce the ability to taste bitterness. The effect on sweetness lasts for hours, and it suppresses the sweetness of saccharin as well as that of sugar. Crystalline sucrose (sugar) taken into the mouth has the feeling of slowly dissolving sand. Gymnemic acid is made up of four components, gymnemic acids A_1, A_2, A_3, and A_4 (Stöcklin *et al.* 1967).

FLAVOR RESTORATION AND DETERIORATION

It has been found possible in certain fruits and vegetables to restore or improve flavors partially lost in processing by means of enzyme preparations. This work was done originally on cabbage with the idea of restoring the flavor of dehydrated cabbage for military uses (Hewitt *et al.* 1956).

The development of off-flavors in foods can result in considerable economic loss. Sometimes only a small amount of deterioration in a food results in a lowering of the grade, but in other cases the product is no longer saleable. The causes for this situation are varied.

Fishy off-flavors in animal products can be traced to diets of fish meal and fish oil. It has been found also that the quantity of highly unsaturated fatty acids in the carcass fat, containing three or more double bonds in the molecule, is proportional to the fishy flavor of the cooked meat. In some instances, fishy flavors could be caused by amines such as trimethylamine. Fish flavor in butter can result from trimethylamine hydrolyzed from lecithin.

A visceral or gamey off-flavor can be detected in cooked meat or poultry held in the uneviscerated condition after slaughter. Off-flavors in poultry and meat products can frequently be traced to rancidity of the fatty materials. This is basically oxidative rancidity. Peroxide values, carbonyl values, and thiobarbituric acid (TBA) values are used in the determination of fat rancidity.

Unblanched frozen peas and spinach held in storage at −18°C yield fatty material which is high in liberated fatty acids and shows variable quantities of peroxides as well as carbonyls. These values parallel the development of off-flavors, which have been described as hay-like.

Several types of off-flavor can develop in milk. Raw milk can develop rancid off-flavor due to the liberation of fatty acids because of the action of natural lipases on the fat in the milk. Oxidized off-flavor can develop in milk spontaneously, but this varies considerably with milks from different sources. The phospholipids connected with the fat globule membrane are probably the site of this off-flavor development. Exposure of milk to sunlight brings about the formation of an off-flavor which is said by some to resemble cabbage and burned feathers. Experimental work has shown that this "sunlight flavor" as it is often called is produced by a reaction involving methionine and riboflavin in the presence of sunlight with the formation of methional (Patton 1954; Patton and Josephson 1953). Homogenized milk is more subject to this off-flavor development than unhomogenized milk (Stull 1953). Incandescent and fluorescent light can bring about these changes also. Actually, the effect of sunlight on milk is twofold, an oxidized flavor which is caused by lipid oxidation, and the "cabbage and burned feathers," which develops in the milk proteins. Patton (1954) has suggested that casein might be the main source of this change.

An important flavor defect in soybeans is a raw bean flavor that is not found in the intact raw bean. This off-flavor develops when the bean is macerated and has been identified as ethyl vinyl ketone by vapor phase chromatography and mass spectrometry (Mattick and Hand 1969).

Another important point in connection with flavor is that a few compounds have the ability to cover flavor defects. One of these compounds is sugar. It can cover up defects in wines. Dry wines must be of good quality because of the lack of sugar would cause any poor qualities to stand out.

A similar situation is the ability of salt to conceal, at least partially, slight flavor defects in butter. It is for this reason that unsalted butter must be of the highest grade.

SUMMARY

Flavor is a very important characteristic of food or food products. It should be more important than appearance.

Four basic tastes are recognized, sweet, sour, salty, and bitter.

The most recent explanation of the sweet taste is the AH,B theory. AH can represent any of the following: OH, an imine group, or a methine group if the conditions are right. The compound must have an electronegative atom A, which is usually oxygen or nitrogen. To this atom is necessarily attached a proton by a single covalent bond. A second electronegative atom B must be within a distance of about 3 Å of the AH proton, usually O or N. The sweetness or lack of it for various compounds is thus explained. This theory explains the sweetness of an extremely sweet compound like saccharin.

A modification of the AH,B theory for sweet taste has been suggested to explain bitterness.

Volatile substances are responsible for the odor part of flavor. The determination of volatiles both qualitatively and quantitatively has been greatly aided by the development of gas chromotography and mass spectrometry. The combined unit of VPC and MS gives the best results. If further work is necessary, infrared spectroscopy and nuclear magnetic resonance can be used.

There has been much speculation concerning the mechanism by which one perceives odor. One which seems to fit the conditions is that worked out by Amoore, who said "the odds are astronomical that a strong correlation does exist between molecular size and shape and odor quality." Two other theories seem to have merit.

The various compounds involved in odor are discussed. The chemical classes of these compounds are important and are given consideration. These include fatty acids, alcohols, esters, aldehydes including the important ring compounds, ketones, terpene alcohols, lactones, phenols, and finally sulfur compounds.

Compounds that bring about flavor enhancement are an interesting development. A Japanese chemist discovered the importance of monosodium glutamate in this respect. Other compounds are now known, including some that enhance sweetness.

Compounds that have the ability to change taste are curiosities that have food applications in the tropical countries.

Off-flavors are important because they must be corrected or avoided when possible.

BIBLIOGRAPHY

AMOORE, J. E. 1952. The sterochemical specifications of human olfactory receptors. Perfum. Essent. Oil Rec. *43,* 321–323.

AMOORE, J. E. 1963. Stereochemical theory of olfaction. Nature *198,* 271–272.

AMOORE, J. E. 1967A. Specific anosmia: a clue to the olfactory code. Nature *214,* 1095–1098.

AMOORE, J. E. 1967B. Stereochemical theory of olfaction. *In* Symposium on Foods: The Chemistry and Physiology of Flavors. H. W. Schultz, E. A. Day, and L. M. Libbey (Editors). AVI Publishing Co., Westport, CT.

AMOORE, J. E. 1970. Molecular Basis of Odor. Charles C. Thomas, Publishers, Springfield, IL.

AMOORE, J. E., JOHNSTON, JR., J. W., and RUBIN, M. 1964. The stereochemical theory of odor. Sci. Am. *210* (2), 42–49.

AMOORE, J. E., and VENSTROM, D. 1966. Sensory analysis of odor qualities in terms of the stereochemical theory. J. Food Sci. *31,* 118–128.

BEATTY, R. M., and CRAGG, L. H. 1935. The sourness of acids. J. Am. Chem. Soc. *57,* 2347–2351.

BEIDLER, L. M. 1954. A theory of taste stimulation. J. Gen. Physiol. *38,* 133–139.

BEIDLER, L. M. 1957. Facts and theory on the mechanism of taste and odor perception. *In* Chemistry of Natural Food Flavors. Quartermaster Food and Container Institute for the Armed Forces, Chicago, IL.

BIRCH, G. G. and LEE, C. 1971. Chemical basis of sweetness in model sugars. *In* Sweetness and Sweeteners. G. G. Birch (Editor). Applied Science Publishers, Ltd., London.

BRICOUT, J., VIANI, R., MÜGGLER-CHAVAN, F., MARION, J. P., REYMOND, D., and EGLI, R. H. 1967. On the composition of the aroma of tea. II. Helv. Chim. Acta *50,* 1517–1522. (French)

COHN, G. 1914. The Organic Flavor Materials. Siemenroth, Berlin, Germany. (German)

CROCKER, E. D. 1948. Meat flavor and observations on the taste of glutamate and other amino acids. *In* Monosodium Glutamate—A Symposium. Quartermaster Food and Container Institute for the Armed Forces, Chicago, IL.

DASTOLI, F. R., and PRICE, S. 1966. Sweet-sensitive protein from bovine taste buds: Isolation and assay. Science *154,* 905–907.

DASTOLI, F. R., LOPEIKES, D. V., and DOIG, A. R. 1968. Bitter sensitive protein from porcine taste buds. Nature *218,* 884–885.

DAVIES, J. T. 1962. The mechanism of olfaction. Symp. Soc. Exp. Biol. *16,* 170–179.

DAVIES, J. T. 1965. A theory of quality of odors. J. Theoret. Biol. *8,* 1–7.

DAVIES, J. T., and TAYLOR, F. H. 1959. The role of adsorption and molecular morphology in olfaction. The calculation of olfactory thresholds. Biol. Bull. *117.*

DAY, E. A. 1966. Role of milk lipids in flavors of dairy products. *In* Flavor Chemistry. R. F. Gould (Editor). Advances in Chemistry Ser. 56. American Chemical Society, Washington, DC.

DAY, E. A. 1967. Cheese flavor. *In* Symposium on Foods: The Chemistry and Physiology of Flavors. H. W. Schultz, E. A. Day, and L. M. Libbey. (Editors). AVI Publishing Co., Westport, CT.

DÖVING, K. B. 1967. Problems in the physiology of olfaction. *In* Symposium on Foods: The Chemistry and Physiology of Flavors. H. W. Schultz, E. A. Day, and L. M. Libbey (Editors). AVI Publishing Co., Westport, CT.

DRAVNIEKS, A. 1966. Current status of odor theories. *In* Flavor Chemistry. R. F. Gould (Editor). Advances in Chemistry Ser. 56. American Chemical Society, Washington, DC.

DRAVNIEKS, A. 1967. Theories of olfaction. *In* Symposium on Foods: The Chemistry and Physiology of Flavors. H. W. Schultz, E. A. Day, and L. M. Libbey (Editors). AVI Publishing Co., Westport, CT.

FLAMENT, I., WILLHALM, B., and STOLL, M. 1967. Research on aromas. Cocoa aroma. III. Helv. Chim. Acta *50,* 2233–2243. (French)

FORSS, D. A. 1969. Role of lipids in flavors. J. Agric. Food Chem. *17,* 681–685.

FORSS, D. A., DUNSTONE, E. A., RAMSHAW, E. H., and STARK, W. 1962. The flavor of cucumbers. J. Food Sci. *27,* 90–93.

FURIA, T. E. and BELLANCA, N. 1971. Revisors and translators. *In* Femaroli's Handbook of Flavor Ingredients. The Chemical Rubber Co., Cleveland, OH.

GOLD, H. J. and WILSON, C. W. III. 1963. The volatile flavor substances of celery. J. Food Sci. *28,* 484–488.

GOLDMAN, I. M., SEIBL, J., FLAMENT, I., GAUTSCHI, F., WINTER, M., WILLHALM, B., and STOLL, M. 1967. Research on aromas. Coffee aroma. II. Pyrazines and pyridines. Helv. Chim. Acta *50,* 694–705. (French).

GOULD, R. F. Editor. 1966. Flavor Chemistry. Advances in Chemistry Ser. 56. American Chemical Society. Washington, DC.

HALL, L. A. 1948. Protein hydrolyzates as a source of glutamate flavors. *In* Monosodium Glutamate—A Symposium. Quartermaster Food and Container Institute for the Armed Forces, Chicago, IL.

HALL, R. L. 1968. Food flavors: benefits and problems. Food Technol. *22,* 1388–1392.

HARPER, R., BATE-SMITH, E. C., and LAND, D. G. 1968. Odour Description and Odour Classification. J. A. Churchill Ltd., London.

HEWITT, E. J., MACKAY, D. A. M., KONIGSBACHER, K. and HASSELSTROM, T. 1956. The role of enzymes in food flavors. Food Technol. *10,* 487–489.

HODGE, J. E. 1964. *In* Chemistry and Physiology of Flavors. H. W. Schultz, E. A. Day, and L. M. Libbey (Editors). AVI Publishing Co., Westport, CT.

HOFFMANN, G. 1961. 3-*cis*-hexenal, the "green" reversion flavor of soybean oil. J. Am. Oil Chem. Soc. *38,* 1–3.

HORNSTEIN, I., and TERANISHI, R. 1967. The chemistry of flavor. Chem. Eng. News *45* (15), 92–108.

HOROWITZ, R. M., and GENTILI, B. 1969. Taste and structure of phenolic glycosides. J. Agric. Food Chem. *17,* 696–700.

JENNINGS, W. G., and SEVENANTS, M. R. 1964. Volatile esters of Bartlett pears. III. J. Food Sci. *29,* 158–163.

JOHNSON, J. A., ROONEY, L., and SALEM, A. 1966. Chemistry of bread flavor. *In* Flavor Chemistry. R. F. Gould (Editor) Advances in Chemistry Ser. 56. American Chemical Society. Washington, DC.

JONES, R. N. 1969. Meat and fish flavors, significance of ribomononucleotides and their metabolites. J. Agric. Food Chem. *17,* 712–716.

JURIENS, G., and OELE, J. M. 1965. Determination of hydroxy-acid triglycerides and lactones in butter. J. Am. Oil Chem. Soc. *42,* 857–861.

KEENEY, P. G., and PATTON, S. 1956. The coconut-like flavor defect of milk fat. I. Isolation of the flavor compound from butter oil and its identification as δ-decalactone. J. Dairy Sci. *39,* 1104–1113.

KEFFORD, J. F. 1959. The chemical constituents of citrus fruits. *In* Advances in Food Research, Vol. 9. C. O. Chichester, E. M. Mrak, and G. F. Stewart (Editors). Academic Press, New York.

KIRIMURA, J., SHIMIZU, A., KIMIZUKA, A., NINOMIYA, T., and KATSUYA, N. The contribution of peptides and amino acids to the taste of foodstuffs. J. Agric. Food Chem. *17,* 689–695.

KUBOTA, T. and KUBO, I. 1969. Bitterness and chemical structure. Nature *223,* 97–99.

KULKA, K. 1967. Aspects of functional groups and flavor. J. Agric. Food Chem. *15,* 48–57.

KUNINAKA, A. 1966. Recent studies of 5′ nucleotides as new flavor enhancers. *In* Flavor Chemistry. R. F. Gould (Editor) Advances in Chemistry Ser. 56. American Chemical Society, Washington, DC.

KURIHARA, K., and BEIDLER, L. M. 1968. Taste-modifying protein from miracle fruit. Science *161,* 1241–1243.

KURIHARA, K., and BEIDLER, L. M. 1969. Mechanism of the action of taste-modifying protein. Nature *222,* 1176–1179.

KUSHMAN, L. J., and BALLINGER, W. E. 1968. Acid and sugar changes during ripening in Wolcott blueberries. Proc. Am. Soc. Hort. Sci. *92,* 290–295.

LINKO, Y.-Y., JOHNSON, J. A., and MILLER, B. S. 1962. The origin and fate of certain carbonyl compounds in white bread. Cereal Chem. *39,* 468–476.

MARION, J. P., MÜGGLER-CHAVAN, F., VIANI, R., BRICOUT, J., REYMOND, D., and EGLI, R. H. 1967. The composition of cocoa aroma. Helv. Chim. Acta *50,* 1509–1516. (French)

MASON, M. E., JOHNSON, B., and HAMMING, M. 1966. Flavor components of roasted peanuts. Some low molecular weight pyrazines and a pyrrole. J. Agric. Food Chem. *14,* 454–460.

MATTICK, L. R., and HAND, D. B. 1969. Identification of a volatile component in soybeans that contributes to the raw bean flavor. J. Agric. Food Chem. *17,* 15–21.

MATTICK, L. R., PATTON, S., and KEENEY, P. G. 1959. The coconut-like flavor defect of milk. III. Observations on the origin of a δ-decalactone in fat containing dairy products. J. Dairy Sci. *42,* 791–798.

MITCHELL, J. H., JR., LEINEN, N. J., MRAK, E. M., and BAILEY, S. D. 1957. Chemistry of Natural Food Flavors. National Academy of Sciences, National Research Council for Quartermaster Food and Container Institute for the Armed Forces, Washington, DC.

MONCRIEFF, R. W. 1951. The Chemical Senses. Leonard Hill Ltd., London.

MONCRIEFF, R. W. 1966. Odour Preferences. Leonard Hill Ltd., London.

NAVES, Y. R. 1957. The relationship between the stereochemistry and odorous properties of organic substances. *In* Molecular Structure and Organoleptic Quality. Soc. Chem. Ind. Monogr. 1. Society of Chemical Industry, London.

N.Y. ACADEMY OF SCIENCES. 1964. Recent advances in odor: Theory, measurement, and control. Ann. N.Y. Acad. Sci. *116,* 357–746.

OERTLY, E., and MYERS, R. G. 1919. A new theory relating constitution to taste. Preliminary paper. Simple relations between the constitution of aliphatic compounds and their sweet taste. J. Am. Chem. Soc. *41,* 855–876.

OUGH, C. S. 1963. Sensory examination of four organic acids added to wine. J. Food Sci. *28,* 101–106.

PANGBORN, R. M. 1963. Relative taste intensities of selected sugars and organic acids. J. Food Sci. *28,* 726–733.

PATTON, S. 1954. The mechanism of sunlight flavor formation in milk with special reference to methionine and riboflavin. J. Dairy Sci. *37,* 446–452.

PATTON, S. 1963. Volatile acids and the aroma of cheddar cheese. J. Dairy Sci. *46,* 856–858.

PATTON, S. 1964. Flavor thresholds of volatile fatty acids. J. Food Sci. *29,* 679–680.

PATTON, S. 1964. Volatile acids of Swiss cheese. J. Dairy Sci. *47,* 817–818.

PATTON, S., and JOSEPHSON, D. V. 1953. Methionine—origin of sunlight flavor in milk. Science *118,* 211.

RIZZI, G. P. 1967. The occurrence of simple alkylpyrazines in cocoa butter. J. Agric. Food Chem. *15,* 549–551.

ROGERS, J. A. 1966. Advances in spice flavor and oleoresin chemistry. *In* Flavor Chemistry, R. F. Gould (Editor). Advances in Chemistry Ser. 56. American Chemical Society, Washington, DC.

SAGHIR, A. R., MANN, L. K., BERNHARD, R. A., and JACOBSEN, J. V. 1964. Determination of aliphatic mono- and disulfides in *Allium* by gas chromatography and their distribution in the common food species. Proc. Am. Soc. Hort. Sci. *84,* 386–389.

SEIFERT, R. M., BUTTERY, R. G., GUADAGNI, D. G., BLACK, D. R., and HARRIS, J. G. 1970. Synthesis of some 2-methoxy-3-alkyl-pyrazines with strong bell pepper-like odors. J. Agric. Food Chem. *18,* 246–249.

SHALLENBERGER, R. S., and ACREE, T. E. 1967. Molecular theory of sweet taste. Nature *216,* 480–482.

SHALLENBERGER, R. S., and ACREE, T. E. 1969. Molecular structure and sweet taste. J. Agric. Food Chem. *17,* 701–703.

SJÖSTRÖM, L. B. 1972. The Flavor Profile. A. D. Little Inc., Cambridge, MA.

SOLMS, J. 1969. The taste of amino acids, peptides, and proteins. J. Agric. Food Chem. *17,* 686–688.
SOLMS, J. 1971. Nonvolatile compounds and the flavor of foods. *In* Gustation and Olfaction. G. Ohloff and A. F. Thomas (Editors). Academic Press, New York.
SOLMS, J., VUATAZ, L., and EGLI, R. H. 1965. The taste of L and D amino acids. Experientia *21,* 692–694.
STARK, W. and FORSS, D. A. 1962. A compound responsible for metallic flavor in dairy products. I. Isolation and identification. J. Dairy Res. *29,* 173–180.
STEVENS, K. L., BOMBEN, J., LEE, A., and McFADDEN, W. H. 1966. Volatiles from grapes. Muscat of Alexandria. J. Agric. Food Chem. *14,* 249–252.
STÖCKLIN, W., WEISS, E., and REICHSTEIN, T. 1967. Gymnemic acid, the antisaccharid principle of *Gymnema sylvestra* R. Br. Isolation and identification. Helv. Chim. Acta *50,* 474–490. (German)
STOLL, M. 1957. Facts old and new concerning relationships between molecular structure and odour. *In* Molecular Structure and Organoleptic Quality. Soc. Chem. Ind. Monogr. 1. Society of Chemical Industry, London.
STONE, H., and OLIVER, S. M. 1969. Measurement of the relative sweetness of selected sweeteners and sweetener mixtures. J. Food Sci. *34,* 215–222.
STULL, J. W. 1953. The effect of light on activated flavor development and on the constituents of milk and its products: A review. J. Dairy Sci. *36,* 1153–1164.
TARR, H. L. A. 1966. Flavor of flesh foods. *In* Flavor Chemistry. R. F. Gould (Editor). Advances in Chemistry Ser. 56. American Chemical Society, Washington, DC.
TERANISHI, R. 1971. Odor and molecular structure. *In* Gustation and Olfaction. G. Ohloff and A. F. Thomas (Editors). Academic Press, New York.
TERANISHI, R., HORNSTEIN, I., ISSENBERG, P., and WICK, E. L. 1971. Flavor Research—Principles and Techniques. Marcel Dekker, Inc., New York.
TSUZUKI, Y. 1948. Sweet taste and chemical constitution. Chem. Chem. Ind. Jpn *1,* 32–40.
VON SYDOW, E. 1971. Flavor—A chemical or psychophysical concept? Food Technol. *25,* 40–44.
WRIGHT, R. H. 1954. Odour and molecular vibration. I. Quantum and thermodynamic considerations. J. Appl. Chem. (London) *4,* 611–615.
WRIGHT, R. H. 1957. Odour and molecular vibration. *In* Molecular Structure and Organoleptic Quality. Soc. Chem. Ind. Monogr. 1. Society of Chemical Industry, London.
WRIGHT, R. H. 1964. Odour and molecular vibration: The far infrared spectra of some perfume chemicals. Ann. N.Y. Acad. Sci. *116,* Art. 2, 552–558.
WRIGHT, R. H. and MICHELS, K. M. 1964. Evaluation of far infrared relations to odor by standard similarity method. Ann. N.Y. Acad. Sci. *116,* Art. 2, 535–551.
WUCHERPFENNIG, K. 1969. Acids—A quality determining factor in wine. Dtsch. Wein Ztg. *30,* 836–840. (German)

11

Natural Colors

Chemistry of Natural Coloring Matters
Caramels and Melanoidins
Color Determination
Summary
Bibliography

Pigments are very important in foods and food products—they impart eye appeal. Unfortunately, many people have a tendency to "eat with their eyes," and it is possible to satisfy many people more because of appearance than flavor.

The bulk of the natural coloring matters in foods fall into the following four categories.

1. **Tetrapyrrole structures: chlorophylls, hemes, and the bile pigments**
2. **Isoprenoid structures: carotenoids**
3. **Benzopyran structures: anthocyanins and flavonoids**
4. **Betacyanins: betanin of the table beet root**

Some pigments occurring in commonly available foods are as follows.

1. **Green vegetables: chlorophylls, carotenoids, and less often, flavonoids**
2. **Root vegetables: carotenoids, betacyanins, anthocyanins, and flavonoids**
3. **Fruits: carotenoids, anthocyanins, flavonoids, and others**
4. **Cereals: carotenoids, caramels[1]**
5. **Syrups: caramels,[1] melanoidins,[2] and others**

[1] **Caramels are formed by heat.**
[2] **Melanoidins can be classed as artifacts as can caramels. Both are included here for convenience.**

6. Meat: hemes, sometimes bilins
7. Fish: hemes, carotenoids
8. Egg yolks: carotenoids
9. Crustaceans: carotenoids
10. Dairy products: carotenoids

The most important natural pigments from higher plant sources are the chlorophylls and the anthocyanins. The former are responsible for the green color of plants, but more importantly act as the catalytic agent in the photosynthetic process, which is essential to the production of food. The latter, the anthocyanins, are phenolic glycosides and are soluble in water.

CHEMISTRY OF NATURAL COLORING MATTERS

Tetrapyrrole Structures (Fig. 11.1)

Heme pigments. The heme pigments found in meat and fish are structurally similar to the chlorophylls. The basic difference is that iron replaces magnesium in the molecule. Also, some of the attached groups are different. The pigment of blood is hemoglobin and in muscle it is myoglobin. Both have the same prosthetic group, but the size of the attached protein is not the same. Under natural conditions, the hemes and the chlorophylls are combined with proteins.

Both hemoglobins and myoglobins can take up oxygen forming oxy compounds; this reaction is reversible. Oxymyoglobin is responsible for the bright red color of fresh meat the dark surface of fish muscle from which the skin has been removed. The iron in these compounds remains in the ferrous state. According to Brown and Tappel (1958) discoloration to brown takes place in two stages.

MbO_2	$\rightleftharpoons$	Mb	$\rightleftharpoons$	Metmb
Oxymyoglobin		**Myoglobin**		**Metmyoglobin**
Fe^{2+} (red)		**Fe^{2+} (purplish red)**		**Fe^{3+} (brownish)**

According to Tressler and Evers (1957), the darkening of bone in frozen chicken is caused by the formation of methemoglobin from the hemoglobin leached from the bone marrow. When MbO_2 is heated in the presence of oxygen, the protein (globin) is denatured, and the iron is converted from the ferrous to the ferric state. The hemichrome compound formed is tan colored. When reducing agents are present, the ferric iron is reduced to ferrous, according to Tappel (1957A, B), and the hemochrome formed is pink. The release of sulfhydryl (—SH) groups as a result of denaturation of the protein is responsible for the reducing conditions. Nitrite reacts with myoglobin, forming nitrosomyoglobin which is red but not highly stable. On heating it is changed to nitrosohemochrome, which is more stable and is the pigment found in cured meats.

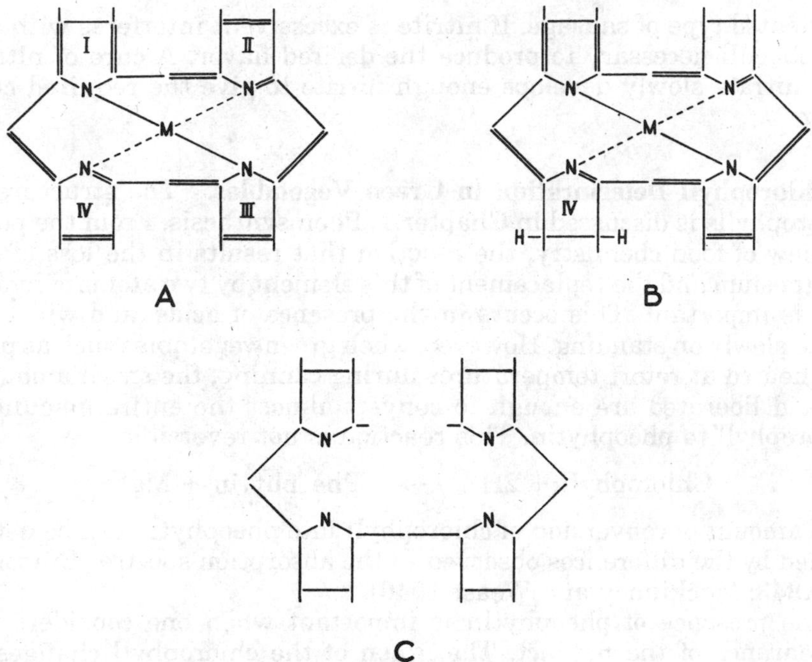

FIG. 11.1. Basic structures for tetrapyrrole pigments. (A) Structure for hemes. (B) Structure for chlorophylls. (C) Structure for bile pigments.
From Mackinney and Little (1962).

Nitrite levels are important in developing color in cured meats. When the amount used is too small, the result is weak color together with inferior light stability. If too great a quantity is used, oxidation of MbO_2 results, giving rise to greenish-gray compounds. It has been found that under high temperatures or high acid conditions, the nitrites can react with certain amino acids forming compounds known as nitrosamines. Nitrosamines are considered to be one of the most potent carcinogenic substances known to man. Nitrosamines can be formed within the highly acid conditions of the stomach by eating meat products containing nitrites. Whether this is a problem in the normal diet is not known, but is currently under investigation. Statistical analysis of populations eating large quantities of nitrite-cured meat correlate with high percentages of cancer. The FDA will probably ban the use of nitrite unless a way is found to prevent the reaction with certain amino acids (Labuza 1974).

The bilins develop only as breakdown products of the hemes. Their presence shows unnecessary loss of the required red which in fresh meat is the oxymyoglobin or in cured meat the denatured MbNO hemochrome. The grayish-green color formation is most acute in the fermented type of sausage. If nitrite is excessive it interferes with the lactobacilli necessary to produce the desired flavor. A cure of nitrite and nitrate slowly develops enough nitrite to give the required cure color

Chlorophyll Deterioration in Green Vegetables. The structure of chlorophylls is discussed in Chapter 1, Photosynthesis. From the point of view of food chemistry, the reaction that results in the loss of the magnesium and the replacement of this element by two atoms of hydrogen is important. This occurs in the presence of acids, and will take place slowly on standing. However, when green vegetables such as peas are heated at retort temperatures during canning, the small amounts of acid liberated are enough to convert almost the entire amount of chlorophyll to pheophytin. This reaction is not reversible.

$$\text{Chlorophyll} + 2H^+ \rightarrow \text{Pheophytin} + Mg^{2+}$$

The amount of conversion of chlorophyll into pheophytin can be determined by the differences observed in the absorption spectra (Dutton *et al.* 1943; Mackinney and Weast 1940).

The presence of pheophytin is important when one considers the appearance of the product. The green of the chlorophyll changes to olive-green and to olive-brown as the quantity of pheophytin increases. These changes take place in underblanched or raw vegetables which have been held in frozen storage. The impairment of the color parallels an impairment of flavor—the vegetables develop haylike flavors and become undesirable for food. These flavor changes, however, are not caused by pheophytin.

In the formula for chlorophyll, pyrrole ring IV has a side chain that is an ester of propionic acid and phytol alcohol. This can be hydrolyzed by chlorophyllase with the liberation of the free acid. Such changes can take place under canning plant conditions. Rather recent work tends to indicate that there is no apparent difference in the rates of conversion of these chlorophyll compounds into magnesium-free compounds.

The chlorophyll in dehydrated vegetables tends to bleach when the product is exposed to light over periods of time. This fading of the color of the vegetable is basically a photo-oxidation.

During the ripening process in fruits the chlorophylls decrease while frequently carotenes tend to increase. The carotenes become visible as the chlorophylls decrease. The ratio of carotenes to xanthophylls also shows an increase.

Carotenoids

The term *carotenoids* is derived from carotene, the main coloring matter in the root of the carrot. Carotenoids are found throughout the plant world, in leaves, red or yellow fruits, flowers, and roots. They are not, however, synthesized by animals. Animals must ingest carotenoids in order to get those necessary (mostly β-carotene) for vitamin A formation. They are absorbed either unchanged or transformed in the intestine.

Usually only a few micrograms dry weight of carotenoid are present in vegetable material. In the carrot, however, the dry weight amounts to about 1 mg per gram of material. Carotene as found in plants is ordinarily made up of a mixture of α-carotene, β-carotene, and γ-carotene, of which β-carotene is the most important and present to the extent of about 85% of the total carotene.

Two basic types of carotenoids are usually present in natural food colors: the carotene type and the lycopene type. Lycopene gives the red color to the red tomato and the red-fleshed watermelon. In the red tomato, the percentage of lycopene is very high. It frequently forms a part of the carotenoid in apricots and peaches. Apricots may have as much as 10% of the pigment as lycopene.

Crustaceans contain astaxanthin which is pink, and red peppers contain capsorubin which is deep red, both of which are examples of carotenoids with structures different from the carotene and lycopene types. When crustaceans such as crabs and lobsters are plunged into boiling water, the color changes from dark blue or blue-gray to red. The reason for this is that during life the carotenoid is attached to a protein. Heat destroys this combination and the color of the free carotenoid can appear.

Carotenoid Structure. Lycopene is completely aliphatic and lacks the ionone ring or rings found in the corresponding carotenes. Both end rings of β-carotene and of β-ionone structure; in α-carotene, one is α-ionone and the other is β-ionone. The β-ionone ring is necessary for the compound to be converted into vitamin A. In all of these compounds, alternate double and single bonds in conjugation make up the chromatophoric grouping. At least seven double bonds in conjugation are necessary before any amount of yellow color is perceptible.

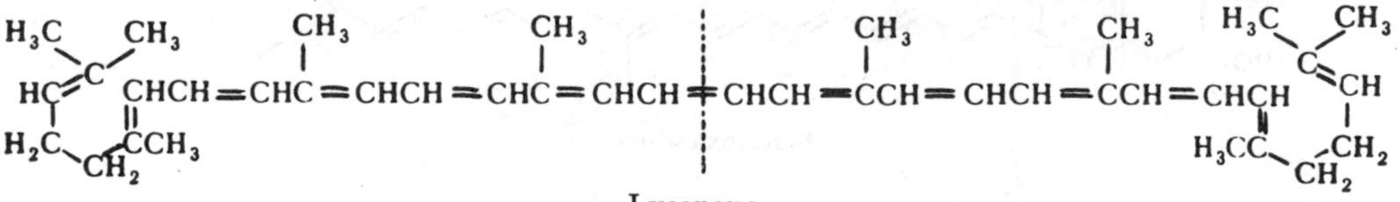

```
H3C  CH3        CH3          CH3       :      CH3          CH3     H3C  CH3
   \C/           |            |        :       |            |         \C/
HC=   CHCH=CHC=CHCH=CHC=CHCH=+CHCH=CCH=CHCH=CCH=CHCH           CH
 |     ||                              :                       ||       |
H2C  CCH3                              :                     H3CC     CH2
   CH2                                 :                         CH2
```

Lycopene

The larger number of carotenoids occurring naturally have the trans structure. The trans structure has greater stability, and the extended all-trans form has the deepest color. The structures shown are extended all-trans configuration, the form usually found in food materials. However, a few mono-cis and di-cis forms are encountered. Isomerization of the all-trans form may account for some color loss. Such isomerization can be increased by the presence of light, acid, or heat. The figure of the carotenoid molecule is V-shaped if it contains a central trans-cis rotation. An all-cis form is hypothetical and would be crumpled.

Astaxanthin of the crustaceans and capsorubin of red peppers have carbonyl groups with the carbon to oxygen double bond in conjugation with the chain of carbon to carbon double bonds. Their importance is limited.

Carotenoid Retention. The following procedures should be followed to retain carotenoids in food and food products: (1) storage at low tem-

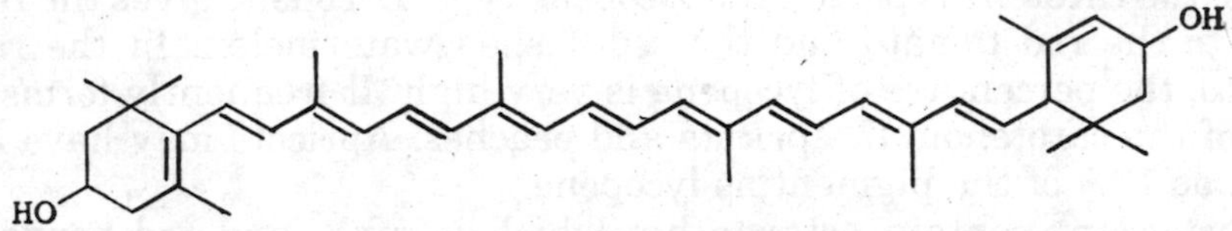

Lutein

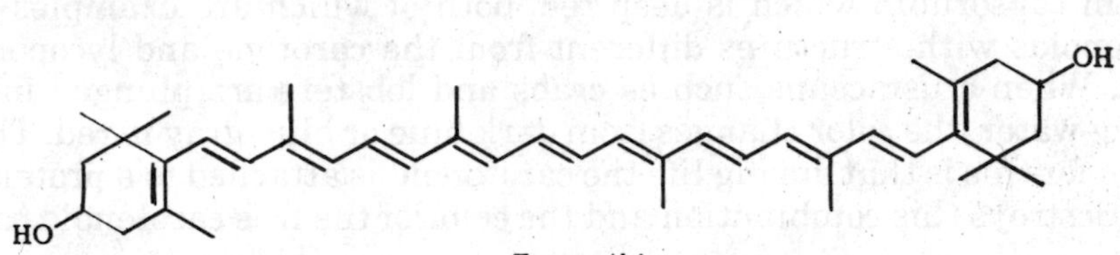

Zeaxanthin

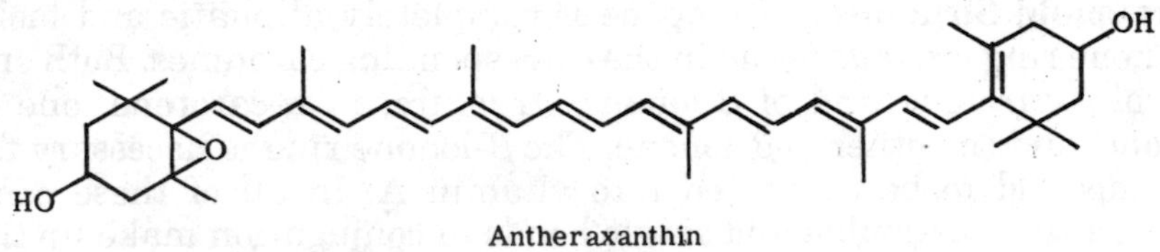

Antheraxanthin

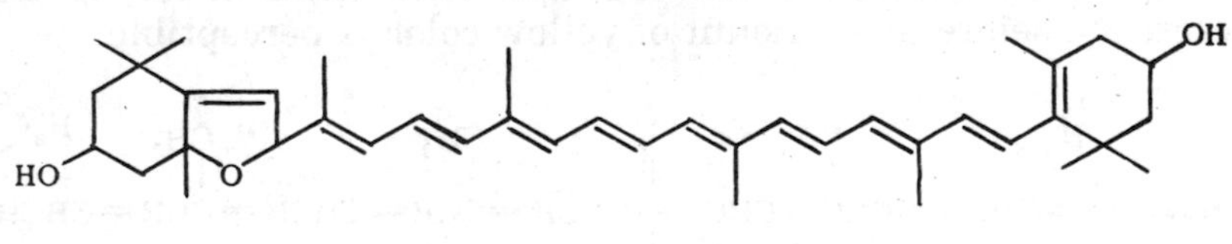

Mutatoxanthin

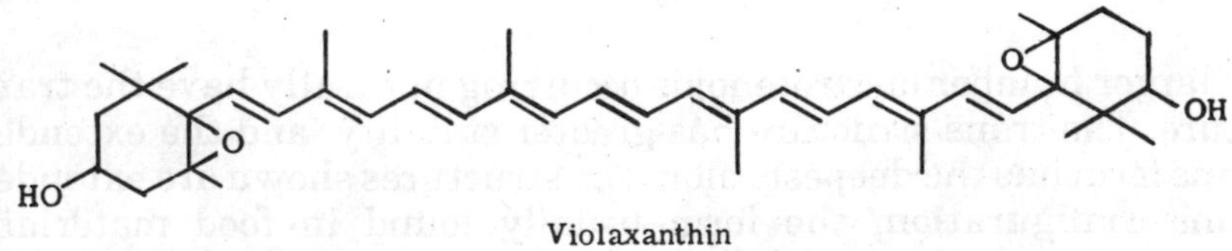

Violaxanthin

FIG. 11.2. Structural formulas of common fruit carotenoids.
From Borenstein and Bunnell (1966). *Used with permission of Academic Press, copyright owner.*

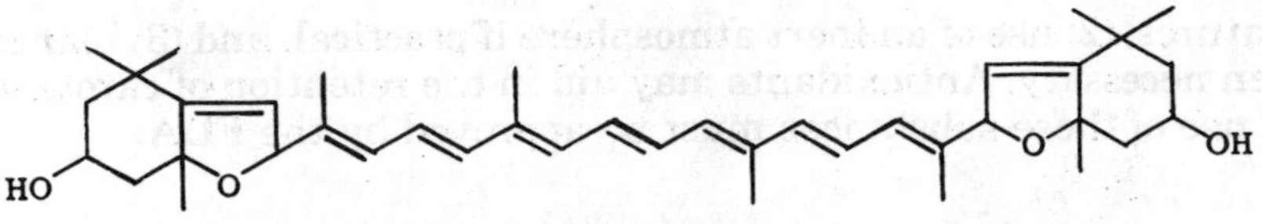

HO O O OH

Auroxanthin

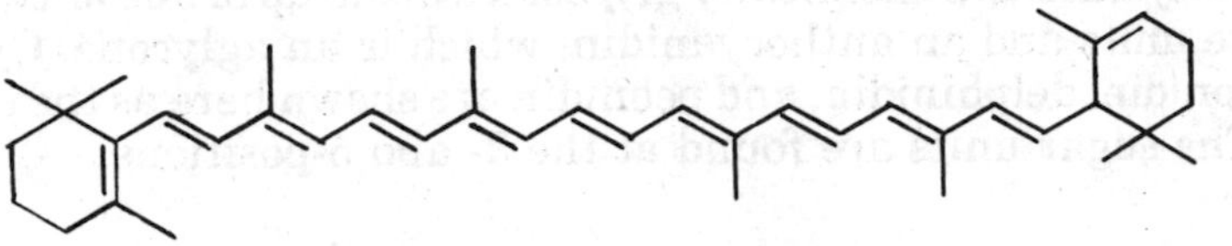

α-Carotene

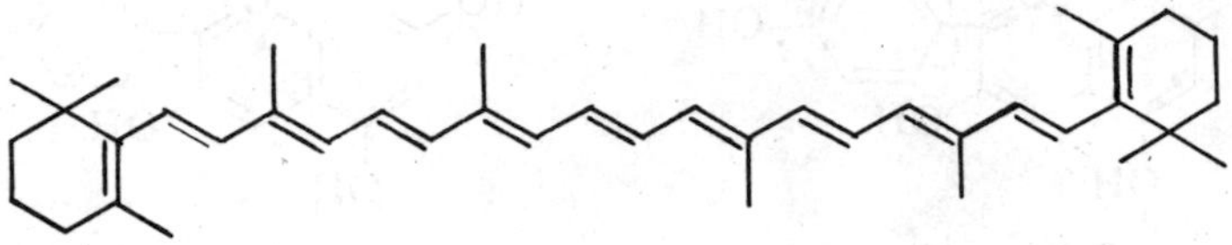

β-Carotene

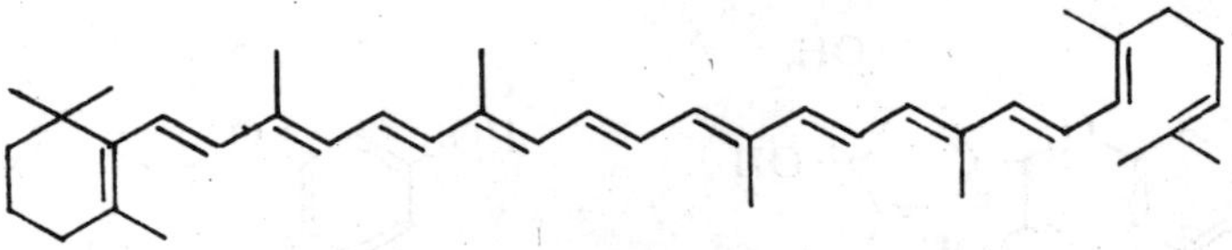

γ-Carotene

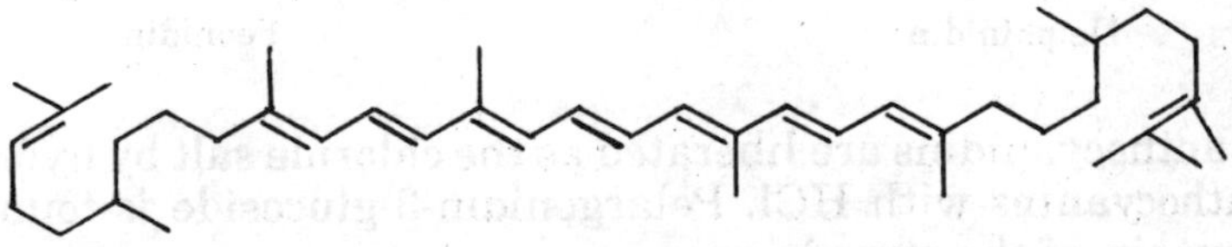

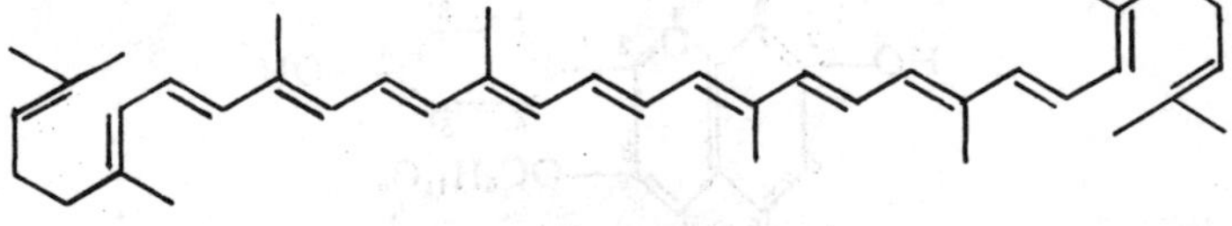

ζ-Carotene

Lycopene

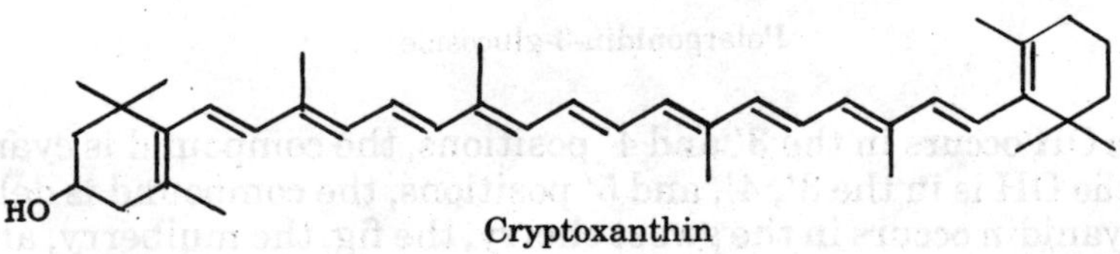

HO **Cryptoxanthin**

FIG. 11.2. (Continued)

peratures, (2) use of an inert atmosphere if practical, and (3) blanching when necessary. Antioxidants may aid in the retention of carotenoids, but use of these substances must be approved by the FDA.

Anthocyanins and Flavonoids

Anthocyanins are chemically glycosides made up of one or two carbohydrate units and an anthocyanidin, which is an aglycone. Cyanidin, pelargonidin, delphinidin, and peonidin are shown here as the oxonium ions. The sugar units are found at the 3- and 5-positions.

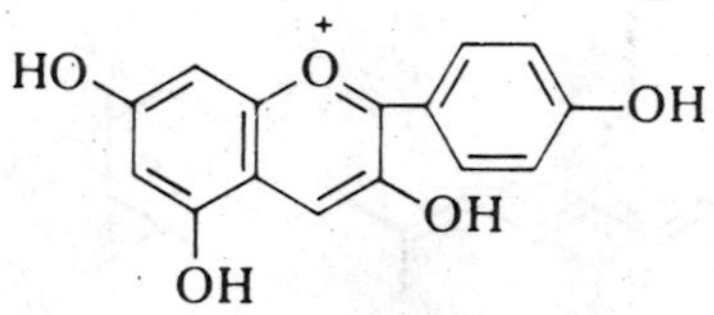

Pelargonidin

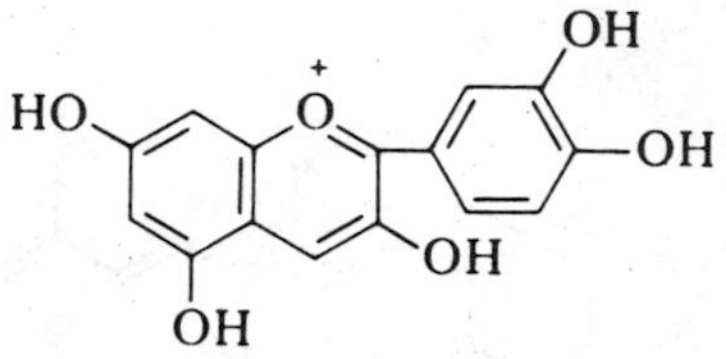

Cyanidin

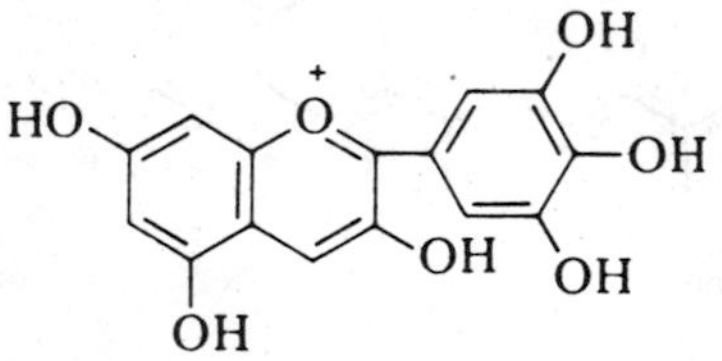

Delphinidin

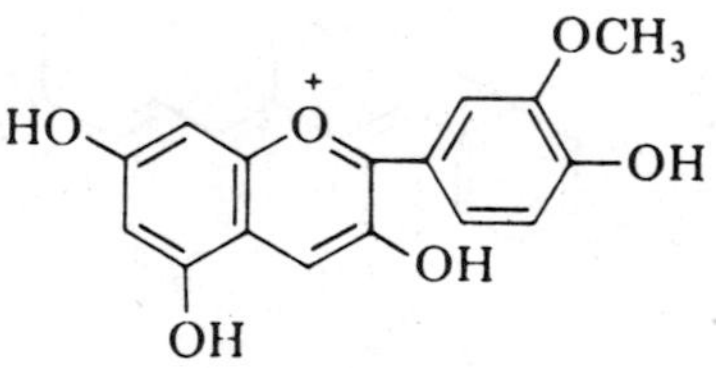

Peonidin

The anthocyanidins are liberated as the chloride salt by hydrolysis of the anthocyanins with HCl. Pelargonidin-3-glucoside is found in the anthocyanin of the strawberry.

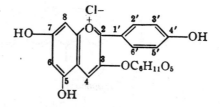

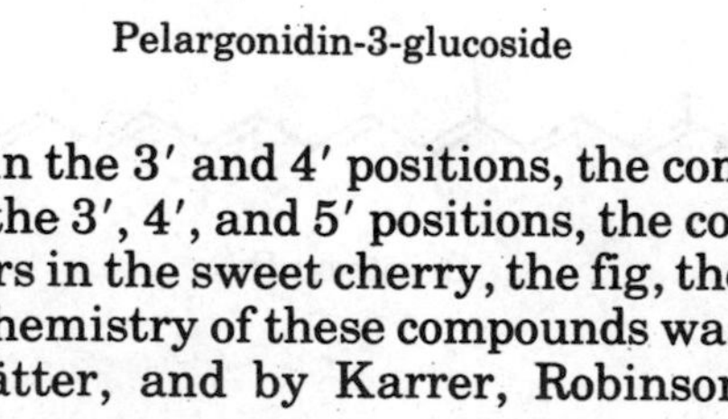

Pelargonidin-3-glucoside

If the OH occurs in the 3′ and 4′ positions, the compound is cyanidin, and if the OH is in the 3′, 4′, and 5′ positions, the compound is delphinidin. Cyanidin occurs in the sweet cherry, the fig, the mulberry, and the blood orange. The chemistry of these compounds was established by the research of Willstätter, and by Karrer, Robinson, and others. The anthocyanidins are altered in color by changes of pH and can serve as indicators. Aside from contributing color they do not seem to have any other function in plants. The depth of color in the fruit depends on the quantity of anthocyanin present.

Table 11.1 gives anthocyanidins found in some fruits and vegetables.

Several papers have been published describing the anthocyanins found in various fruits, including those found in the cranberry (Zapsalis and Francis 1965), the Red Delicious apple (Sun and Francis 1968), the Bing cherry (Lynn and Luh 1964), and Cabernet Sauvignon grapes (Somaatmadja and Powers 1963), as well as those in sour cherries, grapes, and strawberries. For example, Zapsalis and Francis (1965) found the following anthocyanins in the cranberry: cyanidin-3-monogalactoside, peonidin-3-monogalactoside, cyanidin-3-monoarabinoside, and peonidin-3-monoarabinoside, indicating that the color comprises several compounds.

Markakis *et al.* (1957) found that oxygen and ascorbic acid can be detrimental to the anthocyanin found in strawberries. When nitrogen replaced air in the headspace of stored strawberry juice they observed better retention of the anthocyanin.

Daravingas and Cain (1965) found that prolonged storage time and higher storage temperatures reduced the amount of anthocyanins in canned red raspberries. More concentrated syrup and oxygen in the headspace also increased destruction of the color.

Anthocyanins have the ability to form lakes with metals. Lakes are slate-gray or somewhat purple pigments and can be formed in canned foods when the anthocyanin combines with tin.

Jurd (1964) studied the reactions involved in sulfite bleaching of anthocyanins. He showed that this is a reversible process. It does not

TABLE 11.1. Anthocyanidins Found in Some Fruits and Vegetables

Fruit or Vegetable	Anthocyanidin
Apple	Cyanidin
Blueberry	Cyanidin, delphinidin, malvidin, petunidin, and peonidin
Cabbage, red	Cyanidin
Cherry, sweet, Bing	Cyanidin and peonidin
Grape, Concord	Cyanidin, delphinidin, peonidin, malvidin, and petunidin
Grape, European	Malvidin, peonidin, delphinidin, cyanidin, petunidin, and pelargonidin
Peach	Cyanidin
Plum	Cyanidin and peonidin
Pomegranate	Delphinidin
Radish, red	Pelargonidin
Raspberry	Cyanidin
Strawberry, cultivated	Pelargonidin and a little cyanidin

Source: Markakis (1974)

involve hydrolysis of the 3-glycosidic group or reduction of the pigment. Furthermore, it does not involve addition of bisulfite to a ketonic, chalcone derivative. The anthocyanin carbonium ion (R^+) is the reactive species in sulfite decoloration. It reacts with a bisulfate ion to produce a colorless chromen-2 (or 4)-sulfonic acid (R–SO_3H), which is similar in properties and structure to an anthocyanin carbinol base (R–OH). This reaction is shown as follows.

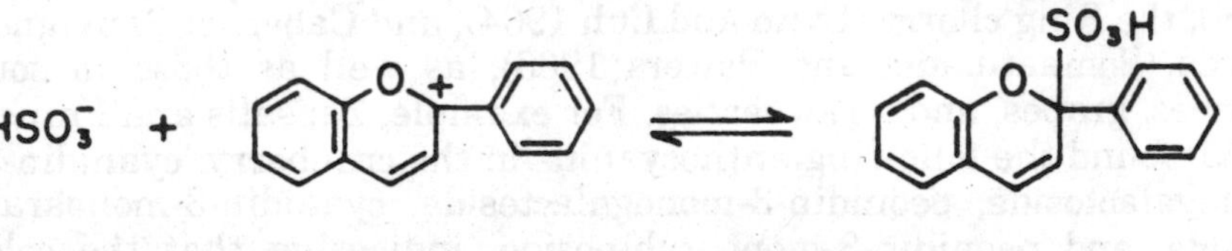

$$HSO_3^- + R^+ \rightleftharpoons R{-}SO_3H$$

Leucoanthocyanins are widely distributed in fruits and vegetables, even though no color can be seen. The color may develop as a result of hydrolysis in the presence of acid. Sometimes color is formed when oxygen is present. Leucoanthocyanins are based frequently on cyanidin. The development of pink color in canned pears, usually as a result of overheating during the cooking stage or too slow cooling after the cooking step of the canning process, may be caused by the change of a leucoanthocyanin to the colored compound.

The flavonoids are not of much importance to the color of foods because they are not strongly pigmented. However, they can, under some conditions, be responsible for off-colors in foods. The basic structure of these compounds is flavone, 2-phenylbenzopyrone. The flavonoids are known also as anthoxanthins.

These compounds occur in the plants as glycosides. Flavone is colorless, but many of the derivatives are yellow in color. These are compounds with hydroxyl or methoxyl groups in the molecule. Flavones have a double bond between carbons 2 and 3. Flavonols have an OH at position 3, while flavanones are saturated at positions 2:3. The flavanonols are saturated at 2:3 with an additional OH at position 3. The isoflavones have the phenol ring at position 3.

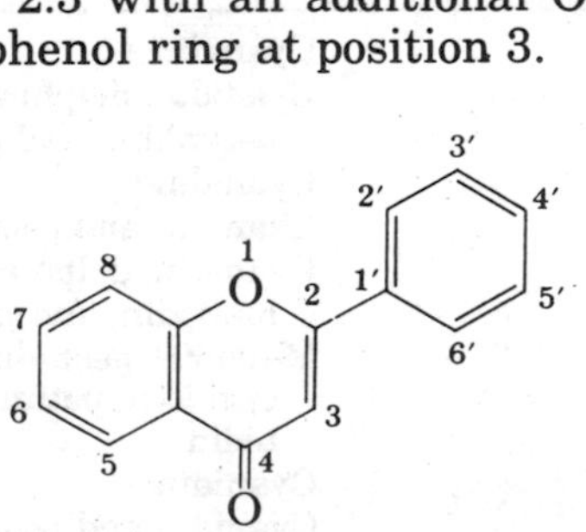

Flavone

Hesperidin, which occurs in citrus fruits, is a flavanone.

Quercetin is a flavonol. Sometimes a yellowish muddy precipitate occurs in canned asparagus which is made up largely of quercetin. It is a matter only of appearance; the product is not harmful in any way.

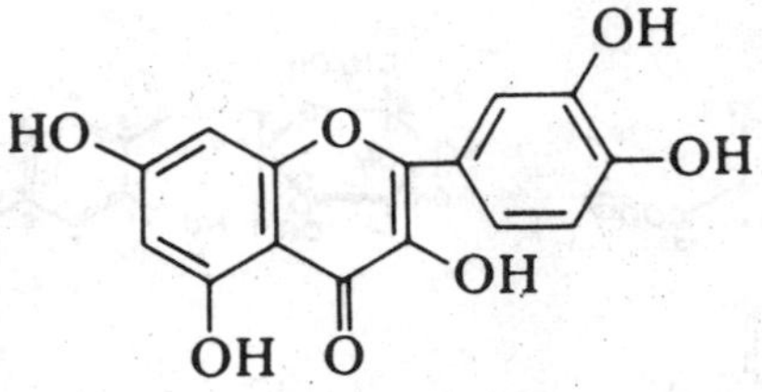

Quercetin

As far as the color of foods is concerned, the anthocyanins are vastly more important than the flavonoids.

Betacyanins

A good source of natural red color for food coloring is the table beet. Von Elbe and Maing (1973) made a study of the color compounds obtained from this vegetable. These colored compounds known as the betalains are made up of red betacyanins and yellow betaxanthins. According to these authors the important betacyanin is betanin (see III, Fig. 11-3). This makes up 75 to 95% of the total coloring matter found in the beet. The other colored compounds are isobetanin (IV), betanidin, isobetanidin, and isobetanin. In addition to these, the sulfate monoesters of betanin and isobetanin are prebetanin and isoprebetanin, respectively. The important yellow color compounds are vulgaxanthin I (V) and vulgaxanthin II (VI). Betanin is the glucoside of betanidin and isobetanin is the C-15 epimer of betanin. Von Elbe *et al.* (1974) studied the effect of pH on the stability of betanin solutions. They concluded that the color of betanin in model systems within certain ranges of pH (between pH 4.0 and pH 6.0) is the most stable. Figure 11.3 shows the visible spectra of betanin at pH 2.0, 5.0, and 9.0. Figure 11.4 shows the degradation rates for betanin in a model system at 100°C and at pH 3.0, 5.0, and 7.0. The thermostability in model systems is dependent on the pH and is greatest in the range of pH 4.0 and 5.0. In beet juice or puree the thermostability is greater than in the model systems. This makes it appear that a protective system is present in the beets. Both air and light have a degrading effect on betanin, which, in addition, is cumulative.

These authors concluded that under suitable conditions betalain pigments lend themselves for use as food colorants.

TABLE 11.2. Average Percentage Retention of Betanin in Beet Puree

Time (min)	Temperature (°C ± 1)			
	102	110	116	129
30	83	68	62	52
45	57	37	35	—
60	46	31	16	—

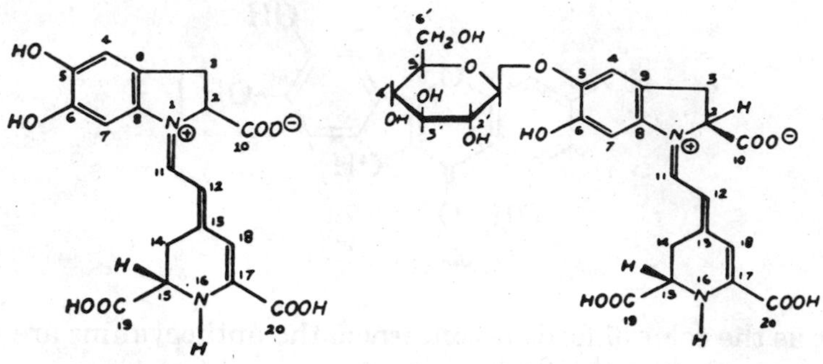

I Betanidin
II Isobetanidin, C-15
Epimer of Betanidin

III Betanin
IV Isobetanin, C-15
Epimer of Betanin

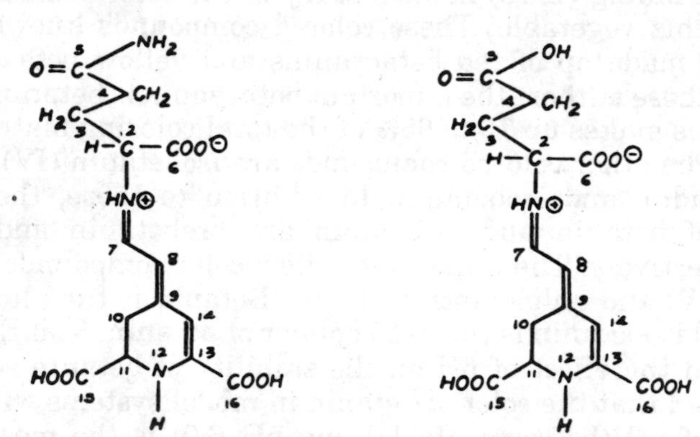

V Vulgaxanthin-I

VI Vulgaxanthin-II

FIG. 11.3. Structure of naturally occurring betalains in red beets.
From von Elbe and Maing (1973).

CARAMELS AND MELANOIDINS

These pigments are important in syrups. Caramels are formed from sugars, mainly sucrose. The concentrated sugar solutions are heated, usually with the addition of ammonia or ammonium salts to speed the process. Caramel coloring is manufactured for use in carbonated beverages, ice creams, and other such products. It can be used, together with certified food colors, to produce a darker chocolate color in bakery products. The melanoidins result from the reaction between basic nitrogenous compounds and reducing sugars, often called the Maillard

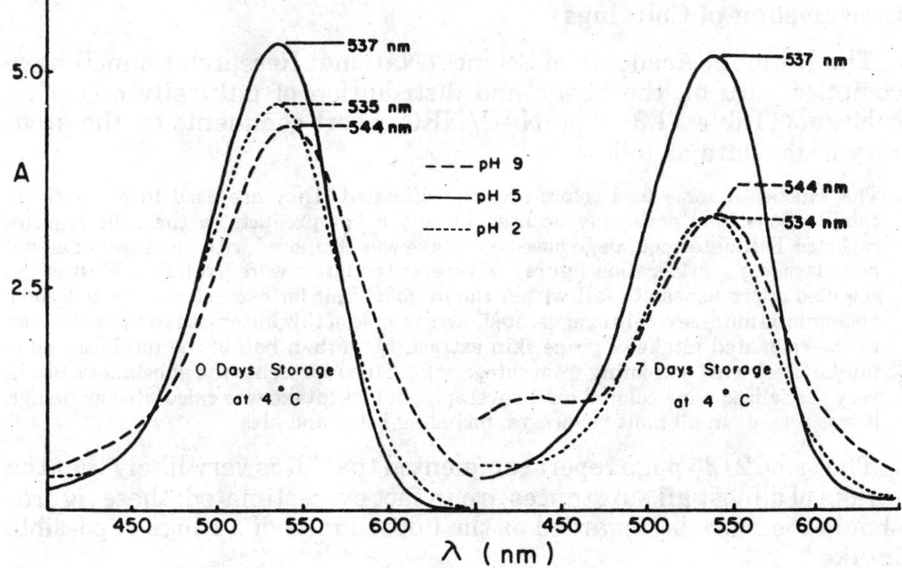

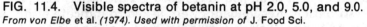

FIG. 11.4. Visible spectra of betanin at pH 2.0, 5.0, and 9.0.
From von Elbe et al. (1974). Used with permission of J. Food Sci.

reaction. These colors are included here because they are not the artificial dye compounds used for the coloring of foods. These compounds will be discussed in Chapter 12, Browning Reactions and Chapter 13, Food Colorings.

COLOR DETERMINATION

The student should consult the table of reactions of some natural colorings given in "Official Methods of Analysis, 11th Edition" (AOAC 1970). Methods for the determination of the natural colorings in foods can also be found in that publication.

A number of methods and instruments are used to measure color in foods. Perhaps the most useful and versatile of the tristimulus instruments is the Hunter Color Difference Meter. It measures values (from lightness to darkness in color), hue, and saturation in the same way as the human eye. It has the advantage of not tiring as does the eye, and it can be used under varying light conditions. In the use of this instrument it is not necessary to alter the sample in any way before reading. One can expose a tomato, a sample of fruit juice, a finely divided powder, a puree, or any other such sample to the instrument and get a value of the color.

Consumption of Colorings

The National Academy of Science / National Research Council have compiled data on the usage and distribution of naturally occurring colorants (Table 11.3). The NAS/NRC report comments on the accuracy of the data as follows:

> The intakes of some food colors are overestimated. They are used in a variety of combinations, and each may be used in only a few products in the subcategories reported. Estimated poundage based on intake was compared with measures of actual poundage (e.g., certification figures) whenever the latter were available. Most of the certified colors appear to fall within the *fivefold* limit for overestimation, but some appear to be more severely exaggerated. An example of this latter situation can be seen in the estimated intake of grape skin extract. More than half of the total estimated intake is reported as coming from category 24. This represents the substance's use in very specialized wine-colored malt beverages, but its intake was calculated as though it was present in all malt beverages, including beers and ales.

The same 2135 page report commented that "It is very likely that the intake of almost all substances are in fact overestimated; these figures should therefore be regarded as the upper limits of a range of possible intake."

The accuracy of data on poundage used, on the other hand, is subject to *under*estimation. Respondents were asked to indicate either actual poundage or a range. If a range was reported, a point one-third the way "up" the range was recorded for totaling purposes. Also, some respondents did not report poundage, and not every food processor responded to the survey. As a result, the report comments that "poundage values must be interpreted only as rough indications of the actual amounts used in food, and in almost all instances will be underestimates. The degree of underestimation will vary with the substance and in some cases may be quite severe."

SUMMARY

Colored compounds are important in foods and food products, mainly because of eye appeal. Most of the natural coloring matters in foods are included in four groups: (1) tetrapyrrole structures (chlorophylls, hemes, and the bile pigments); (2) isoprenoid structures (carotenoids); (3) benzopyran structures (anthocyanins and flavonoids); (4) betacyanins (betanin of the table beet root).

The tetrapyrrole structures—the chlorophylls and the hemes—differ in that magnesium is the metal in the chlorophylls while iron is the metal in the hemes. In addition, some of the attached groups are different. In their naturally occurring state, the hemes and the chlorophylls are combined with proteins.

Hemoglobins and myoglobins can take up oxygen to form oxy compounds. Oxymyoglobin is responsible for the bright red color of fresh meat. The iron remains in the ferrous state. Nitrosohemochrome is the pigment found in cured meats. Nitrite levels are important in developing color in the cured meats. Too little gives a weak color with inferior light stability; too much gives rise to greenish-gray compounds.

When green vegetables are heated at retort temperatures, enough acid is produced to replace the magnesium in the chlorophyll with two atoms of hydrogen, which converts the chlorophyll to pheophytin. As the reaction advances, the bright green of the chlorophyll changes to olive green and then to olive brown.

Carotenoids are found throughout the plant world. They are ingested by animals and used in the formation of vitamin A.

Carotene as found in plants is made up of α-carotene, β-carotene, and γ-carotene. Of these, β-carotene is the most important.

The larger number of carotenoids occurring naturally have the *trans* structure. It has the greatest stability, and the extended all *trans* structure has the deepest color. This is the form usually found in foods.

Anthocyanins are chemically glycosides made up of one or two carbohydrate units and an anthocyanidin. This latter is an aglycone. The sugar units are found at the 3- and 5-positions.

The anthocyanidins are ordinarily liberated as the chlorides by hydrolyzing the anthocyanins with HCl. The anthocyanidins are altered in color by changes of pH and can serve as indicators. The anthocyanins, aside from contributing color, seem to have no other function in plants. In many fruits the color is made up of more than one anthocyanin.

It has been shown that the bleaching of anthocyanins with sulfite is a reversible process. It does not involve hydrolysis of the 3-glycosidic group or reduction of the pigment. Also it does not involve addition of sulfite to a ketonic, chalcone derivative.

Leucoanthocyanins, frequently based on cyanidin, are known to be widely distributed in fruits and vegetables. Color may develop as a result of hydrolysis in the presence of acid. Sometimes color is formed when oxygen is present.

The flavonoids are not of much importance to the color of foods because they are not strongly pigmented. They can be responsible for off-colors in foods. Their basic structure is flavone, 2-phenylbenzopyrone. These compounds occur in plants as glycosides. Flavone is colorless, but many of the derivatives are yellow in color. They are compounds with hydroxyl or methoxyl in the molecule. Different types of these compounds have alterations in the basic structure. The anthocyanins are vastly more important than the flavonoids.

A good source of natural red color for food coloring is the table beet. These colored compounds are known as the betalains and are made up of red betacyanins and yellow betaxanthins. In the beet the important betacyanin is betanin.

A study of the stability of betanin solutions in model systems showed that these solutions are most stable in certain ranges of pH. This is true

TABLE 11.3. Food Color Consumption Data

Colorants from naturally occurring sources	Total U.S. usage in 1976 (lb)	Per capita average daily intake		Distribution of usage among food categories (numbers in boldface refer to food category codes at bottom of this page)
		Age group	mg	
β-apo-8′-carotenal	6,100	6-23 mo	7.7	**8**, 68%; **4**, 26%; **22**, 5%
		6-12 yr	8.6	
		18-44 yr	6.7	
Canthaxanthin	64	6-23 mo	1.3	**21**, 69%; **14**, 11%; **4**, 9%; **10**, 9%; **3**, 1%; **27**, 1%
		6-12 yr	1.4	
		18-44 yr	1.4	
Annatto extract	630,000	6-23 mo	15	**1**, 26%; **10**, 18%; **6**, 12%; **14**, 9%; **15**, 9%; **22**, 6%; **2**, 5%; **4**, 5%; **7**, 4%
		6-12 yr	34	
		18-44 yr	33	
Paprika	7,100,000	6-23 mo	330	**10**, 79%; **6**, 4%; **14**, 4%; **15**, 3%; **4**, 2%; **12**, 2%; **3**, 1% **13**, 1%; **21**, 1%
		6-12 yr	770	
		18-44 yr	1,100	
Paprika oleoresin	500,000	6-23 mo	16	**10**, 47%; **1**, 11%; **4**, 10%; **14**, 7%; **15**, 6%; **45**, 5%; **3**, 4%; **21**, 4%; **6**, 1%
		6-12 yr	36	
		18-44 yr	47	
Turmeric	1,400,000	6-23 mo	23	**1**, 23%; **21**, 21%; **27**, 17%; **10**, 15%; **15**, 9%; **5**, 4%; **13**, 4%; **45**, 3%; **14**, 2%
		6-12 yr	43	
		18-44 yr	46	
Turmeric oleoresin	54,000	6-23 mo	2.6	**21**, 43%; **1**, 35%; **15**, 9%; **22**, 8%; **11**, 3%; **13**, 1%
		6-12 yr	3.5	
		18-44 yr	3.4	

Saffron	890	6-23 mo	2.4	**21**, 46%; **1**, 44%; **10**, 10%
		6-12 yr	3.4	
		18-44 yr	3.4	
Cochineal extract (carmine)	1,600	6-23 mo	11	**1**, 59%; **16**, 31%; **21**, 10%
		6-12 yr	24	
		18-44 yr	25	
Grape skin extract (*Enocianina*)	87,000	6-23 mo	110	**24**, 55%; **23**, 37%; **8**, 8%
		6-12 yr	110	
		18-44 yr	430	
Beet powder	220,000	6-23 mo	92	**23**, 35%; **1**, 31%; **10**, 21%; **20**, 6%; **21**, 3%; **4**, 2%; **15**, 1%
		6-12 yr	160	
		18-44 yr	130	
Carrot oil	10,000	6-23 mo	0.82	**4**, 46%; **27**, 25%; **8**, 20%; **23**, 9%
		6-12 yr	1.3	
		18-44 yr	1.5	

Codes and food categories

01 Baked goods
02 Breakfast cereals
03 Grain products, pasta
04 Fats and oils
05 Milk, milk products
06 Cheese
07 Frozen dairy dessert
08 Fruits, juices
09 Fruit and water
10 Meats, meat products
11 Poultry, poultry products
12 Eggs, egg products
13 Fish, seafood
14 Vegetables, juices
15 Condiments, relish
16 Candy
17 Sugars, frosting
19 Sweet sauces, toppings
20 Gelatins, puddings, custard
21 Soup, soup mixes
22 Snack foods
23 Nonalcoholic beverages
24 Alcoholic beverages
27 Gravies, sauces
28 Dairy product analogs
31 Chewing gum
45 Main dishes, not elsewhere classified
49 Coffee and tea

also for thermostability in model systems. It seems also that a protective system exists in the beet for this compound.

Caramels and melanoidins are pigments important in syrups. The concentrated sugar solutions are heated, usually with the addition of ammonia or ammonium salts to speed the process. The melanoidins are formed from the reaction between basic nitrogenous compounds and reducing sugars, often called the Maillard reaction.

BIBLIOGRAPHY

AMERICAN MEAT INSTITUTE FOUNDATION. Bulletins. Chicago, Illinois.

ANON. 1980. Food colors. A scientific summary by the Institute of Food Technologists Expert Panel on Food Safety and Nutrition and Committee on Public Information. Food Technol. *34*(7), 77–84.

AOAC. 1970. Official Methods of Analysis, 11th Edition. Assoc. of Official Analytical Chemists, Washington, DC.

AOAC. 1980. Official Methods of Analysis, 13th Edition. Assoc. of Official Analytical Chemists, Washington, DC.

BATE-SMITH, E. C. 1954. Flavonoid compounds in foods. *In* Advances in Food Research, Vol. 5, E. M. Mrak, and G. F. Stewart (Editors). Academic Press, New York.

BORENSTEIN, B., and BUNNELL, R. H. 1967. Carotenoids: properties, occurrence, and utilization in foods. *In* Advances in Food Research, Vol. 15. C. O. Chichester, E. M. Mrak, and G. F. Stewart (Editors). Academic Press, New York.

BROWN, W. D., and TAPPEL, A. L. 1958. Oxidative changes in hematin pigments. Wallerstein Lab. Commun. *21,* 299–308.

BROWN, W. D., TAPPEL, A. L., and OLCOTT, H. S. 1958. The pigments of off-color cooked tuna meat. Food Res. *23,* 262–268.

CHEN, S. L., and GUTMANIS, F. 1968. Autoxidation of the extractable color pigments in chili pepper with special reference to ethoxyquin treatment. J. Food Sci. *33,* 274–280.

CHICHESTER, C. O., and McFEETERS, R. 1971. Pigment degeneration during processing and storage. *In* The Biochemistry of Fruits and Their Products. A. C. Hulme (Editor). Academic Press, New York.

CHIRIBOGA, C. D., and FRANCIS, F. J. 1973. Ion exchange purified anthocyanin pigments as a colorant for cranberry juice cocktail. J. Food Sci. *38,* 464–467.

CLYDESDALE, F. M., and FRANCIS, F. J. 1970. Color scales. Food Prod. Dev. *3,* 117, 120, 122, 124–125.

CO, H. J., and MARKAKIS, P. 1968. Flavonoid compounds in the strawberry fruits. J. Food Sci. *33,* 281–283.

CURL, A. L., and BAILEY, G. F. 1956. Carotenoids of aged canned Valencia orange juice. J. Agric. Food Chem. *4,* 159–162.

CURL, A. L., and BAILEY, G. F. 1957. The carotenoids of tangerines. J. Agric. Food Chem. *5,* 605–608.

DARAVINGAS, G., and CAIN, R. F. 1965. Changes in the anthocyanin pigments of raspberries during processing and storage. J. Food Sci. *30,* 400–405.

DIETRICH, W. C. 1958. Determination of the conversion of chlorophyll to pheophytin. Food Technol. *12,* 428.

DUTTON, H. J., BAILEY, G. F., and KOHAKE, E. 1943. Dehydrated spinach. Changes in color and pigments during processing and storage. Ind. Eng. Chem. *35,* 1173–1177.

ELLSWORTH, R. K. 1972. Chlorophyll biosynthesis. *In* Advances in the Chemistry of Plant Pigments, C. O. Chichester (Editor). Academic Press, New York.

EVANS, R. M. 1948. An Introduction to Color. John Wiley & Sons, New York.

FALCONER, M. E., FISHWICK, M. J., LAND, D. G., and SAYER, E. R. 1964. Carotene oxidation and off-flavor development in dehydrated carrot. J. Sci. Food Agric. *15,* 897–901.

FOX, J. B. 1966. The chemistry of meat pigments. J. Agric. Food Chem. *14,* 207–210.

FULEKI, T., and FRANCIS, F. J. 1968. Quantitative methods for anthocyanins. 3. Purification of cranberry anthocyanins. J. Food Sci. *33,* 266–274.

GEISSMAN, T. A., and HINREINER, ELLY. 1952. Theories of the biogenesis of flavonoid compounds. Botan. Rev. *18,* 77–244.

GINGER, I. D., WILSON, G. D., and SCHWEIGERT, B. S. 1954. Biochemistry of myoglobin. Quantitative determination in beef and pork muscle. Chemical studies with purified metmyoglobin. J. Agric. Food Chem. *2,* 1037–1040.

GOODWIN, T. W. 1955. Carotenoids. Annu. Rev. Biochem. *24,* 497–522.

GREENSHIELDS, R. N. 1973. Caramel—Part 2. Manufacture, Composition, and Properties. Process Biochem. *8,* 17–20.

HOROWITZ, R. M. 1961. The citrus flavonoids. *In* The Orange, W. B. Sinclair (Editor). Univ. of California Press, Berkeley, California.

JURD, L. 1964. Reactions involved in sulfite bleaching of anthocyanins. J. Food Sci. *29,* 16–19.

JURD, L. 1972. Recent progress in the chemistry of flavylium salts. *In* Symposia of the Phytochemical Society of North America, Vol. 5, Structural and Functional Aspects of Phytochemistry, V. C. Runeckles and T. C. Tso (Editors). Academic Press, New York.

JURD, L. 1972. Some advances in the chemistry of anthocyanin-type plant pigments. *In* Advances in the Chemistry of Plant Pigments, C. O. Chichester (Editor). Academic Press, New York.

KARRER, P., and JUCKER, E. 1950. Carotenoids. Elsevier Publishing Co., NewYork.

KATZ, J. J. 1972. Chlorophyll function in photosynthesis. *In* Advances in the Chemistry of Plant Pigments, C. O. Chichester (Editor). Academic Press, New York.

LABUZA, T. P. 1974. Food for Thought. AVI Publishing Co., Westport, CT.

LANDROCK, A. H., and WALLACE, G. A. 1955. Discoloration of fresh red meat and its relationship to film oxygen permeability. Food Technol. *9,* 194–196.

LEES, D. H., and FRANCIS, F. J. 1971. Quantitative methods for anthocyanins. 6. Flavonols and anthocyanins in cranberries. J. Food Sci. *36,* 1056–1060.

LUKTON, A., CHICHESTER, C. O., and MACKINNEY, G. 1955. Characterization of a second pigment in strawberries. Nature *176,* 790.

LYNN, D. Y. C., and LUH, B. S. 1964. Anthocyanin pigments in Bing cherries. J. Food Sci. *29,* 735–743.

MACKINNEY, G. 1960. Carotenoids and Vitamin A. *In* Metabolic Pathways, D. M. Greenberg (Editor). Academic Press, New York.

MACKINNEY, G., and CHICHESTER, C. O. 1954. The color problem in foods. *In* Advances in Food Research, Vol. 5, E. M. Mrak and G. F. Stewart (Editors). Academic Press, New York.

MACKINNEY, G., and LITTLE, A. 1962. Color of Foods. AVI Publishing Co., Westport, CT.

MACKINNEY, G., and WEAST, C. A. 1940. Color changes in green vegetables. Ind. Eng. Chem. *32,* 392–395.

MARKAKIS, P. 1974. Anthocyanins. *In* Encyclopedia of Food Technology. A. H. Johnson and M. S. Peterson (Editors). AVI Publishing Co., Westport, CT.

MARKAKIS, P., LIVINGSTON, G. E., and FELLERS, C. R. 1957. Aspects of strawberry pigment degradation. Food Res. *22,* 117–130.

MÖHLER, K. 1974. Formation of curing pigments by chemical, biochemical, or enzymatic reactions. Proc. Int. Symp. on Nitrite in Meat Products. Center for Agricultural Publishing and Documentation, Wageningen, The Netherlands.

MUNSELL, A. H. 1905, 1947. A Color Notation. Munsell Color Co., Baltimore.

NAKAYAMA, T. O. M., and POWERS, J. J. 1972. Absorption spectra of anthocyanin in *vivo. In* Advances in the Chemistry of Plant Pigments, C. O. Chichester (Editor). Academic Press, New York.

PETERSON, R. G., and JOSLYN, M. A. 1960. The red pigment of the root of the beet (*Beta vulgaris*) as a pyrrole compound. Food Res. *25,* 429–441.

PHILIP, T. 1974. Anthocyanins of beauty seedless grapes. J. Food Sci. *39,* 449–451.

PIFFERI, P. G., and CULTRERA, R. 1974. Enzymatic degradation of anthocyanins: The role of sweet cherry polyphenol oxidase. J. Food Sci. *39,* 786–791.

RAMAKRISHNAN, T. V., and FRANCIS, F. J. 19743. Color and carotenoid changes in heated paprika. J. Food Sci. *38,* 25–28.

SAKELLARIADES, H. C., and LUH, B. S. 1974. Anthocyanins in Barbera grapes. J. Food Sci. *39,* 329–333.

SHRIKHANDE, A. J., and FRANCIS, F. J. 1973. Anthocyanin pigments of sour cherries. J. Food Sci. *38,* 649–651.

SHRIKHANDE, A. J., and FRANCIS, F. J. 1973. Flavonol glycosides of sour cherries. J. Food Sci. *38,* 1035–1037.

SIMPSON, K. L. 1972. The biosynthesis of yeast carotenoids. *In* Advances in the Chemistry of Plant Pigments, C. O. Chichester (Editor). Academic Press, New York.

SINGLETON, V. L. 1972. Common plant phenols other than anthocyanins, contributions to coloration and discoloration. *In* Advances in the Chemistry of Plant Pigments, C. O. Chichester (Editor). Academic Press, New York.

SOLBERG, M. 1970. The chemistry of color stability in meat—A review. Can. Inst. Food Technol. J. *3,* 55–62.

SOMAATMADJA, D., and POWERS, J. J. 1963. Anthocyanins. IV. Anthocyanin pigments of Cabernet Sauvignon grapes. J. Food Sci. *28,* 617–622.

SONDHEIMER, E., and KERTESZ, Z. I. 1948. The anthocyanin of strawberries. J. Am. Chem. Soc. *70,* 3476–3479.

SONDHEIMER, E., and LEE, F. A. 1949. Color change of strawberry anthocyanin with D-glucose. Science *109,* 331–332.

SONG, P-S., MOORE, T. A., and SUN, M. 1972. Excited states of some plant pigments. *In* Advances in the Chemistry of Plant Pigments, C. O. Chichester (Editor). Academic Press, New York.

STARR, M. S., and FRANCIS, F. J. 1973. Effect of metallic ions on color and pigment content of cranberry juice cocktail. J. Food Sci. *38,* 1043–1046.

STITT, F., BICKOFF, E. M., BAILEY, G. F., THOMPSON, C. R., and FRIEDLANDER, S. 1951. Spectrophotometric determination of beta carotene stereoisomers in alfalfa. J. Assoc. Off. Agric. Chem. *34,* 460–471.

SUN, B. H., and FRANCIS, F. J. 1968. Apple anthocyanins: identification of cyanidin-7-arabinoside. J. Food Sci. *32,* 647–649.

TAPPEL, A. L. 1957A. Reflectance spectral studies of the hematin pigments of cooked beef. Food Res. *22,* 404–407.

TAPPEL, A. L. 1957B. Spectral studies of the pigments of cooked cured meats. Food Res. *22,* 479–482.

TRESSLER, D. K., and EVERS, C. F. 1957. The Freezing Preservation of Foods, Vol. 1 and 2. AVI Publishing Co., Westport, CT.

URBAIN, W. M., and JENSEN, L. B. 1940. The heme pigments of cured meats. I. Preparation of nitric oxide hemoglobin and stability of the compound. Food Res. *5,* 593–606.

VAN BUREN, J. P., HRAZDINA, G., and ROBINSON, W. B. 1974. Color of anthocyanin solutions expressed in lightness and chromaticity terms. Effect of pH and type of anthocyanin. J. Food Sci. *39,* 325–328.

VON ELBE, J. H., and MAING, I.-Y. 1973. Betalains as possible food colorants of meat substitutes. Cereal Sci. Today *18,* 263–264, 316–317.

VON ELBE, J. H., MAING, I.-Y., and AMUNDSON, C. H. 1974. Color stability of betain. J. Food Sci. *39,* 334–337.

WAGENKNECHT, A. W., LEE, F. A., and BOYLE, F. P. 1952. The loss of chlorophyll in green peas during frozen storage and analysis. Food Res. *17,* 343–350.

WATTS, B. M. 1954. Oxidative rancidity and discoloration in meat. *In* Advances in Food Research, Vol. 5, E. M. Mrak and G. F. Stewart (Editors). Academic Press, New York.

WECKEL, K. G., SANTOS, B., HERNAN, E., LAFERRIERE, L., and GABELMAN, W. H. 1962. Carotene components of frozen and processed carrots. Food Technol. *16*(8), 91–94.

WEEKS, O. B., and ANDREWS, A. G. 1972. Naturally occurring nonapreno and decapreno carotenoids. *In* Advances in the Chemistry of Plant Pigments, C. O. Chichester (Editor). Academic Press, New York.

WHITAKER, J. R. 1972. Principles of Enzymology for the Food Sciences. Marcel Dekker Inc., New York.

WRIGHT, W. D. 1958. The Measurement of Colour. Macmillan Co., New York.

WROLSTAD, R. E., and ERLANDSON, J. A. 1973. Effect of metal ions on the color of strawberry puree. J. Food Sci. *38,* 460–463.

YAMAMOTO, H. Y. 1972. Reaction center chlorophylls. *In* Advances in the Chemistry of Plant Pigments, C. O. Chichester (Editor). Academic Press, New York.

YOKOYAMA, H., GUERRERO, H. C., and BOETTGER, H. 1972. Recent studies on the structures of citrus carotenoids. *In* Advances in the Chemistry of Plant Pigments, C. O. Chichester (Editor). Academic Press, New York.

ZAPSALIS, C., and FRANCIS, F. J. 1965. Cranberry anthocyanins. J. Food Sci. *30,* 396–399.

12

Browning Reactions

Enzymatic Browning
Non-Enzymatic Browning
The Formation of Brown Pigments
Inhibition of Browning
Summary
Bibliography

There are two important forms of browning, enzymatic and non-enzymatic. This color development is usually undesirable, but with a knowledge of the type of reaction involved, it is easier to work out methods for controlling this change. Occasionally this color change is desirable, as in the browning of grapes during drying to produce raisins and during the drying of pruneplums to produce prunes. Long usage of such products has made these changes in color acceptable.

ENZYMATIC BROWNING

A group of enzymes, collectively called "phenolase" is responsible for browning of some cut fruits and vegetables. This group includes such diverse enzymes as phenoloxidase, cresolase, dopa oxidase, catecholase, tyrosinase, polyphenoloxidase, potato oxidase, sweet potato oxidase, phenolase complex.

The use of the term phenolase suffices because studies with the purified or isolated enzyme show that it acts as the catalytic agent in the two different reactions which take place. One of these is the oxidation of *o*-dihydroxyphenols to *o*-quinones, an example of which is the oxidation of catechol to *o*-benzoquinone.

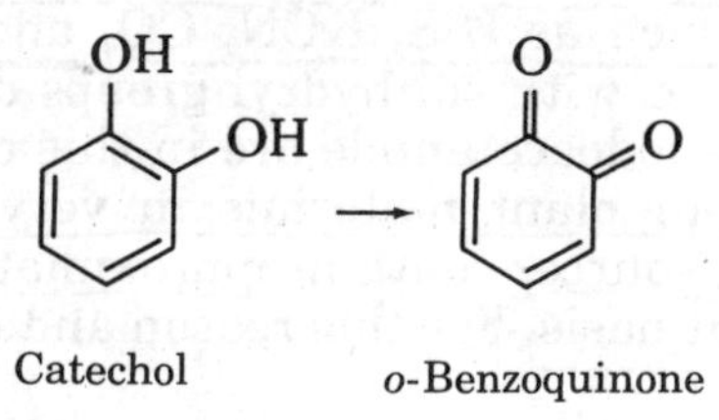

OH, OH → O, O

Catechol → *o*-Benzoquinone

The other example is the hydroxylation of certain monohydroxyphenols to dihydroxyphenols, illustrated by the hydroxylation of *p*-cresol to 3,4-dihydroxytoluene.

OH → OH, OH

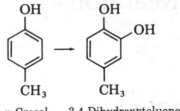

CH_3 CH_3

p-Cresol 3,4-Dihydroxytoluene

When the surfaces of fruits and vegetables are cut, the browning is caused by reactions of the *o*-quinones. These are oxidations that are catalyzed nonenzymatically, followed by polymerization of the oxidation products.

Phenolase is extensively distributed in animals and plants. It is found in such plant materials as squashes, roots, citrus fruits, plums, bananas, peaches, pears, melons, olives, tea, mushrooms, and others. Laccase, another enzyme present in many plant products, has been reported in apples, potatoes, cabbage, and sugar beets. Laccase has not received a great deal of attention as far as foods are concerned, and has been reported to oxidize only polyhydric phenols.

Phenolase has a molecular weight of 128,000 and a copper content of 0.2%, amounting to four molecules of Cu to each enzyme molecule. The freshly prepared enzyme contains copper in the cuprous form, but it slowly oxidizes to the cupric form on aging. This change does not result in the loss of any activity. The apoenzyme, which is free of copper, is not active, but its activity can be restored by the addition of cupric copper. The apoenzyme can be prepared by dialysis of aqueous solutions of the enzyme against dilute solution of potassium cyanide. Phenolase in the pure form is colorless, whereas purified laccase is blue. Concentrated solutions of phenolase are most stable at the neutral pH. Heating for a short time at 60°C inactivates the enzyme. Concentrated solutions of this enzyme in dilute phosphate buffer at a near neutral pH can be held at 1°C or frozen at −25°C without loss of activity for several months. However, it loses activity on prolonged storage. This loss is irreversible and is accompanied by oxidation of the copper from the cuprous to the cupric state. This loss of activity is not caused by oxidation of the cuprous copper. Phenolase is inhibited by substances which form stable complexes with copper such as H_2S, KCN, CO, and *p*-aminobenzoic acid. Reagents which react with sulfhydryl groups do not inhibit the enzyme. Compounds like iodoacetamide are in this class.

Phenolase is present in plant materials in very small amounts. Mushrooms, the richest source, contain approximately 0.03% of the enzyme on the dry weight basis. For this reason and because the plant sources contain the substrates, the preparation of the enzyme is very difficult.

Present in plants are a large number of *o*-diphenolic compounds, which are more or less oxidizable by phenolase. The naturally occurring substrates include 3,4-dihydroxyphenylalanine, the chlorogenic acids, adrenaline, phenylalanine, caffeic and gallic acids, and flavonoids such as daphnetin, and fraxetin. Other substrates that do not occur naturally have been found which are acted upon by this enzyme.

Oxidation of Polyphenols

The mechanism of action of phenolase on *o*-diphenolic compounds is very complicated. Since copper is the prosthetic group of the enzyme, it has been postulated that the activity of phenolase is based on the change of the copper from the cupric to the cuprous state. When the enzyme is isolated, the copper is known to be in the cuprous state. The presence of *o*-dihydroxyphenols brings about the oxidation of the copper to the cupric state. These changes are indicated as follows.

$$4\ Cu^{2+}(\text{enzyme}) + 2\ \text{catechol} \longrightarrow 4\ Cu^{+}(\text{enzyme}) + 2\ o\text{-quinone} + 4\ H^{+}$$

$$4\ Cu^{+}(\text{enzyme}) + 4\ H^{+} + O_2 \longrightarrow 4\ Cu^{2+}(\text{enzyme}) + 2H_2O\ .$$

By losing two electrons and two protons the substrate is oxidized. The copper of the enzyme takes up the two electrons and changes into the cuprous state. The two electrons are rapidly transferred to oxygen. This immediately forms water with the two protons liberated. The enzyme returns to the cupric state and is then ready to repeat the catalytic cycle.

The reaction of phenolase that has been the most extensively studied has been the indirect oxidation of a reducing agent by phenolase with an *o*-dihydroxyphenol. The following equations represent the changes which take place.

$$o\text{-dihydroxyphenol} + \tfrac{1}{2}\ O_2 \xrightarrow{\text{phenolase}} o\text{-quinone} + H_2O$$

$$o\text{-quinone} + RH_2 \longrightarrow o\text{-dihydroxyphenol} + R$$

$$RH_2 + \tfrac{1}{2}\ O_2 \longrightarrow R + H_2O$$

RH_2 is the reducing agent, such as hydroquinone, ascorbic acid, or a reduced phosphopyridine nucleotide. R is the oxidized form, *p*-quinone, dehydroascorbic acid, or oxidized phosphopyridine nucleotide. These reactions continue for a reasonable amount of time if the oxidation of 1 μM of it in the presence of ascorbic acid corresponds to the use of 0.5 μM of O_2. Under some other conditions the reaction can be more complex than is represented here.

Oxidation of Monophenols

The hydroxylation of certain monophenols to *o*-dihydroxyphenols, the second reaction catalyzed by phenolase, is brought about in the same enzyme molecule that produces the oxidation of *o*-dihydroxyphenols. This enzymatic hydroxylation has some very unusual characteristics. The induction period of this reaction is rather long, and it increases with the amount of purification of the enzyme. The usual induction period is a fraction of a second; in this case the period may be several minutes long. However, this induction period can be eliminated or reduced by the addition of a little *o*-dihydroxyphenol, and the rate of oxidation after the induction period becomes linear. Phenolase oxidizes the *o*-dihydroxyphenols at a faster rate than the corresponding monohydroxyphenol. A certain amount of *o*-dihydroxyphenol is always present during the oxidation of monophenols by phenolase. Monophenols can be oxidized to the same potential browning substances as are *o*-dihydroxyphenols.

o-Quinones

o-Quinones are catalytically formed by phenolase and are the precursors of the brown color of certain cut fruits and vegetables. While the *o*-quinones have little color, they are some of the most reactive intermediates found in living matter. Among these reactions is the browning reaction. The formation of the unstable hydroquinone results from the main reaction of the *o*-quinones in the browning reaction. These hydroquinones easily polymerize and are subject to rapid and nonenzymic oxidation, the result of which is a dark brown, slightly soluble polymer.

Simple amines and quinones react with each other readily. An example is the reaction of *o*-benzoquinone with aniline.

o-benzoquinone + 2 aniline ⟶ 4,5-dianilino-1,2-benzoquinone

Benzoquinone reacts with amino acids also.

o-benzoquinone + glycine ⟶ 4-*N*-glycyl-*o*-benzoquinone

The product of this reaction is an intermediate which causes the deamination of glycine and which at the same time forms deep colored pigments. *o*-Quinones formed from *o*-dihydroxyphenols in the presence of phenolase are reacted rapidly with such naturally occurring sulfhydryl compounds as cysteine and glutathione. The result is the formation of characteristic pigments. These reactions take place in addition to the oxidation–reduction reactions.

The enzymatic browning of foods is usually undesirable because it cuts down the acceptability of the food in question for two reasons: (1) the undesirable development of off-color and (2) the formation of off-flavors.

Methods of Control

Boiling or Steaming. When vegetables are packed for freezing preservation, boiling water or steam is used to inactivate the enzymes and control enzymatic browning. This method may not be satisfactory for fruits because of the development of cooked flavor and also a softening of the tissue. It can be used for fruit juices, but these should be held at the proper time-temperature to inactivate the enzymes; usually the minimum gives the best results, cutting down any undesirable changes which might result from excess heating. A temperature of 85 °C could be used because it would cut down the time of exposure.

Sulfur Dioxide. Sulfur dioxide is the chemical inhibitor of phenolase that has been used for years. Very potent as an inhibitor, it is also very inexpensive to use. Sulfur dioxide must be used on the cut tissue or the juice to secure penetration. Since sulfur dioxide also combines with carbonyl compounds, an excess must be added to react with any such compounds present, as well as to inactivate the phenolase. The inhibition of phenolase by SO_2 is not reversible; after removal of the excess SO_2 no regeneration of the phenolase takes place.

The internal atmosphere of the product in question must be considered when using SO_2. Apple slices, for example, have a fair amount of oxygen in the internal tissue, which can cause browning. It is necessary, therefore, that SO_2 penetrate the entire slice, to effectively control browning. Under commercial conditions, the most satisfactory method uses a solution of SO_2 with 2–3% NaCl in a vacuum. The vacuum is subsequently released and SO_2 penetrates the fruit. Better penetration is obtained with free SO_2 than with SO_2 in the form of sodium bisulfite.

Ascorbic Acid. Ascorbic acid can be used to prevent the formation of browning. It reduces the *o*-quinones formed by phenolase to the original *o*-dihydroxyphnolic compounds; this, in turn, prevents the formation of brown substances. Protection against browning lasts as long as any ascorbic acid remains. It works well for peaches frozen in a syrup which contains this vitamin. However, it is not satisfactory for apple slices because of the internal atmosphere of the slices, which, as mentioned before, contains oxygen.

Oxygen can be excluded to control enzymatic browning, usually in connection with other methods. Peaches are packed in containers, covered with syrup that contains ascorbic acid, the oxygen is removed from the headspace and the container hermetically sealed. Apricots do not have much internal oxygen and so may be frozen in friction-lid cans with addition of ascorbic acid to the surface. Adenosine triphosphate (ATP) apparently also forms a reducing agent under these same packing conditions, and it is effective in preventing or reducing enzymatic browning. However, the cost of ATP is too high for commercial feasibility.

Conversion of Natural Substrates. Another method has been developed and it depends on *in situ* changes of the natural substrates of phenolase. The enzyme catechol *o*-methyltransferase is capable of methylating the 3-position of 3,4-dihydroxy aromatic compounds; the *o*-methylation would be irreversible. Caffeic acid yields ferulic acid (Finkle and Nelson 1963).

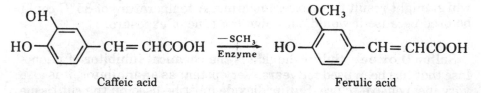

$$\text{HO-C}_6\text{H}_3\text{(OH)-CH=CHCOOH} \xrightarrow[\text{Enzyme}]{-\text{SCH}_3} \text{HO-C}_6\text{H}_3\text{(OCH}_3\text{)-CH=CHCOOH}$$

Caffeic acid — Ferulic acid

Phenolase causes little if any oxidation of monophenolic ferulic acid. Therefore, the *in situ* conversion of all the natural substrates (including the dihydroxyphenolic compounds formed from monophenols as catalyzed by phenolase) could be a way to prevent enzymatic browning. Nelson and Finkle (1964) found that an anaerobic methylation treatment with catechol *o*-methyltransferase system at pH 8 permanently prevents oxidative darkening of apple juice and fruit sections, because it modifies their phenolase substrates. Simple treatment of fruit sections anaerobically at pH 8 results in similar action. This suggests intervention of inherent fruit *o*-methyltransferase. An improved process for the preservation of fresh peeled apples was devised based on the use of small amounts of SO_2 together with the action of inherent *o*-methyltransferase.

Dehydration in Sugar. A method has been developed (Ponting 1960) whereby the fruit is partially dehydrated by reducing to 50% of its original weight by osmosis in sugar or syrup. After draining, the fruit is either frozen or dried further in an air or vacuum dryer. The sugar or syrup inhibits enzymatic browning through the complete dehydration. In addition, it has a protective effect on flavor.

NON-ENZYMATIC BROWNING

Although most non-enzymatic browning in food materials is undesirable because it indicates deterioration in flavor and appearance of the product involved, the development of brown colors in some products is entirely acceptable. Examples of this are the development of brown colors in baked goods during the baking process, in beer, molasses, coffee and substitute cereal beverages, many breakfast foods, and the roasting and other forms of heat preparation of meat. However, the brown colors developing in most other products are not desirable, and methods to prevent or retard such changes are in use. Research is underway and is needed to improve these methods.

Several reactions are known which account partially for these color changes in food products. Sugars are involved. These reactions are called carbonyl-amine or Maillard reactions when they take place in the presence of nitrogenous compounds, particularly when such compounds are primary and secondary amines. They are called caramelization reactions when they take place in the absence of nitrogenous compounds. The initial stages of both of these types of reactions are rather well known. However, the latter stages are not so well known, and will require more research.

Caramelization

When sugars are treated under anhydrous conditions with heat, or at high concentration with dilute acid, caramelization occurs, with the formation of anhydro sugars. Glucose forms glucosan (1,2-anhydro-α-D-glucose) and levoglucosan (1,6-anhydro-β-D-glucose). Since the first of these compounds has a specific rotation of +69°, while the second has −67°, they can be easily distinguished. Yeast will ferment the first compound, but not the second. With similar treatment, fructose gives rise to levulosan (2,3-anhydro-β-D-fructofuranose).

Simultaneous "hydrolysis" and dehydration take place when sucrose is heated at about 200°C, and following this a rapid-dimerization of the products seems to occur. These compounds are characterized by isosacchrosan, which is a sucrose molecule less one molecule of water (Pictet and Stricker 1924). It is not sweet, but mildly bitter.

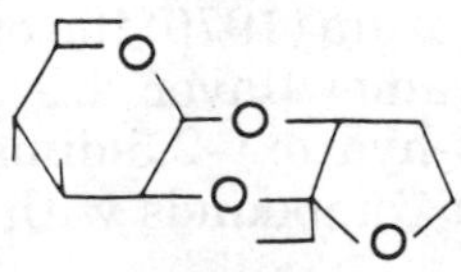

Isosacchrosan

When dilute solutions of reducing sugars are used, the beginning stages of caramelization involve enolization, isomerization, dehydration, and fragmentation. Following this, polymerization reactions take place, which in the end form pigments similar to those formed in more concentrated solutions, or at more elevated temperatures.

Caramels for commercial use are made from glucose syrups, but usually caramelization is the result of reactions that take place when sucrose is heated. There are three stages during this process (at 200°C), during which water is lost and first isosacchrosan and then other anhydrides are formed. The first stage starts with melting of the sucrose, followed by foaming which continues for 35 min. During this period one molecule of water is lost from a molecule of sucrose. The foaming then stops. Shortly after this, a second stage of foaming starts which lasts 55 min. During this stage about 9% of the water is lost, and the compound formed is caramelan, a pigment with the average formula of $C_{24}H_{36}O_{18}$, according to the following equation:

$$2\,C_{12}H_{22}O_{11} - 4\,H_2O = C_{24}H_{36}O_{18}$$

According to Miroshnikova *et al.* (1970) caramelan melts at 138°C, is soluble in water and ethanol, and is bitter in taste. The pigment caramelen is formed during the third stage of foaming which starts after about 55 min. The equation for the formation of this pigment follows:

$$3\,C_{12}H_{22}O_{11} - 8\,H_2O = C_{36}H_{50}O_{25}$$

Caramelen melts at 154°C and is soluble in water. When the heating is continued, the result is the formation of humin, which is an infusible, dark mass with a high molecular weight, and is called caramelin. The formula has been set at about $C_{125}H_{188}O_{80}$ (Janacek 1939). It is obvious, therefore, that a number of compounds are formed during this heating process, and different procedures can result in products of varying composition.

These pigments are colloidal in nature, the isoelectric points of which differ with the method of manufacture used, and the colloids have varying particle sizes. According to Miroshnikova *et al.* (1970), caramel pigments have the following groups: carbonyl, carboxyl, and enolic, together with hydroxyl groups of varying basicity. Plunguian and Hibbert (1935) found that phenolic hydroxyl groups are present. Since iron tends to strengthen the caramel color, the presence of phenolic hydroxyl groups would explain this behavior. It is known that caramelization reactions are qualitatively not the same at different pH values.

According to Jurch and Tatum (1970) the only two degradation products that have typical caramel flavor are aceylformoin (4-hydroxy-2,3,5-hexanetrione) and 4-hydroxy-2,5-dimethyl-3(2*H*)-furanone. In 1967 Hodge suggested that compounds with an odor of caramel have the following groupings:

$$CH_3{-}\bar{C}{=}C(OH){-}\bar{C}{=}O \quad \text{and} \quad CH_3{-}\overset{\overset{\displaystyle O}{\|}}{C}{-}\bar{C}{=}\bar{C}(OH)$$

Furthermore, the structures may be *O*-heterocyclic or alicyclic, or they may have an intramolecular hydrogen-bonded ring:

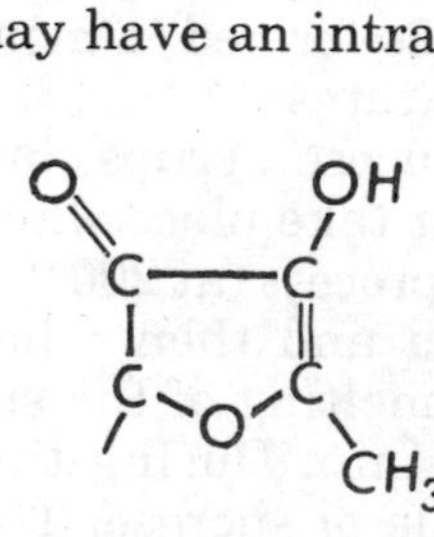

Heterocyclic 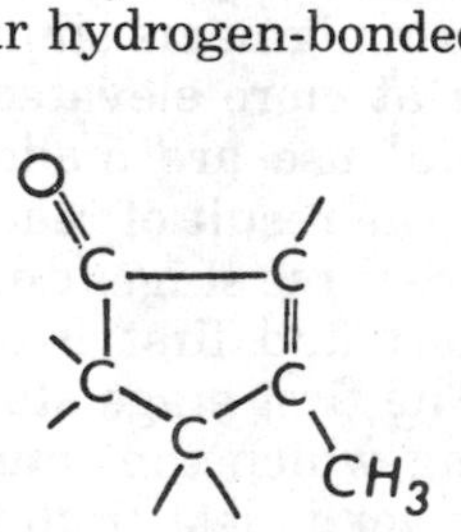 Alicyclic 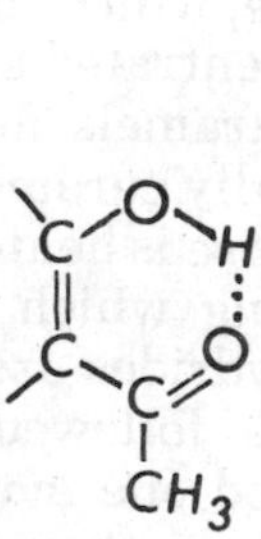 Hydrogen-bonded

In addition to these are analogues of heterocyclic and alicyclic compounds with six-membered rings which behave in the same way.

Ledon and Lananeta (1950) found that the reaction rate for caramelization is ten times greater at pH 8 than at pH 5.9, illustrating the importance of pH on the formation of these pigments.

The Maillard Reaction

The Maillard reaction is the action of amino acids and proteins on sugars. Ammonia (as well as the amino acids) is able to effect this change also. This finding was first published by Maillard in 1912. The carbohydrate must be a reducing sugar because a free carbonyl group is necessary for such a combination. This important reaction is involved in a great many situations in which the browning of foods occurs. The end-product is the melanoidins, which are the brown pigments. The brown color results from the following three stages of development (Fig. 12.1).

- I. Initial stage (colorless, no absorption near ultraviolet)
  - A. Sugar-amine condensation
  - B. Amadori rearrangement
- II. Intermediate stage (colorless or yellow, with strong absorption in near ultraviolet)
  - C. Sugar dehydration
  - D. Sugar fragmentation
  - E. Amino acid degradation
- III. Final stage (highly colored)
  - F. Aldol condensation
  - G. Aldehyde-amine polymerization; formation of heterocyclic nitrogen compounds

Hodge (1953) called the Maillard reaction the carbonyl-amine reaction, because the compounds which react with the amines usually have a carbonyl or a potential carbonyl function. Among the most reactive carbonyl compounds are α,β-unsaturated aldehydes (furaldehyde), and α-dicarbonyl compounds (diacetal and pyruvaldehyde).

$$\begin{array}{c} HC{=}O \\ | \\ COH \\ \| \\ CH \\ | \\ R \end{array} \qquad\qquad \begin{array}{c} HC{=}O \\ | \\ C{=}O \\ | \\ R \end{array}$$

α,β-Unsaturated aldehyde　　α-Dicarbonyl compound

According to Reynolds (1970) the reaction which is reversible, starts between an aldose or ketose sugar and a primary or secondary amine,

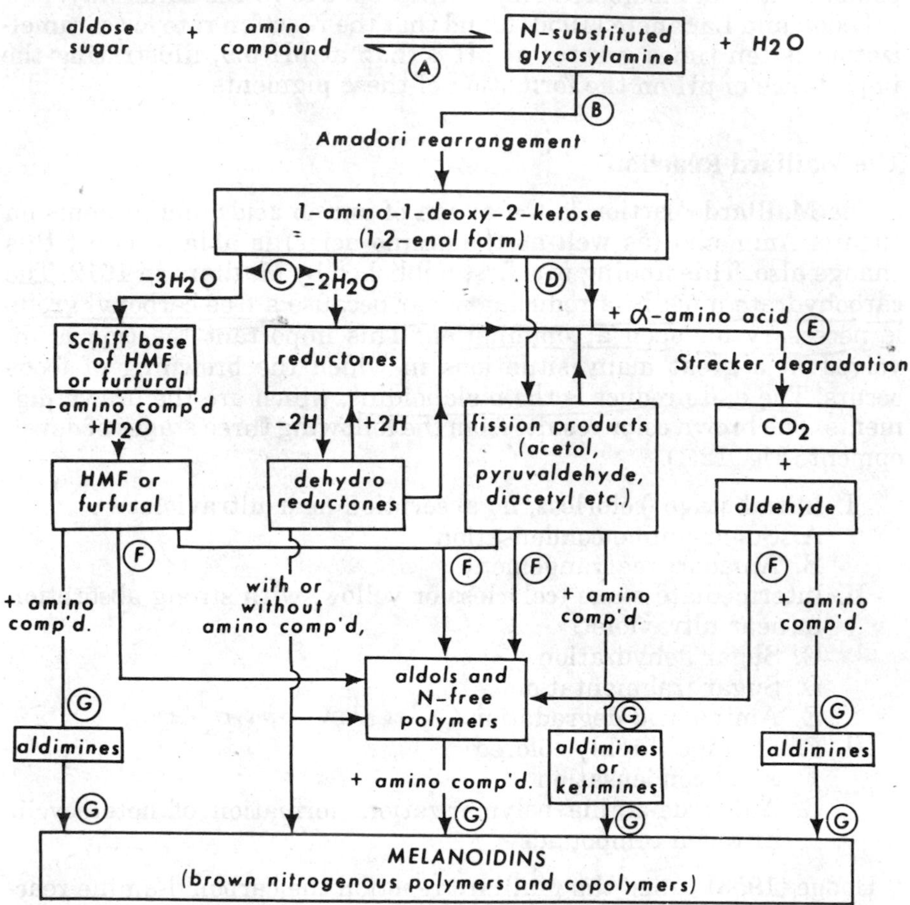

FIG. 12.1. The Hodge scheme. (A) Maillard reaction. (B), Amadori rearrangement. (C) Sugar dehydration. The dehydration of sugar in the sugar–amine browning reaction can take place in two ways. In neutral or acid solutions furfurals are formed. In the dry state or in nonaqueous solvents when amines are present, reductions are formed. (D), Fission products of sugar. (E) Strecker degradation (to aldehydes containing one less carbon than the amino acid, with the liberation of carbon dioxide). (F) Aldol condensation. It is a highly probable reaction in the formation of melanoidins. Nitrogen-free aldols in general are likely to react with amino compounds, alkimines, and ketimines to form nitrogenous melanoidins. (G), Aldehyde–amine polymerization and the formation of melanoidins.

From Hodge (1953). Reproduced with permission of the American Chemical Society.

the product of which is a glycosylamine. The initial product of the reaction between glucose and ammonia is glucosylamine. This rearranges to form 1-amino-1-deoxy-D-fructose in the p esence of an acid catalyst. The yield of glycosylamine is affected by the amount of water present. A substantial amount of this compound is formed when the amount of water present is low. For this reason the carbonyl-amine route may be important in the browning of concentrated and dried foods. The addition of the amine to the carbonyl group after which a molecule of water is eliminated to permit the ring closure is the likely route for the formation of glycosylamine.

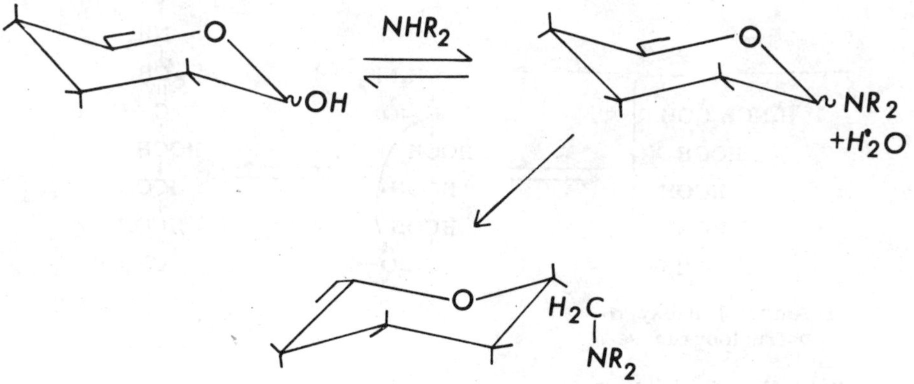

The Amadori Rearrangement

The Amadori rearrangement is the designation of the changes which produce 1-amino-1-deoxy-2-ketone. When glycine and glucose react, 1-deoxy-1-glycino-β-D-fructose is the product.

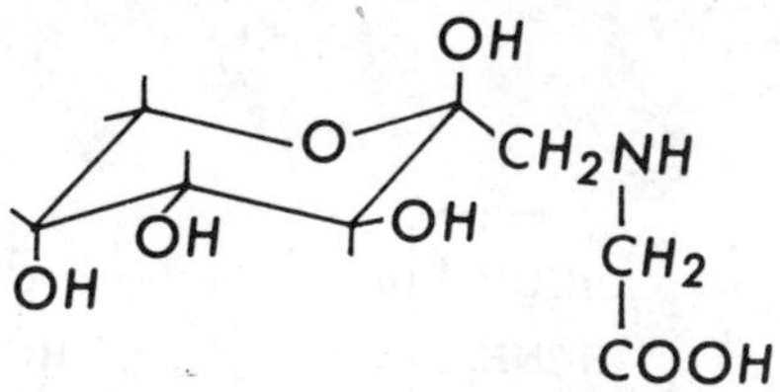

1-Deoxy-1-glycine-β-D-fructose

Free ketoses show less browning activity than the ketosamines. The latter are colorless and rather stable. The yield of these compounds is maximum at a water content of 18%. In dilute solutions the yield is low. Figure 12.2 illustrates the Amadori rearrangement and the Heyns rearrangement. The Heyns rearrangement shows the reaction starting with fructose instead of glucose. Both of these rearrangements bring about the same transformation.

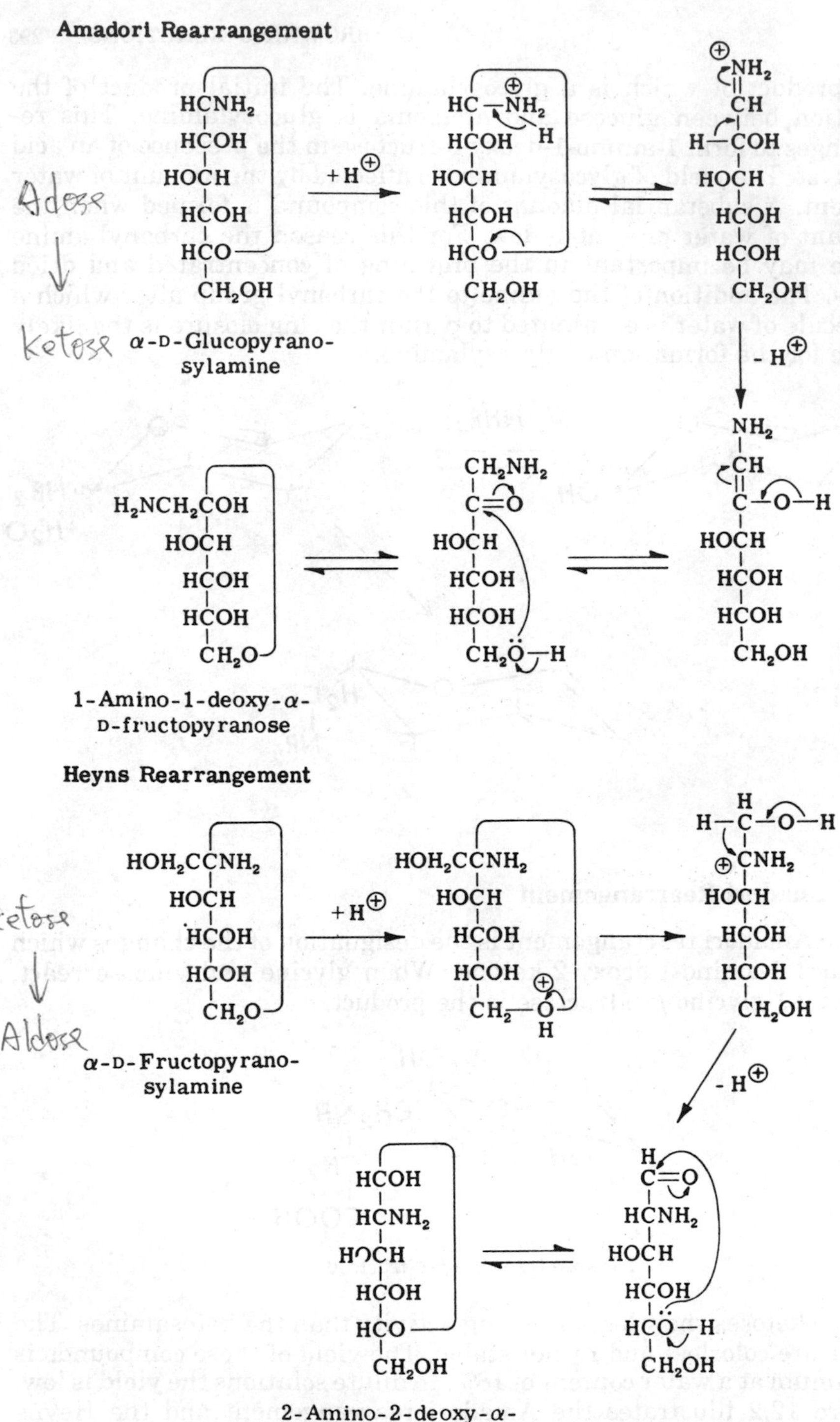

FIG. 12.2. Acid-catalyzed mechanisms for the Amadori rearrangement and Heyns scheme.

From Kort (1970). Reproduced with permission of Academic Press.

According to Reynolds (1970), in the initial stages of the carbonyl-amine reactions secondary products are even more reactive than the primary products. Also, a second mole of aldose can react with a keto-seamine to yield a diketoseamine. A second mole of amine can react with an aldoseamine to form a diamino sugar. The carbonyl-amine reaction in its initial stages is characterized by mole-per-mole addition. However, as the reaction proceeds, the ratio of the reactants consumed changes.

The main pathway for the formation of brown color in foods appears to be degradation and condensation by way of the 1,2-enol forms of the aldose or ketose amines. The aldol condensation mechanism for the α,β-dicarbonyl compounds formed seems to be involved. The precursors of the brown pigments are fluorescent. According to Adhikari and Tappel (1973) a likely structure for a fluorescent compound obtained from a browning system involving glucose and glycine is

$$HO—\overset{\overset{\large O}{\|}}{C}—CH_2—N=CH—CH=\overset{\overset{\large OH}{|}}{C}—$$

The formation of these brown pigments via the carbonyl-amine reactions have similarities to the formation of caramels.

Strecker Degradation

Another step in the formation of brown pigments is the Strecker degradation of the amino acid moiety. This degradation of the alpha amino acids results in aldehydes containing one less carbon atom than the amino acid. The loss of the carbon atom is accounted for by the release of carbon dioxide. In addition to CO_2, it produces carbonyl compounds and amines. Schönberg *et al.* (1948) found that only carbonyl compounds containing the structure

$$—\overset{\overset{\large O}{\|}}{C}—\left[—\overset{|}{C}=\overset{|}{C}—\right]_n—\overset{\overset{\large O}{\|}}{C}—$$

when n is 0 or an integer, are capable of initiating the degradation. It is also necessary that the amino group of the amino acid be in the alpha position to the carbonyl group. The fact that the CO_2 comes from the carboxyl group of the amino acid moiety and not from the sugar was demonstrated by using isotope tracing techniques.

The general equation for the Strecker degradation which takes into consideration the findings of Schönberg *et al.* (1948) is

$$R \cdot CO \cdot CO \cdot R + R'CHNH_2 \cdot COOH \longrightarrow R'CHO + CO_2 + R \cdot CHNH_2CO \cdot R$$

These dicarbonyl compounds are osones and are active agents for Strecker degradation.

Pyrazine compounds with different amounts of substitution are formed in carbonyl-amine reactions and can cause Strecker degradation of the amino acids. 2,5-Dimethylprazine was ioslated from a glucose, glycine reaction mixture by Dawes and Edwards (1966).

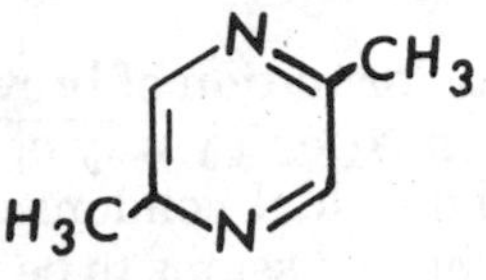

2,5,-Dimethylpyrazine

The events shown in the preceding equations can be stopped if the carbonyl group of the sugar of the first equation is first combined with another compound. This is one of the reasons for the effectiveness of sulfites in the prevention of the formation of the brown colored compounds.

To sum up, the Maillard reaction, which results in the formation of brown nitrogenous polymers and copolymers, involves three stages of two or more steps each. Certain parts of this chain are understood, but the whole is not fully elucidated. The following reactions are now clear from the chemical point of view: (1) the reaction which involves the sugar-amine condensation, (2) the Amadori rearrangement resulting in the keto form, and (3) the Strecker degradation which results in action on the alpha amino acids with the loss of one molecule of CO_2 and the formation of an aldehyde. The products of the first two reactions are colorless, while that of the third (in the second stage) may be either colorless or slightly yellow. It is in the third stage that the really intensely brown colored pigments are formed.

Early workers reported that fructose in the presence of amino acids formed brown color to a greater extent than did glucose. However, later work showed that in buffered mixtures the opposite was true. Also, in unbuffered mixtures in which the sugar was fructose, browning developed faster than when glucose was the sugar. When unbuffered mixtures at 50°C were allowed to set for longer periods of time, considerably more browning took place in the glucose mixtures than when fructose was the sugar involved. Since foodstuffs are ordinarily naturally buffered, the glucose contained in them would brown more readily with α-amino acids than any fructose which might be present.

Brown pigments formed by the reaction between aldoses and amino acids were found to be insoluble in organic solvents but showed variable solubility in water ranging from insoluble to easily soluble. Maple syrup, molasses, and dried apricots that had darkened yielded brown pigments. In each case the pigment had glucose and amino acid residues present. However, the structures of the pigments have not been determined. Reynolds (unpublished data) found that pigments prepared from acetaldehyde and amines and furfural and amines had three to five aldehyde residues to one amine residue. It is likely that the most important precursors of browning reactions in foods are the carbonyl compounds.

A considerable amount of work has been conducted on model systems to obtain clues on intermediates and final products in the development of non-enzymatic brown colors.

Browning and Flavor

Compounds formed as a result of the browning reactions can have an effect of the flavor of the products. Table 12.1 gives some idea of the variety of flavors that might be obtained. Some flavor notes could be the result of the Strecker degradation.

THE FORMATION OF BROWN PIGMENTS

A number of factors can affect the formation of these pigments. Among these are pH, temperature, moisture content, time, concentration, and nature of the reactants. Also, one of these factors may affect another.

The rate of browning increases with rising temperature. In model systems the development increases 2 to 3 times for each 10°C rise in temperature. In natural systems, particularly those high in sugar content, the increase may be faster. Two methods have been used to measure these changes: (1) measure color development and (2) measure the evolution of CO_2.

Although browning reactions usually slow down as the pH decreases until the optimum stability pH for reducing sugars is passed, this is not so important for food products.

Some workers consider that browning in aqueous solution is largely confined to caramelization reactions, while the Maillard reaction is responsible at an alkaline pH or a nearly dry state.

Moisture content seems to have an important effect on the rate of browning. It is quite likely that for moisture contents above 30% a decrease in reaction is caused by dilution. Labuza *et al.* (1970) found that a decrease in reaction at moisture contents below 30% is caused by the intrinsic ability of the sugars to lower water activity.

The development of brown color in dried fruits is largely caused by the reaction between amino acids and glucose, a reaction which is speeded by the presence of organic acids. In the case of sulfited dried apricots, oxidative browning and loss of sulfite was found to occur at moisture contents of 20% or higher and temperatures higher than 43°C. The formation of ketoseamines in dried whole egg or egg white is avoided by fermentation of the glucose before drying.

TABLE 12.1. Flavors of Heated 1:1 α-Amino Acid–Glucose Mixtures

α-Amino Acid	$R \cdot CH(NH_2) \cdot COOH$ where R =	Volatile Strecker Aldehyde RCHO	Aroma Noted by Panel 100°C	Aroma Noted by Panel 180°C	Reference[a]
None			None	Caramel	1
Glycine	H—	Formaldehyde	Caramel	Burnt sugar	1
			Caramel		2
α-Alanine	CH_3—	Acetaldehyde	Caramel, sweet	Burnt sugar	1
α-Aminobutyric	$CH_3 \cdot CH_2$—	Propionic	Caramel	Burnt sugar	1
			Maple syrup		3
Valine	$(CH_3)_2CH$—	Isobutyric	Rye bread	Penetrating chocolate	1
			Fruity aromatic		2
Leucine	$(CH_3)_2CH \cdot CH_2$—	Isovaleric	Sweet, chocolate	Burnt cheese	1
			Toasted, bready		2
Isoleucine	$CH_3 \cdot CH_2 \cdot CH(CH_3)$—	2-Methyl-butanal-(1)	Musty	Burnt cheese	1
			Fruity aromatic		2
Serine	$HO \cdot CH_2$—	Glycolic	Maple syrup		3
Threonine	$CH_3 \cdot CHOH$—	Lactic	Chocolate	Burnt	1
Methionine	$CH_3 \cdot S \cdot CH_2 \cdot CH_2$—	Methional	Potato	Potato	1
Phenylglycine	C_6H_5—	Benzaldehyde	Bitter almond		3
Phenylalanine	$C_6H_5 \cdot CH_2$—	α-Toluic	Violets	Violets, lilac	1
			Rose perfume		2

Tyrosine	$HO \cdot C_6H_4 \cdot CH_2—$	Caramel		1
Proline	N H	Burnt protein	Pleasant, bakery aroma	1
Hydroxyproline	HO— N H	Potato		2
Histidine	N NH	None	Cornbread	1
			Buttery note	2
Arginine	$H_2N \cdot C{:}NH \cdot NH \cdot (CH_2)_2—$		Burnt sugar	1
		Buttery note		2
Lysine · HCl	$H_2N \cdot (CH_2)_4—$	None	Bread-like	1
Aspartic	$HOOC \cdot CH_2—$	Rock candy	Caramel	1
Glutamic	$HOOC \cdot CH_2 \cdot CH_2—$	Caramel	Burnt sugar	1
Glutamine	$H_2N \cdot CO \cdot CH_2 \cdot CH_2—$	Chocolate, pleasant	Butterscotch	1
Cysteine · HCl	$HS \cdot CH_2—$	Sulfide		1
		Meaty		2
Cystine	$—CH_2 \cdot S \cdot S \cdot CH_2—$	Sulfide		1
		Burnt turkey skin		2

Source: Hodge (1967).

[a] Key to reference: 1. Herz and Shallenberger (1960), 2. Kiely *et al.* (1960), 3. Barnes and Kaufman (1947).

Reactions between uronic acids and amines are known to take place, and that solutions of D-galacturonic acid and glycine browned more rapidly at 98 °C than did solutions in which aldose was present in place of D-galacturonic acid. Pectinic acids and α-methyl-D-galacturonic acid methyl ester do not show an effect similar to this; the likely reason is that the galacturonic acid residues are connected with the glycosidic hydroxyl group.

Brown pigments are formed from ascorbic acid when it decomposes under either aerobic or anaerobic conditions. An important point to be noted is that when this takes place, some of the intermediates are reductones, which can interfere with the accurate determination of ascorbic acid. Dehydro-L-ascorbic acid, one of the products resulting from the oxidation of ascorbic acid was found to be broken down to L-xylosone in dilute water solutions of SO_2 (Whiting and Coggins 1960). This compound has been found in apple cider that had been treated with SO_2. Ascorbic acid is important in the browning of systems that contain sugars, amino acids, and ascorbic acid.

Reducing sugars are degraded to furfurals in the presence of either mineral or organic acids. Pentoses yield 2-furfuraldehyde while hexoses yield 5-hydroxymethyl-2-furfuraldehyde, which can break down to levulinic acid, $CH_3COCH_2CH_2COOH$. During the degradation of sugars to furfurals several intermediates are formed. Like most reactions of this character, this tends to be rather complex. The first step of the reaction is the formation of 1,2-enol of the aldose or ketose acted upon. It is known that in the formation of the 1,2-enols, the first step of the degradation of the reducing sugar is more easily accomplished in the presence of alkali than acid. Splitting of the sugar molecule occurs.

```
H—C———C—H
  ||    ||
H—C     C—CHO
   \   /
     O
```

2-Furfuraldehyde

Lipids can be involved in browning reactions. Lipoproteins can be involved in such reactions because their amino groups can act on reducing sugars and aldehydes. Also, when active carbonyl groups are formed in the lipid part, browning can be initiated. A considerable amount of work has been done on the browning of fish oils and fish meals. When browned oils were fractionated, the discoloration was always connected with the oxidized acids.

INHIBITION OF BROWNING

At the present time, the use of sulfur dioxide or sulfites is the most effective method of inhibiting browning in commercial use. When dried apricots and potatoes were analyzed (Legault *et al.* 1949, 1951; Stadtman *et al.* 1946), it was found that in most cases about 50% of the sulfite disappeared from dehydrated vegetables before any significant amount of browning developed. The temperature of storage and the moisture content of the product also influenced the loss of sulfite during storage.

Browning in dehydrated potatoes can be retarded by the use of calcium chloride or, more effectively, by a mixture of calcium chloride and calcium bisulfite. The potatoes are first steam-blanched, and then sprayed with a solution of these compounds before drying. The reason for this effect of calcium chloride has not been fully explained.

SUMMARY

Two important forms of browning are known, enzymatic and nonenzymatic browning. Both are important because this color development is usually undesirable. Control is easier if the type of reaction is known.

Enzymatic browning is brought about by phenolases. A number of diverse enzymes are included. One of these reactions is the oxidation of *o*-dihydroxyphenols to *o*-quinones. The other is the hydroxylation of certain monohydroxyphenols to dihydroxyphenols.

The mechanism of action of phenolase on *o*-diphenolic compounds is very complicated. Copper is the prosthetic group of the enzyme. It has been postulated that the activity of phenolase is based on the change of the copper from cupric to the cuprous state. When the enzyme is isolated the copper is in the cuprous state. The presence of *o*-dihydroxyphenols brings about the oxidation of the copper to the cupric state.

The hydroxylation of certain monophenols to *o*-dihydroxyphenols is the second reaction catalyzed by phenolase.

o-Quinones are catalytically formed by phenolase and are the precursors of the brown color of certain cut fruits and vegetables. *o*-Quinones have little color themselves.

Simple amines and quinones react with each other readily.

The enzymatic browning of foods is usually undesirable because it reduces the acceptability of the food in question because of the development of off-color and the formation of off-flavors.

Methods of control involve boiling water or steam blanching. In addition, sulfur dioxide has been used for a long time. Better penetration of the tissues of apple slices is obtained with free SO_2 than with sodium bisulfite. Another method of control is the use of ascorbic acid, but it is not good for use on apple slices because of the presence of oxygen in the apple. Other methods are known.

Non-enzymatic browning is undesirable in products when it indicates the development of off-color and flavor. However, the development of brown colors in such products as baked goods during the baking process, the brown color of molasses, coffee, breakfast foods, and the heat preparation of meats, are desirable.

These color changes are at least partially accounted for by several reactions. Sugar is involved. When nitrogenous compounds are present the reaction is known as the Maillard reaction. In the absence of nitrogenous substances the changes are known as caramelization. The process of caramelization takes place in several stages and results in water loss from the sugar molecule. Caramelen is the pigment formed during the third stage. Caramelan is formed during the second stage.

The Maillard reaction is the action of amino acids and proteins on sugars. The carbohydrate must be a reducing sugar because a free carbonyl group is necessary. The final brown color results from three stages of development.

The brown color formation proceeds as follows:

I. Initial stage. Sugar-amine condensation. Amadori rearrangement.
II. Intermediate stage.
III. Final stage—highly colored.

The Amadori rearrangement forms 1-amino-1deoxy-2-ketone.

The Strecker degradation of the α-amino acids forms aldehydes containing one less carbon atom than the amino acid. The formation of CO_2 accounts for the loss of the carbon atom.

A number of factors can affect the formation of brown pigments. These include pH, temperature, moisture content, time, concentration, and the nature of the reactants.

At the present time the use of SO_2 or sulfites is the most effective method for control under commercial conditions.

BIBLIOGRAPHY

ADHIKARI, H. R., and TAPPEL, A. L. 1973. Fluorescent products in a glucose–glycine browning reaction. J. Food Sci. *38,* 486–488.

ANET, E. F. L. J. 1964. 3-Deoxyglycosuloses (3-deoxyglycosones) and the degradation of carbohydrates. *In* Advances in Carbohydrate Chemistry, Vol. 19. M. L. Wolfrom and R. S. Tipson (Editors). Academic Press, New York.

BARNES, H. M., and KAUFMAN, C. W. 1947. Industrial aspects of the browning reaction. Ind. Eng. Chem. *39,* 1167–1170.

BAYSE, G. S., and MORRISON, M. 1971. The role of peroxidase in catalyzing oxidation of polyphenols. Biochem. Biophys. Acta *244,* 77–84.

BENDALL, D. S., and GREGORY, R. P. F. 1963. Purification of phenol oxidases. *In* Enzyme Chemistry of Phenolic Compounds. J. B. Pridham (Editor). Macmillan Co., New York.

BOSWELL, J. G. 1963. Plant polyphenol oxidases and their relation to other oxidase systems in plants. *In* Enzyme Chemistry of Phenolic Compounds. J. D. Pridham (Editor). Macmillan Co., New York.

BRUEMMER, J. H., and ROE, B. 1970. Enzymatic oxidation of simple diphenols and flavonoids by orange juice extracts. J. Food Sci. *35,* 116–119.

BURGES, N. A. 1963. Enzymes associated with phenols. *In* Enzyme Chemistry of Phenolic Compounds. J. B. Pridham (Editor). Macmillan Co., New York.

BURTON, H. S., McSWEENY, D. J., and BILTCLIFFE, D. O. 1963. Nonenzymic browning: The role of unsaturated carbonyl compounds as intermediates and of SO_2 as an inhibitor of browning. J. Sci. Food Agric. *14,* 911–920.

CHUBEY, B. B., and NYLUND, R. E. 1969. Surface browning of carrots. Can. J. Plant Sci. *49,* 421–426.

DAWES, I. W., and EDWARDS, R. A. 1966. Methyl-substituted pyrazines as volatile reaction products of heated aqueous aldose–amino acid mixtures. Chem. Ind. (London), 2203.

FINKLE, B. J., and NELSON, R. F. 1963. Enzyme reactions with phenolic compounds: Effect of *o*-methyltransferase on a natural substrate of fruit polyphenol oxidase. Nature *197,* 902–903.

FOSTER, A. B., and HORTON, D. 1959. Aspects of the chemistry of the amino sugars. *In* Advances in Carboyhdrate Chemistry, Vol. 14. M. L. Wolfrom and R. S. Tipson (Editors). Academic Press, New York.

GRIFFITHS, L. A. 1959. Detection and identification of the polyphenoloxidase substrate of the banana. Nature *184,* 58–59.

HERZ, W. J., and SHALLENBERGER, R. S. 1960. Some aromas produced by simple sugar-amino acid reactions. Food Res. *25,* 491–494.

HODGE, J. E. 1953. Dehydrated foods. Chemistry of browning reactions in model systems. J. Agric. Food Chem. *1,* 928–943.

HODGE, J. E. 1955. The Amadori rearrangement. *In* Advances in Carbohydrate Chemistry, Vol. 10. M. L. Wolfrom and R. S. Tipson (Editors). Academic Press, New York.

HODGE, J. E. 1967. Origin of flavor in foods. Nonenzymic browning reactions. *In* Chemistry and Physiology of Flavors. H. W. Schultz, E. A. Day, and L. M. Libbey (Editors). AVI Publishing Co., Westport, CT.

JANACEK, O. 1939. Behavior of sucrose on heating and composition of caramel substances so obtained. Centr. Zukerind. *47,* 647–648.

JOSLYN, M. A., and BRAVERMAN, J. B. S. 1954. The chemistry and technology of the pretreatment and preservation of fruit and vegetable products with sulfur dioxide and sulfites. *In* Advances in Food Research, Vol. 5. E. M. Mrak and G. F. Stewart (Editors). Academic Press, New York.

JOSLYN, M. A., and PONTING, J. D. 1951. Enzyme-catalyzed oxidative browning of fruit products. *In* Advances in Food Research, Vol. 3. E. M. Mrak and G. F. Stewart (Editors). Academic Press, New York.

JURCH, JR., G. R., and TATUM, J. H. 1970. Degradation of D-glucose with acetic acid and methyl amine. Carbohydrate Res. *15,* 233–239.

KERTESZ, D., and ZITO, R. 1962. Phenolase. *In* Oxygenases. O. Hayaishi (Editor). Academic Press, New York.

KERTESZ, Z. I. 1933. The oxidase system of a non-browning yellow peach. N.Y. State Agric. Exp. Stn. Tech. Bull. 219.

KIELY, P. J., NOWLIN, A. C.,and MORIARTY, J. H. 1960. Bread aromatics from browning systems. Cereal Sci. Today *5,* 273–274.

KORT, M. J. 1970. Reactions of free sugars with aqueous ammonia. *In* Advances in Carbohydrate Chemistry and Biochemistry, Vol. 25. M. L. Wolfrom, R. S. Tipson, and D. Horton (Editors). Academic Press, New York.

LABUZA, T. P., TANNENBAUM, S. R., and KAREL, M. 1970. Water content and stability of low-moisture and intermediate-moisture foods. Food. Technol. *24,* 543–555.

LAVOLLAY, J., LEGRAND, G. LEHONGRE, G., and NEUMANN, J. 1963. Enzyme-substrate specificity in potato polyphenol oxidase. *In* Enzyme Chemistry of Phenolic Compounds. J. B. Pridham (Editor). Macmillan Co., New York.

LEDON, A. C., and LANANETA, P. 1950. Spectrophotometric comparison of the factors that affect the formation of colored substances during manufacture of sugar. Bol. Of. Assoc. Tech. Azucar *8,* 547–566.

LEGAULT, R. R., HENDEL, C. E., TALBURT, W. F., and RASMUSSEN, L. B. 1949. Sulfite disappearance in dehydrated vegetables during storage. Ind. Eng. Chem. *41,* 1447–1451.

LEGAULT, R. R., HENDEL, C. E., TALBURT, W. F., and POOL, M. F. 1951. Browning of dehydrated sulfited vegetables during storage. Food Technol. *5,* 417–423.

LERNER, A. B., and FITZPATRICK, T. B. 1950. Biochemistry of melanin formation. Physiol. Rev. *30,* 91–126.

LONG, T. J., and ALBEN, J. O. 1968. Mushroom tyrosinase (polyphenol oxidase). Mushroom Sci. *7,* 69–79.

LUH, B. S., and PHITHAKPOL, B. 1972. Characteristics of polyphenol oxidases related to browning in cling peaches. J. Food Sci. *37,* 264–268.

LÜTHI, H. 1951. Symp. Int. Fruit Juice Union, Paris.

MAILLARD, L. C. 1912. Action of amino acids on sugars. Formation of melanoidins in a methodical way. C.R. Acad. Sci. *154,* 66–68. (French)

MALSTRÖM, B. G., ANDRÉASSON, L.-E., and REINHAMMAR, B. 1975. Copper-containing oxidases and superoxide dismutase. *In* The Enzymes, 3rd Edition, Vol. 12: Oxidation-Reduction Part B. P. Boyer (Editor). Academic Press, New York.

MAPSON, L. W., SWAIN, T., and TOMALIN, A. W. 1963. Influence of variety, cultural conditions, and temperature of storage on enzymic browning of potato tubers. J. Sci. Food Agric. *14,* 673–684.

MASON, H. S. 1955. Comparative biochemistry of the phenolase complex. Adv. Enzymol. *16,* 105–184.

MATHEW, A. G., and PARPIA, H. A. D. 1971. Food browning as a polyphenol reaction. *In* Advances in Food Research, Vol. 19. C. O. Chichester, E. M. Mrak, and G. F. Stewart (Editors). Academic Press, New York.

McSWEENY, D. J., BILTCLIFFE, D. O., POWELL, R. C. T., and SPARK, A. A. 1969. The Maillard reaction and its inhibition of sulfite. J. Food Sci. *34,* 641–643.

MIROSHNIKOVA, Z. P., CHIKIN, G. A., and TOBLINA, N. M. 1970. Chemical nature and the products of sucrose caramelization. Sakh. Prom. *44,* 10–15.

MUNETA, P. 1966. Bisulfite inhibition of enzymic blackening caused by tyrosine oxidation. Am. Potato J. *43,* 397–402.

MUNETA, P., and WALRADT, J. 1968. Cysteine inhibition of enzymatic blackening with polyphenol oxidase from potatoes. J. Food Sci. *33,* 606–608.

NELSON, R. F., and FINKLE, B. J. 1964. Enzyme reactions with phenolic compounds: Effects of *o*-methyltransferase and high pH on the polyphenol oxidase substrates in apple. Phytochemistry *3,* 321–325.

ONSLOW, M. W. 1920. XLV. Oxidizing enzymes. III. The oxidizing enzymes of some common fruits. Biochem. J. *14,* 541–547.

PICTET, A., and STRICKER, P. 1924. Constitution and synthesis of isosacchrosan. Helv. Chim. Acta *7,* 708–713.

PLUNGUIAN, M., and HIBBERT, H. 1935. Studies on lignin and related compounds. XI. The nature of lignite humic acid and of the so-called "humic acid" from sucrose. J. Am. Chem. Soc. *57,* 528–536.

PONTING, J. D. 1960. The control of enzymatic browning in fruit. *In* Food Enzymes. H. W. Schultz (Editor) AVI Publishing Co., Westport, CT.

REYNOLDS, T. M. 1963. Chemistry of nonenzymic browning. I. The reaction between aldoses and amines. *In* Advances in Food Research, Vol. 12. C. O. Chichester, E. M. Mrak, and G. F. Stewart (Editors). Academic Press, New York.

REYNOLDS, T. M. 1965. Chemistry of nonenzymic browning. II. *In* Advances in Food Research, Vol. 14. C. O. Chichester, E. M. Mrak, and G. F. Stewart (Editors). Academic Press, New York.

REYNOLDS, T. M. 1970. Flavors from nonenzymic browning reactions. Food Technol. (Australia), 610–619.

SCHÖNBERG, A., and MOUBACHER, R. 1952. The Strecker degradation of α-amino acids. Chem. Rev. *50,* 261–277.

SCHÖNBERG, A., MOUBACHER, R., and MOSTAFA, A. 1948. Degradation of α-amino acids to aldehydes and ketones by interaction with carbonyl compounds. J. Chem. Soc. 176–182.

SEAVER, J. L., and KERTESZ, Z. I. 1946. The "browning (Maillard) reaction" in heated solutions of uronic acids. J. Am. Chem. Soc. *68,* 2178–2179.

SHANNON, C. T., and PRATT, D. E. 1967. Apple polyphenoloxidase activity in relation to various phenolic compounds. J. Food Sci. *32,* 479–483.

SHUMAKER, J. B., and BUCHANAN, J. H. 1932. A study of caramel color. Iowa State J. Sci. *6,* 367–379.

SIMON M., WAGNER, J. R., SILVEIRA, V. G., and HENDEL, C. E. 1955. Calcium chloride as a non-enzymic browning retardant for dehydrated white potatoes. Food Technol. *9,* 271–275.

STADTMAN, E. R., BARKER, H. A., MRAK, E. M., and MACKINNEY, G. 1946. Storage of dried fruit. Influence of moisture and sulfür dioxide on deterioration of apricots. Ind. Eng. Chem. *38,* 99–104.

STRECKER, A. 1862. Note. On a specific oxidation by means of alloxan. Justus Leibigs Ann. Chem. *123,* 363–365. (German)

SZENT GYÖRGYI, A. 1925. Cell respiration. IV. On the oxidation mechanism of potatoes. Biochem Z. *162,* 399–412. (German)

SZENT GYÖRGYI, A. 1939. Studies on Biological Oxidation and Some of its Catalysts. Williams & Wilkins Co., Baltimore, Maryland.

WHITING, G. C., and COGGINS, R. A. 1960. Formation of L-xylosone from ascorbic acid. Nature *185,* 843–844.

13

Food Colorings

Color Safety and Regulations
Colors Exempt from Certification
Color Analysis and Desired Properties
Summary
Bibliography

While the use of food colorings in some situations may be open to question, there is no disputing the fact that in a number of products they have been used for many years, and their presence has been accepted as desirable. This can be accounted for by the fact that many people eat with their eyes rather than their palates, in many cases a rather unfortunate situation. According to the bulletin *Food Colors* (National Academy of Sciences 1971), the average annual per capita consumption of food in the United States amounts to 645 kg. This contains about 5.5 g of synthetic food colors, the bulk of which (almost 85%) is made up of amaranth (FD&C Red No. 2), tartrazine (FD&C Yellow No. 5), and sunset yellow FCF (FD&C Yellow No. 6). Wine was colored as early as 200 to 300 BC. Many of these early colors were likely mineral pigments, but vegetable or animal colors could have been used. As late as the nineteenth century poisonous metallic compounds were sometimes used to color foods; in 1880 lead chromate was found as a color in candy. There is little doubt, also, that for a long time colorings were used to cover up adulteration.

COLOR SAFETY AND REGULATIONS

After an intensive study to determine safety and suitability, seven dyes were recognized under the Pure Food and Drugs Act of 1906 in the First Food Inspection Decision on dyes issued on July 13, 1907. These dyes were selected after a study of all recorded observations, which showed no unfavorable reports, followed by testing for acute, short-range effects in dogs, rabbits, and human beings. These dyes are shown in Table 13.1. Dyes added to the approved list, 1916 to 1929, are shown in Table 13.2.

Regulations promulgated under the Pure Food and Drugs Act of 1906 helped eliminate undesirable contamination in food dyes because

TABLE 13.1. Seven Dyes Recognized in 1907

Common Name	Later FDA Name[a]
Amaranth	FD&C Red No. 2
Erythrosine	FD&C Red No. 3
Indigo disulfonic acid sodium salt (indigotine)	FD&C Blue No. 2
Light Green SF Yellowish	FD&C Green No. 2[b]
Naphthol Yellow S	FD&C Yellow No. 1[c]
Orange I	FD&C Orange No. 1[d]
Ponceau 3R	FD&C Red No. 1[e]

[a] The FD&C designations were not assigned until 1938, when the Food, Drug, and Cosmetic Act was passed.
[b] Deleted for food, drug, and cosmetic use in 1966 because it was of insufficient economic importance.
[c] Delisted for food use in 1959, but permitted in externally applied drugs and cosmetics under the name Ext. D&C Yellow No. 7.
[d] Delisted for food use in 1956, but permitted in externally applied drugs and cosmetics under the name Ext. D&C Orange No. 3; delisted completely in 1968.
[e] Delisted for food use in 1961, but permitted in externally applied drugs and cosmetics under the name Ext. D&C Red No. 15; delisted completely in 1966.

manufacturing firms could submit to Washington samples of each batch of dye for certification. This was voluntary. However, the Federal Food, Drug, and Cosmetic Act of 1938 made this procedure mandatory, and safety data based on animal tests were supplied to support the continued listing of specific colors under that Act.

In line with the existing requirements, only seven FD&C colors are permitted in the food supply at the present time: FD&C Red No. 3, FD&C Blue No. 2, FD&C Yellow No. 5, FD&C Green No. 3, FD&C Yellow No. 6, FD&C Blue No. 1, and FD&C Red. No. 40. They are listed in the order of their date of approval for food use. Limited use is permitted for two others. Citrus Red No. 2 at a level not to exceed 2 ppm by weight is used to color the skins of oranges not intended or used for further processing. Orange B is permitted at a level which does not exceed 150 ppm on a weight basis. It is used for coloring the surfaces and casings of sausages. However, this color has been taken out of the market by its sole manufacturer.

Lakes

Lakes of the water-soluble FD&C colors, prepared by extending the aluminum or calcium salt of the color on a substratum of alumina, have been permitted for use in coloring nonaqueous food products since 1959. Each lake is considered to be a single color and is listed under the name of the color from which it is formed. Lakes are insoluble pigments which color by dispersion.

Since lake making is very much of an art, manufacturing must be

TABLE 13.2. Ten Dyes Added to the Approved List, 1916 to 1929

Common Name	Later FDA Name	Year Added
Tartrazine	FD&C Yellow No. 5	1916
Sudan I	1[a]	1918
Butter Yellow	1[a]	1918
Yellow AB	FD&C Yellow No. 3[b]	1918
Yellow OB	FD&C Yellow No. 4[c]	1918
Guinea Green B	FD&C Green No. 1[d]	1922
Fast Green FCF	FD&C Green No. 3	1927
Brilliant Blue FCF	FD&C Blue No. 1	1929
Ponceau SX	FD&C Red No. 4[e]	1929
Sunset Yellow FCF	FD&C Yellow No. 6	1929

[a] Withdrawn after having been on the permitted list for about six months because contact dermatitis was observed in up to 90% of the factory workers handling these dyes. There were no reports of any harmful effects in those who consumed foods colored with the dyes. Butter Yellow was found later to be carcinogenic in test animals.
[b] Delisted for food use in 1959, but permitted in externally applied drugs and cosmetics under the name Ext. D&C Yellow No. 9. Delisted completely in 1960.
[c] Delisted for food use in 1959, but permitted in externally applied drugs and cosmetics under the name Ext. D&C Yellow No. 10. Delisted completely in 1960.
[d] Delisted for food, drug, and cosmetic use in 1966 because it was of insufficient economic importance.
[e] Previously permitted only for the coloring of maraschino cherries (no longer permitted).

carefully controlled. While dyes from different manufacturing establishments can be interchanged, shades of difference in lakes can be noticed in products obtained from different manufacturers, even though they contain the same amount of pure dye. The coloring ability of a water-soluble dye is proportional to the dye content of the solutions. This is not the case with lakes. A unit of 20% pure lake is not equivalent in strength of color to two units of 10% pure dye. When a lake is suspended in water, a little bleed of dye into the water results. Lakes are stable at pH 3.5–9.5. Lower or higher pH results in a breakdown of the substratum, liberating the dye.

Amount and Distribution of Color

The data in Table 13.3 show the total annual consumption of food colors and the average daily intake by age of consumer and product category taken from the 1977 Survey of Industry on the Use of Food Additives (NAS/NRC 1979).

This is essentially a survey of a sample of U.S. households to determine the number of times individual food products are consumed in a 2-week period. When this incidence is combined with the average amount of food consumed per eating occasion and the concentration of the particular color used in that food category, it is possible to estimate the per capita consumption of a particular item.

TABLE 13.3. Determining Food Color Consumption Data

Certified colors	Total U.S. ussge in 1976 (lb)	Per capita average daily intake[a] Age group	mg	Distribution of usage among food categories (numbers in boldface refer to food category codes at bottom of this page)
Orange B	3,400	6-23 mo	1.4	**10**, 100%
		6-12 yr	2.1	
		18-44 yr	2.4	
FD&C Blue No. 1	45,000	6-23 mo	2.9	**8**, 58%; **1**, 16%; **23**, 13%; **20**, 4%; **24**, 4%; **16**, 2%; **5**, 1%; **7**, 1%; **10**, 1%
		6-12 yr	4.5	
		18-44 yr	3.6	
FD&C Blue No. 1 aluminum lake[b]	16,000	6-23 mo	0.52	**16**, 43%; **20**, 19%; **31**, 18%; **1**, 10%; **17**, 5%; **23**, 4%; **2**, 1%; **7**, 1%
		6-12 yr	1.0	
		18-44 yr	0.76	
FD&C Blue No. 2	3,600	6-23 mo	0.59	**7**, 28%; **49**, 15%; **31**, 12%; **1**, 11%; **16**, 11%; **23**, 11%; **5**, 5%; **20**, 5%
		6-12 yr	1.6	
		18-44 yr	1.0	
FD&C Blue No. 2 aluminum lake	16,000	6-23 mo	0.35	**16**, 32%; **31**, 27%; **17**, 22%; **1**, 19%
		6-12 yr	0.54	
		18-44 yr	0.49	
FD&C Red No. 3	81,000	6-23 mo	5.9	**10**, 21%; **1**, 20%; **8**, 20%; **2**, 11%; **23**, 7%; **20**, 6%; **7**, 5%; **13**, 3%; **16**, 2%
		6-12 yr	9.9	
		18-44 yr	7.0	
FD&C Red No. 3 aluminum lake	74,000	6-23 mo	1.3	**1**, 46%; **16**, 26%; **20**, 11%; **31**, 8%; **17**, 5%; **22**, 2%; **5**, 1%
		6-12 yr	2.8	
		18-44 yr	2.1	
FD&C Red No. 40	500,000	6-23 mo	19	**8**, 51%; **23**, 19%; **1**, 10%; **20**, 5%; **6**, 2%; **16**, 2%; **19**, 2%; **21**, 2%; **5**, 1%
		6-12 yr	31	
		18-44 yr	26	
FD&C Red No. 40 aluminum lake	11,000	6-23 mo	2.2	**1**, 61%; **16**, 21%; **23**, 10%; **17**, 5%; **31**, 2%; **22**, 1%
		6-12 yr	4.9	
		18-44 yr	3.8	

FD&C Red. No. 40 calcium lake	89	6-23 mo	—	**24**, 100%
		6-12 yr	1.8	
		18-44 yr	2.5	
FD&C Yellow No. 5	420,000	6-23 mo	9.4	**8**, 37%; **1**, 10%; **2**, 9%; **23**, 8%; **14**, 7%; **21**, 5%; **24**, 5%; **16**, 3%; **20**, 3%
		6-12 yr	15	
		18-44 yr	13	
FD&C Yellow No. 5 aluminum lake	160,000	6-23 mo	2.2	**16**, 29%; **1**, 22%; **23**, 12%; **31**, 12%; **20**, 9%; **5**, 5%; **2**, 4%; **17**, 4%; **22**, 3%
		6-12 yr	4.3	
		18-44 yr	3.0	
FD&C Yellow No. 5 calcium lake	150	6-23 mo	0.09	**16**, 100%
		6-12 yr	0.10	
		18-44 yr	0.11	
FD&C Yellow No. 6	400,000	6-23 mo	6.8	**23**, 28%; **1**, 20%; **2**, 11%; **20**, 11%; **10**, 9%; **8**, 5%; **16**, 4%; **7**, 3%; **15**, 2%
		6-12 yr	14	
		18-44 yr	11	
FD&C Yellow No. 6 aluminum lake	72,000	6-23 mo	1.1	**16**, 31%; **17**, 24%; **20**, 13%; **23**, 12%; **1**, 9%; **31**, 5%; **22**, 3%; **2**, 1%; **5**, 1%
		6-12 yr	2.7	
		18-44 yr	1.7	
FD&C Green No. 3	290	6-23 mo	0.68	**8**, 57%; **1**, 21%; **23**, 8%; **16**, 4%; **24**, 4%; **9**, 2%; **20**, 2%; **7**, 1%; **28**, 1%
		6-12 yr	1.1	
		18-44 yr	0.85	

[a] This refers to individuals who indicated that they consume the food in question

[b] The term "lake" refers to insoluble powders prepared by absorption, coprecipitation, or chemical combination of a colorant with an aluminum or calcium compound

Codes and food categories

01 Baked goods
02 Breakfast cereals
03 Grain products, pasta
04 Fats and oils
05 Milk, milk products
06 Cheese
07 Frozen dairy dessert
08 Fruits, juices
09 Fruit and water
10 Meats, meat products
11 Poultry, poultry products
12 Eggs, egg products
13 Fish, seafood
14 Vegetables, juices
15 Condiments, relish
16 Candy
17 Sugars, frosting
19 Sweet sauces, toppings
20 Gelatins, puddings, custard
21 Soup, soup mixes
22 Snack foods
23 Nonalcoholic beverages
24 Alcoholic beverages
27 Gravies, sauces
28 Dairy product analogs
31 Chewing gum
45 Main dishes, not elsewhere classified
49 Coffee and tea

The accuracy of the intakes in the table is limited by two major considerations. One is that an additive may be used in only a few specialized food products within a category, yet it is treated as if it were present in all foods in the category. The second consideration is that some additives are interchangeable, and all of these would be considered present if any were present.

Specifications and Hazards

Two meetings which dealt mainly with food colors were held by the Joint FAO/WHO Expert Committee on Food Additives. The purpose was to evaluate the toxicological hazards attending the use of food colors and to establish specifications of identity and purity for those colors. In all, over 160 food colors that had been used in various countries were studied and placed by the Expert Committee in accordance with its toxicological evaluation, in the following categories:

Category A. These colors were found acceptable for use in food. A maximum acceptable daily intake value was established for each. The Committee emphasized that the assignment of a color to this category should not be interpreted as indicating that further research was unnecessary, but recognized that more work was called for.

Category B. The data available to the Expert Committee for these colors were not wholly sufficient to meet the requirements of Category A.

Category C I. For these colors the available data were inadequate for evaluation, but a substantial amount of detailed information from results of long-term tests was available.

Category C II. The available data on these colors were inadequate for evaluation and virtually no information on long-term toxicity was available. Colors on which there were data from long-term tests for tumor formation unaccompanied by information from other long-term studies were considered as falling within this category.

Category C III. The available data, although inadequate for evaluation, indicated the possibility of harmful effects from these colors.

Category D. Colors for which virtually no toxicological data were available were included in this group.

On the basis of available analytical information, the Committee classified colors also from the standpoint of chemical specifications as follows.

I. Food colors for which the Committee was able to prepare satisfactory specifications.

II. Food colors for which the Committee chose not to prepare specifications, since toxicological studies on substances of defined composition were in progress.
III. Food colors for which the chemical information available to the Committee was inadequate to permit the preparation of fully satisfactory specifications.
IV. Food colors for which the Committee did not attempt to prepare specifications either because toxicological data were totally lacking or because the colors were demonstrably toxic at levels which indicated that their use in foods is undesirable.

Tables 13.4 and 13.5 give an idea of the requirements for listing of a food coloring, and also the acceptable amounts that may be taken daily.

The dyes certified by FD&C are all water soluble and are made up of four types of chemical structure, namely, azodyes, triphenylmethane dyes, fluorescein, and sulfonated indigo. FD&C Red No. 2, FD&C Red No. 4, FD&C Yellow No. 5, and FD&C Yellow No. 6 are azo dyes. FD&C Violet No. 1, FD&C Blue No. 1, and FD&C Green No. 3 are triphenylmethane dyes. FD&C Red No. 3 is a fluorescein dye, and FD&C Blue No. 2 is sulfonated indigo.

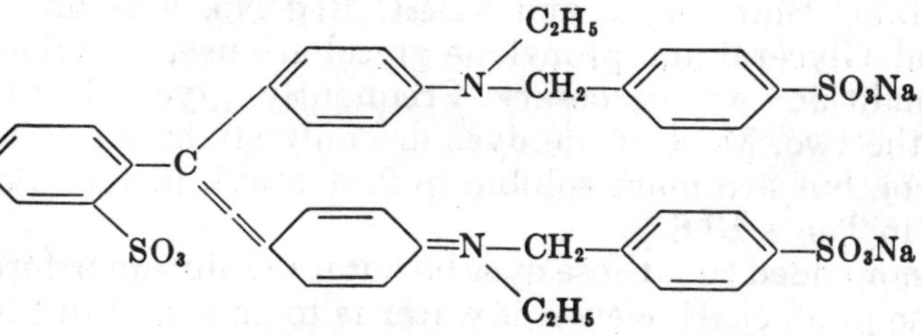

FD & C Blue No. 1

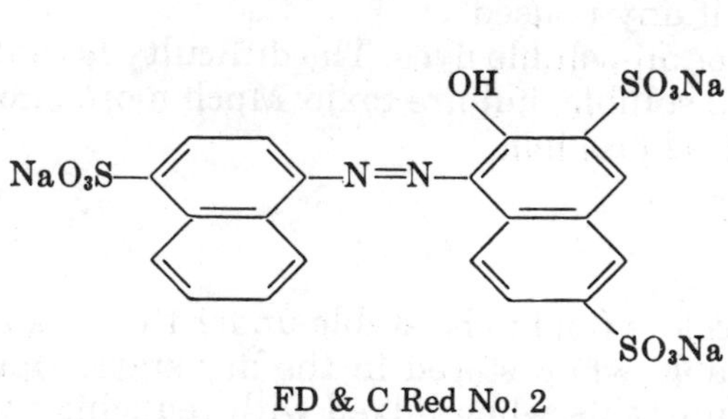

FD & C Red No. 2

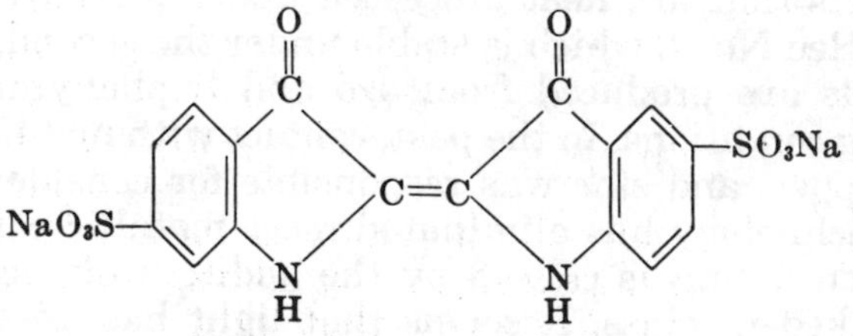

FD & C Blue No. 2 (Sulfonated indigo)

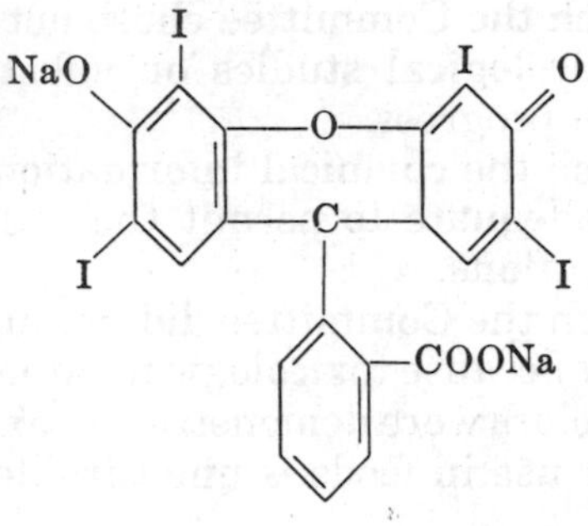

FD & C Red No. 3

Solubility

While FD&C dyes are soluble in water, they are insoluble in most organic solvents. Since the solubility in water of most of the dyes is rather high, solubility is not a problem for general use. Only in the cases of FD&C Blue No. 2 and FD&C Red No. 4 is any difficulty encountered. Glycerol and propylene glycol are used as solvents if anhydrous conditions are necessary. Frequently, glycerol is the better solvent of the two. Most of the dyes are only slightly soluble in 95% ethyl alcohol, but are more soluble in 25% ethyl alcohol. Solubilities are shown in Table 13.6.

It is recommended that these dyes be put into solution before they are added to the product. However, if water is to be added in the process, dry color may be used and solution will be effected by the water and also by the heat if any is used.

A need exists for oil-soluble dyes. The difficulty is that unsulfonated dyes tend to be oil soluble, but are toxic. Much more effort will have to be expended on this problem.

Stability

Certified food colors tend to be stable under the bulk of conditions of use. They are stable when stored in the dry state. However, most of these colors are unstable when mixed with reducing agents and with protein products that are heat processed under pressure (in retorts), except FD&C Red No. 3, which is stable under these conditions. Colorless compounds are produced from azo and triphenylmethane dyes under reducing conditions. In the past, contact with metals such as tin, aluminum, copper, and zinc was responsible for considerable fading, but present technology has eliminated most metallic contamination. Currently, such fading is caused by the addition of ascorbic acid to beverages packed in glass. It seems that light has an effect on this reaction.

TABLE 13.4. Classification of Some U.S. Food Colors by the Joint FAO/WHO Expert Committee

Common Name	FD&C No.	Classification	
		Chemical[a] Specification	Toxicological[a] Category
Amaranth	Red No.	I	A
Annato, bixin and norbixin		I	A
β-Apo-8′-carotenal		I	A
β-Apo-8′-carotenoic acid, methyl or ethyl ester		I	A
Beet red and betain		III	[b]
Benzyl Violet 4B	Violet No. 1[c]	I	C III
Brilliant Blue FCF	Blue No. 1	I	A
Canthaxanthine		I	A
Caramel		IV	D
Carbon Black		II	[b]
Carotene (natural)		III	[b]
β-Carotene (synthetic)		I	A
Citrus Red No. 2	Citrus Red. No. 2[d]	I	E
Cochineal and carminic acid		III	[b]
Erythrosine	Red No. 3	I	A
Fast Green FCF	Green No. 3	I	A
Indigotine	Blue No. 2	I	A
Iron oxides		III	[b]
Ponceau SX	Red No. 4[e]	IV	E
Riboflavin		I	A
Saffron, corcin and crocetin		III	[b]
Sunset Yellow FCF	Yellow No. 6	I	A
Tartrazine	Yellow No. 5	I	A
Titanium dioxide		I	A
Turmeric (curcumin)		I	A
Ultramarine		III	[b]
Xanthophylls		III	[b]

[a] See text for explanation of chemical and toxicological category.
[b] No attempt was made at a toxicological evaluation because of the lack of knowledge of composition and the paucity of biological studies.
[c] A petition for the permanent listing in the United States is under study by the FDA.
[d] Use in the United States is limited to the coloring of skins of oranges not intended or used for further processing, at a level not exceeding 2 ppm by weight.
[e] Use in the United States was limited to the coloring of maraschino cherries at a level not exceeding 150 ppm by weight (no longer permitted).

These dyes are usually stable to light, the only ones causing trouble under certain conditions are FD&C Red No. 3 and FD&C Blue No. 2. Red No. 3 tends to be unstable to light when used in coatings and Blue No. 2 fades when used in solutions. Canned products containing certified food colors and ascorbic acid are relatively stable.

TABLE 13.5. Acceptable Daily Intakes (ADI) of Some U.S. Food Colors as Established by the FAO/WHO Expert Committee on Food Additives

Common Name	FD&C No.	Acceptable Daily Intake for Man (mg/kg Body Weight)		
		Unconditional[a]	Conditional[b]	Temporary[c]
Amaranth	Red No. 2	0–1.5		
Annatto, bixin and norbixin				0–1.25[d]
β-Apo-8′-carotenal		0–2.5[e]	2.5–5.0[e]	
β-Apo-8′-cartenoic acid methyl or ethyl ester		0–2.5[e]	2.5–5.0[e]	
Brilliant Blue FCF	Blue No. 1	0–12.5		
Canthaxanthin		0–12.5	12.5–25.0	
β-Carotene (synthetic)		0–2.5[e]	2.5–5.0[e]	
Erythrosine	Red No. 3			0–1.25[f]
Fast Green FCF	Green No. 3	0–12.5		
Indigotine	Blue No. 2			0–2.5[g]
Riboflavin		0–0.5		
Sunset Yellow FCF	Yellow No. 6	0–5.0		
Tartrazine	Yellow No. 5	0–7.5		
Titanium dioxide		[h]		
Turmeric (curcumin)				0–0.5[i]

Source: National Academy of Sciences (1971).
[a] An unconditional ADI was allocated only to those substances for which the biological data available included either the results of adequate short-term toxicological investigations or information on the biochemistry and metabolic fate of the compound or both.
[b] A conditional ADI was allocated for specific purposes arising from special dietary requirements.
[c] A temporary ADI was allocated when the available data were not fully adequate to establish the safety of the substance, and it was considered necessary that additional evidence be provided within a stated period of time. If the further data requested do not become available within the stated period, it is possible that the temporary ADI will be withdrawn by a future Committee.
[d] Further work required by June 1972: metabolic studies on the major carotenoids of annatto.
[e] Expressed as total carotenoids by weight.
[f] Further work required by June 1972: studies on the metabolism in several species and preferably in man and elucidation of the mechanism underlying the effect of this color on plasma-bound iodine levels.
[g] Further work required by June 1974: 2 yr study in a non-rodent mammalian species.
[h] Not limited except by good manufacturing practice.
[i] Further work required by June 1974: studies on the metabolism of curcumin and a 2 yr study in a non-rodent mammalian species.

Corrosion

Corrosion in canned beverages caused by the use of certain of the certified food colors has been found to take place. The following FD&C dyes have caused some trouble: FD&C Red No. 2, Yellow No. 5, and

Yellow No. 6. The amount of corrosion seems to be directly proportional to the amount of dye used. These are all azo dyes, and it seems that only azo dyes are involved in this problem. Use of 50 ppm or less of these dyes will give the product a reasonable shelf-life. Under these conditions, it is necessary sometimes to improve the color by the addition of some caramel.

COLORS EXEMPT FROM CERTIFICATION

This group of color additives is made up of coloring materials largely of natural origin. These, together with a few which are synthetic, are listed permanently as exempt from certification. The list of these substances in Table 13.7 is rather large because the Color Additives Amendment defined color additive as "a dye, pigment, or other substance made by a process of synthesis or similar artifice, or extracted, isolated, or otherwise derived, with or without intermediate or final change of identity, from vegetable, animal, mineral or other source," that, "when added or applied to a food, drug, or cosmetic, or to the human body or any part thereof, is capable (alone or through reaction with other substance) of imparting color thereto." A number of these are not important.

Carotenoids

These compounds have been discussed under plant pigments in Chapter 11 and specifically, β-carotene, in Chapter 8. Some of these

TABLE 13.6. Solubilities of Certified Food Colors

FD&C Name (Common Name)	Solubilities in Various Solvents (g/100 ml)			
	Water	EtOH 25%	Glycerol	Propylene Glycol
Red No. 2 (amaranth)	20.0	7.0	18.0	1.0
Red No. 3 (erythrosine)	9.0	8.0	20.0	20.0
Red No. 4 (Ponceau SX)[a]	11.0	1.4	5.8	2.0
Yellow No. 6 (Sunset Yellow FCF)	19.0	10.0	20.0	2.2
Yellow No. 5 (tartrazine)	20.0	12.0	18.0	7.0
Green No. 3 (Fast Green FCF)	20.0	20.0	20.0	20.0
Blue No. 1 (Brilliant Blue FCF)	20.0	20.0	20.0	20.0
Blue No. 2 (indigotine)	1.6	0.5	1.0	0.1

Source: National Academy of Sciences (1971).
[a] No longer permitted.

compounds have been synthesized for use as coloring compounds. Originally, the carotenoids were made for coloring by extraction from natural sources—β-carotene was obtained from carrots. This compound has a special value in that it can be transformed in the body to vitamin A and because of this, it is particularly adapted for use in foods. The difficulty involved in the use of these compounds is that they tend to be rather unstable and relatively insoluble. β-Carotene tends to deteriorate in air and light. Oil solutions are stable, so carotene use is to a large extent in fatty foods such as margarine and oils. β-Carotene, in the form of a micro-pulverized suspension, has the advantage of rapidly dissolving in the fatty material under the heat conditions of the process of manufacture. For coloring water-base foods, Bunnell *et al.* (1958) described a water-dispersible beadlet to color such products as puddings, ice cream, and citrus beverages.

TABLE 13.7. Color Additives Permanently Listed for Food Use Exempt from Certification

Color	Use Limitation[a]
Algae meal, dried	For use in chicken feed to enchance the yellow color of chicken skin and eggs
Annatto extract	
β-Apo-8′-carotenal	Not to exceed 15 mg/lb, or pint, of food
Beets, dehydrated (beet powder)	
Canthaxanthin	Not to exceed 30 mg/lb, or pint, of food
Caramel	
β-Carotene	
Carrot oil	
Cochineal extract; carmine	
Corn endosperm oil	For use in chicken feed to enhance the yellow color of chicken skin and eggs
Cottonseed flour, partially defatted, cooked, toasted	
Ferrous gluconate	For coloring ripe olives
Fruit juice	
Grape skin extract	For coloring beverages
Iron oxide (synthetic)	For coloring pet food, not to exceed 0.25% by weight of the food
Paprika	
Riboflavin	
Saffron	
Tagetes meal and extract (aztec marigold)	For use in chicken feed to enhance the yellow color of chicken skin and eggs
Titanium dioxide	Not to exceed 1% by weight of the food
Turmeric; turmeric oleoresin	
Ultramarine blue	For coloring salt intended for animal feed, not to exceed 0.5% by weight of the salt
Vegetable juice	

Source: National Academy of Sciences (1971).

[a] Unless otherwise indicated, the color may be used for the coloring of food generally in amounts consistent with good manufacturing practice.

β-Carotene takes the place of FD&C Yellow No. 3 and FD&C No. 4 which were delisted. It is more expensive, however, and the need exists for less expensive oil-soluble dyes. The carotenoids resist reduction by ascorbic acid, a property of value in the manufacture of beverages.

Annatto Extract, Bixin

The pulp enclosing the seeds of the tree *Bixa orellana,* found in Central America, produces the annatto colors. Bixin, the principal compound in this coloring material, is a carotenoid.

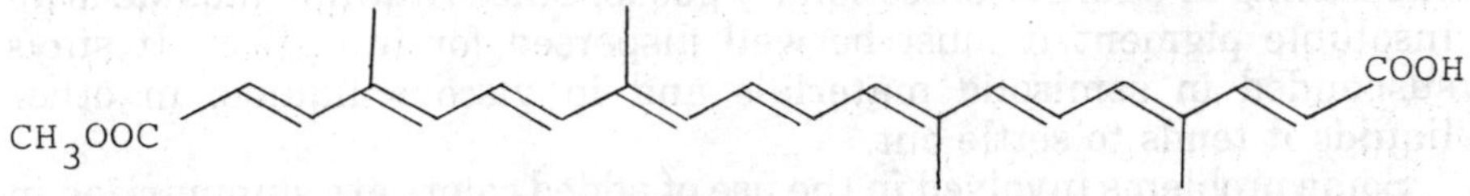

COOH

CH_3OOC

Bixin

The pigment norbixin is prepared by saponification of bixin, during which the methyl ester group splits off. Norbixin is water soluble whereas bixin is oil soluble: the aqueous solution can be used for coloring ice cream; the oil-soluble bixin is used for butter, margarine, salad dressings, and for oil for use on popcorn. Bixin has become more important since the delisting of the certified food colors soluble in oil and has the advantage of being lower in price than β-carotene.

Cochineal

Cochineal is made up of the dried and pulverized bodies of the female insect *Coccus cacti.* This insect lives on a particular variety of cactus and contains an active coloring matter in the form of carminic acid which is extracted by water and forms bright red crystals.

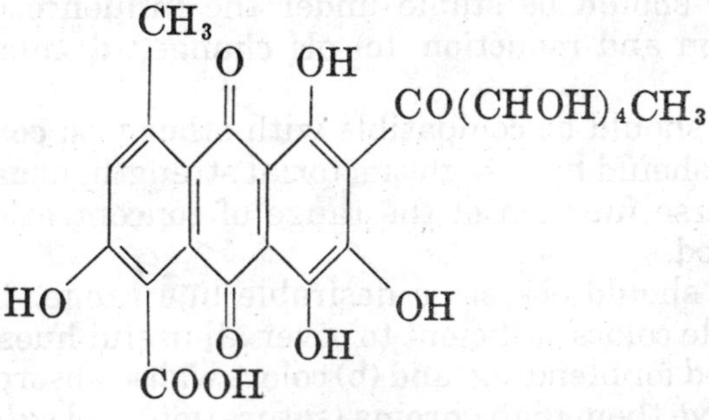

CH_3 O OH $CO(CHOH)_4CH_3$

HO OH

O OH

COOH

Carminic acid

The aluminum lake of carminic acid is known as carmine, and of course, like all lakes, is a pigment. Carmine is used to color coatings pink and can be used in pressure-heated (retorted) protein materials because it is stable under these conditions.

Turmeric

Turmeric is the ground dried rhizome or bulbous root of *Curcurma longa,* a plant of the ginger family. It is used in prepared mustard for flavor but mainly for color.

Titanium Dioxide

This compound is a white pigment and is of particular value because of its effective covering power. It is mainly used in icings and as a subcoating in panned confectionary goods. Since titanium dioxide is an insoluble pigment it must be well dispersed for full effect. It stays suspended in semisolid materials and in viscous liquids; in other liquids it tends to settle out.

Some problems involved in the use of added colors are summarized in Table 13.8.

COLOR ANALYSIS AND DESIRED PROPERTIES

Analytical methods for food color additives are available from the AOAC (1980; see also Furia 1968).

The booklet on food colors of the National Academy of Sciences (1971) lists the following criteria in a rough order of importance.

1. Food colors must be safe for human beings at levels that can reasonably be expected to be consumed when they are properly used for the purpose intended.
2. At the level used, the color must either be tasteless and odorless (as with the certified synthetics), or its organoleptic properties must be inoffensive and must blend well with those of the food it colors (as with paprika, saffron, and turmeric).
3. A color should be stable under the influence of (a) light, (b) oxidation and reduction, (c) pH change, (d) microbiological attack.
4. A color should be compatible with other food components.
5. A color should have high tinctorial strength, usually revealed as an inverse function of the range of concentrations required to color food.
6. A color should possess a desirable hue range. This implies (a) available colors sufficient to cover all useful hues without excessive need for blending, and (b) colors whose absorption characteristics give them high chroma (saturation) and *value* (brightness), i.e., strong colors that contain little black (minimum general absorption).
7. A color should be highly soluble in water and other inexpensive acceptable polar food grade solvents (e.g., alcohol, propylene glycol).

TABLE 13.8. Problems with Dyes

Problem	Cause
Precipitation from color solution or colored liquid food	Exceeded solubility limit Insufficient solvent Chemical reaction Low temperatures, especially for concentrated color solution
Dulling effects instead of bright, pleasing shades	Excessive color Exposure to high temperatures
Speckling and spotting during coloring of bakery and confectionary products	Color not completely dissolved while making solution Employed liquid color containing sediment Attempted dispersion in an aqueous color solution in products containing excessive fat
Fading due to light	Colored products not protected from sunlight
Fading due to metals	Colored solutions or colored products were in contact with certain metals (Zn, Sn, Al, etc.) during handling, dissolving, or storing
Fading due to microorganisms	Color-preparing facilities not thoroughly cleansed to avoid contaminating reducing organisms
Fading due to excessive heat	Processing temperature too high
Fading due to oxidizing and reducing agents	Contacted color with oxidizers such as ozone or hypochlorites or reducers as SO_2 and ascorbic acid
Fading due to strong acids or alkalies	Presence of such strong chemicals during the coloring of certain foods
Fading due to retorting with protein material	Color is unstable under these conditions
Poor shelf-life with colored canned carbonated beverages	Used an excessive amount of certified azo-type dye

Source: Furia (1968). Reproduced with permission of The Chemical Rubber Company.

8. Solubility in edible fats and oils is a desirable property, particularly for certain hues.
9. A color should be easily dispersible, if it is not soluble.
10. A color should be inexpensive in terms of the cost to achieve the desired color level.

SUMMARY

Seven dyes were recognized under the Pure Food and Drugs Act of 1906. Since that time, dyes have been added to and deleted from the approved list. The Federal Food, Drug, and Cosmetic Act of 1938 made it mandatory for firms to submit to Washington samples of each batch of dye for certification.

Lakes of the water-soluble colors are prepared by extending the aluminum or calcium salt of the color on a substratum of alumina. Lakes are insoluble pigments which color by dispersion.

Fairly large quantities of colors are used in various foods. The Government is especially concerned about the safety of various colors for such use. An Expert Committee studied a number of food colors for the FDA and prepared a set of categories for colors depending on the toxicological evaluation. Acceptable daily intakes of U.S. food colors were established by the committee.

The various synthetic colors have one of four chemical structures: azo dyes, triphenyl methane dyes, fluorescein, and sulfonated indigo.

The solubility of the various colors is important since it affects their use. Another important factor is stability. Corrosion caused by dyes is also of much importance.

A group of color additives, largely natural in origin, are permanently exempt from certification. These include such compounds as the carotenes, caramels, and grape skin extract.

The booklet on food colors of the National Academy of Sciences (1971) lists desirable properties food colors should have.

BIBLIOGRAPHY

ANON. 1968. Certified color industry committee. Guidelines for good manufacturing practice. Food Technol. *22,* 946–949.

ANON. 1980. Food colors. A scientific summary by the Institute of Food Technologists Expert Panel on Food Safety and Nutrition and Committee on Public Information. Food Technol. *34* (7), 77–84.

AOAC. 1980. Official Methods of Analysis, 13th Edition. Assoc. of Official Analytical Chemists, Washington, DC.

BAUERNFEIND, J. C., SMITH, E. G., and BUNNELL, R. H. 1958. Coloring fat-base foods with β-carotene. Food Technol. *12,* 527–535.

BUNNELL, R. H., DRISCOLL, W.,and BAUERNFEIND, J. C. 1958. Coloring water-base foods with β-carotene. Food Technol. *12,* 536–541.

FURIA, T. E. 1968. Handbook of Food Additives. The Chemical Rubber Co., Cleveland, Ohio.

GRAICHEN, C., and MOLITOR, J. C. 1963. Determination of certifiable FD&C color additives in foods and drugs. J. Assoc. Off. Agric. Chem. *46,* 1022–1029.

HESSE, B. C. 1912. Coal-tar colors used in food products. Bureau of Chemistry. U.S. Dept. Agric. Bull 147. Washington, DC.

NATIONAL ACADEMY OF SCIENCES. 1971. Food Colors. Committee on Food Protection, Food and Nutrition Board, Division of Biology and Agriculture, National Research Council, Washington, DC.

14

Alcoholic Fermentation

Wine
Beer and Brewing
Distilled Products
Vinegar
Summary
Bibliography

Alcoholic beverages have one point in common. They all depend on the process of fermentation—the conversion of hexose sugars into alcohol and carbon dioxide. This is indeed a very important process and is basic to all of the industries involved. Alcoholic fermentation is an anaerobic process carried on by living yeast cells. The cells absorb the simple sugars, which in turn, are broken down in a series of successive changes in which action by oxidizing and reducing enzymes within the cell takes place. The final result is the formation of ethyl alcohol and carbon dioxide accompanied by the liberation of some energy in the form of heat. This process does not account for all of the original energy in the sugars, however, because a part of it is used for multiplication and metabolism of the yeast cells, and the process, therefore, changes compounds of higher energy to those of lower energy. Buchner (1897) showed that a cell-free extract from yeast cells induced fermentation, demonstrating that the enzymes and not the cells themselves produce the chemical action.

The overall chemical reaction was first described by Gay-Lussac in the beginning of the nineteenth century

$$C_6H_{12}O_6 \longrightarrow 2C_2H_5OH + 2CO_2$$

Pasteur showed that a number of by-products present in the fermented solution were not accounted for by Gay-Lussac's equation. Gay-Lussac's equation turned out to be an overall general equation which showed the starting materials and the end-products. This equation is the summation of a large number of reactions which take place during the formation of the two main products, ethyl alcohol and carbon dioxide. Twenty-two enzymes are necessary for the glycolytic cycle in alcoholic fermentation, together with at least six coenzymes, and magnesium and potassium ions. The reactions are given in Fig. 14.1.

The glycolytic sequence shows the complexity of this system, and also how it is possible for glycerol, lactic acid, and acetaldehyde to build up as by-products. The glyoxylate cycle probably produces most of the succinic acid. Since the entire sequence of events requires the presence of acetaldehyde, in the initial stage (induction phase) hexose phosphate is converted to 3-phosphoglycerate and α-glycerophosphate, and the α-glycerophosphate is then converted to glycerol. There is a delay at the beginning in the entire sequence of reactions because of the absence of acetaldehyde. The 3-phosphoglycerate is converted to pyruvate, which in turn, is decarboxylated to yield acetaldehyde. As acetaldehyde accumulates, it acts as the hydrogen acceptor and reacts with the NADH (reduced coenzyme I) to give ethyl alcohol. Earlier in the process the hydrogen acceptor is dihydroxyacetone phosphate. When the reaction is established, that is, when the steady state is reached, the

1. Glucose[1] $\xrightarrow[\text{(Mg}^{2+}\text{, ATP} \rightarrow \text{ADP)}]{\text{(hexokinase)}}$ glucose-6-phosphate

2a. Glucose-6-phosphate $\xrightarrow[\text{(NADP} \rightarrow \text{NADPH + H}^{4}\text{)}]{\text{(glucose-6-phosphate dehydrogenase)}}$ 6-phosphogluconate → hexose monophosphate shunt system

2. Glucose-6-phosphate $\xrightleftharpoons{\text{(phosphohexoisomerase)}}$ fructose-6-phosphate

2b. Fructose $\xrightarrow[\text{(Mg}^{2+}\text{, ATP} \rightarrow \text{ADP)}]{\text{(hexokinase)}}$ fructose-6-phosphate

3. Fructose-6-phosphate (Neuberg ester) $\xrightarrow[\text{(Mg}^{2+}\text{, ATP} \rightarrow \text{ADP)}]{\text{(phosphofructokinase)}}$ fructose-1,6-diphosphate[2]

4. Fructose-1,6-diphosphate[2] (Harden-Young ester) $\xrightleftharpoons[\text{(Zn}^{2+}\text{, Co}^{2+}\text{, Fe}^{2+}\text{, or Ca}^{2+}\text{)}]{\text{(aldolase)}}$ D-glyceraldehyde-3-phosphate + dihydroxyacetone phosphate

5. D-Glyceraldehyde-3-phosphate (Fischer-Baer ester) $\xrightleftharpoons{\text{(triosphosphate isomerase)}}$ dihydroxyacetone-phosphate

5a. Dihydroxyacetone-phosphate $\xrightleftharpoons[\text{(H}^{+}\text{ + NADH} \rightarrow \text{NAD)}]{(\alpha\text{-glycerolphosphate dehydrogenase)}}$ α-L-glycerol phosphoric acid

5b. α-L-Glycerol phosphoric acid $\xrightleftharpoons{\text{(phosphatase)}}$ glycerol

6. D-Glyceraldehyde-3-phosphate + H_3PO_4 $\xrightleftharpoons[\text{(NAD} \rightarrow \text{NADH} \rightarrow \text{+ H}^{+}\text{)}]{\text{(triosphosphate dehydrogenase)}}$ 1,3-diphosphoryl-D-glycerate

7. 1,3-Diphosphoryl-D-glycerate $\xrightleftharpoons[\text{(Mg}^{2+}\text{, ADP} \rightarrow \text{ATP)}]{\text{(phosphorylglyceryl kinase)}}$ 3-diphosphoryl-D-glycerate

formation of ethanol predominates and only minor quantities of glycerol are formed. If the acetaldehyde is acted on by sulfite, and therefore, removed from the reaction, the induction phase goes ahead, and glycerol is formed. If the concentration of sulfur dioxide in acid solution is high, the main products are glycerol, carbon dioxide, and acetaldehyde, while alcohol is produced in small quantity. However, if the sulfite solution is alkaline, acetaldehyde, alcohol, carbon dioxide, and glycerol are all formed.

The practical yields of alcohol from these reactions are from 90 to 95% of theory. According to the Gay-Lussac equation if a 100% yield were possible, 92 gm of alcohol would be obtained from 180 gm of glucose. This theoretical yield cannot be obtained because some of the glucose is used by the yeast cells for growth and some is converted into small quantities of other carbon compounds.

8. 3-Diphosphoryl-D-glycerate $\underset{\text{(2,3-diphosphoryl-D-glycerate)}}{\overset{\text{(phosphyrlglyceryl mutase)}}{\rightleftharpoons}}$ 2-phosphoryl-D-glycerate

9. 2-Phosphoryl-D-glycerate $\underset{(Mg^{2+})}{\overset{\text{(phosphoenolpyruvic transphorylase)}}{\rightleftharpoons}}$ phosphorylenolpyruvate[3]

10. Phosphorylenolpyruvate $\underset{(Mg^{2+},\ K^{+},\ ADP \rightarrow ATP)}{\overset{\text{(phosphoenolpyruvic transphosphorylase)}}{\rightleftharpoons}}$ pyruvate

11. Pyruvate $\underset{\text{TPP}}{\overset{\text{(carboxylase)}}{\rightleftharpoons}}$ acetaldehyde + CO_2

11a. Pyruvate $\underset{(NADH + H^{+} \rightarrow NAD)}{\overset{\text{(lactic dehydrogenase)}}{\rightleftharpoons}}$ lactic acid

12. Acetaldehyde $\underset{(NADH + H^{+} \rightarrow NAD)}{\overset{\text{(alcohol dehydrogenase)}}{\rightleftharpoons}}$ ethanol

ADP, ATP. Di- and triphosphates of adenosine.

NAD$^+$, NADH. Oxidized and reduced nicotinamide adenine dinucleotide. (NAD is also called coenzyme I or DPN).

NADP, NADPH. Oxidized and reduced nicotinamide adenine dinucleotide phosphate. (NADP is also called coenzyme II or TPN).

TPP. Thiamin pyrophosphate.

[1] Starch is converted to glucose 1-phosphate (the Cremer–Cori ester) with phosphoric acid and phosphorylase. Glucose 1-phosphate plus the enzyme phosphoglucomutase and magnesium ions is converted to glucose-1,6-diphosphate.

[2] Fructose 1,6-diphosphate is also converted to fructose 6-phosphate in the presence of fructose diphosphatase, magnesium ions, and water.

[3] Phosphorylenolpyruvate plus phosphorylenolpyruvate enolase and ITP and TTP (inosine di- and triphosphates) may also produce oxaloacetate.

FIG. 14.1. Chemical reactions in alcoholic fermentation. From Amerine *et al.* (1972).

WINE

Wine is the fermented juice of fruit, usually of the grape, and has been made and used from very early times. The most important grape species is *Vitis vinifera,* but some others, notably *V. labrusca* and hybrids of it are used in the Eastern United States and in Canada.

A variety of wines exist. Still wines, which do not retain the carbon dioxide of the fermentation process are the usual table wines, and are available as red, white, and rosé. The red wines are colored by the pigments found in the skins of the grapes. Rosé (pink) wines are made by removing the skins during fermentation to reduce the amount of color extraction. White wines are made from white grapes or from red grapes from which the juice has been pressed before fermentation.

European wines are often designated by the name of the area or district in which they are produced such as Burgundy, Bordeaux, Moselle, etc.

Sparkling wines are those which have been given a secondary fermentation with the result that they contain carbon dioxide under pressure. Champagne is the prime example of this type of wine.

The quality and character of wine is the result of the following factors: (1) the composition of the fruit from which it is made, (2) the fermentation conditions used, and (3) the changes that take place during storage.

As the grape ripens it progresses from a hard, green berry to one that is turgid and easily crushed when fully ripe. The principal sugars are dextrose and levulose, otherwise known as glucose and fructose. Dextrose predominates in the unripe fruit, but levulose becomes equal to or exceeds dextrose in the ripe and overripe fruit. The reducing sugar content of unfermented musts is from 12 to 27%. Small amounts of sucrose and pentoses are found also.

Acidity decreases during ripening of grapes. The predominant acids are malic and tartaric, and, while there are differences among varieties, the percentage of malates and tartrates decrease as ripening progresses. On the single berry basis, however, the tartrate content stays relatively constant while the malate decreases, but not so much as it does on the percentage basis. This is because the grape increases in size during growth and maturation. Minor amounts of other acids are present. The changes in relative amounts of free acids, acid salts, and salts during ripening are important, and are related to observed changes in pH, which steadily increases during this period. The pH during ripening may change from 2.8 or lower to values above 4.0. This change is caused by the translocation of potassium and other cations into the fruit.

Volatile compounds develop during the growing period. Of all the esters present in the essence of the Concord grape, methyl anthranilate and ethyl acetate are present in the largest quantity. Methyl anthranilate increases during ripening. Linalool and geraniol are the main components of muscat aroma. The presence of other compounds, extremely small in amount but powerful in odor, is suspected in the muscat grape. While a larger number of compounds have been identified as being present in volatiles from grapes, the knowledge of the composition of these aromas is far from complete.

Also present in grapes are tannins, pectins, and anthocyanins which are discussed later under wine composition.

Grapes contain amino acids, proline and threonine being the most abundant, and ammonium salts, as well as mineral salts containing K, Mg, Zn, Co, Ca, I, S, P, as well as others.

The Fermentation of Must

For white wine the must is prepared by pressing the grapes and then fermenting it for 10–15 days at 10°–21°C. The process for the preparation of red wine depends on the extraction of the pigments from the skins. This can be accomplished by heating the grapes before pressing or by fermenting on the skins. In the latter method the grapes are crushed, and the fermentation is carried on in the presence of the skins. The fermentation, at 25°C, is continued for 5–7 days or 10–15 days, depending on the wine being produced. Following the fermentation, the wine is racked from the vat to remove the sediment and skins. The grape skins are then pressed and the resulting pressed wine may be mixed with the wine obtained by racking. The wine is allowed to continue to ferment until only traces of sugar remain. It is then matured for 1–2 years; a shorter period is used for light fruity wines, longer periods for heavy-bodied, high tannin wines. During this period of maturation, racking is employed to remove the wine from its sediment.

Saccharomyces cerevisiae, the important wine yeast, is used frequently to inoculate the musts at the start of fermentation. The yeasts naturally present on the grapes can do the job, but in many wineries the inoculation method is standard practice.

Dextrose is ordinarily more rapidly fermented by wine yeasts than is levulose. However, *S. elegans,* the strain used for sauternes, has the ability to ferment levulose more rapidly. It has been suggested that perhaps levulose can permeate the cell walls of this strain more quickly.

Antiseptics

Sulfur dioxide is the principal antiseptic used in the manufacture of wine to eliminate unwanted organisms and to prevent oxidation. Used for this purpose since the Middle Ages, the sulfur dioxide is added usually before inoculation, often at levels of 50–100 ppm. It is soluble in water and in solution is able to combine with water to form sulfurous acid. This acid is capable of ionizing to yield H^+ and HSO_3^- ions. HSO_3^- can ionize to form H^+ and SO_3^- ions and is capable of forming S_2O^{2-}

with the liberation of water. These are all forms of free sulfur dioxide. The bisulfite ion, HSO_3^-, has the ability to react with such substances as aldehydes, ketones, proteins, pectic substances, and sugars with free aldo or keto groups. Under these conditions, bisulfite addition compounds are formed. In this form the sulfur dioxide is said to be fixed or bound. The general equation for this is

$$\mathrm{R{-}\overset{\displaystyle H}{\overset{|}{C}}{=}O + H{-}O{-}\underset{\displaystyle O}{\underset{\|}{S}}{-}O^- \longrightarrow H{-}O{-}\underset{\displaystyle R}{\underset{|}{\overset{\displaystyle H}{\overset{|}{C}}}}{-}\underset{\displaystyle O}{\underset{\|}{\overset{\displaystyle O}{\overset{\|}{S}}}}{-}O^-}$$

Bilsulfite reacts preferentially with acetaldehyde, but if added in amounts large enough, it will react with dextrose, and in smaller amounts with levulose. This can happen in musts, but wines, according to Joslyn (1952), seldom have enough sulfurous acid in excess of acetaldehyde to combine with sugars.

The antiseptic power of sulfur dioxide depends not only on the forms present but also on the kind of microorganisms and their activity. Molds and many wild yeasts are very sensitive to sulfur dioxide and most microorganisms are restrained at a level of 100 ppm of SO_2. It has been shown that some bacteria are sensitive to rather small amounts of sulfur dioxide. It is possible, however, for yeasts to be conditioned to grow and ferment in an environment high in this substance. That sulfur dioxide is less effective when the yeasts are conducting a rapid fermentation is probably due to the fixation of some of the sulfur dioxide by the acetaldehyde as this compound is produced. This is perhaps one reason that a single addition of sulfur dioxide is more effective than the same quantity added in smaller amounts during fermentation. The effect of this single-addition method is a high content of fixed sulfur dioxide. Sulfur dioxide causes not only an accumulation of acetaldehyde but may effect increased glycerol formation. It prevents malolactic fermentation, but this may not be desirable under all conditions. This fermentation will be discussed under fixed acids.

Temperature of Fermentation

Temperature is an important consideration in the fermentation of musts. It has been found that high temperatures (above 25.6°C) yield products of lower quality. Wines of greater aroma and keeping qualities are produced by cool fermentations. The use of cool fermentation for the production of white table wines has certain advantages. These are (1) reduced activity of wild yeasts and bacteria, (2) larger alcohol yield, (3) smaller losses of aromatic principles, (4) larger residual carbon dioxide, and (5) lower residual bitartrate.

In the case of red wines the effect of temperature has a different consideration—it is necessary to extract color. Amerine and Ough (1957) found better flavor and color for Pinot Noir fermentations at

22.8°–25.6°C than at 10°–15.6°C. It is likely that this is true for the fermentation of most red grapes.

Wine Composition

Ethyl Alcohol. Twelve to 14% ethyl alcohol is usually found in table wines and 16–21% in fortified wines.

Other Alcohols. Methyl alcohol is not a product of alcoholic fermentation, but is mainly formed by hydrolysis of pectins naturally present. It is present in quantities from a trace to 0.635 g/liter.

The following higher alcohols are found in wines, 1-propanol, 1-butanol, 2-butanol, 2-methyl-1-propanol, 2-methyl-1-butanol, 3-methyl-1-butanol, 1-pentanol, and 1-hexanol. These are collectively known as fusel oil and are produced ordinarily by the deamination and decarboxylation of amino acids. It appears, however, that not all higher alcohols are formed in this way. Castor and Guymon (1952) found that their formation paralleled the formation of ethyl alcohol. Guymon and Heitz (1952), as well as others, found that red wines contain slightly more higher alcohols than white wines. These authors noted the sensory importance of higher alcohols in wines; very low concentrations play a desirable rôle in sensory quality. It has been noted that considerable difference exists in the ability of yeasts to produce higher alcohols. The amounts in table wines vary from 0.14 to 0.42% and in dessert wines from 0.16 to 0.90%. The higher amount in the latter comes from the brandy used to fortify them.

Glycerol

Pasteur noted glycerol as a by-product of alcoholic fermentation. No constant ratio of alcohol to glycerol has been established because the amount produced varies with different conditions. Most of it, however, develops in the early stages of fermentation and is considered by most enologists to be of sensory importance because of its slightly sweet taste. It is probably of little importance in sweet wines.

2,3-Butylene Glycol, Acetoin, and Diacetyl

These three compounds are formed during alcoholic fermentation, but their origin is not clear. The amount of 2,3-butylene glycol present in various wines ranges from about 0.1 to 1.6 g/liter, but the average is 0.4–0.9 gm. The amount of 2,3-butylene glycol present does not appear to be related to quality because it has almost no odor and a slight bitter taste which would be masked by the glycerol normally present in amounts 10 to 20 times greater.

Acetoin does have an odor but the amounts found in wines are small, from 2 to 84 mg/liter. Guymon and Crowell (1965) found an increase in acetoin up to the middle of the fermentation to a maximum of 25–100 mg/liter, after which it nearly disappears. This explains the higher amounts in fortified dessert wines in which the fermentation is stopped about midway.

Diacetyl has a pronounced buttery odor and in a few cases may have sensory importance. Dittrich and Kerner (1964) found that normal wines contain an average of 0.2 mg/liter of this compound. Wines having more than 0.89 mg of diacetyl per liter have a sour milk odor.

Acetaldehyde

Acetaldehyde has been discussed earlier as a by-product ordinarily of alcoholic fermentation. Acetaldehyde is present in newly fermented wines below 75 mg/liter, which is of little importance from the sensory point of view. Further, many of the wines have sulfur dioxide added to fix the acetaldehyde. Spanish sherries, however, on the average have over 200 ppm of this compound, primarily the result of the oxidation of ethyl alcohol by film yeasts during aging. While the presence of acetaldehyde is desirable in sherry wine, other factors are involved in the quality of this wine.

Hydroxymethylfurfural

This compound is formed when fructose is heated in acid solutions. Its presence in wines is an indication that they have been heated during processing. Heating of port wine in Portugal is forbidden and a very small amount of this substance is permitted. A large number of California dessert wines have been found to contain hydroxymethylfurfural, an indication of heating during processing.

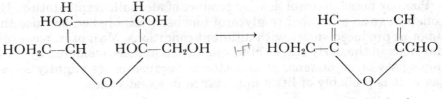

H H
HOC—COH
HOH₂C—CH HOC—CH₂OH → HOH₂C—C CCHO
O O
HC═CH

Fructose → Hydroxymethylfurfural

Ethyl Acetate and Other Esters

Ethyl acetate is the principal ester in wine. It is believed that the importance of the other esters in the odor of normal wine is exaggerated because they are not present in large enough quantities to add significantly to odor. Ethyl acetate could be important if it were present in amounts below 200 mg/liter; quantities higher than this level seem to give an undesirable flavor—a spoiled character—to the wine. Neutral and acid esters occur in wine. Total esters in various wines were found to vary between about 200 and 400 mg/liter, calculated as ethyl acetate. Ports and sherries had higher values. The volatile neutral esters were found to average from 70 to 200 mg/liter calculated as ethyl acetate. Sherry had 344 mg/liter of these same esters.

Fixed Acids

The important fixed (nonvolatile) acids present in wines are tartaric and malic acids. Of lesser importance are citric, succinic, and lactic acids. Small amounts of other fixed acids are known to be present. Fixed acids are important because of the sour taste they impart and because they protect the wine from spoilage and maintain color. The amounts of fixed acids in the finished wine are somewhat smaller than the amounts present in the must from which it was made.

While hydrogen ion content causes the acid taste, this actual taste is influenced by the relative amounts of several acids, the buffer capacity of the wine, and the sugar content. The acid taste of wines is caused mainly by acid salts because most of the acids in wines are partially neutralized. Unless a fairly large amount of the acid is combined with minerals the resulting taste is usually too acid. Amerine *et al.* (1965) found that for the same titratable acidity, the acids in order of sourness were malic, tartaric, citric, and lactic acids. At the same pH, the order was malic, lactic, citric, and tartaric acids. The conclusion: both pH and titratable acidity are important in determining sensory response to sourness.

Tartaric acid decreases mainly by deposition as potassium acid tartrate, commonly known as cream of tartar. This acid salt is present in grapes as a supersaturated solution and, because it is less soluble in alcohol, it tends to precipitate during and after fermentation. Wines contain only dextrorotatory acid, but Cambitzi (1947) found optically inactive racemic calcium tartrate in old wine deposits, an indication of autoracemization.

Grapes grown in warmer countries are lower in acids than those grown in colder regions. The malo-lactic fermentation, which is brought about by lactic acid bacteria, is important as a biological means of acid reduction in areas that produce high-acid grapes. The following equations show the changes which take place.

$$\underset{\text{Malic acid}}{\begin{matrix} CH_2COOH \\ | \\ CHOHCOOH \end{matrix}} \xrightarrow[-CO_2,\ -H_2]{\text{NAD, } Mn^{2+}\text{, malic dehydrase}} \underset{\text{Pyruvic acid}}{\begin{matrix} CH_3 \\ | \\ COCOOH \end{matrix}}$$

$$\xrightarrow[+H_2]{NADH_2\text{, lactic dehydrase}} \underset{\text{Lactic acid}}{\begin{matrix} CH_3 \\ | \\ CHOHCOOH \end{matrix}}$$

About 10–30% of the malic acid has been found to disappear during fermentation.

Grapes of the native American varieties do not have the sugar / acid balance of the *vinifera* varieties. Federal regulations permit the addition of water, sugar, or a combination of these, within limits, to correct this condition. This is known as amelioration, and it is permitted to reduce the total acidity to 5 parts per thousand. It may not, however, exceed 35% of the total volume of the material ameliorated regardless of the total acidity of the wine or juice.

Fixed acids are found in wines in the following amounts. Tartrates calculated as tartaric acid in German white wines, average 0.20 g / 100 ml with a range of 0.14–0.37. French white wines average 0.28 g / 100 ml, range 0.15–0.37. Portuguese red averages 0.19 g / 100 ml, with a range of 0.16–0.32. Malic acid in white wine averages about 0.20 g / 100 ml. Bordeaux white averages 0.14 g / 100 ml. Range in French wines for this acid, 0.067–0.268 g / 100 ml. Citric acid is present in wines in small amounts. About 0.01– 0.03 g / 100 ml of wine is usually found. Succinic acid is a product of alcoholic fermentation, and according to Amerine (1954) it is a general rule that 1% as much succinic acid is present as is alcohol by volume. Lactic acid is a constant by-product of alcoholic fermentation and is present in amounts from 0.04 to 0.75 g / liter.

Volatile Acids

The determination of volatile acidity expressed as acetic acid as an indication of spoilage is one of the standard procedures of modern enology and has become a part of the legal requirements for wine standardization. The quantity of acetic acid formed during alcoholic fermentation is small, usually less than 0.030 g / 100 ml. Bacterial action before, during, and after fermentation may lead to much higher amounts. The spoiled vinegary character of such wines is the reason for the adoption of legal standards. Reduction of high volatile acidity by neutralization is not practical because the fixed acids would be neutralized also. Careful wine makers are able to obtain a product with less than 0.030 g / 100 ml of volatile acids calculated as acetic acid and should not exceed 0.100 g during aging. The legal maximum limit for volatile acidity in California in white wines is 0.110 g / 100 ml and 0.120 g for red wines, exclusive of sulfur dioxide.

Sugars

Very small amounts of levulose and dextrose are present in table wines, the lower the amounts the drier the wines. Sucrose is not naturally found in wines. When added to sweeten wines it hydrolyzes rapidly. If it is to be detected, the test must be carried out not later than a few days after addition. Dessert (sweet) wines contain 10–14% sugar.

The nonfermentable residual reducing material in wines is largely pentoses in concentrations of 0.01–0.20 g/100 ml. The best data indicate that 0.04–0.13 g of arabinose is present in 100 ml of wine.

Pectins

The precipitation of pectin occurs naturally during alcoholic fermentation with the result that the finished wines have 0.3–0.5 g of pectin and gums per 100 ml. The gums of wine are usually arabans, anhydrates of arabinose, and galactans. The mucilaginous materials are known as glucosans, anhydrides of dextrose, or dextrans. Dextrans hinder natural clarification and make filtration difficult. Pectins can be removed with pectic enzymes, added either before or after fermentation.

Nitrogen

The importance of nitrogenous components of wine cannot be overemphasized because they seem to play a role in clarification, in bacterial development, and possibly in odor development. The many different kinds of nitrogenous material in wine include proteins, peptones, polypeptides, amides, amino acids, and ammonia. Only very small amounts of ammonia are found in wine because much of it is utilized by the yeasts in alcoholic fermentation. Alanine is required by lactic acid bacteria. European data (Amerine 1954) gave averages of 0.10–0.77 g of total nitrogen per liter of wine.

Tannins

The usual permanganate oxidation shows that red wines contain 0.1–0.3 g of tannin in 100 ml of wine. The tannin content decreases during the aging of wine. This decrease is caused by a variety of reactions, including combination with aldehydes and precipitation with proteins.

There is little bitter or astringent taste in white wines because of the low tannin content—0.05% or less. Experienced tasters, more able to detect tannin than inexperienced, seem to like wines with higher tannin content.

The following phenolic compounds were found in wine: *d*-catechin, *l*-epicatechin, *l*-epigallocatechin, gallic acid, protocatechuic acid, and *cis*- and *trans*-coumaric acid. The following polyphenols have been found: chlorogenic acid, *cis*- and *trans*-caffeic acids, and probably esculetin. Isochlorogenic acid has been reported, but is now known to be a complex mixture of closely related compounds. Shikimic acid has been found in young wines and in green grapes, whereas quinic acid is in musts and in young and old wines. *p*-Coumarylquinic acid is known to occur in wines.

Color

While a considerable amount of work was done on the anthocyanin pigments in grapes before the general use of chromatographic procedures, progress concerning the identification of structure has been greater since the use of the newer techniques became more widespread. The reason is that pigment mixtures in grapes are rather complex. Current data indicate that *Vitis vinifera,* the European grape, does not contain diglucoside pigments, but that probably most or all of the others do. This means that when diglucosides are found in wine, grape varieties other than *vinifera* have been used. At the present time five anthocyanins are found in *Vitis vinifera,* namely, cyanidin, peonidin, delphinidin, petunidin, and malvidin. Pelargonidin seems to be absent from *Vitis* species. The only sugar moiety known conclusively to occur in the grape anthocyanin molecule is glucose. It is possible to detect the addition of wine from diglucoside-bearing hybrid grapes to the extent of 1%. This analysis is used in France to detect adulteration of pure *vinifera* varietals with the cheaper hybrid wines.

After several years of aging of wine, some anthocyanins are lost as a result of passage to a colloidal state and precipitation, but not all are lost in this process.

The browning of white wines is largely caused by the oxidation of phenolic compounds such as catechins and leucoanthocyanins. The speed with which this change in color takes place depends on the temperature, the amount of dissolved oxygen in the wine, and the amount of phenolic compounds present. Peterson and Caputi (1967) found that two types of browning are probably involved in the development of this color in wines: one oxidative and the other non-oxidative. Caputi and Peterson (1965) found that the content of phenol absorbents varied considerably as regards their effect on browning of white wine. The variety of grape from which the wine is made is important. Some wines brown even though oxygen is excluded. More information is available in the article by Singleton and Esau (1969).

Oxygen

This element is very important as far as the handling of wine is concerned. However, not much data are available on the oxygen content of wines nor on the effects of reactions at different temperatures in different wines. Work on the changes in the oxidation–reduction potential of wines has shown that during fermentation there is a rapid decrease in potential from approximately +0.4 to 0.1 V; there is, however, much variation among musts. It is thought that this decrease is associated with the reduction of quinones during fermentation due to glutathione. Changes in the oxidation–reduction potential are caused by a variety of factors, such as, method of fermentation, composition of the media, degree and temperature of aeration, and the presence of sulfur dioxide. A reduction of potential takes place in sparkling wines during the secondary fermentation. Wines stored in the bottle for several years show a decrease in potential. According to Joslyn (1949) California red table wines usually have a lower potential while aging in wood than do white wines. Wines stored in wood show an increase in potential on racking if air is present.

Mineral Substances

Anions. Phosphates are important in fermentation and, of course, are found in the finished wines. Fifty to 900 mg of phosphate is found in a liter of new wine, of which only about 10–20% is present as organic phosphate such as glycerophosphates. The phosphate content of wine is increased if the method of fermentation is carried out in the presence of the skins.

Sulfates are present in wines as K_2SO_4 and are limited in many countries to about 2–3 g/liter. Sometimes H_2S and mercaptans are found in young wines and arise from sulfur used in mildew control on the grapes in the vineyard. It is for this reason that late spraying for mildew should be avoided. When 5 mg of free sulfur is present, it is difficult to remove by aeration of hydrogen sulfide or mercaptans. Some yeasts can liberate H_2S during fermentation.

The French and Swiss limit the chlorine as chloride to 0.607 g/liter. The presence of bromine in amounts higher than 3 mg/liter of wine indicates the use of monobromacetic acid as an antiseptic, which is illegal in the United States and a number of other countries.

Cations. ***Aluminum.*** Aluminum is found as a component of wines and is present usually to the extent of about 1–3 mg and not more than 15 mg/liter. Red wines contain more aluminum than white wines.

Calcium. Calcium in wines comes from the grapes, soil, calcium-containing filter aids, and concrete tanks. These concrete tanks can be treated to avoid much of this pick-up. Calcium present in wines is deposited as calcium tartrate less in wines with pH below 3.7 than in those with a higher pH. De Soto and Warkentin (1965) recommended that high pH wines be acidified to bring the pH down. The difficulty with excess calcium tartrate is that it takes a long time to settle out.

Copper and Iron. Copper is necessary for the progress of fermentation, and therefore, it is important. Only small amounts of copper are present in new wines as a part of the oxidation–reduction system. Commercial wines were found to average 0.25 mg of copper per liter. The recommended legal maximum is 1 mg/liter. Excessive copper content is caused mostly by contamination, usually with copper equipment. Copper and iron are of interest because they can lead to cloudiness in wine and to flavor impairment. They are, of course, important in oxidation–reduction reactions.

Iron is present in wines up to about 50 mg / liter, but sometimes more can be found. Most California wines contain less than 10 mg / liter.

Lead. Little lead is found in normal musts and wines and that little rapidly declines during storage because of the insolubility of lead tartrate. The legal limit varies from 0.2 mg / liter in Great Britain to 3.5 mg in Switzerland.

Potassium. Potassium makes up about 75% of the total cation content of wine. It is involved in alcoholic fermentation and also in bitartrate stability. Average potassium values vary from 0.36 to 1.1 g / liter of wine. Potassium is effective in the separation of some of the tartaric acid from wine in the form of potassium bitartrate.

BEER AND BREWING

Brewing is the manufacture of beer and other malt beverages such as ale and stout. Basically, the process involves the germination of barley, after which it is carefully dried. The resulting product is known as "malt." It is during this process of manufacture that more hydrolytic enzymes, particularly α-amylase, proteolytic enzymes, and also more β-amylase are formed. In some countries only the malted barley is used for the production of beer, but in the United States another starch source such as corn or rice is used in addition. The reason is that American barley is higher in protein than European, and since the excess protein tends to precipitate out, the starch from the added grains corrects this situation. The grain is treated with hot water (60° to 65°C) to allow the amylases from the malt convert the starch into maltose, dextrins, and soluble starch. Some manufacturers use a series of three different ascending temperatures for this part of the beer-making process. After this step is completed the mixture is filtered, and the resulting filtrate is known as "wort."

The next step is the boiling of the wort with hops which achieves three things: (1) enzymes are destroyed, ending their activity and at the same time sterilizing the wort, (2) flavoring components of the hops are extracted, (3) coagulable proteins are deposited. The flavoring substances in hops consists of tannin, bitter acids and resins, and essential oils.

In the United States a bottom strain of *Saccharomyces cerevisiae* is used. The fermentation is carried out at 3.3°–7°C for 7–12 days. After this the beer is held in lager or cellar fermentation at about 0°C for several weeks or months, and it is chill-proofed with enzymes to remove proteins that would precipitate at lower temperatures and cause a haze. This is necessary for beer consumed in the United States because it is served cold. However, where beer is consumed at room temperature, it is unnecessary. After chill-proofing the beer, it is bottled, canned, or barreled for shipment.

DISTILLED PRODUCTS

Whiskey, brandy, vodka, rum, and gin are all considered distilled products. The basic chemistry of the fermentation in each of these beverages is the same as that for the manufacture of wine or beer, in that yeasts convert the sugars to carbon dioxide and alcohol. For the fermentation of worts made from grain, it is necessary to utilize as much of the carbohydrate as possible by employing yeast strains able to degrade the dextrins left after the action of the α- and β-amylases on the starch. These strains of *S. cerevisiae* are somewhat more tolerant of alcohol. They seem also to have the ability to produce some of the higher alcohols that affect the flavor of the product.

Whiskeys are manufactured from grains as the raw material, usually rye, corn, barley, and wheat. Rye and malt made from rye and barley are the main ingredients of rye whiskey. Bourbon is made from corn and barley or wheat malt, and another grain, frequently rye. The mash must contain a minimum of 51% corn.

The flavor and odor of whiskey are influenced by the raw materials used, the method of fermentation, the distillation process, and the aging in charred white-oak barrels. The determination of the ester content of a whiskey can be indicative of the age because esters increase during storage. Volatile acids also increase during storage. The largest increase in acids, esters, and color takes place during the first 6 months, although increases continue during further storage (Valear and Frazier 1936).

Brandy is made by distilling fermented fruit juice—wine. The bulk of it is from grapes, and the continuous still is extensively employed for this purpose. Brandy is usually aged in charred white-oakbarrels.

Rum is the distillate from fermented molasses syrup or from the syrup of cane sugar. In the United States it is made from black-strap molasses. Different types of stills can be used for this process. The product is usually aged in charred white-oak barrels. This aging process improves the flavor, aroma, and color, and mellows the product.

Gin is prepared by distillation from a mash of grain, mainly barley, to which juniper berries and other aromatics have been added as flavoring materials.

Vodka, first made in Russia, is a distilled alcoholic liquor produced from grain or, in some cases, from potatoes. It is purified so as to be substantially tasteless. Vodka is not aged and is colorless when sold.

VINEGAR

Vinegar is nonalcoholic, but its production involves an alcoholic fermentation step. It is produced from fermented ciders, wines, or alcoholic substrates by any of the numerous species of the genus

Acetobacter, for example, *A. aceti.* The acetic acid bacteria convert the alcohol into acetic acid by the process of oxidation. This enables the bacteria to obtain the energy necessary to carry on their life processes. These reactions proceed as follows.

$$2C_2H_5OH + O_2 \longrightarrow CH_3CHO + 2H_2O$$

$$2CH_3CHO + O_2 \longrightarrow 2CH_3COOH$$

Much of the vinegar made in the United States is from fermented apple juice and is known as cider vinegar. Vinegar made by acetic acid fermentation of dilute distilled alcohol is termed distilled or white vinegar. Wine vinegar is also made.

The solution used in the manufacture of vinegar should range from about 10 to 13% of alcohol in order to yield a satisfactory product. If the concentration of alcohol is too high the alcohol is not completely oxidized to acetic acid, and if the concentration is extremely low, the acetic acid and esters are lost by oxidation.

The oxidation is often carried out on a support material. Beechwood shavings have been preferred, but less costly substitute materials are now being considered. Any such materials must provide large surface area exposed to oxygen to speed the oxidation process. The alcoholic liquid is percolated through this support material.

Equipment has been designed for the manufacture of vinegar by submerged oxidation fermentation, the "submerged method." This is used extensively at the present time for commercial purposes.

The finished vinegar must contain at least 4% of acetic acid (4 g / 100 ml).

SUMMARY

Alcoholic beverages depend on the process of alcoholic fermentation—the conversion of hexose sugars into alcohol and carbon dioxide. Alcoholic fermentation is an anaerobic process in which the simple sugars are broken down in a series of successive changes by oxidizing and reducing enzymes within the yeast cell. Buchner showed that a cell-free extract from yeast cells induced fermentation, which demonstrated that the enzymes and not life itself were necessary to produce the chemical action. Gay-Lussac wrote the overall basic chemical equation. However, Pasteur showed that a number of by-products, present in the fermented solution, were not accounted for by Gay-Lussac's equation. Many enzymes are necessary to bring about the changes in alcoholic fermentation. Other substances are required also. This process is a complex system which explains why such compounds as glycerol, lactic acid, and acetaldehyde build up as by-products. The equations for the various reactions are shown in the text.

Wines are of growing importance in this country. A number of different wines are produced by the process of fermentation.

The principal sugars in the grapes are dextrose and levulose. As the ripening process continues, the quantity of levulose present in the grape berry increases until it is equal to or exceeds that of dextrose. Other changes take place in the chemical composition of the grape berry during ripening. Volatile compounds develop during the growing period.

The wine is prepared by the fermentation of must. Since red wines depend on the extraction of pigment from the skin of the grapes, the processes for the pressing of the grapes for white wines is somewhat different from that used for the preparation of red wines. Antiseptics, mostly SO_2, are necessary in the manufacture of wine. Temperature of fermentation is very important.

A number of different compounds are found in wines. Many of these compounds are formed during the fermentation process, while others are formed by such changes as the hydrolysis of pectins which are naturally present.

Volatile acids, largely acetic, are an indication of spoilage.

The color of grapes is caused by anthocyanins. The browning of white wines is caused by the oxidation of phenolic compounds such as catechins and leucoanthocyanins.

Potassium, present in wine, is involved in alcoholic fermentation, and also in bitartrate stability. Potassium is effective in the separation of some of the tartaric acid from wine in the form of potassium bitartrate.

Beer and distilled products are based chemically on fermentation.

Vinegar is nonalcoholic, but a fermentation step is necessary in its manufacture. It is produced from fermented ciders, wines, or alcoholic substrates. Species of the genus *Acetobacter* are used for this purpose. These bacteria convert the alcohol into acetic acid by oxidation.

ACKNOWLEDGMENT

The author acknowledges use of information on wine from "The Technology of Wine Making," 3rd Edition, 1972 by M. A. Amerine, H. W. Berg, and W. V. Cruess.

BIBLIOGRAPHY

AMERINE, M. A. 1954. Composition of wines. I. Organic constituents. *In* Advances in Food Research, vol. 5, E. M. Mrak and G. F. Stewart (Editors). Academic Press, New York.

AMERINE, M. A., BERG, H. W., and CRUESS, W. V. 1972. Technology of Wine Making, 3rd Edition. AVI Publishing Co., Westport, CT.

AMERINE, M. A., and OUGH, C. S. 1957. Controlled fermentations. III. Am. J. Enol. *8,* 18–30.

AMERINE, M. A., ROESSLER, E. B., and OUGH, C. S. 1965. Acids and the acid taste. I. The effect of pH and titratable acidity. Am. J. Enol. Viticult. *16,* 29–37.

BIDAN, P., and ANDRÉ, L. 1958. Amino acid content of various wines. Ann. Inst. Natl. Rech. Agron. Ser. E. *7,* 403–432. (French)

BUCHNER, E. 1897. Alcoholic fermentation without yeast cells. Chem. Ber. *30,* 117–124. (German)

BURKHARDT, R. 1965. Detection of *p*-coumarylquinic acid in wines and behavior of depsides during after-care. Mitt. Klosterneuburg Ser. A. *15,* (2), 80–86. (German)

CAMBITZI, A. 1947. The formation of racemic calcium tartrate in wines. Analyst *72,* 542–543.

CAPUTI, JR., A., and PETERSON, R. G. 1965. The browning problem in wines. Am. J. Enol. Viticult. *16,* 9–13.

CASTOR, J. G. B., and GUYMON, J. F. 1952. On the mechanism of formation of higher alcohols during alcoholic fermentation. Science *115,* 147–149.

CORSE, J., LUNDIN, R. E., and WAISS, JR., A. C. 1965. Identification of several components of isochlorogenic acid. Phytochemistry *4,* 527–529.

DESOTO, R., and WARKENTIN, H. 1955. Influence of pH and total acidity on calcium tolerance of sherry wine. Food Res. *20,* 301–309.

DITTRICH, H. H., and KERNER, E. 1964. The lactic acid flavor in wine caused by diacetyl and its removal. Wein-Wiss. *19,* 528–535. (German)

GUYMON, J. F., and CROWELL, E. A. 1965. The formation of acetoin and diacetyl during fermentation, and the levels found in wines. Am. J. Enol. Viticult. *16,* 85–91.

GUYMON, J. F., and HEITZ, J. E. 1952. The fusel oil content of California wines. Food Technol. *6,* 359–362.

HENNIG, K., and BURKHARDT, R. 1960. Detection of phenolic compounds and hydroxy acids in grapes, wines, and similar beverages. Am. J. Enol. Viticult. *11,* 64–79.

JOSLYN, M. A. 1949. California wines. Oxidation-reduction potentials at various stages of production and aging. Ind. Eng. Chem. *41,* 587–592.

JOSLYN, M. A. 1952. Chemistry of sulfite addition products. Proc. Am. Soc. Enol. *1952* 59–68.

KIELHÖFER, E., and WÜRDIG, G. 1960A. Sulfurous acid bound aldehydes in wine. I. Acetaldehyde formation by enzymic and nonenzymic oxidation of alcohol. Weinberg Keller 7, (1), 16–22. (German)

KIELHÖFER E., and WÜRDIG, G. 1960B. Sulfurous acid bound aldehydes in wine. II. Acetaldehyde formation by fermentation. Weinberg Keller *7* (2), 50–61. (German)

KLEYN, J. and HOUGH, J. 1971. The microbiology of brewing. Annu. Rev. Microbiol. *25,* 583–608.

KOCH, J., and BRETTHAUER, G. 1960. The glucose-fructose relation of table wines in regard to the various cellar-technical precautions. Z. Lebensm. Untersuch. Forsch. *112,* 97–105. (German)

LEE, C. Y., SHALLENBERGER, R. S., and VITTUM, M. T. 1970. Free sugars in fruits and vegetables. N.Y. State Agric. Exp. Stn. Food Life Sci. Bull. 1.

MÜNZ, T. 1963. Potassium buffering in must and wine. Wein-Wiss. *18,* 496–503. (German)

PASTEUR, L. 1860. I. Report on alcoholic fermentation. Ann. Chim. Phys., *58,* 323–363. (French)

PASTEUR, L. 1860. II. That which happens to yeast during the alcoholic fermentation of beer. Ann. Chim. Phys. Ser. 3, *58,* 364–426. (French)

PASTEUR, L. 1861. Plant physiology. Experience and new views on the nature of fermentations. Compt. Rend. *52,* 1260–1264. (French)

PEDERSON, C. S. 1971. Microbiology of Food Fermentations. AVI Publishing Co., Westport, CT.

PETERSON, R. G., and CAPUTI, JR., A. 1967. The browning problem in wines. II. Ion exchange effects. Am. J. Enol. Viticult. *18,* 105–112.

POMERANZ, Y., and SHANDS, H. L. 1974. Gibberellic acid in malting of oats. J. Food Sci. *39,* 950–952.

PRESCOTT, S. C., and DUNN, C. G. 1959. Industrial Microbiology. McGraw-Hill Book Co., New York.

RANKINE, B. C. 1963. Nature, origin, and prevention of hydrogen sulfide aroma in wines. J. Sci. Food Agric. *14,* 79–91.

RIBÉREAU-GAYON, P., and STONESTREET, E. 1964. The constitution of tannins of grapes and wine. C. R. Seances Acad. Agric. Fr. *50,* 662–670. (French)

ROSE, A. H. 1961. Industrial Microbiology. Butterworths, London.

SALLER, W. 1955. The Improvement of Quality of Wines and Musts by Means of Chilling. Sigurd Horn Verlag., Frankfurt. (German)

SARLES, W. B., FRAZIER, W. C., WILSON, J. B., and KNIGHT, S. G. 1950. Microbiology, General and Applied. Harper and Bros., New York.

SINGLETON, V. L., and ESAU, P. 1969. Phenolic Substances in Grapes and Wine and Their Significance. Academic Press, New York.

TCHELISTCHEFF, A. 1948. Comments on cold fermentation. University of California Wine Technol. Conf. Aug. 11–13. Davis, CA.

VALAER, P., and FRAZIER, W. H. 1936. Changes in whisky stored for four years. Ind. Eng. Chem. *28,* 92–105.

VILLFORTH, F., and SCHMID, W. 1953. Higher alcohols in wine. I. Wein- Wiss. Beih. Fachz. Deut. Weinbau. *7,* 161–170. (German)

VILLFORTH, F., and SCHMID, W. 1954. Higher alcohols in wine. II. Wein- Wiss. Beih. Fachz. Deut. Weinbau. *8,* 107–121. (German)

15

Baked Products

Flour
Leavening
Summary
Bibliography

Baked products, bread, cakes, cookies, pies, and other such items have had a long and interesting history. It is quite likely that the ancient Egyptians were the first to discover leavening action involving (unknown to them) yeast and then to improve their bread by using the leavening process. While their method for making bread was rather advanced, their knowledge of the scientific principles explaining this process was doubtlessly small.

Bread and many other baked products have flour, water, yeast, and salt as basic ingredients. Two of these ingredients, flour and leavening, are responsible for the most important characteristics of the finished products.

FLOUR

It was learned in very early times that grains could be milled to yield substances (flours) usable in the preparation of food products. The milling of grains is a physical process, during which the endosperms are broken apart. In the case of wheat it is essential that all the bran and germ possible is removed to obtain maximum yields of white flour. Most of the mills yield about 80% in this respect. The patent flours are separated first, and the remaining clear flour is separated for use with rye flour. Bran and germ remain at the end of the milling process.

Bread flour is made from hard spring and winter wheats because these wheats mill well and yield flour that has good quality protein and furnishes strong, elastic doughs. Breads made from these flours have good volume, grain, and texture and under a wide range of conditions. Table 15.1 shows the variations one can expect to find in the composition of wheat flours.

The hard spring wheat flours with a protein range of 12.5–12.8% are used mainly for bread, hamburger buns, soft rolls, and other such items and are known ordinarily as bakers' patents. The higher protein spring straight grades containing from 12.8 to 13.1% protein are used for such items as Kaiser rolls, and the spring clear grades are used mostly with rye flour for rye bread. Winter wheat patent flours are used to make bread and buns. Soft wheat flours are used for cake, cookies, and pastry products. The soft wheats are all low in protein and in water-absorption capacity. They are undesirable for commercial bread production, but are especially desirable for cakes. Tables 15.3 and 15.4 give the comparative composition of wheat and various flours.

Amber durum wheat flour is used mainly for the manufacture of alimentary pastes—macaroni, spaghetti, and noodles.

Rye flour is used principally for the making of rye bread. In the United States it is mixed with white flour but in some European countries it is used alone. Since rye flour doughs are very weak they cannot be washed to separate the gluten. Several grades of rye flour are produced. Table 15.5 gives the chemical composition of some of them.

Compounds in Flour Related to Baking Quality

Proteins. Wheat flour has a singular property that distinguishes it from other cereal flours: it is able to form an elastic dough when mixed with the right amount of water. This dough can hold gas, and it has the further property of setting to a spongy structure when it is baked. The result is the well-known bread of today. Protein is the main component of wheat flour responsible for these properties. When treated with water, the bulk of the protein remains undissolved, but it tends to hydrate and form a coherent mass, the gluten. The coiled structure of the protein molecules probably is responsible for their behavior.

Early work on protein (Osborne 1907) indicated that wheat proteins consisted of glutenin and gliadin in about equal amounts, together with about 20% of an albumin, a globulin, and a small amount of proteose. These last three are water soluble or soluble in dilute salt solution. However, later work indicated that gluten of wheat is made up of a large mixture of proteins of varying molecular size (McCalla and Gralén 1942). This mixture can be separated into many fractions that

TABLE 15.1. The Composition of Wheat Flours

Component	Minimum (%)	Maximum (%)
Protein	7.5	15.0
Carbohydrates as starch	68.0	76.0
Fat	1.0	1.5
Fiber	0.4	0.5
Ash	0.3	1.0

Source: Schopmeyer (1960).

TABLE 15.2. Comparative Composition of Wheat and Flour Grades[a]

Type of Wheat	Protein Content of Wheat at 14% Moisture	Grade of Flour	Extraction of Flour Based on Wheat (%)	Ash in Flour at 14% Moisture (%)	Protein in Flour at 14% Moisture (%)
Soft red	9.8	Straight	72	0.38	8.20
		Short patent	35	0.30	7.50
		Baker's patent	65	0.34	7.90
		Clear	10	0.68	9.90
Hard spring	13.50	Straight	72	0.47	12.90
		Short patent	35	0.39	12.00
		Medium patent	60	0.41	12.20
		Long patent	65	0.43	12.50
		First clear	10	0.70	15.50
		Second clear	4	1.20	16.50
Hard winter	12.90	Straight	72	0.47	12.10
		Medium patent	60	0.41	11.70
		Long patent	68	0.45	12.00
		Second clear	4	1.20	16.00

Source: Schopmeyer (1960).
[a] The protein contents of these different wheats show considerable variation, and those of the corresponding flours show similar variation. Some variation can be seen also in the ash contents. These values will vary from season to season because of differences in growing conditions.

differ progressively and systematically both in chemical and physical properties.

Halton and Scott-Blair (1937) stated that wheat flour dough contains protein chains that act like coiled springs, accounting for its elastic behavior. Since the linkages between the chains are not of the same strength at all points, some of them break almost immediately when the dough is extended, causing permanent deformation of flow, while others, remaining intact, maintain the rigid structure of the dough. Adjustments of this character in the protein network must occur in conjunction with a starch–water mixture. While this mixture is basically fluid, it does have some rigid characteristics, which, in turn, complicate the situation and prevent the full relaxation of even the particular protein units that otherwise would be able to effect true elastic recovery.

Swanson (1938) considered the formation of gluten in wheat flour doughs as related to the hydration of the molecules in the gluten. When flour is first mixed with water, the proteins are arranged in a heterogeneous manner. By the action of the mixers, the gluten particles tend to be oriented in a parallel position. The high speed commercial mixers cause a pulling action on the dough that results in a rather parallel arrangement of the protein strands. At this point the dough becomes smooth, showing that the mixing is sufficient. In this state, according to Swanson, the dough shows greatest resistance to pull and also its

TABLE 15.3. Composition and Uses of Hard Spring Wheat Flours

Protein (%)	Ash (%)	Bakery Products in Which the Flour Is Used
12.5–12.8	0.42–0.44	White pan bread (both sponge and straight dough)
		Specialty and variety breads
		Hamburger and wiener rolls
		Yeast raised sweet goods
		Soft rolls (Parker House type)
		Doughnuts, yeast raised
		Hard rolls and hearth bread
12.8–13.1	0.46–0.48	White pan bread (both sponge and straight dough)
		Hard rolls and hearth bread
		Specialty and variety breads
13.5	0.48	Hard rolls and hearth bread
		Blender flour and coarse flour breads
15.0	0.48	High gluten flour for Kaiser rolls
		Italian, French, and Jewish hearth breads
15.0–15.5	0.70–0.74	Carrier flour used in making whole wheat, rye, specialty and variety breads
		Blender flour used in mixes with rye flour for rye bread

Source: Harrel (1959).

greatest elasticity, brought about by the fact that the largest number of gluten coils are able to resist elongation and to straighten back after such elongation. If the mixing is continued beyond this stage the dough breaks down, and it becomes soft and sticky. Doughs that have not been too overmixed will recover when allowed to rest. This is thought to be caused by a film of water surrounding each gluten filament, brought about by the overmixing. These overmixed doughs tend to stick to surfaces through this water film because the forces which attract water to other surfaces are the same as those which hold the filaments together.

The mixing required to develop different flours to the desired degree is not always the same, probably because the rates of combination or hydration are not the same for all flours. Differences in molecular structure and the way the proteins are bound together in the gluten strands may have an influence also.

Using starch gel electrophoresis, Elton and Ewart (1960) observed eight bands that were present in glutens from four different wheat samples, but the intensity of corresponding bands was not the same in all cases. Woychik (1964) also using starch gel electrophoresis observed eight migrating bands from gluten and one stationary band, which was thought to be the glutenin fraction. More work of this nature must be done before any final conclusions can be reached.

Holme and Briggs (1959) showed that when gliadin has 10–50% of the amide nitrogen removed, provided that the conditions are such that no peptide bonds are split, it shows solubility in water at neutral pH. These authors showed also that precipitated deaminated gliadin no longer has the characteristic hydration capacity of normal gliadin. Cunningham *et al.* (1955) showed that insoluble protein of wheat had the highest amide content of the four grains, wheat, barley, rye, and oats. Barley and rye were found to be intermediate in these grains. The elasticity and cohesiveness of the gluten of these grains declined in that order.

Beckwith *et al.* (1966) found a small amount of high molecular weight protein in gliadin from wheat. Meredith *et al.* (1960) concluded that the unique properties of wheat dough seem to be related to the structure of the gel protein and especially to its content of glutamine, proline, lysine, and alanine.

McDermott and Pace (1959) using the Swan (1957) technique, which splits the peptide and protein chains at the cystine or cysteine residues, found that when they used cupric and sulfite ions to split the disulfide bonds in the protein of flour to form sulfur–sulfon derivatives, the protein was solubilized, meaning that the insoluble fraction is partly

TABLE 15.4. Composition and Uses of Soft Wheat Flours

Protein (%)	Ash (%)	Bakery Products in Which the Flour Is Used
7.5–8.0	0.32–0.35	Angel food cakes High ratio cakes and cookies (no spread) (Unbleached) High ratio spead cookies
8.0–8.5	0.35–0.38	Layer cakes and pound cakes Slightly rich wire-cut cookies (no spread) (Unbleached) Cracker topping
9.0–9.5	0.39–0.42	Unbleached flour General line cookies Pie crust Blender flour in yeast raised sweet goods Sugar cones Doughnuts Bleached No spread ice box cookies Loaf cakes Low ratio cup cakes Lunch box cake items
9.5–10.0	0.45–0.48	General purpose flour used much same as 0.39–0.42 ash flour

Source: Harrel (1959).

TABLE 15.5 Composition of Grades of Rye Flour

Component	White Rye Flour (%)	Medium (%)	Dark (%)
Moisture (maximum)	14.5	14.5	14.5
Ash	0.58–0.78	1.11–1.39	2.05–2.83
Protein	7.9–9.1	10.1–12.8	13.7–16.2

Source: Schopmeyer (1960).

dependent on the presence of disulfide cross-linked structure for integrity and rigidity.

Using a sedimentation technique, it was found that the glutenin fraction is made up of a group of compounds with molecular weights in the millions (Anon. 1961). When the disulfide bonds are split by forming the S-sulfonic acid group, or by cleaving them with performic acid, material is produced which seems to be homogeneous with a molecular weight of 20,000. It was concluded that these basic units of polypeptides with a molecular weight of 20,000 are the building blocks of gluten united by cross-linking through disulfide bonds. It can be seen that differences in the number of these basic units, or in the location and number of disulfide bonds linking them, would have a great effect on the rheological properties of gluten. Hird *et al.* (1968) concluded that the glutathione levels in flour are sufficiently large to alter the rheological properties of the dough. Neilsen *et al.* (1968) found that by cleaving the disulfide bonds of the two fractions obtained from classical wheat gliadin by gel-filtration chromatography, more evidence was obtained that the structure and properties of these proteins differ.

Woychik *et al.* (1964) found that reduction of the disulfide bonds of wheat gliadin and glutenin followed by starch gel electrophoresis showed the presence of some components which may be common to both. Intramolecular disulfide bonding was found to be limited in the gliadin fraction.

Woychik and Huebner (1966) separated the γ-gliadin from whole gliadin of Ponca wheat flour. They purified it by ion exchange chromatography and continuous flow paper curtain electrophoresis. End-group analysis indicated that it is a single polypeptide chain.

It is obvious that research has produced a great deal of new knowledge. However, much remains to be done.

Carbohydrates. The carbohydrates of wheat are largely starch, together with small amounts of dextrins and sugars, cellulose (fiber), and gums. Koch *et al.* (1951), making use of paper chromatography on aqueous extracts from bakers' patent flour, found the following sugars: glucose, 0.91%; fructose, 0.02%; sucrose, 0.10%; maltose, 0.08%; melibiose, 0.18%; and raffinose, 0.07%.

Fatty Substances. Studies have been made using gas–liquid chromatography on the fatty acid composition of the fatty material extracted from wheat flour. Thin layer chromatography, as well as other techniques have been employed. Tables 15.6 and 15.7 give results on fatty acid determinations, and Table 15.8 lists the components of wheat flour nonpolar lipids.

Deterioration of the fatty lipid material is closely connected with the decline of flavor quality in four and parallels an increase in the fatty acid content of the fats. Oxidative as well as hydrolytic rancidity are both parts of the picture.

The Effect of Lipids in Bread Making. The amount of lipid in flour is small; wheat flour contains 0.5–3.0% lipid material. However, considerable research demonstrates the importance of the lipid fraction in flour, small as it is, in the crumb grain quality and the volume of the finished loaf.

Mecham and Weinstein (1952) studied lipid binding in doughs and found that when flours are treated with water or made into doughs, a considerable amount of the water-extractable lipid becomes bound. Lipid binding in doughs is decreased by salt. It was found also that a polyoxyethylene-type softener had a similar effect. Lard seems to decrease slightly the phospholipid binding.

Daftary *et al.* (1968) subfractionated the nonpolar lipids into tri-, di-, and monoglycerides. The polar lipids were fractionated into phospholipids and glycolipids. Free polar lipids were found to increase the loaf volume substantially, although the increase was smaller if bound polar lipids were also added. Total free lipids containing a mixture of polar and nonpolar lipids (1:3) produced bread of lesser quality than polar lipids by themselves. The volume of the bread was decreased by nonpolar lipids and the crumb grain was impaired. These effects were counteracted by the addition of polar lipids, suggesting that galactosyl glycerides increased the loaf volume of bread made from petroleum-ether-extracted flours considerably more than phospholipids.

TABLE 15.6. The Fatty Acids of Wheat Flour Lipids (Range for Samples Grown in 1959 to 1962)

Acid	Range (%)	Mean (%)
Palmitic	16.7–24.1	19.5
Stearic	0.3–1.2	0.6
Oleic	7.0–14.5	9.0
Linoleic	60.1–69.8	66.4
Linoleinic	1.9–4.4	2.9

Source: Fisher *et al.* (1966). Reproduced with permission of Dr. N. Fisher of the Flour Milling and Baking Research Association, Chorleywood, Rickmansworth, Herts, England.

Pomeranz *et al.* (1965) studied polar versus nonpolar wheat flour lipids in breadmaking. These authors found that by adding up to 3 g of vegetable shortening for each 100 g of a hard winter wheat flour composite, the crumb grain was improved, and loaf volume increased from 802 to 948 ml. They also found that the addition of 0.5 g of polar flour lipids to bread without vegetable shortening increased the volume of the loaf strikingly. On the other hand, no significant increase in volume resulted from the use of 0.5 g of nonpolar lipids.

Mann and Morrison (1974) extracted lipid material from flour and flour–water doughs with water-saturated *n*-butanol and noted the extent of oxidation and other changes that resulted from the mixing and resting of the dough. They found that the unextracted flour lipids, amounting to 13% of the total lipids, were not oxidized as a result of the mixing of the dough. Changes in the extracted lipids were limited to free fatty acids and monoglycerides, and these showed losses of 18:2 and 18:3 as a result of aerobic dough mixing, caused probably by lipoxygenase activity during and immediately after the mixing period. Since recoveries of free fatty acids, aside from 18:2 and 18:3 were constant, it seems that no lipolysis of glycerolipids, nor general oxidation or degradation of the free fatty acids took place. It also seems that lipoxygenase did not affect the small amount of 18:2 that was present in the "free" free fatty acid extracted by petroleum ether. The process of dough mixing resulted in the binding of considerable amounts of the nonpolar lipids and almost all of the polar lipids. As regards the free fatty acid components, binding was found not to be selective. The only lipid class more largely bound in anaerobic dough than in aerobic was the triglyceride.

Pomeranz et al. (1968B) studied natural and modified phospholipids effects on bread quality. It was found that alcohol-soluble phosphatides containing a 2:1 mixture of phosphatidyl choline and phosphatidyl

TABLE 15.7. Fatty Acid Composition (Weight %) of Wheat Flour Nonpolar Lipids[a]

Fatty Acid	TG	1,3-DG	1,2-DG	1-MG 2-MG	FFA	SE	ESG	EMGDG
12:0					0.2		0.4	0.1
14:0	0.1	0.3	0.2	0.2	0.3	2.5	0.7	0.1
16:0	16.7	17.1	14.8	22.0	13.5	58.2	52.0	24.1
16:1					0.3		1.5	0.4
18:0	1.2	2.7	1.4	1.3	1.4	1.8	5.1	1.8
18:1	13.9	12.2	12.9	8.7	9.3	5.8	6.1	7.2
18:2	63.7	64.4	64.4	62.4	68.0	27.7	28.2	61.6
20:0				1.2				
18:3	4.3	3.3	6.3	4.2	6.5	2.6	3.0	3.6

Source: MacMurray and Morrison (1970). Reproduced with permission of Dr. T. A. MacMurray of H. J. Heinz Co. Ltd. Food Research Laboratory, Hayes Park, Middlesex, England.

[a] Key to abbreviations: TG = Triglyceride, DG = diglyceride, MG = monoglyceride, FFA = free fatty acid, SE = steryl ester, ESG = esterified (6=*O*-acyl) steryl glucoside, EMGDG = esterified (6-*O*-acyl) monogalactosyl diglyceride.

TABLE 15.8. Composition of Wheat Flour Non-Polar Lipids

Lipid	Weight (%)
Steryl ester	14.7
Triglyceride	40.9
1,2-Diglyceride	12.2
1,3-Diglyceride	11.8
Free fatty acid	13.7
Free sterol	4.1
Monoglyceride	2.6

Source: MacMurray and Morrison (1970). Reproduced with permission of Dr. T. A. MacMurray of H. J. Heinz Co. Ltd. Food Research Laboratory, Hayes Park, Middlesex, England.

ethanolamine had the greatest effect in improving crumb grain and loaf volume in bread baked from untreated and petroleum-ether-extracted flour without added shortening. It was noted also that 0.5% alcohol-soluble phospholipids could substitute for both 0.8% free flour lipids and 3% shortening when added to petroleum-ether-extracted flours. Loaf volume of bread was usually correlated with softness retention.

Pomeranz and Wehrli (1969) showed that improvement of bread-baking qualities of synthetic glycosylglycerides depended on the composition, including carbohydrate and lipid composition, in which the chain length and degree of fatty acid unsaturation are important. It was found that glycolipids with the fatty acid attached directly to the sugar moiety were as useful as improvers as glycosylglycerides in restoring loaf volume of bread made with shortening and petroleum-ether-defatted flour. The authors noted that the sugar moiety can be replaced by other polar groups, as is the case with lecithin. Both phosphatidyl serine and crude lecithin were able to restore loaf volume sustantially, indicating that ionic charges, negative or positive, do not inhibit the restoring effect.

Pomeranz *et al.* (1966D) studied the effect of lipids on bread baked from various types of flours. They found that crumb grains improved and loaf volumes increased from 187 to 195 ml when 3 g of vegetable shortening was added to 100 g of flour. They noted that the improvement was striking with additions of up to 1.5 g of shortening, and then only slight up to 4.5 g of shortening. Adding 0.5 g of polar lipids separated from six flours to a composite hard red winter wheat flour was almost equal to the 3 g of shortening. The increase and crumb grain improvement resulted also in retardation of crumb firming during storage.

Pomeranz *et al.* (1968A) investigated reconstitution and properties of defatted flours. In this study bread was baked from defatted flours, with and without the addition of 3% commercial vegetable shortening.

TABLE 15.9. Mineral Content of Wheat and Mill Products Calculated as the Elements from Analysis of the Ash (in Parts per Million of Product, Dry Basis)

Element	Wheat	Patent Flour	Clear Flour	Low-grade Flour	Total Mill-run Middlings	Bran	Germ
Total ash	20,500	4,820	8,040	14,620	47,620	67,480	50,410
Magnesium	1,898	308	624	1,327	4,546	7,166	3,801
Calcium	452	180	227	376	1,115	1,158	692
Phosphorus	4,440	1,162	1,910	3,511	10,446	15,208	12,533
Potassium	2,370	552	875	1,533	5,633	7,098	5,542
Zinc	100	40	48	129	319	562	420
Iron	31	8	11	22	71	95	68
Manganese	24	2	5	12	48	112	67
Copper	6	2	2	4	13	14	9
Aluminum	3	0.6	2	7	8	27	25

Adapted from Sullivan and Near (1929).

Bread baked from these flours showed that added shortening improved loaf volume and crumb grain. The addition of shortening to the dough formula impaired the crumb grain of bread baked from defatted flours. When shortening was added to the dough formula for breads baked from defatted strong flours a decrease in the volume resulted. However, in the case of defatted poor flours the volume increased under comparable conditions. Tests with strong flours showed that the shortening response was fully restored by reconstruction with free lipids. It was further demonstrated that at least half of the quality of free lipid originally present in the flour was necessary to obtain the original loaf volume.

Ash. The grade of flour determines the approximate amount of ash it contains, usually 0.3–1%. Table 15.9 shows the mineral components of wheat and mill products.

Bleaching and Maturing Agents

Freshly milled wheat flour has a slight yellowish color caused by the presence of carotenoid pigments from the grain. Bleaching is used in the United States to produce a whiter flour and maturing agents are used to improve the baking qualities of the dough made from it. However, it should be noted that the bleaching of flour is not a universal practice, and in many countries it is considered undesirable. While this chemical treatment does not chemically change much of the essential unsaturated fatty acids, it does to a certain extent, and at the same time it has a marked effect, apparently, on the natural antioxidant substances in the flour. It is for these reasons that questions have been raised concerning the advisability of the continued bleaching of flour. Unbleached flours and certain breads made of unbleached flour are available on the American market. Benzoyl peroxide is widely used as a bleaching agent for flour. Since the bleaching agents act chemically on the carotenoids in the flour, the benzoyl peroxide must decompose to affect its full bleaching action; this process takes several hours. The bleaching effect can be produced also by certain enzyme preparations. A product made from soybeans, which contains lipoxidase, will, under proper conditions, bleach flour.

Maturing agents or dough improvers are also used, including chlorine dioxide and potassium bromate. The use of these substances is controlled by government regulations. Chlorine dioxide, the maturing agent, acts almost instantly on the flour. The use of this agent results in better handling properties of the dough. The loaf volume of the finished product is said to be better. Chlorine gas acts as an improver as well as a bleach, and is used in cake flour. Its use results in improvement of cake-baking qualities.

Presently, bread flour is treated with 1.8 g of chlorine dioxide per sack. The data given in Table 15.10 were designed to show that the so-called bread improvers and, of course, bleaches in the quantities used have little effect on the essential acids. However, one must conclude that fair amounts of linoleic and linolenic acids were lost, approximately 10% linoleic acid, and 30% linolenic acid when treated with 3.5 g of chlorine dioxide per sack and held for 39 days. It is unfortunate that figures for the 1.8 g chlorine dioxide per sack were not included.

Work has been done on the effect of oxidizing agents on doughs. Jørgensen (1939) considered that they acted on the proteolytic enzymes. However, Freilich and Frey (1939) believed that the components other than enzymes are acted on. Sullivan *et al.* (1940) stated the maturing agents act on the sulfhydryl (S-R) groups in the gluten protein, oxidizing them to more complex groups, such as $R(SO_2)R$, RSO_3H.

TABLE 15.10. Gas-liquid Chromatography Analysis of the Chlorine Dioxide-treated Flour After Treatment with Chlorine Dioxide

Fatty Acid Found as % Methyl Esters	Untreated	After 5 Days		After 39 Days	
		3.5 g/sack	35.0 g/sack	3.5 g/sack	35.0 g/sack
Palmitic	15.4	16.3	20.3	21.3	19.8
Oleic	16.2	16.5	16.8	17.1	7.5
Linoleic	63.2	63.4	60.1	56.4	14.5
Linolenic	5.3	3.8	2.6	3.7	Nil
Undetermined	0.0	0.0	0.2	1.5	58.2

Source: Daniels *et al.* (1960). Reproduced with permission of Dr. N. W. R. Daniels of Spillers Limited, Research and Technology Centre, Station Road, Cambridge, England.

Bread Flavor

According to Johnson and El-Dash (1969) it is during the baking process that free amino acids interact with reducing sugars. This produces many nonvolatile and volatile compounds that contribute to the flavor of bread. It seems that bread flavor is quite complex, and variation in this quality is brought about by formulation, fermentation, and the baking itself. The final bread flavor is produced during the baking process and is influenced by the formulation and fermentation before the baking of the bread. The effect of nonvolatile substances on the fermentation and baking may have a direct or indirect influence on the final flavor of the bread.

According to Johnson *et al.* (1966) experiments with pre-ferments, dough, bread, and oven vapors show that more than 70 different organic compounds have been identified. These compounds are made up of carbonyls, alcohols, organic acids, and esters and are formed during fermentation and baking. These two steps are necessary to produce desirable flavor. Reactions involving reducing sugars and amino groups are active in crust browning. Bread crumbs have smaller amounts of carbonyl compounds than the crust. Staling of bread is accompanied by gradual loss of carbonyl compounds from the crust.

LEAVENING

The common leavening agents are yeast, baking powders, and air. In some cookies ammonium carbonate and ammonium bicarbonate are used. For certain products, sodium bicarbonate alone is employed.

Yeast

The action of yeasts on sugars has already been discussed in Chapter 14 under fermentation. In raising bread dough we have another application of this phenomenon; however, the important product is the carbon dioxide formed by fermentation, rather than the alcohol desirable in the preparation of alcoholic beverages. When used as leavening in dough, bakers' yeast acts predominantly via the fermentation route with only 8% of the energy released as heat. If the oxidative pathway predominated, considerably more heat would be liberated. When yeast is placed in dilute sugar solution and adequately aerated, the sugar is utilized exclusively by the aerobic or oxidative pathway. In the absence of oxygen, or with greater sugar concentrations, the fermentative pathway manifests itself.

$$\text{Oxidation: } C_6H_{12}O_6 \longrightarrow 6CO_2 + 6H_2O; \quad -\Delta F = 686 \text{ kcal}$$
$$\text{Fermentation: } C_6H_{12}O_6 \longrightarrow 2CO_2 + 2C_2H_5OH; \quad -\Delta F = 54 \text{ kcal}$$

According to Matz (1960), high levels of sugar, as are commonly used in sweet doughs, affect the ability of bakers' yeast to ferment rapidly. This type of inhibition is caused by high osmotic pressure. When added to a dough system to provide the same osmotic pressure glucose also inhibits the gas formation. It should also be noted that certain metals and other substances, such as copper, cadmium, mercury, formaldehyde, and chlorine can poison the yeast if a sufficient amount is present. Poisoned yeast will not act.

For growth and fermentation involving bakers' yeast the most favorable temperature range is 29° – 32° C. Table 15.11 shows the relative production of gas at different temperatures and gives information concerning the relative changes in gas production from 17.8° tp 46.7°C. The best dough temperature according to Rumsey (1959) is between 25.6° amd 26.7°C.

There are two principal results of yeast action in panary fermentation. (1) The formation and migration of carbon dioxide in a network of cellular compartments, occupying about 120 $in.^3$ per pound loaf, which lightens and raises the dough, and which, in turn, greatly improves the palatability of the finished product. (2) At the same time certain compounds are produced, namely, alcohols, aldehydes, ketones, and acids which contribute to the odor and flavor of the bread.

The activity of yeast also alters the physical properties of the dough, particularly gluten elasticity because of the forceful stretching actions formed by the diffusion and accumulation of CO_2 throughout the mass of the dough. Active dry yeast performs leavening activities with vigor about equal to compressed yeast, used in the following ratios: 1:2.5 to 1:2.0, active dry yeast to compressed yeast, respectively.

The origin of gas cells in doughs was studied by Baker and Mize (1941). These authors considered five hypothetical sources of gas cells: (1) the gas in the endosperm particles, (2) the gas space between the endosperm particles, (3) the air the mixing beats into the dough and subdivides it to produce bubbles of small size, (4) the gases liberated by the yeast starting new bubbles around the organisms, and (5) the folding, punching, rolling, moulding, and twisting of the dough breaking the gas bubbles to increase their number. One of the results of this is that carbon dioxide in aqueous solution leaves the yeast cells. This solution releases gas by diffusion into air bubbles incorporated during the mixing of the dough. Baker (1941) was able to separate intact gas

TABLE 15.11. Relative Gas Produced at Different Temperatures

Temperature (°C)	Relative CO_2 Evolved[a]
17.8	59
30.8	100
39.0	89
44.0	82
46.7	78

Adapted from Peppler (1960).
[a] After 90 min.

cells from dough as thin transparent protein bubbles by diluting doughs with brine. The wall of these bubbles was found to be glutinous in character and made up (on the dry basis) of about 45% protein and 25% starch.

Baker (1941) made observations on fermenting dough, and he concluded that there was a tendency for the gluten and starch to separate during fermentation and for the gluten to form transparent cells. He concluded further from this that starch in the flour is not necessary to cell formation in doughs and that this film is drawn to the surface because the gas nucleus from which the bubbles originate is a glutinous core. As the bubble expands, the required amount of gluten to satisfy its surface needs is drawn from the starch–gluten matrix of the endosperm material. The properties that enable this to occur may be controlled by the viscosity and fluidity of the gluten and the amount of adhesion of the gluten to starch.

Matz (1960) stated that from these observations that intensive mixing of cake batter is necessary to incorporate minute air bubbles into the batter, which can later expand during baking into fine cells required for the texture of the finished cake. The quantity of gas necessary for the expansion could come from the following sources: (1) chemical leavening added to the mixture, (2) expansion of occluded air bubbles at the baking temperature, and (3) steam formed during baking.

Baking Powder

Baking powders are usually used in cake mixes and are of three kinds: (1) phosphate baking powders, (2) baking powders containing aluminum compounds, called alum baking powders, and (3) the tartrate baking powders. In the past quite a controversy existed between the makers of alum baking powders and the makers of the tartrate-containing product, and involved possible harm from extended use of aluminum containing products, and also from residual (bitter) flavors from them. The sale of baking powders containing aluminum salts is forbidden in European countries, but no such prohibition exists in the United States.

Cream of tarter, the several calcium phosphate salts, together with sodium acid pyrophosphate, and the sodium aluminum salts are used in baking powder to liberate CO_2 from the other active ingredient, sodium bicarbonate. Sodium bicarbonate is especially desirable for use because it is low in cost, is highly pure in the commercial form, does not contribute an undesirable taste to the finished products, is easily handled, and is nontoxic. According to the definition of baking powder by the Federal Government, it is a "leavening agent produced by the mixing of an acid-reacting material and sodium bicarbonate with or without starch or flour, which yields not less than 12% of available CO_2." It is further stated that "the acid reacting materials in baking powders are (1) tartaric acid or its acid salts, (2) acid salts of phosphoric acid, (3) compounds of aluminum, or (4) any combination in substantial proportions of these." While it is true that the stated requirement of 12% available CO_2 could be met by using 23% of sodium bicarbonate in the formula, 26–30% is used to allow for loss of gas in storage.

Ammonium carbonate and bicarbonate mentioned before liberate carbon dioxide, ammonia, and water when heated, and since they decompose completely into gaseous products, they leave no residue.

$$NH_4HCO_3 \longrightarrow CO_2 + NH_3 + H_2O$$

Sodium bicarbonate is used in certain mixtures which contain ingredients able to react with this compound to liberate carbon dioxide.

SUMMARY

The leavening prodess for the making of bread has a long history. It was discovered by the ancient Egyptians. Both flour and the leavening process contribute the most important characteristics to the finished products.

Bread flours are made from hard spring and winter wheats. Breads made from these flours have good volume, grain, and texture. Soft wheat flours are used for cake, cookies, and pastry products. Rye flour is used mostly for rye bread. Rye flour doughs are very weak, compared with those of wheat flour doughs.

Wheat flour has the special property to form an elastic dough when mixed with the right amount of water. The proteins, carbohydrates, and fats present in the wheat contribute to the baking quality of the flour. The lipids present have been shown to be especially important in the quality of the finished bread.

The bleaching of white flour is practiced in the United States to produce a whiter flour. However, this treatment of flour in many countries is not considered desirable. This treatment bleaches the carotenoid pigments found in the grain. Questions have been raised concerning the advisability of the continued bleaching of flour. Benzoyl peroxide is widely used as a bleaching agent for flour.

Dough improvers are chlorine dioxide and potassium bromate. The use of these compounds is regulated by the government. Chlorine gas acts as an improver as well as a bleach and is used for cake flour.

Bread flavor seems to be developed by formulation, fermentation, and the baking itself. This flavor is quite complex and is made up of a number of different organic compounds which include several different types of compounds. Bread staling is accompanied by gradual loss of carbonyl compounds from the crust.

The common leavening agents are yeast, baking powders, and air.

Yeast brings about another application of the phenomenon of fermentation. In the raising of bread dough, however, the important product is not the alcohol as in alcoholic fermentation, but the CO_2. It is this gas that is effective in the raising of bread. The most favorable temperature for fermentation with bakers' yeast is 29°–32°C. The activity of the yeast alters the physical properties of the dough, particularly gluten elasticity. The separation of intact gas cells from dough as thin, transparent protein bubbles was noted.

Baking powders are of several types and their action is the result of liberation of CO_2 from $NaHCO_3$, which is one of the ingredients of these powders. The liberation of CO_2 produces the leavening action.

ACKNOWLEDGMENT

The author would like to acknowledge use of some information in this chapter that previously appeared in "Bakery Technology and Engineering," 1st Edition (1960) and 2nd Edition (1972) by Samuel A. Matz.

BIBLIOGRAPHY

ANON. 1961. Wheat gluten surrenders four more proteins. Zone electrophoresis in starch-gel brings protein total to nine, gives new clues to gluten's elasticity. Chem. Eng. News *39*, No. 23, 44–45.

ARUNGA, R. O., and MORRISON, W. R. 1971. The structural analysis of wheat flour glycerolipids. Lipids *6*, 768–776.

BAKER, J. C. 1941. The structure of the gas cell in bread dough. Cereal Chem. *18*, 34–41.

BAKER, J. C., and MIZE, M. D. 1941. The origin of the gas cell in bread dough. Cereal Chem. *18*, 19–34.

BECKWITH, A. C., NIELSEN, H. C., WALL, J. S., and HUEBNER, F. R. 1966. Isolation and characterization of a high molecular-weight protein from wheat gliadin. Cereal Chem. *43*, 14–28.

BOHN, R. M. 1957. Biscuit and Cracker Production. American Trade Publishing Co., New York.

CHIU, C. M., and POMERANZ, Y. 1966. Changes in extractability of lipids during bread making. J. Food Sci. *31*, 753–758.

CHIU, C. M., and POMERANZ, Y. 1967. Lipids in wheat kernels of varying size. J. Food Sci. *32*, 422–425.

CHIU, C. M., POMERANZ, Y., SHOGREN, M., and FINNEY, K. F. 1968. Lipid binding in wheat flours. Varying in bread-making potential. Food Technol. *22*, 1157–1162.

CHUNG, O., FINNEY, K. F., and POMERANZ, Y. 1967. Lipids in flour from gamma-irradiated wheat. J. Food Sci. *32*, 315–317.

CLAYTON, T. A., and MORRISON, W. R. 1972. Changes in flour lipids during the storage of wheat flour. J. Sci. Food Agric. *23*, 721–736.

CUNNINGHAM, D. K., GEDDES, W. F., and ANDERSON, J. A. 1955. Preparation and chemical characteristics of the cohesive proteins of wheat, barley, rye, and oats. Cereal Chem. *32*, 91–106.

DAFTARY, R. D., and POMERANZ, Y. 1965A. Changes in lipid composition in maturing wheat. J. Food Sci. *30*, 577–582.

DAFTARY, R. D., and POMERANZ, Y. 1965B. Storage effects in wheat. Changes in lipid composition in wheat during storage deterioration. J. Agric. Food Chem. *13*, 442–446.

DAFTARY, R. D., WARD, A. B., and POMERANZ, Y. 1966. Distribution of lipids in air-fractionated flours. J. Food Sci. *31*, 897–901.

DAFTARY, R. D., POMERANZ, Y., SHOGREN, M., and FINNEY, K. F. 1968. Functional bread-making properties of wheat flour lipids. 2. The role of flour lipid fractions in breadmaking. Food Technol. *22*, 327–330.

DANIELS, D. G. H. 1960. Changes in the lipids of flour induced by treatment with chlorine dioxide or chlorine, and on storage. J. Sci. Food Agric. *11*, 664–670.

DANIELS, R. 1970. Modern Breakfast Cereal Processes. Noyes Data Corp., Park Ridge, NJ.

DANIELS, N. W. R., RUSSELL-EGGITT, P. W., and COPPOCK, J. B. M. 1960. Studies on the lipids of flour. I. Effect of chlorine dioxide treatment on the essential fatty acids. J. Sci. Food Agric. *11*, 658–664.

DANIELS, N. W. R., WOOD, P. S., RUSSELL-EGGITT, P. W., and COPPOCK, J. B. M. 1970. Studies on the lipids of flour, V. Effect of air on lipid binding. J. Sci. Food Agric. *21*, 377-384.

ELTON, G. A. H., and EWART, J. A. D. 1960. Starch-gel electrophoresis of wheat proteins. Nature *187*, 600–601.

FEDERAL TRADE COMMISSION. 1927. Report of the trial examiner in Docket No. 540. Royal Baking Powder Co.

FEDERAL TRADE COMMISSION. 1928. The truth about baking powder. Records in Docket No. 540. Calumet Baking Powder Co.

FISHER, N., BELL, B. M., and RAWLINGS, C. E. B. 1973. Lipid binding in flour, dough, and bread. J. Sci. Food Agric. *24*, 147–155.

FISHER, N., BELL, B. M., RAWLINGS, C. E. B., and BENNET, R. 1966. The lipids of wheat. III. Further studies of the lipids of flours from single wheat varieties of widely varying baking quality. J. Sci. Food Agric. *17*, 370–382.

FREILICH, J., and FREY, C. N. 1939. Dough oxidation and mixturing studies. I. The action of potassium bromate in dough. Cereal Chem. *16*, 485–494.

GRAVELAND, A. 1970. Enzymic oxidations of linoleic acid and glycerol-1-monolinoleate in doughs and flour-water suspensions. J. Am. Oil Chem. Soc. *47*, 352–361.

GRAVELAND, A. 1970. Modification of the course of the reaction between wheat flour lipoxygenase and linoleic acid due to adsorption of lipoxygenase on glutenin. Biochem. Biophys. Res. Commun. *41*, 427–434.

GRAVELAND, A. 1973. Analysis of lipoxygenase nonvolatile reaction products of linoleic acid in aqueous cereal suspensions by urea extraction and gas chromatography. Lipids *8*, 599–605.

GRAVELAND, A. 1973. Enzymatic oxidation of linolenic acid in aqueous wheat flour suspensions. Lipids *8*, 606–611.

HALTON, P., and SCOTT-BLAIR, G. W. 1936. Physical properties of flour doughs in relation to their bread-making qualities. J. Phys. Chem. *40*, 561–580.

HALTON, P., and SCOTT-BLAIR, G. W. 1937. A study of some physical properties of flour doughs in relation to their bread-making qualities. Cereal Chem. *14*, 201–219.

HARREL, C.G. 1959. Manufacture of prepared mixes. *In* The Chemistry and Technology of Cereals as Food and Feed, S. A. Matz (Editor). AVI Publishing Co., Westport, CT.

HIRD, F. J. R., CROKER, I. W. D., and JONES, W. L. 1968. Low molecular weight thiols and disulphides in flour. J. Sci. Food Agric. *19*, 602–604.

HOLME, J., and BRIGGS, D. R. 1959. Studies on the physical nature of gliadin. Cereal Chem. *36*, 321–340.

JOHNSON, J. A., and EL-DASH, A. A. 1969. Role of nonvolatile compounds in bread flavor. J. Agric. Food Chem. *17*, 740–746.

JOHNSON, J. A., ROONEY, L., and SALEM, A. 1966. Chemistry of bread flavor. *In* Flavor Chemistry. (Editor). Advances in Chemistry Ser. 56. American Chemical Society, Washington, DC.

JØRGENSEN, H. 1939. Further investigations into the nature of the action of bromates and ascorbic acid on the baking strength of wheat flour. Cereal Chem. *16*, 51–60.

KIM, S. K. and APPOLONIA, B. L. 1977A. Bread staling studies. I. Effect of protein content on staling rate and bread crumb pasting properties. Cereal Chem. *54*, 207–215.

KIM, S. K., and APPOLONIA, B. L. 1977B. Bread staling studies. II. Effect of protein content and storage temperature on role of starch. Cereal Chem. *54*, 216–224.

KIM, S. K., and APPOLONIA, B. L. 1977C. Bread staling studies. III. Effect of pentosans on dough, bread, and bread staling rate. Cereal Chem. *54*, 225–229.

KOCH, R. B., GEDDES, W. F., and SMITH, F. 1951. The carbohydrates of *Gramineae*. I. The sugars of the flour of wheat (*Triticum vulgare*). Cereal Chem. *28*, 424–430.

KOYANAGI, Y., TAKANO, H., TAKAHASHI, T., and TANAKA, Y. 1979. Effect of ascorbic acid on bread making. II. The improvement of rheological properties of dough quality by several chemicals. Shokuhin Sago Kenkyusho Kenkyu Hokoku *34*, 29–34. Chem. Abstr. *94*, 82439c (1981).

LIN, F. M., and POMERANZ, Y. 1968. Characterization of water-soluble wheat flour pentosans. J. Food Sci. *33*, 599–606.

MACMURRAY, T. A., and MORRISON, W. R. 1970. Composition of wheat-flour lipids. J. Sci. Food Agric. *21*, 520–528.

MANN, D. L. and MORRISON, W. R. 1974. Changes in wheat lipids during mixing and resting of flour-water doughs. J. Sci. Food Agric. *25*, 1109–1119.

MATZ, S. A. 1960. Bakery Technology and Engineering, 1st Edition. AVI Publishing Co., Westport, CT.

MATZ, S. A. 1972. Bakery Technology and Engineering, 2nd Edition. AVI Publishing Co., Westport, CT.

MECHAM, D. K., and WEINSTEIN, N. E. 1952. Lipid binding in doughs. Effects of dough ingredients. Cereal Chem. *29*, 448–455.

McCALLA, A. G., and GRALÉN, N. 1942. Ultracentrifuge and diffusion studies on gluten. Can. J. Res. *20*, 130–159.

McDERMOTT, E. E., and PACE, J. 1959. Extraction of the total protein from wheaten flour in the form of soluble derivatives. Nature *184*, 546–547.

MEREDITH, P., SAMMONS, H. G., and FRAZER, A. C. 1960. Examination of wheat gluten by partial solubility methods. I. Partition by organic solvent. J. Sci. Food Agric. *11*, 320–328.

MEREDITH, P., SAMMONS, H. G., and FRAZER, A. C. 1960. Examination of wheat gluten by partial solubility methods. II. Partition by dilute formic acid. J. Agric. Food Chem. *11*, 329–337.

MORRISON, W. R., and MANEELY, E. A. 1969. Importance of wheat lipoxidase in the oxidation of free fatty acids in flour-water systems. J. Sci. Food Agric. *20*, 379–381.

NEILSEN, H. C., BECKWITH, A. C., and WALL, J. S. 1968. Effect of disulfidebond cleavage on wheat gliadin fractions obtained by gel filtration. Cereal Chem. *45*, 37–47.

OSBORNE, T. B. 1907. The proteins of the wheat kernel. Carnegie Inst. Washington Publ. *84*.

PEPPLER, H. J. 1960. Yeast. *In* Bakery Technology and Engineering, S. A. Matz (Editor). AVI Publishing Co., Westport, CT.

POMERANZ, Y. 1965. Isolation of proteins from plant material. J. Food Sci. *30*, 823–827.

POMERANZ, Y. 1966. The role of the lipid fraction in growth of cereals, and in their storage and processing. Wallerstein Lab. Commun. *29*, 17–26.

POMERANZ, Y., and HAYES, E. R. 1968. Hydrogenated corn oil's effect on bread baked from flours of single wheat varieties. Food Technol. *22*, 1446–1448.

POMERANZ, Y., and WEHRLI, H. P. 1969. Synthetic glycosylglycerides in bread making. Food Technol. *23*, 1213–1215.

POMERANZ, Y., RUBENTHALER, G. L., and FINNEY, K. F. 1965. Polar vs. nonpolar wheat flour lipids in bread-making. Food Technol. *19*, 1724–1725.

POMERANZ, Y., CHUNG, O., and ROBINSON, R. J. 1966A. The lipid composition of wheat flours varying widely in bread-making. J. Am. Oil Chem. Soc. *43*, 45–48.

POMERANZ, Y., RUBENTHALER, G. L., and FINNEY, K. F. 1966B. Evaluation of the effects of proteolytic enzymes on bread flour properties. Food Technol. *20*, 327–330.

POMERANZ, Y., RUBENTHALER, G. L., and FINNEY, K. F. 1966C. Studies on the mechanism of the bread-improving effect of lipids. Food Technol. *20*, 1485–1488.

POMERANZ, Y., RUBENTHALER, G. L., DAFTARY, R. D., and FINNEY, K. F. 1966D. Effects of lipids on bread baked from flours varying widely in bread-making potentialities. Food Technol. *20*, 1225–1228.

POMERANZ, Y., SHOGREN, M., and FINNEY, K. F. 1968A. Functional bread-making properties of wheat flour lipids. I. Reconstitution studies and properties of defatted flours. Food Technol. *22*, 324–327.

POMERANZ, Y., SHOGREN, M., and FINNEY, K. F. 1968B. Natural and modified phospholipids effects on bread quality. Food Technol. *22*, 897–900.

POMERANZ, Y., TAO, R. P. C., HOSENEY, R. C., SHOGREN, M. D., and FINNEY, K. F. 1968C. Evaluation of factors affecting lipid binding in wheat flours. J. Agric. Food Chem. *16*, 974–978.

PYKE, M. 1958. The technology of yeast. *In* The Chemistry and Biology of Yeasts, A. H. Cook (Editor). Academic Press, New York.

ROBINSON, R. J. , LORD, T. H., JOHNSON, J. A., and MILLER, B. S. The aerobic microbiological population of pre-ferments and the use of selected bacteria for flavor production. Cereal Chem. *35*, 295–305.

RUMSEY, L. A. 1959. Commercial baking procedures. *In* The Chemistry and Technology of Cereals as Food and Feed, S. A. Matz (Editor). AVI Publishing Co., Westport, CT.

SCHOPMEYER, H. H. 1960. Flour. *In* Bakery Technology and Engineering, S. A. Matz (Editor). AVI Publishing Co., Westport, CT.

SHOUP, F. K., POMERANZ, Y., and DEYOE, C. W. 1966. Amino acid composition of wheat varieties and flours varying widely in bread-making potentialities. J. Food Sci. *31*, 94–101.

SOCIETY OF CHEMICAL INDUSTRY. 1962. Recent Advances in Processing Cereals. S.C.I. Monograph No. 16. Gordon and Breach, Science Publishers, New York.

SULLIVAN, B., and NEAR, C. 1927. The ash of hard spring wheat and its products. Ind. Eng. Chem. *19*, 498–501.

SULLIVAN, B., and HOWE, M. 1937. Isolation of glutathione from wheat germ. J. Am. Chem. Soc. *59*, 2742–2743.

SULLIVAN, B., and HOWE, M. 1938. Lipids of wheat flour. I. The petroleum ether extract. Cereal Chem. *15*, 716–720.

SULLIVAN, B., HOWE, M., SCHMALZ, F. D., and ASTLEFORD, G. R. 1940. The action of oxidizing and reducing agents on flour. Cereal Chem. *17*, 507–528.

SWAM, J. M. 1957. Thiols, disulphides, and thiosulphates: Some new reactions and possibilities in peptide and protein chemistry, Nature *180*, 643–645.

SWANSON, C. O. 1938A. The colloidal structure of dough as a means of interpreting quality in wheat flour. Cereal Chem. Suppl. *15*.

SWANSON, C. O. 1938B. Wheat Flour Quality. Burgess Publishing Co., Minneapolis, MN.

TAO, R. P. C., and POMERANZ, Y. 1967. Water soluble pentosans in flours varying widely in bread-making potential. J. Food Sci. *32*, 162–168.

TAO, R. P. C., and POMERANZ, Y. 1968. Functional bread-making properties of wheat flour lipids. 3. Effects of lipids on rheological properties of wheat flour doughs. Food Technol. *22*, 1145–1149.

WEHRLI, H. P., and POMERANZ, Y. 1969. Synthesis of galactosyl glycerides and related lipids. Chem. Phys. Lipids *3*, 357–370.

WOYCHIK, J. H., and HUEBNER, F. R. 1966. Isolation and partial characterization of wheat gamma-gliadin. Biochem. Biophys. Acta *127*, 88–93.

WOYCHIK, J. H., BOUNDY, J. A., and DIMLER, R. J. 1961. Amino acid composition of proteins in wheat gluten. J. Agric. Food Chem. *9*, 307–310.

WOYCHIK, J. H., HUEBNER, F. R., and DIMLER, R. J. 1964. Reduction and starch-gel electrophoresis of wheat gliadin and glutenin. Arch. Biochem. Biophys. *105, 151–155.*

16

Milk and Milk Products

Milk Composition
Rancidity and Off-Flavor in Milk
Cheese and Cheese Chemistry
Fermentation in Milk
Milk Products Other Than Cheese
Determination of Fat in Dairy Products
Total Solids Analysis
Summary
Bibliography

The cow is the most efficient producer of milk. Cow's milk is defined as "the lacteal secretion practically free from colostrum, obtained by the complete milking of one or more healthy cows, which contains not less than 8¼% of milk-solids-not-fat and not less than 3¼% of milkfat" (Federal Security Agency 1951). The average gross composition of cow's milk is recorded as follows (Watt and Merrill 1963): water, 87%; fat, 3.6%; lactose, 4.9%; proteins, 3.5%; and ash (minerals), 0.7%.

MILK COMPOSITION

The gross composition of milk does not accurately indicate the complete composition, but is intended for commercial evaluation. For such use, total fat and solids are the most important.

Milk contains the following:

1. Water
2. Sugars
   - Lactose (the main carbohydrate in milk)
   - Glucose
   - Galactose
3. Proteins

   The proteins of milk were identified according to the classical nomenclature as casein, 80%; and lactalbumin and lactoglobulin, the serum proteins, comprising the other 20%. However, it is now known that these are mixtures of many proteins. Free amino acids, peptides, and other forms of nitrogen have been found in milk (Jennes *et al.* 1956).

4. Lipids
   Fats
   Phospholipids
   Sterols
5. Enzymes

Aldolase	Lactoperoxidase
Amylase	Phosphatase
Catalase	Protease
Esterase	Ribonuclease
Galactase	Xanthine oxidase
Lipase	

6. Vitamins
   The Table 16.1 lists vitamins in fresh milk.
7. Pigments
   Carotene
   Riboflavin
8. Minerals
   Table 16.2 shows the often-repeated statement that milk is a good source of calcium. In addition, the phosphorus and potassium available in this food is shown.

   A number of trace elements (Table 16.3) measured in parts per million include aluminum, arsenic, barium, boron, bromine, chromium, cobalt, copper, iodine, iron, lead, lithium, manganese, molybdenum, rubidium, selenium, silicon, silver, strontium, tin,

TABLE 16.1 Vitamins in Fresh Milk

	mg/100 ml	Range
Vitamin A[a]	159	136–176
Carotenoids	0.030	0.025–0.060
Vitamin D[b]	2.21	0–10.9
Vitamin E	0.100	0.02–0.18
Vitamin K	0.00467	0.0–0.0160
Vitamin C	2.09	1.57–2.75
Biotin	0.003	0.0012–0.0060
Choline	13.7	4.3–28.5
Folacin	0.0059	0.0038–0.0090
myo-Inositol (Total)[c]	11.0	6.0–18.0
Niacin	0.09	0.03–0.20
Pantothenic acid	0.34	0.26–0.49
Riboflavin	0.17	0.08–0.26
Thiamin	0.04	0.02–0.08
Vitamin B_6	0.06	0.02–0.08
Vitamin B_{12}	0.00042	0.00024–0.00074
p-Aminobenzoic acid	0.01	0.004–0.015

Source: Hartman and Dryden (1965). Reproduced with permission of American Dairy Science Assoc.
[a] Expressed as IU/100 ml (0.048 mg/100 ml).
[b] Expressed as IU/100 ml.
[c] About three-quarters of it exists as free inositol.

TABLE 16.2. Average Values for Milk Salt Constituents

Constituent	Content in Whole Milk (mg/100 ml)	Number of Samples
Calcium	123	824
Magnesium	12	759
Phosphorus	95	829
Sodium	58	491
Potassium	141	472
Chlorine	119	1579
Sulfur	30	80
Citric acid	160	307

Source: Webb *et al.* (1974).

titanium, vanadium, and zinc. They are present usually in very small amounts. Cobalt is essential for vitamin B_{12}; copper is essential for the formation of hemoglobin; manganese is an integral part of the liver enyzme arginase; molybdenum is a part of the enzyme xanthine oxidase; and zinc is a constituent of the enzyme carbonic anhydrase. Radionuclides are present in milk usually in extremely small traces.

9. Miscellaneous
   Gases—CO_2, N_2, and O_2
   Miscellaneous flavor compounds are present according to the animal feeds used.

Carbohydrates

Whereas lactose is the major carbohydrate in milk, others are present; for example, free glucose and galactose are components of fresh cow's milk. The small amount of data available indicate the presence of about 7 mg glucose in 100 ml of milk (Anantakrishnan and Herrington 1948; Honer and Tuckey 1953) and about 2 mg galactose in the same volume of milk. Also present are phosphate ester of glucose, galactose, and lactose in addition to carbohydrate–protein combinations that can be hydrolyzed to one or more of the following: mannose, galactose, fucose, glucosamine, galactosamine, or neuraminic acid. Oligosaccharides have been found in milk, seven of which were investigated by Trucco *et al.* (1954), which on hydrolysis were found to be composed of two, three, or four of the following constituents: lactose, glucose, galactose, neuraminic acid, mannose, and acetylglucosamine.

Lactose

Lactose is a normal component of cow's milk to the extent of 4.4–5.2%, an average of 4.8% of anhydrous lactose. Ordinarily, it amounts to 50–52% of the total solids of skimmed milk. It is found in one of two crystalline forms, alpha hydrate and beta lactose, or as an amorphous mixture of α- and β-lactose (lactose glass).

Lactose is a disaccharide and it yields D-glucose and D-galactose on hydrolysis. It is designated also as 4-*O*-β-D-galactopyranosyl-D-glucopyranose, and it is found in both the alpha and beta forms (Fig. 16.1). Lactose is able to reduce Fehling's solution.

α-Lactose hydrate crystals are found in a number of different shapes depending on the conditions under which crystallization takes place. The form of the crystal of β-anhydride when crystallized from water is an uneven-sided diamond, but curved, needle-like prisms when crystallized from alcohol.

The alpha hydrate (anhydrous weight basis) has an optical rotation in water of $[\alpha]_D^{20} = 89.4°$ and a melting point of 201.6° C. When the beta form is crystallized from aqueous solutions above 93.5° C the crystals formed are anhydrous, have a specific rotation of $[\alpha]_D^{20} = 35.0°$, and a melting point of 252.2°C.

TABLE 16.3. Trace Elements in Cow's Milk (μg/liter)

Element	Cow Receiving Normal Ration	Cow Receiving Supplement of Element in Ration
Aluminum	460	810
Arsenic	50	450
Barium	Qualitative	—
Boron	270	660
Bromine	600	Increases
Bromine (coastal area)	2800	—
Cadmium	26	No Increase
Chromium	15	—
Cobalt	0.6	2.4
Copper	130	No Increase
Fluorine	150	Increases
Iodine	43	Up to 2700
Iron	450	No Increase
Lead	40	Increases
Lithium	Qualitative	—
Manganese	22	64
Molybdenum	73	371
Nickel	27	No Increase
Rubidium	2000	Increases
Selenium (non-seleniferous area)	40	Increases
Selenium (seleniferous area)	Up to 1270	—
Silicon	1430	No Increase
Silver	47	—
Strontium	171	—
Tin	Qualitative	—
Titanium	Qualitative	—
Vanadium	0.092	—
Zinc	3900	5100

Source: Webb *et al.* (1974).

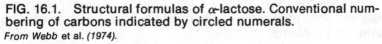

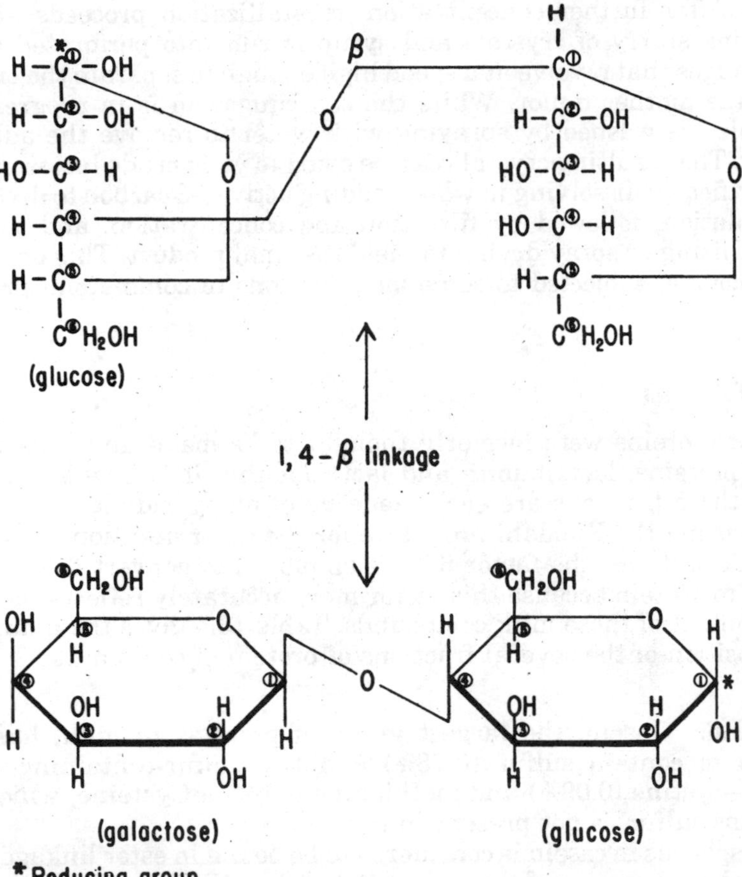

FIG. 16.1. Structural formulas of α-lactose. Conventional numbering of carbons indicated by circled numerals.
From Webb et al. *(1974).*

When either form is dissolved in water there is a gradual shift from one form to the other until equilibrium is established. During this process, mutarotation takes place with the result that the final rotation is $[\alpha]_D^{20} = 55.3°$ at equilibrium. This is a mixture of 37.3% of the alpha form and 62.7% of the beta form.

β-Lactose has been found to be sweeter than α-lactose. Little advantage is obtained, however, by using β-lactose for sweetness because the small difference is quickly eliminated by mutarotation.

Lactose is manufactured at present largely from cheese whey. In the basic process for the preparation of lactose, whey is treated with lime, and then heated and filtered to remove the proteins and calcium phosphate. The clear solution is concentrated to about 30% solids and refiltered, to remove any proteins and salts which have further separated. After further concentration, crystallization proceeds and the resulting slurry of crystals and syrup is run into perforated basket centrifuges that revolve at a speed high enough to separate the crystals from the mother liquor. While the centrifugation is in progress, the crystals are washed by spraying with water to remove the adhering liquor. The resulting crystals can be dried to yield crude lactose, or can be purified by dissolving in water, adding activated carbon to decolorize the solution, followed by filtration and concentration, and then recrystallizing or spray drying to yield the final product. This procedure is, of course, subjected to some modifications in commercial practice.

Milk Proteins

Milk proteins were formerly thought to be made up of casein and whey proteins, lactalbumin and lactoglobulin; it is now known that these three fractions are each made up of many individual proteins. When using the Kjeldahl procedure for the determination of nitrogen in milk proteins, the factor 6.38 is employed to convert the nitrogen found to protein because this factor more accurately reflects the nitrogen content of these milk compounds. Table 16.4 gives the amino acid composition of the several fractions of protein of cow's milk.

Casein. Casein, the largest group of proteins in milk, has been shown to contain sulfur (0.78%) and two sulfur-containing amino acids—cystine (0.09%) and methionine (0.69%). Cysteine, which also contains sulfur, is not present in casein.

Phosphorus in casein is considered to be bound in ester linkages with the hydroxyl groups of threonine and serine. Work on the ester linkages indicates that the probable bond is the *O*-monophosphate ester linkage and that phosphorus is bonded the same way in whole casein, α-casein, and β-casein.

Much work has been done on the structure of casein; the isolation of phosphopeptides and phosphopeptones provides valuable information in this regard. Following the partial hydrolysis of α-casein, Hipp *et al.* (1957) separated phosphoserine, phosphoserylglutamic acid, phosphoserylalanine, and phosphoserylphosphoserine by means of ion exchange chromatography. Relatively large electrophoretically homogeneous phosphopeptone was isolated from tryptic digests of β-casein by Peterson *et al.* (1958). The phosphopeptone had a molecular weight of approximately 3000 and comprised 24 amino acid residues of 10 different amino acids and five phosphoric acid groups most likely attached to the four serine and one threonine residues in the molecule. This phosphopeptone seems to account for almost all of the phosphorus of β-casein. Schormüller *et al.* (1961) compiled the results of other investigations of phosphopeptones prepared from α- and β-caseins.

TABLE 16.4. Amino Acid Composition of Cow's Milk Proteins (g/100 g Protein)

Constituent	α-Lactalbumin	β-Lactoglobulin	Blood Serum Albumin	Immune Globulin	Casein	α-Casein	α_{s1}-Casein B[a]	κ-Casein	β-Casein	γ-Casein
Total N	15.9	15.6	16.1	15.3–16.1	15.6	15.5	15.4	15.3	15.3	15.4
Total P	0.0	0.0	0.0	0.0	0.9	0.99	1.05	0.16	0.61	0.11
Total S	1.9	1.6	1.9	1.0	0.8	0.72	0.68	0.70	0.86	1.03
Glycine	3.2	1.4	1.8	5.2	2.0	2.3	2.9	1.2	1.6	1.5
Alanine	2.1	7.0	6.3	4.8	3.2	3.8	3.4	5.4	2.0	2.3
Valine	4.7	6.1	5.9	9.6	7.2	6.3	5.5	6.3	10.2	10.5
Leucine	11.5	15.5	12.3	9.6	9.2	7.9	9.4	6.1	11.6	12.0
Isoleucine	6.8	6.9	2.6	3.0	6.1	6.4	6.1	7.1	5.5	4.4
Proline	1.5	5.1	4.8	10.0	10.6	7.5	8.3	11.0	15.1	17.0
Phenylalanine	4.5	3.5	6.6	3.9	5.0	4.6	5.6	3.9	5.8	5.8
Tyrosine	5.4	3.7	5.1	6.7	6.3	8.1	7.7	7.6	3.2	3.7
Tryptophan	(7.0) 5.3	2.7	0.58	2.7	1.7	2.2	1.7	1.0	0.83	1.2
Serine	4.8	4.0	4.2	11.5	6.3	6.3	7.1	5.0	6.8	5.5
Threonine	5.5	5.0	5.8	10.5	4.9	4.9	2.5	6.7	5.1	4.4
Cystine + Cysteine	6.4[b]	3.4	6.5	3.2	0.34	0.43	0.0	1.2[b]	0.0	0.0
Methionine	0.95	3.2	0.81	0.9	2.8	2.5	3.2	1.7	3.4	4.1
Arginine	1.2	2.8	5.9	4.1	4.1	4.3	4.4	4.0	3.4	1.9
Histidine	2.9	1.6	4.0	2.1	3.1	2.9	3.2	2.4	3.1	3.7
Lysine	11.5	11.8	12.8	6.8	8.2	8.9	8.7	6.5	6.5	6.2
Aspartic acid	18.7	11.4	10.9	9.4	7.1	8.4	8.5	7.7	4.9	4.0
Glutamic acid	12.9	19.3	16.5	12.3	22.4	22.5	24.3	19.8	23.2	22.9
Amide N	1.4	1.1	0.78	—	1.1	1.6	1.3	1.9	1.6	1.6

Source: Webb *et al.* (1974).

[a] Based on amino acid sequence of protein.

[b] No cysteine present.

Casein Fractions. In processing involving the use of skimmilk, it is highly important that the unique properties of the casein–protein complex be retained. From the chemical point of view, skimmilk is in the class of lyophilic colloids. The reason is that the protein complexes, which make up the dispersed phase, do not spontaneously coagulate, are in the right size range, and interact with and are stabilized by the solvent. Although the protein complex in milk can be separated from the liquid phase by centrifugation, it is unaffected by the gravity of the earth.

Casein is not a single protein. The results of research have indicated that it is made up of a number of different proteins that can be separated and have different properties. Casein has been separated into three main components, namely α_{s1}-casein, β-casein, and κ-casein. Of these α_{s1}-casein occurs in the largest amount, about 50%, is the best characterized protein present in the casein system, and has a molecular weight of 23,600. The molecule is a single-chain polypeptide with 199 amino acid residues together with 8 phosphate residues present as the phosphomonoesters of serine. It has been found that α_{s1}-casein is insoluble under normal conditions of temperature, pH, and ionic strength.

The protein occurring in the next largest amount is β-casein, about 33%. It is a molecule made up of a single chain, has a molecular weight of 24,500, and has five phosphoserine residues. The β-casein molecule lacks a secondary structure, which can be explained partially by the even distribution of proline. As in the case of α_{s1}-casein, β-casein is insoluble at room temperature. Ca^{2+} occurs at concentrations below those found in milk.

κ-Casein is present to the extent of about 15% and differs from the other two in that it is soluble over a very broad range of calcium-ion concentrations. Because of this calcium solubility, Waugh and von Hippel (1956) gave it the role of casein micelle stabilization. In addition, it is the protein of the casein group that can most easily be cleaved by rennin. It seems that κ-casein is important to the micelle structure since it stabilizes the calcium-insoluble α_{s1}- and β-caseins and is the principal site for action by rennin. The products formed are called para-kappa and macropeptide. κ-Casein is the only major component of the casein complex that contains cystine. The primary structure of κ-casein is not well characterized at present. It is, also, the only major component of the casein complex that contains carbohydrate. It is, therefore, a glycoprotein. All the carbohydrate connected with κ-casein is bound to the macropeptide. This is the very soluble part when rennin hydrolyzes the molecule.

Casein Micelles. According to Davies and White (1960), the total calcium content of skimmilk amounts to 30 mmoles/liter. On the other hand, these authors found only ~2.9 mmoles in the serum, prepared either by ultrafiltration or by centrifugation of skimmilk. It seems, therefore, that 90% or more of the calcium of skimmilk is connected in some way with the casein micelles.

TABLE 16.5. Average Composition of Warm Skimmilk Protein

	g/100 g Milk[a]	% Total Protein[a]
Colloidal casein	2.36	74
Serum casein	0.26	8
β-Lactoglobulin	0.29	9
α-Lactalbumin	0.13	4
Bovine serum albumin	0.03	1
Total immunoglobulins	0.06	2
Other proteins	0.06	2

[a] All values normalized to 3.2 g total protein/100 g milk.

Two distinct forms of ions are associated with the casein micelle. One of these is an outer system which, according to Boulet *et al.* (1970), is likely in the form of a charged double layer and also an inner system not washed away easily. While it is true that the casein micelle is an extremely porous and well-solvated system, and it is not unexpected that there would be occlusion of ions within this network, some actual complex formation might take place between the colloidal calcium phosphate and the casein. While it is not established at the present time, it seems that conditions in milk would permit the formation of an amorphous-calcium phosphate-caseinate complex. Indications are that such a complex is connected with maintaining the structural security of the casein micelle. The knowledge of the structure of the casein micelle is not complete, and many possible models have been suggested.

The casein micelles, the proteins in the dispersed phase of milk, account for 74% of the total protein in the skimmilk.

Casein micelles can be considered as swollen, microscopic polyelectrolyte gels containing more than 66% water, of which about one-quarter is bound chemically. Electron microscopy has shown that most of the micelle particles are about 1300 Å in size. The colloidal dispersion in milk is very stable at extremes of temperature and concentration. It substantially recovers its normal dispersion after freezing, desiccation, or concentration. This is important in processing from the point of view of the food technologist.

Whey Proteins. The bulk of the proteins in whey are in the groups known collectively as lactalbumin and lactoglobulin. A water-insoluble crystalline protein in salt-free water was prepared from the classical lactalbumin fraction (Palmer 1934). Named β-lactoglobulin, it is the most abundant of the whey proteins and is a simple protein made up of only amino acids. Unlike casein, β-lactoglobulin contains free sulfhydryl groups in the form of cysteine residues that seem to have a part in the formation of "cooked" flavor when milk is heated (Hutton and Patton 1952).

Present in amounts up to 50% of the non-casein protein of skimmilk, β-lactoglobulin has a monomer molecular weight of 18,000, but in this form it is found only below pH 3.5 and above pH 7.5. Between pH 3.5 and pH 7.5 it exists as a dimer and has a molecular weight of 36,000. A rather complete amino acid sequence for this molecule is known (Frank and Braunitzer 1967). The denaturation of β-lactoglobulin is influenced by the following factors: heat, increased calcium ion concentration, and pH above 8.6.

The other important protein in whey proteins is α-lactalbumin. Second in concentration to β-lactoglobulin, it contains no sulfhydryl groups, but has a high content of cystine.

α-Lactalbumin is the best-characterized protein of milk. It comprises up to 25% of the whey proteins, and ~4% of the total proteins of milk. α-Lactalbumin is a rather stable compound, has four disulfide cross-links, and a monomer molecular weight of 14,000. The complete amino acid sequence of this compound is known.

Of the other proteins present in the serum, serum albumin and immunoglobulin together with enzymes are found in skim milk to the extent of about 4% of total milk proteins.

Solubility. Since solubility of casein refers to the solubility of chemically unaltered hydrogen caseinate, this solubility occurs in solutions having the same isoelectric point as casein; considered usually pH 4.6, but possibly altered by neutral salts in solution. Casein solubility in water was found to 0.05 g/l at 5° C and 0.11 g/l at 25° C (Gordon and Whittier 1965).

Casein solutions rotate the plane of polarized light to the left. A solution of casein in an 8.3% solution of orthophosphoric acid showed a rotation of −86.6°.

Milk Lipids

Milk fat is composed mainly of triglycerides of fatty acids. Table 16.6 shows the composition of cow's milk lipids. The combined fatty acids in milk fat is made up of at least 60 of these acids. Most of them are present in rather small amounts. However, a few are present in significant quantities as shown in Table 16.7. It should be noted that in sheep's and goat's milk the capric and caprylic acids are higher. These acids have an important role in the flavor of cheese made from the milk of these two animals. In goat's milk the amount of capric acid is about 7.9% versus 3.04% for cow's milk and about 2.7% caprilic acid versus 1.06% for cow's. The manner in which the fatty acids of milk fat are associated in the triglycerides has not been entirely worked out.

Phospholipids of Milk. Tables 16.8 and 16.9 show the lipid content of milk and some milk products, and the classes of phospholipids present.

TABLE 16.6. Composition of Bovine Milk Lipids

Class of Lipid	% Total Milk Lipids
Triglycerides of fatty acids	95–96
Diglycerides	1.26–1.59
Monoglycerides	0.016–0.038
Keto acid glycerides (total)	0.85–1.28
Ketonogenic glycerides	0.03–0.13
Hydroxy acid glycerides (total)	0.60–0.78
Lactonogenic glycerides	0.06
Neutral glyceryl ethers	0.016–0.020
Neutral plasmalogens	0.04
Free fatty acids	0.10–0.44
Phospholipids (total)	0.80–1.00
Sphingolipids (less sphingomyelin)	0.06
Sterols	0.22–0.41
Squalene	0.007
Carotenoids	0.0007–0.0009
Vitamin A[a]	0.0006–0.0009
Vitamin D	0.00000085–0.0000021
Vitamin E	0.0024
Vitamin K	0.0001

Source: Webb *et al.* (1974).
[a] Based on the free alcohol.

RANCIDITY AND OFF-FLAVOR IN MILK

In the dairy industry rancidity in milk is considered to be the result of the presence of free fatty acids, specifically the lower volatile acids, that have been hydrolytically separated from milk fat by the lipases normally present in milk. This causes a flavor change, which is undesirable because it affects the salability of milk.

Another form of deterioration in milk and milk products results from autoxidation. The use of inert gas or vacuum packing of dry milks and refrigeration for butter is necessary to retard or prevent lipid deterioration. Some of the reactions taking place have been discussed in Chapter 5.

The following factors contribute to the development of off-flavor in dairy products—oxygen, light, and metals. Of the metals, cupric and ferrous ions are particularly active (the cupric ion is the stronger of the two), and nickel is somewhat less active.

Ascorbic acid (vitamin C) also has an effect in the development of off-flavors. The oxidation of ascorbic acid in fluid milk has been found to be an essential link in the chain of reactions that result in an oxidized flavor. This is perhaps caused by the H_2O_2 released under certain conditions during the oxidation of ascorbic acid (Chapon and Urion 1960).

TABLE 16.7. Acids Found in Large Quantities in Milk

Fatty Acid	Amount
Saturated	
Butyric acid (tetramoic)	2.79%
Caproic acid (hexanoic)	2.34
Caprylic acid (octanoic)	1.06
Capric acid (decanoic)	3.04
Lauric acid (dodecanoic)	2.87
Myristic acid (tetradecanoic)	8.94
Palmitic acid (hexadecanoic)	23.8
Stearic acid (octadecanoic)	13.2
Unsaturated:	
Oleic acid (9:10 octadecanoic)	29.6[a]
Linoleic acid (octadecadienoic)	2.11

Source: Kurtz (1965).
[a] Includes cis and trans isomers.

Saturated and unsaturated aldehydes in extremely small concentrations, can cause off-flavors described as follows: cardboard, fishy, oily, cucumber, tallowy, shrimp-like, green leaves, nutty, orange oil, and paint-like. Patton *et al.* (1959) found that 2,4-decadienal gives a deep fat or oil fried flavor at levels of 0.5 parts per billion. Carbonyls in autoxidized dairy products are shown in Table 16.10.

The development of some off-flavors and browning in dairy products is a result of protein–carbohydrate changes. The two main reactants involved in the browning of dairy products are lactose and casein, although under some conditions the whey proteins are involved. The Maillard reaction is the most important. A little browning, however, is also accomplished by the caramelization of lactose. The protein–carbohydrate changes are responsible for the formation of substances with undesirable flavors. This problem develops particularly in concentrated and dry forms of milk when held for rather long periods in storage.

TABLE 16.8. Phospholipid Content (Percentage) of Milk and Milk Products

Product	Phospholipids in Product	Fat in Product	Phospholipids in Fat
Whole milk	0.0337	3.88	0.87[a]
Skimmilk	0.0169	0.090	17.29
Cream	0.1816	41.13	0.442
Buttermilk	0.1819	1.94	9.378
Butter	0.1872	84.8	0.2207

Source: Webb *et al.* (1974).
[a] The reported value of 0.0869 appears to be a misprint.

TABLE 16.9. Classes of Phospholipids in Milk Fat

Phospholipid Class	Proportion of Total Phospholipids			
	Deutsch *et al.* (1952) (mole %)	Rhodes and Lea (1958) (mole %)	Koops (1958) (mole %)	Smith and Freeman (1959) (wt. %)
Phosphatidyl choline	33	33	30	32
Phosphatidyl ethanolamine	38	29	30	35
Phosphatidyl serine		10	10	
Phosphoinositides	—	5	6	Present
Plasmologens	—	3.3	3	—
Sphingomylein	23	19	25	24
Cerebrosides	—	—	—	6
Lysolecithin	—	—	—	Present
Unidentified	—	—	3–4	Present

Source: Kurtz (1965).

CHEESE AND CHEESE CHEMISTRY

Milk-Clotting Enzymes

Until recently, the enzyme used for the commercial manufacture of cheese was rennin, the gastric enzyme of the calf in the form of a crude extract, powder or paste. There is no longer enough of this enzyme available to satisfy the needs of the cheese industry and this has necessitated the use of other enzymes to take care of these needs. Rennin is still a very desirable enzyme for this purpose, and it is used as the standard of evaluation for other milk-clotting enzymes.

Proteolytic Enzymes. Most of the proteolytic enzymes will clot milk if the conditions are right. These enzymes can be obtained from plant sources, bacteria, fungi, and animal organs. The plant proteases have been rather unsatisfactory because most of them have very strong proteolytic properties, with the following undesirable results: reduced yields of cheese, pasty-bodies cheese, and bitter flavors. Extracts from the flower petals of *Cunara cardunculus* (cardoon) are used in Portugal for the preparation of Serra cheese, which is made from sheep's milk. However, when these extracts were used for other kinds of cheese, the results were not satisfactory.

Fungi. Milk-clotting enzymes are produced from *Endothia parasitica* (EP Protease). These enzymes are used in the manufacture of Emmentaler (Swiss) cheese. In this process, high cooking temperature (51.7°–54.4° C) is used and results in a cheese of high quality. It is likely that the enzyme is destroyed during the cooking process, and therefore, cannot affect the cheese during storage.

TABLE 16.10. Carbonyls Identified in Autoxidized Dairy Products

Product	Methyl Ketone	Alkanal	Alk-2-enal	Alk-2,4-dienal
Phospholipids of butter	—	C_2 to C_{18}[a]	C_5 to C_{15}	C_8, C_9
Butteroil	C_3, C_5, C_7, C_9, C_{11}, C_{13}, C_{15}	C_1 to C_{10}	C_4 to C_{11}	C_7
Butteroil	C_5, C_7	C_2 to C_9	C_5 to C_{10}	C_7, C_{10}
Fishy, butteroil	C_3, C_5, C_7, C_9, C_{11}	C_3 C_5 to C_{10}	C_3, C_5 C_6, C_8, C_9	C_7
Tallow, painty butteroil	C_7	C_5 to C_{10}	C_5 to C_{10}	C_7
Butteroil	C_3, C_5, C_7, C_9, C_{11}, C_{13}, C_{15}	C_1 to C_3 C_5 to C_{10}	C_4 to C_{11}	C_7, C_9 C_{10}, C_{11}
Skimmilk	—	C_2, C_6	C_4 to C_{11}	C_6 to C_{11}
Nonfat dry-milk	C_3, C_4	C_1, C_2 C_6 to C_{10} C_{12}, C_{14} Methylpropanal 3-methylbutanal	—	—
Dry whole milk	C_3, C_4, C_5, C_6, C_7, C_9, C_{11}, C_{13}, C_{15}	C_1 to C_3 C_5 to C_{10} C_{12}	C_5 to C_{11}	—
Washed Cream	C_7	C_5 to C_9	C_5 to C_9	C_7

Source: Parks (1965).
[a] Includes aldehydes released from plasmalogens.

Mucor pusillus var. Lindt produces a milk-clotting enzyme known as MP protease. This enzyme has been used as a rennin substitute in a number of cheeses and has given satisfactory results. Some workers found bitterness in cheese made with several microbial proteases including MP protease, but not in all cases. However, other workers prepared normal cheese including brick, Cheddar, mozzarella, and Parmesan with MP protease, and found bitterness in Cheddar only after 14 months of storage.

A protease obtained from *Mucor miehei* and known as MM protease, was tested in Europe and has been approved for use in the United States. MM protease was found to be completely satisfactory for Emmentaler. In tests made with Edam, Tilsit, and butter cheeses the results showed that the yields and qualities of the experimental and control cheeses made with rennet were similar. Sternberg (1971) was able to prepare MM protease in crystalline form, indicating that the enzyme is not metal-dependent and also that SH groups or serine were not associated with the active site. He found the enzyme to have a broad stability maximum between pH 4.0 and 6.0. Ottesen and Rickert (1970) showed the great stability of MM protease. Over 90% of the activity was retained after 8 days of incubation at 38° C and between pH 3.0 and 6.0. When incubated at pH 6 for 11 hr in 8 *M* urea, the enzyme lost no activity.

Embden-Meyerhof glycolysis scheme whereby glucose of glycogen is converted to lactic acid by animal muscle. Elucidation of these enzymatic pathways was the first important advance in modern metabolic biochemistry and involved many of the best researchers of the period. Research continues on specific mechanisms, physical chemistry, and interactions of the various steps even though most of the enzymes have been crystallized and the sequence of reactions is clear.

Biochemistry

The reactions of alcoholic fermentation are depicted in Fig. 5.1. The hexose (6-carbon) sugars being fermented are isomerized if necessary and phosphorylated to fructose-1,6-diphosphate which is split into two triose units. The triose units are converted to pyruvic acid and this is decarboxylated to acetaldehyde. The decarboxylation step is irreversible, as is hexokinase and, although the other steps theoretically are reversible, in practice the energy input necessary is too high for significant reversal of the phosphofructokinase or phosphopyruvic transphosphorylase steps. Sugar synthesis in the grape vine does use some of the same enzymes in the reverse direction, but avoids those reactions excessively endergonic by alternative routes.

The acetaldehyde is reduced to ethanol by accepting hydrogen from reduced nicotinamide adenine dinucleotide (NADH), the heat-stabile coenzyme with alcohol dehydrogenase. During the initial induction phase of alcoholic fermentation no acetaldehyde is present and 3-phosphoglycerate is converted to glycerol. This is one example of the diversion of sugar to products other than ethanol that explains why the Gay-Lussac equation represents only the theoretical maximum yield. As acetaldehyde accumulates it becomes the hydrogen acceptor (in place of dihydroxyacetone phosphate) and reacts with NADH to produce ethyl alcohol. During the rest of a normal fermentation this process predominates and little glycerol is formed. If acetaldehyde is not available (when bound with excess sulfite, for example) glycerol is produced instead of ethanol.

In the presence of a high concentration of sulfur dioxide in acid solution, acetaldehyde, carbon dioxide, and glycerol are the primary products and alcohol a by-product. If the sulfite solution is alkaline, acetaldehyde, glycerol, alcohol, and carbon dioxide are all produced. Other types of fermentation have been reported. See Amerine (1965). The glycolytic sequence clearly shows the complexity of the system. It also shows how glycerol and acetaldehyde may accumulate as by-products.

Wine fermentations ordinarily occur in a more or less complicated mixture of many different microorganisms. Certain lactic acid bacteria,

1. Glucose $\xrightarrow[\text{(Mg}^{++}\text{, ATP} \rightarrow \text{ADP)}]{\text{(hexokinase, EC 2.7.1.1)}}$ glucose-6-phosphate

1a. Fructose $\xrightarrow[\text{(Mg}^{++}\text{, ATP} \rightarrow \text{ADP)}]{\text{(hexokinase, EC 2.7.1.1)}}$ fructose-6-phosphate

2. Glucose-6-phosphate $\xrightleftharpoons{\text{(phosphoglucoisomerase, EC5.3.1.9)}}$ fructose-6-phosphate

3. Fructose-6-phosphate $\xrightarrow[\text{(Mg}^{++}\text{, ATP} \rightarrow \text{ADP)}]{\text{(phosphofructokinase, EC 2.7.1.11)}}$ fructose-1,6-diphosphate

4. Fructose-1,6-diphosphate $\xrightleftharpoons[\text{(Zn}^{++}\text{, Co}^{++}\text{, Fe}^{++}\text{, Ca}^{++}\text{, or K}^{+}\text{)}]{\text{(fructose diphosphate aldolase, EC 4.1.2.13)}}$ D-glyceraldehyde-3-phosphate + dihydroxyacetone phosphate

. D-Glyceraldehyde-3-phosphate $\xrightleftharpoons{\text{(triosephosphate isomerase, EC 5.3.1.1.)}}$ dihydroxyacetone phosphate

a. Dihydroxyacetone phosphate $\xrightleftharpoons[\text{(H}^{+}\text{ + NADH} \rightarrow \text{NAD}^{+}\text{)}]{\text{(glycerolphosphate dehydrogenase, EC 1.1.1.8)}}$ glycerol-3-phosphate

b. Glycerol phosphate $\xrightarrow{\text{(phosphatase)}}$ glycerol + H_3PO_4

. D-Glyceraldehyde-3-phosphate+H_3PO_4 $\xrightleftharpoons[\text{(NAD}^{+} \rightarrow \text{NADH + H}^{+}\text{)}]{\text{(3-phosphoglyceraldehyde dehydrogenase, EC 1.2.1.12)}}$
1,3-diphosphoryl-D-glycerate

1,3-Diphosphoryl-D-glycerate $\xrightleftharpoons[\text{(Mg}^{++}\text{, ADP} \rightarrow \text{ATP)}]{\text{(phosphorylglycerate kinase, EC 2.7.2.3)}}$ 3-phosphoryl-D-glycerate

3-Phosphoryl-D-glycerate $\xrightleftharpoons[\text{(2,3-diphosphoryl-D-glycerate)}]{\text{(phosphorylglyceromutase, EC 2.7.5.3)}}$ 2-phosphoryl-D-glycerate

2-Phosphoryl-D-glycerate $\xrightleftharpoons[\text{(Mg}^{++}\text{)}]{\text{(enolase, EC 4.2.1.11)}}$ phosphorylenolpyruvate

Phosphorylenolpyruvate $\xrightarrow[\text{(Mg}^{++}\text{, K}^{+}\text{, ADP} \rightarrow \text{ATP)}]{\text{(pyruvate kinase, EC 2.7.1.40)}}$ pyruvate

Pyruvate $\xrightarrow[\text{TPP}]{\text{(pyruvate decarboxylase, EC 4.1.1.1)}}$ acetaldehyde + CO_2

. Pyruvate $\xrightleftharpoons[\text{(Zn}^{++}\text{, NADH + H}^{+} \rightarrow \text{NAD}^{+}\text{)}]{\text{(lactic dehydrogenase, EC 1.1.1.27)}}$ lactic acid

Acetaldehyde $\xrightleftharpoons[\text{(NADH + H}^{+} \rightarrow \text{NAD}^{+}\text{)}]{\text{(alcohol dehydrogenase, EC 1.1.1.1)}}$ ethanol

P, ATP. Di- and triphosphates of adenosine.
D^{+}, NADH. Oxidized and reduced nicotinamide adenine dinucleotide. (NAD^{+} was also called oenzyme I or DPN).
P. Thiamin pyrophosphate.

. 5.1. CHEMICAL REACTIONS IN ALCOHOLIC FERMENTATION

Animal Sources. ***Porcine Pepsin.*** This enzyme has been recommended as a satisfactory substitute for part of the rennet in making a number of varieties of cheese. Earlier it was reported that Cheddar cheese made with pepsin developed bitterness, but this has not been substantiated.

The hog stomach mucosa secretes pepsin as a catalytically inactive pepsinogen that has a molecular weight of 40,400. Pepsinogen was found to be stable in slightly alkaline or neutral solutions. Below pH 5.0 pepsin catalyzes the conversion of pepsinogen to pepsin. Pepsin is a single chain protein (van Vunakis and Herriott 1957), and optical rotary dispersion indicates very little helical coiling even though a globular particle is produced by folding. It is unlikely that hydrogen bonds play an important role in stabilizing the molecule. Three disulfide bonds are present in pepsin, and at least one of these is not vital to enzymatic activity. It is known also that a high number of ionizable groups are present in pepsin which contribute electrostatic charges.

In the cheese-making process, pepsin activity is reduced or eliminated subsequent to or during the coagulation of the milk. Experiments have determined that the bulk or maybe all of the porcine pepsin used as a coagulant is inactivated during the manufacture of Cheddar cheese. Mikelsen and Ernstrom (1972) found the mixtures of porcine pepsin and rennet between pH 5.0 and 6.0 were stable, but above that pH pepsin activity was lost. This loss was found to be caused entirely by pepsin instability. Below pH 5.0 pepsin destroyed rennin activity.

Bovine Pepsin. According to Kassel and Meitner (1970), all the bovine pepsins are about one-third as active as porcine pepsins on the synthetic substrate acetylphenylalanyl diiodotyrosine and about two-thirds as active on hemoglobin substrate. They have milk-clotting action at pH 5.3.

According to Green (1972) Cheddar cheese made from bovine and swine pepsins were only slightly inferior in quality and intensity of Cheddar cheese flavor to the rennet cheeses. The ratios of milk-clotting activity to general proteolytic activity were high for rennet and bovine pepsin and low for swine pepsins. Bovine mucosa gave low milk-clotting activities compared with calf stomach.

According to Fox and Walley (1971) the analysis of Cheddar cheese produced using equivalent amounts of commercial Hansen's rennet, commercial Hansen's half and half (rennet/pig pepsin), and bovine pepsin showed no important differences in losses of fat and protein in whey, rate and extent of soluble protein formation, or electrophoretic patterns of ripened cheese. Development of pH 4.6 soluble nitrogen was slower in cheese made from bovine pepsin. Electrolysis of proteolysis products showed that bovine pepsin produced a well-resolved peptide not produced by either of the other two coagulants. No modification of the manufacturing process is necessary. No signs of bitterness were detected in the cheeses.

TABLE 16.11. Typical Analyses of Cheese

Type	Cheese	Moisture (%)	Fat (%)	Protein (%)	Fat in Dry-matter (%)	Salt (%)	Ash (%)	Lactose (%)	Calcium (%)	Phosphorus (%)
Soft-unripened Low -fat	Cottage	79.0	0.4	16.9	1.9	1.0	0.8	2.7	0.09	0.05
	Creamed cottage	78.3	4.2	13.6	19.3	1.0	0.8	3.3	0.09	0.05
	Quarg	72.0	8.0	18.0	28.5			3.0	0.30	0.35
	Quarg (high fat)	59.0	18.0	19.0				3.0	0.30	0.35
Soft-unripened high	Cream	51	37	8.8	75.5	1.0	1.2	1.5–2.1	0.08	0.06
	Neufchatel	55	23	18.0	51.1	1.0	2.0			
Soft-ripened by surface bacteria	Limburger	46	27	21.5	50.0	2.0	3.6	0–2.2	0.5	0.4
	Liederkranz	52	28	16.5	58.3	1.5	3.5	0	0.3	0.25
		53	25.5	16.8	54.2	1.7	3.9			
Soft-ripened by external molds	Camembert	51	26	20.0	53.0	2.5	3.8	0–1.8	0.6	0.5
	Brie	45	30	21.6	54.5	2.0	4.0	0–2.0	0.6	0.4
Soft-ripened by bacteria, preserved by salt	Feta	57	24	20	55.8	5.0				
	Domiati	55	25	20.5	55.5	4.8				
Semi-soft, ripened by bacteria with surface growth	Brick	42	31	21	53.4	2.0	4.2	0–1.9	0.6	0.4
	Muenster	44	28	25	50.0	1.8			0.5	0.35

Semi-soft, ripened by internal molds	Blue	41.5	30.5	21.5	52.1	4.0	6.0	0–2.0	0.7	0.5
	Roquefort	40.0	31.0	22.0	50.1	4.2	6.0		0.65	0.45
	Gorgonzola	36.0	32.0	26.0	50.0	2.4	5.0			
Hard, ripened by bacteria	Cheddar	37.0	32.0	22	50.8	1.6	3.7	0–2.1	0.7	0.5
	Colby	39.0	31.0	21	50.8	1.7	3.6	0	0.7	0.5
Hard, ripened by eye-forming bacteria	Swiss	37	28	27.5	44.4	1.3	3.8	0–1.7	1.0	0.6
	Edam	39	25	28.0	40.9	2.0	4.4	0–1.0	0.75	0.45
	Gouda	36.5	29	25.0	45.6	1.7		0–1.0	0.60	0.38
Very hard, ripened by bacteria	Parmesan	30.0	26.0	36.0	37.1	1.8	5.1	0–2.9	1.1	0.8
	Romano	32.0	30.0		44.1	4.6	5.4	0		
Pasta filata (stretch cheese)	Provolone	38.0	28	28	45.1	3.0	4.0	0	0.7	0.6
	Mozarella	53	18	22	38.3	1.0		0.3		
Low-fat or skim milk cheese (ripened)	Euda	56.5	6.5	30.0		2.6		1.0		
	Sapsago	37.0	7.4	41.0		4.5				
Whey cheese	Ricotta	72.0	10.0	12.5	35.7	1.2	3.6	3.0		
	Primost	13.8	30.2	10.9	35.0	—		36.6		
Process cheese	Process Cheddar	39.5	31.5	22.2	52.0	1.7	4.9	0	0.7	0.7
	Process cheese food	43.0	24.0	20.5	42.1	1.0		7.0	0.6	0.6
	Process cheese spread	48.5	21.5	16.0	41.1	1.0		7.0	0.8	0.8

Source: Webb *et al.* (1974).

As a result of the practice of extracting the stomachs from older calves and adult cattle, the use of bovine pepsin as a coagulant has increased.

Rennet and Rennin. As was mentioned earlier, calves' stomachs (vells) are the source of rennin. Crude rennet extract contains active rennin together with an inactive precursor known as prorennin. When acid is added to the extract, the conversion of prorennin to rennin is facilitated, which in turn, produces an extract of maximum activity. The pH recommended for this purpose is 5.0. Below this value the stability of rennin is poor in the presence of sodium chloride, and yields are, therefore, lower.

Rennin has been prepared in crystalline form, as block-shaped crystals or needles. It has been observed that needle-shaped crystals change into rectangular plates during storage at −15°C (Foltmann 1959). Ernstrom (1958) worked out a good method for the preparation of these crystals. Crystalline rennin has been shown to be heterogeneous by a number of methods which included polyacrylamide gel electrophoresis (Asato and Rand, 1971). Prorennin is heterogeneous also. Two minor and two major rennins and one minor and two major prorennins are also present. Commercial rennet extracts contain substantial amounts of sodium chloride. It is necessary to activate them slowly at pH 5.0 to obtain maximum yields (Rand and Ernstrom 1964). These workers noted that at pH 5.0 the activation of prorennin was predominantly autocatalytic, especially in the presence of 1.7 *M* sodium chloride. However, Foltmann (1966) showed that the course of the activation at pH 4.7 was not purely autocatalytic.

During the activation of prorennin, the peptides are split from the N-terminal end of prorennin and at the same time the molecular weight is reduced from about 36,000 to 31,000. From the structural data obtained prorennin must be a single-chain protein which contains three disulfide bridges that remain in the rennin part of the molecule after activation. The amino acid sequence in the core of the rennin molecule is not known, but that of the N-terminal end is known, as are those of the region of each sulfide bridge. Prorennin has been found to be stable at room temperature from pH 5.3 to 9.0. The optimum stability for rennin is between pH 5.3 and pH 6.3 (Foltmann 1959).

The fact that rennin can be digested by other proteolytic enzymes is very important because of the commercial availability of blends of rennin with other milk-clotting enzymes such as porcine pepsin, which was found to be most stable at pH 5.5 (Mickelson and Ernstrom 1972).

An investigation of the rennin clotting of milk under the electron microscope was carried out by Hostettler and Imhoff (1951, 1955). They noticed that clotting started clumping of the nearly spherical caseinate particles to form agglomerates, which changed by continued cross-linking into a network of fibrous structures. At the end of the final coagulum was an irregular structure of paracaseinate particles in a three-dimensional threadlike network. The milk fat globules and whey were within the paracasein structure.

Coagulation

Cheese is composed of the casein and usually almost all of the fat, insoluble salts, and colloidal materials, along with part of the serum of the milk used. This serum contains lactose, whey proteins, soluble salts, vitamins, and other milk components. The preparation of cheese does not eliminate any of the principal milk solids; they are made available in a concentrated form less subject to spoilage. However, the proportional interrelationships of all the basic components of milk are changed. The coagulation of the milk is affected by the addition of rennet, or by the acid formed by bacteria, or by both. A starter, in the form of a bacterial culture is usually added. The coagulate of the casein brings about curdling, entraping as much as 90% of the fat. The curd is then cut into cubes and the excess moisture is expelled by the development of acid, by the continued action of rennet, by a moderate or high degree of heating, by stirring, or, with harder varieties, by pressing the curd. The curd is then shaped into forms, the surface is salted if salt has not already been added to the curd in the vat, and the cheese is stored in curing rooms at controlled temperature and humidity. More than 400 varieties of cheeses are known, but there are actually only about 14 distinct types of cheese in existence. The variations in cheese are the result of the type and composition of milk used, substances added, and the process of manufacturing and curing.

In cheese making, the coagulum formed by rennet has elasticity, and shrinkage takes place with the separation of whey. This shrinking proceeds with the development of acidity. When coagulation is produced by the addition of acid alone, the curd is not elastic. It is, however, gelatinous and fragile and tends to shatter more and contract less than when formed by rennet. The coagulum from rennet envelops with it most of the fat and the insoluble salts of the milk; however, when the curd is formed by means of acid instead of rennet, the insoluble salts become soluble by action ofthe acid and stay mostly in the whey. About one-quarter of the phosphorus is in organic combinations and because calcium becomes soluble more quickly than the phosphorus and other components of the ash, the Ca/P ratio and the percentage of calcium are rather low in cheese made under conditions of high acidity during drainage, especially if the acid curd is washed.

Cheddar cheese ordinarily retains about 60–65% of the calcium and about 50–60% of the phosphorus from the milk (Irvine *et al.* 1945A; Mattick 1938; McDowall and Dolby 1935). However, higher values for calcium and lower for phosphorus have been found. Retention of these elements in soft cheeses, cottage, Neufchâtel, and cream cheese, is considerably smaller, ordinarily 20% for calcium and 37% for phosphorus (McCammon *et al.* 1933). Cottage cheese was shown to contain an average of about 0.91% of calcium and 0.18% of phosphorus, but variation was noted. Stilton cheese showed 0.20–0.258% of calcium and 0.247–0.34% of phosphorus (Mattick 1938).

Hard cheeses contain about 30–40% moisture, while semisoft cheeses contain about 39–50% moisture, and soft cheeses from 50–75% moisture, although some varieties may be as high as 80%.

The temperature of cogulation is important and, although temperature may vary for different kinds of cheese, for most optimum temperature is between 22° and 35° C. The temperature of coagulation for soft cheese is ordinarily lower than that for hard cheese. The milk used for Cheddar cheese is treated with rennet, or "set," at 30° or 31° C. The amount of rennet added and the temperature used control the formation of acid to yield the necessary type of curd.

It has been suggested that the control of the development of acidity has more influence on the quality of the cheese than any other factor. It has been found that acidity can be controlled by using pasteurized milk, by varying the proportion of the starter in accordance with the characteristics of the milk, and by allowing the milk to ripen for 1 hr after the addition of the starter. Wilson *et al.* (1945) found that a pH value between 5.40 and 5.50 at the time of milling the curd is desirable for making the best cheese. According to these authors, Cheddar cheese of inferior quality is obtained when the hydrogen-ion concentration or the titratable acidity is excessive at any given stage during manufacturing from raw or pasteurized milk. According to Hood and Gibson (1948), acid and bitter flavors found frequently in raw milk cheese are usually a result of overdevelopment of acidity during the manufacturing process.

Heating or Cooking

In the manufacture of Swiss cheese, less time is consumed and higher heating is used than for Cheddar (between 51° and 54° C for Swiss and about 38° C for Cheddar). The heating step accomplishes several things: it speeds the expulsion of whey from the curd, alters the texture to increase compactness, and reduces the number of openings, increases the elasticity, and modifies the bacterial flora. The acidity increases only slightly, from pH 6.57 (average) to 6.48 (average), with at least part of the lowering of the pH caused by physicochemical rather than bacterial factors (Frazier *et al.* 1934). Acidity develops rapidly while the cheese cools and drains on the press, primarily from the activity of *Streptococcus thermophilus* (Frazier *et al.* 1935A.). After a few hours, acidity increases from the activity of *Lactobacillus bulgaricus* or *L. lactis,* closely related types (Tittsler *et al.* 1954). Cheese of good quality reaches a range of acidity between pH 5.0 and 5.3, or better between pH 5.10 and 5.15, in 1 day (Burkey *et al.* 1935; Frazier *et al.* 1935B). Since the formation of propionic acid is the primary cause of production of the eyes in Swiss cheese, it is necessary that this step proceed normally. When the pH is higher than 5.3 at 21 hr, the cheese ordinarily develops too many eyes.

Curing of Cheese

Curing is the methods of handling—care and treatment from manufacturing to marketing—including control of temperature, humidity, and sanitation.

Wilson *et al.* (1941) reported that cheese curd stored from 3 to 4 months at 10° C prior to low-temperature storage had better quality than cheese stored continuously for 6 months at 1° C. They found also, that cheese made properly from good quality milk could be cured at temperatures up to 10° C with likely development of good flavor, but cheese made from milk of poor quality needed a lower curing temperature. Wilson *et al.* (1945) found that for Cheddar cheese made from pasteurized milk, a curing temperature up to 10° C and perhaps somewhat higher insured suitable ripening, but a temperature as low as 1° C is too low for ripening but suitable for storing ripened cheese and retarding the formation of undesirable flavors.

Cheese Ripening. Ripening of cheese consists of chemical and physical changes which take place in cheese while curing. Proteins, fat, and lactose are the important components of milk that show physical and chemical changes during the ripening process. Rapid ripening is promoted by high moisture content, the flavor is improved by added salt, and at the same time biological activity is retarded. The formation of biacetyl is assisted by the presence of citric acid. The important chemical changes taking place according to Davis (1941) include (1) fermentation of lactose to lactic acid, some small amounts of acetic and propionic acids, and carbon dioxide, (2) breakdown of proteins, and (3) some breakdown of fat with liberation of fatty acids. These changes are largely enzymatic.

The chemical and physical changes taking place in Cheddar cheese during ripening cause the cheese to change from a tough and curdy state to one soft and mellow. During this process, the insoluble nitrogenous components undergo change to soluble forms. In the course of this progressive proteolysis, the paracasein and the lesser proteins are slowly converted to simpler nitrogenous compounds, including proteoses, peptones, amino acids, and ammonia.

A considerable amount of work has been done on the flavor of Cheddar and other cheeses. A number of keto acids and also neutral carbonyl compounds were found in Cheddar and other cheeses. It should be noted that for each neutral carbonyl, a corresponding intermediate precursor in the group of acidic carbonyls is possible. Pyruvic and α-ketoglutaric acids were found to be important components of all varieties. α-Acetolactic acid, which is the immediate precursor of acetylmethylcarbinol, was found in all cheeses with the exception of Camembert cheese. Day *et al.* (1960) identified 10 volatile carbonyls positively, and four tentatively, in Cheddar cheese. This completely reproduced mixture gave a noticeable but not typical, cheese-like odor. A synthetic mixture of carbonyls with acetic, butyric, caproic, capric, and 3-mercaptopropionic acids yielded a more cheese-like odor. However, this aroma was still not complete. Patton *et al.* (1958) found that of the compounds identified in Cheddar cheese aroma, dimethyl sulfide was the only one considered to be obviously and directly important to cheese aroma, with the possible exception of diacetyl and 3-hydroxybutanone. Other workers (Kristoffersen and Gould 1960) found that the concentrations of ammonia, free amino acids, and free fatty acids in Cheddar cheese increased continuously during the ripening process, but the concentration of hydrogen sulfide fluctuated. Definite differences have been found in the amounts of hydrogen sulfide and free fatty acids in commercial Cheddar cheese made from raw milk and Cheddar cheese made from pasteurized milk.

Salt is added to cheese for two main purposes: to control ripening and to improve flavor. Experiments have shown that salt has a desiccating effect also, increasing firmness especially in rind formation, which in turn makes the cheese more stable in handling. However, the amount of salt is important. Tustin (1946) found that too little salt in Cheddar cheese causes a weak and pasty body, abnormal ripening, and increased shrinkage in curing, while too much salt causes a dry, brittle body along with cracking of the rind. It has been found that oversalting of Swiss cheese causes slow and insufficient formation of the eyes. In addition, other defects including discoloration and mold damage are possible. The interested student should read the review of the literature relating to the chemical and microbiological aspects of the ripening of Cheddar cheese (Marth 1963).

FERMENTATION IN MILK

In the manufacture of cheeses, lactic acid bacteria are extensively used as starters to begin and control acid formation, and also to initiate the necessary lactic acid fermentation. These bacteria are also beneficial in the manufacture of butter. They inhibit the growth of undesired organisms and assist in the development of desirable aroma and flavor. Lactic acid bacteria work similarly in the manufacture of sour cream, margarine, and cultured buttermilk. Whereas other organisms may be present, the predominant organisms present in the lactic starter are *Streptococcus lactis* and/or *S. cremoris.* They produce large amounts of dextrolactic acid together with lesser amounts of acetic acid.

A small amount of biacetyl (0.05 mg or less per 100 g) makes a contribution to the characteristic flavor of Cheddar cheese. More biacetyl than this, however, can cause a flavor defect. *Streptococcus lactis* and *Lactobacillus citrovorum* produce diacetyl when present in the same culture. Both diacetyl and larger amounts of acetylmethylcarbinol are formed, but neither compound is stable in the culture and tends to be reduced to 2,3-butylene glycol. The formation of 2,3-butylene glycol is assisted by optimum conditions for the growth of the culture.

MILK PRODUCTS OTHER THAN CHEESE

The manufacture of most of the commercial milk products is the result of separation of one or more of the components from the milk; one of these components is separated and either it or the remaining mixture or both may be commercially important. Since this separation is not ordinarily complete, most of the products made from milk contain some or all of its original components. The cream separator, however, leaves the bulk of the fat in the cream and most of the other compounds in the skimmed milk.

Butter

Butter is the fatty material from whole milk; it must contain not less than 80% milk fat. The usual composition is fat, 80.5%; moisture, 16.5%; curd, 0.7%; and salt, 2.3%. Salt is omitted from unsalted (sweet) butter.

Butter in some form has been made since the dawn of history. The basis of butter production is mechanical agitation to destabilize the liquid-phase emulsion. The churning is started by agitation together with the incorporation of many small air bubbles. Small clumps, which are "centered" by liquid fat exuded from the fat globules, are formed when partially denuded fat globules collect at the fat/plasma interface. The air bubbles collapse when a part of this hydrophobic liquid fat spreads over their surface. The monomolecular layer of film fat is dispersed in the plasma phase as colloidally dispersed fat particles. This accounts for about 25% of the lipids found in buttermilk. The gathering together of the fat globules, prompted by the forces of churning and refloatation proceeds until butter granules are produced. A considerable amount (about one-half) of the fat-globule membrane material goes into the buttermilk.

Continuous buttermaking processes are widely used because of greater manufacturing efficiencies. The Cherry-Burrell and Creamery Package processes are American processes. The Cherry-Burrell process reheats and agitates the first cream to obtain partial destabilization of the lipid phase. A reseparation step achieves complete destabilization. In the Creamery Package process, homogenization is used to destabilize the reseparated cream. In either process the lipid and aqueous phases are separated. It is necessary, therefore, to be cautious in the remixing step to obtain homogeneity in the final product. A series of chill workers are fed the standardized mixture. After this it goes to working units where the congealed mixture is given the proper consistency, which means a complete absence of globular fat in the continuous phase and by large crystals or crystal aggregates.

Consistency of Butter. The rheological properties of butter include texture, structure, firmness, and hardness. Important to obtaining the desired qualities are thermal and mechanical treatment of the butter, composition of the milkfat, and temperature treatment of the cream.

Evaporated Milk

Evaporated milk is prepared by removing part of the water from whole milk by the use of vacuum treatment—2 $^{1}/_{10}$ lb of fresh milk is the equivalent of 1 lb of evaporated milk. It must contain no less than 7.9% milk fat and also not less than 25.9% of milk solids.

Dry Whole Milk

Few standards exist for dry whole milk. In the making of this product the milk is first evaporated under vacuum so that it is concentrated to about one-third of the original. After evaporation, it is spray-or drum-dried. An average composition is 2% moisture, 27.5% fat, and 26.4% protein. One pound of this dried product is usually equivalent to 8 lb of fresh milk.

Sweetened Condensed Milk

Sweetened condensed milk is prepared by adding about 18 lb of sugar to 100 lb of milk. The mixture is then concentrated by vacuum. Since sugar in this strength is a preservative, the product is canned without sterilization. According to Federal standards, this product should contain not less than 8.5% milk fat and not less than 28.0% total milk solids.

Malted Milk

The preparation of malted milk involves concentrating a mixture of milk and an extract made from a mash of ground barley malt and wheat flour to produce a solid which is ground to powder. Malted milk powder contains not less than 7.5% of milk fat nor more than 3.5% moisture. An average composition is 2.6% moisture, 8.3% fat, 14.7% protein, 70.8% carbohydrate, and 3.6% ash. The carbohydrate is made up of 20% lactose, 50.5% maltose and dextrin, and 0.3% fiber.

Cream

Cream is a mixture of milk fat with some of the other solids and water from the milk. Fat in cream varies from 11.5% in "half and half" to 36% in heavy cream. A product made for use in manufacturing is known as plastic cream and contains 80% fat. The designation of the cream is by the amount of fat it contains—for example, 18% cream contains 18% milk fat.

Skimmilk

While there is no Federal standard for skimmilk, the percentage of all the components of skimmilk except fat are proportionally larger than in whole milk from which it is made. A usual composition (Bell and Whittier 1965) is water, 90.5%; fat, 0.1%; protein, 3.6%; lactose, 5.1%; and ash, 0.7%.

Nonfat Dry Milk

This product is obtained when water and fat are removed from milk and is made up of milk proteins, lactose, and mineral substances, all in the same relative proportions as in milk. One pound of nonfat dry milk is approximately equal to 11 lb of skimmilk. The average composition is moisture, 3.0%; fat, 0.8%; protein, 35.9%; lactose, 52.3%; and ash, 8.0%.

Buttermilk

Two kinds of buttermilk are available: one is the byproduct resulting when cream is churned to make butter, and the other is made from skimmilk which is acted upon by a culture of lactic acid bacilli (usually called cultured buttermilk). These products are, therefore, similar to skimmilks except that they contain lactic acid, 0.1% in buttermilk from the churning process to 0.9% in some cultured buttermilks. The average composition for cultured buttermilk follows: water, 90.5%; fat, 0.1%; protein, 3.6%; lactose, 4.3%; lactic acid, 0.8%; and ash, 0.7%.

Frozen Desserts

This group of products includes ice cream, frozen custard, milk ice, and sherbet. Most states require not less than 10% of milk fat in plain ice cream, while required milk fat in nut, fruit, or chocolate ice cream is 8%. Frozen custard has egg yolk in addition. Total milk solids in ice cream are usually 20%. Sherbet contains about 4% milk solids, 30% sugar, and some citric acid. Ice milk varies from 2 to 7% milk fat, but aside from this it is much like ice cream.

DETERMINATION OF FAT IN DAIRY PRODUCTS

Originally the fat content of a sample of milk was determined by extraction with ethyl ether with the Soxhlet extractor, a time-consuming process. Various modifications of parts of this method were tried, principally involving the drying of the sample previous to the extraction. After 12–16 hr, the extracted material was dried and weighed.

The Röse–Gottlieb method of wet ether extraction is employed for fat determination and is quite satisfactory. It can be used for milk samples or for such products as evaporated milk; however, for evaporated milk, water must be added so that satisfactory extraction can be obtained. Ammonium hydroxide is added to neutralize the acids and facilitate extraction, and then the sample is treated with ethyl alcohol, ethyl ether, and petroleum ether. The alcohol facilitates extraction, and the petroleum ether is added to cut down the quantity of alcohol and water in the ether layer which holds the fat. Originally, the Röhrig tube was the equipment used, but this has since been replaced with the Mojonnier flask, specially shaped, which is more accurate, and can be centrifuged to break emulsions.

The Babcock test, devised for rapid commercial fat determinations in milk and cream, uses sulfuric acid to destroy interfering substances in the sample, while leaving the fat unharmed. The whole is centrifuged, the fat is brought up into the calibrated neck of the Babcock bottle, and the determination is read directly in percentage of fat.

TOTAL SOLIDS ANALYSIS

Total solids can be obtained by drying, but conditions must be carefully controlled if accurate results are to be obtained. The reasons for this are varied. (1) If the fat undergoes some hydrolysis, some volatile fatty acids could be lost. (2) If heating continues too long, oxidation of unsaturated fatty acids could take place which would increase the weight. Furthermore, this oxidation could result in the loss of some volatile oxidation products. (3) Changes which affect the weight in proteins can take place. (4) Finally, lactose heated together with phosphates can result in a water loss because of interactions between the two compounds. For these reasons, the rigid standard conditions for the determination of total solids is necessary.

SUMMARY

Milk is quite complex in chemical composition and contains a large number of substances.

The disaccharide lactose is the main carbohydrate found in milk. It can be manufactured from cheese whey. While β-lactose is sweeter than α-lactose, it is not advantageous to use it because the small difference is quickly eliminated by mutarotation.

Milk proteins are made up of three fractions. However, each of these fractions comprises many individual proteins. Casein is the largest group of proteins in milk.

Phosphorus in casein is considered to be bound in ester linkages with the hydroxyl groups of threonine and serine. Work on the ester linkages indicates that the probable bond is the *O*-monophosphate ester linkage and that phosphorus is bonded the same way in whole casein, α-casein, and β-casein.

Casein is not a single protein. It is made up of three main components, α_{s1}-casein, β-casein, and κ-casein. α_{s1}-Casein is found in the largest amount. κ-Casein is important to the micelle structure since it stabilizes the calcium-insoluble α_{s1}- and β-caseins and is the principal site for action by renin.

The casein micelles, the proteins in the dispersed phase of milk, account for 74% of the total protein in skimmilk.

Other important proteins are present in whey proteins. These are β-lactoglobulin and α-lactalbumin.

Milk lipids are composed mainly of triglycerides of fatty acids. The combined fatty acids in milk fat is made up of at least 60 of these acids. Most of them are present in rather small amounts. In sheep's and goat's milk capric and caprylic acids are higher than in cow's milk. These acids have an important role in the flavor of cheese made from the milk of these two animals.

Hydrolytic and oxidative rancidity in milk is undesirable since either effects its salability. The following factors contribute to the development of off-flavor in dairy products: oxygen, light, and metals.

The oxidation of ascorbic acid (vitamin C) in fluid milk has been found to be an essential link in the chain of reactions which result in an oxidized flavor. This may be caused by H_2O_2 released under certain conditions during the oxidation of ascorbic acid.

Aldehydes, saturated or unsaturated, in extremely small concentrations, can cause a variety of off-flavors.

The Maillard reaction is the most important in the development of browning in dairy products.

Milk-clotting enzymes are very important in the making of cheese. Rennin was the enzyme used for this purpose, but now, because of the shortage of rennin, other enzymes are being used.

The coagulation process is important in the manufacture of cheese. Cheese is composed of the casein and usually almost all of the fat, insoluble salts, and colloidal materials, along with part of the serum of the milk used. The preparation of cheese does not eliminate any of the principal milk solids.

In cheese making the coagulum formed by rennet has elasticity, and shrinkage takes place with the separation of whey.

Processing procedures vary for different kinds of cheese.

Curing and cheese ripening are of importance in preparing the cheese for market.

Biacetyl amounting to 0.05 mg or less per 100 g makes a contribution to the characteristic flavor of Cheddar cheese. More of it, however, causes a flavor defect.

A number of dairy products other than cheese are important.

ACKNOWLEDGMENT

The author would like to acknowledge "Fundamentals of Dairy Chemistry," 1st edition, 1965, by B. H. Webb and A. H. Johnson, and 2nd edition, 1974 by B. H. Webb, A. H. Johnson, and J. A. Alford, from which some information in this chapter was adapted.

BIBLIOGRAPHY

ADACHI, S., and PATTON, S. 1961. Presence and significance of lactulose in milk products: A review. J. Dairy Sci. *44*, 1375–1393.

ALAIS, C., and JOLLÈS, P. 1961. Comparative study of the caseinoglycopeptides obtained after rennin digestion of the caseins of the milk of cow, sheep, and goat. II. Study of the non-peptidic part. Biochem. Biophys. Acta *51*, 315–322. (French)

ANANTAKRISHNAN, C. P., and HERRINGTON, B. L. 1948. Trace sugars in milk. Arch. Biochem. *18*, 327–337.

ANDERSON, D. F., and DAY, E. A. 1966. Quantitation, evaluation, and effect of certain microorganisms on flavor components of blue cheese. J. Agric. Food Chem. *14*, 241–245.

AOAC. 1980. Official Methods of Analysis, 13th Edition. Assoc. of Official Analytical Chemists, Washington, DC.

ASATO, N., and RAND, JR., A. G. 1971. Resolution of prorennin and rennin by polyacrylamide gel electrophoresis. Anal. Biochem. *44*, 32–41.

AVIS, P. G., BERGEL, F., and BRAY, R. C. 1955. Cellular constituents. The chemistry of xanthine oxidase. Part I. The preparation of a crystalline xanthine oxidase from cow's milk. J. Chem. Soc. 1100–1110.

BASSETTE, R., and KEENEY, M. 1960. Identification of some volatile carbonyl compounds from nonfat dry milk. J. Dairy Sci. *43*, 1744–1750.

BELL, R. W., and WHITTIER, E. O. 1965. The composition of milk products. *In* Fundamentals of Dairy Chemistry, 1st Edition, B. H. Webb and A. H. Johnson (Editors). AVI Publishing Co., Westport, CT.

BERRIDGE, N. J. 1943. Pure crystalline rennin. Nature *151*, 473–474.

BOULET, M., YANG, A., and RIEL, R. R. 1970. Examination of the mineral composition of the micelle of milk by gel filtration. Can. J. Biochem. *48*, 816–822.

BURKEY, L. A., SANDERS, G. P., and MATHESON, K. J. 1935. The bacteriology of Swiss cheese. IV. Effect of temperature upon bacterial activity and drainage in the press. J. Dairy Sci. *18*, 719–731.

CHAPON, L., and URION, E. 1960. Ascorbic acid and beer. Wallerstein Lab. Commun. *23*, 38–44.

COHN, E. J. 1922. Studies in the physical chemistry of the proteins. I. The solubility of certain proteins at their isoelectric points. J. Gen. Physiol. *4*, 697–722.

COLE, W. C., and TARASSUK, N. P. 1946. Heat coagulation of milk. J. Dairy Sci. *29*, 421–429.

CORBIN, E. A., and WHITTIER, E. O. 1965. Composition of milk. *In* Fundamentals of Dairy Chemistry, 1st Edition, B. H. Webb and A. H. Johnson (Editors). AVI Publishing Co., Westport, CT.

CRANE, J. C., and HORRALL, B. E. 1943. Phospholipids in dairy products. II. Determination of phospholipids and lecithin in lipids extracted from dairy products. J. Dairy Sci. *26*, 935–942.

DAVIES, D. T., and WHITE, J. C. D. 1960. The use of ultrafiltration and dialysis in isolating the aqueous phase of milk and in determining the partition of milk constituents between aqueous and disperse phases. J. Dairy Res. *27*, 171–190.

DAVIS, J. G. 1941. Enzymes in cheese. Chem. Ind. 259–265.

DAY, E. A., and ANDERSON, D. F. 1965. Gas chromatographic and mass spectral identification of natural components of the aroma fraction of blue cheese. J. Agric. Food Chem. *13*, 2–4.

DAY, E. A., and LIBBEY, L. M. 1964. Cheddar cheese flavor: gas chromatographic and mass spectral analyses of the neutral components of the aroma fraction. J. Food Sci. *29*, 583–589.

DAY, E. A., and LILLARD, D. A. 1960. Autoxidation of milk lipids. I. Identification of volatile monocarbonyl compounds from autoxidized milk fat. J. Dairy Sci. *43*, 585–597.

DAY, E. A., FORSS, D. A., and PATTON, S. 1957. Flavor and odor defects of γ-irradiated skimmilk. I. Preliminary observations and role of volatile carbonyl compounds. J. Dairy Sci. *40*, 922–931.

DAY, E. A., FORSS, D. A., and PATTON, S. 1957. Flavor and odor defects of γ-irradiated skimmilk. II. Identification of volatile components by gas chromatography and mass spectrometry. J. Dairy Sci. *40*, 932–941.

DAY, E. A., BASSETTE, R., and KEENEY, M. 1960. Identification of volatile carbonyl compounds from Cheddar cheese. J. Dairy Sci. *43*, 463–474.

DAY, E. A., LINDSAY, R. C., and FORSS, D. A. 1964. Dimethyl sulfide and the flavor of butter. J. Dairy Sci. *47*, 197–198.

DEUTSCH, A., MATTSSON, S., and SWARTLING, P. 1958. Composition of the phospholipid fraction of butter with special regard to the polyenoic fatty acids. Milk Dairy Res. Rept. 54.

DORNER, W., and WIDMER, A. 1931. Rancidity of milk due to homogenization. Lait *11*, 545–567.

DUTHRIE, A. H., JENSEN, R. G., and GANDER, G. W. 1961. Interfacial tensions of lipolyzed milk fat-water systems. J. Dairy Sci. *44*, 401–406.

EL-NEGOUMY, A. M., MILES, D. M., and HAMMOND, E. G. 1961. Partial characterization of the flavors of oxidized butteroil. J. Dairy Sci. *44*, 1047–1056.

ERNSTROM, C. A. 1958. Heterogeneity of crystalline rennin. J. Dairy Sci. *41*, 1663–1670.

FEDERAL SECURITY AGENCY, Public Health Service, Washington, DC. 1951. Public Health Bull. 220.

FOLTMANN, B. 1959. Studies on rennin. II. On the crystallization, stability, and proteolytic activity of rennin. Acta Chem. Scand. *13*, 1927–1935.

FOLTMANN, B. 1966. A review on prorennin and rennin. C. R. Trav. Lab. Carlsberg *35*, 143–231.

FORSS, D. A., PONT, E. G., and STARK, W. 1955. Further observations on the volatile compounds associated with oxidized flavor in skimmilk. J. Dairy Res. 22, 345–348.

FORSS, D. A., DUNSTONE, E. A., and STARK, W. 1960. Fishy flavor in dairy products. II. The volatile compounds associated with fishy flavor in butterfat. J. Dairy Res. *27*, 211–220.

FORSTER, T. L., BENDIXEN, H. A., and MONTGOMERY, M. W. 1959. Some esterases of cows' milk. J. Dairy Sci. *42*, 1903–1912.

FOX, P. F., and WALLEY, B. F. 1971. Bovine pepsin. Preliminary cheese-making experiments. Ir. J. Agr. Res. *10*, 358–360.

FRANK, G., and BRAUNITZER, G. 1967. The primary structure of β-lactoglobulins. Z. Physiol. Chem. *348*, 1691–1692. (German)

FRAZIER, W. C., SANDERS, G. P., BOYER, A. J., and LONG, H. F. 1934. The bacteriology of Swiss cheese. I. Growth and activity of bacteria during manufacturing processes in the Swiss cheese kettle. J. Bacteriol. *27*, 539–549.

FRAZIER, W. C., JOHNSON, JR., W. T., EVANS, F. R., and RAMSDELL, G. A. 1935A. The bacteriology of Swiss cheese. III. The relation of acidity of starters and of pH of the interior of Swiss cheeses to quality of cheeses. J. Dairy Sci. *18*, 503–510.

FRAZIER, W. C., BURKEY, L. A., BOYER, A. J., SANDERS, G. P., and MATHESON, K. J. 1935B. The bacteriology of Swiss cheese. II. Bacteriology of the cheese in the press. J. Dairy Sci. *18*, 373–387.

GADDIS, A. M., ELLIS, R., and CURRIE, G. T. 1961. Carbonyls in oxidizing fat. V. The composition of neutral volatile monocarbonyl compounds from autoxidized oleate, linoleate, linolenate esters, and fats. J. Am. Oil Chem. Soc. *38*, 371–375.

GORDON, W. G., and WHITTIER, E. O. 1965. Proteins in milk. *In* Fundamentals of Dairy Chemistry, 1st Edition, B. H. Webb and A. H. Johnson (Editors). AVI Publishing Co., Westport, CT.

GORDON, W. G., and ZIEGLER, J. 1955. Amino acid composition of crystalline α-lactalbumin. Arch. Biochem. Biophys. *57*, 80–86.

GORDON, W. G., SEMMETT, W. F., CABLE, R. S., and MORRIS, M. 1949. Amino acid composition of α-casein and β-casein. J. Am. Chem. Soc. *71*, 3293–3297.

GORDON, W. G., SEMMETT, W. F., and BENDER, M. 1950. Alanine, glycine, and proline of casein and its components. J. Am. Chem. Soc. *72*, 4282.

GORDON, W. G., SEMMETT, W. F., and BENDER, M. 1953. Amino acid composition of γ-casein. J. Am. Chem. Soc. *75*, 1678–1679.

GOULD, I. A., and TROUT, G. M. 1936. The effect of homogenization on some of the characteristics of milk fat. J. Agric. Res. *52*, 49–57.

GREEN, M. M. 1972. Assessment of swine, bovine, and chicken-pepsins as rennet substitutes for Cheddar cheese-making. J. Dairy Res. *39*, 261–273.

GREENBANK, G. R. 1954. The xanthine oxidase (Schradinger's enzyme) content of dairy products and its inhibition by heat. J. Dairy Sci. *37*, 644.

HANKINSON, C. L. 1943. The preparation of crystalline rennin. J. Dairy Sci. *26*, 53–62.

HANSEN, R. G., and CARLSON, D. M. 1956. An evalution of the balance of nutrients in milk. J. Dairy Sci. *39*, 663–673.

HANSEN, R. G., POTTER, R. L., and PHILLIPS, P. H. 1947. Studies on proteins from bovine colostrum. II. Some amino acid analyses of a purified colostrum pseudoglobulin. J. Biol. Chem. *171*, 229–232.

HARPER, W. J., and HUBER, R. M. 1956. Some carbonyl compounds in raw milk. J. Dairy Sci. *39*, 1609.

HARTMAN, A. M., and DRYDEN, L. P. 1965. American Dairy Science Assoc., Champaign, Ill. Also personal communication (1972).

HERB, S. F., MAGIDMAN, P., LUDDY, F. E., and RIEMENSCHNEIDER, R. W. 1962. Fatty acids of cows' milk. B. Composition by gas-liquid chromatography aided by other methods of fractionation. J. Am. Oil Chem. Soc. *39*, 142–146.

HILL, R. J., and WAKE, R. G. 1969. Amphiphile nature of κ-casein as the basis for its micelle stabilizing property. Nature *221*, 635–670.

HIPP, N. J., GROVES, M. L., and MCMEEKIN, T. L. 1957. Phosphopeptides obtained by partial acid hydrolysis of α-casein. J. Am. Chem. Soc. *79*, 2559–2565.

HIPP, N. J., BASCH, J. J., and GORDON, W. G. 1961. Amino acid composition of α_1-, α_2-, and α_3-caseins. Arch. Biochem. Biophys. *94*, 35–37.

HOLM, G. E., WRIGHT, P. A., and DEYSHER, E. F. 1936. The phopholipids of milk. IV. Their chemical nature and their distribution among some milk products. J. Dairy Sci. *19*, 631–639.

HONER, C. J., and TUCKEY, S. L. 1953. Chromatographic studies of reducing sugars, other than lactose, in raw and autoclaved milk. J. Dairy Sci. *36*, 559.

HOOD, E. G., and GIBSON, C. A. 1948. The problem of bitter flavour in Cheddar cheese. Can. Dairy Ice Cream J. *27*(11), 45–47.

HOSTETTLER, H., and IMHOFF, K. 1951. Electron optical investigations on the fine structure of milk. Milchwissenschaft *6*, 351–354; 400–402. (German)

HOSTETTLER, H., and IMHOFF, K. 1955. Gel produced by rennet and cheese curd. I. Basic conditions for the production of a gel by rennet. Schweiz. Milchztg. Suppl. *31*, 144–248. (German)

HUTTON, J. T., and PATTON, S. 1952. The origin of sulfhydryl groups in milk proteins and their contributions to "cooked" flavor. J. Dairy Sci. *35*, 699–705.

IRVINE, O. R., BRYANT, L. R., SPROULE, W. H., JACKSON, S. H., CROOK, A., and JOHNSTONE, W. M. 1945A. The retention of nutrients in cheese making. I.

The retention of calcium, phosphorus, and riboflavin in Cheddar cheese made from raw milk. Sci. Agric. *25*, 817–832.

IRVINE, O. R., BRYANT, L. R., SPROULE, W. H., JACKSON, S. H., CROOK, A., and JOHNSTONE, W. M. 1945B. The retention of nutrients in cheese making. II. The effect of pasteurization of the milk upon retention of calcium, phosphorus, and riboflavin in Cheddar cheese, Sci. Agric. *25*, 833–844.

IRVINE, O. R., BRYANT, L. R., SPROULE, W. H., JACKSON, S. H., CROOK, A., and JOHNSTONE, W. M. 1945C. The retention of nutrients in cheese making. III. The calcium, phosphorus, and riboflavin contents of cream, cottage, brick, and blue cheese. Sci. Agric. *25*, 845–853.

JENNES, R., LARSON, B. L., McMEEKIN, T. L., SWANSON, A. M., WHITNAH, C. H., and WHITNEY, R. McL. 1956. Nomenclature of the proteins of bovine milk. J. Dairy Sci. *39*, 536–541.

JOLLÈS, J., JOLLÈS, P., and ALAIS, C. 1969. Present knowledge concerning the amino-acid sequence of cow κ-casein. Nature *222*, 668–670.

KASSEL, B., and MEITNER, P. A. 1970. Bovine pepsinogen and pepsin. *In* Methods in Enzymology, Vol. 19, Gertrude E. Perlman, and L. Lorand (Editors). Academic Press, New York.

KEENEY, M., and DAY, E. A. 1957. Probable role of the Strecker degradation of amino acids in development of cheese flavor. J. Dairy Sci. *40*, 874–876.

KOOPS, J. 1958. The composition of the phosphatides of butter made from ripened cream. Neth. Milk Dairy J. *12*, 226–237.

KRISTOFFERSEN, T., and GOULD, I. A. 1960. Cheddar cheese flavor. II. Changes in flavor quality and ripening products of commercial Cheddar cheese during controlled curing. J. Dairy Sci. *43*, 1202–1215.

KRISTOFFERSEN, T., GOULD, I. A., and HARPER, W. J. 1959. Cheddar cheese flavor. I. Flavor and biochemical relationships of commercial Cheddar cheese. Milk Prod. J. *50*, 14–21.

KURTZ, F. E. 1965. Lipids of milk: Composition and properties. *In* Fundamentals of Dairy Chemistry, 1st Edition, B. H. Webb, and A. H. Johnson (Editors). AVI Publishing Co., Westport, CT.

KURTZ, F. E., JAMIESON, G. S., and HOLM, G. E. 1934. The lipids of milk. I. The fatty acids of the lecithin-cephalin fraction. J. Biol. Chem. *106*, 717–724.

LEA, C. H., MORAN, T., and SMITH, J. A. B. 1943. The gas-packing and storage of milk powder. J. Dairy Res. *13*, 162–215.

LIBBEY, L. M., and DAY, E. A. 1963. Methyl mercaptan as a component of Cheddar cheese, J. Dairy Sci. *46*, 859–861.

MACY, I. G., KELLY, H. J., and SLOAN, R. E. 1953. The composition of milks. A compilation of the comparative composition and properties of human, cow, and goat milk, colostrum, and transitional milk. Natl. Res. Council–Natl. Acad. Sci. (U.S.) Publ. 254.

MARTH, E. H. 1963. Microbiological and chemical aspects of Cheddar cheese ripening. A review. J. Dairy Sci. *46*, 869–890.

MATTICK, E. C. V. 1938. The calcium and phosphorus contents of some types of British cheese at various stages during manufacture and ripening. J. Dairy Res. *9*, 233–241.

McCAMMON, R. B., CAULFIELD, W. J., and KRAMER, M. M. 1933. Calcium and phosphorus of cheese made under controlled conditions. J. Dairy Sci. *16*, 253–263.

McDOWALL, F. H., and DOLBY, R. M. 1935. Studies on the chemistry of Cheddar-cheese making. I. The mineral content of cheese and whey. J. Dairy Res. *6*, 218–234.

McKENZIE, H. A. 1967. Milk proteins. *In* Advances in Protein Chemistry, Vol. 22, C. B. Anfinsen, M. L. Anson, J. T. Edsall, and F. M. Richards (Editors). Academic Press, New York.

MIKELSEN, R., and ERNSTROM, C. A. 1972. Effect of pH on the stability of rennin-porcine pepsin blends. J. Dairy Sci. *55*, 294–297.

NITSCHMANN, H., and HENZI, R. 1959. Rennet and its action on the casein of milk. XIII. Research on the peptide set free by rennin action. Helv. Chem. Acta *42*, 1985–1995.

OTTESEN, M., and RICHERT, W. 1970. Isolation and partial characterization of an acid protease produced by *Mucor miehei*. C. R. Trav. Lab. Carlsberg *37*, 301–325.

PALMER, A. H. 1934. The preparation of a crystalline globulin from the albumin fraction of cow's milk. J. Biol. Chem. *104*, 359–372.

PARKS, O. W. 1965. Lipids of milk: Deterioration. Part II. Autoxidation. *In* Fundamentals of Dairy Chemistry, 1st Edition, B. H. Webb and A. H. Johnson (Editors). AVI Publishing Co., Westport, CT.

PARKS, O. W., and PATTON, S. 1961. Volatile carbonyl compounds in stored dry whole milk. J. Dairy Sci. *44*, 1–9.

PATTON, S. 1963. Volatile acids and the aroma of Cheddar cheese. J. Dairy Sci. *46*, 856–858.

PATTON, S., FORSS, D. A., and DAY, E. A. 1956. Methyl suflide and the flavor of milk. J. Dairy Sci., *39*, 1469–1470.

PATTON, S., WONG, N. P., and FORSS, D. A. 1958. Some volatile components of Cheddar cheese. J. Dairy Sci. *41*, 857–858.

PATTON, S., BARNES, I. J., and EVANS, L. E. 1959. *n*-Decca-2,4-dienal, its origin from linoleate and flavor significance in fats. J. Am. Oil Chem. Soc. *36*, 280–283.

PERTZOFF, V. 1927. The effect of temperature upon some of the properties of casein. J. Gen. Physiol. *10*, 961–985.

PETERSON, R. F., NAUMAN, L. W., and McMEEKIN, T. L. 1958. The separation and amino acid composition of a pure phosphopeptone prepared from β-casein by the action of trypsin. J. Am. Chem. Soc. *80*, 95–99.

PIEZ, K. A., DAVIE, E. W., FOLK, J. E., and GLADNER, J. A. 1961. β-Lactoglobulins A and B. I. Chromatographic separation and amino acid composition. J. Biol. Chem. *236*, 2912–2916.

POLIS, B. D., and SHMUKLER, H. W. 1950. Aldolase in bovine milk. J. Dairy Sci. *33*, 619–622.

POLIS, B. D., and SHMUKLER, H. W. 1953. Crystalline lactoperoxidase, I. Isolation by displacement chromatography. II. Physicochemical and enzymatic properties. J. Biol. Chem. *201*, 475–500.

RAND, A. G., JR., and ERNSTROM, C. A. 1964. Effect of pH and sodium chloride on activation of prorennin. J. Dairy Sci. *47*, 1181–1187.

RHODES, D. N., and LEA, C. H. 1958. The composition of the phospholipids of cow milk. J. Dairy Res. *25*, 60–69.

RYLE, A. P. 1970. The porcine pepsins and pepsinogens. *In* Methods in Enzymology, Vol. 19, Gertrude E. Perlmann and L. Lorand (Editors). Academic Press, New York.

SCHORMÜLLER, J. 1968. The chemistry and biochemistry of cheese ripening. *In* Advances in Food Research, Vol. 16, C. O. Chichester, E. M. Mrak, and G. F. Stewart (Editors). Academic Press, New York.

SCHORMÜLLER, J., BELITZ, H. D., and BACHMAN, E. 1961. Phosphates and organic phosphorus compounds in foods. X. Phosphopeptides from enzymic hydrolyzates of α- and β-casein. Z. Lebensm. Untersuch. -Forsch. *115*, 402–409.

SMITH, L. M., and FREEDMAN, N. K. 1959. Analysis of milk phospholipids by chromatography and infrared spectrophotometry. J. Dairy Sci. *42*, 1450–1462.

STEIN, W. H., and MOORE, S. 1949. Amino acid composition of β-lactoglobulin and bovine serum albumin. J. Biol. Chem. *178*, 79–91.

STERNBERG, M. Z. 1971. Crystalline milk-clotting protease from *Mucor miehei* and some of its properties. J. Dairy Sci. *54*, 159–167.

SVEDBERG, T. 1930. Ultracentrifugal dispersion determinations of protein solutions. Kolloid Z. *51*, 10–24. (German)

THURSTON, L. M., and BARNHART, J. L. 1935. A study of the relation of materials adsorbed on the fat globules to the richness of flavor of milk and certain milk products. J. Dairy Sci. *18*, 131–137.

TITTSLER, R. P., WOLK, J., and HARGROVE, R. E. 1954. Effects of temperature and pH on the survival and growth of Swiss cheese bacteria. J. Dairy Sci. *37*, 638.

TRUCCO, R. E., VERDIER, P., and REGA, A. 1954. New carbohydrate compounds from cow milk. Biochem. Biophys. Acta *15*, 582–583.

TUSTIN, E. B. 1946. Effect of varying amounts of salt on the quality of Cheddar cheese. Natl. Butter Cheese J. *37*(9), 44–45.

U.S. DEPT. OF AGRIC. 1969. Cheese varieties and descriptions. Agric. Handbook 54.

U.S. DEPT. OF AGRIC. 1971. Federal and State standards for the composition of milk products. Agric. Handbook 51.

U.S. FOOD AND DRUG ADMIN. 1959. Cheeses and cheese products, definitions and standards. Federal Food, Drug, and Cosmetic Act, Part 19, Title 21, with revisions.

VAN, DUIN H. 1958. Carbonyl compounds in butter. II. Preliminary investigation into the autoxidation products of phosphatides from butter. Neth. Milk Dairy J. *12*, 81–89.

VAN VUNAKIS, H., and HERRIOTT, R. M. 1957. Structural changes associated with the conversion of pepsinogen to pepsin. II. The N-terminal amino acid residues of pepsin and pepsinogen, the amino acid composition of pepsinogen. Biochem. Biophys. Acta *23*, 600–608.

WARNER, R. C., 1944. Separation of α- and β-casein. J. Am. Chem. Soc. *66*, 1725–1731.

WARNER, R. C., and POLIS, E. 1945. On the presence of a proteolytic enzyme in casein. J. Am. Chem. Soc. *67*, 529–532.

WATT, B. K., and MERRILL, A. L. 1963. Composition of foods. U.S. Dept. Agric., Agric. Handbook 8.

WAUGH, D. F., and GILLESPIE, J. M. 1958. Annu. Meeting Am. Chem. Soc. (Sept). Abstr.

WAUGH, D. F., and VAN HIPPEL, P. 1956. κ-casein and the stabilization of casein micelles. J. Am. Chem. Soc. *78*, 4576–4582.

WEBB, B. H., and JOHNSON, A. H. 1965. Fundamentals of Dairy Chemistry, 1st Edition. AVI Publishing Co., Westport, CT.

WEBB, B. H., JOHNSON, A. H., and ALFORD, J. A. 1974. Fundamentals of Dairy Chemistry, 2nd Edition. AVI Publishing Co., Westport, CT.

WHITE, J. C. D., and DAVIES, D. T., 1958. The relation between the chemical composition of milk and the stability of the caseinate complex. I. General introduction, description of samples, methods, and chemical composition of samples. II. Coagulation by ethanol. III. Coagulation by rennet. IV. Coagulation by heat. J. Dairy Res. *25*, 236–296.

WHITNEY, R. M. 1958. The minor proteins of bovine milk. J. Dairy Sci. *41*, 1303–1323.

WHITTIER, E. O., and WEBB, B. H. 1950. By-products from Milk. Reinhold Publishing Co., New York.

WILSON, H. L. 1942. U.S. Dept. Agr. Bur. Dairy Ind. BDIM-947.

WILSON, H. L., HALL, S. A., and JOHNSON, JR., W. T. 1941. Relationship of curing temperatures to quality of American Cheddar cheese. J. Dairy Sci. *24*, 169–177.

WILSON, H. L., HALL, S. A., and ROGERS, L. A. 1945. The manufacture of Cheddar cheese from pasteurized milk. J. Diary Sci. *28*, 187–200.

17

Coffee

Composition of the Green Coffee Bean
Changes during Roasting
Roasted Coffee
Coffee Products
Summary
Bibliography

Coffee is used extensively as a beverage and is also employed as a flavoring material. Its use probably originated in Ethiopia, and from there it spread to Egypt and Arabia before the thirteenth century. Coffee is said to have been first sold in Amsterdam in 1640. A little later it was regularly delivered in Amsterdam, and it had begun to be used in England, France, and Italy. It was cultivated in the Western Hemisphere in the eighteenth century.

Coffee is used for its flavor and aroma, as well as for the stimulating effects of the caffeine it contains, unless the caffeine has been removed. As a beverage it contributes little to nutriton if cream has not been added to it.

The coffee tree belongs to the madder family, the genus *Coffea,* and the species most commonly cultivated is *Coffea arabica.* Perhaps 90% of the world production comes from this species, with about 9% of *Coffea canephora* and about 1% or less *Coffea liberica* accounting for the remaining production. The most common variety of *C. canephora* is *robusta,* a low-priced coffee used in England, France, Belgium, and Portugal as ground coffee; it finds extensive application also in the manufacture of decaffeinated coffee. About 85% of the coffee from Africa is of this variety.

The coffee tree grows up to 20 ft high but can be controlled under commercial conditions by pruning. The fruits of the coffee tree are called cherries and are about the size and color of cranberries. The seed is the commercially important part; one to three seeds may be found in a single fruit. Generally, the best coffee is grown at high altitudes.

Coffee is prepared from the fruit by either of two methods. The first of these is the so-called washed or wet process, in which the ripe fruit is harvested and put through a pulping machine to remove most of the soft outer pulp, leaving an exposed, slippery layer of mucilage. The mucilage is removed by a water wash, the beans are dried to about 12%

moisture, and then the outer layers are removed by hulling. This produces the high quality green coffee of commerce.

The second method is the dry or natural method. The fruit is left on the tree beyond the fully ripe stage, partially dried before harvesting, and then collected and dried to about 12% moisture. After this, the outer layers are removed by hulling, resulting in the finished product. This method is cheaper, but the quality of the coffee is usually not so high as that made by the washing process.

COMPOSITION OF THE GREEN COFFEE BEAN

Like all natural entities, coffee beans are rather complex in chemical composition. The green bean contains a variety of carbohydrates making up about 60% of its composition, some protein, and some oil, together with organic acids, caffeine, other compounds, and mineral substances. Samples of different varieties, as well as those harvested in different seasons of the year and areas of production, show some variance in chemical composition. Table 17.1 gives the composition of green coffee beans, together with information as to the solubility of the various materials.

Carbohydrates

As shown in the Table 17.1, a small amount of reducing sugars is present. French workers studying water-soluble oligosaccharides in green coffee found sucrose as well as two galactosides of sucrose, raffinose and stachyose.

Wolfrom and Patin (1965) found that the green coffee bean contains 50–60% of polysaccharides. The constituent sugars of these polysaccharides are mainly D-mannose, together with L-arabinose, D-galactose, and D-glucose. Most of the L-arabinose and D-galactose in the green bean are water-soluble arabinogalactan, ratio 2:5. Dilute acid removes the L-arabinose from this polysaccharide, leaving a galactan. Wolfrom and Patin (1964) showed in addition that the principal polysaccharide found in the green bean is a hard, insoluble $(1 \rightarrow 4)$-β-D-mannan. When this compound is removed, what remains is a glucan demonstrated to be cellulose comprising about 5% of the green coffee bean.

Thaler and Arneth (1968A) studied polysaccharides in green Arabica coffee. Two complexes in raw coffee that amounted to about 3.4% (dry weight basis) were separated and found to be composed of mannoarabinogalactans in which galactan was the main constituent. One was soluble in cold water, the other only in hot water. In addition, a so-called holocellulose, which contained cellulose together with galactan and mannan, was found. A fourth complex made up of galactan, araban, and mannan, of which galactan made up about 70% while mannan was present in small amount, was obtained.

Nitrogenous Substances

The protein content of green coffee is about 14%. The following free amino acids were reported in green coffee: aspartic acid, 0.33%; serine, 0.12%; asparagine, 0.3%; glutamic acid, 0.49%; proline, 0.14%; glycine, 0.02%; alanine, 0.24%; valine, 0.02%; isoleucine, 0.03%; leucine, 0.03%; tyrosine, 0.04%; phenylalanine, 0.081%; γ-aminobutyric acid, 0.3%; lysine, 0.04%; histidine, 0.04%; and arginine, 0.04%. Thaler and Arneth (1969) reported the presence also of threonine and methionine, and found that the proteins of green coffee vary with the type of coffee.

TABLE 17.1. Chemical Composition of Green Coffee Beans[a]

Classes and Components	Water Solubility	Percentage of Green Coffee		
		Total	Compound or Group	Soluble
Carbohydrates		60		
Reducing sugars	Soluble		1.0	
Sucrose	Soluble		7.0	
Pectins	Soluble		2.0	
			10.0	10
Starch	Easily solublized		10.0	
Pentosans	Easily solublized		5.0	
			15.0	—
Hemicelluloses	Hydrolyzable		15.0	—
Holocelluloses	Non-hydrolyzable fiber		18.0	
Lignin	Non-hydrolyzable fiber		2.0	
			20.0	
Oils	Insoluble	13		—
Protein (N × 6.25)	Depends on amount denatured	13		4
Ash as oxide	Depends on amount hydrolyzed	4		2
Non-volatile acids				
Chlorogenic[b]	Soluble		7.0	
Oxalic	Soluble		0.2	
Malic	Soluble		0.3	
Citric	Soluble		0.3	
Tartaric	Soluble		0.4	
			8.2	8
Trigonelline	Soluble	1		1
Caffeine (Arabica 1.0, Robusta 2.0%)	Soluble	1		1
		100		26

Source: Sivetz (1963), with data averaged from Elder (1949), Lockhart (1957), Mabrouk and Deatherage (1956), Merritt *et al.* (1957); Moores and Heininger (1951), and Winton and Winton (1939), and others.

[a] Approximate composition based on dry weight.

[b] Probably a mixture of chlorogenic, isochlorogenic and neochlorogenic acids.

In the surface wax of the green coffee bean Harms and Wurziger (1968) found hydroxytryptamides composed of 48% arachidic acid-5-(hydroxy)-tryptamide, 48% behenic acid-5-(hydroxy)-tryptamide, and 4% lignoceric acid-5-(hydroxyl)-tryptamide.

Lipid Material

Kaufmann and Schickel (1965) found the lipid content of green Colombia coffee as follows; phystosterin esters, hydrocarbons, and waxes, 2.0%; triglycerides, 81.30%; diterpen-fatty acid esters (wax alcohols) 15.90%; free sterols, 0.39%; and diterpenes, 0.15%.

Among the reported unsaponifiable materials in coffee oil are γ-sitosterol, stigmasterol, dehydrositosterol, together with the two diterpene compounds, cafestol ($C_{20}H_{28}O_3$) and kahweol. The structure of cafestol was worked out by Finnegan and Djerassi (1960). Cafestol is the principal component of the unsaponifiable material in coffee oil and occurs in the form of fatty acid ester. Kahweol, present in small quantity along with cafestol and extracted with it, differs form cafestol because it contains two additional double bonds. Kahweol can be obtained from cafestol by reduction with nickel or palladium or with nascent hydrogen.

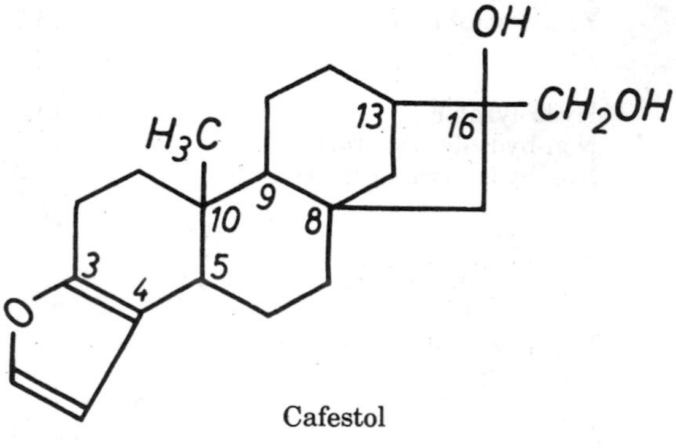

OH
CH2OH
H3C
13
16
9
10
8
3
5
4
O

Cafestol

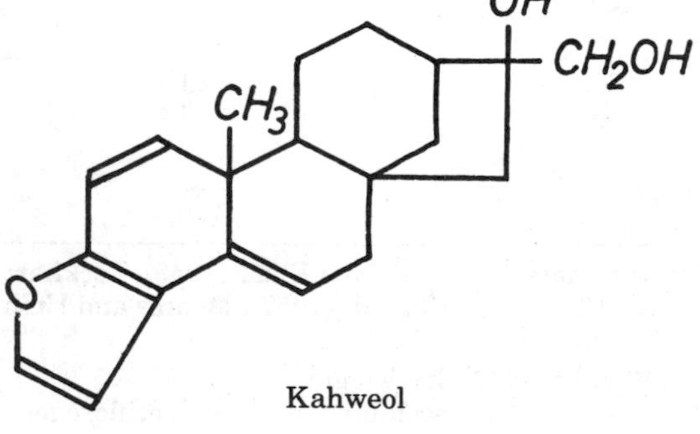

OH
CH2OH
CH3
O

Kahweol

Kaufmann and Hamsagar (1962A) found the following fatty acids in the saponified lipid material from green Brazilian coffee beans: linoleic acid, 39.0%; oleic acid, 17.2%; palmitic acid, 25.3%; stearic acid, 13.1%; arachidic acid, 4.2%; and behenic acid, 1.0%. In addition to these, Barbiroli (1966) has reported myristic, palmitoleic, and linolenic acids.

Other Compounds

Corse *et al.* (1965) showed that isochlorogenic acid of green coffee is a mixture of closely related compounds: 4,5-dicaffeoylquinic acid, 3,4-dicaffeoylquinic acid, 3,5-dicaffeoylquinic acid, and a fourth compound which seems to be a part of the mixture, 3′-methyl ethers of 3,5-dicaffeoylquinic acid.

Also present in microgram amounts in green coffee are chlorogenic acid, neochlorogenic acid, and scopoletin, $C_{10}H_8O_4$ (7-hydroxy-6-methoxy coumarin).

Aflatoxins, which are toxins produced by certain molds when moisture is present, have been found in very small quantities in green coffee.

CHANGES DURING ROASTING

Carbohydrates

Most of the sucrose in coffee is caramelized during the roasting process. Feldman *et al.* (1969) found that among the carbohydrates, sucrose and arabinogalactan are decomposed, depending on the degree of roast. Also, while soluble mannan markedly increases with the roast, the holocellulose's mannose decreases.

Thaler and Arneth (1969) found that during roasting of Arabica coffee, the higher polymer carbohydrates showed thermal losses, particularly at the beginning of the roasting process. Cellulose and araban showed decreases, mannan was rendered soluble and became part of the water-soluble polysaccharides. Galactan, however, was included in the solubilized polysaccharides.

Nitrogenous Substances

Thaler and Gaigl (1963), after investigating the effect of roasting on the nitrogen compounds in coffee, found that arginine is completely destroyed in roasting, and cystine is extensively so. However, some of the other amino acids, alanine, glutamic acid, leucine, isoleucine, phenylalanine, proline, and valine are increased during roasting.

Feldman *et al.* (1969) examined the chemical composition of green and roasted coffee and the changes that take place during the roasting process. They noted that of the free and combined amino acids, arginine, cystine, lysine, and serine undergo destruction during roasting.

Numerous investigators found that nicotinic acid increases in coffee during roasting. Barbiroli (1965) reported that when the roasting of coffee is done under constant conditions, the transformation of trigonelline to nicotinic acid is about the same for different types of coffee (not greater than 5–10%). However, not enough of this vitamin (nicotinic acid) is formed to satisfy the nutritional needs. Thiamin is destroyed by roasting. The small amounts of other vitamins in coffee are too small to have any dietary significance.

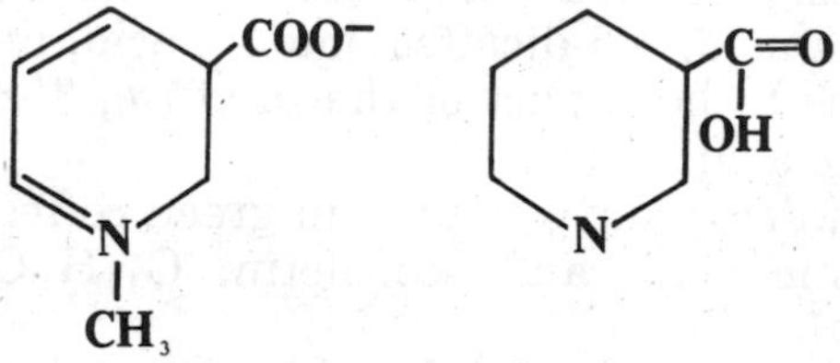

Trigonelline Nicotinic acid

Lipid Material

According to Kaufmann and Schickel (1965), the lipid material contained in coffee decreases during roasting. Colombian beans showed a triglyceride decrease from 81.3% of total lipid to 78.75%, while the triglyceride of the lipid of decaffeinated Colombian coffee was reduced to 76.0% in the same process. Similar decreases were found in the other components of the lipid material.

Kaufmann and Hamsagar (1962B) found that during the roasting process a complex change takes place in the diterpene ester. One of the results is a decrease of about 2% in the unsaponifiable material. Also, a simply hydrolytic cleaving takes place with the liberation of diterpene.

Other Changes

Color, Odor, and Flavor Volatiles. During the roasting process about 16% of the moisture is lost, and it seems that the development of most of the odor, color, and volatile flavor components start after the water has been substantially driven off. Many of the volatiles are not released until near the end of the roast, making the quality of the roast very important. These characteristics are discussed under coffee volatiles.

Little *et al.* (1958) showed that development of color is rather constant during the latter part of the roasting period.

Acids. Deshusses (1961) determined that about 135–220 mg of formic acid per 100 g of coffee forms during roasting. Feldman *et al.*

(1969) showed that volatile, nonvolatile, and phenolic acids decreased as the time of roast was increased. During roasting a large percentage of chlorogenic acid is destroyed (only 13% is retained), isochlorogenic acid is completely destroyed, and neochlorogenic acid showed a 67% retention. (Corse *et al.* 1970). They stated that no compound has been isolated from roasted coffee that is definitely derived from chlorogenic acid in the roasting process. This contradicts other published reports.

ROASTED COFFEE

Roasted coffee contains some unchanged substances as well as some formed during roasting, including the volatiles so important to the final aroma and flavor of the brewed coffee. Table 17.2 shows the general composition of roasted coffee, and the nature of these substances as to solubility. Table 17.3 shows the composition of soluble

TABLE 17.2. Chemical Composition of Soluble and Insoluble Portions of Roast Coffee[a b]

Chemical Compound or Class	Solubles (%)	Insolubles (%)
Carbohydrates (53%)		
Reducing sugars	1–2	—
Caramelized sugars	10–17	7–0
Hemicellulose (hydrolyzable)	1	14
Fiber (not hydrolyzable)	—	22
Oils	—	15
Proteins (N × 6.25) amino acids are soluble	1–2	11
Ash (oxide)	3	1
Acids, nonvolatile		
Chlorogenic	4.5	—
Caffeic	0.5	—
Quinic	0.5	—
Oxalic, malic, citric, tartaric	1.0	—
Acids, volatile	0.35	—
Trigonelline	1.0	
Caffeine (Arabicas 1.0, Robustas 2.0%)	1.2	
Phenolics (estimated)	2.0	—
Volatiles		
Carbon dioxide	trace	2.0
Essense of aroma and flavor	0.04	—
Total	27-35	73–65

Source: Sivetz (1963) and data was averaged from Elder (1949), Lockhart (1957), Mabrouk and Deatherage (1956), Merritt *et al.* (1957), Winton and Winton (1939) and others.

[a] Approximate, dry basis.

[b] Among the volatiles may be found acids, amines, sulfides, carbonyls (aldehydes and ketones), and other chemical groups. Nonvolatiles may contain carbohydrates, proteins, oils phospholipids, acids, minerals and others.

materials extracted from roasted coffee, together with similar information on the grounds that remain. The amounts of soluble materials that can be extracted depend on several factors, including the size to which the coffee is ground, the length of the extraction period or whether the extraction is complete or partial, and the temperature of the water used for the extraction.

Table 17.4 shows the mineral composition of coffee.

Caffeine

About 1% caffeine is found in Arabica coffee and about 2% in Robusta coffee. Cups of these coffees contain about 100 and 200 mg of caffeine, respectively. Caffeine is an alkaloid (a basic nitrogenous plant product), is white and crystalline, and is soluble in organic solvents other than petroleum ether, in which it is insoluble. Slightly soluble in cold water, 2% at 20°C and 18% at 80°C, caffeine has a slightly bitter flavor and no odor. It sublimes at 178°C and is physiologically active as a stimulant.

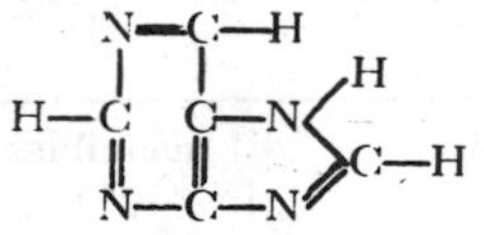

Purine 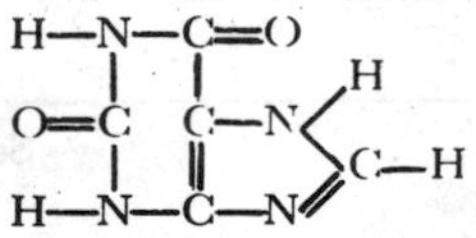 | Xanthine (dioxypurine) 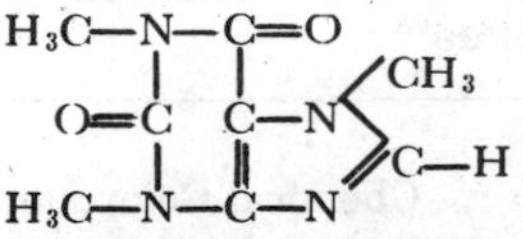 | Caffeine (trimethylxanthine)

The structural formulae show the relationship among the three compounds, purine, xanthine, and caffeine. Xanthine and caffeine are purine bases—xanthine is 2,6-dioxypurine, whereas caffeine is 1,3,7-trimethylxanthine. Caffeine could be called 1,3,7-trimethyl-2,6-dioxypurine, since it is a trimethyl derivative of xanthine. Caffeine is related to theobromine, the alkaloid found in cacao. Theobromine is a dimethyl derivative of xanthine. Some caffeine occurs in cacao along with theobromine and is also found in tea.

The caffeine content of roasted coffee is higher than the caffeine content of green coffee, regardless of the fact that a small amount of caffeine is lost by sublimation during the roasting process. This higher caffeine content is accounted for by the 16% loss of weight in the coffee that occurs during the roasting process, a much greater percentage than the amount of caffeine lost. Caffeine lost by sublimation accumulates in the roaster stacks.

Very small amounts of carcinogenic substances have been found in roasted coffee (Fritz 1968; Grimmer 1968; Grimmer and Hildebrandt 1966).

Volatiles

As has already been mentioned, odor and flavor are developed during the roasting of the coffee bean. The quality of the roast is, therefore,

TABLE 17.3. Chemical Composition of Coffee Solubles and Spent Grounds (Insoluble)[a]

Chemical Compound or Class	Soluble (%)	Spent Grounds (%)
Carbohydrates (3–5% reducing sugars)	35.0	65
Browning complexes	15.0	—
Oils and fatty acids	0.2	18
Proteins (amino acids and complexes)	4.0	15
Ash (oxide)	14.0	Fraction of 1%
Acids, nonvolatile		
Chlorogenic	13.0	—
Caffeic	1.4	—
Quinic	1.4	—
Others	3.0	—
Trigonelline	3.5	Few tenths %
Caffeine		
Arabicas	3.5	Few tenths %
Robustas	7.0	Few tenths %
Phenols (estimated)	5.0	Few tenths %
Volatiles		
Before drying—acids and essence	1.1	Nil
After drying	Nil	Nil
Total	100.0	98+

Source: Sivetz (1963). Averaged and calculated from data of Elder (1949), Lockhart (1957), Mabrouk and Deatherage (1956), Merritt *et al.* (1957), Winton and Winton (1939), and others.

[a] Approximate dry basis.

TABLE 17.4. Estimated Coffee Ash Distribution

	Green Coffee	Roast Coffee	Soluble Powder	Dry Spent Grounds
Dry weight relations, dry basis	1.176	1.000	0.380	0.620
Percentage ash content, dry basis	4.00	4.71	10.00	1.47
Weight ash per unit weight roast coffee dry basis		0.0471	0.0380	0.0091

Mineral Oxide	Green, Roast Ash (%)	Solubles Ash: Total (%)	Solubles Ash: Soluble (%)	Grounds Ash: Total (%)	Grounds Ash: Grounds (%)
K_2O	62.5	52.0	75.59	10.5	33.65
P_2O_5	13.0	3.0	4.36	10.0	32.05
CaO	5.0	2.0	2.90	3.0	9.62
MgO	11.0	8.0	11.63	3.0	9.62
Fe_2O_3	1.0	0.4	0.58	0.6	1.92
Na_2O	0.5	0.4	0.58	0.1	0.32
SiO_2	1.0	—	—	1.0	3.21
SO_3	5.0	2.0	2.90	3.0	9.61
Cl	1.0	1.0	1.46	—	—
	100.0	68.8	100.0	31.2	100.0

Source: Sivetz (1963).

very important. Volatiles are the primary factor in the flavor of freshly brewed coffee and are difficult to achieve and retain. A considerable amount of work has been done on this difficult problem, starting originally with Reichstein and Staudinger (1955).

There are difficulties in solving this problem (Gianturco *et al.* 1963). (1) The aroma complex lacks chemical stability, and (2) the roasted coffee volatile compounds are present in extremely low concentrations, 80% of which comprise 17 compounds, and 90% of the aroma mixture contains 30 compounds. Since more than 300 compounds are known to be in the mixture, it can be seen that the largest number are present in very small amounts.

Table 17.5 lists some of the compounds found in coffee aroma; many other compounds are also present. The current techniques of gas liquid chromatography and mass spectroscopy make identification and determination easier and more accurate.

Other workers have investigated the composition of coffee aroma. Gianturco *et al.* (1966) listed 85 volatile compounds in roasted coffee, 54

TABLE 17.5. Analysis of Coffee Aroma Essence[a]

	Mol Wt	Amt. Present (%)	Boiling Point °C	Relative Flavor Importance[b]
Acetaldehyde	44	19.9	20.8	1
Acetone	58	18.7	56.2	2
Diacetyl	86	7.5	88.0	1
n-Valeraldehyde	86	7.3	103.0	2
2-Methylbutyraldehyde	86	6.8	92.0	2
3-Methylbutyraldehyde	86	5.0	92.5	2
Methylfuran	82	4.7	63.0	2
Propionaldehyde	58	4.5	48.8	2
Methyl formate	60	4.0	32.0	2
Carbon dioxide	44	3.8	−78.0	—
Furan	68	3.2	31.4	1
Isobutyraldehyde	72	3.0	63.0	1
Pentadiene (isoprene)	68	3.0	33.0	2
Methylethyl ketone	72	2.3	79.6	2
C_4-C_7 paraffins and olefins	—	2.0	—	2
Methyl acetate	74	1.7	57.0	2
Dimethyl sulfide	62	1.0	37.3	1
n-Butyraldehyde	72	0.7	74.7	1
Ethyl formate	74	0.3	54.0	2
Carbon disulfide	76	0.2	46.0	2
Methyl alcohol	32	0.2	64.7	3
Methyl mercaptan	48	0.1	6.0	1
Thiophene	84	0.1	84.1	—

Source: Zlatkis and Sivetz (1960). Reproduced with permission of the Institute of Food Technologists.

[a] Coffee aroma essence represents about 200 ppm of the roasted coffee bean.

[b] 1—large; 2—medium; 3—small. These results were obtained by the use of gas chromatography and mass spectrometry.

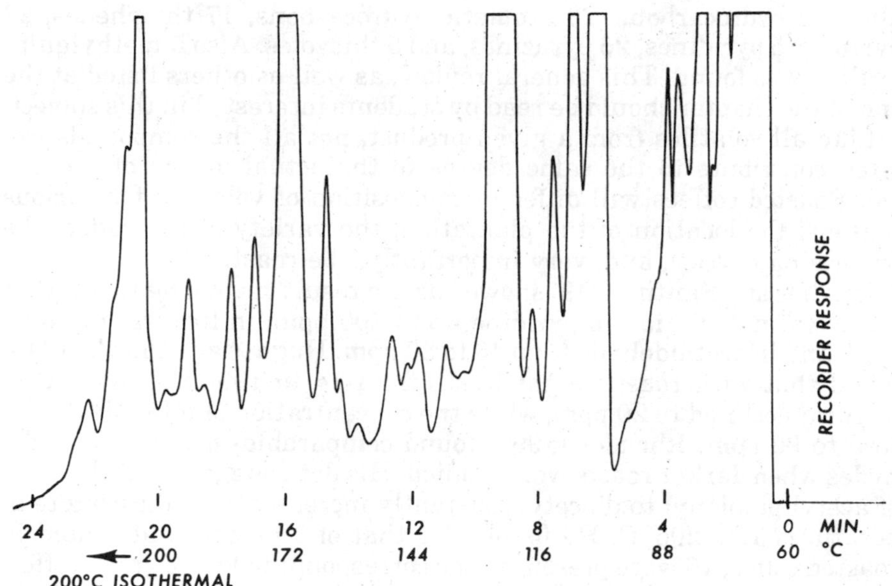

FIG. 17.1. Volatiles from Colombian coffee beans. This profile chromatogram was obtained under laboratory conditions. The peaks here have not been identified. Many compounds are overlapping in each peak. This work was done to obtain a general idea of the volatiles evolved in the effluent gas during a normal roasting cycle, and no attempt was made to identify compounds, or whether variation of conditions would change it.
Reproduced with permission of Blaw-Knox Food and Chemical Equipment, Inc., Jabez-Burns Division, one of the White Consolidated Industries.

of which were newly identified. Stoll *et al.* (1967) worked on coffee concentrate and identified 202 compounds of which 154 were identified for the first time. These same workers and others (Goldman *et al.* 1967) studied the pyrazines and pyridines of coffee, identified 25 compounds, and found that about 10 others are present. Figure 17.1 is a chromatogram of volatiles from Colombian coffee beans.

Work completed thus far indicates that approximately 50% of the volatiles are aldehydes, 20% are ketones, 8% are esters, approximately 7% are heterocyclic, approximately 2% dimethyl sulfide, together with smaller amounts of other organic compounds and sulfides. Since volatile organic compounds containing sulfur have considerable odor, they are usually important in such flavors.

Walter and Weidemann (1969) in a general review of coffee flavor compounds say that the number of known compounds that comprise coffee flavor has risen to over 300. These are made up of the following groups: 22 alcohols, 15 ethers, 30 aldehydes, 64 ketones, 25 diketones, 30 esters, 26 phenols, 41 sulfur compounds, 58 furan compounds. In addition, 3 nitriles, 16 carboxylic acids, 17 aliphatic hydrocarbons, 3 alicyclic hydrocarbons, 20 aromatic hydrocarbons, 17 thiophenes, 23 pyrroles, 5 pyridines, 26 pyrazines, and 5 thiazoles. Also 5-methylquinoxaline was found. This general review, as well as others listed at the end of the chapter should be read by students interested in this subject.

Like all volatiles from a given product, not all the compounds isolated contribute to the same degree to the actual aroma of the coffee. Roasted coffees will differ in composition of volatiles for various reasons: the location of the plantation, the variety of the coffee, the processing history, and, very importantly, the roast.

Hughes and Smith (1949) showed as the result of varying roasts that there was a variation in pyridine up to 200 ppm, in furfural up to 80 ppm, and in acetaldehyde from 40 to 80 ppm. Hughes and Smith (1949) found that with roast weight losses from 14 to 19%, the amount of furfural declined to 20 ppm, while the concentration of total aldehydes rose to 90 ppm. Rhoades (1960) found comparable increases in aldehydes when darker roasts were studied. His data suggest that the ratio of acetyl propionyl to diacetyl constantly increases with roasting temperatures over 205° C. He found also that of 19 compounds found in roasted coffee, 16 were present in the corresponding lots of green coffee, but usually in much smaller amounts.

Of this list of carbonyls and alcohols, only diacetyl, butyraldehyde, and valeraldehyde are important contributors to the flavor of coffee. The others contribute a general fruity-to-sweet background aroma and smooth flavor to the cup of coffee.

Aldehydes

Acetaldehyde. A volatile substance, CH_3CHO contributes to coffee aroma and flavor. This compound has an irritating odor, but in the small quantities present it is evidently a necessary part of the coffee aroma. According to Sivetz (1963) it can be detected when coffee is being ground. Rhoades (1958) states that commercial instant coffees are high in acetaldehyde content compared with roast coffee; the content found in instant coffee was 100 ppm.

Propionaldehyde. CH_3CH_2CHO has one more carbon atom in the chain than acetaldehyde, and about 200 ppm is present in the volatiles from roast coffee and to a lesser extent in soluble coffees. It is present in coffee odor and flavor, but it is probably not an important factor.

Butyraldehyde. The two butyraldehydes, normal and *iso* [CH_3CH_2-CH_2CHO and $(CH_3)_2CHCHO$], seem to be important as components of the aroma of coffee. In many natural mixtures the ratio of isobutyraldehyde to normal is about 5:1. Taken together, they are present in the volatiles of roast coffees to the extent of about 30 ppm, at which level Sivetz (1963) notes the odor and taste are detectable. It is his opinion that soluble coffee flavor is benefited very much by the retention of a few parts per million of butyraldehyde. Low concentration of these aldehydes contribute to the coffee flavor.

Valeraldehyde. The valeraldehydes, normal (straight chain), isovaleraldehyde, and 2-methyl isomers, constitute about 20% of the coffee essence.

$CH_3—CH_2—CH_2—CH_2—CHO$
n-Valeraldehyde

$(CH_3)_2CH \cdot CH_2CHO$
Isovaleraldehyde

$CH_3CH_2(CH_3)CHCHO$
2-Methyl butyraldehyde

The two iso forms are present in approximately equal amounts, about 60 ppm in roast coffee and something less in soluble coffee. Sivetz (1963) considers them part of the coffee flavor. Zlatkis and Sivetz (1960) and Rhoades (1960) have found varying amounts of valeraldehydes in roast coffee.

Furfural. Like the other volatiles, furfural is produced during roasting. Furfural is formed by acid hydrolysis during the commercial distillation of pentosans, and is generated in coffee during the roasting process by the same reaction. Hughes and Smith (1949) found 90 ppm at 15.7% roasting loss, 22 ppm at 18.2%, and 11 ppm at 19% roasting loss. In addition, they found that after 6 weeks the roast coffee did not lose furfural.

```
H—C——C—H
  ||    ||
H—C    C—CHO
   \  /
    O
```
Furfural

Samples of stale roast coffee held for more than 2 years were found to contain about 50 ppm of furfural, and soluble coffee contained 100–200 ppm.

Ketones

Acetone. CH_3COCH_3 comprises approximately 20% of the coffee essence. While the aroma of acetone tends to be sweetish and pungent, it is said to give smoothness to the coffee when small amounts are added. Rhoades (1958) and Zlatkis and Sivetz (1960) found about 50 ppm of acetone in both roast and instant coffees. Rhoades (1960) and Hughes and Smith (1949) found higher concentrations in the darker roasted coffees.

Diacetyl. About 20 ppm $CH_3COCOCH_3$ is found in roast coffee, as reported by Rhoades (1960) and Zlatkis and Sivetz (1960). Diacetyl has an odor and flavor that is sweet and buttery, and it is considered an important component of the flavor of coffee. It is used, incidentally, to strengthen the butter flavor in such products as baked goods and margarine. Hughes and Smith (1949) found that the flavor of diacetyl can be noticed at 1–2 ppm in roast coffee. In roast coffee two months old about half of the diacetyl is reduced to methylacetylcarbinol.

Methylacetylcarbinol. $CH_3CHOHCOCH_3$, known also as acetoin, is used as a butter flavor at not more than 100 ppm. As might be expected, when it is oxidized it forms diacetyl.

Methyl Ethyl Ketone. This ketone is found in roasted coffee to the extent of 10 to 20 ppm, and it possesses an odor similar to that of acetone.

Alcohols

Methyl Alcohol. A few parts per million of this compound is found in roast coffee, as determined by Zlatkis and Sivetz (1960). According to Sivetz (1963) it can be tasted at this concentration.

Esters

Methyl formate, ethyl formate, and methyl acetate are among those esters present in coffee essence. Methyl formate has been reported to be 10–20 ppm in roast coffee, or about 4% of the coffee essence. Ethyl formate amounts to about 2 ppm in roast coffee or about 0.3% of coffee essence, and methyl acetate about 2–5% of coffee essence, or 10–20 ppm of roast coffee. These esters are of pleasant aroma, and it is considered that each contributes a type of dark roast flavor.

Heterocyclic Compounds

Furan. Furan (b.p. 32°C) and methyl furan (b.p. 63°C) were found by various workers in coffee essence in amounts of 3–5%, which is equivalent to 10–20 ppm in roast coffee. The odor of furan is listed as being not unpleasant, while methyl furan is a little less pleasant. Furfural boils at 162°C and has a sweet, pleasant odor in dilute concentrations. Furfural has been already mentioned under the aldehydes.

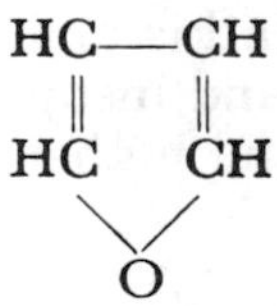

```
HC——CH
‖    ‖
HC   CH
  \ /
   O
```

Furan

Pyrrole. This compound has an NH substituted for the O in furan (see Chapter 5). It has a boiling point of 131°C, has a strong, nauseating odor, and is present in coffee aroma to the extent of 0.5%, or about 2 ppm according to Merritt *et al.* (1957).

Pyridine. A six-membered heterocyclic compound, pyridine is produced from trigonelline during the roasting of coffee.

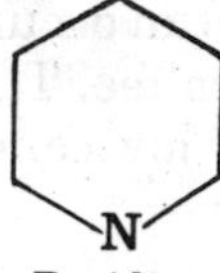

Pyridine

It was mentioned earlier that trigonelline also yields nicotinic acid during the roasting process. Pyridine boils at 115°C, has a repulsive odor, a sharp taste, and about 200 ppm is found in roast coffee. Highly roasted coffee is noted for its high pyridine content. Pyridine's possible involvement in coffee staling is discussed later in this chapter. The threshold of pyridine odor is about 3 ppb in air. Research by Goldman *et al.* (1967) on pyrazines and pyridines should be consulted for further information.

Sulfur Compounds

Many volatile compounds that have disagreeable odors in concentrated form lend very pleasing and desirable aromas when the concentration is very low, and this is often the case with volatile sulfur compounds. These compounds occur in coffee aroma in very small amounts and probably are derived from proteins containing sulfur.

Thiophene. Another compound with a five-membered ring, thiophene is similar in composition to furan, but the oxygen atom is replaced by sulfur. It boils at 84°C and was found by Zlatkis and Sivetz (1960) to be in coffee essence to the extent of about 0.1% or 1 to 2 ppm in roast coffee. Although thiophene is a sulfur compound, it does not have much odor.

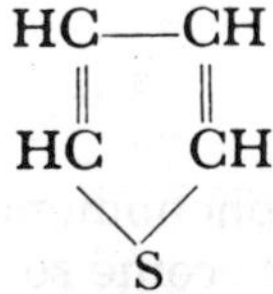

Thiophene

Dimethyl Sulfide. $(CH_3)_2S$ has a boiling point at 38°C. Zlatkis and Sivetz (1960) found 4 ppm in roasted coffee or about 1% of the recovered coffee essence. The odor of dimethyl sulfide can be detected in parts-per-billion concentrations. This compound is very strong at higher concentrations, but in the amounts found in coffee aroma it gives a very desirable character to the coffee grown in high altitudes like Colombia.

Methyl Mercaptan. CH_3SH has a boiling point at 8°C and has a very strong odor at rather high concentrations. Methyl mercaptan has a threshold of detection in water of about 2 ppb, while that of dimethyl sulfide is 12 ppb. Methyl mercaptan occurs in coffee essence at the 0.1% level and about 3 ppm in roast coffee. This mercaptan odor is a part of coffee aroma and flavor, with the low concentration an important factor in its acceptability.

Hydrogen Sulfide. H_2S is said to be present in very small amounts in roast coffee, but it is of no importance in coffee aroma and flavor.

Carbon Disulfide. CS_2 has a boiling point of 46°C and 3% is present in coffee essence, and about 1 ppm is roasted coffee. It seems to add something to coffee aroma.

Ammonia and Volatile Amines

Volatile amines and ammonia, formed during roasting, especially in the dark roasts, have a recognizable odor. Ammonia, methylamine, dimethylamine, and trimethylamine are products of coffee roasting. Trimethylamine has a fishy odor, which can be recognized in dark roast coffees.

Caramelized Sugars

Caramelized sugars represent about 4% of solubilized substances from roast coffee and 14% in soluble coffee. Most of the increased solubles yield are carbohydrates from darker roast coffees. Under the conditions of roasting the reducing sugars and 8% of the sucrose in the green coffee undergo pyrolysis to give some of the desired coffee odors and aldehydes.

Staling

Staling is a very important phenomenon in coffee storage. As a matter of fact many people have become so accustomed to drinking the beverage made from stale coffee that they have developed a taste for it.

At one time it was thought that staling was caused by the development of rancidity in coffee oil. However, staling is now considered to be the result of a reduction in or changes in the composition of the volatile compounds in roasted coffee. Many compounds in the essence of coffee are easily oxidized, as well as volatile. If a flavorful aromatic substance is lost or chemically altered, it is likely that the aroma of the coffee, and therefore, the flavor, will be lower in quality. Aldehydes are easily oxidized, and so is dimethyl sulfide Other compounds are also subject to change.

It has been thought possible that the undesirable odor of stale coffee is at least partially caused by the loss of flavorful volatiles, leaving behind pyridine, which in turn, could materially affect the odor of the product. As has already been pointed out, the odor threshold of pyridine is about 3 ppb in air.

Moisture content is probably also involved in the staling of coffee. Normally fresh-roasted coffees have very little moisture, but some vacuum-packed coffees have up to 3% moisture, causing them to stale faster after opening for use. A tray spread with ground coffee containing 1% moisture will stale appreciably in 1 hour when exposed to the air.

It has been said that the best vacuum-packed coffee is never so good in flavor as the day it was placed in the can. Some staling is caused by the small amount of oxygen left in the can, and after it is opened, staling continues. It is possible to retard the staling of coffee by storing in airtight containers at −1°C or lower. Vacuum pack or nitrogen gas pack provide the best practical protection available for coffee if staling is to be minimized.

COFFEE PRODUCTS

Decaffeinated Coffee

Because of the stimulating effect of caffeine, coffee is undesirable as a beverage at night for many people. Around the beginning of this century, a process to remove most of the caffeine was devised in Germany, and this process was given patent protection in the United States in 1908. According to this patent, the green beans are steamed until they contained 21% moisture. The beans are then treated with water, and if necessary, with ammonia or acid, and then solvent-extracted with trichloroethylene, choloroform, or benzene. The extracted beans are steamed to get rid of the excess solvent, and then dried.

Many other patents for the decaffeination of coffee have been issued. Among them is one issued to General Foods in this country, which uses water to extract 98% of the caffeine in an 8 hour extraction time. Decaffeinated beans are roasted in the usual fashion.

The natural flavor of the coffee is modified by the decaffeination processes and some of the peak flavor notes are lost. For this reason, the tendency is to use cheaper coffees. Another advantage of the use of the Robusta coffees is that double the amount of caffeine is contained in this variety. The caffeine is recovered and sold as a by-product. When possible, beans are decaffeinated in the coffee-producing countries. The green beans must be extracted wet and shipped dry, and if they are decaffeinated before shipment, only one drying is necessary.

Instant Coffee

Instant coffee, also known as soluble coffee, is made up of coffee solids that are completely soluble in water. Prepared by a process involving the use of percolation with water to extract the ground coffee, the solution is then dried by one of several methods. (1) In spray drying, the extracted liquid is forced through nozzles into a stream of hot air or inert gases, the result of which is almost instantaneous drying. The dried material collects in a suitable chamber. (2) Vacuum drying on a moving belt yields a dried product in the form of flakes which are very soluble, even in cold water. (3) Freeze drying, otherwise known as lyophilization, depends on the removal of water by evaporation in a high vacuum and at very low temperatures. This method yields a product which is closer to brewed coffee than the others, but it is expensive to produce. Most instant coffee is produced by spray drying.

Coffee Substitutes

In some of the coffee-consuming countries, chicory is blended with coffee. This reduces the cost of the product, but alters the flavor somewhat. In France, this combination is extensively used. Chicory and many types of roasted cereals are used as coffee substitutes by themselves.

SUMMARY

Like all natural substances, coffee beans are rather complex in chemical composition. The green bean contains about 60% carbohydrates, with a high amount of polysaccharides.

The lipid material of green coffee is quite high in triglycerides. The saponified lipid material from green Brazilian coffee beans is quite high in linoleic, oleic, palmitic, and stearic acids.

Changes take place during roasting. Most of the sucrose in coffee is caramelized during the roasting process. Mannan increases while the holocellulose mannose decreases.

Nitrogenous substances show some changes as a result of roasting. Arginine is completely destroyed during roasting and cystine extensively so, lysine and serine also show losses. Thiamine is destroyed during roasting, but nicotinic acid is increased, though not enough to satisfy nutritional needs.

Lipid material decreases during roasting.

During the roasting process 16% of the moisture is lost, and the development of the bulk of the odor, color, and volatile flavor components start after the water is driven off. A large group of volatiles are released at the end of the roast.

About 1% caffeine is found in Arabica coffee and about 2% in Robusta. This roughly equals about 100 mg and 200 mg of caffeine in a cup of coffee of each of these varieties, respectively. Caffeine is a purine base.

The number of known compounds that comprise coffee flavor is now over 300. About 50% of the volatiles are aldehydes, 20% are ketones, 8% are esters, and approximately 7% are heterocyclic, 2% dimethylsulfide, together with smaller amounts of other organic compounds and sulfides. Diacetyl, butyraldehyde, and valeraldehyde make important contributions to the flavor of coffee. It is thought that acetaldehyde contributes to coffee aroma and flavor. Methyl formate, ethyl formate, and methyl acetate are present in coffee essence. Some sulfur compounds present in coffee essence doubtless make a contribution to its quality.

It is possible that the undesirable odor of stale coffee is at least partially caused by the loss of flavorful volatiles.

BIBLIOGRAPHY

BARBIROLI, G. 1965. Free and combined amino acids in some types of green and roasted coffee. Rass. Chim. *17,* 220–225. (Italian)

BARBIROLI, G. 1966. Transformation of fatty substances in coffee during industrial roasting. Quad. Merceol. *5,* 223–235. (Italian)

BÜCHI, G., DEGEN, P., GAUTSCHI, F., and WILLHALM, B. 1971. Structure and synthesis of kahweofuran, a constituent of coffee aroma. J. Org. Chem. *36,* 199–200.

CORSE, J., LAYTON, L. L., and PATTERSON, D. C. 1970. Isolation of chlorogenic acid from roasted coffee. J. Sci. Food Agric. *21,* 164–168.

CORSE, J., LUNDIN, R. E., and WAISS, A. C. Jr. 1965. Identification of several components of isochlorogenic acid. Phytochemistry *4,* 527–529.

DESHUSSES, J. 1961. Formic acid content of roasted coffee, chicory, soluble coffee extracts, and coffee substitutes. Mitt. Geb. Lebensmittelunters. Hyg. *52,* 428–430. (German)

ELDER, L. W. 1949. Coffee. *In* Encyclopedia of Chemical Technology, Vol. IV, R. E. Kirk, and D. F. Othmer (Editors). Wiley (Interscience) Publishers, New York.

FELDMAN, J. R., RYDER, W. S., and KUNG, J. T. 1969. Importance of nonvolatile compounds to the flavor of coffee. J. Agric. Food Chem. *17,* 733–739.

FINNEGAN, R. A., and DJERASSI, C. 1960. Terpenoids. XLV Further studies on the structure and absolute configuration of cafestol. J. Am. Chem. Soc. *82,* 4342–4344.

FRITZ, W. 1968. Formation of carcinogenic hydrocarbons during thermal treatment of foods. II. Roasting of coffee beans and coffee substitutes. Nahrung *12,* 799–804. (German). Chem. Abstr. *71,* 2301p (1969).

GIANTURCO, M. A., GIAMMARINO, A. S., and PITCHER, R. G. 1963. The structures of five cyclic diketones isolated from coffee. Tetrahedron *19,* 2051–2059.

GIANTURCO, M. A., GIAMMARINO, A. S., FRIEDEL, P., and FLANAGAN, V. 1964. The volatile constituents of coffee. IV. Furanic and pyrrolic compounds. Tetrahedron *20,* 2951–2961.

GIANTURCO, M. A., GIAMMARINO, A. S., and FRIEDEL, P. 1966. Volatile constituents of coffee. V. Nature *210,* 1358.

GOLDBLATT, L. A. 1969. Aflatoxin. Scientific Background, Control, and Implications. Academic Press, New York.

GOLDMAN, I. M., SEIBL, J., FLAMENT, I., GAUTSCHI, F., WINTER M., WILLHALM, B., and STOLL, M. 1967. Research on aromas. XIV. Coffee aroma. II. Pyrazines and pyridine. Helv. Chim. Acta *50,* 694–705. (French)

GRIMMER, G. 1968. Carcinogenic hydrocarbons in the human environment. Dtsh. Apoth. Ztg. *108,* 529–533. (German)

GRIMMER, G., and HILDEBRANDT, A. 1966. Hydrocarbons in the surroundings of man. IV. Polycyclic hydrocarbon content of coffee and tea. Dtsh. Lebensm. Rundsch. *62,* 19–21. (German)

HARMS, U., and WURZIGER, J. 1968. Carboxylic acid 5-hydroxytryptamides in coffee beans. Z. Lebensm. Unters. -Forsch. *138,* 75–80. (German)

HUGHES, E. B., and SMITH, R. F. 1949. Volatile constitutents of roasted coffee. J. Soc. Chem. Ind. *68* 322–327. (London)

KAUFMANN, H. P., and HAMSAGAR, R. S. 1962A. To the knowledge of the lipids of coffee beans. I. The fatty acid esters of cafestol. Fette, Seifen, Anstrichm. *64,* 206–217. (German)

KAUFMANN, H. P., and HAMSAGAR, R. S. 1962B. To the knowledge of the lipids of coffee beans. II. The change of the lipids during coffee roasting. Fette, Seifen, Anstrichm. *64,* 734–738. (German)

KAUFMANN, H. P., and SCHICKEL, R. 1963. To the knowledge of the lipids of coffee beans. IV. Further research on the behavior of the lipids during coffee roasting. Fette Seifen, Anstrichm. *65,* 1012–1016. (German)

KAUFMANN, H. P., and SCHICKEL, R. 1965. To the knowledge of the lipids of coffee beans. V. The lipid content of coffee as prepared for drinking. Fette, Seifen, Anstrichm. *67,* 115–120. (German)

KAUFMANN, H. P., and SEN GUPTA, A. K. 1964. On the lipids of coffee beans. V. The triterpenes and hydrocarbons. Fette, Seifen, Anstrichm. *66,* 461–466. (German)

KLEIN, P. 1962. Caffeine sublimation in vacuum-packed coffee. Food Technol. *16,* 96–98.

LEHMANN, G., and MARTINOD, P. 1965. On the occurrence of theobromine in coffee. Z. Physiol. Chem. *341,* 155–156. (German)

LITTLE, A. C., CHICHESTER, C. O., and MACKINNEY, G. 1958. On the color of coffee. II. Food Technol. *12,* 505–507.

MABROUK, A. F., and DEATHERAGE, F. E. 1956. Organic acids in brewed coffee. Food Technol. *10,* 194–197.

MERRITT, M. C., and PROCTOR, B. E. 1959. Effect of temperature during the roasting cycle on selected components of different types of whole bean coffee. Food Res. *24,* 672–680.

MERRITT, C., SULLIVAN, J. H., and ROBERTSON, J. H. 1957. Volatile components of coffee aroma. Anal. Rept. 12. U. S. Army Quartermaster Corps., Natick, MA.

MOORES, R. G., and GRENINGER, DOROTHY M. 1951. Determination of trigonelline in coffee. Anal. Chem. *23,* 327–331.

PETEK, F., and DONG, T. 1961. Separation and study of two α-galactosidases of coffee beans. Enzymologia *23,* 133–142. (French)

RADTKE, R., SPRINGER, R., MOHR, W., and HEISS, R. 1963. Research on the chemical processes during the aging of roastd coffee. I. Gas-chromatographic analysis of the readily volatile aroma components. Z. Lebensm. Unters. -Forsch. *119,* 293–302. (German)

REICHSTEIN, T., and STAUDINGER, H. 1955. The aroma of coffee. Coffee Tea Ind. *78,* No. 6, 91.

REYMOND, D., CHAVAN, F., and EGLI, R. H. 1963. Gas-chromatographic analysis of the highly volatile constituents of roasted coffee. *In* Recent Advances in Food Science, Vol. 3., J. Leitch and D. Rhodes (Editors) Butterworths, London.

RHOADES, J. W. 1958. Sampling method of analysis of coffee volatiles by gas chromatography. Food Res. *23,* 254–261.

RHOADES, J. W. 1960. Analysis of the volatile constituents of coffee. J. Agric. Food Chem. *8,* 136–141.

SCOTT, P. M. 1968. Note on analysis of aflatoxins in green coffee. J. Assoc. Off. Anal. Chem. *51,* 609.

SIVETZ, M. 1963. Coffee Processing Technology, Vol. 2. AVI Publishing Co., Westport, CT.

SIVETZ, M., and FOOTE, H. E. 1963. Coffee Processing Technology. Vol. 1. AVI Publishing Co., Westport, CT.

STAUDINGER, H., and REICHSTEIN, T. 1929. Artificial coffee aroma. U.S. Pat. 1,696,419. Dec. 25.

STOFFELSMA, J., and PYPKER, J. 1968. Some new constituents of roasted coffee. Rec. Trav. Chim. Pays-Bas *87,* 241–242.

STOFFELSMA, J., SIPMA, G., KETTENES, D. K., and PYPKER, J. 1968. New volatile components of roasted coffee. J. Agric. Food Chem. *16,* 1000–1004.

STOLL, M., WINTER, M., GAUTSCHI, F., FLAMENT, I., and WILHALM, B. 1967. Research on aromas. On the aroma of coffee. I. Helv. Chim. Acta *50,* 628–694. (French)

THALER, H. 1964. Research on coffee and substitute coffee. X. On cell-wall carbohydrate of the roasted coffee bean. Z. Lebensm. Unters. -Forsch. *125,* 369–375. (German)

THALER, H., and ARNETH, W. 1968A. Research on coffee and substitute coffee. XI. Polysaccharides of the green beans of *Coffea arabica.* Z. Lebensm. Unters. -Forsch. *138,* 26–35. (German)

THALER, H., and ARNETH, W. 1968B. Research on coffee and coffee substitutes. XII. Polysaccharides of roasted Arabica-coffee. Z. Lebensm. Unters. -Forsch. *138,* 137–145. (German)

THALER, H., and ARNETH, W. 1969. Studies on coffee and coffee substitutes. XIII. Behavior of polysaccharide-complexes of raw Arabica-coffee during roasting. Z. Lebensm. Unters. -Forsch. *140,* 101–109. (German)

THALER, H., and GAIGL, R. 1963. Research on coffee and coffee substitutes. VIII. The behavior of nitrogenous substances on the roasting of coffee. Z. Lebensm. Unters.-Forsch. *120,* 357–363. (German)

VIANI, R., MÜGGLER-CHAVAN, P., REYMOND, D., and EGLI, R. H. 1965. On the composition of the aroma of coffee. Helv. Chim. Acta *48,* 1809–1815. (French)

WALTER, W., GRIGAT, H. G., and HUEKESHOVEN, J. 1970. On free amino acids in green coffee. Naturwissenschaften *57,* 246–247. (German)

WALTER, W., and WEIDEMANN, H. L. 1969. Coffee flavor compounds. Z. Ernaehrungswiss. *9,* 123–147. (German)

WEIDEMANN, H., and MOHR, W. 1970. Specificity of roasted coffee aroma. Lebensm.-Wiss. Technol. *13*(2), 23–32. (German)

WINTON, A. L., and WINTON, K. B. 1939. Analysis of Foods, Vol. IV. John Wiley & Sons, New York.

WOLFROM, M. L., and PATIN, D. L. 1964. Coffee constituents. Isolation and characterization of cellulose in the coffee bean. J. Agric. Food Chem. *12,* 376–377.

WOLFROM, M. L., and PATIN, D. L. 1965. Carbohydrates of the coffee bean. IV. An arabinogalactan. J. Org. Chem. *30,* 4060–4063.

WOLFROM, M. L., LAVER, M. L., and PATIN, D. L. 1961. Carbohydrates of the coffee bean. II. Isolation and characterization of a mannan. J. Org. Chem. *26,* 4533–4535.

ZLATKIS, A., and SIVETZ, M. 1960. Analysis of coffee volatiles by gas chromatography. Food Res. *25,* 395–398.

18

Tea

Methods of Preparation
Chemical Components of Tea
Chemistry of Tea Manufacture
Tea Aroma
Summary
Bibliography

Tea is one of the world's most extensively used beverages. The use of tea started many centuries ago in the country of its origin, China. Some authorities believe that its use as a beverage goes back more than 5000 years, although tea leaves may have been used earlier as a drug.

The plant belongs to the Theaceae family, the genus is *Cameillia* and the species either *sinensis* or *assamica*. Some authors use *Thea* as the genus name rather than *Camellia*. *Camellia sinensis* is the Chinese variety, while *Camellia assamica* is the Assam variety. A number of hybrids of the tea plant are used extensively in commercial production. The plant is an evergreen shrub, and it is pruned so that the average height is about 3–5 ft to facilitate harvesting.

The important areas of production are largely in Asia, namely, China, India, Sri Lanka, and Japan and Taiwan. Tea is also produced extensively in the southwestern part of the Soviet Union. World tea production is about 2.4 billion pounds annually. The highest quality of tea is grown in high altitude gardens. The part of the plant harvested for tea manufacture is the growing shoot tips, including the second or third leaf, called the tea flush.

There are three main classes of tea on the market. The first is black tea, the production of which involves a fermentation step. The second is Oolong tea, a partially fermented type of tea. The last is green tea which is prepared without fermentation. Oolong tea comes largely from Taiwan and is said to be made for the American market. Many variations and grades of these classes are known.

METHODS OF PREPARATION

Black Tea

Black tea is produced as the result of four basic steps: withering, rolling, fermenting, and firing.

Withering. The tea leaves are harvested by plucking from the shrubs, and after harvesting are spread on trays to wither. As much as 30 sq ft of area is needed to wither 1 lb of fresh tea flushes. Withering must be conducted as a slow and even process, or good tea will not be obtained. The usual time period is 18 hr, but this may be shorter depending on the weather. When the natural conditions are unfavorable, heat must be applied to speed the withering. Artificial withering equipment has been developed for use in the USSR to decrease the withering time to 6–8 hr (Bokuchava and Skobeleva 1969). The fresh tea leaf contains about 75–80% moisture, and when withering is complete, the final moistuer content differs depending on the aera of production, ranging from 58 to 68%. In northeast India, moisture content changes from 77%, the normal moisture content of the fresh leaves, to 68% during the withering process, about a 9% loss. In this condition the flushes can be rolled without undue breaking of the leaves.

Rolling. Following withering, the leaves are rolled, distorting the shape. This twisting of the leaves injures the leaf cells so the juices with the enzymes exude. The combination of the flavonols and catechol oxidase results in the start of the fermentation process. In short, the rolling step replaces respiration, and other normal biochemical processes in the leaf, with fermentation, which is fundamental to the production of black tea and is, in reality, a process of enzymic oxidation. It is not analogous to the anaerobic fermentation that produces alcohol.

Different methods are used in the rolling process. In India (Assam and Darjeeling), rolling is carried on for periods of 30 min. This is usually done three times, giving a total roll of 1½ hr. At the end of the first roll, the finer particles of leaf are separated, and since they are sufficiently damaged can be sent to fermentation. Although fermentation starts in the roll process, there is not enough time in this step to complete the process, and the next step, fermentation, must be used.

Fermentation. The fermenting leaves are spread out on suitable equipment in the fermentation room at room temperature. The relative humidity of the fermentation room is held as near 100% as possible. This step usually requires about 3 hr, but in some instances, longer time may be required. During this period the leaves change from green to a coppery-red color and the odor changes from grassy to that of fermented tea. The pigments formed are products of tannin oxidation, and are the cause of the color of the prepared tea infusion which is used as the beverage. Another result is the change in the taste from bitter to pleasant and not so astringent.

Firing. Firing ends the fermentation because it stops enzyme activity, and so stops the biological processes. It is an important step for this reason. In addition, firing reduces the moisture content of the tea to about 3%. The temperature used for this treatment is 82°–93°C, but the Russians use a temperature range of 90°–95°C in their process. During firing some volatile materials are lost, principally volatile flavoring compounds, and leaf proteins are coagulated and become insoluble.

A special thermal process has been developed in Russia for the manufacture of black tea (Bokuchava and Skobeleva 1969) that is said to shorten the fermentation process and apply the thermal treatment to the underfermented tea. This treatment consists of holding the underfermented tea at a temperature of 50°–65°C for a period of 2–5 hr depending on the size and moisture content of the lot being treated. This method increases the concentration of extractives, tannin, catechins, and other substances. One of the Russian factories using this process reported an increase in tannins and catechins of about 25%.

Russian workers stress the value of so-called vitamin P in tea. They contend that tea catechins have high vitamin P activity. This view is not accepted in the West. While a protective effect on capillary resistance is associated with these compounds and other flavonoids, it is thought that it is not a specific physiological function. It is held wiser, therefore, not to consider this as vitamin activity.

After firing, the final sorting and packing occurs and the tea is separated into different grades.

Oolong Tea

The Oolong teas are semifermented. After they are harvested, but before they are panned, the leaves are slightly withered, which permits a small amount of fermentation. The leaves are spread in bamboo baskets and held in the shade for about 5–6 hr at temperatures of about 28.3°–29.4°C. At this point the leaves have changed color and give off an aroma akin to apples. The fermentation is stopped by drying for 10 min in a hot pan at about 204°C. Much agitation is used in this step to avoid scorching. The tea is then fired. The first firing lasts about 3 hr; the second or last firing lasts from about 5 to 12 hr, at a temperature of about 100°C. The longest firing time is used for the highest quality teas.

Green Tea

Green tea is made by omitting the withering and fermentation steps. The plucked leaves are immediately steamed or heated in pans to inactivate the enzymes, which hinders blackening. After this they are rolled until they are dried. The leaves are then further dried to about 2% moisture content. Some dried green teas are rotated in a drum together with French chalk or soapstone as a polishing material to improve the appearance of the product.

CHEMICAL COMPONENTS OF TEA

Table 18.1 gives the general composition of tea and Table 18.2 lists the phenolic compounds present in freshly harvested tea flush. Approximately 30% of the dry weight is made up of these important compounds.

"Tea tannins" was the term formerly used for the compounds in tea that are made up of a number of phenolic substances. These compounds belong to four basic groups: flavanols (catechins), flavonols and flavonol glycosides, leucoanthocyanins, and depsides. The difference between flavanols (catechins) and flavonols is that the former are reduced forms while the flavonols are the oxidized forms. This is illustrated as follows with corresponding parts of the catechin and flavone molecules.

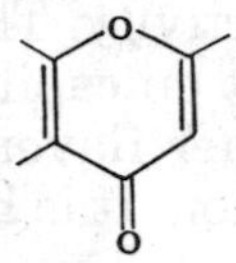

Flavone

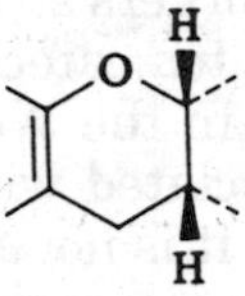

Catechin

Depsides are formed by interaction between the carboxyl group of a phenolic acid with the hydroxy group of another. An example of a depside is theogallin (galloylquinic acid) (see also Table 18.2).

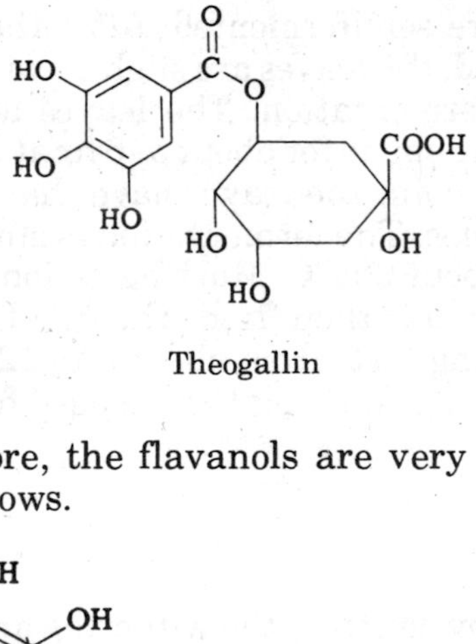

Theogallin

As mentioned before, the flavanols are very important, and their structures are as follows.

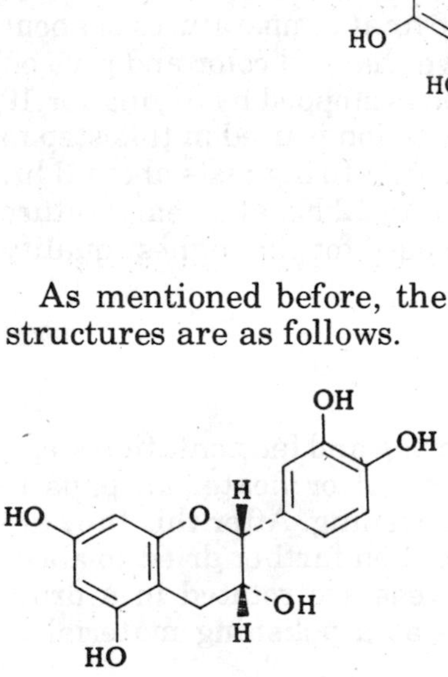

(−)-Epicatechin

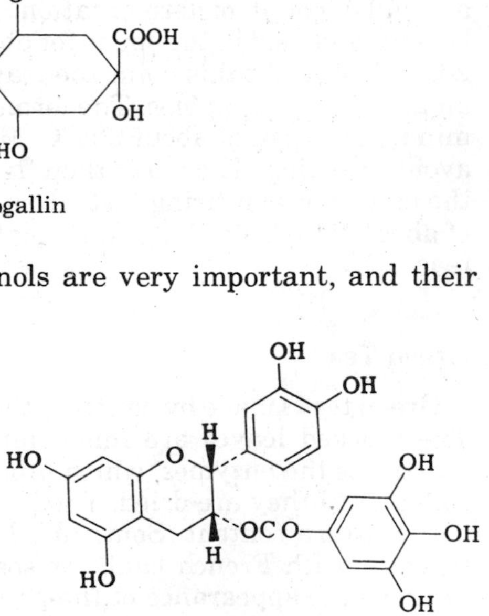

(−)-Epicatechin gallate

TABLE 18.1. Approximate General Analysis of Tea

Component	Fresh Flush	Black Tea[a]	Black Tea Brew[b,c]
Protein	15[c]	15	Trace
Fiber	30	30	0.0
Pigments	5	5	Trace
Caffeine	4	4	~3.2
Polyphenols, simple	30	5	4.5
Polyphenols, oxidized	0	25	15.0
Amino acids	4	4	3.5
Ash	5	5	4.5
Carbohydrates	7	7	4.0
Volatile compounds	0.01	0.01	0.01

Source: Sanderson (1972). Reproduced with permission of Academic Press, New York.

[a] Many of these components have been altered during the black tea manufacturing process, and they would not necessarily be recognizable as the component in the fresh flush from which they were derived.

[b] These are the substances extracted in a normal 5-min extraction in a teapot.

[c] Values are expressed as percentage of whole tea dry weight.

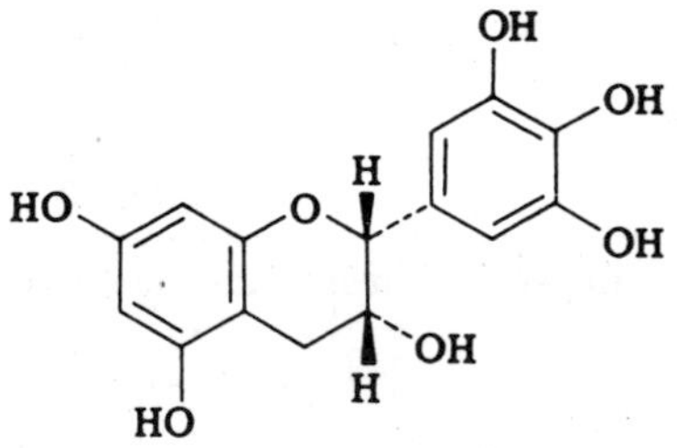

(−)-Epigallocatechin

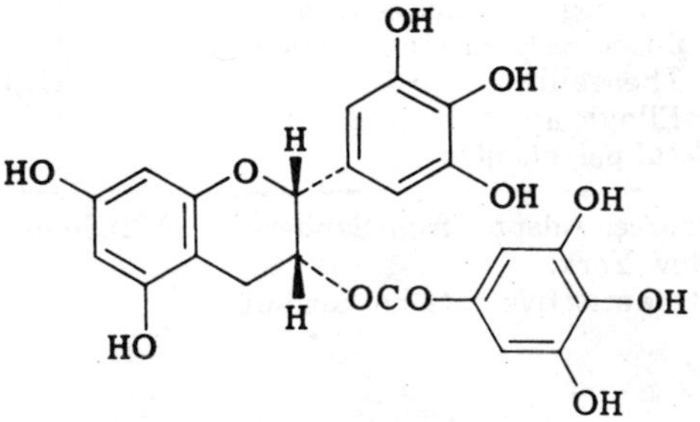

(−)-Epigallocatechin gallate

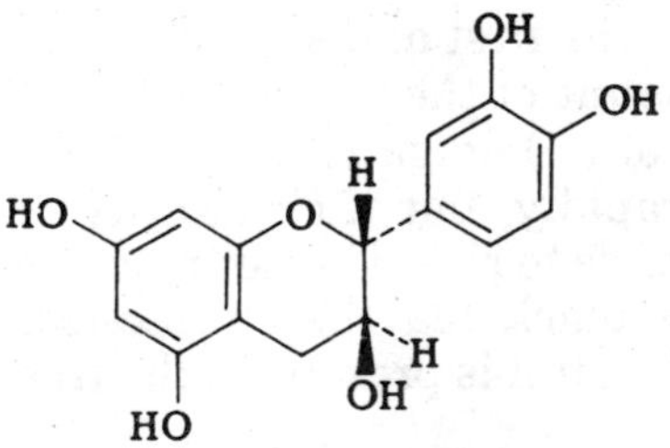

(+)-Catechin

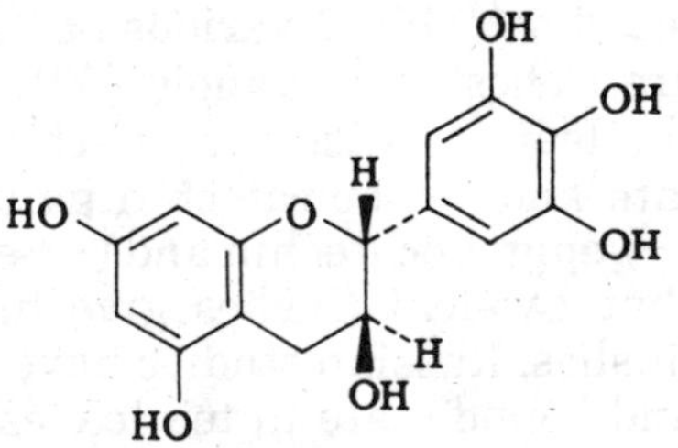

(+)-Gallocatechin

An inspection of the structural formulas shows that those flavanols of the tea leaf that are esterified with gallic acid frequently have this union in the 3-position.

TABLE 18.2. Phenolic Compounds Found in Fresh Tea Flush

	Molecular Formula	Molecular Weight	Amount Found in Flush (% dry wt)
Flavanols			
(−)-Epicatechin	$C_{15}H_{14}O_6$	290	1–3
(−)-Epicatechin gallate	$C_{22}H_{18}O_{10}$	442	3–6
(−)-Epigallocatechin	$C_{15}H_{14}O_7$	306	3–6
(−)-Epigallocatechin gallate	$C_{22}H_{18}O_{11}$	458	9–13
(+)-Catechin	$C_{15}H_{14}O_6$	290	1–2
(+)-Gallocatechin	$C_{15}H_{14}O_7$	306	3–4
Flavonols and flavonol glucosides			
Quercetin	$C_{15}H_{10}O_7$	302	—[a]
Kaempferol	$C_{15}H_{10}O_6$	286	—
Quercetin 3-rhamnoglucoside (rutin)	$C_{27}H_{30}O_{16}$	610	—
Kaempferol 3-rhamnoglucoside	$C_{27}H_{30}O_{15}$	594	—
Quercetin 3-rhamnodiglucoside	$C_{33}H_{40}O_{21}$	772	—
Kaempferol 3-rhamnodiglucoside	$C_{33}H_{40}O_{20}$	756	—
Leucoanthocyanins			2–3
Acids and depsides			~5
Gallic acid	$C_7H_6O_5$	170	—[a]
Chlorogenic acids (4 isomers)	$C_{16}H_{18}O_9$	354	—
p-Coumarylquinic (4 isomers)	$C_{16}H_{18}O_8$	338	—
Theogallin	$C_{14}H_{15}O_{10}$	343	~1
Ellagic acid	$C_{14}H_6O_8$	302	—
Total polyphenols			25–30

Source: Adapted from Sanderson (1972). Reproduced with permission of Academic Press, New York.
[a] Quantitative data not available.

Kursanov and Brovchenko (1950) showed that (−)-epigallocatechin gallate is the flavanol found in the greatest quantity in all parts of the tea shoot. Of the various parts of the shoot, the first and second leaves are richest in flavanols. With age, the amount of these compounds in the leaf decreases. In effect, the quantity of (−)-epigallocatechin gallate and (−)-epicatechin gallate decline rapidly, but the quantity of (−)-epigallocatechin and (−)-epicatechin tends to rise. The net result is that two-leaf flushes give higher quality black tea than three-leaf flushes. Russian studies have shown that inositol is present in the free and bound state in tea leaves.

Sanderson (1966) found the specific activity of 5-dehydroshikimate reductase in crude preparations of tea flush to be 13 to 90 times larger than that found in any other organism. This activity, together with the quantity of shikimic acid found, indicate an active shikimic acid pathway in the biosynthesis of flavanols in the tea flush. This is in agreement with the very high amount of flavonoids (including flavanols) present in these tissues. Experimental work by Patschke and Grisebach (1965) indicates that 4,2′,4′,6′-tetrahydroxychalcone-2′-glucoside is a precursor of (−)-epicatechin in tea leaves. Flavanols of tea are synthesized in the leaves of the plant, and immature leaves are more active in this way than mature leaves (Sanderson and Sivapalan 1966B; Zaprometov 1958).

Analytical Procedures

For methods for the determination of flavanols in tea extracts, consult Pierce *et al.* (1969) and Collier and Mallows (1971).

Enzymes

The most important enzyme in the tea flush is catechol oxidase. It acts in the oxidation of tea catechins in the fermentation step of the manufacture of black tea. Since steam treatment of tea leaves made fermentation impossible, it was obvious that this change was brought about by an enzyme or enzymes. Srerrangachar (1939, 1943) showed this enzyme to be a polyphenol oxidase that contains copper. For more information, read these articles, as well as the one by Roberts (1958A).

Catechol oxidase was long thought to be insoluble in water. Modern experimental methods, however, indicate that it is indeed soluble (Sanderson 1964A). Wickremasinghe *et al.* (1967) found that this enzyme occurs in the epidermis and around the vascular bundles of the tea leaf.

In addition to catechol oxidase, peroxidase, chlorophyllase, pectin methyl esterase, peptidase, phosphatases, 5-dehydroshikimate reductase, and leucine transaminase, are present in tea leaves. The function of pectin methyl esterase is the demethylation of pectins during tea manufacture. Peptidase breaks down the proteins during withering. The action of 5-dehydroshikimate reductase is the biosynthesis of tea polyphenols and that of leucine transaminase is the biosynthesis of terpenes. The function of peroxidase, chlorophyllase, and phosphatases is unknown.

AMINO ACIDS

A number of free amino acids have been reported in the tissues of the fresh tea flush: aspartic acid, glutamic acid, asparagine, alanine, β-alanine, glycine, serine, threonine, lysine, glutamine, histidine, arginine, tyrosine, valine, leucine, isoleucine, proline, phenylalanine, and tryptophan. In addition to these amino acids, theanine is also found. It is present in the largest amount, about 1.2% on the dry weight basis of the flushes (Cartwright *et al.* 1954). It is thought that these amino acids are responsible for the formation of tea aroma during the process of manufacture.

COOH
|
HCNH$_2$
|
CH$_2$
|
CH$_2$
|
$O=C-NHC_2H_5$

Theanine

Purine Bases

Caffeine is the most important purine base in tea, about 3–5%. Theobromine and theophyllin are present also, but in much smaller quantities, 0.17% for theobromine and 0.013% for theophylline in black tea (Michl and Haberler 1954).

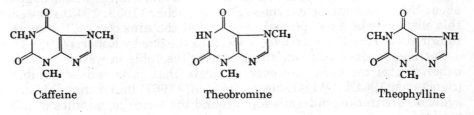

Caffeine Theobromine Theophylline

Carbohydrates

The sugars found in the fresh tea flush are glucose, fructose, and sucrose, together with traces of raffinose and stachyose. Pectin, other extractable polysaccharides, and crude fiber are also present. There has been some disagreement in the literature on the exact compounds and quantities occurring in tea. Starch is reported as being present by some workers but absent by others. This may be partially caused by differences in variety, growing conditions, and methods of determination.

Japanese workers found that fructose, glucose, arabinose, sucrose, maltose, raffinose, and stachyose were present in black tea (Mizuno and Kimpyo 1955). Table 18.3 shows Russian data for sugars in black tea manufactured by the new Russian technology and by classic technology. From these and other studies, the Russian workers concluded that as a result of the processing, the content of sucrose decreases and that of the monosaccharides increases.

Chlorophyll and Carotenoid Pigments

According to Sanderson (1972), chlorophyll *a* and *b* and the carotenoids are the important pigments in the fresh tea flush. Co and Sander-

TABLE 18.3. Sucrose, Glucose, and Fructose in Tea (% Dry Matter)

Sugars	Manufacture of Black Tea	
	New Technology	Classic Technology
Sucrose	0.17	0.09
Glucose	0.46	0.78
Fructose	0.40	0.48
Total sugar content	1.03	1.35

Source: Bokuchava and Skobeleva (1969). Reproduced with permission of Academic Press, New York.

son (1970) found 1.4 mg of chlorophyll in 1 g, dry weight, of fresh tea flush. During the manufacture of black tea, a considerable amount of the chlorophyll is changeed to degradation products, specifically pheophytin *a*. However, the commercial sample they used also showed the formation of pheophytin *b* and pheophorbide. The chlorophyll *b* in their fresh tea flush was 481 μg/g, and the chlorophyll *a* content was 955 μg/g. At the end of the processing, after firing and drying, the amounts of chlorophyll *b* decreased to 315 μg/g and chlorophyll *a* to 650 μg/g. At the same time the pheophytin *a* increased to 234 μg/g. Russian workers showed considerable loss in the yellow pigments during the manufacture of black tea (Bokuchava and Skobeleva 1969; Nikolaishvili and Adeishvili 1966; see also Sanderson *et al*. 1971).

Organic Acids

The following organic acids have been found in fresh tea flush: oxalic, malic, citric, succinic, and isocitric acids. Quinic and shikimic acids have been found also. Small amounts of aconitic acid have been found in extracts from black tea, formed as a result of the oxidation of theaflavin gallate, (−)-epigallocatechin gallate, and gallic acid (Roberts and Russell 1957).

Mineral Substances

A variety of minerals are found in the tea leaf. Copper is extremely important because it is a constituent of tea catechol oxidase which is vital in the manufacture of black tea. It has been found that approximately 12–18 ppm of copper (dry weight) in the tea leaf is necessary to produce enough catechol oxidase for the fermentation step to make black tea (Child 1955).

The tea plant is an accumulator of aluminum, but the role of this metal in the biochemistry of the plant is not clear (Chenery 1955; Jurd 1962).

Volatiles

This subject is discussed in the next section.

CHEMISTRY OF TEA MANUFACTURE

In studying Table 18.1, one can see that the important chemical change in the manufacturing process for black tea is that from simple polyphenols to oxidized polyphenols. Other changes may not be obvious, but it is easy to see the shift in color of the prepared infusions of green tea to the reddish-brown of black tea infusions. The fact that no protein material is found in the brew made from black tea indicates that though initially insoluble, it became so in the process of manufacture.

Withering

Not too much is known concerning the biochemistry of withering; however, there is an increase in the permeability. According to Sanderson (1964C), this increased permeability is important in the mixing of the cell contents when the leaves are rolled. In this step the catechol oxidase and the tea catechins, which are found in different locations of the tea leaf tissue, are brought together so that fermentation can start.

Fermentation

This is the most important step that takes place in the manufacture of black tea. It allows the catechins of the tea leaf to be oxidized. As a result of these reactions, two main fractions of pigments, theaflavins and thearubigins, are formed. These fractions are mixtures rather than pure compounds. Other compounds are also formed, both colored and colorless. To understand the connection between the formation of substances that produce the color of the infusion of black tea and the oxidation of tea catechins, Roberts (1958A) used a model tea fermentation system.

It has been shown that all tea catechins are oxidized during the fermentation of tea (Dzhemukhadze *et al.* 1957, 1964). However, according to Pierce *et al.* (1969B) the gallocatechins are oxidized in preference to the catechol-catechins. These findings were later confirmed (Pierce *et al.* 1969A). This agrees with Roberts (1957) concerning the relative oxidation potentials of these compounds.

According to Sanderson (1964C), the extent of the oxidation of the catechins depends on the fermentation conditions used. It involves also the efficiency of the rolling process, as well as the time used for the oxidation, the state of the activity of the catechol oxidase found in the tea flush, and the catechins' relative oxidation potentials (Robert 1957; Sanderson 1964B, C; Takeo 1966).

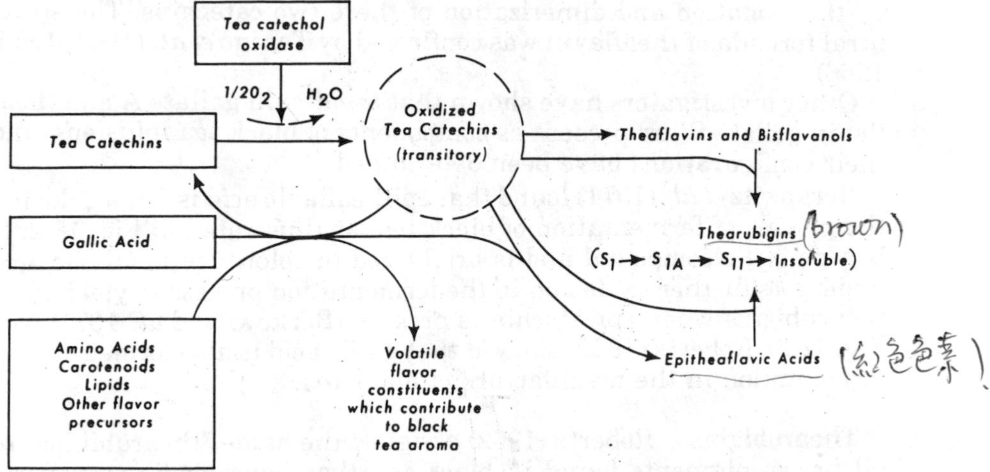

FIG. 18.1. Proposed scheme for tea fermentation. Materials present in fresh (withered) tea flush are enclosed in rectangles. The amount of any of the above materials present in black tea would be dependent on the conditions of black tea.
From Sanderson (1972). Reproduced with permission of Academic Press.

Figure 18.1 shows a proposed scheme for tea fermentation. It gives a graphic representation of the changes that take place during the manufacture of black tea and the origins of the compounds present in the finished product.

Theaflavins. In the mixture of theaflavins, a pure theaflavin has been separated. Compounds of theaflavin exist in the mixture also.

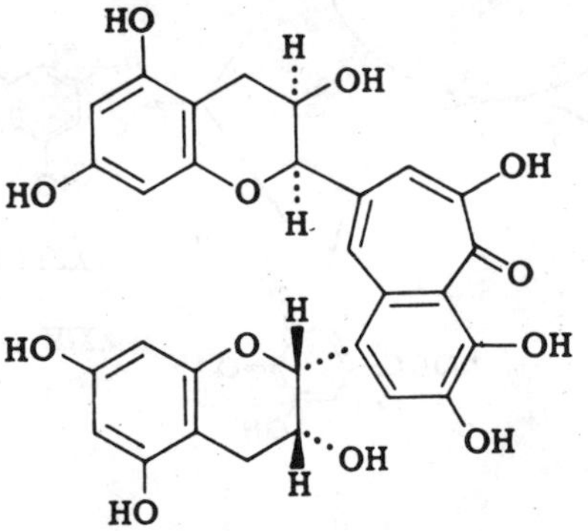

Theaflavin

Takino and Imagawa (1964) showed that the pigment formed from the two catechins, (−)-epicatechin and (−)-epigallocatechin, was the same as the theaflavin extracted from black tea. Theaflavin is formed by the oxidation and dimerization of these two catechins. The structural formula of theaflavin was confirmed by Takino *et al.* (1964, 1965, 1966).

Other investigators have shown that theaflavin gallate A and theaflavin gallate B are present as components of black tea infusions, and their configurations have been determined.

Berkowitz *et al.* (1971) found that epitheaflavic acid is formed during the process of fermentation of black tea, as illustrated in Fig. 18.2. It is a phenolic compound and is bright red in color. Epitheaflavic acid undergoes further oxidation in the fermentation process to yield other thearubigins when epicatechin is present (Berkowitz *et al.* 1971) (see Fig. 18.3). Roberts (1962) showed that gallic acid is also formed during fermentation in the manufacture of black tea.

Thearubigins. Roberts (1962) assigned the name "thearubigins" to all brown pigments found in black tea that have acidic properties. Thearubigins are made up of a group of phenolic compounds that are heterogeneous in composition. Roberts and Smith (1963) worked out a procedure for the assay of tea for these compounds, which is also used as a method for their separation. According to these authors thearubigins make up about 9–19% of black tea—this is a large percentage of

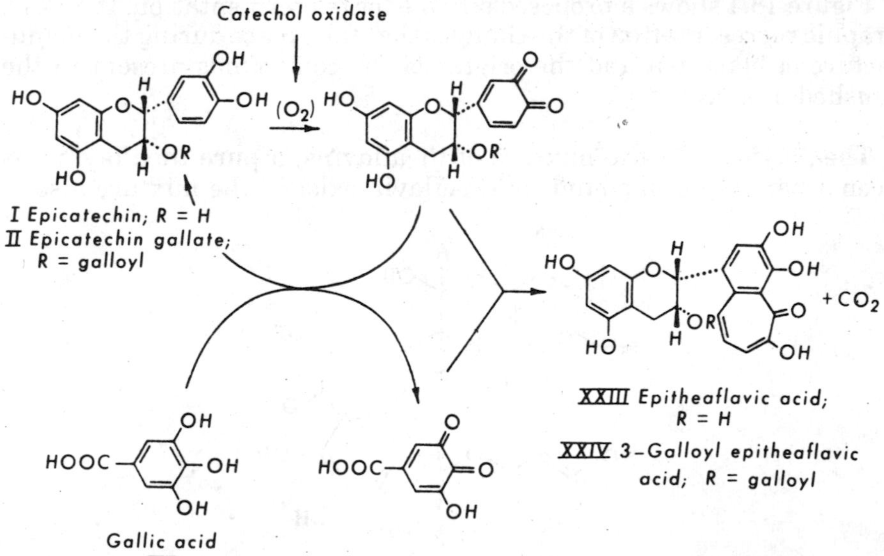

FIG. 18.2. Formation of epitheaflavic acids in tea fermentation system. Note dependence of gallic acid oxidation on (−)-epicatechin (gallate) oxidation, and formation of (3-galloyl)-epitheaflavic acid from oxidized (−)-epicatechin (gallate) and oxidized gallic acid.
From Sanderson (1972). Reproduced with permission of Academic Press.

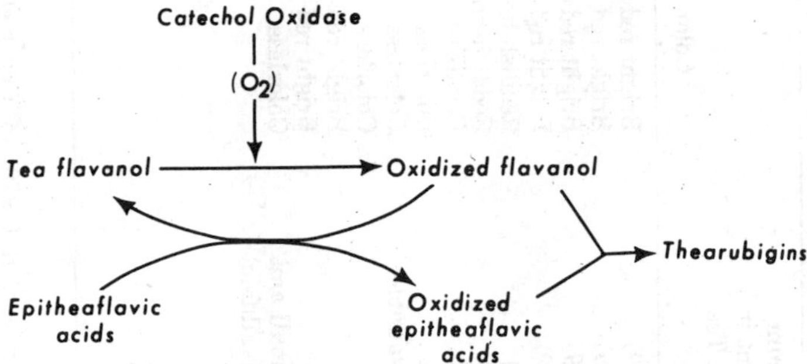

FIG. 18.3. Formation of thearubigins from epitheaflavic acids in tea fermentation system. Note dependence of epitheaflavic acids transformation on oxidation of tea flavanols.
From Sanderson (1972). Reproduced with permission of Academic Press.

the solids (30–60%) found in an infusion of black tea—and are quite largely responsible for its color.

Roberts (1962) found that during the fermentation of tea, theaflavins initially increase, but then eventually decrease. However, the thearubigins continue to increase throughout the entire fermentation. Sanderson *et al.* (1972) found that thearubigins are formed as a result of the oxidation of any one of the catechins of tea, or a combination of them.

Sanderson *et al.* (1972), using model systems, found that different thearubigins are formed when different combinations of catechins are the starting material. Also, the quantity of thearubigins formed increased, but at the same time their solubility lessened with lengthening of the period of oxidation. This decreased solubility is most likely the result of an increase in the molecular weights and also the more complex composition of the thearubigins which are in turn probably caused by oxidative polymerization between the polyphenolic compounds present. Considerable work has been done on this problem, but much more is left to be accomplished.

Table 18.4 gives an excellent summary of the phenolic compounds that form during tea fermentation.

The only known enzymic oxidations taking place during the fermentation step in the manufacture of black tea are those of the tea catechins acted on by catechol oxidase. It is likely that these actions bring about the conditions for the subsequent secondary oxidations which take place.

Tea Cream. According to Roberts (1962, 1963) the precipitate that forms in brewed tea when it cools, does so naturally, and is known as tea cream. It is made up mainly of thearubigins, theaflavins, and caffeine, present in about 66 : 17 : 17 parts, respectively. This precipitation causes problems in the making of instant tea.

TABLE 18.4. Phenolic Compounds Known to Form During Tea Fermentation

Compound	Molecular Formula	Molecular Weight	Best Present Structure Assignment[a] (References)	Amount Present in Black Tea (%)	Color
Theaflavin	$C_{29}H_{24}O_{12}$	564	*2, 10*	0.28–1.63 (*9*)	Bright red
Theaflavin gallate A	$C_{36}H_{28}O_{16}$	716	*4, 5*	0.28–1.63 (*9*)	Bright red
Theaflavin gallate B	$C_{36}H_{28}O_{16}$	716	*4, 5*	0.28–1.63 (*9*)	Bright red
Theaflavin digallate	$C_{43}H_{32}O_{20}$	868	*4, 5*	0.28–1.63 (9)	Bright red
Thearubigins—S_{I}	—	700 to	*3*	5.1–14.8 (*9*)	Reddish-brown
Thearubigins—S_{IA}	—	40,000 (*1*)	*3*	5.1–14.8 (*9*)	Reddish-brown
Thearubigins—S_{II}	—		*3*	5.1–14.8 (*9*)	Reddish-brown
Bisflavanol A	$C_{44}H_{34}O_{22}$	914	*7*	Small Amt.	Colorless
Bisflavanol B	$C_{37}H_{30}O_{18}$	762	*7*	(No data available)	Colorless
Bisflavanol C	$C_{30}H_{26}O_{14}$	610	*7*		Colorless
Epitheaflavic acid	$C_{21}H_{16}O_{10}$	428	*4, 6*	Trace (*1*)	Bright red
3-Galloylepitheaflavic acid	$C_{28}H_{20}O_{14}$	580	Unconfirmed		Bright red
Tricetinidin	$C_{15}H_{11}O_{6}$	287	Unconfirmed	Trace (*8*) Small amt. no data available)	Colorless
Ellagic acid	$C_{14}H_{6}O_{8}$	302			
Unknown Q ("Q complex")	Thought to be a mixture of 11 and 12 and others		Unconfirmed	Trace (*8*)	
Unknown R			Unconfirmed	Trace (*8*)	
Unknown Z			Unconfirmed	Trace (*8*)	

Source: Sanderson (1972). Reproduced with permission of Academic Press, New York.

Key to references: (*1*) Berkowitz *et al.* (1971); (*2*) Brown *et al.* (1966); (*3*) Brown *et al.* (1969A, 1969B); (*4*) Bryce *et al.* (1970); (*5*) Coxon *et al.* (1970A); (*6*) Coxon *et al.* (1970B); (*7*) Ferretti *et al.* (1968); (*8*) Roberts (1962); (*9*) Roberts and Smith (1963); (*10*) Takino *et al.* (1965, 1966).

TEA AROMA

The aromas of green and black tea develop during the manufacturing process, and they both, of course, differ from each other and from the aroma of the fresh tea flush. The odor of black tea develops during the entire production process. A variety of odors can be noticed during production, but the usual odor of black tea is achieved as a result of the final or firing stage of the process.

It is only since the advent of gas chromatography and other techniques capable of analyzing very small samples that really satisfactory results have been possible. However, much remains to be done, particularly in regard to the pathways by which these compounds are formed. It is now known that about 140 compounds are present in black tea aroma, and many more compounds are likely to be found. Further, some compounds present in very small quantities may significantly influence the aroma of tea.

Studies were conducted by Yamanishi *et al.* (1966) on the development and changes in volatile compounds during the preparation of black tea. The withering stage produced the largest increases in the following compounds: hexyl alcohol, nerol, and *trans*-2-hexenoic acid. Other compounds to show increases during this period included *trans*-2-hexenol, linalool oxide, *n*-valeraldehyde, capronaldehyde, *n*-heptanal, *trans*-2-hexenal, *trans*-2-octenal, benzaldehyde, phenylacetaldehyde, *n*-butyric, isovaleric, *n*-caproic, *cis*-3-hexenoic, and salicylic acids, and *o*-cresol. The following compounds decreased considerably during this period: *cis*-2-pentenol, linalool, geraniol, benzyl alcohol, phenylethanol, and acetic acid. During the fermentation step most of the aroma components increased, but increases for the following were larger than usual: 1-pentene-3-ol, *cis*-2-penteneol, benzyl alcohol, *trans*-2-hexeneal, benzaldehyde, *n*-caproic, *cis*-3-hexenoic, and salicylic acids. As a result of firing, rather large decreases were noticed in most alcohols, carbonyl, and phenolic compounds. Acetic, propionic, and isobutyric acids also showed considerable increases.

Saijo and Kuwabara (1967A) found that black tea leaves had more aldehydes and acids, whereas the fresh tea leaves from which they were made were relatively richer in alcohols.

Other workers (Yamanishi *et al.* 1968) found that Ceylon tea flushes grown in high altitudes yielded high quality black teas with larger amounts of linalool, linalool oxides, *cis*-jasmone, and geraniol than the lower quality black teas produced from tea flushes grown in lower altitudes.

For a list of the compounds comprising tea aroma, the student is advised to consult Table 13 of Sanderson (1972).

Flavoring components of manufactured green tea were studied by Yamanishi *et al.* (1966B). The essential oil from manufactured green tea was separated into carboxylic, phenolic, carbonyl, and alcoholic fractions. The components detected were isobutyl alcohol, butyl alcohol, 1-penten-3-ol, isoamyl alcohol, *trans*-2-hexenal, amy alcohol, *cis*-2-pentenol, hexanol, *cis*-3-hexenol, *trans*-2-hexenol, linalool oxide (*trans,* furanoid), linalool oxide (*cis,* furanoid), linalool, benzaldehyde, linalool oxide (*cis,* pyranoid), nerol, geraniol, benzyl alcohol, phenylethanol, and several unidentified compounds.

Yamanishi *et al.* (1970) worked on the medium- and high-boiling flavor components of green tea. They found the following compounds: 4-ethylguaiacol, carinenol, nerolidol, 3,7-dimethyl-1,5,7-octatriene-3-ol, 3,5-octadiene-2-one, α-ionone, β-ionone, indole, and dibutylphthalate. The first six of these compounds, together with linalool, acetophenone, and three unidentified compounds seemed to form the flavor characteristic of green tea.

Compounds Concerned with Aroma Development

A number of substances have been shown or suggested as being involved in the formation of compounds in tea aroma.

Amino Acids. Certain of the amino acids are known to be effective in the development of aroma compounds. Valine gives rise to isobutyraldehyde, leucine to isovaleraldehyde, isoleucine to 2-methylbutanal, methionine to methional, phenylalanine to phenacetaldehyde, glycine to formaldehyde, and alanine to acetaldehyde (Co and Sanderson 1970; Nakabayashi 1958; Saijo and Takeo 1970B). These aldehydes are produced via the Strecker degradation. Using model systems it has been further determined that both tea enzymes and tea catechins are essential for the formation of aldehydes in these reactions, but some other oxidizing agents, such as dehydroascorbic acid (Co and Sanderson 1970), can bring about the development of aldehydes from amino acids by the Strecker degradation.

Carotenes. There is a considerable decrease in the carotenoids of the fresh tea flush during its preparation to yield black tea (Sanderson *et al.* 1971). The largest decrease takes place during the first part of the fermentation step, and oxidative degradation of β-carotene results in the formation of β-ionone together with some other unidentified volatile and nonvolatile compounds. The oxidation of tea catechins is essential before the breakdown of β-carotene can take place. It seems quite possible that other carotenoids could undergo oxidative degradation to yield volatile compounds that could be a part of the aroma of tea.

Other Compounds. It is indeed probable that other compounds are present in the harvested tea flush, which during processing could produce compounds present in the aroma of black tea. Such things as tea lipids could perhaps yield these compounds. In addition, alcohols could be oxidized to aldehydes and acids. These are yet to be studied.

SUMMARY

Three main types of tea are produced: black, Oolong, and green teas.

Black tea is produced as the result of four basic steps: withering, rolling, fermenting, and firing. Each of these steps is important for the production of good black tea.

The rolling step injures the leaf cells so that the fermentation step can begin. This is brought about by combination of the flavonols and the catechol oxidase. This process, therefore, is a process of enzymatic oxidation. Fermentation is completed in the fermentation step. Enzymatic activity is stopped in the firing step.

Oolong tea is semi-fermented. This is accomplished by a shortened withering step.

A mixture of a number of phenolic compounds are present in tea. The four basic groups are flavanols (catechins), flavonols and flavonol glycosides, leucoanthocyanins, and depsides. About 30% of the dry weight of the fresh tea flush is made up of these important compounds. (−)-Epigallocatechin gallate is the quantitatively important flavanol in all parts of the tea shoot. Inositol is present in the free and bound state in tea leaves.

The most important enzyme in the tea flush is catechol oxidase.

Among the amino acids present, theanine is present in the largest amount.

Caffeine is the most important purine base present in tea, from 3 to 5%.

It was shown that the pigment formed from the two catechins, (−)-epicatechin and (−)-epigallocatechin, is the same as the theaflavin extracted from black tea.

Epitheaflavic acids are formed during the process of fermentation of black tea. These compounds are bright red in color.

Thearubigins, found in black tea, are a group of phenolic compounds that are heterogeneous in composition.

Black tea leaves have more aldehydes and acids, whereas the fresh tea leaves from which they were made are relatively richer in alcohols.

The carotenes and amino acids are concerned with aroma development. About 140 compounds are present in black tea aroma.

BIBLIOGRAPHY

BERKOWITZ, J. E., COGGON, P., and SANDERSON, G. W. 1971. Formation of epitheaflavic acid and its transformation to thearubigins during tea fermentation. Phytochemistry *10,* 2271–2278.

BOKUCHAVA, M. A., and SKOBELEVA, N. I. 1969. The chemistry and biochemistry of tea and tea manufacture. *In* Advances in Food Research, Vol. 17, C. O. Chichester, E. M. Mrak, and G. F. Stewart (Editors). Academic Press, New York.

BRADFIELD, A. E., and BATE-SMITH, E. C. 1950. Chromatographic behaviour and chemical structure. II. The tea catechins. Biochim. Biophys. Acta *4,* 441–444.

BRADFIELD, A. E., and PENNEY, M. 1948. The catechins of green tea. II. J. Chem. Soc. 2249–2254.

BRADFIELD, A. E., PENNEY, M., and WRIGHT, W. B. 1947. The catechins of green tea. I. J. Chem. Soc. 32–36.

BROWN, A. G., EYTON, W. B., HOLMES, A., and OLLIS, W. D. 1969A. Identification of the thearubigins as polymeric proanthocyanidins. Nature *221,* 742–744.

BROWN, A. G., EYTON, W. B., HOLMES, A., and OLLIS, W. D. 1969B. The identification of the thearubigins as polymeric proanthocyanidins. Phytochemistry *8,* 2333–2340.

BROWN, A. G., FALSHAM, C. P., HASLAM, E., HOLMES, A., and OLLIS, W. D. 1966. Constitution of theaflavin. Tetrahedron Lett. *11,* 1193–1204.

BRYCE, T., COLLIER, P. D., FOWLIS, I., THOMAS, P. E., FROST, D., and WILKINS, C. K. 1970. Structures of the theaflavines of black tea. Tetrahedron Lett. *32,* 2789–2792.

CARTWRIGHT, R. A., and ROBERTS, E. A. H. 1954. The sugars of manufactured tea. J. Sci. Food Agric. *5,* 600–601.

CARTWRIGHT, R. A., ROBERTS, E. A. H., and WOOD, D. J. 1954. Theanine, an amino-acid *N*-ethyl amide present in tea. J. Sci. Food Agric. *5,* 597–599.

CARTWRIGHT, R. A., ROBERTS, E. A. H., FLOOD, A. E., and WILLIAMS, A. H. 1955. The suspected presence of *p*-coumarylquinic acids in tea, apple, and pear. Chem. Ind. 1062–1063.

CHENERY, E. M. 1955. A preliminary study of aluminum and the tea bush. Plant Soil *6,* 174–200.

CHILD, R. 1955. Copper: Its occurrence and role in tea leaf. Trop. Agric. (St. Augustine) *32,* 100–106.

CO, H., and SANDERSON, G. W. 1970. Biochemistry of tea fermentation: Conversion of amino acids to black tea aroma constituents. J. Food Sci. *35,* 160–164

COLLIER, P. D., and MALLOWS, R. 1971. The estimation of flavanols in tea by gas chromatography of their trimethylsilyl derivatives. J. Chromatogr. *57,* 29–45.

COXON, D. T., HOLMES, A., and OLLIS, W. D. 1970A. Isotheaflavin. A new black tea pigment. Tetrahedron Lett. *60,* 5241–5246.

COXON, D. T., HOLMES, A., and OLLIS, W. D. 1970B. Theaflavic and epitheaflavic acids. Tetrahedron Lett. *60,* 5247–5250.

COXON, D. T., HOLMES, A., OLLIS, W. D., and VORA, V. C. 1970C. The constitution and configuration of the theaflavin pigments of black tea. Tetrahedron Lett. *60,* 5237–5240.

DZHEMUKHADZE, K. M., SHAL'NEVA, G. A., and MILESHKO, L. F. 1957. Catechol changes in tea fermentation. Biokhimiya *22,* 836–840. (Russian) Chem. Abstr. *52,* 14023a (1958).

DZHEMUKHADZE, K. M., BUZUN, G. A., and MILESHKO, L. F. 1964. Enzymic oxidation of catechols. Biokhimiya *29,* 882–888. (Russian) Chem. Abstr. *62,* 1893h (1965).

EDEN, T. 1965. Tea, 2nd Edition. Longmans, Green and Co., London.

FERRETTI, A., FLANAGAN, V. P., BONDAROVICH, H. A., and GIANTURCO, M. A. 1968. The chemistry of tea. Structures and compounds A and B of Roberts and reactions of some model compounds. J. Agric. Food Chem. *16,* 756–761.

GONZALEZ, J. G., COGGON, P., and SANDERSON, G. W. 1972. Biochemistry of tea fermentation: Formation of *t*-2-hexenal from linolenic acid. J. Food Sci. *37,* 797–798.

GREGORY, R. P. F., and BENDALL, D. S. 1966. The purification and some properties of the polyphenol oxidase from tea (*Camellia sinensis* L.). Biochem. J. *101,* 569–581.

HAINSWORTH, E. 1969. Tea. *In* Encyclopedia of Chemical Technology, Vol. 19. John Wiley & Sons, New York.

JURD, L. 1962. Special properties of flavonoid compounds. *In* Chemistry of Flavonoid Compounds, T. A. Geissman (Editor). Macmillan Co., New York.

KHAREBAVA, L. G., DARASELIYA, Z. G., GUGUSHVILI, T. A., and STARODUBTSEVA, V. P. 1969. Volatile substances in a tea shoot. Subtrop. Kul't. No. 2. 65–69. (Russian) Chem. Abstr. *72,* 120224j (1970).

KURSANOV, A. L., and BROVCHENKO, M. I. 1950. Tannins of various organs of the tea plant. Biokhim. Chain. Proizvod. No. 6, 53–69. (Russian) Chem. Abstr. *46,* 2631a (1952).

KURSANOV, A. L., DZHEMUKHADZE, K. M., and ZAPROMETOV, M. N. 1947. Condensation of tea leaf catechols during oxidation. Biokhimiya *12,* 421–436 (Russian) Chem. Abstr. *43,* 728 (1949).

LI, L. P., and BONNER, J. 1947. Experiments on the localization and nature of tea oxidase. Biochem. J. *41,* 105–110.

MICHL, H., and HABERLER, F. 1954. On the determination of purines in caffeine containing drugs. Monatsch. Chem. *85,* 779–795. (German)

MILLIN, D. J., CRISPIN, D. J., and SWAINE, D. 1969. Nonvolatile components of black tea and their contribution to the character of the beverage. J. Agric. Food Chem. *17,* 717–722.

MIZUNO, T., and KIMPYO, T. 1955. Carbohydrates of tea. I. The kinds of carbohydrates in black tea. J. Agric. Chem. Soc. Jpn. *29,* 847–852.

NAKABAYASHI, T. 1958. Formation mechanism of black tea aroma. V. The precursor of the volatile carbonyl compounds. Nippon Nogei Kagaku Kaishi *32,* 941–945. (Japanese) Chem. Abstr. *55,* 1577a (1961).

NEISH, A. C. 1960. Biosynthetic pathways of aromatic compounds. Annu. Rev. Plant Physiol. *11,* 55–80.

NIKOLAISHVILI, D. K., and ADEISHVILI, N. I. 1966. Chromatographic study of the quantitative changes in tea leaf pigments during production of black tea. Byull. Vses. Nauch.-Issled. Inst. Chain. Promsti. No. 2, 57–60. (Russian) Chem. Abstr. *69,* 66240j (1968).

OTA, I., NAKATA, N., and WADA, K. 1970. Volatile components of black tea. I. Difference of volatile components in essential oil of fresh tea leaves. Chagyo Gijutsu Kenkyu No. 39, 53–62. (Japanese) Chem. Abstr. *73,* 97544y (1970).

PATSCHKE, L., and GRIESBACH, H. 1965. 2′,4,4′,6′-tetrahydroxychalone 2′-β-D-glucoside-^{14}C as a precursor of tea catechols. Z. Naturforsch. Teil B *20,* 399. (German)

PIERCE, A. R., GONZALEZ, J. G., and SANDERSON, G. W. 1969A. Changes in tea flavanols during tea fermentation. Unpublished observations.

PIERCE, A. R., GRAHAM, H. N., GLASSNER, S., MADLIN, H., and GONZALEZ, J. G. 1969B. Analysis of tea flavanols by gas chromatography of their trimethylsilyl derivatives. Anal. Chem. *41,* 298–302.

ROBERTS, E.A.H. 1942. The chemistry of tea-fermentation. Adv. Enzymol. *2,* 113–133.

ROBERTS, E. A. H. 1957. Oxidation-reduction potentials in tea fermentation. Chem. Ind. (London) 1354–1355.

ROBERTS, E. A. H. 1957. Oxidative condensation of flavanols in tea fermentation. Chem. Ind. (London) 1355–1356.

ROBERTS, E. A. H. 1958A. The phenolic substances of manufactured tea. II. Their origin as enzymic oxidation products in fermentation. J. Sci. Food Agric. *9,* 212–216.

ROBERTS, E. A. H. 1958B. The chemistry of tea manufacture. J. Sci. Food Agric. *9,* 381–390.

ROBERTS, E. A. H. 1962. Economic importance of flavonoid substances: Tea fermentation. *In* The Chemistry of Flavonoid Compounds, T. A. Geissman (Editor). Pergamon, London.

ROBERTS, E.A.H. 1963. The phenolic substances of manufactured tea. X. The creaming down of tea liquors. J. Sci. Food Agric. *14,* 700–705.

ROBERTS, E. A. H., CARTWRIGHT, R. A., and OLDSCHOOL, M. 1957. The phenolic substances of manufactured tea. I. Fractionation and paper chromatography of water-soluble substances. J. Sci. Food Agric. *8,* 72–80.

ROBERTS, E. A. H., CARTWRIGHT, R. A., and WOOD, D. J. 1956. The flavanols of tea. J. Sci. Food Agric. 7, 637–646.

ROBERTS, E. A. H., and MYERS, M. 1958. Theogallin, a polyphenol occurring in tea. II. Identification as a galloylquinic acid. J. Sci. Food Agric. *9,* 701–705.

ROBERTS, E. A. H., and MYERS, M. 1959A. The phenolic substances of manufactured tea. IV. Enzymic oxidations of individual substrates. J. Sci. Food Agric. *10,* 167–172.

ROBERTS, E. A. H., and MYERS, M. 1959B. The phenolic substances of manufactured tea. VI. The preparation of theaflavin and of theaflavin gallate. J. Sci. Food Agric. *10,* 176–179.

ROBERTS, E. A. H., and RUSSELL, G. R. 1957. Oxidation of gallic acid and gallic acid esters to aconitic acid. Chem. Ind. (London) 1598–1599.

ROBERTS, E. A. H., and SMITH, R. F. 1963. The phenolic substances of manufactured tea. IX. The spectrophotometric evaluation of tea liquors. J. Sci. Food Agric. *14,* 689–699.

ROBERTS, E. A. H., and WOOD, D. J. 1950. The fermentation process in tea manufacture. II. Oxidation of substrates by tea oxidase. Biochem. J. *47,* 175–186.

SAIJO, R., and KUWABARA, Y. 1967A. Volatile flavor of black tea. I. Formation of volatile components during black tea manufacture. Agric. Biol. Chem. *31,* 389–396.

SAIJO, R., and KUWABARA, Y. 1967B. Volatile flavor of black tea. II. The differences in the volatile components among tea seasons and tea varieties. Chagyo Gijutsu Kenkyu *35,* 88–92. (Japanese) Chem. Abstr. *68,* 77047g. (1968).

SAIJO, R., and TAKEO, T. 1970A. Production of phenylacetaldehyde from L-phenylalanine in tea fermentation. Agric. Biol. Chem. *34,* 222–226.

SAIJO, R., and TAKEO, T. 1970B. Formation of aldehydes from amino acids by tea leaves extracts. Agric. Biol. Chem. *34,* 227–233.

SANDERSON, G. W. 1964A. Extraction of soluble catechol oxidase from tea shoot tips. Biochim. Biphys. Acta *92,* 622–624.

SANDERSON, G. W. 1964B. Changes in the level of polyphenol oxidase activity in tea flush on storage after plucking. J. Sci. Food Agric. *15,* 634–639.

SANDERSON, G. W. 1964C. Theory of withering in tea manufacture. Tea Qt. *35,* 146–163.

SANDERSON, G. W. 1966. 5-Dehydroshikimate reductase in the tea plant (*Camellia sinensis* L.). Properties and distribution. Biochem. J. *98,* 248–252.

SANDERSON, G. W. 1968. Change in cell membrane permeability in the tea flush on storage after plucking and its effect on fermentation in tea manufacture. J. Sci. Food Agric. *19,* 637–639.

SANDERSON, G. W. 1972. The chemistry of tea and tea manufacturing. *In* Recent Advances in Phytochemistry, Vol. 5, V. C. Runeckles and T. C. Tso (Editors). Academic Press, New York.

SANDERSON, G. W., and GRAHAM, H. W. 1973. Formation of black tea aroma. J. Agric. Food Chem. *21,* 576–585.

SANDERSON, G.W., and SELVENDRAN, R. R. 1965. The organic acids in tea plants. A study of the non-volatile organic acids separated on silica gel. J. Sci. Food Agric. *16,* 251–258.

SANDERSON, G. W., and SIVAPALAN, K. 1966A. Translocation of photosynthetically assimilated carbon in tea plants. Tea Qt. *37,* 140, 142–153.

SANDERSON, G. W., and SIVAPALAN, K. 1966B. Effect of leaf age on photosynthetic assimilation of carbon dioxide in tea plants. Tea Qt. *37,* 11–26.

SANDERSON, G. W., CO, H., and GONZALEZ, J. G. 1971. Biochemistry of tea fermentation: The role of carotenes in black tea aroma formation. J. Food Sci. *36,* 231–236.

SANDERSON, G. W., BERKOWITZ, J. E., CO, H., and GRAHAM, H. N. 1972. Biochemistry of tea fermentation: Products of the oxidation of tea flavanoles in a model tea fermentation system. J. Food Sci. *37,* 399–404.

SATO, S., SASAKURA, S., KOBAYASHI, A., NAKATANI, Y., and YAMANISHI, T. 1970. Flavor of black tea. VI. Intermediate and high boiling components of the neutral fraction. Agric. Biol. Chem. *34,* 1355–1367.

SREERANGACHAR, H. B. 1939. The endo-enzyme in tea fermentation. Curr. Sci. *8,* 13–14.

SREERANGACHAR, H. B. 1943. Studies on the "fermentation" of Ceylon tea. 6. The nature of the tea-oxidase system. Biochem. J. *37,* 661–667.

STAHL, W. H. 1962. The chemistry of tea and tea manufacturing. *In* Advances in Food Research, Vol. 11, C. O. Chichester, E. M. Mrak, and G. F. Stewart (Editors). Academic Press, New York.

TAKEI, S., SAKTO, Y., and ONO, M. 1935. Odoriferous substances of green tea. VI. Constituents of tea oil. Bull. Inst. Phys. Chem. Res. (Tokyo) *14,* 1262–1274. (Japanese) Chem. Abstr. *30,* 6889[5] (1936).

TAKEO, T. 1966. Tea leaf polyphenol oxidase. III. Changes of polyphenol oxidase activity during black tea manufacture. Agric. Biol. Chem. *30,* 529–535.

TAKINO, Y., and IMAGAWA, H. 1964. Crystalline reddish-orange pigment of manufactured black tea. Agric. Biol. Chem. *28,* 255–256.

TAKINO, Y., IMAGAWA, H., HORIKAWA, H., and TANAKA, A. 1964. The mechanism of the oxidation of tea leaf catechins. III. Formation of a reddish-orange pigment and its spectral relation to some benzotropolone derivatives. Agric. Biol. Chem. *28,* 64–71.

TAKINO, Y., FERRETTI, A., FLANAGAN, V., GIANTURCO, M., and VOGEL, M. 1965. Structure of theaflavin, a polyphenol of black tea. Tetrahedron Lett. *45,* 4019–4025.

TAKINO, Y., FLANAGAN, V., GIANTURCO, M. A., and VOGEL, M. 1967. Spectral evidence for the structure of three flavanotropolones related to theaflavin, an orange-red pigment of black tea. Can. J. Chem. *45,* 1949–1956.

WICKREMASINGHE, R. L., ROBERTS, G. R., and PERERA, B. P. M. 1967. The localization of the polyphenol oxidase of tea leaf. Tea Qt. *38,* 309–310.

YAMANISHI, T., KIRIBUCHI, T., SAKAI, M., FUJITA, N., IKEDA, Y., and SASA, K. 1963. The flavor of green tea. V. Examination of the essential oil of the tea-leaves by gas-liquid chromatography. Agric. Biol. Chem. *27,* 193–198.

YAMANISHI,T., KIRIBUCHI, T., MIKUMO, Y., SATO, H., OMURA, A., MINE, A., and KURATA, T. 1965. The flavor of green tea. VI. Neutral fraction of the essential oil of tea leaves. Agric. Biol. Chem. *29,* 300–306.

YAMANISHI, T., KOBAYASHI, A., SATO, H., NAKAMURA, H., OSAWA, K., UCHIDE, A., MORI, S., and SAIJO, R. 1966A. Flavor of black tea. IV. Changes in flavor constituents during the manufacture of black tea. Agric. Biol. Chem. *30,* 784–792.

YAMANISHI, T., KOBAYASHI, A., UCHIDA, A., and KAWASHIMA, Y. 1966B. Flavor of green tea. VII. Flavor components of manufactured green tea. Agric. Biol. Chem. *30,* 1102–1105.

YAMANISHI, T., WICKREMASINGHE, R. L., and PERERA, K. P. W. C. 1968. Quality and flavor of tea. III. Gas-chromatographic analyses of the aroma complex. Tea Qt. *39,* 81–86.

YAMANISHI, T., NOSE, M., and NAKATANI, Y. 1970. Flavor of green tea. VIII. Flavor constituents in manufactured green tea. Agric. Biol. Chem. *34,* 599–608.

YAMANISHI, T., KOSUGE, M., TOKITOMO, Y., and MAEDA, R. 1980. Flavor constituents of Pouchong tea and a comparison of aroma pattern with jasmine tea. Agric. Biol. Chem. *44,* 2139–2142.

ZAPROMETOV, M. N. 1958. The site of the formation of catechins in the tea plant. Fiziol. Rast. *5,* 51–61. (Russian) Chem. Abstr. *52,* 10298a (1958).

ZAPROMETOV, M. N. 1962. The mechanism of catechol biosynthesis. Biokhimiya *27,* 366–376. (Russian) Chem. Abstr. *57,* 2601f (1962).

19

Cocoa and Chocolate

Introduction
Manufacture of Cocoa and Chocolate
Chemical Composition
Changes during Manufacture of Cocoa and Chocolate
Volatiles and Chocolate Flavor
Chocolate and Cocoa Products
Summary
Bibliography

INTRODUCTION

Cacao originated in the Americas, probably in the jungles of the Orinoco and Amazon valleys. Cultivated by the Aztecs of Mexico and by the Mayas of Central America, cacao was being used by them at the time of the discovery of America by Columbus and the conquest of Mexico by Cortés. Columbus brought some cacao beans back to Europe. A beverage called *chocolatl* made from these beans was consumed in large amounts by the Emperor Montezuma. The basic concoction comprised cakes made from the fluid mass of ground chocolate, to which corn and spices as well as vanilla were added. Parts of the cakes were beaten with water and the other ingredients were added resulting in a final product that had the approximate consistency of honey. The Spaniards were said to have been attracted to it only when it was sweetened.

The botanical name, *Theobroma cacao* Linn., is the species most usually cultivated for commercial purposes (although a number of others are known) and belongs to the Sterculiaceae family. It is believed that *T. pentagona* was the species cultivated by the Aztecs at the time of the conquest. There are two main groups of *Theobroma cacao*, namely, Criollo and Forastero. A third form, Trinitario, is, in reality, a hybrid of these two. Forastero is the commercially important group at the present time. However, the other two varieties are used also, and the beans produced command a higher price because of their flavor qualities. Criollo and Trinitario are used in the manufacture of some dark eating chocolates. They are of little importance in the manufacture of milk chocolate.

The cacao tree grows to 15–25 ft high and has evergreen leaves. The tree takes from 8 to 10 years to reach full development although fruits are produced after about 3 years of growth. The flowers of the tree are found on the trunk or the main branches and appear throughout the year. They are white or pale pink in color and are odorless. The ripe cacao pods (the fruit of the tree) are 6–8 in. long and are made up of a rather thick husk in which are found from 20 to 40 seeds surrounded by a mucilaginous pulp. The seeds are the cacao beans. The pods take between 5 to 6 months to mature and when ripe weigh approximately 1 lb.

The cacao tree requires a tropical climate, with humidity and warm temperature necessary for its successful cultivation. The ideal mean temperature is 26.6° C with a range of 18.3°–35° C. Whereas it can be grown at almost sea level, the higher quality product is grown at higher altitudes. In Sri Lanka choice cacao is grown at an elevation of over 1400 ft. It is possible, under certain conditions, to grow cacao trees with irrigation.

MANUFACTURE OF COCOA AND CHOCOLATE

Fermentation

After the fully mature pods are harvested, the pods are cut open, the beans are removed and subjected to fermentation, a curing process. Great care is necessary in the fermentation process to get a satisfactory product. Cacao fermentation is an art rather than a strict science.

Fermentation of the separated beans is accomplished by either of two methods. These methods were designed to provide the necessary heat and to make possible the draining of the sweatings (the liquefaction of the adhering pulp) from the beans. Liquefaction occurs mainly during the first day of fermentation. During this time the temperature rises steadily. It is essential to use sufficient beans so that the heat produced by the process is not lost, but raises the temperature above that of the ambient temperature. In the first method, used by the small farmers, 500–600 lb of beans are piled on banana or plantain leaves in the form of a flat cone. The top is covered with more of these leaves. The second method, the box method, or modifications of it, is used by the large plantations. This method is better because it permits a more even fermentation. The boxes used are 3-ft cubes, which hold about 500 lbs of the beans and have holes or slats in the bottoms to allow the liquefied pulp to drain off. The beans are transferred to other boxes of the same type every two days to permit aeration and to equalize the temperature. The filling of the boxes must be reasonably uniform, and the mixing must be done at regular intervals. It takes about 3 days to attain the temperature of 45° C, and when it reaches approximately 50° C this temperature is maintained until the fermentation is finished. This temperature is necessary to complete the fermentation. Tests showed that all the beans were still alive after 20 hr, but not after 68 hr. In the interim, both types of beans were found. After approximately 160 hours, the air starts to penetrate the cotyledons and browning, which is caused by the action of polyphenol oxidase, begins. These changes are normal.

Roberts (1959) has stated that the fermentation process can be classed as an anaerobic phase, while the later drying stage is oxidative. Powell (1959) has disputed the need for the first or anaerobic phase; however, anaerobic conditions do exist during the time that the pulp is undergoing liquefaction.

Fermentation results in beans that have the required composition to permit the preparation of an acceptable cocoa or chocolate.

Drying

After fermentation, the beans are dried to about 6–8% moisture content. During this process about a 40% loss takes place. This step is necessary so that the beans can be stored without mold formation. The most satisfactory temperature for this purpose is between 45° and 60° C, and it can be done either in the sun or artificially. Too rapid drying at the beginning of the process is undesirable because sufficient oxidation of the tannin components will not take place. After the beans are dried they are ready for shipment.

Roasting

The beans are cleaned before being roasted, during which process broken beans, dirt and other foreign matter are removed. The roasting process is necessary to develop aroma and flavor, and at the same time it reduces the amount of moisture present and loosens the shell for easy removal.

There are a number of different designs for roasters, but all of them employ the basic fundamentals necessary for the satisfactory roasting of the beans. Higher temperatures are used for roasting beans that are to be used in the manufacture of cocoa than for those to be used in the production of chocolate. For cocoa, the temperatures used are usually from 116° to 121° C, whereas for chocolate production the temperatures are from 99° to 104° C. Beans that are to be used mainly for the production of cocoa butter used in the manufacture of milk chocolate are roasted at lower temperatures. In this case, the beans are just warmed to loosen the shell. They yield a cocoa butter that has the desired mildness essential for this purpose. The time used for roasting, depending on the equipment used and the size of the batch, may be from 15 to 70 min.

Winnowing

After roasting, the shells of the beans are removed using the winnowing equipment, which separates the cotyledons (nibs) from the shells and the germs.

CHEMICAL COMPOSITION

Unfermented Cacao Beans

Ripe fresh cacao beans contain about 36% moisture and 30% fat. The remainder is made up of sugars, starch, and other carbohydrate material, phenolic compounds, and protein substances, together with the purine bases, theobromine and caffeine.

Most of the cells of the cotyledons are of two types of parenchyma cells. The more numerous of these are small cells in which protoplasm and fat droplets and starch and aleurone grains are found. Enzymes are also present in these cells (Forsyth 1955; Forsyth and Quesnel 1963). Scattered among the first type are larger cells, polyphenol storage cells, containing all the purine bases and phenols, but no protein or fat, nor do they contain enzymes.

A method to separate these two types of cells was developed by Brown (1954) that involves the use of light petroleum in a sedimentation procedure. This method has been helpful in research on the chemistry of these cells.

Polyphenols

The polyphenols of cacao are very important compounds, since they are involved in the flavor and color of the finished chocolate and cocoa. According to Forsyth and Quesnel (1963), the three important groups of phenols present are anthocyanins, leucocyanidins, and catechins, and they are present in the following approximate percentages: anthocyanins, 4%; leucocyanidins, 58%; and catechins, 37%. (−)-Epicatechin comprises over 90% of the catechins. This catechin was first found in cacao in 1932 (Freudenberg *et al.* 1932). Forsyth (1952A, 1955) found the other catechins to be (+)-catechin, (+)-gallocatechin, and (−)-epigallocatechin. Forsyth and Quesnel (1957) reported that 3-β-D-galactosidyl cyanidin and 3-α-L-arabinosidyl cyanidin salts are the anthocyanin pigments present.

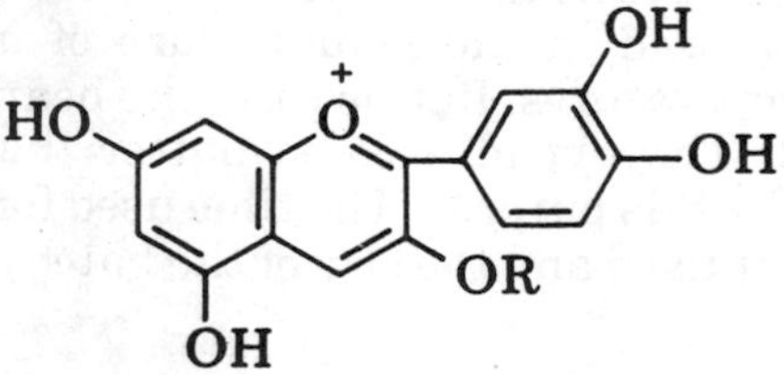

Anthocyanin structure in cacao. R = arabinose or galactose

The anthocyanins of cacao are crystalline, somewhat soluble in water, and easily dissolved by methyl and ethyl alcohol. They are insoluble in ether, chloroform, petroleum ether, and benzene. Anthocyanins are purple in neutral aqueous solution. Weak alkaline solutions are violet, but turn to green in the presence of strong alkali. Acid solutions are pink, but become colorless if they are diluted. It is obvious that these color changes, the result of variations in the pH of the medium, account for the changes in color in the beans while they are undergoing fermentation.

Griffiths (1958) determined that *p*-coumarylquinic acid, also a polyphenol compound is present in the cotyledons of the cacao bean.

The polyphenol composition of the storage cells is given in Table 19.1. Figure 19.1 shows a chromatogram of these various compounds.

Enzymes and Amino Acids

The bulk of the protein work on cacao beans reported in the literature has to do with enzymes. Among the enzymes identified are β-galacosidase (anthocyanase), α-amylase (diastase), pectinesterase, polygalacturonidase, cellulase, proteinase, polyphenol oxidase, peroxidase, and catalase. Also, an enzyme is known in the unfermented bean which is able to split 3-α-L-arabinosidyl cyanidin chloride and 3-β-D-galactosidyl cyanidin chloride to yield sugar and cyanidin. This enzyme is a β-galactosidase (Forsyth and Quesnel 1963).

The unfermented cotyledons contain the following amino acids: aspartic acid, glutamic acid, asparagine, glutamine, alanine, leucine, proline, valine, and α-aminobutyric acid (Schormüller and Winter 1959).

TABLE 19.1. Composition of Polyphenol Storage Cells and of Dry Cotyledons

	Polyphenol Storage Cells (%)	Dry Cotyledons (%)	Total Polyphenol (%)
Catechins	25.0	3.0	37.6
Leucocyanidin 1	14.0	1.7	21.0
Leucocyanidin 2,3	7.0	0.8	10.5
Anthocyanins	3.0	0.4	4.5
Polymeric leucocyanidins	17.5	2.1	26.3
Total polyphenols	66.5	8.0	99.9
Theobromine	14.0	1.7	
Caffeine	0.5	0.1	
Free sugars	1.6		
Cell wall and starch	3.0		
Unknown[a]	14.4		
Total	100.0		

Source: Forsyth and Quesnel (1963). Reproduced with permission of Wiley (Interscience)
[a] Would include glycosidic residues on the polymeric leucocyanidins.

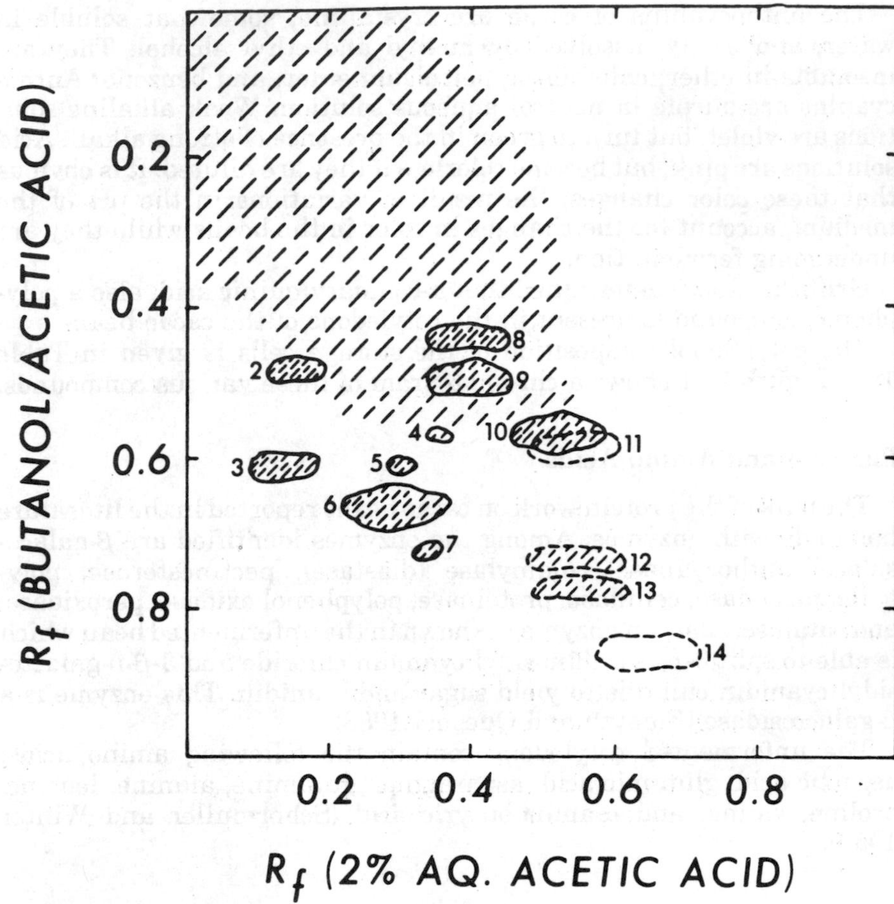

FIG. 19.1. Chromatogram of polyphenol compounds from an extract of Cacao in 80% aqueous methanol containing approximately 1% HCl. Shaded areas react with ferricyanide-Fe^{3+} reagent. Spots with the dashed outline are fluorescent. (1) See Fig. 19.2. The chromatogram is first run in 2% aqueous acetic acid and then in butanol-acetic acid-water (4:1:5). Identification: (2) Cyanidin-3-galactoside, (3) cyanidin-3-arabinoside, (4) (−)-epigallocatechin, (5) (+)-gallocatechin, (6) (−)-epicatechin, (7) (+)-catechin, (8) leucocyanidin L_3, (9) leucocyanidin L_2, (10) leucocyanidin L_1, (11) unknown, (12, 13) possibly neochlorogenic and chlorogenic acids, (14) p-coumaryl-quinic acid. Area in top left corner is polymeric leucocyanidin.
From Forsyth and Quesnel (1963). Reproduced with permission of John Wiley & Sons.

Carbohydrates

Quantitative work on the polyphenol storage cells shows the following results: sucrose, 1%; galactose and glucose as a unit, 0.3%; and arabinose, 0.3%. Starch in the unfermented dry cotyledon is present to the extent of 5 to 9%. Stachyose and raffinose have been found in the parenchymatous cells and pectic substances and cellulose have also been detected.

Other Substances

Theobromine. Cacao is the most important source of theobromine. Theobromine is present in the unfermented dry cotyledons, and analytical results show that about 1 to 2% is present. Theobromine is a white crystalline powder that has the property of subliming at 290° C. It is sparingly soluble in alcohol and water when cold, but is more soluble when the solvent is heated. It is soluble in chloroform, but only slightly soluble in benzene, ether, and petroleum ether. Theobromine is a weak base, and the aqueous solution has a bitter taste. The structural formula is given in Chapter 18.

Caffeine. Caffeine is also found in the cotyledons, but the quantity present is small.

CHANGES DURING MANUFACTURE OF COCOA AND CHOCOLATE

Changes During Fermentation

A number of organisms are responsible for changes which occur during the fermentation of the cacao. This sequence is as follows: yeasts, followed by lactic acid bacteria, followed by acetic acid bacteria, followed finally by spore-forming bacilli. The changing conditions during the fermentation process are instrumental in the determination of this sequence. While it is true that the seeds are sterile while they are in healthy pods, contamination results when they are withdrawn from the pods and placed in piles or boxes.

The first change to take place in the fermenting pulp is, as one would expect, the fermentation by yeasts and the formation of alcohol from the sugars. The pulp contains 10–13% sugars, about two-thirds of which are monosaccharides; the remainder is sucrose. Bacteria then work on the alcohol to produce acetic acid. It is quite likely that other reactions, such as pectin breakdown, take place at the same time. The fermentation process should not be unnecessarily delayed because other organisms could cause undesirable changes, mainly putrefaction.

Fermentation causes considerable change in the beans. The almost neutral sap of the fresh beans gradually becomes acid during fermentation, and in the case of the beans with purple anthocyanins in the cotyledons (Forastero and hybrid beans), the purple color changes to bluish violet and then to red violet with further development of acidity. Pure Criollo beans contain no pigments because of the presence of colorless leucoanthocyanins.

An important and necessary change that takes place during the fermentation process is the death of the beans. They lose their ability to germinate and lose by diffusion the substances in the polyphenol storage cells. Important contributors to this are the development of acetic acid and ethanol, together with the accompanying increase in temperature. The development of acetic acid is the principal cause of the death. The ability to germinate ends when the temperature reaches 43°–44° C. At the same time the semipermeability of the cell membrane is lost. Water is absorbed, by the dead beans, and this makes possible the movement of the polyphenols from the storage cells to the smaller cells containing the enzymes. The presence of alcohol and acetic acid help to bring about the solution of the polyphenolic substances and this in turn makes possible the movement of the polyphenols. Water is not present in sufficient amounts to bring this about alone. Enzyme action is directed principally at sugars, polyphenols, and proteins.

Chatt (1953) and Forsyth and Quesnel (1963) have noted that if germination is permitted, the finished chocolate products will be bitter, an undesirable effect.

Chemical Changes During Fermentation. The *fat* (cocoa butter) is not changed during the curing process. Using paper chromatography, Forsyth (1952B) found that the anthocyanins do undergo rapid destruction during the fermentation process. The reduction in quantity of the catechins and simple leucocyanidins was likely the result of exuding into the testa. It has long been considered that purple-colored cured beans are of lower quality. The pH drops when the acid works into the bean and the anthocyanin pigments continue to take on the pseudo form which is colorless. Also, enzymatic hydrolysis takes place as the fermentation progresses. When conditions of fermentation are such that complete hydrolysis of the anthocyanins takes place, the finished chocolate has good flavor.

Forsyth *et al.* (1958) believe that some tanning of the proteins of cacao takes place during the curing process. The proteins become water insoluble, directly proportional to the accompanying developing insolubility of the leucoanthocyanins. The proteins and the leucoanthocyanins appear to be firmly united.

Figure 19.2 shows the tannin compounds in the fully cured cacao bean. Bracco *et al.* (1969) found that the amount of catechins significantly decreased during fermentation. On the other hand, noncatechinic tannins showed only small variations They suggested that a ratio of catechins to soluble tannins be used as an index of fermentation since it decreased regularly during this operation. They found also that the ratio of soluble nitrogen to total nitrogen could be used as a fermentation index, confirming Rohan and Stewart (1967A, B, or C?). In addition the ratio of glucose and fructose to sucrose correlated with fermentation. They concluded that well-fermented cacao from the Ivory Coast would show a catechin index of about 0.20, a nitrogen index of

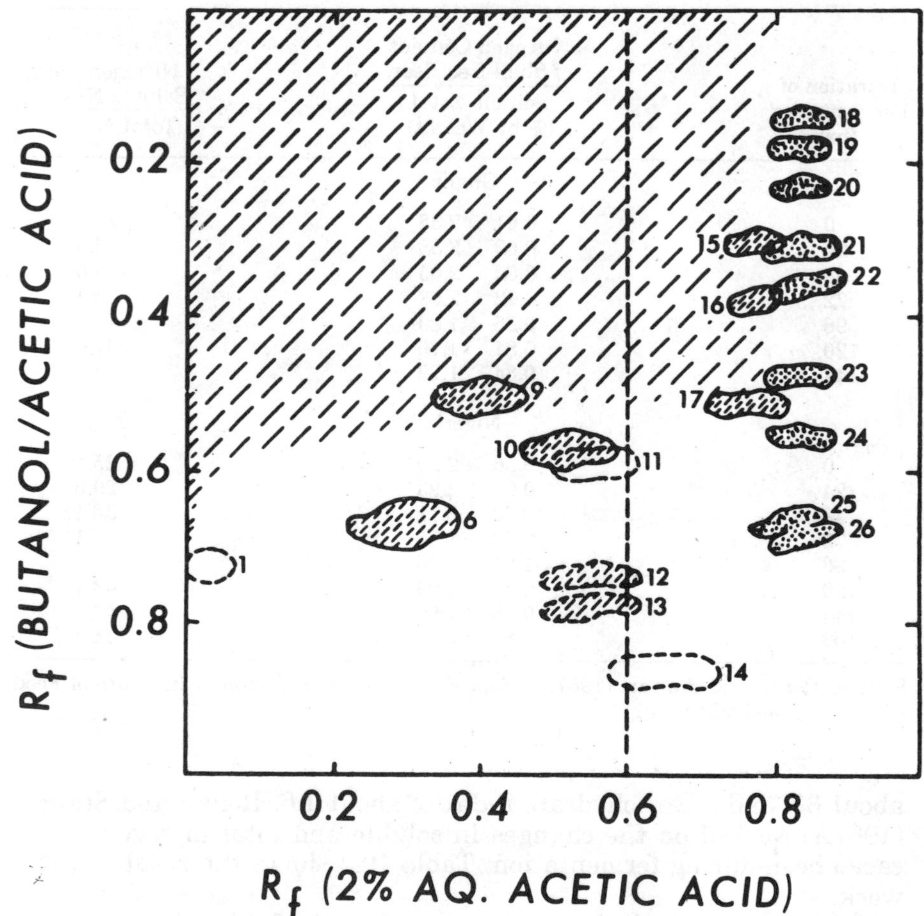

FIG. 19.2 Chromatogram of an extract of fully cured cacao in 80% aqueous methanol containing approximately 1% HCl. Areas with dashed shading react with the ferricyanide-Fe^{3+} reagent. Areas with dotted shading give mauve colors with diazotized p-nitroaniline followed by Na_2CO_3. Spots with a dashed outline are fluorescent. The chromatogram is run first in 2% aqueous acetic acid and then in butanol-acetic acid-water (4:1:5). Identification: (1) Unknown, (6) (−)-epicatechin, (9) leucocyanidin L_2, (10) leucocyanidin L_1, (11) unknown, (12, 13) possibly neochlorogenic and chlorogenic acids, (14) p-coumaryl-quinic acid, (15, 16) presumed X substances of Schubiger et al. (1957), (17) unknown (also reacts gray-mauve with diazotized *p*-nitroaniline, (18–26) unknown. Spots 17–26 occur in traces in the unfermented bean.
From Forsyth and Quesnel (1963). Reproduced with permission of John Wiley & Sons.

TABLE 19.2. Changes in Soluble and Total Nitrogen of Ghanan and Nigerian Cacao Beans During Fermentation

Duration of Fermentation (hr)	Nitrogen Content of Shell-free Bean Soluble (% by Weight)	Total	Nitrogen index $\frac{\text{Soluble N}}{\text{Total N}} \times 100$
	Ghana		
0	0.61	2.38	25.6
24	0.63	2.38	26.5
48	0.69	2.05	33.6
72	0.72	2.09	36.4
96	0.85	1.80	47.2
120	0.81	1.70	47.6
144	0.84	1.79	46.9
	Nigeria		
0	0.56	2.20	25.4
24	0.66	2.23	29.6
48	0.76	2.11	36.1
72	0.98	1.95	50.1
96	1.01	1.96	51.5
120	0.84	1.95	43.1
144	0.78	1.85	42.1
168	0.81	2.11	38.4

Source: Rohan and Stewart (1967A). Reproduced with permission of Institute of Food Technologists.

about 33, and a carbohydrate index of about 1.6. Rohan and Stewart (1967A) worked on the changes in soluble and total nitrogen in the cacao bean during fermentation. Table 19.2 shows the results of this work.

Holden (1959) studied the enzyme changes in fermenting cacao. He found that enzymes were inactivated sooner at the top of the heap than in the center. At the top the temperature rose much more rapidly. The concentration of enzymes gradually decreased while the beans were dying until, after three days of fermentation, little activity could be detected. However, Forsyth *et al.* (1958) reported that the enzymes as well as the protein became insoluble in water by phenolic tannin. The enzymes in this insoluble condition retain however, a slight amount of activity. This remaining action of polyphenol oxidase is responsible for the retarded browning that occurs later in the drying stage.

Rohan and Stewart (1967A) showed that, during the fermentation of cacao beans, amino acids are formed at a rate similar to that known for flavor and aroma production. That these compounds play a likely role as precursors of aroma is given support as a result of this investigation.

Rohan and Stewart (1967B) also noted that during fermentation the sucrose is hydrolyzed to glucose and fructose at a rate so as to include these reducing sugars among other substances as precursors of chocolate aroma and flavor (see Table 19.3).

According to Humphries (1944B) about 40% of the theobromine is lost from the cotyledons during fermentation.

TABLE 19.3. Changes in Sugars of Ghana and Nigerian Cacao Beans During Fermentation

Duration Fermentation (hr)	Reducing Sugars % of Cotyledons		Total Sugars % of Cotyledons		(Reducing Sugars / Total Sugars) × 100	
	Ghana	Nigeria	Ghana	Nigeria	Ghana	Nigeria
0	0.12	0.33	1.44	2.20	11.0	14.9
24	0.24	0.67	1.37	1.98	17.5	33.8
48	0.62	1.22	1.38	1.79	49.0	68.0
72	0.90	1.28	1.31	1.72	65.0	74.5
96	1.04	1.34	1.25	1.57	83.0	85.4
120	1.22	1.22	1.22	1.40	100	87.1
144	0.94	1.11	1.01	1.28	93.9	87.1
168	—	1.16	—	1.46	—	80.0

Source: Rohan and Stewart (1967B). Reproduced with permission of Institute of Food Technologists.

Changes During Drying

It is during the drying stage that important formation of brown colors typical of chocolate takes place. When fermentation is complete the leucocyanidins and the (−)-epicatechin remain either free or in combination with peptides. On completion of drying only traces of these substances can then be detected. These compounds cannot be acted on by the polyphenol oxidase because the fermentation proceeds under anaerobic conditions. During the drying process, oxygen is available and the brown substances can be formed. Griffiths (1957) was able to demonstrate that the important substrate for polyphenol oxidase is epicatechin. However, the others constribute also.

The following aromatic acids were found (Quesnel 1965; Quesnel and Roberts 1963) in fermented and dried cacao beans: *o*-hydroxyphenylacetic acid, phenylacetic acid, *p*-hydroxyphenylacetic acid, *p*-hydroxybenzoic acid, vanillic acid, protocatechuic acid, syringic acid, *p*-coumaric acid, ferulic acid, phloretic acid, and aesculetin. The presence of phenylacetic acid, however, was not conclusively proved.

Bracco *et al.* (1969) noted that the catechins significantly decreased particularly during the period of sun drying of the beans.

Changes During Roasting

Table 19.4 gives the results of the analysis of cocoa nibs showing the effects of roasting on the general composition.

TABLE 19.4. Analysis of Anhydrous Cocoa Nibs Showing the Effects of Roasting

										Constants of Ether Extract (Fat)		
	Nibs in Bean (%)	Protein (%)	Theo-bromine (%)	Caffeine (%)	Total Nitrogen (%)	Fat (%)	Starch (%)	Crude Fiber (%)	Ash (%)	m.p. (°C)	Refractive Index	Iodine Number
Raw (fermented)	85.8	11.99	1.09	0.44	2.38	54.22	7.31	2.01	3.21	33.0	1.4576	36.33
Under roasted	87.6	12.36	0.98	0.43	2.41	54.08	7.96	2.16	3.29	32.5	1.4576	35.69
Medium roasted	88.0	12.01	1.06	0.43	2.38	53.63	7.70	2.82	3.26	32.7	1.4576	35.61
Over roasted	87.7	12.26	0.98	0.38	2.38	53.15	7.78	2.93	3.33	32.5	1.4576	35.66

Source: Winton *et al.* (1902).

Pinto and Chichester (1966) report that the important changes during the roasting of cacao beans are the destruction of the reducing sugars and amino acids, after which volatile carbonyl compounds are formed. Most of the compounds formed during roasting are produced by oxidative deamination of free amino acids. These authors further suggest that action between the amino acids and the reducing sugars forms products of the Amadori rearrangement. As the heating is continued, these products decompose to form aldehydes and other compounds from the sugars. Tables 19.5 and 19.6 show the changes in amino acids and reducing sugars as the result of roasting.

Rohan (1964) showed that an aqueous methyl alcohol extract of fermented cacao beans gave a chocolate odor when heated. A similar extract from unfermented beans did not give such an aroma. This extract contained amino acids, sugars, and flavonoids.

Reineccius *et al.* (1972) determined that pyrazines are generated in cacao during roasting in amounts of 142–698 μg/100 g of beans. These authors noted that the greatest potential for the generation of pyrazines was in beans from the countries where traditional fermentation was practiced. Those found in the largest amount included tetramethylpyrazine, trimethylpyrazine, and a group of pyrazines under one peak that was a mixture of 2-ethylpyrazine, 2,5-dimethylpyrazine and 2,6-dimethylpyrazine.

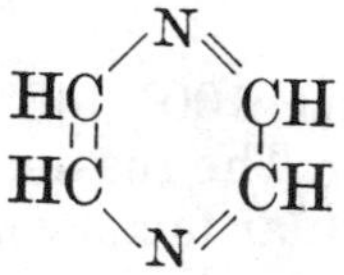

```
      N
   ⁄    ⸗
HC        CH
 ‖         |
HC        CH
   ⸌    ⁄⁄
      N
```

Pyrazine (1,4-Diazine)

Bailey *et al.* (1962) pointed out the probable importance of the Strecker degradation in the formation of aldehydes. They found isovaleraldehyde, isobutyraldehyde, propionaldehyde, methyl alcohol, acetaldehyde, methyl acetate, *n*-butyraldehyde, and diacetyl. These compounds were found in both unroasted and roasted beans in the usual order of relative descending amounts, isovaleraldehyde being in the largest amount. Eight more compounds were found in small amounts. These compounds were also found in the raw bean volatiles, also in smaller amounts. The same components but in different ratios were found in the several varieties studied.

Rohan and Stewart (1967C) found a curious temperature effect in connection with the Strecker degradation of amino acids by reducing sugars in cacao beans, and to such an extent that it could affect the flavor of the final product. A model amino acid reducing sugar system was studied and the results support the idea that a relationship exists among temperature of the reaction, extent of the degradation of amino acid, and the formation of volatile flavors during roasting. A sample of Accra cacao beans was roasted under factory conditions using three

TABLE 19.5. Amino Acids Present in Unroasted and Roasted Bahia (Good Quality) Cacao Beans[a]

	Roasting Time (min)		
Amino Acid	0	30	45
Lysine	14.15	12.54	10.50
Histidine	4.93	4.55	3.43
Arginine	14.76	12.68	12.39
Aspartic acid	9.54	10.03	8.38
Threonine	15.57	13.28	10.41
Serine	13.10	12.94	11.96
Glutamic acid	16.76	9.50	6.55
Proline	15.78	12.38	12.38
Glycine	8.16	7.77	7.36
Alanine	32.82	32.38	28.30
Valine	17.94	17.02	15.37
Methionine	2.04	1.99	1.94
Isoleucine	11.97	11.74	9.78
Leucine	39.01	36.84	30.91
Tyrosine	13.88	12.13	10.58
Phenylalanine	26.24	23.87	20.10
Ammonia	30.08	29.18	37.77

Source: Pinto and Chichester (1966). Reproduced with permission of Institute of Food Technologists.

[a] Given in mg/g of beans.

different temperatures of roast, 100°, 140°, and 180° C for 100 min, 80 min, and 30 min, respectively. The roasted beans were then made into plain chocolate using the regular factory method, and after 3 weeks of storage were sent out for organoleptic scoring. The beans heated at 100° C were scored as being weakest in chocolate flavor. However, they were described as fruity, something akin to Arriba cocoa. Chocolate from beans heated at 180° C was not a great deal better. The highest score was given to the chocolate processed at 140° C. It seems, therefore, that formation of flavor volatiles during the roasting of cocoa beans is related to the temperature of reaction and the extent of amino acid degradation.

TABLE 19.6. Changes in the Reducing Sugars During Roasting of Bahia Cacao Beans (Good Quality)

	Roasting Time (min)			
	0	15	30	45
Reducing sugars (mg dextrose per gm of beans)	17.62	12.22	8.04	3.94

Source: Pinto and Chichester (1966). Reproduced with permission of Institute of Food Technologists.

Some of the fermented cacao beans were ground, extracted with 80% methanol, and then filtered to get rid of extraneous material. The methyl alcohol was removed in a rotary evaporator, and the residue was dialyzed. The diffusate was freeze-dried, and the residue developed the chocolate aroma on heating at 140° C. Different samples of the residue were heated at various temperatures and analyzed for reducing sugars and amino acids, as illustrated in Table 19.7.

Similar experiments making use of the polycomponent synthetic mixture are shown in Table 19.8. The fact that a real chocolate aroma was not produced by the synthetic mixture indicates that the problem of the production of chocolate flavor is far from solved.

TABLE 19.7 Destruction of Amino Acids and Reducing Sugars on Heating Aroma-Precursor Extract

Heating Conditions					
Temp (°C)	Temp (min)	Amino Acid Destruction (%)	Reducing Sugar Destruction (%)	Total Sugar Destruction (%)	Aroma of Product
50	30	6.2	—	—	—
50	60	8.9	27.0	—	—
50	120	8.9	41.1	—	—
100	30	18.8	27.5	35.8	cocoa
100	60	29.6	56.6	79.6	cocoa
100	120	30.9	81.4	22.8	cocoa
120	30	43.0	84.1	38.9	cocoa
120	60	54.4	100	79.2	cocoa
120	120	54.4	93.5	91.2	cocoa
150	10	58.8	—	—	cocoa
150	30	81.7	—	—	cocoa
150	60	85.8	—	—	—
150	120	85.1	—	—	—

Source: Rohan and Stewart (1967C). Reproduced with permission of Institute of Food Technologists.

Keeney (1972) reviewed the various interactions in chocolate flavor and noted that the raw beans may have the following flavors: acid-like, musty, astringent, bitter, nutty, unclean, and perhaps somewhat chocolate-like, depending on the beans. Different batches are not all the same. During the roasting process some volatile materials present in the raw beans are lost and the final chocolate flavor is developed. Roasting reduces the moisture in the beans from 6 to about 1%.

Conching, which is the mixing or rolling of the chocolate mass, has a mellowing effect on chocolate flavor. Basically, this process is mechanical in nature, and probably involves changes in surface phenomena.

TABLE 19.8. Destruction of Amino Acids and Reducing Sugars on Heating Polycomponent Synthetic Mixture

Heating Conditions				
Temp (°C)	Time (min)	Amino Acid Destruction (%)	Reducing Sugar Destruction (%)	Aroma of Product
100	30	15.6	17.9	aldehydic
100	60	17.6	30.1	aldehydic
100	120	21.1	41.0	aldehydic
120	15	40.2	—	aldehydic
120	30	41.3	31.0	meaty
120	60	41.5	50.8	burnt chocolate
120	120	41.2	67.8	burnt chocolate
150	30	73.0	82.1	burnt chocolate
150	60	77.0	100	—
150	120	79.3	—	—

Source: Rohan and Stewart (1967C). Reproduced with permission of Institute of Food Technologists.

During the roasting process nonenzymatic browning takes place. Ammonia nitrogen is higher in beans that have been through the fermentation process.

VOLATILES AND CHOCOLATE FLAVOR

There is still much work to be done in the area of chocolate flavor. The final chocolate flavor is a combination of many compounds, both somewhat fixed and volatile. It is decidedly complex. Bainbridge and Davies (1912) prepared volatile oil from 2000 g of lightly roasted Arriba cocoa beans from the Arriba district of Ecuador. This cacao is not well cured. A yield of 24 ml was obtained, was fractionally distilled, and the first fractions were found to be rich in esters. The middle distilled fractions were rich in *d*-linalool, probably owing to incomplete curing. The esters, when saponified, gave mixtures of amyl alcohol with acetic, propionic, octoic, and nonoic acids.

Since then many volatile compounds, have been isolated. In addition, cured beans extracted with chloroform have yielded an odoriferous resinous product which chromatographic studies have shown is made up of aromatic acids and carbonyl compounds.

Van Pragg *et al.* (1968) first reported that "5-methyl-2-phenyl-2-hexenal possesses a deep, bitter, consistant cocoa note." A table identifying compounds in various fractions of cocoa nibs distillate is included in their article. Among these compounds are a number of pyrazines, methyl sulfide and other sulfur compounds, alcohols, aldehydes, ketones, 5-methyl-2-phenyl-2-hexenal.

The volatile compounds of roasted cocoa were investigated by van der Wal *et al.* (1971), and they found and identified 181 compounds, of which 112 were detected for the first time. The authors prepared a synthetic mixture and, although it had a cocoa-like odor, it lacked the strong odor of the original. There was no problem distinguishing the two. It is quite likely that important compounds are yet to be found. Work of this character has been possible only since the advent of gas chromatography, infrared and mass spectrometry. Earlier methods were not capable of such sensitive precision. According to Keeney (1972) more than 300 volatile compounds have been found in roasted cocoa beans.

CHOCOLATE AND COCOA PRODUCTS

Chocolate is made by grinding the hot nibs to a dark brown fluid known as chocolate liquor. There is a large amount of fat, or cocoa butter, in chocolate liquor, about 55% of the nib. The cooled and molded liquor is sold as unsweetened chocolate.

Cocoa is made by subjecting the chocolate liquor to pressure in the hydraulic press and removing a part of the fat, which is known as prime cocoa butter. Another method for the preparation of prime cocoa butter from the nibs was described earlier under chocolate manufacture. After the cocoa butter has been separated from the chocolate liquor by pressure, a hard cake which contains about 22 to 25% cocoa butter remains. This is ground and sieved to yield cocoa powder. Cocoa powder results in a more satisfactory beverage than chocolate because of the lower fat content. The prime cocoa butter is used mainly for the manufacture of chocolate products, particularly for milk chocolate, and for pharmaceutical purposes. The preparation of cocoa powder was first done by van Houten in 1826.

Alkalized or Dutch Process cocoa was first developed by van Houten in 1828. This process involves the use of potassium or sodium carbonate to improve the dispersability of the chocolate, the color, and the flavor, although the latter improvement has been disputed. Other alkaline substances can also be used. The maximum amount permitted for use is 2.5–3 parts of potassium carbonate, or equivalent, to 100 parts of nib. The variations in color produced are doubtless caused by modification of the polyphenolic substances present; the cocoa particles also show some swelling. Machinery is now available to alkalize the nibs, dry them, and roast them in a continuous process. The color and flavor are affected by the temperature of the roast. Low temperatures yield reddish colors, whereas high roasting temperatures yield dark brown colors and strong flavors. Alkalinization temperatures must be 80°–85°C if good flavors are to be obtained.

Definitions

According to United States law (Food and Drug Admin. 1964) sweet chocolate contains not less than 15% by weight of chocolate liquor. This is calculated as follows:

$$\text{Wt. of chocolate liquor} - (\text{Wt. of cacao fat} + \text{alkali and seasonings}) \times 2.2 \div \text{Wt. of finished chocolate} \times 100$$

Bitter sweet chocolate is sweet chocolate which contains not less than 35% by weight of chocolate liquor.

Milk chocolate, sweet milk chocolate, and milk chocolate coatings are the solid or semiplastic foods made from chocolate liquor (with or without the addition of cacao fat) and one or more of the optional dairy ingredients, sweetened with one of the optional saccharin ingredients or combinations of these. It may be spiced, flavored, or otherwise seasoned with one or more of the optional permitted ingredients. The finished milk chocolate contains not less than 3.66% by weight of milk fat, not less than 12% by weight of milk solids, and not less than 10% by weight of chocolate liquor.

It is required by law in Germany that sweet chocolate, milk chocolate, and chocolate coating be labeled with the percentage of chocolate liquor contained therein. German milk chocolate contains 28 or 31% chocolate liquor, the 28% usually sells for a little lower price. Other products in Germany such as cream bitter (43% chocolate liquor) are higher in chocolate liquor. Since these products are higher in chocolate liquor than the corresponding American products, they have a stronger chocolate flavor and are less sweet. Mandatory labeling of chocolate products as to chocolate liquor content in the United States would be desirable.

Analytical Procedures

Methods for the analysis of the cacao bean and its products are given in Official Methods of Analysis (AOAC 1980).

SUMMARY

Cacao beans are removed from the pods after harvest and are subjected to a fermentation process. This prepares them for drying and roasting to produce the cocoa and chocolate of commerce.

The compounds found in unfermented cacao beans affect the final product. Polyphenols of cacao are very important compounds. They are involved in the flavor and color of the finished product. The three important groups of phenols present are anthocyanins, leucocanidins, and catechins. It is known that (−)-epicatechin comprises over 90% of the catechins. Two anthocyanin pigments, 3-β-D-galactosidyl cyanidin and 3-α-L-arabinosidyl cyanidin are present. They are purple in neutral aqueous solution and change to different colors with changes in the pH. This accounts for the changes in the color of the beans during fermentation.

Most of the protein work reported on cacao beans has to do with enzymes, a number of which have been identified.

Theobromine is the important purine base present. Some caffeine is present in the bean also.

An important change during fermentation is the loss of the ability of the beans to germinate. This is caused by the development of acetic acid in the beans.

Amino acids increase during fermentation, supporting the idea that amino acids are involved in the production of aroma.

During the drying stage, the formation of brown colors typical of chocolate are formed. This is because the leucocyanidins and the (−)-epicatechin are acted on by polyphenol oxidase in the presence of oxygen. During fermentation an anaerobic condition exists.

During the roasting of cacao beans, destruction of reducing sugars and amino acids takes place after which volatile carbonyl compounds are formed.

An aqueous methyl alcoholic extract of fermented cacao beans yields a chocolate odor when heated. A similar extract from unfermented beans gives no such odor.

Formation of flavor volatiles during roasting of cacao beans is related to the temperature of reaction and the extent of amino acid degradation.

More than 300 volatile compounds have been found in roasted cacao beans.

BIBLIOGRAPHY

ANTENER, I. 1958. The vitamin B-complex in cocoa beans. Mitt. Gebiete Lebensm. Hyg. *4*, 127–139.

AOAC. 1980. Official Methods of Analysis, 13th Edition. Assoc. of Official Analytical Chemists, Washington, DC.

BAILEY, S. D., MITCHELL, D. G., BAZINET, M. L., and WEURMAN, C. 1962. Studies on the volatile components of different varieties of cocoa beans. J. Food Sci. *27*, 165–170.

BAINBRIDGE, J. C., and DAVIES, S. H. 1912. The essential oil of cocoa. J. Chem. Soc. Trans. 2209–2221.

BRACCO, U., GRAILHE, N., ROSTAGNE, W., and EGLI, R. H. 1969. Analytical evaluation of cocoa curing in the Ivory Coast. J. Sci. Food Agric. *20*, 713–717.

BROWN, H. B. 1954. Separation of pigment cells of cacao. Nature *173*, 492.

BROWN, H. B. 1957. Chocolate manufacture with special reference to cacao constituents. Chem. Ind. (London) 680–685.

CHATT, E. M. 1953. Cocao. Economic Crops, Vol. III. Wiley (Interscience) Publishers, New York.

DIMICK, P. S., and HOSKIN, J. M. 1981. Chemico-physical aspects of chocolate processing. A review. Can. Inst. Food Sci. Technol. J. *14*, 269–282.

FOOD AND DRUG ADMIN. 1964. Cacao products. Definitions and standards under the Federal Food, Drug, and Cosmetic Act. U.S. Department of Health, Education, and Welfare. Food, Drug and Cosmetic Regulations, Part 14, Dec. Amendment: Fed. Regist., Mar. 19, 1968, 33 F.R. 4656.

FORSYTH, W. G. C. 1949. Method for studying the chemistry of cacao fermentation. Nature *164*, 25–26.

FORSYTH, W. G. C. 1952A. Cacao polyphenolic substances. 1. Fractionation of the fresh bean. Biochem. J. 51, 511–516.

FORSYTH, W. G. C. 1952B. Cacao polyphenolic substances. 2. Changes during fermentation. Biochem. J. *51*, 516–520.

FORSYTH, W. G. C. 1955. Cacao polyphenolic substances. 3. Separation and estimation on paper chromatograms. Biochem. J. *60*, 108–111.

FORSYTH, W. G. C., and QUESNEL, V. C. 1957. Cacao glycosidase and colour changes during fermentation. J. Sci. Food Agric. *8*, 505–509.

FORSYTH, W. G. C., and QUESNEL, V. C. 1963. The mechanism of cacao curing. Adv. Enzymol. *25*, 457–492.

FORSYTH, W. G. C., and ROBERTS, J. B. 1958. Cacao "leucocyanidin." Chem. Ind. (London) 755.

FORSYTH, W. G. C., QUESNEL, V. C., and ROBERTS, J. B. 1958. The interaction of polyphenols and proteins during cacao curing. J. Sci. Food Agric. *9*, 181–184.

FREUDENBERG, K., COX, R. F. B., and BRAUN, E. 1932. The catechin of the cacao bean. J. Am. Chem. Soc. *54*, 1913–1917.

GRIFFITHS, L. A. 1957. Detection of the substrate of enzymic browning in cacao by a post-chromatographic enzymatic technique. Nature *180*, 1373–1374.

GRIFFITHS, L. A. 1958. Phenolic acids and flavonoids of *Theobroma cacao,* L. Separation and identification by paper chromatography. Biochem. J. *70*, 120–125.

HOLDEN, M. 1959. Processing of raw cocoa. III. Enzymic aspects of cocoa fermentation. J. Sci. Food Agric. *10*, 691–700.

HUANG, H. T. 1956. The kinetics of the decolorization of anthocyanins by fungal "anthocyanase." J. Am. Chem. Soc. *78*, 2390–2393.

HUMPHRIES, E. C. 1944A. Studies on tannin compounds. 1. Changes during autolysis of minced cacao bean. Biochem. J. *38*, 182–187.

HUMPHRIES, E. C. 1944B. Some problems of cacao fermentation. Trop. Agric. (Trinidad) *21*, 166–169.

KEENEY, P. G. 1972. Various interactions in chocolate flavor. J. Am. Oil Chem. Soc. *49*, 567–572.

KNAPP, A. W. 1937. Cacao Fermentation. Bale, Sons and Curnow, London.

LAWRENCE, W. J. C., PRICE, J. R., ROBINSON, G. M., and ROBINSON, R. 1938. A survey of anthocyanins. V. Biochem. J. *32*, 1661–1667.

LEVANON, Y., and ROSSETINI, S. M. O. 1965. A laboratory study of farm processing of cocoa beans for industrial use. J. Food Sci. *30*, 719–722.

MARION, J. P., MÜGGLER-CHAVAN, F., VIANI, R., BRICIUT, J., REYMOND, D., and EGLI, R. H. 1967. On the composition of the aroma of cacao. Helv. Chim. Acta *50*, 1509–1516. (French)

MINIFIE, B. S. 1970. Chocolate, Cocoa, and Confectionery: Science and Technology. AVI Publishing Co., Westport, CT.

PINTO, A., and CHICHESTER, C. O. 1966. Changes in the content of free amino acids during roasting of cocoa beans. J. Food Sci. *31*, 726–732.

POWELL, B. D. 1959. Cocao fermentation: A note on the anaerobic and aerobic phases. Chem. Ind. (London), 991–992.

QUESNEL, V. C. 1965. Chloroform-extractable aromatic acids of cacao, J. Sci. Food Agric. *16*, 596–599.

QUESNEL, V. C., and ROBERTS, J. B. 1963. Aromatic acids of fermented cacao. Nature *199*, 605–606.

REINECCIUS, G. A., KEENEY, P. G., and WEISSBERGER, W. 1972. Factors affecting the concentration of pyrazines in cocao beans. J. Agric. Food Chem. *20*, 202–206.

ROBERTS, J. B. 1959. Cacao fermentation: The existence of anaerobic cotyledons during fermentation. Chem. Ind. (London), 1410–1411.

ROELOFSEN, P. A. 1958. Fermentation, drying, and storage of cacao beans. *In* Advances in Food Research, Vol. 8, E. M. Mrak and G. F. Stewart (Editors). Academic Press, New York.

ROHAN, T. A. 1958. Processing raw cocoa. II. Uniformity in heap fermentation and development of methods for rapid fermentation of West African Amelonado cocoa. J. Sci. Food Agric. *9*, 542–551.

ROHAN, T. A. 1963. Precursors of chocolate aroma. J. Sci. Food Agric. *14*, 799–805.

ROHAN, T. A. 1964. The precursors of chocolate aroma: A comparative study of fermented and unfermented cocoa beans. J. Food Sci. *29*, 456–459.

ROHAN, T. A., and CONNELL, M. 1964. The precursors of chocolate aroma: A study of the flavonoids and phenolic acids. J. Food Sci. *29*, 460–463.

ROHAN, T. A., and STEWART, T. 1967A. The precursors of chocolate aroma: Production of free amino acids during fermentation of cocoa beans. J. Food Sci. *32*, 395–398.

ROHAN, T. A., and STEWART, T. 1967B. The precursors of chocolate aroma: Production of reducing sugars during fermentation of cocoa beans. J. Food Sci. *32*, 399–402.

ROHAN, T. A., and STEWART, T. 1967C. Precursors of chocolate aroma: Studies in the degradation of amino acids during the roasting of Accra cocoa beans, J. Food Sci. *32*, 625–629. (German with English Summary).

SCHORMÜLLER, J., and WINTER, H. 1959. Amino acid content of the proteins of the cacao bean. Nahrung *3*, 187–192.

SCHUBIGER, G. F., ROESCH, E., and EGLI, R. H. 1957. Contribution to the knowledge of the cacao polyphenols and related substances through application of spectrophotometry, chromatography and electrophoresis. Fette Seifen Anstrichm. *59*, 631–636. (German)

VAN DER WAL, B., KETTENES, D. K., STOFFELSMA, J., SIPMA, G., and SEMPER, T. J. 1971. New volatile components of roasted cocoa. J. Agric. Food Chem. *19*, a76–280.

VAN ELZAKKER, A. H. M. and VAN ZUTPHEN, H. J. 1961. Research on cacao aroma with aid of gas chromatography. Z. Lebensm. Untersuch.-Forsch. *115*, 222–226. (German)

VAN PRAAG, M., STEIN, H. S., and TIBBETTS, M. S. 1968. Steam volatile aroma constituents of roasted cocoa beans. J. Agric. Food Chem. *16*, 1005–1008.

VITZTHUN, O. G., WERKHOFF, P., and HUBERT, P. 1975. Volatile components of roasted cocoa: Basic fraction. J. Food Sci. *40*, 911–916.

WINTON, A. L., SILVERMAN, M., and BAILEY, E. M. 1902. The effects of roasting on the chemical composition of cocoa beans. Conn. Agr. Exp. Stn. Rep. 265–269.

20

Meat and Meat Products

Muscle Composition
Changes in Muscle after Slaughter and during Processing
Effect of Ionizing Radiation
Summary
Bibliography

Knowledge of animal muscle and the process by which it is converted to food has developed considerably. Much of this progress has resulted from the great strides that have been made in the basic understanding of proteins. Basic work of this character has already been discussed in Chapter 6 on proteins. The electron miscroscope has been instrumental in the understanding of the structure of muscles. Along with proteins, fats in meat are important from the nutritional as well as functional standpoint.

MUSCLE COMPOSITION

Structure of Muscle Fiber

Skeletal muscle is made up of long, cross-striated muscle fibers. The fiber includes the myofibrils, the sarcoplasmic matrix together with some small structural elements including the sarcoplasmic reticulum, the mitochondria, and the nucleus. The myofibrils, much smaller units that make up the fiber, can be seen at a magnification of ×2000. Electron microscopic study has shown the sarcoplasmic reticulum to be made up of a series of minute tubes, which may be involved in chemical control of the function of the muscle. The skeletal muscle mitochondria have a fine membranous internal structure. Each fiber is surrounded by a sheath known as the sarcolemma, which is composed of a double membrane. The sarcoplasm is a liquid that surrounds the myofibrils.

Accounting for approximately 80% of the fiber volume, the partially crystallized protein gel of myofibrils is the contractile substance. The protein concentration is between 15 and 20%. Because the protein of the myofibrils lacks homogeneity, an optical effect of cross-striation—an alternating sequence of anisotropic (A) and isotropic (I) bands—occurs (Fig. 20.1) (DuPraw 1968). The A-band is made up of thick filaments, which are parallel and arranged in an hexagonal system.

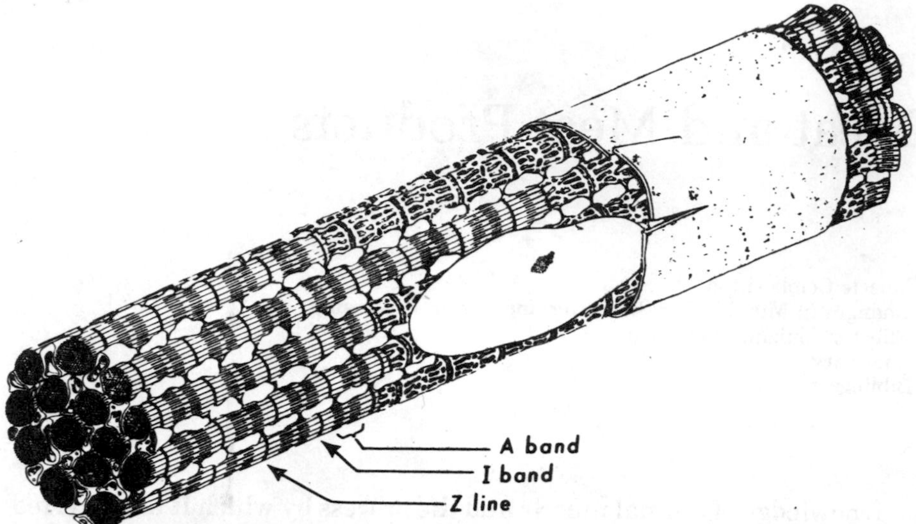

FIG. 20.1. A dissected skeletal muscle fiber, showing cylindrical myofibrils surrounded by mitochondria and enclosed in a sarcolemma. Individual myofibrils are organized as a series of sarcomeres 2–3 μm in length. A nucleus is seen in a typical position just beneath the sarcolemma. In the right part of the fiber, the sarcoplasmic reticulum surrounding the single fibrils, is not removed. Band and Z line designations are added.
From DuPraw (1968). Reproduced with permission of Academic Press.

The I-band comprises thin filaments which are bisected by the Z-lines. The thick filaments and the muscle protein myosin are identical and account for about 38% of the total muscle protein. The thin filaments contain largely F-actin (about 14%) and also tropomyosin. These proteins are involved in muscular contraction and so are called "contractile proteins." Figures 20.2 and 20.3 further illustrate these structures (Hamm 1970).[1]

Red and White Muscle Fibers. Muscles, according to Dubowitz (1966) are divided into two fiber types: type I, rich in dehydrogenases and poor in phosphorylase, and type II, the reverse. Fibers of intermediate activity also exist. These are also known as red (Type I) and white (Type II) fibers. It has been shown by enzymes studies that other fibers also exist.

The existence of red and white muscle fibers has long been known. As a rule, red fibers are smaller than white fibers. Myoglobin, the principal muscle pigment, is present in larger quantities in red than in

[1]This discussion is a very brief resumé and shows the complexity of skeletal muscle comprising the meat of commerce. It does not pretend to be complete and for further information the interested student is advised to consult several of the books on meat science.

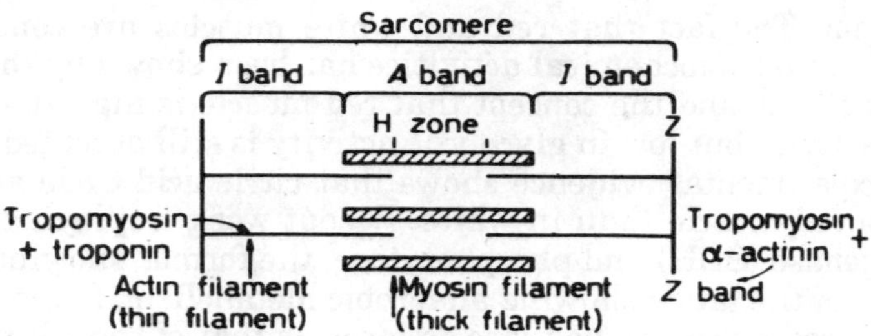

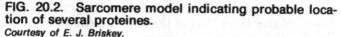

FIG. 20.2. Sarcomere model indicating probable location of several proteines.
Courtesy of E. J. Briskey.

white muscle. More soluble protein is found in white muscle than in red. Red fibers, unlike white fibers, are rich in mitochondria. Mitochondria are formed structures that contain insoluble enzymes involved in respiration and oxidative phosphorylation. George and Naik (1958) think that high mitochondrial content, high fat content, together with high lipase activity, the presence of myoglobin, and larger surface area of the narrow red fibers, combine to make a good oxidative system. Red muscle is for aerobic metabolism, while white muscle is for glycolytic

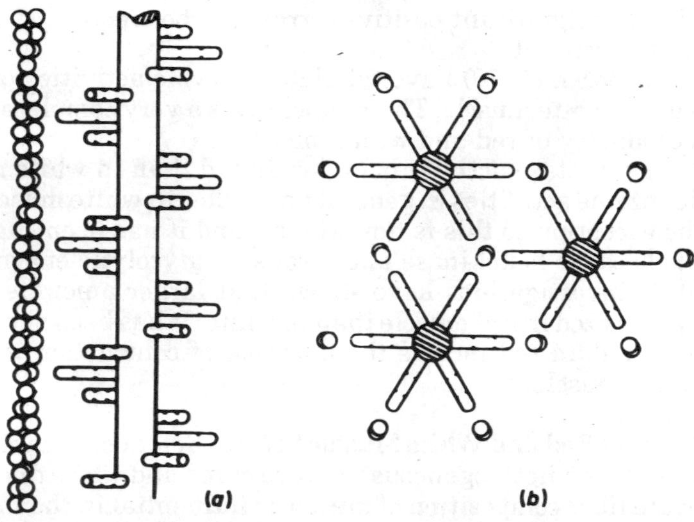

FIG. 20.3. Six thin actin fliaments are arranged around one thick myosin filament. Both kinds of filaments are connected by cross-linkages. (a) The double-stranded beaded structure of actin, the pitch of the spiral being about 350 Å, and the six staggered rows of feet on the myosin filament. (b) The alignment of one actin filament, opposite each row of feet, in cross section.

metabolism. The fact that red and white muscles are considerably different in their biochemical activities has been shown by their enzymatic reactions, and the concept that red muscle is high in oxidative enzyme activity but low in glycolytic activity is still accepted. A great deal of experimental evidence shows that citric acid cycle activity is higher in red muscle than in white. Recent work employing succinic dehydrogenase (SDH) and phosphorylase, the former showing aerobic metabolism, the latter showing anaerobic metabolism, found the SDH activity of red muscle to be twice as large as that of the white. At the same time the phosphorylase activity of the white muscle was found to be three times greater than that of the red muscle. RNA concentration was found to be higher and more active in red muscle than in white.

Ashmore, *et al.* (1972) worked on muscles from the chick, made up mainly of αR, βR, or αW fibers and compared them in connection with whole muscle succinic dehydrogenase (SDH) activity, mitochondrial protein yield, mitochondrial SDH activity, and the oxidation of α-glyceral phosphate and β-hydroxybutyrate by the isolated mitochondria. They found that total SDH activity, which is higher in αR than in βR, is basically the result of higher specific activity of SDH of the mitochondria than to more mitochondria. They concluded that αR and βR fibers are basically not the same and that studies involving "red" muscles must take into consideration the varying amounts of αR and βR fibers.

El-Badawi and Hamm (1970) made a study of pork and beef muscle and found a very significant positive correlation between SDH activity and myoglobin content.

Beatty and Bocek (1970) have tabulated enzyme activities of mammalian red and white muscle. Their paper gives a very complete review of the biochemistry of red and white muscle.

Glycogen synthetase activity is higher in red than in white muscle. Glycolytic enzyme activities in general are higher in white muscle than in red. The exception to this is hexokinase, and it has been suggested that this is because hexokinase is not solely a glycolytic enzyme.

Laboratory investigations have shown that larger amounts of protein are synthesized in red muscle than in white. It has been shown also that in mammalian red muscle the duration of contraction is longer than in white muscle.

Importance of Red and White Muscle to Meat Science. Mammalian skeletal muscle is a heterogeneous mixture of red and white fibers. The red and white fiber composition of muscle is influential in the interpretation of biochemical studies on the use of muscles as food (Cassens and Cooper 1971).

Beecher *et al.* (1965) made the first study on pig muscle with this in mind. They used a number of different muscles; those having the smallest to the largest content of red fibers. The following factors were studied: SDH activity, myoglobin content, fat content, pH fall postmortem, glycogen and lactate levels, and postrigor sarcomere length. Those muscles that were made up of more than 40% red fibers were considered red muscles; those with less than 30% red fibers were rated as white muscles. This study was continued using red and white parts of the semitendinosus muscle of the pig (Beecher *et al.* 1965). It was found that the myoglobin level percentage of red fibers and SDH activity were two times larger in the red part, but the quantities of ADP and P_i (inorganic phosphate) were similar in both. Amounts of iron and zinc were larger in the red part, but calcium, nickel, boron, and potassium were about the same in both. Sodium was lower and phosphorus, sarcoplasmic nitrogen, and lipid were higher in the white.

In general, a high content of myoglobin is related to high succinic dehydrogenase activity. It seems therefore that myoglobin insures high oxygen tension for the muscular oxidase system. On the other hand, muscles with low myoglobin content but high succinic dehydrogenase activity could have superior muscle circulatory systems.

At the time of death the lactic acid concentrations were about the same in all muscles. However, values determined after 24 hr tended to be a little higher in white muscles.

The light part of the semitendinosus muscle was found to contain more lipid and more sarcoplasmic nitrogen than was found in the dark part.

From these data it can be inferred that the white and red portions within the semitendinosus muscle have properties of a physical and chemical nature that are not far from those of uniform white and uniform red muscles. A third paper in this group (Beecher *et al.* 1968) continued this study on postmortem metabolism in red and white muscle.

According to Beatty and Bocek (1970), in the present state of knowledge it is advisable to correlate function with results obtained by histochemical and biochemical means only in a general way as far as red and white muscle are concerned. However, the scientific evidence obtained strongly supports the idea that red muscles (or predominately red) have higher rates of oxidative metabolism and are best able to carry on continued activity and the extended process of energy production. Sudden spurts of activity are best taken care of when white muscle predominates because its glycolytic activity is higher.

Meat Proteins

The proteins in meat contribute a great deal to the maintenance of life and health. A number of workers have contributed to the knowledge of meat proteins, including Szent-Györgyi (1951).

According to Bendall (1964), meat proteins, because of the special functions they have in the animal, must have rather unusual properties. The proteins must be able to contract and relax rapidly. In addition, these proteins must have a complex system of enzymes to furnish the needed energy. Further, they must have the network of connective tissue to hold them in place to sustain the tension of their movements, and a nervous system and a supply of blood as well.

Proteins of the Skeletal Muscle. The myofibrillar proteins are actin, myosin, tropomyosin, troponin A, troponin B, α-actinin, and β-actinin. α-Actinin has two components, which have been extensively studied because they are important in muscular contraction; recent work has shown that they undergo changes during postmortem storage. One of these is the alteration of the actin–myosin interaction, which results in a considerable increase in the activity in Mg^{2+}-modified actinomyosin adenosine triphosphatase.

According to Seifer and Gallop (1966) the myofibril proteins of the skeletal muscle of the rabbit consist of 21% actin, 54% myosin, 15% tropomyosin B, and 10% other proteins including such compounds as α-actinin and β-actinin.

When the water-soluble sarcoplasmic proteins are removed from the muscle tissue by extraction with a salt solution of low ionic strength, the contractile proteins of the myofibrils and the stroma proteins of the connective tissue remain behind. When myofibrils and stroma proteins are extracted by salt solutions of high ionic strength, the result is a viscous solution which contains the myosin and actin, as well as actomyosin, a complex of actin and myosin. It is possible to separate pure myosin by step-wise lowering the ionic strength.

The compounds present in the largest amounts in skeletal muscle are water (about 75%) and protein (about 20%). The remaining 5% is made up of fat, carbohydrate, non-protein aqueous extractives, and minerals. Of the protein fraction about 65% comprises myosin and actin. Myosin is the bulk of this fraction, and it falls under the classification of globulins, since it is soluble in aqueous salt solution. If this solution is heated or diluted, a gel is formed.

Soluble in salt solution, actin is able to combine with myosin to form actomyosin. The actomyosin complex serves as the contractile material in muscle. Connective tissue comprises about 10–15% of the protein fraction; the other proteins present include albumins, globulins, and respiratory proteins.

Myosin. This protein was shown by Singer and Barron (1944) to be a sulfhydryl enzyme, adenosine triphosphatase (ATPase), which can be denatured rather easily by heat and by freeze-drying.

Myosin has a molecular weight of about 500,000 and is a thread-shaped molecule, with a head and a tail as illustrated in Fig. 20.4. This value for the molecular weight represents a fair average based on published information, rather than an exact figure, since variation exists in the values available (Dreizen *et al.* 1966; Frederiksen and Holtzer 1968; Keilley and Harrington 1960). Myosin behaves as an enzyme (ATPase), and it is able to break down adenosine triphosphate (ATP) to adenosine diphosphate and inorganic phosphate (P_i). In this

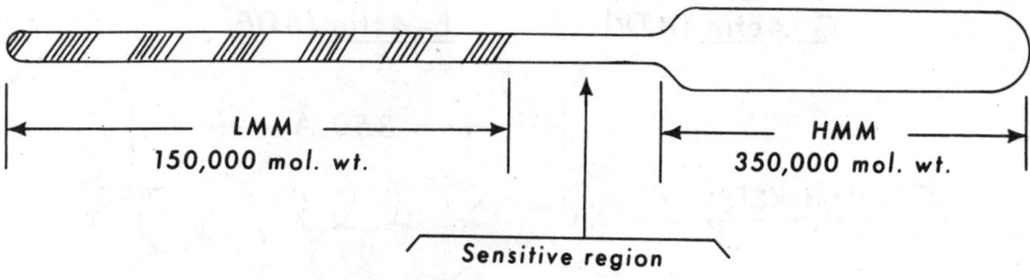

FIG. 20.4. Myosin molecule and portion of thick filament.
Courtesy of E. J. Briskey.

action, the terminal phosphate is split off, and energy is released. Of vast importance in muscular contraction, this enzyme action is stimulated by calcium ions but inhibited by magnesium. Actin in the presence of myosin alters this ATPase activity. Under these circumstances, magnesium ions present in low ionic strength no longer inhibit but strongly activate this activity, with the result that the final effect is stronger than that with calcium ions. In this case, actinomyosin ATPase takes the place of myosin ATPase.

Actually, myosin as ordinarily prepared is made up of two proteins, true myosin and actin (Bendall 1964; Schramm and Webber 1942). It is possible by means of proteolytic enzymes (trypsin and chymotrypsin) to split myosin itself into heavy meromyosin (H-meromyosin, HMM) with a molecular weight of about 380,000 and light meromyosin (L-meromyosin, LMM) with a molecular weight of about 120,000. Of these two fragments, only HMM possesses enzyme activity, which means that only this part is able to bind with actin. Aside from contractibility these two compounds have all the properties of the original myosin. Light meromyosin retains the solubility properties of the original myosin. Figures 20.4 and 20.5 show the structures of actin and myosin. Figure 20.6 shows the changes in the position of filaments during extension and contraction.

Actin. It was found that actin was a major component of muscle as a result of the discovery that certain extracts of muscle that contained myosin were unable to hold their gel-like consistency when they were treated with ATP, but others were not so affected. Actin has a molecular weight of about 60,000 and occurs in two forms. One is clear in

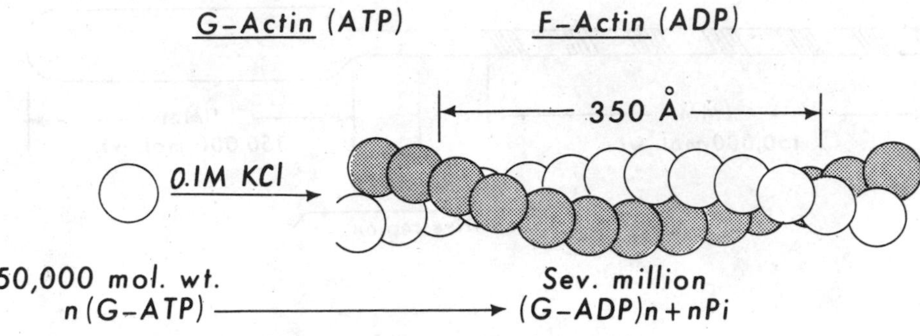

FIG. 20.5. Characteristics of G- and F-actin.
Courtesy of Nature.

solution and is known as globular or G-actin. In this form it has been found to be made up of separate beads (Fig. 20.5). It can be polymerized into the F-form which is fibrous, when neutral salts are added. This takes place simultaneously with enzymatic splitting of ATP to ADP and P_i. In the F-form the beads arrange side by side and two such strings are wound together as a double helix. Polymerized F-actin is generally held to be linear aggregate of G-actin units; there are, however, some irregularities in this explanation. Only F-actin can interact with myosin to form a colloidal system, the condition of which is strongly affected by ATP.

Actomyosin. Actomyosin results when solutions of F-actin and myosin are mixed together. This solution is highly viscous. When ATP or P_i is added to this solution, a drop in viscosity results because of the dissociation of the actomyosin into myosin and actin. As mentioned earlier, the actomyosin complex, under the right conditions, is contractile. The addition of ATP under specific conditions to actomyosin in the gel form causes the protein system to contract. Actomyosin does not exist in resting muscle. It comprises about 80% of the structural protein that can be extracted from contracted muscle. Very small amounts of actin will cause solutions of myosin to contract.

When actin and myosin are together in the presence of ATP and Mg they contract and remain so while they hydrolyze the ATP until it is used up. When the ATP no longer exists in the system, a stable complex forms between actin and myosin, actomyosin. This is a state of rigor. Actin and myosin are dissociated when in a state of rest. This can be maintained under special physiological conditions involving pH and ionic strength, and the presence of the two proteins, troponin and tropomyosin. Relaxation necessitates a combination of troponin and tropomyosin. Also, they have been shown to bind to actin and not to myosin. The binding of ATP to myosin is not inhibited by tropomyosin or troponin. Of the contractile proteins, myosin has the only available binding site for ATP. Most of the muscle protein used as food is actomyosin which has been formed during rigor by the polymerization of actin and myosin. The probability is that tropomyosin regulates the interaction between myosin and actin in the presence of ATP in molecular contraction.

It is possible to produce a cycle of contraction, rigor, relaxation, and rigor and then contraction again artificially by the addition of ATP.

Tropomyosin and Troponin. Tropomyosin accounts for about 10–12% of the total myofibrillar proteins. While it is very close to myosin in amino acid composition, its molecular weight amounts to about only 50,000. Its aqueous solution is very viscous, but this viscosity is effectively decreased by the addition of a little salt. Tropomyosin is highly α-helical, does not seem to have any enzyme activity, and in solution fails to combine with actin or myosin. It is resistant to denaturation.

Tropomyosin B is a protein complex of tropomyosin and troponin. Troponin, is very important physiologically: it is Ca^{2+} receptive and is involved in the Ca^{2+} sensitivity of the actomyosin system together with tropomyosin. The troponin–tropomyosin complex is considered effective in triggering the action of Ca^{2+} on muscular contraction.

Fatty Acids as a Source of Energy. Striated muscle burns fatty acids by preference. The availability of fatty acids to muscle and the general

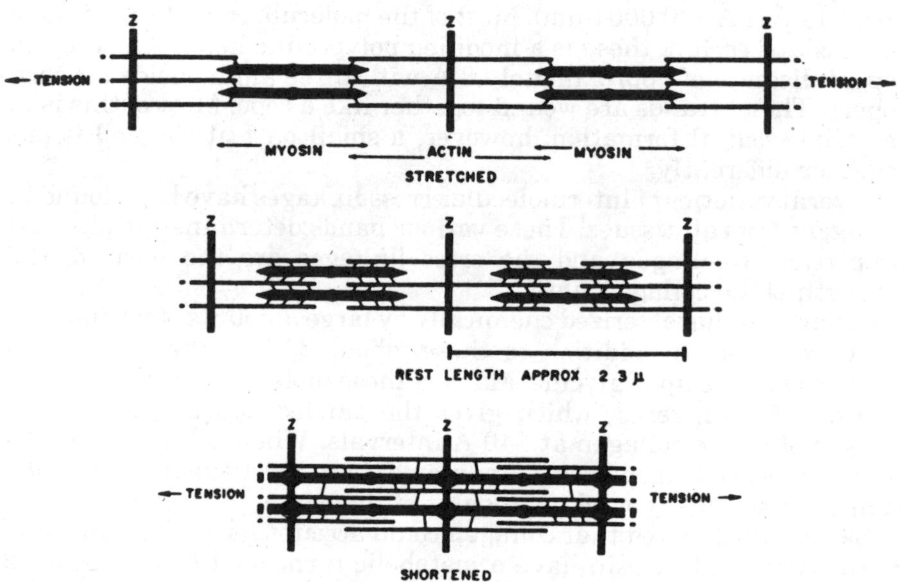

FIG. 20.6. Schematic diagram of contraction of skeletal muscle showing some completed cross-bridges at various lengths.
Courtesy of E. J. Briskey.

pathways used are known, as well as the medium through which they are carried to the muscle for use (Havel 1970). In the skeletal muscles, which contain both red and white fibers, the necessary function of making available additional ATP is provided by glycogenolysis. ATP is an energy source also. It is needed for continued activity when the work loads are high, in such situations where the fatty acid uptake is insufficient to cover the need.

Connective Tissue. Collagen and elastin are the two main proteins of connective tissue, the former being quantitatively the more important. Another protein, reticulin, is often mentioned, but it is so similar to collagen that many authors regard them as identical. For this reason, it will not be discussed further. Collagen is found extensively in animals. It is important in the tendons that attach the muscle to bone. Extensive work on collagen and elastin has been done by Piez (1966) and Partridge (1962, 1966). It should be noted that the texture of connective tissue tends to render it as undesirable in muscle tissue used as food.

Collagen. White connective tissue contains as its major component the protein collagen. The exact molecular weight of the collagen molecule is not known, but it is close to 300,000. The molecule is represented as a cylinder and has an extimated length of 2990 Å and a diameter of about 14 Å (1 Å = 0.0001 μm). Most of the molecule is made up of three strands and each of these is a modified polyproline helix, which has a repeat distance of approximately 3 Å with three amino acids for each repeat. These strands are wound together like a rope. Most of this is in the triple helical formation; however, a small part at the end is put together differently.

Several varieties of intermolecular cross linkages have been found in collagens from all tissues. These various bands determine the physical properties of collagen and the cross linkages are the basis of the strength of the collagen fibril.

Collagen is characterized chemically by large amounts of proline and hydroxyproline. In addition to these, about 33% of the amino acid residues is made up of glycine. Many of these molecules are set parallel and quarter-staggered, which gives the banded appearance characteristic of native collagen at 640 Å intervals. When collagen is denatured, soluble collagen breaks up into its three constituent polypeptide chains, or strongly bound aggregates of them.

Bailey (1972) noted that collagen could account for the toughness of meat. Actin and myosin have a metabolic turnover time of about 12 days. The turnover time for collagen, however, is very long, which, in turn, gives the cross links opportunity to accumulate and to become stabilized. This could account for the developing toughness of meat with increasing age of the animal.

The total amount of collagen in the meat of animals does not of necessity correlate with the age of the animals. However, as an animal gets older an increase in the cross links between the collagen molecules does occur, and the solubility in salt solution decreases. This is explained by this alteration in the collagen, which in turn, is responsible for the increase in toughness. Another thought to explain the effects of collagen on increasing toughness of the meat with advancing age of the animal was suggested by Cormier *et al.* (1971). Collagen is closely associated with polysaccharides of the connective tissue. The decrease of polysaccharides with advancing age may be a principal factor in the increased insolubility of collagen. The polysaccharides could play a vital role in plasticizing the collagen fibers, which in turn, could explain their contribution to meat tenderness.

Satterlee (1971) studied the activity of porcine pancreatic collagenase on connective tissue and found that both porcine and bovine pancreas extracts had collagenolytic activity of pH 5.5 and 37° C. This capacity to break down connective tissue was caused by a combination of collagenolytic and elastolytic activity.

As the temperature increases, so does the degree of solubility of collagen. According to various workers, collagen A shortens at about 60° C and is converted to collagen B. When collagen B is heated in water, swelling and softening results. Finally, it is converted into gelatin, but only after heating a long time at 100° C. This conversion takes place rapidly if the collagen is heated under pressure at temperatures ranging from 115° to 125° C. Ritchey *et al.* (1963) made a study of the conversion of collagen to gelatin during cooking. Gelatin is a very unusual protein. Proline and hydroxyproline make up over 30% of gelatin protein. The unusual properties of gelatin make it important in household and laboratory uses.

Elastin. Elastin, the protein found in animal connective tissue in small amounts, tends to be rubbery. Fairly large amounts of elastin are found in the muscles of the rump of beef animals, but in other muscles it is present in rather small quantities. In order to have the necessary rubbery properties, the protein chains are coiled in random fashion for most of their length. They are kinetically free and at intervals they are cross-linked by bonds that are thermally stable. The net result is rubber-like elasticity. Chemical proof exists for the presence of covalent cross links in elastin. Elastin must be in an environment containing about 40% water for its rubbery qualities to be evident. It is likely that a tendency to form an α-helix with internal hydrogen bonds is reduced or prevented by the amino acid sequence. This protein contains a fair amount of proline among its amino acids. Two previously unknown amino acids have been shown to be present in elastin—desmosine and isodesmosine. The presence of desmosine cross links renders elastin's solubility very low.

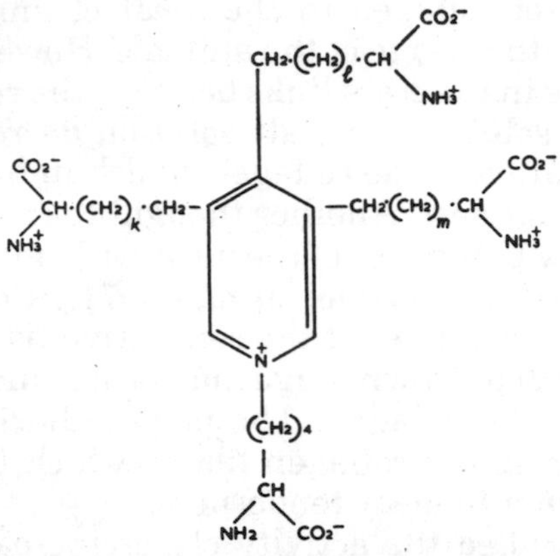

Desmosine (proposed partial structure)

Both desmosine and isodesmosine are tetracarboxylic-tetraamino acids. They also contain quaternary pyridinium rings, but reversible oxidation–reduction reactions do not take place. This is different from derivatives of nicotinic acid. The pyridine nucleus of these two amino acids is destroyed as a result of mild oxidation with alkaline ferricyanide, releasing lysine as well as some other ninhydrin-reacting compounds.

Neither collagen or elastin contain all of the essential amino acids, and so are not complete in nutritive value. However, they are excellent sources of some amino acids. The undesirable feature of these proteins is that they impart toughness to the muscle.

Sarcoplasm. The intracellular fluid of muscle is known as sarcoplasm, and it is from this fluid that the sarcoplasmic proteins are obtained. Most of the soluble proteins contained in sarcoplasm are enzymes, one of which, glyceraldehyde phosphate dehydrogenase (GAPDH), is found in rather large amounts in muscles such as longissimus dorsi. Others are also present in rather large amounts; however, it is most likely that many hundreds of these enzymes exist in very small quantities. Among those in larger amounts are enolase, triose phosphate isomerase, phosphoglucose isomerase, creatine kinase, aldolase, phosphoglucomutase, and pyruvate kinase.

Meat Pigments

The color of meat is of importance commercially because the purchaser so often judges the quality of the cut by the appearance.

According to Fox (1966) "the chemistry of the color of meat is the chemistry of the heme pigments, myoglobin and hemoglobin, which, insofar as meat color is concerned, are identical in their reactions."[2] Myoglobin is the principal source, although not the entire, source of the

[2] Used with permission of the American Chemical Society.

color of meat. Hemoglobin comprises 20–30% of the meat pigment present and sometimes more. While it is indeed true that reactions of these two pigments are identical in most of the cases, several important reactions proceed at different rates, including denaturation, reaction with nitrite, and autoxidation. It is for this reason that the amounts and reactions of myoglobin and hemoglobin must be considered if color changes or intensity in meat are to be correlated with pigment concentration.

Figure 20.7 shows the heme pigment reactions of meat and meat products.

Myoglobin. The myoglobin molecule is similar to that of hemoglobin of blood, except that it contains only one heme group and hemoglobin contains four. In the older literature, one finds myoglobin written as muscle hemoglobin. This has been discarded, and in all modern literature the term myoglobin is used (Theorell 1932). Ginger *et al.* (1954) found, as one might suspect, that myoglobin is higher in beef than in pork. In beef versus light-colored pork, the ratio is 4.7 to 1. In dark-colored pork muscle the ratio was found ot be 2.6 to 1. Lewis and

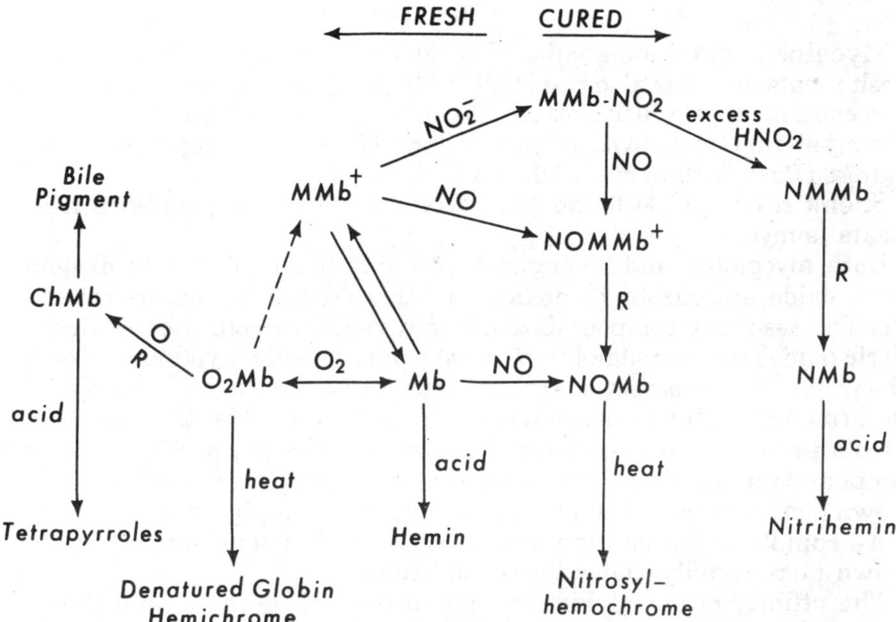

FIG. 20.7. Heme pigment reactions of meat and meat products. ChMb, cholemyoglobin (oxidized porphyrin ring); O_2Mb, oxymyoglobin (Fe^{2+}); MMb, metmyoglobin (Fe^{3+}); Mb, myoglobin (Fe^{2+}); MMb·NO_2, metmyoglobin nitrite; NOMMb, nitrosylmetmyoglobin; NOMb, nitrosylmyoglobin; NMMb, nitrimetmyoglobin; NMb, nitrimyoglobin, the latter two being reaction products of nitrous acid and the heme portion of the molecule; R, reductants; O, Strong oxidizing conditions.

From Fox (1966). Reproduced with permission of the American Chemical Society.

Schewigert (1955) determined that beef myoglobin in crystalline form is electrophoretically hetercgeneous.

$CH_3C{=}CCH{=}CH_2$
HC—C, N, C=CH
CH_3C—C, $C{=}CCH_3$
N—Fe—N
C—C, $C{=}CCH{=}CH_2$
$HOOCCH_2CH_2$, HC—C, N, C=CH
$HOOCCH_2CH_2$—$C{=}CCH_3$

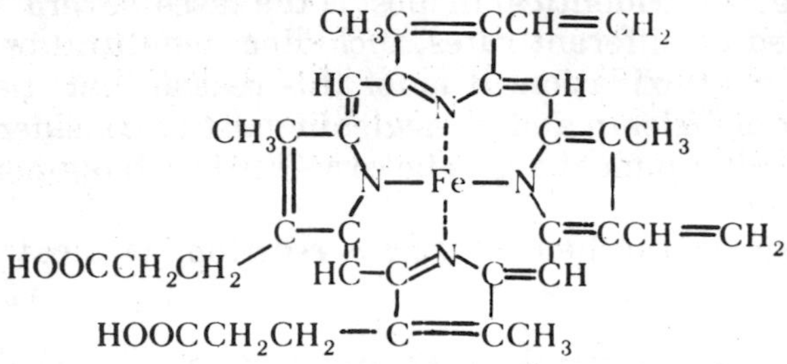

Heme (Ferroprotoporphyrin)

Heme can combine with globin at near neutral pH to yield hemoglobin.

CHANGES IN MUSCLE AFTER SLAUGHTER AND DURING PROCESSING

Color

Myoglobin and Hemoglobin. Myoglobin is important in the color of fresh meats, but it is also responsible for the color of cured meats when it is combined with nitric oxide. The nitric oxide red pigment changes during storage to brown and gray colors. Heat and/*or* light speed this change (Ramsbottom *et al.* 1951; Watts 1954).

Shenk *et al.* (1934) found that over 90% of the pigments in fresh meats is myoglobin.

Both myoglobin and hemoglobin are able to combine with oxygen, nitric oxide, and carbon monoxide, and these combinations are reversible. The resulting compounds are bright red oxymyo(or hemo)-globin, nitric oxide myo(hemo)globin, and carbon monoxide myo(hemo)globin. The iron in these compounds stays in the ferrous state. The corresponding brown met-pigments are formed in each case when the compound loses an electron and oxidation is the result. The oxy and nitric oxide compounds must dissociate to myoglobin before the oxidation to the brown ferric compound can take place. Fresh meat pigments dissociate more rapidly at lower than atmospheric oxygen tensions and would brown more rapidly under these conditions.

The affinity of myoglobin for oxygen is much greater than that of hemoglobin. Kiese and Kaeske (1942) found that on exposure to atmospheric oxygen, myoglobin oxidizes 14 to 16 times as quickly as hemoglobin to form the brown ferric compound.

The bright red color of fresh meat is the result of oxygenated heme pigments at the surface, which are exposed to the oxygen of the air; the purplish red of the interior is caused by the reduced pigments. Brown, gray, and green discolorations of meat are abnormal. Two kinds of oxidative changes are the principle cause of the formation of these off-colors. The brown oxidation products, metmyoglobin and methemoglobin, are the ones usually found. They are produced by oxidation of the iron from the ferrous to the ferric condition (Brooks 1938). The second is the result of the action of oxygen on the heme structure. Ascorbic acid, useful under some conditions in meats, can reduce methemoglobin.

Sometimes an objectionable green color appears in meat, caused by pigments that are the result of transformations in the heme pigments (Brooks 1938). In these instances the prophyrin ring is acted on, and frequently the point of attack is the α-methene bridge. The result is destruction of the double bond at this point which, in turn, breaks the succession of the series of conjugated double bonds that make up the porphyrin ring, and destroys the resonance structure. The ring is not necessarily broken at this point because other green compounds are known with the porphyrin rings intact. In addition, green verdohemes can be formed by splitting out the α-methene carbon atom with the resulting opening of the prophyrin ring. It is possible for other methene bridges to be acted upon in the same way and so form pyrrole fragments.

The development of green colors in meat can also result from the generation of H_2S or H_2O_2. Under these conditions it is likely that bacteria are involved. In the case of fresh meat, the formation of H_2O_2 would not be a problem because of the presence of catalase in the meat, because catalase decomposes H_2O_2. In the case of cured meats, no catalase is present and in the presence of H_2O_2, rapid greening results.

Nitric acid myoglobin and the corresponding hemo compound form in meat during the curing process. They retain their bright red colors on heat denaturation, accounting for the fact that corned beef and other cured meats are red colored during and after cooking.

According to Greenberg *et al.* (1943), the nitrite in the curing medium reacts with oxyhemoglobin to form methemoglobin as shown in the following reaction.

$$NO_2^- + 2HbO_2 \rightarrow NO_3^- + Hb_2\,O + O_2$$

When the globin is denatured, even partially, the bonds connecting the heme with the globin are weakened. Under these conditions the heme is no longer able to combine reversibly with oxygen to form the bright red oxy compound. In place of this it is rapidly oxidized to the ferric form which is brown. The red color of the cured meat pigment, nitric oxide myo(hemo)globin, is not destroyed, even though the denaturation of the globin is complete and irreversible.

Color and Meat Packaging. The limited color stability of fresh meat limits central meat-packing operations. This, in turn, raises the cost of handling of meat. If it were possible to increase the color stability of fresh meat, a savings could be passed on to the consumer.

pH

Bate-Smith and Bendall (1956) found it essential to have high levels of creatine phosphate and ATP in the muscles initially to slow the rate decline in the pH of muscle after death. If meat of top quality is to be produced, the rate of destruction of ATP in the muscle must be kept to a minimum, or the resynthesis of ATP must be sufficient to meet the needs. In addition, a satisfactory balance of creatine phosphate must be maintained.

Another point to be considered in connection with pH change in the muscle is the nutritional condition of the animal, since the amount of glycogen in the muscle is instrumental in effecting the amount of postmortem change in the pH.

Meat with a high final pH completely loses its commercial acceptability for several reasons. Among them are reduced penetration to salts in curing, dryness, apparent toughness, and the cosmetic effect of dark cutting meat which is objectionable to many. However, Penny *et al.* (1964) found that a superior freeze-dried product results when meat of high ultimate pH is used.

Rigor Mortis

It has been suggested that the important factor in turning muscle into a cut of meat to be used as food is the degradation of ATP that takes place during the period starting with death of the animal to the disappearance of rigor mortis. This indicates the end of post-morten glycolysis due to the exhaustion of the ATP.

It is generally agreed that the important chemical changes which take place between the time of death and the start of rigor mortis are connected with a lowering of glycogen content, pH, ATP, and creatine phosphate, together with larger amounts of lactic acid, ammonia, and decomposition products of ATP, and finally with the formation of actomyosin which results in a tough, inflexible muscle.

The extreme changes that take place in meat while it passes into rigor mortis are the well-known stiffening and loss of extensibility, the increase in acidity caused by the formation of lactic acid by means of the anaerobic glycolytic cycle, and finally the loss of water-holding capacity. Rigor mortis results from the formation of cross links between the actin and myosin filaments of the myofibril while at the same time the ATP supply of the muscle is exhausted after death. This gives an inextensible and rigid structure caused by the shortening of the muscle and accounts for the rigidity of rigor. According to Stromer and Goll (1967) the myofibrils of the beef muscle are supercontracted after 24 hr.

Szent-Györgyi (1951, 1960) found originally that an actomyosin thread shortens when put in contact with ATP. Immediately he saw the connection between this and the contraction of muscle. During contraction, the hydrolyzed ATP is restored by the Lohman reaction, in which phosphocreatine (CP) phosphorylates the ADP and releases creatine (C), according to the following reaction.

$$ADP + CP \rightleftharpoons ATP + C$$

According to Marsh (1966), if bound calcium is contained in the contractile protein of muscle actomyosin, then in the presence of ATP the muscle contracts. On the other hand, it relaxes with ATP if part of the calcium is removed. The relaxing factor controls the amount of calcium in actomyosin. This is done by regulating the amount of Ca^{2+} available. Research on the relaxing factor could lead to a more complete understanding of rigor mortis.

Bendall (1951) agrees with the idea that rigor mortis and muscular contraction are the result of the same mechanism. There is no general agreement concerning the chemical changes involved in the resolution of rigor mortis. While a great deal of work on muscle has been done with the rabbit, Marsh (1954) did work with beef muscle on the onset of rigor mortis. He found that the results were similar to those obtained with rabbit muscle.

The most apparent change obvious on visual inspection is that of color of porcine muscle. In this case the change is from relatively dark red to a rather lighter grayish pink. Rigor morits and other post mortem changes are due mainly to glycolytic chemical changes. This amounts to the change of glycogen to lactic acid, and it can be noted by monitoring the decrease in pH. High temperature and low pH occurring together soon after death can be the cause of the loss of color and extensive lowering of water-binding capacity.

Briskey (1964) published a good review on the status and associated studies of pale, soft, exudative (PSE) porcine musculature. He noted that the muscles of living hogs are moderately dark in color, the appearance is dry, and the texture is firm. As the animal is prepared for processing and is processed, many biochemical, physical, and physiological changes take place that have an important affect on the muscles. If these changes occur within 0.5–1.5 hr at an elevated temperature and produce acidic conditions, the final result is muscles with pale color, soft texture, and an excessively exudative appearance. This condition is highly undesirable since it results in significant economic losses. Excessively rapid rates of glycolysis and low pH values at rather high body temperatures are connected with the production of this condition. Briskey stated that muscles that were normal at death, but had pH values significantly lower 40 min after death, developed the undesirable PSE condition. The pH values of the muscles had reached in many instances 5.0 or lower while the muscle temperatures held to 36° to 41°C.

Some pigs are known to be susceptible to stress, and postmortem glycolysis of their muscle goes on very rapidly, resulting in pale, soft, and exudative muscle, which is undesirable for use as food. Cooper *et al.* (1969) found that many of the red fibers in stress-susceptible animals were not really red fibers. Such fibers had phosphorylase and adenosine triphosphatase activity as well as positive tests for the oxidative enzyme reactions, and were therefore, intermediate fibers. Cassens *et al.* (1969) found a large fiber, neither red nor white, in pig muscle. They occur in less than 1% of the fibers, but frequently in muscles of stress-susceptible animals. They give a negative reaction for phosphorylase and a positive for adenosine triphosphatase.

Günther and Schweiger (1966) investigated the feeding of sucrose to animals after exhaustive transportation. A rest of a few hours after long distance transport, followed by sucrose feeding, restored the exhausted animals not only to normal glycolysis but to normal water-holding capacity and consistency of the muscle. The sucrose feeding decreased the pH values and increased lactate levels in the muscles investigated. By this procedure some characteristics of the quality of the meat were improved.

A number of authors found that fat, protein, and moisture contents of normal and PSE muscles are not consistently different. Much work remains to be done on this problem.

Tenderness

When muscle tissue is cooked immediately after death of the animal but before rigor mortis starts, it is tender because the important proteins of the myofibrils, actin and myosin, are not connected. After the start of rigor mortis, the length of the sarcomere decreases, the muscle becomes inextensible, and it is tough when cooked. During the process of aging the length of the sarcomere increases, the muscle becomes pliable again, and increasingly tender on cooking.

It would seem, on the face of it, that the improvement in tenderness connected with aging could be caused by changes in one or a combination of the following: (1) connective tissue changes, (2) dissolution of actomyosin, (3) an increase in the hydration of the proteins, (4) proteolysis.

It is quite probable that the most important factor in tenderness–toughness in meat is the degree of cross-linking in the actomyosin. This is a much more important factor than the amount of collagen and contraction during rigor.

Muscle Proteins

Muscle protein is very important in a number of processed foods. In a wiener, which contains a considerable amount of water-soluble and salt-soluble proteins, the resulting texture is materially affected by these proteins. The temperature, rate, and extent of postmortem decrease in pH influence the characteristics of the muscle proteins and those of the food products into which they are made. As the pH falls, the adenosine triphosphatase activity, the solubility of the salt-soluble protein, as well as the solubility of the water-soluble protein, and the heat gelling properties are reduced.

Two especially important properties of muscle proteins from a practical standpoint are (1) water-holding capacity and (2) fat-emulsifying capacity.

Water-Holding Capacity. Water-holding capacity is intimately related to tenderness, color, and taste and is affected by the treatment of the animal before slaughter. Also, it influences the quality of meat during most of the processing operations. These include storage, aging, grinding, curing, salting, heating, drying, freezing, and thawing. The binding of water within the muscle is brought about by the muscle proteins. Tenderness, texture, and other factors including shrinkage on cooking are connected with the condition of hydration of muscle proteins.

Bound Water. Although meat contains about 75% water, only about 4–5% of it is tightly bound (bound water) to the muscle proteins. The changes in the water-holding capacity that take place during storage and processing are affected by the immobilization of the physicochemically free water within the microstructure of the tissue. The myosin and actin of the myofibrillar proteins are responsible for binding hydration water and for immobilization of free water in meat.

Within the protein network there exists a continuous transition from the bound water to the unbound or loose water. The free or unbound water within the tissue is affected by the spatial molecular arrangement of the myofibrillar proteins. If the network of proteins is tightened (that is, by drying or freezing, in rigor mortis, protein denaturation by heating, effect of cations), the bound water decreases and the amount of unbound water increases. Loosening of the protein structure by salting or aging has the opposite effect. The attraction or repulsion between the charged groups of adjacent protein molecules or by linking or loosening the cross linkages between the peptide chains may cause such structural changes. The water-holding capacity is strongly influenced by pH and by ions of such salts as NaCl. It is possible to explain this by changes in the electrical charges of the myofibrillar proteins.

The salts dissolved in the sarcoplasma decidedly influence the water-holding capacity of meat. Approximately half of the water-holding capacity of beef muscle is caused by the effect of the sarcoplasmic ions on myofibrillar proteins; this effect can occur during storage and thawing with a resulting loss of meat juice.

Fat-Emulsifying Capacity. The second item of practical importance, fat-emulsifying capacity, is essential to the preparation of meat emulsions. The meat emulsion is a muscle–fat–water system in which some of the muscle protein dissolves, especially when salt is present. Under these conditions the protein exerts an emulsifying effect by forming a thin layer around the fat droplets. In a sausage mixture the sarcoplasmic proteins are likely to have a dominating effect because more of them will be in solution than myofibrillar proteins. However, the emulsifying ability of muscle proteins decreases as in the following order: G-actin, myosin, actomyosin, sarcoplasmic proteins, F-actin.

Oxidation and Rancidity

A discussion of oxidative rancidity in fats is given in Chapter 5 and will not be repeated here. However, oxidative rancidity in meats is important because it limits the salability of meat so affected.

According to Gunstone and Hilditch (1945) the methylene group between two double bonded carbon atoms is considerably more active to oxidative assault than carbon atoms adjacent to a single double bond. As a result, linoleic acid with one active methylene group can oxidize 10–12 times as quickly as oleic acid, while linolenic acid with two labile carbon atoms is able to oxidize at twice the speed of linoleic acid. Hydroperoxides, which are intermediates, are formed and they are unstable and break down, giving rise to a number of products, some of which are the cause of the developing rancidity. The start of rancidity is controlled by the varying amounts of α-tocopherol, a natural antioxidant.

While it is known that pork fat contains more unsaturated fatty acids than lamb or beef fats, actual amounts found depend on the diet of the animal. This is particularly true for hogs than for the other two species. Hogs have a tendency to the deposition of fats from the diet. When added to the diet, tocopherol tends to be deposited in the animal fat, but not in amounts large enough to improve effectively the storage life of the meat because much of the tocopherol administered is eliminated rather than being deposited in the fat.

Although unsaturated fats and heme pigments in meat are oxidized separately, a coupled reaction involving both of these can in turn, speed color loss and rancidity. This is a catalytic effect. Robinson (1924) studied the activity of hemes as catalysts on the uptake of oxygen by linseed oil. She found that hemin, hemoglobin, and methemoglobin, when used in equivalent concentrations, had approximately the same effect in speeding the reaction.

The destruction of porphyrin takes place during its extended reaction with unsaturated fats. These unsaturated acids were more unsaturated than oleic acid. Inorganic iron was released and color faded. It is only in heterogeneous systems that this reaction has been found to occur, and it has been concluded that enzymes are not involved in this reaction. It is likely that this reaction is involved in the deterioration of ground meats held in freezing storage. Dark meat of turkey becomes rancid considerably faster than the white meat (Klose *et al.* 1950).

If muscle extracts or meat are heated so as to coagulate the myoglobin or hemoglobin, the catalytic effect, as far as the activity on the oxidation of fat is concerned, is destroyed.

According to Lea (1937) increasing pH within the range normally found in meat lowers the rancidity of muscle extracts.

The main cause of deterioration in quality of meat and meat products is lipid oxidation, and this can take place in the stored triglycerides or in the phospholipids of the tissues. The essential prooxidants involved in tissue lipid oxidation are the ferric heme pigments. Lipid and pigment oxidation are closely related, and lipid oxidation is thought to be promoted by ferric hemes.

Fresh meat as well as many processed meat products are subject to lipid oxidation. Oxidative deterioration can take place in raw meat during storage under refrigeration, and during frozen storage as well.

Phospholipids are susceptible because of their high content of unsaturated fatty acids. The phospholipids of beef contain fatty acids, 19% of which have four or more double bonds. Phosphatidylethanolamine of pork belly muscle was found to contain rather high amounts of linoleic and arachidonic acids. This may account for the development of oxidative deterioration in this material. Hornstein *et al.* (1961) found that extracts containing total lipid and phospholipids prepared from beef and pork developed rancid odors faster when exposed to air than did corresponding samples of neutral fat. This suggests that the phosphlipids assist in the development of poor flavor, especially in very lean meat. Lipid oxidation takes place very slowly if the pigment of the meat is in the pink cured ferrous form. The pigment is converted to the brown ferric form during storage of cured meat and this, in turn speeds the oxidation.

El-Gharbawi and Dugan (1965) showed that the lipids extracted from freeze-dried raw beef undergo oxidation in two stages: (1) the phospholipids are oxidized first, and (2) then the neutral fats are oxidized. They found, in addition, that the loss of unsaturated fatty acids was greater in the phospholipids than in the neutral fat.

Muscles that have pH values in the upper normal range (normal pH of meat is 5.2 to 6.6) fade least on freezer storage. The reason is that at the highest pH values the reaction between unsaturated fat and myoglobin is inhibited. The oxidation of the fresh meat pigments is accelerated if the pH values are in the lower part of the range. In general, pork to be cured by conventional methods is in the lower end of the pH range.

Salt has an accelerating action on the development of rancidity, as has been noted by various investigators. The condition needed to bring this about is the contact of the salt with the fat over an extended surface. Copper and iron also have been shown by a number of workers to have an accelerating effect on the oxidation of fats or foods containing fats. Although other metals can do this, copper and iron are the most likely to contaminate foods. Smoke renders fleshy foods less suceptible to rancidity.

Action of Light

In the presence of oxygen, light speeds all oxidative changes in meats. Also, discoloration by light is brought about in cured meats more readily than in fresh (Ramsbottom 1951). It takes several days for lighting used in display cases to bring about any noticeable discoloration in fresh meats, but cured meats show evidence of fading in about an hour under the same conditions. Ultraviolet light, which can denature proteins, discolors fresh meat pigments to which it is exposed.

Antioxidants

In order to prevent the development of rancidity, a number of compounds are used. Most of them are phenolic compounds. Those which have approval for use in lard include gum guaiac, 0.1%; butylated hydroxy-anisole (BHA), 0.02%; propyl gallate, 0.01%; and the naturally occurring tocopherols, 0.03%. These compounds lengthen the induction period in the process of oxidation of fats, probably by absorbing the activating energy of the peroxides in the fat. This, of course, would break the chain reactions that result in the oxidation of anitoxidants (Filer *et al.* 1944; Lowey *et al.* 1969; Mahan and Chapman 1953).

Another group of so-called "synergistic" antioxidants improve the shelf-life of animal fats in combination with antioxidants, although they have no effect as protectants when used alone with fats. Compounds that improve the ability of the phenolic antioxidants to retard rancidity include phosphoric acid, 0.005%; citric acid, 0.005%; thiodipropionic acid, 0.01%, and esters of this acid, and lecithin in any quantity. Aside from the use of lecithin, these are patented processes. The action of these compounds is not entirely clear. A number of other compounds have also been found to possess this property.

Ascorbic acid when added to meats show irregularities in any action against rancidity. When the content of tocopherol is low, ascorbic acid is able to speed the oxidation of fat. Also, it is able to have an inhibitory action on fat oxidation in the presence of polyphosphate or other metal chelating agents. (see Scarborough and Watts 1949 and Lehmann and Watts 1951).

The use of antioxidants with cooked meats that are to be dehydrated or frozen works quite well. The storage life of such products can be lengthened effectively.

Effects of Heat on Meat Products

Great changes in the muscle proteins result from the heating of meat. These changes are particularly important because heat treatment (cooking) usually precedes use. Changes in the myofibrillar proteins cause shrinkage of the tissue and release of juice (Rogers *et al.* 1967). There is a decrease in the solubility of myofibrils and sarcoplasmic proteins with increasing duration and temperature of heating.

The changes that take place in the heating of meat are most profound in the muscles themselves. The adenosine triphosphate activity of myosin is destroyed by heating for short periods of time at 37°C. However, the tissue changes that take place between 40° and 60°C are more unusual. From the food scientist's point of view it is of greater importance to study the muscle tissue rather than the individually separated muscles. This is because of the different behavior under some conditions of the separates muscles as compared with that in the tissues. The pH of muscle tissue rises during heating (Hamm and Detherage 1960). According to Hamm (1966) this could be caused by hydrogen bonding or charge changes, or perhaps both, taking place in the myofibrillar proteins. Also, Hamm (1936A) believes that the isoelectric point of actomyosin is related approximately to the pH at which the water-holding capacity is minimal; this is 5.0 in the native muscle. The likelihood is that shrinkage of tissues and release of juice by heating up to 40°C are caused by some unfolding of the peptide chains. Hamm and Deatherage (1960) found that "acid stable" cross linkages are formed only at temperatures above 50°C. Rising temperature increases the formation of protein network linked by such bonds.

During the heating and denaturing of muscle tissue, unfolding of the peptide helix appears to be closely connected with coagulation. The unfolding is complete above 70°C. Many other changes take place during this process. One of these is the release of reactive sulfhydryl groups which are concealed inside the folded structure of the protein. Proteins migrating in an electric field with the greatest velocity (anodic and cathodic) are the most quickly denatured (Grau and Lee 1963).

The formation of H_2S seems to be the most important result of the destruction of sulfhydryl groups (Hamm 1961). Hydrogen sulfide starts to form from myofibrils at around 80°C with an exponential increase as the temperature goes up. Also as the heating time increases, so does the formation of H_2S. Most of the H_2S formed during heating comes from the structural proteins, and not from cysteine or glutathione, which are water-soluble substances. This amounts to 18.9 ± 0.4 μg from a gram of protein. This is substantially the quantity from heating the myofibrils.

The physical and chemical changes in muscle occur in distinct steps during heating.

From 20° to 30°C. The colloidal chemical properties and the ion-binding muscle proteins show no alteration in this range of temperature. At 30°C however, the enzyme activity (ATPase) of myosin is reduced.

From 30° to 50°C. The alterations in the myofibrillar proteins include an affect on the rigidity and water-holding capacity of the muscle, the pH, and ability of muscle proteins to bind Ca^{2+} and Mg^{2+}. These changes are twofold: (1) the peptide chains unfold and (2) rather unstable cross linkages develop that bring about a tighter structural network at the isoelectric pH. The enzyme activity (ATPase) is lost and denaturation of a little of the sarcoplasmic proteins takes place also. Most of these changes take place between 40° and 50°C.

From 50° to 55°C. The myofibrillar proteins rearrange and this results in a slowing of the water-holding capacity changes, pH, and protein bound Ca^{2+} and Mg^{2+}. The building of new cross linkages is started; these cannot be split by weak acid or base. Sarcoplasmic proteins continue to be denatured.

From 55° to 80°C. The bulk of the changes which started between 40° and 50°C continue, but the amount of these changes is less. When the temperature of 65°C has been reached, coagulation has taken place in the bulk of the globular and myofibrillar proteins. Temperatures in the neighborhood of 63°C shrink the collagen and at higher temperatures it may be converted, in part, to gelatin.

Above 80°C. The oxidation of the sulfhydryl groups of actomyosin resulting in the formation of disulfide bonds starts between 70° and 90°C, and continues with rising temperature. As the temperature goes above 90°C, hydrogen sulfide separates from the sulfhydryl groups of actomyosin. Continued heating to 120°C produces more changes in the muscle proteins with a decrease in protein-bound Ca^{2+} and Mg^{2+}. The flavor of canned meat has been found to correlate significantly with H_2S and other volatile sulfur compounds that have formed. At about 90°C, Maillard reactions begin and continue with length of heating time and increasing temperature. The browning of meat has been found to be caused principally by the reaction between carbohydrates of muscle and the amino acid groups. According to Bowers *et al.* (1968) browning increases with the amount of reducing sugar in meat. An increase in tenderness results from the conversion of collagen to gelatin. Elastin cannot be denatured by heat and remains in its original native form after cooking.

While normal methods of cooking may have some influence on the digestibility of meat proteins, they do not greatly change the amino acid content of meat.

EFFECT OF IONIZING RADIATION

Irradiation studies on meat and meat products got started about the time of World War II. Such treatment has been found to increase storage life. Changes in proteins are caused by ionizing radiation and are controlled by the dose of the radiation and by the nature of the proteins. Treatment with a dose of 5 Mrad (about the dose necessary for microbial sterilization) results in a definite loss of water-holding capacity. Since many enzyme proteins are relatively stable against radiation, they require more than 5 Mrad before they are inactivated. There can be an effect on the decomposition of the amino acids, but the damage was found to be very small (Johnson and Moser 1967).

Meat Tenderizers

Proteolytic enzymes have been used for a long time to increase the tenderness of meat. Papain is the enzyme most often used, but others are effective also. Among these are the proteolytic enzymes, ficin and bromelain. Ficin and Bromelain are of plant origin and can hydrolyze the peptide bonds of meat proteins. Such tenderizers are being used commercially, and they are used either singly or in mixtures of one or more.

Smith *et al.* (1973) found that antemortem injection of proteolytic enzyme into bullocks resulted in a significant increase in tenderness of the flesh and yielded steaks that were more comparable with those from the carcasses of steers.

Meat Flavor

Much attention has been focused on efforts to understand the flavors formed in meat as a result of the various ways of preparation. Much better methods of investigation now are available. In this work, compounds formed are identified and the importance of each in the final flavor is evaluated.

Jones (1969) studied the natural flavors of ribomononucleotides and their breakdown products in meat and fish flavors, both alone and in mixtures. While attention has been given to the flavor-enhancing properties of some of these compounds, evidence is available that nucleotide derivatives are active in flavor-precursing systems. The taste and flavor can be affected in several ways by nucleotides and their breakdown products. They are precursors of the odor of flesh food, and in some concentrations in muscle they have tastes of their own, although these may be subject to some alteration, and in addition they have the ability to enhance flavors. Concentrations of 5′-ribomononucleotides and their break-down products, especially hypoxanthine, can be used as indices of freshness and quality.

Malbrouk *et al.* (1969A, B) identified several polar lipids in the nonaqueous beef flavor components. These included phosphatidylcholine, lysophosphatidylcholine, phosphatidylethanolamine, phosphatidylserine, sphingomyelin, and phosphatidylinositol. In addition, sulfatide was identified tentatively. These authors found that the dry substance of aqueous meat extract prepared from lyophilized meat following removal of petroleum ether extractable lipids accounted for 6.5% of the fresh weight. Finally, fractions separated from this with high meaty aroma amounted to 1.1% based on fresh weight.

Sink and Smith (1972) studied the increase in carbonyls during the aging of beef. This was done because of the importance of these compounds in flavor. Longissimus dorsi muscle was held at 3°C and dramatic increases in total carbonyls and also in monocarbonyls were obtained. Methyl ketones made up the largest group of monocarbonyls and increased five-fold after 14 days of storage. Flavor significance, however, was not reported.

Cho and Bratzler (1970) investigated the use of sodium nitrite in curing pickle for pork and its effect on the flavor of the finished product. The use of this compound resulted in a product with a greater "cured pork" flavor than when it was omitted from the pickling solution. In all cases, NaCl was used in this solution. Smoking the final product did not mask this effect.

Volatile Flavors

The study of the volatile flavors of meat has made much progress because of the advances made in analytical techniques, specifically gas chromatography and mass spectroscopy. The idea that beef flavor was the result of the combination of carbonyls, H_2S—NH_3, prevalent until about 1966, does not provide the answer.

Hirai *et al.* (1973) in a study of the compounds present in the volatile flavor from boiled beef, isolated 54 compounds, no one of which had a typical aroma of boiled beef, suggesting that boiled beef aroma is a very complex mixture of compounds. In this investigation it was found that three unsaturated alcohols were present. These included 1-octen-3-ol, 1-penten-3-ol, and 2-hexen-1-ol, which had previously been found in other materials. Among the carbonyls, diacetyl and acetoin have previously been reported as components of meat flavor. These were found by Harai *et al.* (1973) in relatively large amounts. Because acetoin can easily be converted to diacetyl in the presence of air, the rather large amount of acetoin shown to be present could be a good source of diacetyl. Two heterocyclic compounds, 2,5-dimethyl-1,3,4-trithiolane, with sulfure in the molecule, and 2,4,5-trimethyl-3-oxazoline, with nitrogen, were found earlier to be components of volatiles of boiled beef. However, their odor in the pure form was not characteristic of that of boiled beef either. In addition, two furane compounds, 2-methyltetrahydrofuran-3-one and 2-pentylfuran, that had been isolated from other substances previously, and of two heterocyclic compounds which contained sulfur were found. One of these that had a spicy meat odor was thiophen-2-carboxaldehyde; the other was 5-thiomethylfurfural.

Tonsbeek *et al.* (1971, 1968) made a study of the compounds occurring in beef broth. They isolated 2-acetyl-2-thiazoline, 4-hydroxy-5-methyl-3(2*H*)-furanone and its 2,5-methyl homolog from beef broth.

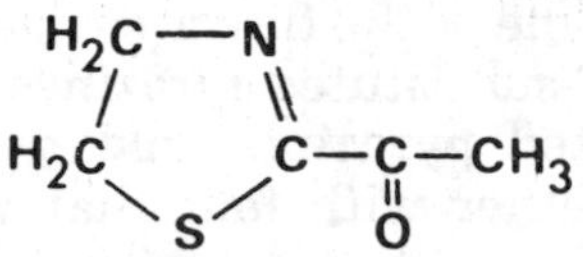

2-Acetyl-2-thiazoline

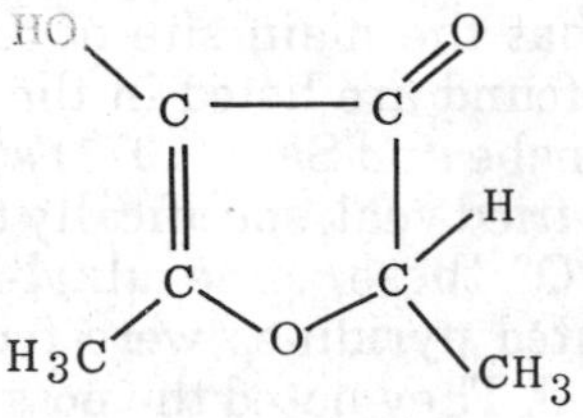

4-Hydroxy-2,5-dimethyl-3(2H)-furanone)

Tonsbeek *et al.* (1969) studied further the natural precursors of 4-hydroxyl-5-methyl-3(2*H*)-furanone. They found that the precursors of this compound are ribose-5-phosphate, and pyrrolidone carboxylic acid or taurine or both. It was shown by Peer and van den Ouweland (1968) that 4-hydroxy-5-methyl-3-(2*H*)-furanone can be prepared from D-ribose-5-phosphate in concentrated buffer systems at pH 5.5. However, in a diluted solution such as beef broth, this reaction requires the presence of the compounds containing nitrogen, pyrrolidone carboxylic acid, or taurine.

Brinkman *et al.* (1972) analyzed the volatiles of the headspace of beef broth, and found the following: 3-methyl-butanol, benzaldehyde, and 3,5-dimethyl-1,2,4-trithiolane (*cis* and *trans*). All have been found previously as components of beef flavor. In addition, 2,3-pentanedione, methional, 1-methylthioethanethiol, and 2,4,6-trimethylperhydro-1,3,5-dithiazine (thialdine) were not previously known as beef flavor components. 1-Methylthioethanethiol, which has an odor of fresh onions, had not been previously described in the literature. Evidence indicates that the trithiolanes and the thialdine are formed largely during the isolation procedure.

Schutte and Koenders (1972) set forth systematic reactions for the formation of the flavor compound 1-methylthioethanethiol in beef broth from components known to be present in beef. It was found that 1-methylthioethanethiol is produced when ethanal, methanethiol, and hydrogen sulfide are heated in aqueous solution at pH 6. These immediate precursors are generated in turn and under the same conditions from alanine, methionine, and cysteine in the presence of a Strecker degradation agent such as pyruvaldehyde. The formation of methanethiol from methional, the initial degradation product of methionine, was found to be the weakest link in the reaction sequence; for this reason, it was studied in some detail.

Liebich *et al.* (1972) studied the volatile components of roast beef by gas–liquid chromatography and mass spectrometry. They found that alkanals, alk-2-enals, and alka-2,4-dienals are the major volatile components. Also, 3-hydroxy-2-butanone and γ-butyrolactone were present in high concentration. Other classes of components include 2-*n*-alkylfurans, 2-alkanones, 3-alkanones, 2,3-alkadiones, pyrazines, primary and secondary alcohols, acids, γ- and δ-lactones, alkanes, aromatic compounds, sulfur compound, and acetylpyrrole. It was also found that the main site of the aldehydes is the fat. Tables of compounds found are listed in the article.

Watanabe and Sato (1971) studied some of the flavor components of shallow fried veal, specifically the alkyl-substituted pyrazines and pyridines. Of these, nine alkyl-substituted pyrazines and two alkyl-substituted pyridines were found, together with four that were not identified. They noted the possibility of a reaction between sugars and amino acids to produce a specific flavor in beef. They identified methylpyrazine, 2-ethylpyridine, 2,5-dimethylpyrazine, 2,6-dimethylpyrazine, 2,3-dimethylpyrazine, 2-ethyl-5-methylpyrazine, trimethylpyrazine, 2,5-dimethyl-3-ethylpyrazine, tetramethyl-pyrazine, 2,6-diethyl-3-methylpyrazine, and 2-pentylpyradine. These authors had previously identified four of these compounds in beef fat flavor.

Persson and von Sydow (1973A, B) made a study of the aroma of canned beef. The headspace gas contained a number of volatiles—95 of these were identified and were made up of 21 sulfur compounds, 12 aldehydes, 16 ketones, 8 alcohols, and 11 furans. Among these, ethylene sulfide, propylene sulfide, 2-methyl furan, and some thiophene derivatives had not been previously reported in beef aroma, but some had been found in chicken aroma. Of the new compounds, ethylene sulfide and some of the furans could perhaps be of importance in the aroma. Another compound, 2-methyl butanal, which had been reported previously and found in this investigation, could also be one of the more important compounds. Another previously reported compound, 3,5-dimethyl-1,2,4-trithiolane was present in the head space gas, but only in very low concentration (0.84–5.00 ppb v/v) and therefore probably not too important.

The following sulfur compounds were identified in the headspace of canned beef: hydrogen sulfide, methyl mercaptan, ethyl mercaptan, dimethyl sulfide, methyl ethyl sulfide, ethylene sulfide, dimethyl disulfide, thiophene, 2-methyl thiophene, and 3,5-dimethyl-1,2,4-trithiolane. The first four compounds of this list were reported in the following concentrations (in ppb) 1100, 1800, 190, and 600.

Hydrogen sulfide was found to be present in amounts up to 6000 times the threshold of value, $(CH_3)_2S$ up to 700 times, methyl mercaptan up to 2200 times, and ethyl mercaptan up to 300 times. Of the ketones, only 2,3-butanedione and 2,3-pentanedione contributed to the aroma.

Simulated Meats

Interest is developing in products that can take the place of meat in the diet. These products are made from vegetable proteins textured to resemble meat. Soybeans are an important raw material used for this purpose. Derivatives of soybeans have been used in the Orient for hundreds of years.

The techniques necessary for the spinning of the protein were made practical in the early thirties. The proteins are extracted with weak alkali, and after refining, yield a bland material which is about 96% pure. This is dispersed, and then spun into fibers by forcing through spinerettes, into a coagulating bath in which the fibers are formed. Treatment can be varied so as to produce the filament desired. The fibers are formed into bundles, these bundles are tied together, treated with a binding agent such as egg albumin, and then flavored and colored and treated to resemble existing products. If a product such as bacon is desired, smoking can be done (Thulin and Kuramoto 1967; Lockmiller 1972).

Simulated meats can be made to contain a minimum of 50% protein, which can include the essential amino acids, vitamins, and minerals. These products are a good source of protein in such cases where the use of meat must be avoided. Soy protein in particularly valuable because of its nutritional value, as well as its economy.

Sausage Production

Sausage manufacture has a long history and considerable research has gone into the improvement of sausage products. Meat emulsions are very important in their preparation.

Hansen (1960) discussed the fundamental structure of sausages and the emulsion formation in them. Sausages are made up of very finely comminuted meat components which are thoroughly dispersed in an emulsion of fat and water. Water and salt-soluble proteins from the meat make up a protein matrix that surrounds the emulsified fat globules. These proteins are able to form a rather solid membrane since they are denatured on contact with the fat. The relationship between the concentration of protein and capacity for emulsification is curvilinear. Swift *et al.* (1961) showed that during emulsification, 84% of the original protein is taken out of solution. This further substantiates the idea that the emulsion is made up of globules of fat enveloped by a matrix of protein.

In the preparation of meat for sausage, temperature of chopping is very important. Temperatures for this operation should not exceed 30°C, usually they are somewhat lower. An inverse linear relationship exists between the emulsifying ability of the protein and the temperature attained during emulsification. This linear relationship was shown by Swift *et al.* (1961) to be between the temperature limit of 15°–45°C.

Separation of fat from sausage has not been completely explained. In experiments on batches of sausage it was found that fat separation occurred during cooking only when the beef was incompletely comminuted. More work must be done on this phase of the work (Sulzbacher 1973).

The article by Saffle (1968) goes into considerable detail on the subject of meat emulsions and is suggested reading for interested students.

Modern American practice in sausage making is based on the use of frozen beef. However, grinding immediately after slaughter is done in Germany frequently. Work done in Hamm's Laboratory shows chemical changes that take place in the meat as a result of this treatment.

Beef for sausage manufacture is frequently ground and salted immediately after slaughter (Hamm and van Hoof 1971). This makes available the high water-holding capacity of the tissue before the start of rigor mortis. The beef must be salted to avoid the breakdown of large amounts of ATP and glycogen. In the experimental work, the effect of grinding on the rate of biochemical changes that occur in the tissue postmortem were followed. These included changes in ATP, ADP, AMP, IMP (inosine monophosphate), glycogen, inosine, glucose 6-phosphate, lactate, and pH during the first 72 hr at +4°C. Results were obtained on intact muscle as well, for comparative purposes.

While grinding of longissismus dorsi muscle within an hour after death of the animal shows an acceleration of glycolysis, the final amounts of glycogen, lactate, and pH after 24 or 72 hr were the same in both the minced and the intact muscle. After 12 hr postmortem the ground sample had more glucose 6-phosphate than did the intact muscle. During storage of either intact or ground muscle, 55–70% of the glycogen used was converted to lactate. An accelerated hydrolysis of ATP and ADP resulted from the grinding of the pre-rigor muscle, which speeded the increase in the concentration of IMP. After 72 hr, however, the same amount of IMP was found in both samples, and the ATP in both samples was entirely hydrolyzed.

This work indicates that when beef is ground before the start of rigor mortis there is a more rapid conversion of ATP to IMP and hypoxanthine and of glycogen to lactate. As soon as ATP is used up, a maximum concentration of IMP is attained. A faster increase in the concentration of hypoxanthine takes place in the ground muscle than in the intact. Inosine behaves in just the opposite manner.

Analysis of Meat and Meat Products

Information on the analysis of meat and meat products can be found in Chapter 24 of Official Methods of Analysis (AOAC) (1980).

SUMMARY

The structure of muscle fiber is of importance in meat. Muscles are divided into two fiber types. Type I is red muscle fibers and type II is white. Red muscle is rich in dehydrogenases and poor in phosphorylase while white muscle is the reverse. Myoglobin, the principal muscle pigment, is present in larger amounts in the red muscle. More soluble protein is found in white muscle.

Red muscle is for aerobic metabolism, while white muscle is for glycolytic metabolism. Succinic acid dehydrogenase activity in red muscle is twice as large as in which muscle. The phosphorylase activity of white muscle was found to be three times that of the red muscle.

The scientific evidence strongly supports the idea that red muscles (or predominantly red) have higher rates of oxidative metabolism and are best able to carry on continued activity and long drawn out production of energy. Sudden spurts of activity are best taken care of when white muscle predominates because glycolytic activity is higher.

The proteins of meat are very important from the point of view of the food value of meat.

The myofibrillar proteins are actin, myosin, tropomyosin, troponin A, troponin B, α-actinin, and β-actinin. α-Actinin contains two components which are important in muscular contraction.

The compounds present in the largest amount are water (about 75%) and protein (about 20%). Of this protein fraction about 65% comprises myosin and actin. Myosin behaves as an enzyme, and it is able to break down adenosine triphosphate.

Actomyosin results when solutions of F-actin and myosin are mixed together. This solution is highly viscous. When ATP or P_1 is added to this solution viscosity decreases because of the dissociation of the actomyosin into myosin and actin. Actomyosin, under the right conditions, is contractile. It does not exist in resting muscle.

Most of the muscle protein used as food is actomyosin. It is formed during rigor by the polymerization of actin and myosin.

Straited muscle burns fatty acids by preference as a source of energy. ATP is an energy source also and is used when fatty acid is insufficient to cover the need.

Collagen and elastin are the two main proteins of connective tissue. The former is quantitatively more important. The texture of connective tissue tends to render it undesirable in muscle tissue used as food. Collagen could account for the toughness of meat.

The polysaccharides could play a vital role in plasticizing the collagen fibers, which in turn, could explain their contribution to meat tenderness.

Elastin, the protein found in animal connective tissue in small amounts, tends to be rubbery. Fairly large amounts of elastin are found in the muscles of the rump of beef animals, but in other muscles it is present in rather small amounts. Elastin must be in an environment containing about 40% of H_2O for its rubbery qualities to be effective.

The undesirable feature of both collagen and elastin is that they impart toughness to the muscle.

Myoglobin is the principal source of color in meat, but hemoglobin may make up 20–30% of it. Myoglobin contains only one heme group while hemoglobin contains four.

The role of myoglobin in the color of fresh meats is significant. Also, it forms the color of cured meats when it is combined with nitric oxide. The nitric oxide red pigment can change during storage to brown and gray colors.

Heme is the prosthetic group of both myoglobin and hemoglobin. The affinity of myoglobin for xoygen is much greater than that of hemoglobin. The bright red color of fresh meat is the result of oxygenated heme pigments at the surface which are exposed to the oxygen of the air. The purplish red of the interior is caused by the reduced pigments. Brown, gray, and green are the three abnormal discolorations of meat.

In the red color of cured meat, the pigment, nitric oxide myo-(hemo)globin is not destroyed, even though the denaturation of the globin is complete and irreversible.

It has been found to be essential to have high levels of creatine phosphate and ATP in the muscles initially in order to obtain a slow rate of decline in the pH of muscle after death. If top quality of meat is to be produced, the rate of destruction of ATP in the muscle must be kept to a minimum. Meat with high final pH is undesirable because of reduced penetration of salts in curing, dryness, and the cosmetic effect of dark cutting meat, which is undesirable.

Important chemical changes take place between the time of death and the start of rigor mortis.

The contraction of muscle is a very important phenomenon. Rigor mortis and muscular contraction are the result of the same mechanism.

Rigor mortis and other post mortem changes are mainly due to glycolytic chemical changes. This amounts to the change of glycogen to lactic acid. This is determined by monitoring the decrease in pH. High temperature and low pH occurring together soon after death can cause the loss of color and extensive lowering of water-binding capacity.

If bound calcium is contained in the contractile protein of muscle actomyosin, in the presence of ATP the muscle contracts. It relaxes with ATP if some of the calcium is removed.

Pale, soft, exudative muscle is not desirable in pork because it results in significant economic losses.

It is quite probable that the most important factor for tenderness-toughness in meat is the degree of cross linking in the actomyosin.

Two very important properties of muscle proteins from a practical standpoint are (1) water-holding capacity and (2) fat-emulsifying capacity. Tenderness, texture, and shrinkage on cooking are connected with the condition of hydration of muscle proteins. The water-holding capacity is strongly influenced by pH and by ions of salts such as NaCl.

The second item of practical importance, fat-emulsifying capacity, is essential to the preparation of meat emulsions. The meat emulsion is a muscle–fat–water system in which some of the muscle protein dissolves, especially when salt is present. The protein exerts an emulsifying effect by forming a thin layer around the fat droplets. This is important in the manufacture of sausage.

Oxidative rancidity in meats is important because it limits the salability of meat so affected.

Although unsaturated fats and heme pigments in meat are oxidized separately, a coupled reaction which involves both of these entitles, can in turn, speed color loss and rancidity. This is a catalytic effect. The activity of hemes as catalysts on the uptake of oxygen by linseed oil was studied. It was found that hemin, hemoglobin, and methemoglobin when used in equivalent concentrations had approximately the same effect in speeding the reaction.

The destruction of porphyrin takes place during the period of its extended reaction with unsaturated fats. It is likely that this reaction is involved in the deterioration of ground meats held in frozen storage.

Phospholipids are susceptible to oxidation because of their high content of unsaturated fatty acids.

In the presence of oxygen, light speeds all oxidative changes in meats.

To prevent the development of rancidity, a number of compounds, mostly phenols, are used.

During the heating and denaturing of muscle tissue, unfolding of the peptide helix appears to be closely connected with coagulation. Unfolding is complete above 70°C. Proteins that migrate with the greatest velocity in an electric field are the most quickly denatured.

One of the changes taking place during heating is the release of sulfhydryl groups. H_2S forms from myofibrils at around 80°C with exponential increase as the temperature goes up. Most of this comes from the structural proteins rather than from cysteine or glutathione.

Work has been done on the volatile flavors of meat, as well as on other forms of meat flavor.

BIBLIOGRAPHY

ANSON, M. L., and PADER, M. 1959. Method for preparing a meat-like product. U.S. Patent 2,879,163.

ASHMORE, C. R., TOMPKINS, C, and DOERR, L. 1972 Comparative aspects of mitochondria isolated from αW, αR, and βR muscle fibers of the chick. Exp. Neurol. *35*, 413–420.

ASSOC. OF OFFICIAL ANALYTICAL CHEMISTS. 1980. Official Methods of Analysis, 13th Edition. Assoc. of Official Analytical Chemists, Washington, DC.

BAILEY, A. J. 1972. The basis of meat texture. J. Sci. Food Agric. *23*, 995–1007.

BAILEY, M. E., FRAME, R. W., and NAUMANN, H. D. 1964. Studies of the photooxidation of nitrosomyoglobin. J. Agric. Food Chem. *12*, 89–93.

BATE-SMITH, E. C., and BENDALL, J. R. 1949. Factors determining time course of rigor mortis. J. Physiol. *110*, 47–65.

BATE-SMITH, E. C., and BENDALL, J. R. 1956. Changes in muscle after death. Br. Med. Bull. *12*, 230–235.

BATZER, O. F., SANTORO, A. T., and LANDMANN, W. A. 1962. Identification of some beef flavor precursors. J. Agric. Food Chem. *10*, 94–96.

BEATTY, C. H., and BOCEK, R. M. 1970. Biochemistry of the red and white muscle. *In* Physiology and Biochemistry of Muscle as Food, Vol. 2, E. J. Briskey, R. G. Cassens, and B. B. Marsh (Editors). University of Wisconsin Press, Madison, WI.

BEATTY, C. H., PETERSON, R. D., BASSINGER, G. M., BOCEK, R. M., and YOUNG, M. M. 1966. Major metabolic pathways for carbohydrate metabolism of voluntary skeletal muscle. Am. J. Physiol. *210*, 404–410.

BEECHER, G. R., CASSENS, R. G., HOEKSTRA, W. G., and BRISKEY, E. J. 1965. Red and white fiber content and associated post-mortem properties of seven porcine muscles. J. Food Sci. *30*, 969–976.

BEECHER, G. R., KASTENSCHMIDT, L. L., CASSENS, R. G., HOEKSTRA, W. G., and BRISKEY, E. J. 1968. A comparison of the light and dark portions of a striated muscle. J. Food Sci. *33*, 84–88.

BENDALL, J. R. 1951. The shortening of rabbit muscle during *rigor mortis*: Its relation to the breakdown of adenosine triphosphate and creatine phosphate and to muscular contraction. J. Physiol. (London) *114*, 71–88.

BENDALL, J. R. 1960. Post mortem changes in muscle. *In* Structure and Function of Muscle, Vol. III, G. H. Bourne (Editor). Academic Press, New York.

BENDALL, J. R. 1964. Meat proteins. *In* Symposium on Foods: Proteins and Their Reactions, H. W. Schultz and A. F. Anglemier (Editors). AVI Publishing Co., Westport, CT.

BENDALL, J. R. 1966. The effect of pre-treatment of pigs with curare on the post-mortem rate of pH fall and onset of *rigor mortis* in the musculature. J. Sci. Food Agric. *17*, 333–338.

BENDALL, J. R. 1967. The elastin content of various muscles of beef animals. J. Sci. Food Agric. *18*, 553–558.

BENDALL, J. R. 1969. Muscles, Molecules, and Movement. Heinemann Educational Books, London.

BENDALL, J. R. 1972. Consumption of oxygen by the muscles of beef animals and related species, and its effect on the colour of meat. I. Oxygen consumption in pre-rigor muscle. J. Sci. Food Agric. *23*, 61–72.

BENDALL, J. R., and TAYLOR, A. A. 1972. Consumption of oxygen by the muscles of beef animals and related species. II. Consumption of oxygen by post-rigor muscle. J. Sci. Food Agric. *23*, 707–719.

BENDALL, J. R., and VOYLE, C. A. 1967. A study of the histological changes in growing muscles of beef animals. J. Food Technol. *2*, 259–283.

BENDALL, J. R., and WISMER-PEDERSEN, J. 1962. Some properties of the fibrillar proteins of normal and watery pork muscle. J. Food Res. *27*, 144–159.

BERMAN, M. D., and SWIFT, C. E. 1964. Meat curing. The action of NaCl on meat electrolyte binding. J. Food Sci. *29*, 182–189.

BOGNAR, A. 1971. Effect of heat treatment on the amino acid content of beef. Ernaehr. Umsch. *18*, No. 5, 200–204. (German)

BOWERS, J. A., HARRISON, D. L., and KROPF, D. H. 1968. Browning and associated properties of porcine muscle. J. Food Sci. *33*, 147–151.

BOYER, R. A. 1954. High protein food product and process for its preparation. U.S. Patent 2,682,446.

BODRERO, K. O., PEARSON, A. M., and MAGEE, W. T. 1981. Evaluation of the contribution of flavor volatiles to the aroma of beef by surface response methodology. J. Food Sci. *46*, 26–31.

BRAUNITZER, G., HILZE, K., RUDOLFF, V., and HILSCHMANN, N. 1964. The hemoglobins. *In* Advances in Protein Chemistry, Vol. 19, C. B. Anfinsen, M. L. Ansan, J. T. Edsall, and F. M. Richards (Editors). Academic Press, New York.

BRINKMAN, H. W., COPIER, H., de LEUW, J. J. M., and TJAN, S. B. 1972. Components contributing to beef flavor. Analysis of headspace volatiles of beef broth. J. Agric. Food Chem. *20*, 177–181.

BRISKEY, E. J. 1964. Etiological status and associated studies of pale, soft, exudative porcine musculature. *In* Advances in Food Research, Vol. 13, C. O. Chichester, E. M. Mrak, and G. F. Stewart (Editors). Academic Press, New York.

BRISKEY, E. J., and FUKAZAWA, T. 1971. Myofibrillar proteins of skeletal muscle. *In* Advances in Food Research, Vol. 19, C. O. Chichester, E. M. Mrak, and G. F. Stewart (Editors). Academic Press, New York.

BROOKS, J. 1938. Color of meat. Food Res. *3*, 75–78.

CARMICHAEL, D. J., and LAWRIE, R. A. 1967. Bovine collagen. 1. Changes in collagen solubility with animal age. J. Food Technol. *2*, 299–311.

CASSENS, R. G. 1966. General aspects of postmorten changes. *In* Physiology and Biochemistry of Muscle as a Food, E. J. Brikey, R. G. Cassens, and J. C. Trautman (Editors). University of Wisconsin Press, Madison, WI.

CASSENS, R. G., and COOPER, C. C. 1971. Red and white muscle. *In* Advances in Food Research, Vol. 19, C. O. Chichester, E. M. Mrak, and G. F. Stewart (Editors). Academic Press, New York.

CASSENS, R. G., COOPER, C. C., and BRISKEY, E. J. 1969. The occurrence and histochemical characterization of giant fibers in the muscle of growing and adult animals. Acta Neuropathol. *12*, 300–304.

CHO, I. C., and BRATZLER, L. J. 1970. Effect of sodium nitrite on flavor of cured pork. J. Food Sci. *35*, 668–670.

COHEN, E. H. 1966. Protein changes related to ham processing temperatures. I. Effect of time-temperature on amount and composition of soluble proteins. J. Food Sci. *31*, 746–750.

COOPER, C. C., CASSENS, R. G., and BRISKEY, E. J. 1969. Capillary distribution and fiber characteristics in skeletal muscle of stress-susceptible animals, J. Food Sci. *34*, 299–302.

CORMIER, A., WELLINGTON, G. H., and SHERBON, J. W. 1971. Epimysial connective tissue polysaccharides of bovine semimembranosus muscle and alterations in their type with age and sex differences. J. Food Sci. *36*, 199–205.

DAVIES, R. E. 1963. A molecular theory of muscle contraction: calcium dependent contractions with hydrogen bond formation plus ATP-dependent extensions of part of the myosin-actin cross-bridges. Nature *199*, 1068–1074.

DICKERSON, J. W. T., and WIDDOWSON, E. M. 1960. Chemical changes in skeletal muscle during development. Biochem. J. *74*, 247–257.

DONNELLY, T. H., RONGEY, E. H., and BARSUKO, V. J. 1966. Protein composition and functional properties of meat. J. Agric. Food Chem. *14*, 196–200.

DRAUDT, H. N. 1963. The meat smoking process: A review. Food Technol. *17*, 1557–1562.

DREIZEN, P., HARTSHORNE, D. J., and STRACHER, A. 1966. The subunit structure of myosin. 1. Polydispersity in 5 *M* guanidine. J. Biol. Chem. *241*, 443–448.

DUBOWITZ, V. 1966. Enzyme histochemistry of developing human muscle. Nature *211*, 884–885.

DUBOWITZ, V., and EVERSON PEARSE, A. G. 1960. Reciprocal relationship of phosphorylase and oxidative enzymes in skeletal muscle. Nature *185*, 701–702.

DUPRAW, E. J. 1968. Cell and Molecular Biology. Academic Press, New York.

EBASHI, S., and KODAMA, A. 1966. Interaction of troponin with F-actin in the presence of tropomyosin. J. Biochem. (Tokyo) *59*, 425–426.

EBASHI, S., KODAMA, A., and EBASHI, F. 1968. Troponin. I. Preparation and physiological function. J. Biochem. (Tokyo) *64*, 465–477.

EBASHI, S., EBASHI, F., and KODAMA. A. 1967. Troponin as the Ca^{++}-receptive protein in the contractile system. J. Biochem. (Tokyo) *62*, 137–138.

EITENMILLER, R. R., KOEHLER, P. E., and REAGAN, J. O. 1978. Tyramine in fermented sausages: factors affecting formation of tyramine and tyrosine decarboxylase. J. Food Sci. *43*, 689–693.

EL-BADAWI, A. A., and HAMM, R. 1970. Relation between enzyme activity and myoglobin content in the muscles of slaughtered animals. Fleischwirtschaft *50*, 966–967. (German)

EL-GHARBAWI, M. I., and DUGAN, JR., L. R. 1965. Stability of nitrogenous compounds and lipids during storage of freeze-dried raw beef. J. Food Sci. *30*, 817–822.

ENDO, M., NONOMURA, Y., MASAKI, T., OHTSUKI, I., and EBASHI, S. 1966. Localization of native tropomyosin in relation to striation patterns. J. Biochem. (Tokyo) *60*, 605–608.

ENGELHARDT, W. A., and LJUBIMOWA, M. N. 1939. Myosine and adenosinetriphosphatase. Nature *144*, 668–669.

FANELLI, A. R., ANTONINI, E., and CAPUTO, A. 1964. Hemoglobin and myoglobin. *In* Advances in Protein Chemistry, Vol. 19, C. B. Anfinsen, M. L. Anson, J. T. Edsall and F. M. Richards (Editors). Academic Press, New York.

FILER, L. J., JR., MATTIL, K. F., and LONGENECKER, H. E. 1944. Antioxidant losses during the induction period of fat oxidation. Oil Soap *21*, 289–292.

FOX, JR., J. B. 1966. The chemistry of meat pigments. J. Agric. Food Chem. *14*, 207–210.

FREDERIKSEN, D. I., and HOLTZER, A. 1968. The substructure of the myosin molecule. Production and properties of the alkali subunits. Biochem. *7*, 3935–3950.

GAUTHIER, G. F. 1970. The ultrastructure of three fiber types in mammalian skeletal muscle. *In* Physiology and Biochemistry of Muscle as a Food, Vol. 2, E. J. Briskey, R. G. Cassens, and B. B. Marsh (Editors). University of Wisconsin Press, Madison, WI.

GEORGE, J. C., and NAIK, R. M. 1958. Relative distribution and chemical nature of the fuel store of the two types of fibers in the pectoralis major muscle of the pigeon. Nature *181*, 709–711.

GEORGE, P., and STRATMANN, C. J. 1952. The oxidation of myoglobin to metmyoglobin by oxygen. 2. The relation between the first order rate constant and the partial pressure of oxygen. Biochem. J. *51*, 418–425.

GERGELY, J. 1964. The relaxing factor of muscle. Fed. Proc. *23*, 885–886.

GINGER, I. D., and SCHWEIGERT, B. S. 1954. Chemical studies with purified metmyoglobin. J. Agric. Food Chem. *2*, 1037–1040.

GINGER, I. D., WILSON, G. D., and SCHWEIGERT, B. S. 1954. Biochemistry of myoglobin. Quantitative determination in beef and pork muscle. J. Agric. Food Chem. *2*, 1037–1038.

GOLL, D. E., HENDERSON, D. W., and KLINE, E. A. 1964. Post-mortem changes in physical and chemical properties of bovine muscle. J. Food Sci. *29*, 590–596.

GOLL, D. E., ARAKAWA, H., STROMER, M. H., BUSCH, W. A., and ROBSON, R. M. 1970. Chemistry of muscle proteins as a food. *In* Physiology and Biochemistry of Muscle as a Food, Vol. 2, E. J. Briskey, R. G. Cassens, and B. B. Marsh (Editors). University of Wisconsin Press, Madison, WI.

GORBATOV, V. M., KRYLOVA, N. N., VOLOVINSKAYA, V. P., LYASKOVSKAYA, YU. N., BAZAROVA, K. I., KHLAMOVA, R. I., and YAKOVLEVA, G. YA. 1971. Liquid smokes for use in cured meats. Food Technol. *25*, 71–77.

GORDZIEVSKII, L. N. 1967. Autolytic processes in frozen meat during storing. Izv. Vyssh. Uchebn. Zaved. Pishch. Tekhnol. No. 2, 40–43. (Russian) Chem. Abstr. *67*, 20715y (1967).

GRAU, R., and LEE, F. A. 1963. On the influence of temperature on the behavior of bovine sarcoplasm. Naturwissenschaften *50*, 379. (German)

GREENBERG, L. A., LESTER, D., and HAGGARD, H. W. 1943. The reaction of hemoglobin with nitrite. J. Biol. Chem. *151*, 665–673.

GREGORY, J. F., and KIRK, J. R. 1977. Interaction of pyridoxal and pyridoxal phosphate with peptides in a model food system during thermal processing. J. Food Sci. *42*, 1554–1557.

GUNSTONE, F. D., and HILDITCH, T. P. 1945. The union of gaseous oxygen with methyl oleate, linoleate, and linolenate. J. Chem. Soc. 836–841.

GÜNTHER, H., and SCHWEIGER, A. 1966. Changes in the concentration of lactic acid and free sugars in post-mortem samples of beef and pork muscle. J. Food Sci. *31*, 300–308.

HAMM, R. 1958. Adenosinetriphosphoric acid and its importance for meat quality. II. Function of in post-morten changes of muscle. Fleischwirtschaft *10*, 80–87. (German)

HAMM, R. 1958. On the biochemistry of the salting of meat. Z. Lebensm. Unters.-Forsch. *107*, 1–15. (German)

HAMM, R. 1960. Biochemistry of meat hydration. *In* Advances in Food Research, Vol. 10, C. O. Chichester, E. M. Mrak, and G. F. Stewart (Editors). Academic Press, New York.

HAMM, R. 1963A. The water imbibing power of foods. Recent Adv. Food Sci. *3*, 218–229.

HAMM, R. 1963B. The microstructure of muscle and its relation to the water-binding capacity of meat. Fleischwirtschaft *15*, 298–301.

HAMM, R. 1966. Heating of muscle systems. *In* Physiology and Biochemistry of Muscle as a Food, E. J. Briskey, R. G. Cassens, and J. C. Trautman (Editors). University of Wisconsin Press, Madison, WI.

HAMM, R. 1970. Properties of meat proteins. *In* Proteins as Human Food, R. A. Lawrie (Editor). AVI Publishing Co., Westport, CT.

HAMM, R. 1972. Colloid chemistry of meat. Water-binding Capacity of Muscle Protein in Theory and Practice. Paul Parey, Berlin. (German).

HAMM, R., and DEATHERAGE, F. E. 1960. Changes in hydration, solubility, and charges of muscle proteins during heating of meat. Food Res. *25*, 587–610.

HAMM, R., and VAN HOOF, J. 1971. Influence of grinding of beef muscle on the break down of adenosine triphosphate and glycogen post mortem. Z. Lebensm. Unters. -Forsch. *147*, 193–200. (German)

HANSEN, L. J. 1960. Emulsion formation in finely comminuted sausage. Food Technol. *14*, 565–569.

HARRINGTON, W. F., and VON HIPPEL, P. H. 1961. The structure of collagen and gelatin. *In* Advances in Protein Chemistry, Vol. 16, C. B. Anfinsen, M. I. Anson, and K. Bailey (Editors). Academic Press, New York.

HASSELBACH, W., and MAKINOSE, M. 1961. The calcium pump of the granules of the muscle and their dependence on adenosine triphosphate hydrolysis. Biochem. Z. *333*, 518–528. (German)

HAUROWITZ, F., SCHWERN, P., and YENSON, M. M. 1941. Destruction of hemin and hemoglobin by the action of unsaturated fatty acids and oxygen. J. Biol. Chem. *140*, 353–359.

HAVEL, R. J. 1970. Lipid as an energy source. *In* Physiology and Biochemistry of Muscle as a Food, Vol. 2, E. J. Briskey, R. G. Cassens, and B. B. Marsh (Editors). University of Wisconsin Press, Madison, WI.

HAWRYSH, Z. J., BERG, R. T., and HOWES, A. D. 1975. Eating quality of mature marbled beef. Can. Inst. Food Sci. Technol. J. *8*, 30–34.

HEDRICK, H. B., BRADY, D. E., and TURNER, C. W. 1957. The effect of antemortem stress on postmortem beef carcass characteristics. Am. Meat Inst. Proc. Conf. Res. *9*, 9–14.

HERRING, H. K., CASSENS, R. G., and BRISKEY, E. J. 1965. Sarcomere length of free and restrained bovine muscles at low temperature as related to tenderness. J. Sci. Food Agric. *16*, 379–384.

HERRING, H. K., CASSENS, R. G., and BRISKEY, E. J. 1967. Factors affecting the collagen solubility in bovine muscles. J. Food Sci. *32*, 534–538.

HERRING, H. K., CASSENS, R. G., FUKAZAWA, T., and BRISKEY, E. J. 1969. Studies on bovine natural actomyosin. 2. Physicochemical properties and tenderness of muscle. J. Food Sci. *34*, 571–576.

HERZ, K. O., and CHANG, S. S. 1970. Meat flavor. *In* Advances in Food Research, Vol. 18, C. O. Chichester, E. M. Mrak, and G. F. Stewart (Editors). Academic Press, New York.

HILL, F. 1966. The solubility of intramuscular collagen in meat animals of various ages. J. Food Sci. *31*, 161–166.

HIRAI, C., HERZ, K. O., POKORNY, J., and CHANG, S. S. 1973. Isolation and identification of volatile flavor compounds in boiled beef. J. Food Sci. *38*, 393–397.

HODGE, A. J., and PETRUSKA, J. A. 1963. Recent studies with the electron microscope on ordered aggregates of the tropocollagen macromolecule. *In* Aspects of Protein Structure, G. N. Ramachandran (Editor). Academic Press, New York.

HOEVE, C. A. J., and FLORY, P. J. 1958. The elastic properties of elastin. J. Am. Chem. Soc. *80*, 6523–6526.

HORNSTEIN, I., CROWE, P. F., and SULZBACHER, W. L. 1960. Constituents of meat flavor: Beef. J. Agric. Food Chem. *8*, 65–67.

HORNSTEIN, I., CROWE, P. F., and HEIMBERG, M. J. 1961. Fatty acid composition of meat tissue lipids. J. Food Sci. *26*, 581–586.

HUXLEY, A. F., and HUXLEY, H. E. 1964. A discussion on the physical and chemical basis of muscular contraction. Proc. R. Soc. London Ser. B *160*, 433–452.

HUXLEY, H. E. 1965. Structural evidence concerning the mechanism of contraction in striated muscle. *In* Muscle, W. M. Paul *et al.* (Editors). Pergamon Press, Oxford.

HUXLEY, H. E., and HANSON, J. 1960. Molecular basis of contraction in cross-striated muscles. *In* Structure and Function of Muscle, Vol. 1, G. H. Bourne (Editor). Academic Press, New York.

IGENE, J. O., KING, J. A., PEARSON, A. M., and GRAY, J. I. 1979. Influence of heme pigments, nitrite, and non-heme iron on development of warmed-over flavor (WOF) in cooked meat. J. Agric. Food Chem. *27*, 838-842.

ISHLER, N. H., MACALLISTER, R. V., and SZCZESNIAK, A. S. 1963. Process for preparing a food product having a fibrous texture and the resulting product. U.S. Patent 3,093,483.

JENSEN, L. B., and URBAIN, W. M. 1936. Bacteriology of green discoloration in meats and spectrophotometric characteristics of the pigments involved. Food Res. *1*, 263–273.

JOHNSON, B. C., and MOSER, K. 1967. *In* Radiation Preservation of Foods. R. F. Gould (Editor). Adv. Chem. Ser. 65. American Chemical Society, Washington, DC.

JONES, N. R. 1969. Meat and fish flavors. Significance of ribomononucleotides and their metabolites. J. Agric. Food Chem. *17*, 712–716.

KARMAS, E., and DI MARCO, G. R. 1970A. Dehydration thermoprofiles of amino acids and proteins. J. Food Sci. *35*, 615–617.

KARMAS, E., and DI MARCO, G. R. 1970B. Denaturation thermoprofiles of some proteins. J. Food Sci. *35*, 725–727.

KENDREW, J. C. 1963. Myoglobin and the structure of proteins. Crystallographic analysis and data-processing techniques reveal the molecular architecture. Science *139*, 1259–1266.

KIELLEY, W. W., and HARRINGTON, W. F. 1960. A model for the myosin molecule. Biochim. Biophys. Acta *41*, 401–421.

KIESE, M., and KAESKE, H. 1942. Derivatives of myoglobin. Biochem. Z. *312*, 121–149. (German)

KLOSE, A. A., HANSON, H. L., and LINEWEAVER, H. 1950. The freezing preservation of turkey meat steaks. Food Technol. *4*, 71–74.

KOMINZ, D. R. 1966. Interactions of calcium and native tropomyosin with myosin and heavy meromyosin. Arch. Biochem. Biophys. *115*, 583–592.

KOMINZ, D. R., and MARUYAMA, K. 1967. Does native tropomyosin bind to myosin? J. Biochem. (Tokyo) *61*, 269–271.

LANDMANN, W. A., and BATZER, O. F. 1966. Influence of processing procedures on the chemistry of meat flavors. J. Agric. Food Chem. *14*, 210–214.

LAWRIE, R. A. 1952. Biochemical differences between red and white muscle Nature *170*, 122–123.

LAWRIE, R. A. 1966. Meat Science. Pergamon Press, Oxford.

LAWRIE, R. A. 1966. Metabolic stresses which affect muscle. *In* Physiology and Biochemistry of Muscle as a Food, E. J. Briskey, R. G. Cassens, and J. C. Trautman (Editors). University of Wisconsin Press, Madison, WI.

LAWRIE, R. A 1968. Chemical changes in meat due to processing—A review. J. Sci. Food Agric. *19*, 233–240.

LEA, C. H. 1937. Influence of tissue oxidases on rancidity; oxidation of fat of bacon. J. Soc. Chem. Ind. *56*, 376–380T.

LEE, S. H., CASSENS, R. G., WINDER, W. C., and FENNEMA, O. R. 1978. Factors affecting the formation of nitrate from added nitrite in model systems and cured meat products. J. Food Sci. *43*, 673–676.

LEMBERG, R., and LEGGE, J. W. 1949. Hematin Compounds and Bile Pigments. Wiley (Interscience) Publishers, New York.

LEMBERG, R., and LEGGE, J. W. 1950. Pyrrole pigments. Annu. Rev. Biochem. *19*, 431–452.

LEWIS, U. J., and SCHWEIGERT, B. S. 1955. Biochemistry of myoglobin. III. Homogeneity studies with crystalline beef myoglobin. J. Biol. Chem. *214*, 647–655.

LIEBICH, H. M., DOUGLAS, D. R., ZLATKIS, A., MÜGGLER-CHAVAN, F., and DOUZEL, A. 1972. Volatile components in roast beef. J. Agric. Food Chem. *20*, 96–99.

LIN, T.-S. and HULTIN, H. O. 1978. Glutathione peroxidase of skeletal muscle. J. Food Biochem. *2*, 39–47.

LISTER, D., SAIR, R. A., WILL, J. A., SCHMIDT, G. R., CASSENS, R. G., HOEKSTRA, W. G., and BRISKEY, E. J. 1970. Metabolism of striated muscle of stress-susceptible pigs breathing oxygen or nitrogen. Am. J. Physiol. *218*, 102–107.

LOCKER, R. H. 1960. Degree of muscular contraction as a factor in tenderness of beef. Food Res. *25*, 304–307.

LOCKER, R. H., and HAGYARD, C. J. 1963. A cold shortening effect in beef muscles. J. Sci. Food Agric. *14*, 787–793.

LOCKMILLER, N. R. 1972. What are textured protein products? Food Technol. *26*, No. 5, 56–58.

LOGAN, J. R., and MEDVED, E. 1966. Effects of prior knowledge about composition on scoring of simulated meats. Food Techol. *20*, 675–676.

LOVE, J. D., and PEARSON. A. M. 1971. Lipid oxidation in meat and meat products—a review. J. Am. Oil Chem. Soc. *48*, 547–549.

LOWEY, S., SLAYTER, H. S., WEEDS, A. G., and BAKER, H. 1969. Substructure of the myosin molecule. I. Subfragments of myosin by enzymic degradation. J. Mol. Biol. *42*, 1–29.

LUNDBERG, W. O., DOCKSTADER, W. B., and HALVORSON, H. O. 1947. The kinetics of the oxidation of several antioxidants in oxidizing fats. J. Am. Oil Chem. Soc. *24*, 89–92.

MACLEOD, G., and SEYYEDAIN-ARDEBILI, M. 1981. Natural and simulated meat flavors (with particular reference to beef). CRC Crit. Rev. Food Sci. Nutr. *14*, 309–437.

MAHON, J. H., and CHAPMAN, R. A. 1953. The relative rates of destruction of propyl gallate and butylated hydroxyanisole in oxidizing lard. J. Am. Oil Chem. Soc. *30*, 34–39.

MALBROUK, A. F., JARBOE, J. K., and O'CONNOR, E. M. 1969A. Water-soluble flavor precursors of beef. Extraction and fractionation. J. Agric. Food Chem. *17*, 5–9.

MALBROUK, A. F., O'CONNOR, E. M., and JARBOE, J. K. 1969B. Nonaqueous beef flavor components. Composition of petroleum ether-extractable intramuscular polar lipids. J. Agric. Food Chem. *17*, 10–14.

MARSH, B. B. 1954. Rigor mortis in beef. J. Sci. Food Agric. *5*, 70–75.

MARSH, B. B. 1966. Relaxing factor in muscle. *In* Physiology and Biochemistry of Muscle as a Food, E. J. Briskey, R. G. Cassens, and J. C. Trautman (Editors). University of Wisconsin Press, Madison, WI.

MARSH, B. B., and LEET, N. G. 1966. Studies in meat tenderness. III. The effects of cold shortening on tenderness. J. Food Sci. *31*, 450–459.

MARUYAMA, K., and EBASHI, S. 1970. Regulatory proteins of muscle. *In* Physiology and Biochemistry of Muscle as a Food, Vol. 2, E. J. Briskey, R. G. Cassens, and B. B. Marsh (Editors). University of Wisconsin Press, Madison, WI.

McCANCE, R. A., and WIDDOWSON, E. M. 1956. Metabolism, growth, and renal function of piglets in their first days of life. J. Physiol. *133*, 373–384.

McLOUGHLIN, J. V. 1969. Relationship between muscle biochemistry and properties of fresh and processed meats. Food Manuf. *44*, No. 1, 36–40.

McLOUGHLIN, J. V., and GOLDSPINK, G. 1963. Post-mortem changes in the colour of pig Longissimus dorsi muscle. Nature *198*, 584–585.

MELLO, F. C., JR., FIELD, R. A., and RILEY, M. L. 1978. Effect of age and anatomical location on composition of bovine bone. J. Food Sci. *43*, 677–679.

MIN, D. B. S., INA, K., PETERSON, R. J., and CHANG, S. S. 1979. Preliminary identification of volatile flavor compounds in the neutral fraction of roast beef. J. Food Sci. *44*, 639–642.

MIYAKE, M., and TANAKA, A. 1971. Studies on utilization of amino acids for food-stuffs. 5. Basic amino acids as a meat improving agent. J. Food Sci. *36*, 874–876.

MÖHLER, K. 1970. Formation of curing pigment in beef muscle. I. Oxydation of muscle pigment and nitrite. Z. Lebensm. Unters. -Forsch. *142*, 169–179. (German)

MOODY, W. G., and CASSENS, R. G. 1968. Histochemical differentiation of red and white muscle fibers. J. Animal Sci. *27*, 961–968.

MORRISON, D. R., and CAMPBELL, A. M. 1971. Phospholipids as related to total lipid and DNA light and dark portions of porcine semitendinosus muscle. J. Food Sci. *36*, 1103–1104.

NEEDHAM, D. M. 1960. Biochemistry of muscular action. *In* Structure and Function of Muscle, Vol. II, G. H. Bourne (Editor). Academic Press, New York.

NEWBOLD, R. P. 1966. Changes associated with rigor mortis. *In* Physiology and Biochemistry of Muscle as a Food, E. J. Briskey, R. G. Cassens, and J. C. Trautman (Editors). University of Wisconsin Press, Madison, WI.

NEWBOLD, R. P., and SCOPES, R. K. 1971. Post-mortem glycolysis in ox skeletal muscle: Effect of adding nicotinamide-adenine dinucleotide to diluted mince preparations. J. Food Sci. *36*, 215–218.

NONOMURA, Y. 1967. The physicochemical properties of α-actin. J. Biochem. (Tokyo) *61*, 796–802.

PARTRIDGE, S. M. 1962. Elastin. *In* Advances in Protein Chemistry, Vol. 17, C. B. Anfinsen, M. I. Anson, K. Bailey, and J. T. Edsall (Editors). Academic Press, New York.

PARTRIDGE, S. M. 1966. Elastin. *In* Physiology and Biochemistry of Muscle as a Food, E. J. Briskey, R. G. Cassens, and J. C. Trautman (Editors). University of Wisconsin Press, Madison, WI.

PAUL, P. C., BUCHTER, L., and WIERENGA, A. 1966. Solubility of rabbit muscle proteins after various time-temperature treatments. J. Agric. Food Chem. *14*, 490–492.

PEER, H. G., and VAN DEN OUWELAND, G. A. M. 1968. Synthesis of 4-hydroxy-5-methyl-2,3-dihydrofuran-3-one from D-ribose 5-phosphate. Rec. Trav. Chim Pays-Bas *87*, 1017–1020.

PENNY, I. F., VOYLE, C. A., and LAWRIE, R. A. 1964. Some properties of freeze dried pork muscles of high or low ultimate pH. J. Sci. Food Agric. *15*, 559–565.

PERRY, S. V., and CORSI, A. 1958. Extraction of proteins other than myosin from the isolated rabbit myofibril. Biochem. J. *68*, 5–12.

PERSSON, T., and VON SYDOW, E. 1973A. Aroma of canned beef: Gas chromatographic and mass spectrometric analysis of the volatiles. J. Food Sci. *38*, 377–385.

PERSSON, T., and VON SYDOW, E. 1973B. Aroma of canned beef: Sensory properties. J. Food Sci. *38*, 386–392.

PIEZ, K. A. 1966. Collagen. *In* Physiology and Biochemistry of Muscle as a Food, E. J. Briskey, R. G. Cassens, and J. C. Trautman (Editors). University of Wisconsin Press, Madison, WI.

RAMSBOTTOM, J. M., GOESER, P. A., and SHULTZ, H. W. 1951. How light discolors meat: What to do about it. Food Ind. *23*, No. 2, 120–124, 222–223.

RITCHEY, S. J., COVER, S., and HESTETLER, R. L. 1963. Collagen content and its relation to tenderness of connective tissue in two beef muscles. Food Technol. *17*, 194–197.

ROBINSON, M. E. 1924. Haemoglobin and methaemoglobin as oxidative catalysts. Biochem. J. *18*, 255–264.

ROBINSON, R. F. 1972. What is the future of textured protein products? Food Technol. *26*, No. 5, 59–63.

ROGERS, P. J., GOERTZ, G. E., and HARRISON, D. L. 1967. Heat induced changes of moisture in turkey muscles. J. Food Sci. *32*, 298–304.

SAFFLE, R. L. 1968. Meat emulsions. *In* Advances in Food Research, Vol. 16, C. O. Chichester, E. M. Mrak, and G. F. Stewart (Editors). Academic Press, New York.

SANDERSON, A., PEARSON, A. M., and SCHWEIGERT, B. S. 1966. Effect of cooking procedure on flavor components of beef. Carbonyl compounds. J. Agric. Food Chem. *14*, 245–247.

SANDERSON, M., and VAIL, G. E. 1963. Fluid content and tenderness of three muscles of beef cooked to three internal temperatures. J. Food Sci. *28*, 590–599.

SATO, K., HEGARTY, G. R., and HERRING, H. K. 1973. The inhibition of warmed-over flavor in cooked meats. J. Food Sci. *38*, 398–403.

SATTERLEE, L. D. 1971. Effect of porcine pancreatic collagenase on muscle connective tissue. J. Food Sci. *36*, 130–132.

SCHRAMM, G., and WEBER, H. H. 1942. Monodisperse myosin solutions. Kolloid Z. *100*, 242–247. (German)

SCHUTTE, L., and KOENDERS, E. B. 1972. Components contributing to beef flavor. Natural precursors of 1-methylthio-ethanethiol. J. Agric. Food Chem. *20*, 181–184.

SEIFTER, S., and GALLOP, P. M. 1966. The structure of proteins. *In* The Proteins, 2nd Edition, Vol. IV, H. Neurath (Editor). Academic Press, New York.

SHENK. J. H., HALL, J. L., and KING, H. H. 1934. Spectrophotometric characteristics of hemoglobins. 1. Beef blood and muscle hemoglobins. J. Biol. Chem. *105*, 741–752.

SHULTS, G. W., RUSSEL, D. R., and WIERBICKI, E. 1972. Effect of condensed phosphates on pH, swelling and water-holding capacity of beef. J. Food Sci. *37*, 860–864.

SILBERSTEIN, D. A., and LILLARD, D. A. 1978. Factors affecting the autoxidation of lipids in mechanically deboned fish. J. Food Sci. *43*, 764–766.

SINGER, T. P., and BARRON, E. S. G. 1944. Effect of sulfhydryl reagents on adenosinetriphosphatase activity of myosin. Proc. Exp. Biol. Med. *56*, 120–124.

SINK, J. D., and SMITH, P. W. 1972. Changes in the lipid soluble carbonyls of beef muscle during aging. J. Food Sci. *37*, 181–182.

SMITH, G. C., WEST, R. L., REA, R. H., and CARPENTER, Z. L. 1973. Increasing the tenderness of bullock beef by use of antemortem enzyme injection. J. Food Sci. *38*, 182–183.

SOLBERG, M. 1970. The chemistry of color stability in meat. A review. Can. Inst. Food Technol. J. *3*, 55–62.

STROMER, M. H., and GOLL, D. E. 1967. Molecular properties of post-mortem muscle. II. Phase Microscopy of myofibrils from bovine muscle. J. Food Sci. *32*, 329–331.

SULZBACHER, W. L. 1973. Meat emulsions. J. Sci. Food Agric. *24*, 589–595.

SUTHERLAND, J. P., GIBBS, P. A., PATTERSON, J. T., and MURRAY, J. G. 1976. Biochemical changes in vacuum packaged beef occurring during storage at 0–2°C. J. Food Technol. *11*, 171–180.

SWIFT, C. E., LOCKETT, C., and FRYAR, A. J. 1961. Comminuted meat emulsions—the capacity of meats for emulsifying fat. Food Technol. *15*, 468–473.

SZCZESNIAK, A. S., and TORGESON, K. W. 1965. Methods of meat texture measurement viewed from the background of factors affecting tenderness. *In* Advances in Food Research, Vol. 14, C. O. Chichester, E. M. Mrak, and G. F. Stewart (Editors). Academic Press, New York.

SZENT-GYÖRGYI, A. 1951. The Chemistry of Muscular Contraction, 2nd Edition. Academic Press, New York.

SZENT-GYÖRGYI, A. G. 1953. Chemical Physiology of Contraction in Body and Heart Muscle. Academic Press, New York.

SZENT-GYÖRGYI, A. G. 1960. Proteins of the myofibril. *In* Structure and Function of Muscle, Vol. II, G. H. Bourne (Editor). Academic Press, New York.

SZENT-GYÖRGYI, A. G. 1966. Nature of actin-myosin complex and contraction. *In* Physiology and Biochemistry of Muscle as a Food, E. J. Briskey, R. G. Cassens, and J. C. Trautman (Editors). University of Wisconsin Press, Madison, WI.

SZENT-GYÖRGYI, A. G., and BORBIRO, M. 1956. Depolymerization of light meromyosin by urea. Arch. Biochem. Biophys. *60*, 180–197.

TARRANT, P. V. 1977. The effect of hot-boning on glycolysis of beef muscle. J. Sci. Food Agric. *28*, 927–930.

THEORELL, A. H. T. 1932. Crystalline myoglobin. I. Crystalization and purification of myoglobin and a preliminary study of its molecular weight. Biochem. Z. *252*, 1–7. (German)

THOMAS, J., ELSDEN, D. F., and PARTRIDGE, S. M. 1963. Degradation products from elastin. Nature *200*, 651–652.

THOMPSON, G. B., DAVIDSON, W. D., MONTGONERY, M. W., and ANGLEMEIER, A. F. 1968. Alternations of bovine sarcoplasmic proteins as influenced by high temperature aging. J. Food Sci. *33*, 68–72.

THULIN, W. W., and KURAMOTO, S. 1967. "Bontrae"—A new meat-like ingredient for convenience foods. Food Technol. *21,* 168–171.

TONSBEEK, C. H. T., PLANCKEN, A. J., and VAN DER WEERDHOF, T. 1968. Components contributing to beef flavor. Isolation of 4-hydroxy-5-methyl-3(2*H*)-furanone and its 2,5-dimethyl homolog from beef broth. J. Agric. Food Chem. *16*, 1016–1021.

TONSBEEK, C. H. T., KOENDERS, E. B., VAN DER ZIJDEN, A. S. M., and LOSEKOOT, J. A. 1969. Components contributing to beef flavor. Natural precursors of 4-hydroxy-5-methyl-3(2H)-furanone in beef broth. J. Agric. Food Chem. *17*, 397–400.

TONSBEEK, C. H. T., COPIER, H., and PLANCKEN, A. J. 1971. Components contributing to beef flavor. Isolation of 2-acetyl-2-thiazoline from beef broth. J. Agric. Food Chem. *19*, 1014–1019.

TRAUB, W., and PIEZ, K. A. 1971. The chemistry and structure of collagen. *In* Advances in Protein Chemistry Vol. 25, C. B. Anfinsen, M. L. Anson, J. T. Edsall, and F. M. Richards (Editors). Academic Press, New York.

TRAUTMAN, J. C. 1966. Effect of temperature and pH on the soluble proteins of ham. J. Food Sci. *31*, 409–418.

TRAUTMAN, J. C. 1966. Role of muscle protein in processed foods. *In* Physiology and Biochemistry of Muscle as a Food, E. J. Briskey, R. G. Cassens, and J. C. Trautman (Editors). University of Wisconsin Press, Madison, WI.

TUOMY, J. M., and LECHNIR, R. J. 1964. Effect of cooking temperature and time on the tenderness of pork. Food Technol. *18*, 219–222.

VAN HOOF, J. and HAMM, R. 1973. Influence of sodium chloride on the breakdown of adenosine triphosphate and glycogen in ground beef muscle *post mortem*. Z. Lebensm. Unters.-Forsch. *150*, 282–295. (German)

VARRINO-MARSTON. E., DAVIS, E., HUTCHISON, T. E., and GORDON, J. 1978. Postmortem aging of bovine muscle: A comparison of two preparation techniques for electron microscopy. J. Food Sci. *43*, 680–683, 688.

WALSH, K. A., and ROSE, D. 1956. Factors affecting the oxidation of nitric oxide myoglobin. J. Agric. Food Chem. *4,* 352–355.

WARRIS, P. D. and RHODES, D. N. 1977. Haemoglobin concentrations in beef. J. Sci. Food Agric. *28*, 931–934.

WASSERMAN, A. E. 1972. Thermally produced flavor components in the aroma of meat and poultry. J. Agric. Food Chem. *20*, 737–741.

WASSERMAN, A. E. 1979. Symp. on meat flavor. Chemical basis for meat flavor: A review. J. Food Sci. *44,* 6–11

WATANABE, K., and SATO, Y. 1971. Some alkyl-substituted pyrazines and pyridines in the flavor components of shallow fried beef. J. Agric. Food Chem. *19*, 1017–1019.

WATTS, B. M. 1954. Oxidative rancidity and discoloration in meat. *In* Advances in Food Research, Vol. 5, E. M. Mrak and G. F. Stewart (Editors). Academic Press, New York.

WATTS, B. M., and PENG, D.-H. 1947. Rancidity development in raw versus precooked frozen pork sausage. J. Home Econ. *39*, 88–92.

WEBER, H. H., and STÖVER, R. 1933. The colloidal behavior of muscle proteins. IV. The weight of particles of muscle protein and the van der Waal active volume of myogen particles. Biochem. Z. *259*, 269–284. (German)

WHITAKER, J. R. 1959. Chemical changes associated with aging of meat with emphasis on the proteins. *In* Advances in Food Research, Vol. 9, C. O. Chichester, E. M. Mrak, and G. F. Stewart (Editors). Academic Press, New York.

WIERBICKI, E., KUNKLE, L. E., and DEATHERAGE, F. E. 1957. Changes in the water-holding capacity and cationic shifts during the heating and freezing and thawing of meat as revealed by a simple centrifugal method for measuring shrinkage. Food Technol. *11,* 69–73.

21

Fruits and Vegetables

Ripening and Postharvest Changes
Storage
Chemistry Involved in the Texture of Fruits and Vegetables
Summary
Bibliography

RIPENING AND POSTHARVEST CHANGES

Changes that occur in fruits and vegetables during the ripening process and after harvest influence their use as foods. Following harvest, the sources of nutrients and water that supplied the growing entities are no longer available. But life processes go on: respiration continues with the uptake of oxygen and loss of carbon dioxide and water. Deterioration accompanies the use of the reserves of respirable materials accumulated during photosynthesis that are no longer replenished. Under these conditions, while new compounds are formed by the continuation of anabolism, the major metabolic activity in this situation is catabolism. Fruit ripening is basically a process of aging. The ripening process is accompanied by developing senescence and then deterioration. The speed with which the reserve materials are used depends on the rate of respiration. In some species, change after harvest takes place much faster than in others. Apples will hold much longer in proper storage than strawberries, raspberries, or peaches. Also, deterioration by the growth of microorganisms in fruits and vegetables held in storage is a prominent factor.

Respiration

Respiration is carried on by all living organisms and is the key activity of the life process. It is basically the transformation of potential energy into kinetic energy and is exothermic—heat is evolved. Respiration makes use of oxygen and eliminates carbon dioxide released as a result of the oxidation of the carbohydrates present in the plant material. Adenosine triphosphate (ATP) is formed and is stored in the cells as available energy. If respiration takes place under anaerobic conditions or in the presence of very limited amounts of oxygen, breakdown of the carbohydrates is not complete, and smaller amounts of ATP are produced. The carbon dioxide and oxygen are the beginning and end substances and in no way indicate the very complex reactions connected with this vital process.

Mitochondria. Mitochondria are considered to be vital components of the protoplast. They appear as small rods, filaments, or granules in the cytoplasm of most plant and animal cells. Mitochondria are composed primarily of protein and lipid.

Much experimental work has been done on the function of mitochondria. They are involved fundamentally in the oxidation of pyruvate through the tricarboxylic acid cycle. Also, inorganic phosphate is converted to adenosine diphosphate (ADP) and ATP. Mitochondria are active also in the uptake of oxygen. This activity probably constitutes an important part of the respiration of cells, since the oxygen uptake is proportional to that of the intact cells.

Lyons and Raison (1970) found that mitochondria from plant tissue that is sensitive to chilling respire at a significantly lower rate below 10°C than mitochondria from tissue resistant to chilling. This is consistent with the idea that lower temperatures have an effect on some physical property of the mitochondrial membranes. Membrane lipids may be involved.

Pathways of Respiration. From the biochemical point of view, two pathways are operational in plant respiration, and therefore, in the ripening process. These are the Embden–Meyerhof–Parnas (EMP) glycolytic–tricarboxylic acid (TCA) cycle, and the pentose pathway. The pentose pathway is a shunt that takes the place of the EMP. The stage of maturity influences the activity of each pathway in ripening. In many plants the larger amount of the respiration takes place via the glycolytic–tricarboxylic acid cycle pathway. In pepper fruits 28–36% of the glucose respires via the pentose pathway (Doyle and Wang 1958), while the remainder followed the glycolytic–tricarboxylic acid pathway.

Embden–Meyerhof–Parnas Glycolytic–Tricarboxylic Acid Pathway. This pathway starts with the breakdown of oligosaccharides and polysaccharides to hexose and hexose phosphates as a result of the activity of hydrolases and phosphorylases.

$$\text{Polysaccharides and oligosaccharides} \longrightarrow \text{hexoses and hexose phosphates}$$

Glucose forms glucose 6-phosphate:

$$\text{Glucose} + \text{ATP} \xrightarrow[\text{Mg}^{2+}]{\text{hexokinase}} \text{glucose 6-phosphate} + \text{ADP} + \text{H}^+ \quad (1)$$

The glucose 6-phosphate undergoes the following reaction:

$$\text{Glucose 6-phosphate (70\%)} \xrightarrow{\text{phosphoglucoisomerase}} \text{fructose 6-phosphate (30\%)} \qquad (2)$$

The fructose 6-phosphate, in turn, reacts with ATP.

$$\text{Fructose 6-phosphate} + \text{ATP} \xrightarrow{\text{phosphofructokinase}} \text{fructose 1,6-diphosphate} + \text{ADP} + H^+ \qquad (3)$$

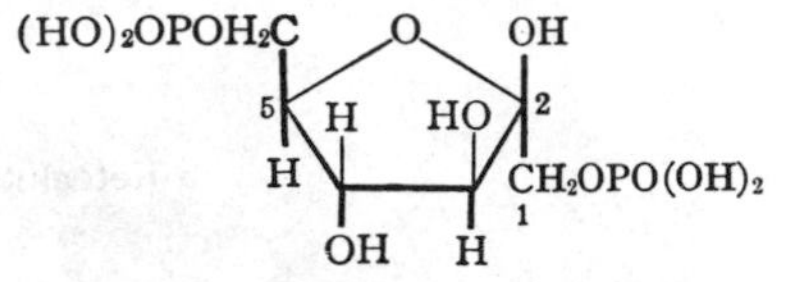

$(HO)_2OPOH_2C$ — O — OH; 5 H; H HO 2; $CH_2OPO(OH)_2$ 1; OH H

Fructose 1,6-diphosphate

The next stage of the EMP pathway is the glycolysis stage. The first reaction is the conversion of fructose 1,6-diphosphate to dihydroxyacetone phosphate. The second is that of dihydroxyacetone phosphate to D-glyceraldehyde 3-phosphate.

$$\text{Fructose 1,6-dephosphate} \xrightleftharpoons{\text{adolase}} \text{dihydroxyacetone phosphate} \xrightleftharpoons[\text{isomerase}]{\text{triosephosphate}} \text{D-glyceraldehyde 3-phosphate} \qquad (4)$$

D-Glyceraldehyde 3-phosphate is in turn converted into phosphoenol pyruvic acid in four steps using four enzymes. Phosphoenol pyruvic acid is converted to pyruvic acid by reacting with ADP.

$$\text{Phosphoenol pyruvic acid} + \text{ADP} \xrightarrow[\text{Mg2}^+]{\text{pyruvate kinase}} \qquad (5)$$

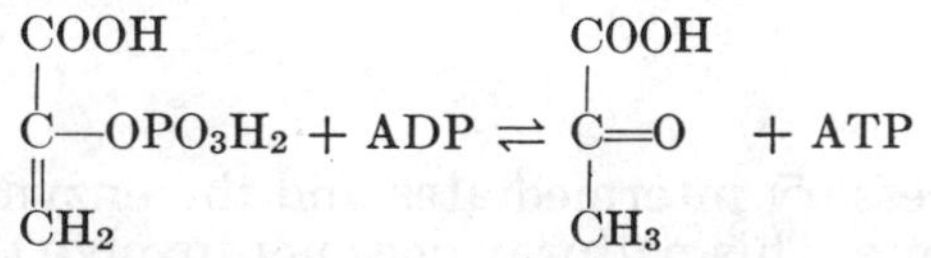

$$\begin{array}{c} COOH \\ | \\ C\text{—}OPO_3H_2 \\ \| \\ CH_2 \end{array} + \text{ADP} \rightleftharpoons \begin{array}{c} COOH \\ | \\ C{=}O \\ | \\ CH_3 \end{array} + \text{ATP}$$

Phosphoenol pyruvic acid Pyruvic Acid

Pyruvate is then converted to acetyl coenzyme A and carbon dioxide. After this, it acts on oxaloacetate in the tricarboxylic acid cycle (see Fig. 21.1). Organic acids in the tricarboxylic (Krebs citric acid) cycle are found in fruits and vegetables.

Pentose Pathway. An alternative to the EMP glycolytic pathway for the degradation of the hexose sugars to yield pyruvate is the pentose

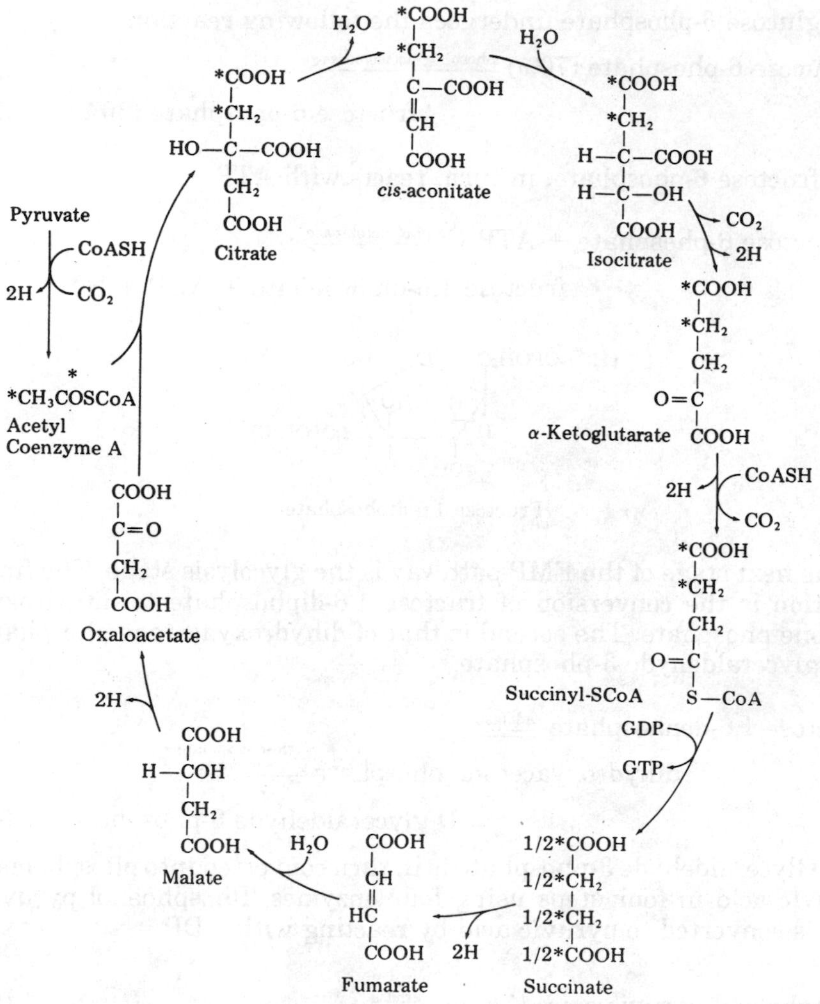

Pyruvate → (CoASH; 2H, CO_2) → *$CH_3COSCoA$ Acetyl Coenzyme A

Oxaloacetate: COOH—C=O—CH_2—COOH

Citrate: *COOH—*CH_2—HO—C—COOH—CH_2—COOH

(H_2O) *cis*-aconitate: *COOH—*CH_2—C—COOH=CH—COOH

(H_2O) Isocitrate: *COOH—*CH_2—H—C—COOH—H—C—OH—COOH

(CO_2, 2H) α-Ketoglutarate: *COOH—*CH_2—CH_2—O=C—COOH

(CoASH; 2H, CO_2) Succinyl-SCoA: *COOH—*CH_2—CH_2—O=C—S—CoA

(GDP → GTP) Succinate: 1/2*COOH—1/2*CH_2—1/2*CH_2—1/2*COOH

(2H) Fumarate: COOH—CH=HC—COOH

(H_2O) Malate: COOH—H—COH—CH_2—COOH

(2H) → Oxaloacetate

FIG. 21.1. Tricarboxylic acid (Krebs citric acid) cycle.

pathway. The necessary intermediates and the enzymes required are present in the plants. This pathway does not involve the enzyme aldolase. In the glycolytic system, DPN is necessary; in the pentose pathway, TPN is present.

In this pathway, glucose 6-phosphate in the presence of triphosphopyridine nucleotide is converted to 6-phosphogluconate plus TPNH. The 6-phosphogluconate in the presence of TPN is converted to ribulose-5-phosphate with the release of CO_2 and the formation of TPNH. Ribulose-5-phosphate is converted to ribose-5-phosphate, and in another reaction ribulose-5-phosphate is converted to xylulose-5-

phosphate. Specific enzymes bring these reactions about. Carbon atoms are transferred from xylulose-5-phosphate to ribose-5-phosphate. The result is the formation of sedoheptulose phosphate and 3-phosphoglyceraldehyde. Finally, fructose 6-phosphate is formed. It is probable that the triose phosphate that is produced is alternatively converted into pyruvate as a result of the glycotic sequence and then to acetyl coenzyme A (CoA) and into the tricarboxylic acid cycle, as does the EMP sequence. In other words, the pentose pathway is another path for the breakdown of hexoses to trioses.

Faust *et al.* (1966) studied the skins of harvested apples over a period of four months to determine the respiratory pathways during aging. It was found that increasing senescence resulted in an increase in the use of the pentose pathway. The evolution of CO_2 from the pentose pathway in senescent apples was directly proportional to the uptake of O_2. Change in cell permeability was indicated by the increase in glucose uptake as the tissue aged.

Respiration Climacteric. Kidd and West (1925, 1930) found a very large increase in the rate of respiration after harvest of apples stored at 22.5°C. They called this the "climacteric rise." (Fig. 21.3). This large increase in CO_2 output, or respiration climacteric, shown by many fruits is a sudden activity. It indicates the point during which growth and development stop and breakdown and senescence begin. The climacteric in fruits may bring about color changes. Bananas change from green to yellow as a result of chlorophyll destruction, while xanthophyll and carotene change very little. Similar changes take place in some varieties of apples and pears, while plums change to red. The breakdown of the cholorphyll allows the yellow pigments that were previously masked to appear, and the formation of other pigments as a result of synthesis. Other changes occur also: starch changes to sugars, softening ensues as a result of changes in the pectic substances, and proteins are modified. In addition, flavor volatiles are synthesized and changes can take place in the acid composition. In some fruits, wax is deposited on the skin during this phase.

The respiration climacteric may be very small or totally absent in some kinds of fruit. Also, some fruits may show a climacteric under certain conditions and not others. A climacteric has been shown for apples that have been left on the tree during ripening. The output of carbon dioxide was stable as long as the fruit remained on the tree; the postclimacteric decline in release of carbon dioxide took place after harvest (Hulme 1954; Kidd and West 1945; Romani and Ku 1966). In just the opposite manner, the avocado does not show a climacteric rise in respiration until it has been separated from the tree (Biale 1960B); the avocado tree probably furnishes a factor that controls this phenomenon.

According to Biale (1960A, B) fruits can be classfied based on which show a climacteric and which do not. Among those that have a climac-

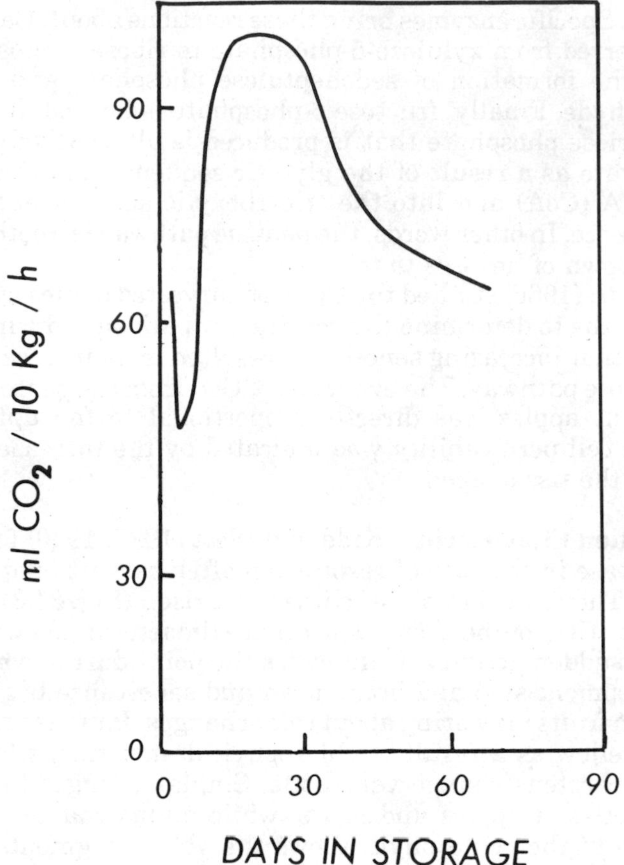

FIG. 21.2. Respiration climacteric in apples.

teric are apples, bananas, peaches, pears, and plums. Some of those which seem to be without a climacteric are citrus fruits, cherries, grapes, and strawberries. Fruits without a climacteric, usually ripen on the tree or vine. Small quantities of ethylene seem to have some effect on these two kinds of respiratory activity. This effect will be fully discussed later. The existence of non-climacteric fruits has been disputed (Pratt and Goeschl 1964). While citrus fruits have been classed as non-climacteric, some experiments have shown that if they are harvested before ripeness is reached, a climacteric rise in respiration is found to occur (Aharoni 1968).

Sacher (1966) made a study of the changes in free space in banana tissue during the respiratory climacteric. As free space increased (about 44 hr before the start of the climacteric), so did the proportion of cells which became entirely permeable to the ambient solution. It was concluded that the beginning of permeability changes marks the start of senescence.

Marks *et al.* (1957) demonstrated that the maximum phosphorylation occurred at the maximum of climacteric respiration. Oxidative phosphorylation is now known to continue throughout the ripening process in fruits. Using ^{32}P tracer methods Marks *et al.* (1957) found that phosphate esterification takes place both in preclimacteric tomato fruits and during the postclimacteric senescent period. Also, inhibition of the formation of high energy phosphate esters results in failure of the fruit to ripen normally. This was the case for tomatoes treated with DNP (2,4-dinitrophenol). Energy is necessary, therefore, to complete the ripening process. Phosphorylative capacity is no longer observed when ripening is complete.

Coupling of oxidation by mitochondria can be retarded by such reagents as DNP. This compound has the ability to cut down the need of respiratory oxidation for coupled phosphorylation.

Esterification of phosphate proceeds considerably faster in climacteric than in preclimacteric tissue (Young and Biale 1967). The ratio of ADP to ATP substantially decreases as ripening advances. Neither uncoupling nor acceptor control accounts for the start of the rise in respiration, but permeability changes during the climacteric are important in the metabolism of fruit during this period.

Chalmers and Rowan (1971) obtained results which indicate that phosphofructokinase is stimulated by the increase in concentration of orthophosphate in the cytoplasm of the fruit, which contributes to the respiration climacteric. Ordinarily, the rate of respiration depends on the presence of such phosphate acceptors as ADP, and an increase in the amount of such substances could account for the climacteric rise in respiration in the ripening process.

It has been suggested that fructose 1,6-diphosphate increases in concentration in the banana which results in the respiration climacteric (Barker and Solomos 1962). Barker (1968) found that glucose 6-phosphate, fructose diphosphate, 3-phosphoglycerate, and phosphoenolpyruvate, which are present in apple tissue, were substantially unchanged as a result of the climacteric rise in respiration. He concluded that the climacteric rise in respiration in apples seems to have a mechanism that is different from that in bananas.

Neal and Hulme (1958) noted that when malate is added to peel disks of Bramley's Seedling apples that are beyond the high point of their climacteric rise in respiration, the result is a strong increase in the evolution of carbon dioxide. Furthermore, a corresponding increase in the uptake of O_2 does not take place. They also discovered that malate decarboxylation, connected with the start of the respiration climacteric in fruit, goes on actively after the climacteric peak is attained. It was suggested that this decarboxylation, although not restricted to malate (even though malate is responsible for the greatest amount of this effect), could cause the development of the respiration climacteric. In preclimacteric fruits the malate effect is small. The authors believed it to be the beginning of senescence.

Flood *et al.* (1960) found that the malate effect has the greatest activity in the epidermal and hypodermal tissues of the apple.

The activity of four hydrolytic enzymes in connection with the rate of respiration during the climacteric was studied by Rhodes and Woolterton (1967). The enzymes were chlorophyllase, lipase, ribonuclease, and acid phosphatase. During the buildup of the respiration climacteric in apples stored at 12°C, the activities of lipase, acid phosphatase, and ribonuclease reached a peak approximately coinciding with the height of the respiration activity and then declining as the rate of respiration fell. The authors determined that acid phosphatase and ribonuclease were present in the soluble protein fractions. The activity of chlorophyllase increased before the start of the respiration rise, and continued after the peak of activity.

Kato and Chachin (1970) found that when mature green tomatoes were irradiated with gamma rays at 250 krad, the ripening was delayed about two days as compared with unirradiated fruit. When 1000 krad was used, a climacteric rise did not take place and the fruit did not ripen normally.

Ethylene and Ripening. Gane (1934) was the first to discover that plants produced ethylene (C_2H_4). Very small amounts of this gas (approximately 1 ppm) are known to act as a stimulant to respiratory activity, and in this activity ethylene influences ripening. According to Biale and Young (1962) climacteric fruits are affected by ethylene in the preclimacteric stage. Non-climacteric fruits after harvest show stimulation by ethylene in all stages of ripeness.

Burg and Burg (1962) concluded as a result of experimental data that ethylene is a ripening hormone; they consider it the cause of ripening and not the effect. Their results show that many fruits produce small amounts of ethylene. Ethylene is the only such compound known to be produced in plants and is present in all plant cells. However, the action of ethylene as a hormone has been disputed (Fidler and North 1971).

Pathways for the Formation of Ethylene. Lieberman and Mapson (1964) were the first to report that methionine could be a starting material for ethylene production. This was discovered while studying a system containing ascorbate, linolenic acid, and copper. It was reported that when tissues are supplied with methionine more ethylene was formed than in the controls. Somewhere along the pathway for the conversion of methionine to ethylene, oxygen is also necessary (Baur *et al.* 1971). However, the method of participation is not yet clearly understood. It is possible that an intermediate that depends on oxygen is formed. Hansen (1942) first showed that oxidative metabolism is involved in the formation of ethylene in pears. This was found also by Burg and Thimann (1961) in apple tissue.

Burg and Clagett (1967) found that methionine metabolism is a major pathway to the formation of ethylene in vegetative tissues, as well as in sections of apple and banana fruits. C-1 of the methionine gives rise to CO_2, while C-3–C-4 give rise to ethylene. The methyl carbon, the sulfur, and the C-2 of the molecule remain in the tissue and are metabolized widely.

$$CH_3SCH_2CH_2CH(NH_2)COOH$$

L-Methionine

It is now generally recognized that methionine is the precursor of ethylene in plants. Other pathways have been suggested (Baur and Yang 1969; Burg and Burg 1965; Jacobsen and Wang 1965) that may produce some ethylene under specified conditions.

Recently, Wardale (1973) studied the effect of phenolic compounds in the tomato on the synthesis of ethylene. It is apparent from this and other work that phenolic compounds are involved in the conversion of methionine to ethylene. In Wardale's work, the ethylene was produced from 4-methylmercaptan-2-oxobutyric acid with peroxidase in the presence of phenolic substances. The effective phenolic compounds present in the tomato fruit were naringenin and *p*-coumaric acid. Other phenolic compounds present included caffeic, ferulic, chlorogenic and sinapic acids. Chlorogenic acid makes up 75% of the phenolic compounds preset in the mature green fruit; in ripe fruit it comprises only 35% of the total phenols present.

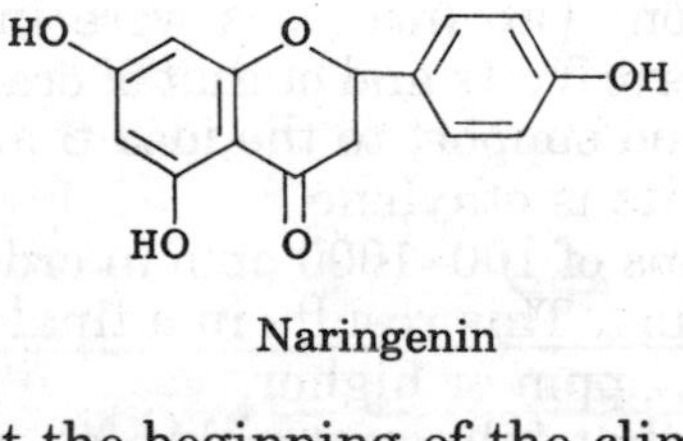

HO O OH HO O

Naringenin

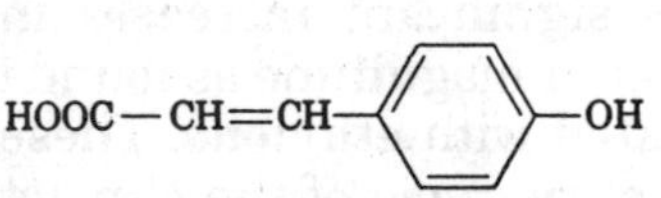

HOOC—CH=CH— —OH

p-Coumaric acid

At the beginning of the climacteric, a large increase in naringenin takes place in the skin. As the concentration of naringenin increases in the skin, so does the formation of ethylene. The phenol compounds present in the skin other than *p*-coumaric acid and naringenin were found to inhibit the formation of ethylene.

When ripening starts in the apple, the increase in lipoxidase activity occurs first, then the release of ethylene, which precedes the respiration climacteric. This increase in the activity of respiration takes place in the presence of a rapid buildup of free and esterified fatty acids. After this, free acids start to disappear and then esterified fatty acids are lost. The respiratory changes during ripening have no effect on the metabolism of triterpenoid components found in the skin.

Ethylene Production. The mechanism necessary for the production of ethylene is available in the fruit from the beginning. According to Abeles (1972) ethylene is produced in all the cells, particularly in higher plants. Ethylene is formed in the fruits and so does not have to be transported from other parts of the plant.

Ethylene to about 0.01 ppm can be found in any plant tissue. The physiological threshold for ripening is about 0.1 ppm of ethylene. However, some fruits cannot start ripening until it is removed from the tree. In such cases something supplied by the tree inhibits the start of ripening, and harvest is necessary to trigger the effect.

Pratt and Goeschl (1969) report that a significant rise (50- to 100-fold) in the rate of formation of ethylene takes place previous to the increase in the rate of respiration. Burg and Burg (1965) showed that ripening can be delayed by removing ethylene from tissues of fruit in the preclimacteric stage.

The use of ethylene with mature but preclimacteric fruits triggers the formation of ethylene in the fruit, and increases it in such fruits. In non-climacteric but mature fruits the small peak forms but drops off again when the ethylene is removed.

Biale *et al.* (1954) found that fruits with a higher climacteric showed moderately high to high rates of ethylene production. Mango was the only expection.

Ethylene production has been reported to cease under anaerobic conditions. The evolution of ethylene after the start of ripening is much larger than the amount necessary to initiate the process of ripening of the same fruit (Reid and Pratt 1970). The respiratory climacteric could be the result of high concentrations of ethylene found in most fruits during the ripening process. The work of Huelin and Barker (1939) was repeated with potatoes but using higher temperatures, and the result was significant increases in respiration. The increases were in the order of magnitude as found in climacteric fruits and of that of oranges treated with ethylene. These results lend support to the idea that the primary cause of the climacteric in fruits is ethylene.

Ethylene is best used in concentrations of 100–1000 ppm in order to speed the rate and uniformity of ripening. This results in a final concentration of ethylene in the fruit of 0.1 ppm or higher.

Wang and Mellenthin (1972) found that fully mature Anjou pears had an internal concentration of ethylene amounting to 0.94 μl/liter when taken out of cold storage. This level of ethylene later dropped, but it remained above the level necessary for fruit softening. Later when the climacteric developed, the ethylene had increased to 0.46 μl/liter. During the climacteric the ethylene content within the fruit rose rapidly. In 11 days, it reached a peak of 40.66 μl/liter. It seems, therefore, that the mechanism responsible for the respiration climacteric must be stimulated by a substantially high level of ethylene. This high concentration of ethylene appears to be necessary for the completion of the climacteric. Other work seems to indicate that maturation is connected with an increase in the sensitivity to ethylene together with the development of the mechanism which brings about ripening. Low temperatures seem to effect the biosynthesis of ethylene in the pear.

McMurchie *et al.* (1972) used propylene on green bananas instead of ethylene and the result was respiration typical of the climacteric pattern. The rise in respiration was evident approximately 10 hr after the start of the treatment. Propylene-treated fruit ripened normally. An endogenous production of ethylene was not discovered until more than 3 hr after the respiratory increase. In normally ripened fruit, however, endogenous production of ethylene and respiration started to increase slowly at least 18 hr before the start of the high climacteric rise in respiration. A rapid rise in ethylene production took place at the same time. There are a number of compounds besides propylene which will do this, but most of them are not related chemically.

Ethylene is not often used commercially because the need usually is to delay ripening, rather than to advance it. However, ethylene is useful and at times must be employed.

Chemical Changes During Ripening

Carbohydrates and Acids. Carbohydrates contribute the major amount of energy needed for respiration. During the climacteric, polysaccharides such as starch change to sugars and, together with acid, form carbon dioxide during respiration. The sugars of apples and pears consist mainly of fructose, with smaller amounts of other sugars present.

De Fekete and Cardini (1964) studied the mechanism of sucrose–starch transfer in the endosperm of sweet corn. Endosperm preparations are able to incorporate radioactivity from [^{14}C]sucrose in the starch granule when ADP or UDP (uridine diphosphate) are present. In order to transfer [^{14}C]glucose from glucose 1-phosphate to the granule Mg^{2+} and ATP must be present. Two possible sequences for the incorporation of glucose into the starch granule are suggested:

Sucrose $\rightleftharpoons$ ADP-glucose (or UDP-glucose) $\rightarrow$ starch

Sucrose $\rightleftharpoons$ UDP-glucose $\rightleftharpoons$ glucose 1-phosphate $\rightleftharpoons$ ADP-glucose $\rightarrow$ starch

According to Pressey (1970) two enzymes are involved in the synthesis of sucrose in potatoes. These are sucrose synthetase and sucrose phosphate synthetase. Sugars increased rapidly when the potatoes were held in cold storage. The sugars produced varied with the variety of potato.

Yamaguchi *et al.* (1967) found that when $^{14}CO_2$ was injected into the cavity of green pepper fruits, most of it disappeared within 1 hr. As much as 70% of the labeled CO_2 reappeared as label in sugars including raffinose. When ripe fruits so treated were held in the dark for 7 hr, approximately 16% of the injected ^{14}C was incorporated. Most of it was found in malic acid, smaller amounts in citric acid, with traces in the amino acid fraction. This suggests that previous to the ripe stage CO_2 is synthesized into sugars. When ripeness is reached the CO_2 goes into acids, although in lesser quantities.

Citrus fruits are usually ripened on the trees and in the mature state contain no starch. Therefore, the sugars present are probably derived from pectins and hemicelluloses, and also from the fruit acid, since the amount of acid in oranges and grapefruits decreases in quantity during the ripening process.

Postharvest apples decrease in titratable acid as ripening progresses. The acid predmoninantly found in mature apple is malic acid. However, in earlier stages L-quinic acid is found. Quinic acid is found in the young fruits of the Bartlett pear and in green peaches.

$$HOOC{-}CH_2{-}CHOH{-}COOH$$

L-Malic acid

Krishnamurthy *et al.* (1971) studied the changes that take place in oxo acids and free amino acids in the mango during the respiratory climacteric. The acids included in the study were α-oxoglutaric, oxaloacetic, pyruvic, aspartic, glutamic, and γ-aminobutyric. Accumulation of oxaloacetate did not take place during ripening process. The amounts of aspartate and glutamate increased for approximately three days after harvest, but at the maximum point of the climacteric their concentration was reduced. During this same period, δ-aminobutyrate increased in amount.

Pectic Substances. Changes in pectic substances affect the texture of fruits. After the harvest of apples, protopectin decreases rather rapidly; at the same time pectin increases. After that, and for a period of time those two forms of pectin remain rather constant in quantity. At the end of the life of the fruit, a final decrease in pectin content takes place. Similar changes occur in the pear, but more rapidly.

Pectinesterase and polygalacturonase are the two pectin enzymes found in fruits. it is likely that a few other enzymes are involved. The ripening process in fruits involves changes in the pectic substances. These changes include loss of methyl groups from the carboxyl groups, deacetylation, and shortening of the chain length of the polymer, which may be the cause of the softening of the tissue as ripening proceeds.

Work on Fuerte avocados and Bartlett pears (McCready and McComb 1954) showed that polygalacturonase is not active in the unripe fruits. On the other hand, the activity of this enzyme is high in the corresponding ripe fruits. It is quite likely that this enzyme is active in cutting down the size of the molecules of pectin in the ripened fruit. McClendon *et al.* (1959) found a very large increase (10 times) in the content of galacturonic acid of apples during the process of ripening. It is likely that much more was formed at this time, but it was in turn metabolized, and therefore, disappeared.

Protein. Only small amounts of protein are found in ripe fruits. A minor increase in the quantity of protein material present takes place in apples, tomatoes, and avocados during the climacteric.

In a study of the intact Bartlett pear, Frenkel *et al.* (1968) found that protein synthesis is necessary for normal ripening. These authors determined that the enzymes needed for ripening are indeed the proteins synthesized during the early ripening process. In order to determine the relationship between protein synthesis and ripening, cycloheximide was used to infilter the fruits. Cycloheximide forcefully inhibits synthesis of protein. The ripening process was stopped when cycloheximide was infiltrated in fruits in their early climacteric stage. However, at later stages it was increasingly less effective. When fruits in the early climacteric stage were treated with cycloheximide a major reduction in the synthesis of ethylene, softening of flesh, and degradation of chlorophyll resulted, indicating that a positive relationship between ripening and the synthesis of proteins exists. However, protein synthesis should not be considered an explanation of the climacteric rise as has been suggested by earlier workers. In addition, it was found that the inhibition of ripening by cycloheximide was not voided by ethylene.

Hulme *et al.* (1968) showed that enzyme systems develop in apple peel during the progress of the respiration climacteric involving the whole fruit. They include incorporation of acetate into lipid, evolution of ethylene, amino acid incorporation into protein, and finally the malate effect, which is the decarboxylation of added malate. Although the formation of ethylene has been thought to be the first step in the sequence of events leading to the climacteric, these authors believe that, as a result of their experiments, an increase in the lipid turnover takes place before the increase in the production of ethylene.

The following enzymes were found (Krishnamurthy 1971) to follow the respiratory pattern of the fruit: malic enzyme (L-malate), malate dehydrogenase (oxaloacetate-decarboxylating) ($NADP^+$) (1.1.1.40), and acetylornithine aminotransferase (2.6.1.11). On the other hand, glutamate decarboxylase (4.1.1.15) gradually increased in activity.

Pigments. According to Miller *et al.* (1940), in lighter colored citrus fruits (i.e., lemons and grapefruit), carotenoid compounds tend to decrease with ripening. In oranges, however, they tend to increase. Knee (1971C) found that total chlorophyll decreased whereas total carotenoids increased during the respiration climacteric in apples.

The problem of the development of yellow color in vegetables held in storage is largely the result of cholorphyll degradation.

In work on Hungarian yellow wax peppers Schanderl and Lynn (1966) studied compounds that appeared at progressive stages of ripening. The stages of ripeness used for analysis were whitish yellow, dark green, olive orange, and the final stage of bright orange. Chlorophylls *a* and *b*, as well as at least five other pink fluorescent compounds, appeared at various stages of ripening. The ripe red pepper contained no pink fluorescent compounds. Pheophytin first appeared in the second stage of ripeness, but this and other green compounds disappeared when the final ripe stage was reached. In this last stage, only two fluorescent compounds remained, but their composition was not clear.

Although much has been accomplished on changes in chemical composition during ripening, much more remains to be done.

STORAGE

A primary objective when storing fruits and vegetables is to retard deterioration by reducing respiration. This is achieved by lowering the temperature. A 10°C decrease in temperature cuts the respiration rate approximately in half, and extends the storage life of the fruits or vegetables. Changing the composition of the surrounding atmosphere is known as controlled-atmosphere storage (CA). High CO_2 and low oxygen are necessary in CA, but low oxygen has less effect than the high concentration of CO_2. The climacteric stage of the fruit at the time of harvest has a definite effect on its behavior in storage. Only preclimacteric fruit can benefit from low temperature and CA storage.

Fidler and North (1967) studied the effect of storage conditions on the respiration of apples. They investigated the relationship between the loss of respirable substrate and the formation of the end products on Cox's Orange Pippin apple. Several different gas mixtures and temperatures were employed in the study. It was found that the acid concentration in the apples decreased logarithmically with the length of storage time for uninjured apples, and acid loss accelerated when low temperature breakdown occurs at the same time. The speed of acid loss is lower when the concentration of oxygen is lowered or when the carbon dioxide concentration is raised. The rate of respiration controls the loss of carbohydrate, and so carbohydrate loss is reduced under conditions of controlled-atmosphere storage. Sorbitol accumulated in apples in storage in progressively larger amounts as the temperature was lowered. When apples were stored in air at 12°C, very little sorbitol was formed, and in the absence of oxygen at temperatures from 0 to 12°C only small amounts were formed.

$$
\begin{array}{c}
CH_2OH \\
| \\
H-C-OH \\
| \\
HO-C-H \\
| \\
H-C-OH \\
| \\
H-C-OH \\
| \\
CH_2OH
\end{array}
$$

D-Sorbitol

In the past, the effect of ethylene on fruits held in CA cold storage was the subject of controversy. However, Blanpied *et al.* (1972) and others found little effect of ethylene up to 1% in CA storage.

The work of McGlasson and Wills (1972) is of interest because it illustrates the effects on the ripening and organic acids of some varieties of bananas when the CO_2, O_2 and N_2 are varied during CA storage. The three different CA concentrations employed were (A) high CO_2: 5%

CO_2, 20% O_2, 20% O_2, 75% N_2; (B) low oxygen: 0% CO_2, 3% O_2, 97% N_2; (C) high CO_2 and low O_2: 5% CO_2, 3% O_2, 92% N_2. They found delay in ripening of the fruit in conditions A, B, and C to be at least 2, 8 and 12 times greater, respectively, than would be found in air. They found further that the three CA streams of gases reduced the uptake of the total oxygen during the period of time before the start of the respiratory climacteric.

Changes in the organic acids of the bananas were found. In the first four days of the treatment, atmosphere-A brought about increases in pyruvate, oxaloacetate, 2-oxoglutarate, glyoxulate, glutamate, aspartate, citrate, and malate, but no increase in succinate. During the same period, atmosphere-B produced greater increases in the 2-oxo acids, while the other acids decreased. Atmosphere-C resulted in smaller increases in pyruvate, 2-oxoglutarate, and glyoxylate, but held down the increase in oxalocetate and further reduced citrate, malate, and aspartate as contrasted with B. The largest changes in the acids took place in the first day.

Chance *et al.* (1958) surmised from theoretical considerations that low oxygen cut down the operations of the Krebs cycle between pyruvate or oxaloacetate and citrate, and between 2-oxoglutarate and succinate. The reduction of malate and the accumulation of glyoxylate in low O_2 would agree with a lowering of the malate synthetase activity. Control points for high CO_2 were not apparent. Significant increases in all the acids except succinate were produced by high CO_2.

The greatest changes found in the organic acids were caused by low oxygen. The concentration of these acids increased in the case of the 2-oxo acids while that of the non-volatile acids decreased. One-day exposure to controlled atmospheres caused the greatest changes in the concentrations of the individual acids.

Controlled Atmosphere Storage of Vegetables

Broccoli. Lebermann *et al.* (1968A) studied the respiration of broccoli shoots and the color of the flower heads when stored in modified atmospheres. Atmospheres containing 2–21% O_2 and 0–20% CO_2 were employed. Progressive increases in CO_2 and decreases of O_2 reduced respiration. The atmosphere containing 20% CO_2 with 21% oxygen inhibited respiration about the same amount as an atmosphere containing 2% O_2 with no additional CO_2. Progressive increases in CO_2 and decreases of O_2 resulted in improved organoleptic color scores and chlorophyll retention. Also, chlorophyll was more effectively retained by a high level of CO_2 than by a low level of O_2. Good color retention resulted from 28 days of storage at 34°C.

Lebermann *et al.* (1968B) noticed a decrease in titratable acid in broccoli stored in atmospheres containing increasing amounts of CO_2. This was reversed by removing the samples to air storage. When stalks stored in high CO_2 were cooked, the result was softer texture and brighter green color.

Lettuce. Temperature of storage is most important if the quality of lettuce is to be maintained (Singh *et al.* 1972). When a controlled atmosphere containing 2.5% O_2 and 2.5% CO_2 (35°F, 1.6°C) was employed in lettuce storage, quality after 39 days was much higher than the control held in normal air. The benefits obtained by storing lettuce in CA include the lowering of decay and russet spotting, together with less of butt discoloration and pink rib.

Potatoes. Harkett (1971) stored potatoes at 1°C for one month in atmospheres containing concentrations of oxygen from 0–100%. Determinations were made of reducing sugar, sucrose, and CO_2 production during this period. Oxygen concentrations of 3% or less resulted in a delay in the accumulation of sugars as contrasted with air. In an atmosphere of N_2 no accumulation occurred; however, CO_2 was evolved at a low rate. When the tubers were held in oxygen concentrations greater than 3% for a period of three weeks, a maximum level of sucrose resulted, after which it decreased slowly. On the other hand, the formation of reducing sugar continued to increase throughout the period. The formation of CO_2 showed a rapid increase for approximately seven days, after which it declined slowly to approximately the initial amount.

Deterioration caused by the growth of microorganisms while the fruits or vegetables are held in storage has practical importance, but is beyond the scope of this book. The interested student is advised to consult sources such as the excellent one by Fidler *et al.* (1973).

Chemical Analysis

The student is advised to consult "Official Methods of Analysis," Chapter 22, Fruits and Fruit Products, and Chapter 32, Vegetable Products, Processed (AOAC 1980).

CHEMISTRY INVOLVED IN THE TEXTURE OF FRUITS AND VEGETABLES

An important characteristic of foods is texture. Control and modification of this characteristic is an objective of modern food technology. Changes in texture during handling and processing must be carefully considered so that the consumer gets a product which, as far as the texture is concerned, is in line with his expectations. This requires a knowledge of the factors which influence texture.

One of the important factors is the chemistry of the compounds that affect texture. This is especially vital when it concerns the changes in texture that may take place during storage, processing, or cooking. The main concern is the components of the cell wall and the middle lamella because these are the compounds that have an important effect on the final texture of the fruit or vegetable when used.

The cell walls of plants are made up of a two phase system. In the higher plants cellulose is found as the skeletal material. It is a partially crystalline phase. The amorphous is made up largely of hemicellulose, pectic substances, and lignins. Some other substances including proteins are found also. The walls of the parenchyma cells of apples are thin and made up of middle lamella and primary cell wall, according to Nelmes and Preston (1968).

Similar percentages of pectic substances, hemicellulose, and cellulose occur in the primary cell wall. The cellulose gives rigidity as well as resistance to tearing. Pectic substances and hemicelluloses give plasticity together with ability to stretch. The middle lamella is an extension of the matrix material found in the primary cell wall, but lacking the cellulose fibrils. Since it is the outermost portion of the plant cell, it plays the primary role in intercellular adhesion. Covalent bonding, hydrogen bonding, and crystallite bonding together with ionic bonding (Albersheim 1974) are thought to make up the bonding among the polymeric cell wall and middle lamella components. Other types of bonding have been suggested. It should be realized that these types of bonding are possibilities rather than established facts.

Reeve (1970) noted that structural features play a vital part in determining texture. Consideration must be given to the relative proportion of the more tender parenchyma tissues to the fibrous collenchyma, sclerenchyma, and xylem tissues. Of importance also are the thickness of the cell wall, the shape and size of the cells, and the volume of intercellular spaces. Thin areas of the cell walls may be sites of easier wall breakage, an example of which is the plasmadesmata of bean parenchyma cells (Grote and Fromme 1978).

Pectic Substances

About one-third of the dry material of the primary cell walls of fruits and vegetables is made up of pectic substances, and it is likely that they comprise a much greater amount of the dry substance of the middle lamella. In this way they contribute to the mechanical strength of the wall and also to the adhesion between cells. These substances are more chemically reactive than other cell wall polymers and are more easily brought into solution. Processes that bring about texture changes, such as cooking. storage, and ripening, are accompanied by significant changes in the characteristics of the pectic substances.

Structurally, the pectic substances are made up of chains of galacturonic acid residues linked by α-(1 $\rightarrow$ 4) glycosidic bonds. Some inserts of rhamnose appear in the uronide chain making up about 2% of the main chain residues. Between 150 and 1500 saccharide units are present in the main chain of solubilized pectic substances. The carboxyl groups of the galacturonic acid residues are extensively esterified with methyl alcohol. This esterification usually amounts to between 50 and 90%. The acetylation of hydroxyl groups of the galacturonic acid residues is often observed, especially in beet and pear pectic substances.

A number of different side chains can be found attached to the main chain residues. Side chains made up of neutral sugars such as galactose, arabinose, and xylose can be quite extensive and may merge with the hemicellulose portion of the cell wall.

Some of the pectic substances of tissues are often found to be water soluble under mild extraction conditions, and these usually have low percentages of neutral sugars and high percentages of methoxylation. Other pectic substances can be put into solution by the use of chelating agents such as sodium hexametaphosphate or ethylenediaminetetraacetic acid. It is possible that these portions may have been insoluble originally because of Ca^{2+} bridges between adjacent pectic polymers. Usually they have lower percentages of methoxylation than the readily soluble pectins of the tissues.

Further portions can be extracted in the soluble form only following drastic treatments such as the use of high temperatures and the presence of acid or alkali. Reasons for this behavior can include salt linkages, covalent and noncovalent bonding to hemicelluloses, and mechanical intermeshing. After such extraction these pectic materials often show evidence of degradation, mainly lower intrinsic viscosity. This indicates lower molecular weights.

Pectic substances have the ability to undergo many chemical modifications. These can occur under the conditions which one finds associated with foods. Under mild alkaline or acid conditions demethoxylation of the esterified carboxyl group takes place. During heating at neutral or alkaline pH splitting of glycosidic linkages between galacturonic acid residues can take place by a process of β-elimination. Under mild acid conditions hydrolytic cleavage of the glycosidic bonds is enhanced. Increased solubility is the result when glycosidic bonds of pectic substances imbedded in the cell wall matrix are split.

It is rather easy to saponify the methoxyl groups in pectin under acid or alkaline conditions. This reaction can be catalyzed by pectin methylesterases. Chemical saponification results in random de-esterification along the polymer chain, while enzymatic de-esterification generates consecutive sequences of de-esterified galacturonate residues. Since depolymerization by β-elimination splitting of the glycosidic bonds is catalyzed by hydroxyl ions (BeMiller and Kumari 1972) it can proceed under non-acid conditions ordinarily found during canning or cooking of vegetables (Keijbets and Pilnik 1974).

In this reaction the glycosidic bond is split along with the formation of a double glycosidic bond between the C-5 and C-4 of the newly formed nonreducing end group. That this reaction takes place can be

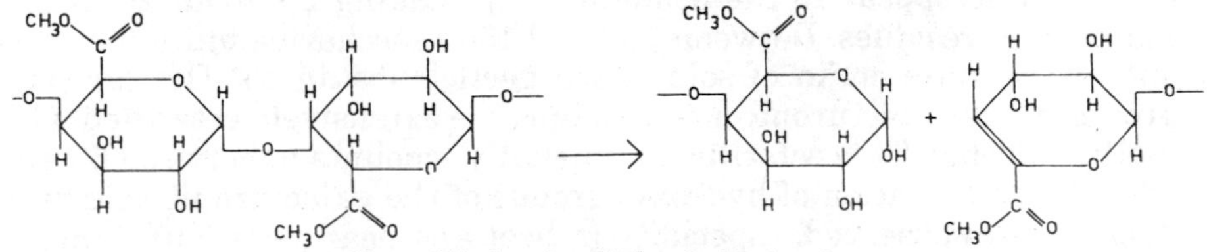

Glycosidic bond split by β-elimination

demonstrated by an increased solubility of pectic materials, by decreased intrinsic viscosity of soluble pectin, by increased optical absorbance at 235 nm wavelength, or by increased color production in the periodate–thiobarbituric acid test (Keijbets 1974). In this reaction it is necessary that the carboxyl group of the residue undergoing β-elimination be esterified since this enhances the electron deficit at the C-5 position. Demethoxylated pectic substances do not undergo the β-elimination reaction and are rather stable when heated under mild alkaline conditions.

Hydroxyl ions speed the reaction as a result of the removal of H^+ from the C-5 position. According to Neukom and Deuel (1958) β-elimination is rapid at room temperatures in alkaline conditions. Keijbets and Pilnik (1974) found that heating speeds the reaction and as a result even at pH 6.1 β-elimination in appreciable amounts takes place in 30 min at 100°C.

A number of other ions are able to enhance the rate of β-elimination. According to Keijbets (1974) and Keijbets and Pilnik (1974), Ca^{2+}, Mg^{2+}, K^+, citrate, malate, and phytate are responsible for a considerable increase in the velocity, particularly at molar concentrations the same or higher than the molar concentration of the pectic materials.

BeMiller (1967) reviewed the acid-catalyzed hydrolytic splitting of glycosidic bonds. These reactions proceed through an initial protonation of the glycosidic oxygen followed by a rate-limiting formation of a cyclic oxocarbonium ion leading to rearrangements which result in the breakage of the disaccharide linkage. The glycoside bonds which join the uronides tend to be rather resistant to this kind of hydrolysis compared with similar bonds of neutral sugar oligosaccharides. BeMiller suggested that this is due to both steric and inductive effects of the carboxyl group at the C-6 of uronides. Thus, it is most likely that glycosidic hydrolysis of pectic substances occurs in neutral sugar side chains or in the main chain at rhamnose units. It should be noted that the degree of polymerization of low ester pectin produced by mild acid hydrolysis is 45 (Rombouts *et al.* 1970), close to that expected if cleavage of the pectin chain had taken place at the rhamnose inserts.

Reviews by Pilnik and Voragen (1970) and Doesburg (1965) show that enzymes capable of catalyzing the demethoxylation and depolymerization of pectic materials are widespread in fruits and vegetables. Almost all the depolymerizing enzymes reported in fruits and vegetables have been described (or assumed) as causing the hydrolytic splitting of the glycosidic bond between uronide residues. Only one, in peas, is described as catalyzing a β-elimination reaction (Albersheim and Killias 1962). Most higher plant pectin depolymerizing enzymes have not been tested by methods that would distinguish between hydrolytic and β-elimination cleavage. This brings up the possibility that β-eliminative pectin depolymerizing enzymes may be more prevalent than it would seem at the present time.

A feature of pectic substances that has an effect on their function in the cell wall matrix is their ability to form gels in the presence of divalent ions or under acid conditions with high concentrations of sugar. Such gelling properties are important commercially and have been studied extensively. Reviews by Feldman and Taüfel (1956) and Neukom (1967) have been published on this subject. Gel formation with divalent or trivalent ions is dependent on the presence of a sufficient percentage of nonesterified carboxyl groups. Gel formation in sugar–acid systems is favored by a high percentage of methoxylation, a low percentage of acetylation, a long polymer chain length, and a low proportion of neutral sugar side chains.

Hemicellulose

Hemicelluloses, according to Isherwood (1970) can be considered as neutral sugar polysaccharides extracted by alkaline solutions from the residue of cell wall material after removal of pectic substances. Together with pectic substances they make up the bulk of the polymeric noncellulosic dry material in the primary cell wall. Although hemicellulose can be roughly divided into some rather distinct types of polymers, there are many variations in the particular architecture of the polymer segments. Thus, according to Northcote (1972) hemicelluloses are thought to show "micro dispersity." This term could be used in connection with pectins also.

In many cases, the principal chain of a particular kind of hemicellulose is made up largely of one type of monosaccharide. One of them may be a galactan, another a xylan, and another a glucan. Recently, Das and Das (1977) isolated a β-D-(1 $\rightarrow$ 4)galactan from garlic. Often these main or interior chains have a preponderance of another kind of sugar as side chains. One has a xyloglucan, where xylose is the side chain, or an araginogalactan, where arabinose is the side chain. Albersheim (1974) proposed a fairly regular system such as this for most of the hemicellulose content of the cell wall.

This view is complicated somewhat by the frequent discovery that many different separable hemicellulose fractions contain four or more types of saccharides in alkaline extracts from cell walls (Knee 1975). There seems to be much crosslinking through covalent and noncovalent bonding, between the rather homogeneously constructed polymer chains. This kind of crosslinking can include pectic substances (Isherwood 1970).

The main or interior chains participate in noncovalent complex formation with the main chains of other polymer molecules. It is particularly important that xylans and glucans have the ability to complex with cellulose. This attaches the cellulose fibrils to the matrix materials. These complexes are rather strong and need drastic action such as 0.5 N NaOH to bring about dissociation.

The amount of water adsorbed by the polymer is affected by the side chains. Arabinose side chains are especially effective in this way (Northcote 1972). Rees and Wight believe that the side chains interfere with noncovalent interchain bonding. On the other hand, the matrix can be further stabilized when a galactose side chain becomes involved in covalent linkages between polymer molecules.

Cellulose

Cellulose is a β-(1 → 4)glucan with 8000–12,000 glucose units per chain. In the primary cell walls it occurs as linear associations of the polymer molecules called fibrils, the diameters of which are about 3.4 nm. Up to about 70% of the fibril volume is crystalline in form. These portions are distributed along the fibrils about 40 nm apart (Preston 1974).

Especially in its crystalline form, cellulose is a rather inert material. The degree of hydration of this substance can be increased by heating (Sterling and Shimazu 1961) and may influence wall properties. Walter *et al.* (1977) found that gels containing 3% apple cellulose, a concentration lower than found in a primary cell wall, are free standing.

A number of special solvents such as cupriethylene diamine (Kertesz *et al.* 1958) and dimethyl sulfoxide paraformaldehyde (Johnson *et al.* 1975) can bring cellulose into solution.

Proteins

As indicated by chemical studies, proteins are always present in isolated cell walls (Mühlethaler 1967; Lamport 1977). Nitrogen organically bound varies from 0.5 to 2.0% of the dry weight. According to Knee (1975), apple has been reported to contain 0.1% of the fresh weight of nitrogen. The predominant amino acid of bound mature runner bean wall protein hydrolysate is hydroxyproline, with serine and proline as other major components (Selvendran 1975).

According to Mühlethaler (1967) wall protein is especially resistant to attack by proteolytic enzymes. However, treatment of cell walls with proteolytic pronase reduces its tensile strength (Preston 1974) and liberates pectic and hemicellulose fragments along with peptides from suitably prepared cell walls (Keegstra *et al.* 1973). These results indicate that the protein is bound to cell wall polysaccharides. This binding is supported by the isolation of cell wall fragments containing galactose-*O*-serine and arabinose-*O*-hydroxyproline glycosidic linkages (Preston 1974; Cho and Chrispeels 1976). It seems that the cell wall matrix contains a protein–hemicellulose network.

Lignin

According to Freudenberg (1968) and Sakakibara (1977) lignin is a very complex polymer derived from phenolic compounds. It has the ability to replace water in the cell wall matrix and, as a result, increases the rigidity and cohesion of the walls. It appears in significant amounts in vegetables during the later stages of cell wall maturity along with the formation of cellulose-rich secondary walls. Lignin is not present or is found only in very small amounts in the cells of fruits and vegetables at the stage when they are edible. An exception is the specialized vascular and structural tissue cells (Nelmes and Preston 1968).

Water

A large portion of young cell walls is water. Its contribution to texture can be very great, which corresponds to its great complexity of behavior. The influence of this compound can be seen by a comparison of the textures of fresh and dehydrated fruits and vegetables. Northcote (1972) made the suggestion that water plays four important functions in the walls. As a part of the matrix gel it is a structural component; it can act as a wetting agent interrupting direct hydrogen bonding between polymers; it can help in stabilizing conformations of polymers; and it acts as a solvent for the presence and transport of salts, low molecular weight organic compounds, and enzymes (see Frank 1965).

Maturation and Storage of Fruits

Totally different changes take place in the textures of fruits and vegetables during maturation and storage. Fruits soften, while vegetables become tougher. These changes involve alterations in their cell wall components. For these reasons it is necessary to consider these foodstuffs separately.

The softening of fruits during the ripening process has for a long time been attributed to changes in the quantity and character of the polysaccharides of the cell wall middle lamella (Pilnik and Voragen 1970; Hulme 1958). The pectic materials particularly have been watched repeatedly. The bulk of the evidence shows that increased solubility of the pectic substance takes place during maturation and softening. The reason for this increased solubility in sound fruit is of continued interest and interpretation. It is quite a complex situation.

According to Lamport (1970) it is reasonable to assume that bond-splitting processes are active in the cell walls of fruit during the entire growing period. The continued growth of the cells in the surface area together with little change in wall dry matter per unit area (Nelmes and Preston 1968) indicates that new polymeric material must be continuously put into the matrix of the wall. A great deal of the physiology of cell wall growth was discussed by Darriel *et al.* (1977). It is possible for the matrix in place to stretch to provide a place for the new material. This stretching is suggested by the requirement for turgor pressure for cell growth and the partial orientation of wall polysaccharides (Morikawa and Sendra 1974). However, stretching can be expected to be limited in a matrix as fully interconnected by covalent bonds as in models put forward by Keegstra *et al.* (1973). The dynamic equilibrium of hemicelluloses (Nelmes and Preston 1968; Labavitch and Ray 1964) and of cell wall galactose (Knee *et al.* 1977) gives the idea that covalent bond splitting does take place. This is strengthened by the identification of a number of enzymes able to hydrolyze the glycosidic linkages of cell wall polymers.

A continuation of glycosidic bond splitting during that part of the fruit life cycle at which incorporation of new saccarides and reformation glycosidic bonds slows down or ceases would decrease the extent of covalent interconnection between the polymer chains of the matrix. Tavakoli and Wiley (1968), Bartley (1976), and Knee *et al.* (1975) found that during apple ripening a loss of cell wall galactose takes place. It is likely that this would weaken the elasticity and resistance to rupture of the cell wall. The extension of such a process to the middle lamella would decrease the strength of intercellular adhesion. Tavakoli and Wiley (1968) showed a high correlation between objective firmness changes and galactose disappearance. Along with the changes in texture, more matrix polymer could be freed from the matrix meshwork to act as soluble material. Buescher *et al.* (1976) showed that inhibition of ripening, as found with some tomato genotypes, decreases the formation of soluble pectic substances.

A preexisting bond-splitting process may be overshadowed by the action of additional hydrolytic enzymes, the activity of which increases as fruits ripen. According to Sacher (1973), many hydrolases, including polygalacturonases, increase with maturation. Most of the cell wall material depolymerizing enzymes that have been examined for this behavior were found to act in similar fashion.

Fruit polygalacturonase has been found in tomatoes, peaches, avocados, pears, and pineapples (Pilnik and Voragen 1970). It was reported as found in some other fruits, but the results have been questioned.

Although hemicellulases have not been so widely found in higher plants, a sufficient number have been described to establish that they can be considered normal fruit enzymes. β-Galactosidase is present in apples (Bartley 1974, 1977) and tomatoes (Wallner and Walker 1975). Matheson and Saini (1977) purified two α-L-arabinofuranosidases and three β-D-galactopyranosidases from germinating lupin cotyledons. Pierrot and van Wielink (1977) found β-galactosidase together with other glycosidases to be located in living cell walls.

Cellulase was found in the avocado by Pesis *et al.* (1978). Its activity appears to be induced by ethylene. Barnes and Patchett (1976) found that cellulase activity was present in ripe strawberries and increased in overripe fruit. Cellulase activity has been studied in tomatoes (Hobson 1968), in dates (Hasegawa and Smolensky 1971), and peaches (Hinton and Pressey 1974).

Hobson concluded that cellulase is not an important factor in the softening of tomato fruit during the ripening period. Its activity in small green fruits was high initially, but it gradually fell during fruit swelling. The activity increased again during incipient ripening, and the rise continued to the full red condition.

Hinton and Pressey (1974) concluded that cellulase is low in immature peaches. It increases during the ripening process. However, the enzyme level in ripe peaches was found to be much lower than that reported for tomatoes (Dickinson and McCollum 1964) and dates (Hasegawa and Smolensky 1971). Peach cellulase was found to be of the same type as that found in tomatoes, and that the largest increase in cellulase activity in peaches occurs before an important change in fruit firmness.

It has been suggested that the increase in cellulase activity during peach ripening acts in conjunction with other hydrolytic enzymes such as the pectic enzymes and contributes to the softening of the peaches during ripening. Since Pressey *et al.* (1971) found that polygalacturonase activity developed during the later stages of peach softening, it seems that cellulase forms prior to full softening.

Hasegawa and Smolensky (1971) believe that softening of fruit during ripening is connected with alterations in the pectic substances as a result of the action of pectic enzymes. This seems to be the case with dates. These authors showed that mature dates have much higher polygalacturonase activity and also that there is a close correlation between polygalacturonase activity and fruit softening. They showed the presence of cellulase in the Deglet Noor dates and, in addition, its activity very much increases during ripening. There is a possible role for cellulase in the softening process.

The bulk of the water-soluble polymeric cell wall material of fruits consists of pectic material (Joslyn 1962; Knee 1973). However, varying amounts of neutral sugar polymers are found also. The preponderance of polygalacturonic material can indicate that such polymers have fewer cross-polymer bonds and/or that such bonds are more rapidly broken.

According to Doesburg (1965) the degree of pectin esterification decreases as the fruit ripens. Reeve (1970) found that an exception to this is the peach. An increase in free pectin carboxyl groups might be expected to increase the importance of calcium as a firmness increasing agent since the result is the formation of calcium pectate. While little evidence is available on maturing fruit, evidence shows that calcium plays a role in firming fruit tissue.

The extraction of cell walls with solutions of calcium binding substances such as oxalate, ethylenediaminetetraacetic acid, or sodium hexametaphosphate, additional pectic materials are brought into solution according to Kertez (1951) and Doesburg (1965). It seems, therefore, that calcium ions in cooperation with pectin have a role in maintaining the integrity of the matrix.

When calcium ions are added to solutions of pectic substances having a low percentage of esterification the result is the formation of a gel. This has an application in the manufacture of low sugar jellies. It has been proposed that the formation of a calcium bridge between the ionized carboxyl group of different polynuronide chains takes place. However, the explanation does not seem to be so simple. Other explanations, which involve the secondary hydroxyl groups and degree of polymerization, may come into play according to Pilnik and Zwiker (1970). The complexing of calcium with pectin can be influenced by other ions, especially K^+. Usually the predominant cation in fruits, K^+ decreases the calcium complex according to Kohn and Furda (1967). Therefore, the salt composition of the cell wall and the metabolic processes affecting the movement of salts (Sacher 1973) can influence texture.

Regardless of the mechanism, the calcium-mediated linking of polyuronide chains can serve to anchor some pectic substances, not covalently bound to the insoluble portions of the matrix, to more firmly bound material and can also increase the overall extent of crosslinking in the matrix.

Addition of calcium to fruits, particularly applies (Betts and Bramlage 1977; Mason *et al.* 1975; Riley and Kohattukudy 1976) was found to increase their firmness and cut down softening during storage. Calcium is used to maintain and toughen the texture of cherries stored in bisulfite brine in the process for making maraschino cherries (Van Buren *et al.* 1967).

The normal level of calcium in edible tissues is rather low. Potatoes have 0.07–0.13% calcium on a dry weight basis (Addiscott 1974), whereas apples have an average of 0.005% on the fresh weight basis (Goodall 1969; Perring 1974).

The modification and control of physiological senescence in fruits is brought about by controlled atmosphere–controlled temperature storage and treatment with ethylene (Smock and Neubert 1950; Sacher 1973). These methods speed or retard ripening and softening in fruits during storage.

Maturation and Storage of Vegetables

Many of the same considerations involved in the chemistry of the texture of fruits apply to vegetables as well. The difference involves the differentiation of tissues, enlargement of cells, and continuation of cell wall growth throughout the stage of edibility. There is an increase of fibrous tissue with maturation. The potato is an exception to this (see Reeve 1970: Warren and Woodman 1974).

The toughening of vegetable tissue in most of its aspects is accounted for by the continued growth and thickening of the cell wall. This eventually involves lignification and formation of secondary cell wall.

Heat Processing and Texture Chemistry

The softening of fruits and vegetables when heated is partly due to the loss of turgor, particularly with leafy vegetables, but also due to a variety of chemical changes in the cell wall matrix polysaccharides. These polysaccharide modifications can be influenced by a number of factors, the most important of these being pH and the amounts and types of salts that are present.

Figure 21.3 illustrates the influence of pH as derived by Doesburg (1961) from studies on a variety of fruits and vegetables. Enhanced softening at low pH has been ascribed to hydrolytic cleavage of glycosidic bonds of neutral sugar components of the cell wall. The enhanced softening at neutral pH has likewise been associated with polymer cleavage, in this case through the β-elimination reaction involving the polyuronides. This has been demonstrated in potatoes by Keijbets and Pilnik (1974). At intermediate pH neither reaction would be expected to proceed at a significant rate.

Warren and Woodman (1974) discussed the hydration of cell wall components that accompanies cooking. The uptake and adsorption of water by polysaccharides can reduce the cohesiveness of the matrix, soften the cell wall, and decrease the intercellular adhesion. Matrix composition can play a role in the degree of hydration; it has been suggested that a high degree of methoxylation of cell wall pectin would increase water uptake (Warren and Woodman 1974).

When an undesirable degree of softness is expected during heat processing, calcium salts are often added before cooking to improve the texture (Collins and Wiley 1963; Doesburg 1965; Loconti and Kertesz 1941; Van Buren 1968; Van Buren and Peck 1963). For practical purposes the amount of calcium salt added is rarely over 0.1% of the fruit or vegetable weight in order to avoid off-flavors. Overhardening could be induced in canned peas when unsoftened water containing 80 ppm

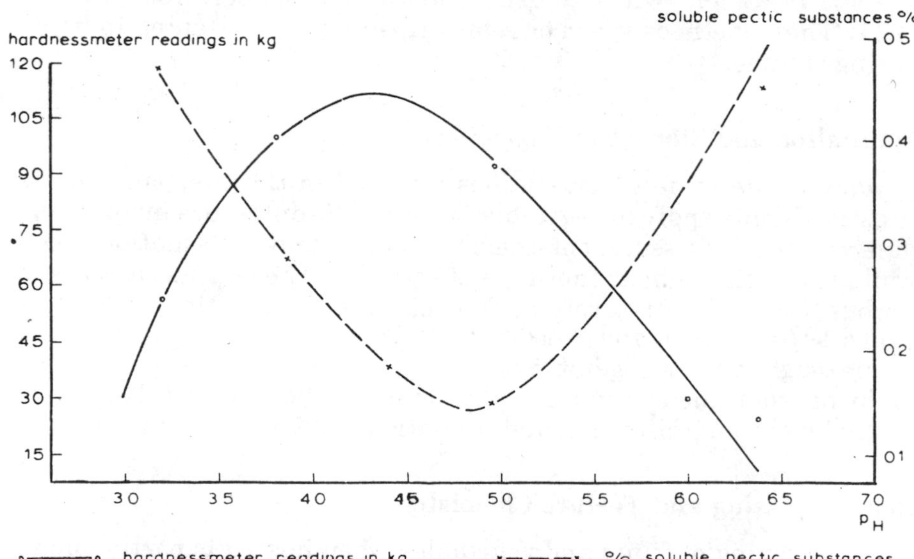

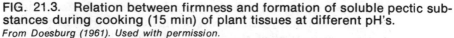

FIG. 21.3. Relation between firmness and formation of soluble pectic substances during cooking (15 min) of plant tissues at different pH's.
From Doesburg (1961). Used with permission.

of calcium is used for the brine (Bigelow and Stevenson 1923). Green beans and beets showed no effect at this calcium level.

Another role for salts in heat-processed foods is their enhancement of β-elimination reactions. Calcium and potassium have been shown to increase β-elimination in boiled potatoes (Keijbets 1974; Keijbets *et al.* 1976). Potassium increased sloughing and pectin solubility. According to Hughes *et al.* (1975) the compressive strength of boiled potato was found to be decreased in the presence of potassium chloride in the cooking solution. Under the same conditions pectin solubility was increased. When salts were removed from raw potatoes (Zaehringer and Cunningham 1971; Davis and LaTourneau 1973), the result was a decrease in softening during cooking.

Appropriate fertilization can alter the concentrations of cations, especially K^+, in vegetables. This, in turn, influences the firmness of the cooked product. This behavior in snap beans is illustrated in Table 21.1 (Peck and Van Buren 1975). According to Peck and MacDonald (1972) potassium can accumulate in other vegetables, such as the beet, in which the level can be as high as 0.6%.

Calcium has two opposing effects on texture. By complexing with pectic substances it firms tissues. It is able to enhance the softening of tissues as a result of the β-elimination reaction. The final result of the addition of calcium, however, has been invariably the firming of the tissue.

In addition to the influence of salts during cooking is the effect when salts are added to cooked materials. Outstanding in this case is the influence on firming by calcium and magnesium salts (Sterling 1968; Doesburg 1965). Monovalent cations have a softening effect. There are several probable explanations of the effects that take place between these salts and pectins, which have a significant amount of free carboxyl groups. (1) Divalent cations increase the ease of the formation of complexes between pectin molecules (Pilnik and Zwiker 1970). (2) Monovalent salts compete with divalent salts for positions in the neighborhood of the dissociated carboxyl groups (Kohn and Furda 1967). (3) H^+ reduces the dissociation of free carboxyl groups diminishing electrostatic repulsion (Doesburg 1965). (4) The effective repulsive forces between the negatively charged dissociated carboxyl groups are decreased by both monovalent and divalent cations. According to Linehan and Hughes (1969) this accounts for the increased firmness brought about by the addition of monovalent cations to tissues which had previously been freed from calcium.

Ca^{2+} and K^+ have more influence on vegetable texture than on fruit texture because the pH of most vegetables is in the range at which free carboxyl groups of pectins are dissociated and can interact with positive ions. A role is played by pH itself in the texture of cooked tissues with firmness increasing as the pH is lowered (Hughes *et al.* 1975).

The salt and pH effects are much more readily noticed with cooked tissue than with fresh (Sterling 1968). Only a slight effect on the tex-

TABLE 21.1. Potassium Fertilization and Texture of Canned Snap Beans

Fertilization Rate KCl (kg/hectare)	K in Pods (% of Fresh Weight)	Cooked Pods[a] Resistance to Compression (kg)
0	0.10	51
67	0.14	43
269	0.19	41
1076	0.25	40

Source: Peck and Van Buren (1975).
[a]Size three pods cooked 20 min at 115°C.

ture was observed by Massey and Woodams (1973) when calcium was added to raw carrots, beets, and potatoes. This is in contrast with the marked effect found on cooked vegetable tissue by Sterling (1968). The breakage of covalent links and the hydration of the wall materials that take place during cooking results in the weakening of other wall matrix interconnect mechanisms. This may increase the relative importance of the ionic mechanisms that involve pectic materials and cations.

A very important reaction connected with the texture of processed fruits and vegetables is the demethoxylation of pectic substances. This reaction is catalyzed by the enzyme pectin methylesterase. Although this enzyme is present in most plant tissue it is rather inactive. However, if the tissue is damaged by bruising, freezing, or heating to 50°–80°C this enzyme becomes active. Conditions such as these frequently take place during processing. According to Bartolome and Hoff (1972) cell membranes might become more permeable to cations during these treatments. This would permit the activation of the cell enzyme (Van Buren *et al.* 1962). A firmer texture of fruits before cooking and of vegetables after cooking is frequently seen as a result of demethoxylation (Van Buren 1974).

Two separate phenomena may be involved in this firming effect. The possibilities and strength of calcium or magnesium binding between pectin polymers are increased by the formation of free carboxyl groups. The suceptibility of pectin to the β-elimination chain-breaking reaction during cooking would be lessened by the decrease in the methyl ester content. Among the fruits and vegetables that have been found to show firmer texture after activation of pectin methylesterase are the following: snap beans (Van Buren *et al.* 1960; Sistrunk and Cain 1960), frozen snap beans (Steinbush 1976), cauliflower (Hoogzand and Doesburg 1961), potato (Bartolome and Hoff 1972), tomatoes (Hsu *et al.* 1965), apple (Wiley and Lee 1970), and cherries (Buch *et al.* 1961; LaBelle 1971; Van Buren 1974).

Sometimes when enzyme reactions are initiated at early stages of processing a loss of pectin results. This was noted in the preparation of tomato paste at low break temperature by Sherkat and Luh (1976).

SUMMARY

Changes in fruits and vegetables during the ripening process and after harvest influence their use as foods. Fruit ripening is basically a process of aging.

Respiration is carried on by all living organisms and is the key activity of the life process. Basically, it is the transformation of potential energy into kinetic energy. Heat is evolved.

Mitochondria are the vital components of the protoplast and are observed as small rods, filaments, or granules in the cytoplasm of most plant cells. They are made up of protein and lipid.

Two biochemical pathways are known in plant respiration. They are involved in plant respiration, and therefore, in the ripening process. These are the Embden–Meyerhof–Parnas glycolytic–tricarboxylic acid cycle and the pentose pathway.

The respiration climacteric is a very great increase in respiration after harvest in such fruits as apples held in storage. This is the climacteric rise in output of CO_2. Some changes take place in the chemical composition of the fruits during the climacteric.

It was shown that oxidative phosphorylation continued throughout the ripening process in fruits. Phosphate esterification takes place in preclimacteric tomato fruits as well as during the postclimacteric senescent period.

Ethylene in very small amounts is important in the ripening of fruits. A major pathway to the formation of ethylene is methionine metabolism.

Since carbohydrates contribute the major amount of energy required for respiration, changes in these compounds take place during ripening.

Pectic changes taking place in fruits during ripening are important because of their effects on texture.

As a result of a study on the Bartlett pear it was found that protein synthesis is necessary for fruit ripening. It was noted that the enzymes needed for ripening are the proteins synthesized during the early part of the ripening process.

The primary objective of storing fruits and vegetables is to retard deterioration by reducing the respiration during storage.

An important characteristic of foods is texture. One of the principal factors is the chemistry of the compounds that affect texture. This is vital when the changes are brought about by storage, processing, and cooking. The components of the cell wall and the middle lamella are most important. Covalent bonding, hydrogen bonding, and crystallite bonding together with ionic bonding are thought to make up the bonding among the polymeric cell wall and middle lamella components. These are not established facts.

About one-third of the dry material of the primary cell walls of fruits and vegetables is made up of pectic substances. Cooking brings about changes in these compounds.

Pectic substances can undergo many chemical modifications. These can occur under the conditions one finds associated with foods.

It is easy to saponify the methoxyl groups in pectin under acid or alkaline conditions. Since depolymerization by β-elimination splitting of the glycosidic bonds is catalyzed by hydroxyl ions, it can proceed under non-acid conditions usually found during canning or cooking of vegetables. In this reaction, the glycosidic bond is split and a double glycosidic bond is formed between the C–5, and C–4 of the newly formed nonreducing end group. Demethoxylated pectic substances do not undergo the β-elimination reaction and are rather stable when heated under mild alkaline conditions.

A feature of pectic substances that has an effect on their function in the cell wall matrix is their ability to form gels in the presence of divalent ions, or under acid conditions with high concentrations of sugar. These properties are of importance commercially.

Hemicelluloses and pectic substances make up the bulk of the polymeric noncellulosic dry material in the primary cell wall. The principal chain of a particular kind of hemicellulose is made up largely of one type of monosaccharide.

It is particularly important that xylans and glucans have the ability to complex with cellulose. This attaches the cellulose fibrils to the matrix materials. These complexes are rather strong.

A large portion of cell walls is water. Its contribution to texture can be very great. Water plays four important functions in the walls.

The softening of fruits during the ripening process has been attributed to changes in the quantity and character of the polysaccharides of the cell wall middle lamella. The evidence shows that increased solubility of the pectic substance takes place during maturation and softening. This is quite a complex situation. A high correlation has been shown between objective firmness changes and galactose disappearance.

Many hydrolases, including polygalacturonases, increase with maturation.

Hemicellulases are considered normal fruit enzymes.

Cellulase activity contributes to the softening of peaches during ripening.

The salt composition of the cell wall and the metabolic processes affecting movement of salts can influence texture.

When an undesirable degree of softness is expected during heat processing, calcium salts are often added before cooking to improve the texture.

ACKNOWLEDGMENT

The author acknowledges use of information in the last section from J. P. Van Buren. 1979. The chemistry of texture in fruits and vegetables. *J. Texture Studies 10,* 1-23.

BIBLIOGRAPHY

ABELES, F. B. 1972. Biosynthesis and mechanism of action of ethylene. Annu. Rev. Plant Physiol. *23*, 259–292.

ADDISCOTT, T. M. 1974. Potassium and the distribution of calcium and magnesium in potato plants. J. Sci. Food Agric. *25*, 1173–1183.

AHARONI, Y. 1968. Respiration of oranges and grapefruits harvested at different stages of development. Plant Physiol. *43*, 99–102.

AHARONI, Y., LATTAR, F. S., and MONSELISE, S. P. 1969. Postharvest response of oranges to ethylene. Plant Physiol. *44*, 1473–1474.

AKAMINE, E. K., and GOO, T. 1973. Respiration and ethylene production during ontogeny of fruit. J. Am. Soc. Hort. Sci. *98*, 381–383.

ALBERSHEIM, P. 1974. The primary cell wall and central control of elongation growth. *In* Plant Carbohydrate Chemistry (J. B. Pridham, ed.), Academic Press, New York.

ALBERSHEIM, P., AND KILLIAS, V. 1962. Studies relating to the purification and properties of pectin transeliminase. Arch. Biochem. Biophys. *97*, 107–115.

ALBERSHEIM, P., NEUKOM, H., and DEUEL, H. 1960. On the formation of unsaturated degradation products by a pectin degrading enzyme. Helv. Chim. Acta *43*, 1422–1426. (German).

AP REES, T., and ROYSTON, B. J. 1971. Control of respiration in disks of carrot storage tissue. Phytochemistry *10*, 1199–1206.

AOAC. 1980. Official Methods of Analysis, 13th Edition. Assoc. of Official Analytical Chemists, Washington, DC.

AURAND, L. W., and WOODS, A. E. 1973. Food Chemistry. AVI Publishing Co., Westport, CT.

BARKER, J. 1968. Studies in the respiratory and carbohydrate metabolism of plant tissues. XXIII. The mechanism of the 'climacteric' rise in respiration of apples. New Phytologist *67*, 213–217.

BARKER, J., and SOLOMOS, T. 1962. Mechanism of the 'climacteric' rise in respiration in banana fruits. Nature *196*, 189.

BARNES, M. F., and PATCHETT, B. J. 1976. Cell wall degrading enzymes and the softening of senescent strawberry fruit. J. Food Sci. *41*, 1392–1395.

BARTLEY, I. M. 1974. β-Galactosidase activity in ripening apples. Phytochemistry *13*, 2107-2111.

BARTLEY, I. M. 1976. Changes in glucans of ripening applies. Phytochemistry *15*, 625–636.

BARTLEY, I. M. 1977. A further study of β-galactosidase activity in apples ripening in stores. J. Exp. Bot. *28*, 943–948.

BARTOLOME, L. G., and HOFF, J. E. 1972. Firming of potatoes: Biochemical effects of preheating. Agric. Food Chem. *20*, 266–270.

BAUR, A. H., and YANG, S. F. 1969. Precursors of ethylene. Plant Physiol. *44*, 1347–1349.

BAUR, A. H., YANG, S. F., PRATT, H. K., and BIALE, J. B. . 1971. Ethylene biosynthesis in fruit tissues. Plant Physiol. *47*, 696–699.

BEEVERS, H. 1960. Respiratory Metabolism in Plants. Harper & Row Biological Monographs, Harper & Row, New York.

BeMILLER, J. N. 1967. Acid-catalyzed hydrolysis of glycosides. Adv. Carbohydr. Chem. *22*, 25–108.

BeMILLER, J. N., and KUMARI, G. V. 1972. β-elimination in uronic acids: Evidence for an ElcB mechanism, Carbohydr. Res. *25*, 419–428.

BEN-YEHOSHUA, S. 1969. Gas exchange, transpiration, and the commercial deterioration in storage of orange fruit. J. Am. Soc. Hort. Sci. *94*, 524–528.

BETTS, H. A., and BARMLAGE, W. J. 1977. Uptake of calcium by apples from postharvest dips in calcium chloride solutions. J. Am. Soc. Hort. Sci. *102*, 785–788.

BIALE, J. B. 1960A. The postharvest biochemistry of tropical and subtropical fruits. *In* Advances in Food Research, Vol. 10. C. O. Chichester, E. M. Mrak, and G. F. Stewart (Editors). Academic Press, New York.

BIALE, J. B. 1960B. Respiration of fruits. *In* Handbuch der Pflanzenphysiologie, Vol. 12, Part 2, W. Ruhland (Editor). Springer Verlag, Berlin and New York.

BIALE, J. B., and YOUNG, R. E. 1962. The biochemistry of fruit maturation. Endeavour *21*, 164–174.

BIALE, J. B., YOUNG, R. E., and OLMSTEAD, A. J. 1954. Fruit respiration and ethylene production. Plant Physiol. *29*, 168–174.

BIALE, J. B., YOUNG, R. E., POPPER, C. A., and APPLEMAN, W. E. 1957. Metabolic processes in cytoplasmic particles of the avocado fruit. 1. Preparative procedure, cofactor requirements, and oxidative phosphorylation. Physiol. Plant. *10*, 48–63.

BIGELOW, W. D., and STEVENSON, A. E. 1923. The effect of hard water in canning vegetables. National Canners Association Research Laboratories Bull. No. 20-L.

BLANPIED, G. D., and HANSEN, E. 1968. The effect of oxygen, carbon dioxide, and ethylene on the ripening of pears at ambient temperature. J. Am. Soc. Hort. Sci. *93*, 813–816.

BLANPIED, G. D., CADUN, O., and TAMURA, T. 1972. Ethylene in apple and pear experimental CA chambers. J. Am. Soc. Hort. Sci. *97*, 204–206.

BOE, A. A. 1971. Ethyl hydrogen 1-propylphosphate as a ripening inductant of green tomato fruit. HortScience *6*, 399–400.

BONNER, J., and VARNER, J. E. 1965. The path of carbon in respiratory metabolism. *In* Plant Physiology, J. Bonner and J. E. Warner (Editors). Academic Press, New York.

BUCH, M. L., SATORI, K. G., and HILLS, C. H. 1961. The effect of bruising and aging on the texture and pectic constituents of canned red tart cherries. Food Technol. *15*, 526–531.

BUESCHER, R. W., SISTRUNK, W. A., and TIGCHELAAR, E. C. 1976. Softening, pectolytic activity, and storage life of rin and nor tomato hybrids. Hort. Sci. *11*, 603–604.

BURG, S. P. 1962. The physiology of ethylene formation. Annu. Rev. Plant Physiol. *13*, 265–302.

BURG, S. P., and BURG, E. A. 1962. Role of ethylene in fruit ripening. Plant Physiol. *37*, 179–189.

BURG, S. P., and BURG, E. A. 1964. Biosynthesis of ethylene. Nature *203*, 869–870.

BURG, S. P., and BURG, E. A. 1965. Ethylene action and the ripening of fruits. Science *148*, 1190–1196.

BURG, S. P.; and BURG, E. A. 1966. Fruit storage at subatmospheric pressures. Science *153*, 314–315.

BURG, S. P. and CLAGETT, C. O. 1967. Conversion of methionine to ethylene in vegetative tissue and fruits. Biochem. Biophys. Res. Commun. *27*, 125–130.

BURG, S. P., and THIMANN, K. V. 1961. The conversion of glucose-C^{14} to ethylene by apple tissue. Arch. Biochem. Biophys. *95*, 450–457.

CARDINI, C. E. LELOIR, L. F., and CHIRIBOGA, J. 1955. The biosynthesis of sucrose. J. Biol. Chem. *214*, 149–155.

CHALMERS, D. J., and ROWAN, K. S. 1971. The climacteric in ripening tomato fruit. Plant Physiol. *48*, 235–240.

CHANCE, B., HOLMES, W., HIGGINS, J., and CONNELLY, C. M. 1958. Localization of interaction sites in multi-component transfer systems: Theorems derived from analogues. Nature *182*, 1190–1193.

CHEN, C.-H., and LEHNINGER, A. L. 1973. Ca^{2+} transport activity in mitochondria from some plant tissues. Arch. Biochem. Biophys. *157*, 183–196.

CHO, Y. P., and CHRISPEELS, M. J. 1976. Serine-O-galactosyl linkages in glycoproteins from carrot cell walls. Phytochemistry *15*, 165–169.

CLIJSTERS, H. 1965. Malic acid metabolism and initiation of the internal breakdown in "Jonathan" apples. Physiol. Plant. *18*, 85–94.

COLLINS, J. L., and WILEY, R. C. 1963. Influence of added calcium salts on texture of thermal-processed apple slices. Maryland Univ. Agric. Ex. Stn. Bull. A–130.

CRAFT, C. C., DUNCAN, G., and FOUSE D. 1968. Respiratory activity of small lemons and lemon tissue as influenced by modified atmospheres. J. Am. Soc. Hort. Sci. *93*, 173–185.

DARRIELL, A. G., SMITH, C. J., and HALL, M. A. 1977. Auxin induced proton release, cell wall structure, and elongation growth. *In* Regulation of Cell Membrane Activities in Plants. E. Marre ad O. Cieferri (Editors). Elsevier-North Holland, Amsterdam.

DAS, N. N., and DAS, A. 1977. Structure of the D-galactan isolated from garlic *(Allium sativum)* bulbs. Carbohydr. Res. *56*, 337–349.

DAVIS, W. C., and LeTOURNEAU, D. J. 1973. Leaching of solutes and sloughing of potato tuber tissue. Am. Potato J. *50*, 35–41.

DeFEKETE, M. A. R., and CARDINI, C. E. 1964. Mechanism of glucose transfer from sucrose into the starch granule of sweet corn. Arch. Biochem. Biophys *104*, 173–184.

DICKINSON, D. B., and McCOLLUM, J. P. 1964. Cellulase in tomato fruits. Nature *203*, 525–526.

DOESBURG, J. J. 1961. Relation between the behavior of pectic substances and changes in firmness of horticultural products during heating. Qual. Plant. Mater. Veg. *8*, 115–129.

DOESBURG, J. J. 1965. Pectic substances in fresh and preserved fruits and vegetables. I.B.V.T. Comm. Nr. 25, Wageningen, The Netherlands.

DOYLE, W. P., and WANG, C. H. 1958. Glucose catabolism in pepper fruit (*Capsicum frutescens* Longum.). Can. J. Bot. *36*, 483–490.

ESKIN, N. A. M., HENDERSON, H. M., and TOWNSEND, R. J. 1971. Biochemistry of Foods. Academic Press, New York.

FAUST, M., CHASE, B. R., and MASSEY, JR., L. M. 1966. Omtogenetic changes in respiratory pathways in Cortland apple skin during storage. Plant Physiol. *41*, 1610–1614.

FELDMAN, G., and TAÜFEL, K. 1956. Chemistry of gelation of pectin substances. Ernährungsforsch. ,*1*, 260–270 (German).

FIDLER, J. C., and NORTH, C. J. 1967. The effect of conditions of storage on the respiration of apples. II. The effect on the relationship between loss of respirable substrate and the formation of end products. J. Hort. Sci. *42*, 207–221.

FIDLER, J. C., and NORTH, C. J. 1970. Sorbitol in stored apples. J. Hort. Sci. *45*, 197–204.

FIDLER, J. C., and NORTH, C. J. 1971. The effect of conditions of storage on the respiration of apples. VI. The effects of temperature and controlled atmosphere storage on the relationship between rates of production of ethylene and carbon dioxide. J. Hort. Sci. *46*, 237–243.

FIDLER, J. C., and NORTH, C. J. 1971. The effect of conditions of storage on the respiration of apples. VII. The carbon and oxygen balance. J. Hort. Sci. *46*, 245–250.

FIDLER, J. C., WILDINSON, B. G., EDNEY, K. O., and SHARPLES, R. O. 1973. The Biology of Apple and Pear Storage. Commonwealth Agricultural Bureaux, Slough, England.

FLOOD, A. E., HULME, A. C., and WOOLTORTON, L. S. C. 1960. The organic acid metabolism of Cox's Orange Pippin apples. I. Some effects of the addition of organic acids to the peel of the fruit. J. Exp. Bot. *11*, 316–334.

FRANCIS, F. J., HARNEY, P. M., and BULSTRODE, P. C. 1955. Color and pigment changes in the flesh of McIntosh apples after removal from storage. Proc. Am. Soc. Hort. Sci. *65*, 211–213.

FRANK, H. S. 1965. The structure of water. Fed. Proc. Suppl. *15*, Pt 3, S1–S11.

FRENKEL, C. 1972. Involvement of peroxidase and indole-3-acetic acid oxidase isozymes from pear, tomato, and blueberry fruit in ripening. Plant Physiol. *49*, 757–763.

FRENKEL, C., KLEIN, I., and DILLEY, D. R. 1968. Protein synthesis in relation to ripening of pome fruits. Plant Physiol. *43*, 1146–1153.

FREUDENBERG, K. 1968. Constitution and biosynthesis of lignin. *In* Molecular Biology, Biochemistry, and Biophysics A. Kleinzeller (Editor), pp. 45–122, Springer Verlag, Berlin and New York.

GALLER, M., and MACKINNEY, G. 1965. The carotenoids of certain fruits (apple, pear, cherry, strawberry), J. Food Sci. *30*, 393–395.

GALLIARD, T., RHODES, M. J. C., WOOLTORTON, L. S. C., and HULME, A.

C. 1968. Metabolic changes in excised fruit tissue. III. The development of ethylene biosynthesis during the ageing of disks of apple peel. Phytochemistry *7*, 1465–1470.

GANE, R. 1934. Production of ethylene by some ripening fruits. Nature *134*, 1008.

GANE, R. 1935. Identification of ethylene among the volatile products of ripe apples. Dept. Sci. Ind. Res., Rep. Food Invest. Board 1934, 122–128. HMSO, London.

GOODALL, H. 1969. The composition of fruits. British Food Manufacturing Industries Res. Assoc. Sci. Tech. Survey, No. 59. Leatherhead, Surrey, England.

GROESCHEL, E. C., NELSON, A. I., and STEINBERG, M. P. 1966. Changes in color and other characteristics of green beans stored in controlled refrigerated atmospheres. J. Food Sci. *31*, 488–496.

GROTE, M., and FROMME, H. G. 1978. Electron microscope studies in cultivated plants. I. Green pods of *Phaseolus vulgaris*. Z. Lebensm. Unters.-Forsch. *166*, 69–73.

HALL, C. B. 1963. Cellulase in tomato fruits. Nature *200*, 1010–1011.

HANSEN, E. 1942. Quantitative study of ethylene production in relation to respiration of pears. Bot. Gaz. (Chicago) *103*, 543–558.

HARKETT, P. J. 1971. Effect of oxygen concentration on the sugar content of potato tubers stored at low temperature. Potato Res. *14*, 305–311.

HANSEN, E. 1966. Postharvest physiology of fruits. Annu. Rev. Plant Physiol. *17*, 459–480.

HARKETT, P. J., HULME, A. C., RHODES, M. J. C., and WOOLTORTON, L. S. C. 1971. The threshold value for physiological action of ethylene on apple fruits. J. Food Technol. *6*, 39–45.

HASEGAWA, S., and SMOLENSKY, D. C. 1971. Cellulase in dates and its role in fruit softening. J. Food Sci. *36*, 966–967.

HEATHERBELL, D. A., and WROLSTAD, R. E. 1971. Carrot volatiles. 2. Influence of variety, maturity, and storage. J. Food Sci. *36*, 225–227.

HINTON, D. M., and PRESSEY, R. 1974. Cellulase activity in peaches during ripening. J. Food Sci. *39*, 783-785.

HOBSON, G. E. 1968. Cellulase activity in tomato. J. Food Sci. *33*, 588–592.

HOOGZAND, C., and DOESBURG, J. J. 1961. Effect of blanching on texture and pectin of canned cauliflower. Food Technol. *15*, 160–163.

HSU, C. P., DESHPANDE, S. N., and DESROSIER, N. W. 1965. Role of pectin methylesterase in firmness of canned tomatoes. J. Food Sci. *30*, 583–588.

HUELIN, F. E., and BARKER, J. 1939. The effect of ethylene on the respiration and carbohydrate metabolism of potatoes. New Phytol. *38*, 85–104.

HUGHES, J. C., FAULKS, R. M., and GRANT, A. 1975. Texture of cooked potatoes: The effect of ions and pH on the compressive strength of cooked potatoes. J. Sci. Food Agric. *26*, 739–748.

HULME, A. C. 1954. Studies on the maturity of apples. Respiration progress curves for Cox's Orange Pippin apples for a number of consecutive seasons. J. Hort. Sci. *29*, 142–149.

HULME, A. C. 1958. Some aspects of the biochemistry of apple and pear fruits. *In* Advances in Food Research, Vol. 8. E. M. Mrak and G. F. Stewart (Editors). Academic Press, New York.

HULME, A. C., RHODES, M. J. C., GALLIARD, T., and WOOLTORTON, L. S. C. 1968. Metabolic changes in excised fruit tissue. IV. Changes occurring in discs of apple peel during the development of the respiration climacteric. Plant Physiol. *43*, 1154–1161.

HULME, A. C., RHODES, M. J. C., and WOOLTORTON, L. S. C. 1971. The relationship between ethylene and the synthesis of RNA and protein in ripening apples. Phytochemistry *10*, 749–756.

ISHERWOOD, F. A. 1970. Hexosans, pentosans, and gums. *In* The Biochemistry of Fruits and Their Products. A. C. Hulme (Editor), Academic Press, New York.

ITOH, T. 1975. Cell wall organization of cortical parenchyma of angiosperms observed by the freeze etching technique. Bot. Mag. *88*, 145–156.

JACOBSEN, D. W., and WANG, C. H. 1965. The conversion of acrylic acid to ethylene in *Penicillium digitalum*. Proc. Plant Physiol. *40*, xix.

JOHNSON, D. C., NICHOLSON, M. D., and HAIGH, F. C. 1975. Dimethyl sulfoxide/paraformaldehyde: A nondegrading solvent for cellulose. Inst. of Paper Chemistry, Tech. Paper Ser. 5, Appleton, WI.

JOSLYN, M. A. 1962. The chemistry or protopectin. *In* Advances in Food Research, Vol. 11, C. O. Chichester, E. M. Mrak, and G. F. Stewart (Editors). Academic Press, New York.

JUNG, D. W. and HANSON, J. B. 1973. Respiratory activation of 2,4-dinitro-phenol-stimulated ATPase activity in plant mitochondria. Arch. Biochem. Biophys. *158*, 139–148.

KATO, K., and CHACHIN, K. 1970. Maturation changes in fruits induced by ionizing radiation. V. Effects of gamma radiation on respiration and ethylene production of tomatoes. Nippon Shokuhin Kogyo Gakkai-Shi *17*, 97–103. (Japanese) Chem. Abstr. *73*, 97579p (1970).

KEEGSTRA, K., TALMADGE, K. W., BAUER, W. D., and ALBERSHEIM, A. 1973. The structure of plant cell walls. III. A model of the walls of suspension-cultured sycamore cells based on the interconnections of the macromolecular components. Plant Physiol. *51*, 188–196.

KEFFORD, J. F. 1959. The chemical constituents of citrus fruits. *In* Advances in Food Research, Vol. 9. E. M. Mrak and G. F. Stewart (Editors). Academic Press, New York.

KEIJBETS, M. J. H. 1974. Pectic substances in the cell wall and the intercellular cohesion of potato tuber tissue during cooking. Thesis. Agric. Univ. Wageningen, The Netherlands.

KEIJBETS, M. J. H., and PILNIK, W. 1974. β-Elimination of pectin in the presence of anions and cations. Carbohydr. Res. *33*, 359–362.

KEIJBETS, M. J. H., PILNIK, W., and VAAL, J. F. A. 1976. Model studies on behavior of pectic substances in the potato during boiling. Potato Res. *19*. 289-303.

KERTESZ, Z. I. 1951. The Pectic Substances. Wiley (Interscience) Publishers, New York.

KERTESZ, Z. I., EUCARE, M., and FOX, G. 1958. A study of apple cellulose. Food Res. *24*, 14–19.

KIDD, F., and WEST, C. 1925. The course of respiratory activity throughout the life of an apple. Dept. Sci. Ind. Res., Rept. Food Invest. Board 1924, 27–32. HMSO, London.

KIDD, F., and WEST, C. 1930. Physiology of fruit. I. Changes in the respiratory activity of apples during their senescence at different temperatures. Proc. R. Soc. London Ser. B *106*, 93–109.

KIDD, F., and WEST, C. 1939. Gas storage of English-grown Doyenne du Comice pears. Dept. Sci. Ind. Res. (Brit.), Rept. Food Invest. Board 1938, 157–160.

KIDD, F., and WEST, C. 1945. Respiratory activity and duration of life of apples gathered at different stages of development and subsequently maintained at constant temperature. Plant Physiol. *20*, 467–501.

KNEE, M. 1971A. Ripening of apples during storage. I. Introduction. J. Sci. Food Agric. *22*, 365–367.

KNEE, M. 1971B. Ripening of apples during storage. II. Respiratory metabolism and ethylene synthesis in Golden Delicious apples during the climacteric, and under conditions simulating commercial storage practice. J. Sci. Food Agric. *22*, 368–371.

KNEE, M. 1971C. Ripening of apples during storage. III. Changes in chemical composition of Golden Delicious apples during the climacteric and under conditions simulating commercial storage practice. J. Sci. Food Agric. *22*, 371–377.

KNEE, M. 1973. Polysaccharides and glycoproteins of apple fruit cell walls. Phytochemistry *12*, 637–653.

KNEE, M. 1975. Soluble and wall-bound glycoproteins of apple fruit tissue. Phytochemistry *14*, 2181–2188.

KNEE, M., FIELDING, A. H., ARCHER, S. A., and LABORDA, F. 1975. Enzymatic analysis of cell wall structure in apple fruit cortical tissue. Phytochemistry *14*, 2213–2222.

KNEE, M., SARGENT, J. A., and OSBORN, D. J. 1977. Cell wall metabolism in developing strawberry fruits. J. Exp. Bot. *28*, 377–396.

KODENCHERY, U. K., and NAIR, M. P. 1972. Metabolic changes induced by sprout inhibiting dose of γ-irradiation in potatoes. J. Agric. Food Chem. *20*, 282–285.

KOHN, R., and FURDA, I. 1967. Interaction of calcium and potassium ions with carboxyl groups of pectin. Collect. Czech. Chem. Commun. *32*, 4470–4484.

KRISHNAMURTY, S., PATWARDHAN, M. V., and SUBRAMANYAM, H. 1971. Biochemical changes during ripening of the mango fruit. Phytochemistry *10*, 2577–2581.

LABAVITCH, J. M., and RAY, P. M. 1974. Relationship between promotion of xyloglucan metabolism and induction of elongation by indoleacetic acid. Plant Physiol. *54*, 499–502.

LABELLE, R. L. 1971. Heat and calcium treatments for firming red tart cherries in a hot fill process. J. Food Sci. *36*, 323–326.

LABELLE, R. L., WOODAMS, E. E., and BOURNE, M. C. 1964. Recovery of Montmorency cherries from repeated bruising. Proc. Am. Soc. Hort. Sci. *84*, 103–109.

LAMPORT, D. T. A. 1970. Cell wall metabolism. Annu. Rev. Plant Physiol. *21*, 235–270.

LAMPORT, D. T. A. 1977. Structure, biosynthesis, and significance of cell wall glycoproteins. *In* Recent Advances in Phytochemistry Vol. II. Plenum Press, New York.

LEBERMANN, K. W., NELSON, A. I., and STEINBERG, M. P. 1968A. Postharvest changes of broccoli stored in modified atmospheres. I. Respiration of shoots and color of flower heads. Food Technol. *22*, 487–490.

LEBERMANN, K. W., NELSON, A. I., and STEINBERG, M. P. 1968B. Postharvest changes of broccoli stored in modified atmospheres. 2. Acidity and its influence on texture and chlorophyll retention of the stalks. Food Technol. *22*, 490–493.

LELOIR, L. F., and CARDINI, C. E. 1955. The biosynthesis of sucrose phosphate. J. Biol. Chem. *214*, 157–165.

LELOIR, L. F., DE FEKETE, M. A. R., and CARDINI, C. E. 1961. Starch and oligosaccharide synthesis from uridine diphosphate glucose. J. Biol. Chem. *236*, 636–641.

LIEBERMAN, M., and KUNISHI, A. T. 1971. An evaluation of 4-*S*-methyl-2-keto-butyric acid as an intermediate in the biosynthesis of ethylene. Plant Physiol. *47*, 576–580.

LIEBERMAN, M., and MAPSON, L. W. 1964. Genesis and biogenesis of ethylene. Nature *204*, 343–345.

LINEHAN, D. J., and HUGHES, J. C. 1969. Texture of cooked potato. III. Intercellular adhesion of chemically treated tuber sections. J. Sci. Food Agric. *20*, 119–123.

LIPTON, W. J. 1972. Market quality of radishes stored in low O_2-atmospheres. J. Am. Soc. Hort. Sci. *97*, 164–167.

LOCONTI, J. D., and KERTESZ, Z. I. 1941. Identification of calcium pectate as the tissue-firming compound formed by treatment of tomatoes with calcium chloride. Food Res. *6*, 499–508.

LOONEY, N. E. 1971. Interaction of ethylene, auxin, and succinic acid-2,2-dimethylhydrazide in apple fruit ripening control. J. Am. Soc. Hort. Sci. *96*, 350–353.

LYONS, J. M., and RAISON, J. K. 1970. Oxidative activity of mitochondria isolated from plant tissues sensitive and resistant to chilling injury. Plant Physiol. *45*, 386–389.

MAPSON, L. W., and ROBINSON, J. E. 1966. Relation between oxygen tension, biosynthesis of ethylene, respiration, and ripening changes in banana fruit. J. Food Technol. *1*, 215–225.

MAPSON, L. W., and WARDALE, D. A. 1972. Role of indolyl-3-acetic acid in the formation of ethylene from 4-methylmercapto-2-oxobutyric acid by peroxidase. Phytochemistry *11*, 1371–1387.

MAPSON, L. W., MARCH, J. F., and WARDALE, D. A. 1969. Biosynthesis of ethylene. 4-Methylmercapto-2-oxobutyric acid: An intermediate in the formation from methionine. Biochem. J. *115*, 653–661.

MAREI, N., and CRANE, J. C. 1971. Growth and respiratory response of fig (*Ficus caria* L. cv. Mission) fruits to ethylene. Plant Physiol. *48*, 249–254.

MARKAKIS, P., LIVINGSTON, G. E., and FELLERS, C. R. 1957. Quantitative aspects of strawberry pigment degradation. Food Res. *22*, 117–130.

MARKS, J. D., BERHLOHR, R., and VARNER, J. E. 1957. Esterification of phosphate in ripening fruit. Plant Physiol. *32*, 259–262.

MASON, J. L., JASMIN, J. J., and GRANGER, R. L. 1975. Softening of McIntosh apples reduced by a post harvest dip in calcium chloride solutions plus thickener. Hort. Sci. *10*, 524–525.

MASSEY, L. M., Jr., and WOODAMS, E. E. 1973. Effect of calcium on the texture profile of irradiated carrots, beets, and potatoes. J. Texture Stud. *4*, 242–247.

MATHESON, N. K., and SAINI, H. S. 1977. α-L-Arabinofuranosidases and β-D-galactosidases in germinating lipin cotyledons. Carbohydr. Res. *57*, 103–116.

McCLENDON, J. H., WOODMANSEE, C. W., and SOMERS, G. F. 1959. On the occurrence of free galacturonic acid in apples and tomatoes. Plant Physiol. *34*, 389–391.

McCREADY, R. M., and McCOMB, E. A. 1954. Pectic constituents in ripe and unripe fruit. Food Res. *19*, 530–535.

McGILL, J. N., NELSON, A. I., and STEINBERG, M. P. 1966. Effects of modified storage atmospheres on ascorbic acid and other quality characteristics of spinach. J. Food Sci. *31*, 510–517.

McGLASSON, W. B., and WILLS, R. B. H. 1972. Effects of oxygen and carbon dioxide on respiration, storage life, and organic acids of green bananas. Aust. J. Biol. Sci. *25*, 35–42.

McMURCHIE, E. J., McGLASSON, W. B., and EAKS, I. L. 1972. Treatment of fruit with propylene gives information about the biogenesis of ethylene. Nature *237*, 235–236.

MEHERIUK, M., and PORRITT, S. W. 1972. Effects of waxing on respiration. ethylene production, and other physical and chemical changes in selected apple cultivars. Can J. Plant Sci. *52*, 257–259.

MEIGH, D. F., JONES, J. D., and HULME, A. C. 1967. The respiration climacteric in the apple. Production of ethylene and fatty acids in fruit attached to and detached from the tree. Phytochemistry *6*, 1507–1515.

MILLER, E. V., WINSTON, J. R., and SCHOMER, H. A. 1940. Physiological studies of plastid pigments in rinds of maturing oranges. J. Agric. Res. *60*, 259–267.

MONRO, J. A., PENNY, D., and BAILEY, R. W. 1976. The organization of primary cell walls of lupin hypocotyl. Phytochemistry 15, 1193–1198.

MOORE, A. E., and STONE, B. A. 1972. Effect of senescence and hormone treatment on the activity of a β-1,3-glucan hydrolase in *Nicotiana glutinosa* leaves. Planta *104*, 93–109.

MORIKAWA, H., and SENDRA, M. 1974. Oriented structure of matrix polysaccharides in an extension growth of *Nitella* cell wall. Plant Cell Physiol. *15*, 1139–1143.

MÜHLETHALER, K. 1967. Ultrastructure and formation of plant cell walls. Annu. Rev. Plant Physiol. *18*, 1–24.

MUNCH-PETERSEN, A., KALCKAR, H. M., CUTOLO, E., and SMITH, E. E. B. 1953. Urdyl transferases and the formation of uridine triphosphate. Enzymic production of uridine triphosphate: Uridine diphosphoglucose pyrophosphorolysis. Nature *172*, 1036–1037.

MURATA, T., TATEISHI, K., and OGATA, K. 1968. Controlled atmosphere storage of fruits and vegetables. Effect of controlled atmosphere storage on the quality of tomatoes at two ripening stages. Engei Gakkai Zasshi *37*, 391–396. (Japanese) Chem. Abstr. *71*, 111620n (1969).

NEAL, G. E., and HULME, A. C. 1958. The organic acid metabolism of Bramley's Seedling apple peel. J. Exp. Bot. *9*, 142–157.

NELMES, B. J., and PRESTON, R. D. 1968. Wall development in apple fruit: A study of the life history of a parenchyma cell. J. Exp. Bot. *19*, 496–518.

NEUKOM, H. 1967. Pectic substances. Kirk-Othmer Encyl. Chem. Technol. *14*, 636–451.

NEUKOM, H., and DEUEL, H. 1958. Alkaline degradation of pectin. Chem. Ind. (London) 683.

NORTHCOTE, D. H. 1958. The cell walls of higher plants: Their composition, structure, and growth. Biol. Rev. *33,* 53–102.

NORTHCOTE, D. H. 1972. Chemistry of the plant cell wall. Annu. Rev. Plant Physiol. *23*, 113–132.

PALMER, J. M. 1966. The influence of growth regulating substances on the development of enhanced metabolic rates in thin slices of beetroot storage tissue. Plant Physiol. *41,* 1173–1178.

PEARSON, J. A., and ROBERTSON, R. N. 1954. The physiology of growth in apple fruits. VI. The control of respiration rate and synthesis. Aust. J. Biol. Sci. *7,* 1–17.

PECK, N. H., and MacDONALD, G. E. 1972. Plant response to concentrated superphosphate and potassium chloride. IV. Table beet (*Beta vulgaris* L.). Search *2* (14), 1–32.

PECK, N. H., and VAN BUREN, J. P. 1975. Plant response to concentrated superphosphate and potassium chloride fertilizers. V. Snap Beans (*Phaseolus vulgaris* var. humilis). Search *5* (2), 1–32.

PERRING, M. A. 1974. The mineral composition of apples. Method for calcium. J. Sci. Food Agric. *25,* 237–245.

PESIS, E., FUCHS, Y., and ZAUBERMAN, G. 1978. Cellulase activity and fruit softening in avocado. Plant Physiol. *61,* 416–419.

PIERROT, H., and VAN WEILINK, J. E. 1977. Localization of glycosidases in the wall of living cells from cultured *Convulvulus arvensis* tissue. Planta *137,* 235–242.

PILNIK, W., and VORAGEN, A. G. J. 1970. Pectic substances and other uronides. *In* The Biochemistry of Fruits and Their Products, Vol. 1 A. C. Hulme (Editor). Academic Press, New York.

PILNIK, W., and ZWIKER, P. 1970. Pektine. Gordian *70,* 202–204, 252–257, 302–305, 343–346.

POLÁCSEK-RACZ, M., and POZSÁR-HAJNAL, K. 1976. Determination of pectin methylesterase, polygalacturanase, and pectic substances in some fruits and vegetables. Acta Aliment. Acad. Sci. Hung. *5,* 189–204.

PRATT, H. K., and GOESCHL, J. D. 1962. A comparison of growth and ripening of honey dew and cantaloupe fruits. Proc. Plant Physiol. *37,* xix.

PRATT, H. K., and GOESCHL, J. D. 1964. Some relationships of ethylene to the respiratory patterns of honey dew melons. Proc. Plant Physiol. *39,* ix.

PRATT, H. K., and GOESCHL, J. D. 1969. Physiological roles of ethylene in plants. Ann. Rev. Plant Physiol. *20,* 541–584.

PRESSEY, R. 1970. Changes in sucrose synthetase and sucrose phosphate synthetase activities during storage of potatoes. Am. Potato J. *47,* 245–251.

PRESSEY, R., and AVANTS, J. K. 1976. Pear polygalacturonases. Phytochemistry *15,* 1349–1351.

PRESSEY, R., HINTON, D. M., and AVANTS, J. K. 1971. Development of polygalacturonase activity and solubilization of pectin in peaches during ripening. J. Food Sci. *36,* 1070–1073.

PRESTON, R. D. 1974. The Physical Biology of Plant Cell Walls. Chapman and Hall, London.

PRIDHAM, J. B. 1974. Plant Carbohydrate Biochemistry. Academic Press, New York.

REES, D. A. 1969. Structure, conformation, and mechanism in formation of polysaccharide gels and networks. Advances in Carbohydrate Chemistry, Vol. *24,* M. L. Wolfrom and R. S. Tipson (Editors). Academic Press, New York.

REES, D. A., and WELSH, E. J. 1977. Secondary and tertiary structure of polysaccharides in solutions and gels. Angew. Chem. Int. Ed. Engl. *16,* 214–224.

REES, D. A., and WIGHT, J. N. 1969. Molecular cohesion in plant cell walls. Biochem. J. *115,* 431–439.

REEVE, R. M. 1970. Relationships of histological structure of texture of fresh and processed fruits and vegetables. J. Texture Stud. *1,* 247–284.

REID, M. S., and PRATT, H. K. 1970. Ethylene and the respiration climacteric. Nature *226,* 976–977.

RHODES, M. J. C., and WOOLTORTON, L. S. C. 1967. The respiration climacteric in apple fruits. The action of hydrolytic enzymes in peel tissue during the climacteric period in fruit detached from the tree. Phytochemistry *6,* 1–12.

RHODES, M. J. C., and WOOLTORTON, L. S. C. 1971. The effect of ethylene on the respiration and on the activity of phenylalanine ammonia lyase in swede and parsnip root tissue. Phytochemistry *10,* 1989–1997.

RILEY, R. G., and KOHATTUKUDY, P. E. 1976. Effect of treatment with calcium ion-containing formulations on the firmness of Golden Delicious apples. Hort. Sci. *11,* 249–251.

ROMBOUTS, F. M., NORDE, W., and PILNIK, W. 1970. A simple method for the determination of the number-average degree of polymerization of pectic substances. Lebensm.-Wiss. Technol. *3,* 94–97.

RONGINE DE FEKETE, M. A., and CARDINI, C. E. 1964. Mechanism of glucose transfer from sucrose into the starch granule of sweet corn. Arch. Biochem. Biophys. *104,* 173–184.

ROBERTSON, R. N., and TURNER, J. F. 1951. The physiology of growth in apple fruits. II. Respiratory and other metabolic activities as functions of cell number and cell size in fruit development. Aust. J. Sci. Res. Ser. B *4,* 92–107.

ROMANI, R. J., and KU, L. 1966. Direct gas chromatographic analysis of volatiles produced by ripening pears. J. Food Sci. *31,* 558–560.

ROMANI, R. J., and YU, I. K. 1967. Mitochondrial resistance to massive irradiation in vivo. III. Suppression and recovery of respiratory control. Arch. Biochem. Biophys. *117,* 638–644.

RYALL, A. L., and LIPTON, W. J. 1979. Handling, Transportation, and Storage of Fruits and Vegetables, 2nd Edition. Vol. 1 Vegetables and Melons. AVI Publishing Co., Westport, CT.

SACHER, J. A. 1962. Relations between changes in membrane permeability and the climacteric in banana and avocado. Nature *195,* 577–578.

SACHER, J. A. 1966. Permeability characteristics and amino acid incorporation during senescence (ripening) of banana tissue. Plant Physiol. *41,* 701–708.

SACHER, J. A. 1973. Senescence and postharvest physiology. Annu. Rev. Plant Physiol. *24,* 197–224.

SAKAKIBARA, A. 1977. Degradation products of protolignin and the structure of lignin. *In* Recent Advances in Phytochemistry Vol. 11. Plenum Press, New York.

SALUNKHE, D. K., and WU, M. T. 1974. Developments in technology of storage and handling of fresh fruits and vegetables. Crit. Rev. Food Technol. *5,* 15–54.

SCHANDERL, S. H., and LYNN, D. Y. C. 1966. Changes in chlorophylls and spectrally related pigments during ripening of *Capsicum fructescens.* J. Food Sci. *31,* 141–145.

SCHWIMMER, S., and ROREM, E. S. 1970. Biosynthesis of sucrose by preparation from potatoes stored in the cold and at room temperature. Nature *187,* 1113–1114.

SELVENDRAN, R. 1975. Cell wall glycoproteins and polysaccharides of parenchyma of *Phaseolus coccineus 14,* 2175–2180.

SHERKAT, F., and LUH, B. S. 1976. Quality factors of tomato paste made at different temperatures. J. Agric. Food Chem. *24,* 1155–1158.

SINGH, B., LITTLEFIELD, N. A., and SALUNKHE, D. K. 1970. Effects of controlled atmosphere (CA) storage on amino acids, organic acids, sugars, and rate of respiration of Lambert sweet cherry fruit. J. Am. Soc. Hort. Sci. *95,* 458–461.

SINGH, B., YANG, C. C., SALUNKHE, D. K., and RAHMAN, A. R. 1972. Controlled atmosphere storage of lettuce. 1. Effects on quality and the respiration rate of lettuce heads. J. Food Sci. *37,* 48–51.

SINGH, B., WANG, D. J., and SALUNKHE, D. K. 1972. Controlled atmosphere storage of lettuce. 2. Effects on biochemical composition of leaves. J. Food Sci. *37,* 52–55.

SISTRUNK, W. A. 1965. Effect of storage time and temperature of fresh snap beans on chemical composition of the canned product. Proc. Am. Sco. Hort. Sci. *86,* 380–386.

SISTRUNK, W. A., and CAIN, R. F. 1960. Chemical and physical changes in green beans during preparation and processing. Food Technol. *14,* 357–362.

SMOCK, R. M., and NEUBERT, A. M. 1950. Apples and Apple Products. Wiley (Interscience) Publishers, New York.

STEINBUCH, E. 1976. The improvement of texture of frozen vegetables by stepwise blanching treatments. J. Food Technol. *11,* 313–315.

STERLING C. 1963. Texture and cell-wall polysaccharides in foods. *In* Recent Advances in Food Science, Vol. 3 J. M. Leitch and D. N. Rhodes (Editors). Butterworths, London.

STERLING, C. 1968. Effect of solutes and pH on the structure and firmness of cooked carrot. J. Food Technol. *3,* 367–371.

STERLING, C., and SHIMAZU, F. 1961. Cellulose crystallinity and the reconstitution of dehydrated carrots. J. Food Sci. *26,* 479–484.

TAIZ, L., and HONIGMAN, W. A. 1976. Production of cell wall hydrolyzing enzymes by barley aleurone layers in response to gibberellic acid. Plant Physiol. *58,* 380–386.

TAVAKOLI, M., and WILEY, R. C. 1968. Relation of trimethylsilyl derivatives of fruit tissue polysaccharides to apple texture. Proc. Am. Soc. Hort. Sci. *92,* 780–787.

VAN BUREN, J. P. 1968. Adding calcium to snap beans at different stages in processing: Calcium uptake and texture of the canned product. Food Technol. *22,* 790–793.

VAN BUREN, J. P. 1974. Heat treatments and the texture and pectins of red tart cherries. J. Food Sci. *39,* 1203–1205.

VAN BUREN, J. P. 1979. The chemistry of texture in fruits and vegetables. J. Texture Stud. *10,* 1–23.

VAN BUREN, J. P., and PECK, N. H. 1963. Effect of calcium level in nutrient solution on quality of snap bean pods. Proc. Am. Soc. Hort. Sci. *82,* 316–321.

VAN BUREN, J. P., MOYER, J. C., WILSON, D. E., ROBINSON, W. B., and HAND, D. B. 1960. Influence of blanching conditions on sloughing, splitting, and firmness of canned snap beans. Food Technol. *14,* 233–236.

VAN BUREN, J. P., MOYER, J. C., and ROBINSON, W. B. 1962. Pectin methylesterase in snap beans. J. Food Sci. *27,* 291–294.

VAN BUREN, J. P., LaBELLE, R. L., and SPLITTSTOESSER, D. F. 1967. The influence of SO_2 level pH and salts on color, texture and cracking of brined Windsor cherries. Food Technol. *21,* 1028–1030.

VENDRELL, M., and McGLASSON, W. B. 1971. Inhibition of ethylene production in banana fruit tissue by ethylene treatment. Aust. J. Biol. Sci. *24,* 885–895.

WADE, N. L., and BRADY, C. J. 1971. Effects of kinetin on respiration, ethylene production. Aust. J. Biol. Sci. *24,* 165–167.

WAGER, H. G. 1964. Physiological studies of the storage of green peas. J. Sci. Food Agric. *15,* 245–252.

WALLNER, S. J., and WALKER, J. E. 1975. Glycosidases in cell wall degrading extracts of ripening tomato fruit. Plant Physiol. *55,* 94–98.

WALTER, R. H., RAO, M. A., VAN BUREN, J. P., SHERMAN, R. M., and KENNY, J. F. 1977. Development and characterization of an apple cellulose gel. J. Food Sci. *42,* 241–243.

WANG, C. Y., and HANSEN, E. 1970. Differential response to ethylene in respiration and ripening of immature 'Anjou' pears. J. Am. Soc. Hort. Sci. *95,* 314–316.

WANG, C. Y., and MELLENTHIN, W. M. 1972. Internal ethylne levels during ripening and climacteric in Anjou pears. Plant Physiol. *50,* 311–312.

WANG, C. Y., and MELLENTHIN, W. M. 1973. Chlorogenic acid levels, ethylene production, and respiration of d'Anjou pears affected with cork spot. HortScience *8,* 180–181.

WANG, C. Y. MELLENTHIN, W. M., and HANSEN, E. 1971. Effect of temperature on development of premature ripening in 'Bartlett' pears. J. Am. Soc. Hort. Sci. *96,* 122–125.

WANG, C. Y., MELLENTHIN, W. M., and HANSEN, E. 1972. Maturation of 'Anjou' pears in relation to chemical composition and reaction to ethylene. J. Am. Soc. Hort. Sci. *97,* 9–12.

WANG, S. S., HAARD, N. F., and DIMARCO, G. R. 1971. Chlorophyll degradation during controlled-atmosphere storage of asparagus. J. Food Sci. *36,* 657–661.

WANKIER, B. N., SALUNKHE, D. K., and CAMPBELL, W. F. 1970. Effects of controlled atmosphere storage on biochemical changes in apricot and peach fruit. J. Am. Soc. Hort. Sci. *95,* 604–609.

WARDALE, D. A. 1973. Effect of phenolic compounds in *Lycopersicon esculentum* on the synthesis of ethylene. Phytochemistry *12,* 1523–1530.

WARREN, D. S., and WOODMAN, J. S. 1974. The texture of cooked potatoes: A review. J. Sci. Food Agric. *25,* 129–138.

WATADA, A. E., and MORRIS, L. L. 1966. Effect of chilling and non-chilling temperatures on snap bean fruits. Proc. Am. Soc. Hort. Sci. *89,* 368–374.

WATADA, A. E., and MORRIS, L. L. 1966. Post-harvest behavior of snap bean cultivars. Proc. Am. Soc. Hort. Sci. *89,* 375–380.

WILEY, R. E., and LEE, Y. S. 1970. Modifying texture of processed apple slices. Food Technol. *24,* 1168–1170.

WISKICH, J. T. 1966. Respiratory control by isolated apple mitochondria. Nature *212,* 641–642.

WORKMAN, M. 1963. Color and pigment changes in Golden Delicious and Grimes Golden apples. Proc. Am. Soc. Hort. Sci. *83,* 149–161.

YAMAGUCHI, M., TIMM, H., CLEGG, M. D., and HOWARD, F. D. 1966. Effect of stage of maturity and postharvest conditions on sugar conversion, and chip quality of potato tubers. Proc. Am. Soc. Hort. Sci. *89,* 456–463.

YAMAGUCHI, M., HOWARD, F. D., HUGHES, D. L., and THOMPSON, R. H. 1967. Carbon dioxide in pepper fruits: its utilization and effect on respiration. Proc. Am. Soc. Hort. Sci. *91,* 428–435.

YANG, S. F. 1967. Biosynthesis of ethylene. Ethylene formation from methional by horseradish peroxidase. Arch. Biochem. Biophys. *122,* 481–487.

YEMM, E. W. 1965. The respiration of plants and their organs. *In* Plant Physiology, Vol. 4A, F. C. Steward (Editor). Academic Press, New York.

YOUNG, E. Y., and BIALE, J. B. 1967. Phosphorylation in avocado fruit slices in relation to respiratory climacteric. Plant Physiol. *42,* 1357–1362.

ZAEHRINGER, M. V., and CUNNINGHAM, H. H. 1971. Potato extractives: Sloughing as related to replacement of anions or cations. Am. Potato J. *48,* 385–389.

Index

A

Acacia, 75
Acetoin, 410
Aconitic acid, 427
Aceylformoin, 290
Acrolein, 248
Adenine (6-aminopurine), 159
Adenosine diphosphate, 59, 191, 324, 479, 506-507, 511
Adenosine triphosphate, 4-5, 163, 191, 324-325, 470-472, 478-479, 492, 505-507, 511, 515
  and muscle contraction, 478-479
ADP, *see* Adenosine diphosphate
Adsorption isotherm, 15-16
Aflatoxins, 401
Agar, 73-74
Aglycone, 268
Albumins, 144
Alcoholic fermentation, 323-339, *see also* Wine; Beer; Distilled Products; Vinegar
Alcohols, 246
Aldoses, 48
Aldehydes, 247
Aldaric acids, 49
Aldonic acids, 48
Algin, 73
Alkalies, action on sugars, 44-47
Allicin, 253
Allyl isothiocyanate, 252
Allyl propyl disulfide, 252
Allyl sulfide (allyl disulfide), 252
Amadori rearrangement, 291-295, 453
Amelioration, 332
Amino acids, 133-143, 170, 434, 445, 454-455, 474, 491
  activation, 163
  essential, 135
Amino acyl-adenosine monophosphate (AMP), 163
Amino acyl-tRNA, 163
Amino sugars, 52
α-Amylase, 179, 192
β-Amylase, 180, 192
Amyloglucosidase, 180, 192
Angular rotation, 37
Anionic minerals, 231
Anisaldehyde, 249
anti-Anisaldehyde oxime, 239
syn-Anisaldehyde oxime, 239
Annato extract, 319
Anthocyanidins, 268-269
Anthocyanins, 268-269
  bleaching, 269-270
  in fruits, 269
Anthrone color reaction, 76
Anticodon, 161
Antioxidants, 108, 118, 120
Apoenzyme, 187
L-Arabinose, 35
Ascorbase, 179
Ascorbate oxidase, 179
Ascorbic acid, 210-212
  acid property, 211
  determination, 212
  peroxide release, 211
  reducing properties, 211
  sources, 210
  structure, 210
  synthesis, 211
Ascorbic acid oxidase, 193
Ash, 352, 364-366, 399, 403, 405, 423, 452
Astaxanthin, 265
Asymmetric carbon atom, 37
ATP, *see* Adenosine triphosphate
ATPase, 468-469
Autoxidation, 108-109, 114-115
  autocatalytic, 99-110

catalysis, light and radiations, 115
fats, carbonyls, 114-115
free radical mechanism, 110-112
non-autocatalytic, 108
of fats, 108
rate, acceleration, inhibition, 108-109
Avidin, 165, 207

B

Bacteria, and photosynthesis, 2
Baked products, 343, *see also* specific items
Baking powder, 356
Beer, hydrolytic enzymes, 336
Beet colors
betalains, 271
betacyanins, 271
betacyanins, red, 271
betanidin, 271, 272
betanin, 271-272
betaxanthins, yellow, 271
isobetanidin, 271-272
isobetanin, 271, 272
vulgaxanthin I, 271-272
vulgaxanthin II, 271-272
Benzaldehyde, 248
o-Benzoquinone, 283
BET isotherm, 16
Bial test for pentoses, 78
Bilins, 263
Biotin, 207
physical properties, 207
sources, 207
Biuret reaction, 144
Bixin, 319
Black tongue, 203-204
Blue-green algae, 2
Bound water, 18, 481
determined by NMR, 19
Brandy 337
Bread, 343
dough temperature, 355
flavor, 354
flour, 352
Brewing, 336
Browning
enzymatic, 283-288
chemical reactions, 283-284
control, 287-288
dehydration in sugar, 288
distribution of phenolase,
oxidation of monophenols, 286
oxidation of polyphenols, 285
pigment formation, 286
o-quinones, 286
reaction of phebolase, 383, 384-385
formation of pigment, 297-300
ascorbic acid, 287, 300
in dried fruits, 297
moisture content and, 297
inhibition of, 300-301
non-enzymatic, 288-297
Amadori rearrangement, 293-295
caramel flavor, 290
caramelization, 289-291
carbonyl compounds as precursors, 291, 297
desirability, 288-289
2,5-dimethylpyrazine, 296
effect on flavor, 297-299
Heyn's rearrangement, 293-294
Hodge scheme, 292
isosacchrosan, 289
lipid involvement, 300
Maillard reaction, 291, 296
moisture control, 297
Strecker degradation, 295-297
and water activity, 17
Butter, 385
consistency, 386
rancidity in, 245
Buttermilk, 385, 387

C

CA, *see* Controlled atmosphere storage
Cacao beans
aldehydes, 453
catechin index, 448
death, 448, 450
changes during drying, 451
during fermentation, 448-451
cured, 449
drying, 443
fermented
aromatic acids, 541
enzyme changes, 450
germination effect, 448
leucocyanidins, 541
roasted
pyrazines in, 453
volatiles in, 456-457

roasting, 443
formation of volatile flavors, 453-455
esters from Arriba beans, 456
d-linalool in volatile oil, 456
5-methyl-2-phenyl-2-hexenal, 456
other volatile compounds, 457
roasting changes, 451-456
amino acids, 453-456
reducing sugars, 453-455
sucrose changes during fermentation, 450-451
Theobroma cacao, 441
Theobroma pentagona, 441
tree, 441-442
winnowing, 444
unfermented
amino acids, 445
anthocyanins, 444-445
carbohydrates, 446-447
chemical composition, 444-447
enzyme changes, 445
polyphenols, 444-445
Cadmium, 233
Cafestol, 400
Caffeic acid, 288
Caffeine, 404-405, 426, 447
Cake batter, 356
Calciferol, 214
Calcium, 228
Canned foods, staining in, 233
Canning equipment, metals from, 233
Cans
detinning, 233
internal corrosion, 233
manufacture, 233
swelling, 233
Capric acid, 372-374
Caproic acid, 246
Caprylic acid, 374
Capsorubin, 265-266
Caramel
color and iron, 290
coloring, use, 272
degradation products, 291
and malanoidins, 272-273
pigments, 290
Caramelization, 289-291
glucose, 289
sucrose, 3 stages, 289
Caramelan, 290
Caramelen, 290
Caramelin, 290
Carbohydrates, 33
classification, 34
digestion of, 76
formation in plants, 34
qualitative tests, 76-78
quantitative tests, 78-79
structure, 39-43
Carbon disulfide, 411-412
Carbon monoxide myo(hemo)globin, 476
Carbonyl-amine reactions, *see* Maillard reaction
Carbonyl compounds, 291
Carbonyls, 383-384
Carboxyl(acid) proteinases, 181
Carminic acid, 319
Carotenes, 213, 265
determination, 214
properties, 213
sensitivities, 213
sources, 214
structural differences, 213
β-Carotene, 213
Carotenoids, 212, 261, 264, 317
retention, 266
Carrageenan, 74
Casein, 368
fractions, 370
micelles, 370
solubility and optical rotation, 372
Cassia, oil, 249
Catalase, 179, 192-193
Catechins, 448
Catechol, 283
Cellobiose, 70
Cellulose, 69-70
beta-linked, 70
Cephalins, 104
Cheddar cheese
flavor, 245, 383-384
from raw and pasteurized milk, 384
rennet-pig pepsin mixture, 377
ripening changes, 384
volatile carbonyls, 383-384
Cheese and Cheese chemistry, 375, *see also* specific cheeses
analysis, 378-379
aroma synthetic, 383-384
calcium and phosphorus retention, 381
clotting enzymes, 375
coagulation, 381-382
Cunara cardunculus, 375
curing, 383

Lactobacillus bulgaricus, 382
Lactobacillus lactis and acidity, 382
manufacture
acid control, 382
acid development control, 382
coagulation temperature, 382
heating step, 382
ripening, 383-384
chemical changes, 383-384
salt, 384
Chelates, 225-227
Chicory, 414
Chitin, 71
Chloride, 231
Cholorogenic acid, 513
Chlorophyll, 1-3, 261, 263-264
complexes, 2
deterioration, 264
found in different plants, 2
chloroplasts, 2
role in photosynthesis, 1, 2
separation of, 2
structure, 3
wavelengths of light absorbed, 2, 3
Chlorophyll *a,* 2, 3
Chlorophyll *b,* 2, 3
Chlorophyll *c,* 2
Chlorophyll *d,* 2
Chocolate, 357, *see also* Cocoa
milk, 458
products, German, 548
sweet, 458
Cholecalciferol, *see* Vitamin D_3
Cholesterol, 107
Choline, 103
Chromdroitin, 49
Chromoproteins, 146
Chymopapain, 181
Cinnamaldehyde (cinnamic aldehyde), 249
Cinnamic acid, 249
Cinnamic alcohol, 249
Cinnamon oil, 249
Citral, 248
Citronellol, 250-251
Climacteric fruits, 509-512
Cobalt, 230
Cobamide coenzyme, 208
Cochineal, 319
Cocarboxylase, 201
Cocoa
alkalized (Dutch Process), 457
analytical procedures, 458
butter, 443, 457
conching, 455
changes during manufacture, 447-456
drying, 451
fermentation 448-451
roasting, 451-456
chemical composition,
amino acids and enzymes, 445
anthocyanins-polyphenols, 444-446
carbohydrates, 446
theobromine and caffeine, 447
manufacture, 442-444
drying, 443
fermentation, 442-443
roasting, 443
winnowing, 444
unfermented beans, 444
Codon, 161
Coenzyme A, 206
Coenzymes, 182
Coffee, 397
acids, 402-403
ash, 405
cafestol, 400
caffeine, 404
caramelized sugars, 401
carbohydrates, 398
carcinogenic substances, 404
changes during roasting, 401-403
chlorogenic acid, 401
Coffee arabica, 397
Coffea canephora, 397
composition of green bean, 398-401
solubles, 405
decaffeinated coffee, 413
3,4-dicaffeoylquinic acid, 401
3,5-dicaffeoylquinic acid, 401
4,5-dicaffeoylquinic acid, 401
fatty acids, 401
high altitude grown, 397
7-hydroxy-6-methoxy coumarin, 401
instant, 414
kahweol, 400
lipids, 400, 402
methods of preparation, 397-398
neochlorogenic acid, 401
nicotinic acid, 402
nitrogenous substances, 399
changes during roasting, 401-402
purine, 404
bases, 404
roasted, chemical composition, 403

robusta coffee, 397
staling, 412-413
moisture and, 413
substitutes, 414
theobromine, 404
thiamin, 402
trigonelline, 402
use, 397
volatiles
alcohols, 410
aldehydes, 408-409
ammonia and volatile amines, 412
analysis, 406
effect of varying roasts, 408
esters, 410
from Colombian coffee, 407
heterocyclic compounds, 410-411
ketones, 409-410
research difficulties, 406
sulfur compounds, 411, 412
xanthine, 404
Collagen, 472-473
Collagenolytic activity, 473
Colloids, 25-32
behavior in an electrical field, 26
breaking of emulsions
double layer theory, *see* Helmholtz-Gouy theory
emulsions, 28-29
foams,
gels, 27
hydrophilic (lyophilic), 25-27
hydrophobic (lyophobic), 26-27
imbibition, 28
protective colloid, 27
salting-in, 26-27
salting out, 26-27
size of particles, 25
syneresis, 27-28
viscosity, 30-31
Colorings, food, 307-322, *see also* specific pigments
acceptable daily intakes, 316
amount and distribution, 309
annato extract, bixin, 319
cochineal, 319
classification of some US food colors, 315
color analysis and desired properties, 321
colors exempt from certification, 317
color safety and regulations, 307
corrosion, 316-317
dyes
chemical structure, 313-314
problems, 321
food color consumption data, 310-311
lakes, 308
permanently listed color additives, 318
solubility, 314
stability, 314
specifications and hazards, 312-313
titaniim dioxide, 320
turmeric, 320
Colors, natural, 261-281
anthocyanins and flavonoids, 268-271
anthocyanidins, 269
betacyanins, 271
caramels and melanoidins, 272-273
carotenoid retention, 266
carotenoid structure, 265-267
carotenoids, 264-265
chemistry, 262
chlorophyll deterioration, 264
consumption, 274
determination, 273
flavone, 270
heme pigments, 262
lycopene, 265
natural, categories, 261
tetrapyrrole structure, 263
Color tests, protein and amino acids, 135
Competitive inhibition, 185
Conalbumin, 165
Copper, 230
dietary sources, 230
Connective tissue, 472
Controlled-atmosphere storage, 518-520
broccoli, 519
lettuce, 520
potatoes, 520
ripening and organic acids, 518-519
p-Coumaric acid, 513
p-Coumarylquinic acid, 445
Covalent bonds, 147
Cream, 386
Cream of tartar, 331
p-Cresol, 283
Cresolase, 283
Cyanidin, 268
Cyanocabalamin, *see* Vitamin B_{12}
Cycloheximide, 517
Cysteine, 138
Cystine, 138
Cytosine, 158

D

Dairy products, fat determination, 387
Danish agar, 75
n-Decyl alcohol, 246
Degradation products, secondary, 113-115
L-Dehydroascorbic acid, 193, 210
Dehydroascorbic reductase, 193
7-Dehydrocholesterol, 107, 214
Delphinidin, 268
Denaturation, 152
Dental caries, 232
Deoxyribonucleic acid, 35, 158
  chains, 159-160
  genetic substance, 159
Deoxy-D-ribose, 52
2-Deoxyribose, 35
Deoxy sugars, 52
Derived lipids, 107
Desmosine, 474
  structure, 474
Desorption isotherm, 15-16
Dextranase, 180
Dextran gels, 155
Dextrorotation, 37
Diacetyl, 249-250, 291, 329-330, 384-385, 406, 409-410
Diastase, 179-180
3,4-Dicaffeoylquinic acid, 401
3,5-Dicaffeoylquinic acid, 401
4,5-Dicaffeoylquinic acid, 401
Dienes, 214-215
Differential thermal analysis, 19
Differential enthalpic analysis, 19
Digestion
  of carbohydrates, 76
  of lipids, 120-121
  of proteins, 157
Dihydroxytoluene, 284
2,5-Dimethylpyrazine, 296
Dimethyl sulfide, 284, 411-412
2,4-Dinitrophenol, 511
Dipalmitoleyl-L-α-glycerylphosphorylcholine, 103
o-Diphenolic compounds, 285
Diphosphopyridine nucleotide, 204
Disaccharides, 53
Distilled alcoholic products, 337
Di- and triglycerides, 94
DNA, *see* Deoxyribonucleic acid
DNP, *see* 2,4-Dinitrophenol
Dopa oxidase, 283
Double layer theory, *see* Helmholtz-Gouy theory
Dough, 346, 348, 355
  elastic properties, 345
  gas cells, 355-356
  lipid binding, 349
  mixing, 346
  overmixing, 346
  oxidizing agents, 353
  rye flour, 344
  wheat type, effect of, 347
DPN, *see* Diphosphopyridine nucleotide
Dyes, problems with use, 321

E

Edman degradation, 148-149
Egg yolk, 165
Egg white, 165
Elaidic acid, 192
Elastin, 472, 473-474
β-Elimination reactions, 522-523, 530-532
Elongation factors, 163
Electron acceptor, 1
Emmentaler (Swiss) cheese, 245, 375, 379
Emulsifying agents, 28
  breaking down, 29
Emulsions, 28-29
Enantiomorphs, 38
1,2-Enediol, 45
2,3-Enediol, 46
Energy, 1
Enolization, 45
Enzymatic browning, 283
Enzyme activators, 189
Enzyme activity
  complexity, 190
  in food processing, 193
  mode of, 186-187
Enzymes
  active sites, 190
  Buchner's work, 177
  catalytic efficiency, 190
  classification, 178-181
  composition, 182
  concentration, 182
  covalent substrate intermediate, 182
  inactivation, 193
  inhibition, 184
  kinetics, 182
  metal containing, 182
  milk clotting animal sources, 375

molecular or molar activity, 187
properties, 182
proteolytic, 375
reaction rate, 187-188, 189
and temperature, 188
regeneration (renaturation), 188
specificity, 186
turnover number, 187
Epocatechin, polyphenol oxidase substrate, 451
Ergo-calciferol, *see* Vitamin D_2
Ergosterol, 107, 214-215
Esterases, 190
Esters, 245-247
ethyl *n*-butyrate, 245
ethyl formate, 245
Ethylene
controlled atmosphere storage, 518
concentration used, 514
formation pathways, 512-513
production, 514
in fruit, 514
and lipid turnover, 517
ripening, 512
Eugenol, 251-252
Evaporated milk, 386
Exo-1,4-glucosidase, 180
Exudate gums, 75

F

Fat
determination in dairy products, 287-288
emulsifying capacity, 481-482
manufacturing, tempering, 98-99
new terminology, 89
and oils, 88-89
in shortening, 98
analysis, 100-101
classification, 99
preparation, 100
Fatty acids, 89-91, 244-246
cis-trans susceptability to autoxidation, 108
cyclic, 93
determination, 94
energy source, 471-472
in edible lipid materials, table, 90
over water, monomolecular layer, 89
saturated, 89
unsaturated, 91
position of double bond, 92
Fehling's method, 47
Fermentable sugars, 39
Fermentation
alcohol yields, 325
alcoholic, 323-341
of milk, 384
temperature, 328
Ferroprotoporphyrin, 476
Ferulic acid, 288
Ficin, 181
Fischer–Tollens formulas, 40
Fish proteins, 168
Flavanols, 270
Flavanones, 270
Flavin adenine dinucleotide, 203
Flavone (2-phenylbenzopyrone), 270
Flavonoids, 268, 270
Flavonols, 270
Flavoprotein, 165, 203
Flavor, 237-256
alcohols, 246
aldehydes, 247-249
bread, 354
basic tastes, 237
defects, concealing, 255
detection, 237
enhancement, 253
esters, 246-247
fatty acids, 245-246
ketones, 249-250
lactones, 251
mint, 250
phenols, 251-252
restoration and deterioration, 254
sulfur compounds, 252
taste and odor, 237
terpene alcohols, 250
and volatiles, 241
Flour, 343
ash, 352
bleaching and maturing agents, 352
carbohydrates, 348
composition, 344-345, 347
compounds related to baking quality, 344
defatted, 351-352
fatty substances, 349
glutathione, 348
hard spring wheat, 343, 346
lipids and bread making, 349-352
proteins, 344-348
effect of cross-linked structure, 347-348

rheological properties,
rye, 344-348
soft wheat, 344, 347
wheat, amber durum, 344
Fluidity, 30
Fluorine, 232
Foams, 29-30
on beer, 29
stability, 29
Folic acid, 208
dietary sources, 208
Formaldehyde, 3-4
Forastero, *see* Cacao
Formic acid, 245
Fructopyranose, 35
Fructose, 36, 39
Fructose 6-phosphate, 4
Fructose 1,6-diphosphate, 324
Fruits, 505-534
chemical changes during ripening, 555
citric acid cycle, 508
ethylene formation, 512-514
and ripening, 512
maturation, 526-529
mitochondria, 506
pathways of respiration, 506-509
respiration climacteric, 509
ripening, 505-517
p-coumaric acid, 513
cycloheximide, 517
methionine, 513
naringenin, 513
pectic substances, 516
pigments, 517
proteins, 516-517
storage, 518-520
texture, 520-532
cellulose, 525
hemicellulose, 524-525
heat processing and, 529-532
lignin, 525-526
pectic substances, 521-524
proteins, 525
water, 526
β-Fructofuranosidase, 180
Fucose, 52
Furcellaran, 75
Furanose ring, 41
Furfural, 44, 409
by degradation of reducing sugars, 300

G

D-Galactosamine, 52
D-Galactose, 25-36, 77
test, 77
Galactosyl glycerides, 349
Galactuoranan, 72
D-Galacturonic acid
Garlic oil, 252
Gas chromatography, 241-242
Gay-Lussac equation, 323
Gelatin, 19, 26
Gels, 27
diffusion in, 27
heat reversible and non-reversible, 27
Geraniol. 251-251, 433
Gin, 337
Gliadin, 347
Globulins, 145
Glucaric acid, 49
Glucitol (sorbitol), 50
D-Gluconic acid, 48
Glucono-δ-lactone, 48
Glucono-γ-lactone, 48
α-D-Glucopyranose, 40
β-D-Glucopyranose, 40
Glucosamine, 52
Glucosamylase, 180
Glucose, 35, 77
and browning, 296
Glucose 6-phosphate, 228, 506-508
Glucuronic acid, 48
Glucuronides, formation, 49
Glutamic acid, 141
Glutathione, 193
Glutelins, 145
Gluten, building blocks, 348
Glutenin, 348
Glyceraldehyde, 38
D-Glyceraldehyde-3-phosphate, 507
Glycerol, 93-94
Glycogen, 70
Glycogen synthetase activity, 466
Glycoproteins, 145
Glycosidases, 179
Glycosides, 51
Glycosylglycerides synthetic, 351
Goiter, 232
Grana, 2
Grape(s)
acidity, 326
anthocyanin in *Vitus vinifera*, 334
Concord, 247

diglucoside pigments, importance, 334
effect of temperature in growth areas, 331
sugar acid balance, 332
volatile compounds, 226
Guanine, 159
L-Guluronic acid, 73
Gum arabic, 75
Gums
exudate, 75
seaweed, 73

H

Haworth formulas, 41-42
Hematin compounds, 117
Hematin-catalyzed peroxidation 117-118
Heme pigments, 262-263
Hemicelluloses, 73, 524
Hemochrome, 262
Hemoglobin, 262, 475-477
coagulation, 482-483
n-Heptylic acid, 246
Hesperidin, 270
Heteropolysaccharides, 56
Hexaphosphoinositol, 228
Hexoses, 35
accumulation in photosynthesis, 5
Heyns rearrangement, 293-294
Histidine, 140
Histones, 145
Hodge scheme, 292
Holoenzyme, 187
Homopolysaccharides, 56
Hops, 336
Humins, 44
Hunter Color Difference Meter, 273
Hydrogen bonding, 11
Hydrogen donor, 1
Hydrogen sulfide, 384, 412, 484-485, 488, 490
Hydrogenation, 121
Hydrolases, 178-179
Hydroperoxide, 108
development curve, 109
dismutation pathways, 113
secondary degradation products, 113-115
3-Hydroxybutanone, 384
Hydroxyl groups, sugars, 50
7-Hydroxy-6-methoxy coumarin, 401
Hydroxymethylfurfural, 330
Hydroxyproline, 140

I

Ice crystal growth, 14
Inbibition, 28
Initiation, 163
meso-Inositol, 209-210
Inositol phosphatides, 105
Interesterification, 122
directed, 122-123
Inulase, 180
Inulin, 70
Inulinase, 180
Invertase, 180
Iodine, 231
Iodine number, 100-101
Ionization radiation, effect, 486-487
β-Ionone, 212
Ionones, 250
Iron, 229
sources, 229
use by the body, 229-230
Iron porphyrin, 229
Iron protein complexes, 229
Irradiation effect, 512
Irreversible inhibition, 186
Isobetanidin, 271
Isobetanin, 271-272
Isodesmosine, 474
Isoelectric focusing, 155
Isoelectric point, 135
Isoeugenol, 252
Isoleucine, 137
Isomaltase, 180
Isomerases, 178
Isomerism, 39, 91-92
cis, 92
trans, 92
Isoprene, 251
Isoprenoid structures, *see* Carotenoids
Isosacchrosan, 289

J

cis-Jasmone, 433

K

Kahweol, 400
δ-Ketoglutaric acid, 383
Ketoses,
conversion to aldoses, 48
hydrogenation, 50
Kwashiorkor, 229

L

Laccase, 284
α-Lactalbumin, 372
Lactic acid, 387
  bacteria, 382, 384, 387
Lactobacillus citrovorum, 384
β-Lactoglobulin, 371-372
Lactones, 251
δ-Lactones, 251
  in butter, 251
α-Lactose, 54, 365-366
β-Lactose, 366-367
Lakes, 308
  variation in shades, 309
Lamellae, 2
Lead in canned foods, 232
Leavening, 354
  agents, 354
Lecithins, 103-104
Leucoanthocyanins, 270
Leucine, 135, 137
Levorotation, 47
Levulose, 36
Ligands, 225
Light
  oxidation, 484
  and radiation, 115
  ultraviolet effects, 115
  visible, effect on peroxides, 115
  wavelengths absorbed by chlorophyll, 2-3
Ligases, 178
Lignins, 525
Limit dextrinase, 180
Linalool, 250
  oxides, 433
Lineweaver-Burk equation, 184
Linoleic acid, 90, 93
Lipids, 87
  in bread making, 349-352
  classification, 87
    scheme, 88
  commercial application, 121-124
  composite, 102-106
  composition, 102
  derived, 107
  nonpolar, 349
  milk, 372
    bovine, 373
  oxidation, 108
    effect of heavy metals, 116
  peroxidation of lipids by hematin compounds, 117-118
  polar, 349
  primary alkoxy radical reactions, 115
  simple, 87
Lipoproteins, 145
Lipovitellenin, 165
Lipovitellin, 165
Lipoxidase, 179
Lipoxigenase, 118, 179
  catalytic action, 118
  use in baking, 118
Livetin, 165
Lobry de Bryn-Alberda van Ekenstein transformation, 45
Lycopene, 205
Lyophilic colloid, 25
Lyophilization, 13
Lyophobic colloid, 26
Lysine, 135, 139
Lysozyme, 165

M

Magnesium, 228
Maillard reaction, 272, 278, 289, 291
  carbonyl-amine reaction, 291
  three stages, 291
  α-β-unsaturated aldehydes and α-dicarbonyl compound, 291
Malic acid, 516
Malt, 336
Maltose, 55
Manganese, 230
Mannose, 35-36
Margarine, 123
Muscle
  connective tissue, 472
    collagen, 472
    elastin, 473
  proteins, 480
    action of light, 484
    antioxidants, 484
    bound water, 481
    effect of heat, 484-486
    fat emulsifying capacity, 481
    oxidation and rancidity, 482
  sarcoplasm, 474
  skeletal muscle
    composition, 463
    effect of slaughter, 476
      abnormal coloration, 477
      browning intensity, 169
      color of cured meats and myoglobin, 476

color and meat packaging, 477
pH, 478
rigor mortis, 478
tenderness, 480
fibers, 464-467
structure, 463
proteins, 468
actin, 469
actomyocin, 470
tropomyocin, 471
troponin, 471
Meat products
analysis of, 492
effects of heat, 484
flavors, 487-488
5′-ribomononucleotides, 487
volatile flavors, 488-490
odor, thiophen-2-carboxaldehyde, 488
pigments, 474
hemoglobin, 475
myoglobin, 475-476
proteins, 467
simulated, 490-491
sausage, 491-492
tenderizers, 487
Melanoidins, 272, 291
Melibiose, 56
Menadione, *see* Vitamin K_3
1-Menthol, 250
1-Menthone, 250
3-Mercaptopropionic acid, 384
Mercury absorption from sea water, 229
meso-Inositol, structure, 209
Meta-saccharinic acid, 46-47
Methemoglobin, 262
L-Methionine, 513
2-Methoxy-4-allylphenol, *see* Eugenol
p-Methoxybenzaldehyde, 249
2-Methoxy-4-propenylphenol, 252
Methyl amyl ketone, 249
Methyl anthranilate, 247
Methyl benzaldehyde, 248
Methyl benzoate, 247
2-Methyl butyraldehyde, 406
2-Methyl-3-difarnesyl-1,4-naphthoquinone, *see* Vitamin K_2
Methyl ethyl ketone, 410
Methyl formate, 410
Methyl lineolate
effect of monochromatic light, 115
peroxides, decomposition curve, 111
Methyl mercaptan, 412
Methyl oleate, 114
5-Methyl-2-phenyl-2-hexenal, 457
2-Methyl-3-phytyl-1,4-naphthoquinone, *see* Vitamin K, 216
Methyl salicylate, 247
Methylacetylcarbinol, 410
Metmyoglobin, 477
Micelles, 26
Michaelis-Menten equation, 184-185
Milk and milk products, 363
analysis, total solids, 388
ascorbic acid and off-flavors, 373
butter, 385-386
buttermilk, 387
carbohydrates, 365
clotting enzymes, animal sources, 376
composition, 363-364
cow's
defined, 363
proteins, amino acid composition, 369
cream, 386
dry whole, 386
evaporated, 386
fatty acids, 374
fermentation, 384
lipids, 372
malted, 386
nonfat dry, 387
off-flavor, 251
phospholipids, 372
proteins, 368
rancidity and off-flavor, 255, 373-374
salt components, 365
skimmilk, 387
sunlight flavor, 255
sweetened condensed, 386
trace elements, 366
vitamins, 364
whole dry, 386
Millon reaction, 144
Minerals, 225
in canned foods, 232-233
content of wheat and mill products, 352
detection and determination, 225
in foods, 226
major amounts, 225, 228
minor amounts, 225, 229
occurrence, 228
in physiological and biochemical reactions, 226
Mitochondria, 506
Molecular weight, 156

Molisch reaction, 76
Molybdenum, 231
 in dental caries, 231
 nitrate hemochrome, denatured, 263
Monoamino-monocarboxylic acid titration curve, 143
Monophenols, oxidation, 286
Monosaccharides, 34, 77
 oxidation, 47
 phenylhydrazine reactions, 50-51
 qualitative paper chromatography, 78
 reduction, 49
MRNA, 160
Mucoproteins, 145
Mucor miehei-MM protease, 376
Mucor pusillus, var. Lindt-MP protease, 376
Muscle "acid stable" cross linkages, 485
Must fermentation, 327
Mutarotation, 39, 40-43
 of fructose, 42
Mustard oil, 252
Myoglobin, 262, 475-477
 affinity for oxygen, 476
 coagulation, 482-483
Myosin 468-471
 molecule, 469
Myosin from fish, 168

N

NADP, *see* Nicotinamide adenine dinucleotide phosphate
Naringenin, 513
Natural substrates, conversion, 288
Neochlorogenic acid, 401
Nerol, 250
Niacin, 203-205
 determination, 205
 sources, 205
Nicol prisms, 36
Nicotinamide, 203-204
 coenzymes, 204
 properties, 204
Nicotinamide adenine dinucleotide, 204
Nicotinamide adenine dinucleotide phosphate, 2, 204
Nicotinic acid, 203
 synthesis, 204
Night blindness, 212, 214
Ninhydrin, 135
Nitric oxide myo(hemo)globin, 476
Nitrite, nitrate cure, 263-264, 477
 and cancer, 263
Nitrosamines, 263
Nitrosohemochrome, 262
Nitrosomyoblobin, 262
Nonalactone, 251
Nonclimacteric fruits, 510
Noncompetitive inhibition, 186
Nonenzymatic browning, *see* Browning, non-enzymatic
Nucleoproteins, 145

O

Odor,
 classes of organic compounds, 244
 theories, 243-244
Off-flavors, influence of aldehydes, 115
Oleic acid, 90-92
Oligo-1,6-glucosidase, 180
Oligosaccharides, 53, 78
Optical activity, 36-37
Optical rotation, 36-37
Osazones, 50
Osmotic pressure, 21
Osones, 50
Ovalbumin, 165
Ovomucin, 165
Ovomucoid, 165
Oxidation
 and heavy metals, 116
 measurement of, 113
 of monoenoic acids, 112
 of polyenoic acids, 112
 and rancidity, 254-255, 482-483
Oxidative phosphorylation, 511
Oxidative rancidity and hematin compounds, 117
Oxidoreductases, 178-179, 192-193
Oxygen, from photosynthesis, 6
Oxygenated heme pigments, 262, 476
Oxymyo(hemo)globin, 262, 476

P

Pale, soft, exudative porcine musculature, 479
Pancreatic amylase, 76
Pantothenic acid, 206-207
 in coenzyme A, 206-207
 properties, 207
 sources, 207

Papain, 181
Papainase, 181
Pectic enzymes and fruit juices, 194
Pectic substances, 521-524
  changes during ripening, 516
Pectin, 49, 71-73, 521-524
  composition, 71-72
Pectin gel formation, 72-73
Pectinase, 180
Pectindepolymerase, 180
Pectinesterases and other pectic enzymes, 179, 192
Pectinmethylesterase, 179
Pelargonidin, 268
Pelargonidin-3-glucoside, 268
Pellagra, 203-205
Pentose pathway, 507-509
Pentoses, 34-35
  regeneration in photosynthesis, 4, 5
Peonidin, 268
Pepsin,
  bovine, 377, 380
  porcine, 377
Pepsin A (pepsin), 181
Pepsin B, 181
Pepsin C, 181
Peptide bond, 134
Peptide hydrolase, 180
Pernicious anemia, 208, 230
Peroxidase, 179, 192-193
Peroxide content of corn oil, 115
Phenolase, 174, 283
  chemical action, 283
  distribution, 284
  inhibition, 284
  naturally occurring substrates, 285
Phenoloxidase, 283
Phenols, 251-252
Phenylalanine, 135, 142
Phenylhydrazones, 50-51
Pheophytin, 3, 264
Phosphatidyl ethanolamine, 105
Phosphatidyl inositol, 105
L-α-Phosphatidyl serine, 104
Phosphoenol pyruvic acid, 507
Phosphoglyceraldehyde, 4
3-Phosphoglyceric acid, 4
Phospholipids, 102-106
  effect on bread quality, 350-351
  and off-flavor, 483
  rancidity, 483
Phosphorus, 232
Phosphorylase, 466
Phosvitin, 165
Photosynthesis, 1
  chemistry, 3-5
  reaction types, 6
  sugar production, 4-5
Photosynthetic process, 3
Photosystem I, 3
Photosystem II, 3
Phytic acid, 232
β-Picoline, 204
Pigment changes and ripening, 517
  cured meats, 162, 477
  in foods, 261
  meat, 474-476
Plasmalogens, 105
Polarimeter, 36
Polygalacturonase, 180, 192
Polymethylgalacturonidases, 192
Polymorphism, 96
Polypeptides, 134
  synthesis, 163
Polyphenoloxidases, 193, 283
Polyphenols, oxidation, 285
Polysaccharides, 56
Pork, curing pickle, 488
Porphin, 116
  complexes, 116
Potassium, 229
Potato oxidase, 283
Porphyrin destruction, 482
Prebetanin, 271
Processing losses, 217
Prolamines, 145
Proline, 140
Propionaldehyde, 408
Propionic acid, 245
  and Swiss cheese, 382
Propylene activity, 514-515
Prorennin, 380
Protamines, 145
Proteases, 134
Proteinases, 77, 192
Proteins, 133-176, 525
  α-helix form of molecule, 151
  amino acid sequence, 148-149
  in baked goods, 344
  in cereals, 168-169
  changes during processing and cooking, 169
    by heat, 484-486
  classified, 144-146

color tests, 135, 143-144
conformation, 157
conjugated, 145-146
defined, 133-134
denaturation, 152
derived proteins, 146
digestion, 157
egg, 165
fibrous, 144, 150
fish, 168
in foods, 165
globular, 144, 150
hydrogen bonds, 150
isoelectric focusing, 155
hydrolysis, 134
insoluble salts, 152
meat, 467-474
molecular weight, 156-157
by gel filtration, 155
myoribrillar, 468
N-terminal amino group, 148
phosphoproteins, 146
preservation by lyophilization, 152
primary derivatives, 146
primary structure, 146
properties, physical and chemical, 152
quaternary structure, 150
ripening, 516-517
separation, 153
ammonium sulfate fractionation, 153
chromatography, 153
electrophoresis, 153-154
gel filtration, 155
isoelectric focuring, 155
organic solvents, 153
polyacrylamide, 155
starch-gel electrophoresis, 154-155
ultracentrifugation, 153-154
secondary derivatives, 146
secondary structure, 150
simple, classified, 144-145
skeletal muscle, 468-472
skimmilk, composition, 371
structural forces, 147
structure, 146-151
synthesis, 158
tertiary structure, 150
from wheat flour, 344
whey, 371-372
Proteolytic enzymes, 156, 194, 375
Provitamins A, 213
Provitamin D, 214
Pteroylglutamin acid, *see* Folic acid
Purine, 159, 404
Pyradoxine, 205
crystals, 205
determination, 206
isolation, 205
properties, 205
synthesis and formula, 205
Pyranose ring, 41
Pyrazine, 453
Pyridine, 411
3-Pyridine carboxylic acid, *see* Niacin
Pyridoxal, 205
Pyridoxal phosphate, coenzyme activity, 205
Pyridoxamine phosphate, enzyme activity, 205
Pyrimidine, 158
Pyrrole, 116
Pyruvaldehyde, 291
Pyruvate kinase, 189
Pyruvic acid, 201

Q

Qualitative tests, carbohydrates, 76-78
Quantasomes, composition, 2
Quercitin, 270-271
L-Quinic acid, 516
o-Quinones, 286

R

Racemic mixture, 37
Raffinose, 55
Raman spectroscopy, 244
Recommended daily allowances, 200-201
Reducing sugars and alkali, 47
Regeneration (renaturation), 188
Reichert-Meissl number, 101
Release factors, 163
Rennet, 375, 380-381
Rennin
clotting, 380-381
substitutes, 375
Respiration, 505-515
changes in oxo and free amino acids, 516
climacteric, 509-512
pentose pathway, 507
tricarboxylic pathway, 506-507
Respiration pattern, enzyme involved, 517
Retinol, *see* Vitamin A

L-Rhamnose, 52
Riboflavin
 assay, 203
 as coenzyme, 202-203, 218
 deficiency, 203
 effect of light, 202
 enzyme activity, 202
 formula and synthesis, 203
 sources, 203
Ribonucleic acids, 158
D-Ribose, 35
Ribose-5-phosphate, 508-509
Ribulose-5-phosphate, 508
Ribulose-1,5-diphosphate, 4
Rickets, 214-215
Rigor mortis, 478-479
 chemical changes, 478
Ring compounds, 248
Ripening
 carbohydrates and acids, 515-516
 chemical changes, 515-517
 postharvest changes, 505

S

Saccharimeter, 36
Saccharinic acids, 46-47, 49
Saccharomyces cerevisiae, 336
Salting in, 27
Salting out, 26-27
Sanger, 148
Saponification, 96
Saponification number, 100
Sarcomere model, 465
Sarcoplasm, 474
 enzymes, 474
Sausage
 color formation in, 263-264
 production, 491-492
Schiff's reagent, 39
Scleroproteins, 145
Scurvy, 210
Seaweed gums, 73
Seliwanoff color test for ketose sugars, 77
Sedoheptulose phosphate, 509
Selenium, 232
Serine, 137
Serine proteinases, 180
Serra cheese, 375
Shadow matching, 245
SH proteinases, 181
Sialic acids, 52
Sinigrin, 252
Skimmilk, 387
Sodium, 229
Solutions, 19
 basis of, 22
 boiling point, 21
 freezing point, 22
 osmotic pressure, 21
D-Sorbitol, 49-50, 518
Sephadex, 155
Sphingolipids, 106
Sphingomyelin, 106
Sphingosine, 106
Spinach quantasome, 2
Squalene, 107
Stachyose, 55-56
Starch, 57-69
 acid modified, 68
 biosynthetic pathways, 58-60
 conversion to glucose, 62-63
 corn, gelatinization temperature, 65
 cross-linked, 68
 effect of various substances on viscosity, 65
 esterification, 67-68
 esters, 68
 fractionation, 63-64
 gelatinization, 64-66
 granule swelling, 65-66
 hot paste viscosity effect, 67-68
 hydrolysis, 60-63
 modified, 67
 moisture absorption, 64
 native, gelatinization characteristics, 65
 phosphates, 69
 potato
  gelatinization starting temperature, 66
  weak internal binding, 66
  synthesis, enzymes involved, 58-60
  technology, 57
  wrinkled pea starch, 57
Stereochemical theory, 244
Steroids, 107
Sterols, 107
Storage, fruits and vegetables, 518
Strecker degradation, 292, 295, 453
 carbonyl structure necessary, 295
 equation with Schonberg information, 295
Streptococcus cremoris, 384
Streptococcus lactis, 384
Subtilisin, 181
Succinic dehydrogenase, 466-467

Sucrase, 180
Sucrose, 53, 78
  feeding exhausted animals, 480
  structure, 53
  transfer mechanism to starch, 515
Sugar acetyls as surfactants, 51
Sugar esters, 51-52
Sugar phosphate esters, 52
Sugars
  hydroxyl groups, 150
  qualitative tests, 77-78
Sulfhydryl groups, meat, 262
Sulfur compounds in flavors, 252-253
Sulfur dioxide, 327-328
Sumner, urease, 177
Supersaturated solutions, 21
Surface tension, 21
Sweet potato oxidase, 283
Syneresis, 27
Synthetases, 178

T

Tartaric acid, 38
Taste, 237
  AH, B unit and, 241
  bitter, 237, 241
  and chemical structure, 237
  salty, 238
  sour, 238
  sweet and chemical structure, 238-240
Tastes, four basic, 237
Tea, 419
  amino acids, 425
  aroma, 433-434
    amino acids, 434
    carotene, 434
    compounds formed, 433-434
    other compounds, 434
  black, 419-421
    carbohydrates, 426
    flavor, 433
    production, 419-421
  Camellia assamica, 419
  Camellia sinensis, 419
  carotenoid pigments, 426-427
  catechins, 422-423
  chemical components, 422-425
  chlorophyll, 426-427
  classes, 419
  cream, 431
  depsides, 422
  enzymes in tea leaves, 425
  (−)-epigallocatechin gallate, 424
  epitheaflavic acids, formation, 430
  fermentation, 420, 428-431
  firing, 420-421
  flavanols, determination in tea extracts, 425
  flavone, 422
  flush
    amino acids, 425
    carbohydrates in fresh, 426
    catechol oxidase, 425
    other enzymes in tea leaves, 425
    organic acids, 427
  green, 421
    volatile compounds, 433-434
  manufacture, chemistry, 428
  methods of preparation, 419-421
  minerals in leaves, 427
  Oolong, 421
  phenolic compounds, 424
    in fermentation, 431-432
  pheophytin formation, 427
  production areas, 419
  purine bases, 426
  rolling, 420
  theoflavins, 429-430, 432
  theanine, 425-426
  thearubigins, 430-431
    formation, 431
  theobromine, 447, 451
  theophylline, 426
  vitamin P, 421
  withering, 420, 428
Tempering, 98
Tenderness, 480
Terpene alcohols, 250
Terpenes, 107
Tetrapyrrole structure, 261-263
Tetrasaccharides, 55-56
Texture
  chemistry of fruits and vegetables, 520
    and heat processing, 529
  changes during maturation and storage, 526
  demethoxylation of pectic substances, 532
  effect of pH, 429-430
  improvement with Ca salts during heating, 530
  and salts, 530-532
Thiamin, *see* Vitamin B_1
Thiaminase, 202
Thiazole ring, 201

Thiochrome, 202
Thiophene, 411
Thixotropy, 27
Threonine, 135, 137
Thymine, 158
Thymol, 252
Thyroxine, 231
Titanium dioxide, 320
α-Tocopherol, 215-216
  properties, 216
  reversible antioxidants, 216
Total extinction, 36
Total solids, analysis in dairy products, 388
Toughness meat, 472-474
Tragacanth, 75-76
Tragacanthic acid, 75
Transferases, 178
Trans structure, stability, 265
Triacylglycerol lipase, 179
Triacylglycerols, 89, 94-95
Tricarboxylic acid (Krebs) cycle, *see* Pentose pathway
Triketohydrindene hydrate, 143
Trigonelline, 402
Trimethylamine, 255
Triphosphopyridine nucleotide, 204
Trisaccharides, 55
tRNA, 161
Tropomyosin, 468, 471
Tropomyosin B, 471
Troponin, 471
Troponin A, 468
Troponin B, 468
α-Trypsin, 180
β-Trypsin, 100
Tryptophan, 135, 142
  niacin precursor, 204
Turmeric, 320
Turnover number, 187-188
Tyrosinase, 283

U

Ultracentrifuge, 153
Uncompetitive inhibition, 186
Unsaturated carbonyl compounds, 193
Uracil, 158
Uronic acid, 48

V

Valeraldehyde, 409
n-Valeric acid, 246
Valine, 135-136
Van der Waals forces, 147
Vanillin, 249, 251-252
Van't Hoff, 39
Veal, flavor components, 490
Vegetable processing and enzymes, 193
Vegetables
  dry blanching, 218
  heat processing and texture chemistry, 529-530
  maturation and storage, 529
  postharvest, 505-517
  storage, 518-520
    texture changes, 520-532
  toughening in, 529
Vibrational theory, 244
Vinegar, 337-338
  manaufacture, 338
  submerged method, 338
Viscosity, 30
  of colloidal dispersions, 31
  determination, 30
  factors affecting viscosity, 31
Vitamin A, 212
  assay, 214
  deficiency effect, 214
  structure, 212
Vitamin A_1 and A_2, structural differences, 212
Vitamin B_1, 199, 201
  deficiency, 199-200, 202
  effect of SO_2 and sulfites, 202
  occurrence, 202
  properties, 201
  quantitative determination, 202
  sources, 202
  structure changes and activity, 201
  structure and synthesis, 200
Vitamin B_2, *see* Riboflavin
Vitamin B_6, *see* Pyridoxine
Vitamin B_{12}, 208
  assay, 208
  sources, 208
Vitamin C, *see* Ascorbic acid
Vitamin D, 214
  assay, 215
Vitamin D_2, 107, 215
  properties, 215

Vitamind D_3, 107, 214-215
Vitamin E, 215-216
  assay, 216
Vitamin K, 217
Vitamin K_2, 217
Vitamin K_3, 217
Vitamins, 199-224
  B group, 199
  fat soluble, 212
  history, 199
  instability, 217-218
  malnutrition, 217
  processing losses, 217-218
  recommended daily allowances, 200-201
  water soluble, 199
Vitus vinifera, 326
Vitus labrusca, and hydrids, 326
Vodka, 337
Volatile substances
  and chocolate flavor, 456-457
  gas chromatography and mass spectrometry, 241
Vulgaxanthin I, 271, 272
Vulgaxanthin II, 271, 272

W

Water, 9, 526
  content of foods, table, 10
  density, 12
  latent heat of fusion, 12
  latent heat of vaporization, 12
  physical properties, 12
  structure, 9, 11
  supercooling, 12
  unusual physical properties, 12
Water activity, 14-18
  and enzymes, 17-18
Water-holding capacity, 481
Waxes, 102
  bloom of fruits, 102
Wheat gluten proteins, 168
Whey proteins, 371
Whiskey, 326-327
Wine, 326-336
  acetaldehyde, 330
  acetoin, 329-330
  acids
    fixed, 331-332
    volatile, 332
    legal maximum limit, 332
  alcohols, other than ethyl, 329
  amelioration, 332
  anthocyanins, aging, 334
  antiseptics, 327-328
  arabinose, 333
  browning, types, 334
  2,3-butylene glycol, 329
  calcium, 335
  color, 334
  composition, 329
  copper and iron, 335-336
  cream of tartar, 331
  dessert, sugar content, 332-333
  diacetyl, 329-330
  ethyl acetate, levels, 330-331
  other esters, 330
  ethyl alcohol, 329
  fixed acids, 331
  glycerol, 329
  grape species, 326
  hydroxymethylfurfural, 330
  lead, 336
  malo-lactic fermentation, 331
  minerals
    anions, 325
    cations, 335-336
    nitrogen, 333
    oxygen, 334-335
    pectins, 333
    phenolic compounds, 333-334
    pentoses, 333
    phosphates, 335
    potassium, 336
    Saccharomyees cerevisiae, 327
    sourness, pH, and titratable acidity, 331
    spoilage, and acids, 332
    sugar, 332
    sulfates, 335
    sulfur compounds, 327-328
    tannins, 333
    white, browning, 334
    yeasts, 327
Wintergreen oil, 247
Wort, 336

X

Xanthine, 404
Xerogel, 27
Xerophthalmia, 212, 214
D-Xylose, 35
Xylulose-5-phosphate, 509

Y

Yeast, 354
action, results of, 355
dry and compressed, 355
fermentation pathway, 354
gas produced, temperature effect, 355

Z

Zeta potential, 26
Zinc, 230
sources, 230
Zwitterion, 135